Cosmetic and Drug Preservation

COSMETIC SCIENCE AND TECHNOLOGY SERIES

Series Editor
ERIC JUNGERMANN
Jungermann Associates, Inc.
Phoenix, Arizona

Volume 1: Cosmetic and Drug Preservation: Principles and Practice, *edited by Jon J. Kabara*

Volume 2: The Cosmetic Industry: Scientific and Regulatory Foundations, *edited by Norman F. Estrin*

Volume 3: Cosmetic Product Testing: A Modern Psychophysical Approach, *by Howard R. Moskowitz*

Other Volumes in Preparation

Cosmetic and Drug Preservation

PRINCIPLES AND PRACTICE

Edited by
JON J. KABARA
Department of Biomechanics
Michigan State University
East Lansing, Michigan

MARCEL DEKKER, INC.　　　　New York and Basel

Library of Congress Cataloging in Publication Data
Main entry under title:

Cosmetic and drug preservation.

(Cosmetic science and technology ; v. 1)
Includes index.
1. Cosmetics--Preservation. 2. Cosmetics--
Microbiology. 3. Drugs--Preservation. 4. Drugs--
Microbiology. I. Kabara, Jon J., [date].
II. Series.
TP983.C774 1984   668'.55   84-15603
ISBN 0-8247-7104-4

COPYRIGHT © 1984 by MARCEL DEKKER, INC.   ALL RIGHTS RESERVED

Neither this book nor any part may be reproduced or transmitted in
any form or by any means, electronic or mechanical, including photo-
copying, microfilming, and recording, or by any information storage
and retrieval system, without permission in writing from the publisher

MARCEL DEKKER, INC.
270 Madison Avenue, New York, New York 10016

Current printing (last digit):
10 9 8 7 6 5 4 3 2

PRINTED IN THE UNITED STATES OF AMERICA

ABOUT THE SERIES

The rapid growth of cosmetic science has made it virtually impossible for a single author or a single book to present a coherent review of the entire field. This series was conceived to permit discussion of the broad spectrum of current knowledge and theories of cosmetic science and technology. The series is made up of a number of books either written by single authors, or edited, with a number of contributors. Well-known authorities from industry, academia, medicine, and the government are participating in writing these books.

Our aim is to cover many facets of advances in the field of cosmetic science. Topics will be drawn from a wide spectrum of disciplines ranging from chemical, physical, analytical, and consumer evaluations to safety, efficacy, and regulatory questions. Organic, inorganic, physical and polymer chemistry, emulsion technology, microbiology, toxicology, dermatology, and more — all play a role in cosmetic science. There is little commonality in the scientific methods, processes, or formulations required for the wide variety of cosmetics and toiletries manufactured. Products range from hair care, oral care, and skin care products to lipsticks, nail polishes, deodorants, powders, aerosol products to over-the-counter products such as antiperspirants, dandruff treatments, antibacterial soaps, acne creams, and suntan lotions. Thus cosmetics represent a highly diversified field with many subsections of science and art, for indeed, even today a lot of art and instinct is used and needed in the formulation and evaluation of cosmetics.

Emphasis is placed on reporting the current status of cosmetic technology and science in addition to historical reviews. The series will include books on preservatives, cosmetic safety, cosmetic product testing, hair and hair care, oral hygiene products, scientific and regulatory foundations of the cosmetic industry, and methods of cutaneous investigation. Contributions range from highly sophisticated and scientific treatises to practical applications and pragmatic presentations. Authors were encouraged to present their own concepts as well as established theories. They have been asked not to shy away from fields which are still in a state of development, nor to hesitate in presenting detailed discussions of their own work. Altogether, we intend to develop in this series a collection of critical surveys by noted experts covering most phases of the cosmetic business.

The first book in this series, entitled *Cosmetic and Drug Preservation: Principles and Practice* edited by Dr. Jon J. Kabara, contains contributions from many experts in the field. It provides an overview of product preservation ranging from the theory of the mechanism of preservative action to practical applications.

Finding effective and safe antimicrobials for a broad line of product categories presents a continuing challenge to chemists in the cosmetic industry. The need to be effective in oil and water systems that contain a myriad of ingredients, ranging from surfactants to proteins, oils, emulsifiers and a host of natural ingredients, often requires special preservative systems. This book includes most of the commercially available preservatives, test procedures, formulations, and safety considerations.

Finally, I want to thank all contributors and editors who are participating in the development of this series, the editorial staff at Marcel Dekker, Inc., and above all, my wife Eva, without whose editorial help and constant support I would never have undertaken this project.

Eric Jungermann, Ph.D.

FOREWORD

It is quite remarkable to review the recent literature in cosmetic preservation and realize how the subject has grown in importance and recognition in the industry.

It was not many years ago that the parabens and formaldehyde were just about the only preservatives used in cosmetics. While these still remain important, a host of additional antimicrobial products have been developed or discovered in response to some very important safety and efficacy requirements.

The major influences that have made cosmetic preservation a very important specialty are the following:

1. *Safety Considerations*: Historically, enough preservative was added to a product after it was formulated to eliminate any microbial growth and then a little extra was added for safety. The result was often levels that could provoke a reaction in consumers, although generally a very small percentage of the population. With the much greater awareness of the consumer reactions to cosmetics, the level of any one preservative is now reduced as low as possible by taking advantage of the synergistic effects of preservative combinations.
2. *Formula Complexity*: The simple formulas of years ago consisted of a few well-known chemical ingredients that were familiar to the formulator and whose performance were well known. Now some very sophisticated formulas with many more ingredients provide a greater challenge to the formulator. Preservative/ingredient interactions are far more common with the wide range of emulsifiers and surfactants now in common use in the industry.
3. *Natural Materials*: The great popularity of natural materials has added immeasureably to the problem of microbial growth in cosmetics. This is partly because many of these products have a heavy microbial load on arrival in the cosmetic factory, and they tend to be

very good growth mediums for bacteria. Many are heat sensitive and, therefore, cannot be sterilized before use. This is probably the major influence in the growing importance of multiple preservative systems.

4. *Consumer Awareness*: The consumer today is less likely to throw away and forget a package that had obviously spoiled en route from the factory to the home. Now complaints are registered both with the Food and Drug Administration and more often with the manufacturer. This has put greater pressure to reduce the costs of handling complaints and of customer dissatisfaction by doing an adequate job of preservation during product formulation.
5. *Product Recall*: The cost of a product recall can be tremendous and can wipe out years of profits from a particular individual product. This potential economic catastrophe justifies a great deal of more careful and extensive preservation testing before marketing.
6. *Resistant Organisms*: As the use of preservatives of all types proliferates throughout laboratories and factories, the development of resistant organisms simply was inevitable. This development has made the microbiologist's life much more difficult and challenging. Much of the impetus to the development of new preservatives and preservative systems has been this problem of traditional preservatives that are no longer as effective as they used to be.

Several proposals for avoiding the use of preservatives entirely have been put forward. Included is an attempt to market cosmetics through channels of distribution that handle refrigerated and frozen products. This is feasible, as long as the consumer recognizes that the product must be kept in a refrigerator at home. Thus far, there seems to be little acceptance for this principle on the part of the consumer.

Another solution to the preservative problem, or at least reducing its complexity, is the single-use package. While there has been some marketing of single-use packaging, this proves to be a relatively expensive method and has also been met with very limited acceptance by the consumer.

As long as there remains a vigorously growing mass market for personal care products, there will be a growing need for new development of antimicrobial agents to preserve a continually wider variety of the protective and decorative products which we call cosmetics and toiletries.

The greatest need is still a better understanding of the dynamics of product spoilage and appreciation of the importance of considering product preservation as an integral part of the product development process.

A complicating factor that is now being discussed in regulatory circles is the need for products to resist contamination during use by the consumer. It is not enough that the product arrives in the consumer's home in a well-preserved condition. After the package is opened, the consumer inevitably introduces microbial contamination. The ability of a product to resist this type of contamination will probably be the biggest challenge facing the cosmetic formulators in the near future.

The chapters of this book provide the most comprehensive collection of information on all aspects of cosmetic preservation that have ever been gathered together in one publication. It is certainly a testimony to the importance of this subject and the need to keep abreast of the latest developments.

Stanley E. Allured
Allured Publishing
Wheaton, Illinois

PREFACE

Chemicals, sterilizing gases, and irradiation have all been used to control microbial growth or promote sterility in cosmetics, toiletries, and pharmaceutical preparations. The application of these diverse methods did not always reflect optimal use levels or guaranteed products free from contamination. Because of these failures and the shrinking list of chemical preservatives available for investigation, there has been increased interest in the science of cosmetic preservation. Such heightened attention reflects the broader spectrum of products available on the market and the greater awareness of consumers to possible toxic or adverse effects. The unwanted effect may be due to toxins from contaminated products, allergic reactions, or toxicity of the preservative itself. As scientists our ability to deal with these issues can only be effective if there is a clear understanding of the basic principles and practices involved in the science of cosmetic formulas. The present book is an attempt to collate such technical information on cosmetic preservation both from world literature as well as from the experiences of experts.

This book is an effort to define the principles and practices currently in vogue in the area of cosmetic and drug preservation. In many cases these principles are useful to all industries where microorganisms are or can become a problem. The material in this book should therefore be of particular value to the microbiologist, formulator, fragrance specialist, and others in the cosmetic and drug industries who may need to understand what is involved in bringing a safe and sterile product to the marketplace.

The chapters comprising this work deal with various facets of cosmetic and drug preservation, ranging from general outlines of principles and concepts to specific, often cookbook-approach, problem-solving instructions. Since there is no one best way to solve problems dealing with cosmetic and drug preservation, the editor tended to give full freedom of expression to the experts in presenting their individual chapters. In some cases this has led to a certain amount of duplication between chapter contents, but with little redundancy.

I gratefully acknowledge receiving permission from Stanley Allured for publishing two papers which appeared in Cosmetics and Toiletries. My thanks to Dr. M. Gay and his FIP committee for graciously allowing reproduction of their Report on the preservation test for pharmaceuticals. Special mention must be given to Ann D. Eschtruth, Ann Pepp, and Sue Pepp, whose secretarial services and support over the past three years lifted a great deal of burden from my shoulders and made the task of editor easier. Added appreciation is extended to Dr. Eric Jungermann and the staff of Marcel Dekker for their faith in believing the venture worthwhile.

Finally, I acknowledge the generous assistance and cooperation of the contributing authors, who were able to submit their manuscripts on time and without the usual "gentle" persuasion. Maybe the threat of meeting my collaborator, the Baron von Richthofen, my Doberman, had a positive influence on their promptness.

It is my hope that the critical reader will call to my attention any errors or shortcomings of the volume. While editorial license was exercised with restraint, the final responsibility for omissions, errors, and so on, for this first edition rests with me alone. Advice useful for the second edition will be most welcome.

Jon J. Kabara

CONTRIBUTORS

Philip A. Berke Sutton Laboratories, Inc., Chatham, New Jersey

Robert L. Bronaugh Division of Toxicology, U.S. Food and Drug Administration, Washington, D.C.

Betty Croshaw* Quality Control Department, The Boots Company PLC, Nottingham, England

Heinz J. Eiermann Division of Cosmetic Toxicology, U.S. Food and Drug Administration, Washington, D.C.

Stig E. Friberg Chemistry Department, University of Missouri-Rolla, Rolla, Missouri

J. R. Furr Welsh School of Pharmacy, University of Wales Institute of Science and Technology, Cathays Park, Cardiff, Wales

M. Gay Hoffman-LaRoche and Co., Ltd., Basel, Switzerland

Thomas E. Haag Drug and Cosmetic Chemicals Division, Mallinckrodt, Inc., St. Louis, Missouri

Allen L. Hall Cincinnati Technical Center, Emery Industries, Inc., Cincinnati, Ohio

J. Roger Hart Organic Chemicals Division, W. R. Grace & Co., Lexington, Massachusetts

Victor R. Holland Quality Control Department, The Boots Company PLC, Nottingham, England

---

*Now retired in Burton Joyce, Nottingham, England

W. B. Hugo Department of Pharmacy, The University of Nottingham, Nottingham, England

Gene A. Hyde Olin Corporation, New Haven, Connecticut

G. P. Jacobs Department of Pharmacy, School of Pharmacy, Hebrew University of Jerusalem, Jerusalem, Israel

Jon J. Kabara Department of Biomechanics, Michigan State University, East Lansing, Michigan

Edward S. Lashen Research Division, Rohm and Haas Company, Spring House, Pennsylvania

F. Lautier Institut d'Hygiène et de Médecine Préventive, Faculty de Médecine, Strasbourg, France

Andrew B. Law Research Division, Rohm and Haas Company, Spring House, Pennsylvania

Donald F. Loncrini Drug and Cosmetic Chemicals Division, Mallinckrodt, Inc., St. Louis, Missouri

O. J. Lorenzetti Dermatology and Research Development, Alcon Laboratories, Inc., Fort Worth, Texas

Joseph M. Madden Division of Microbiology, U.S. Food and Drug Administration, Washington, D.C.

Howard I. Maibach Department of Dermatology, University of California Medical School, San Francisco, California

S. Ross Marouchoc Microbiology Product Development, The Dow Chemical Company, Midland, Michigan

Terence J. McCarthy School of Pharmacy, University of Port Elizabeth, Port Elizabeth, Republic of South Africa

Jack N. Moss Research Division, Rohm and Haas Company, Spring House, Pennsylvania

John D. Nelson, Jr. Olin Corporation, New Haven, Connecticut

Donald S. Orth The Andrew Jergens Company, Cincinnati, Ohio

Malcolm S. Parker Department of Pharmacy, Brighton Polytechnic, Brighton, England

Donald Joseph Reinhardt Biology Department, Georgia State University, Atlanta, Georgia

C. Richards Pharmacy Department, Llandough Hospital, Cardiff, Wales

Marvin Rosen Glyco, Inc., Williamsport, Pennsylvania

William E. Rosen Sutton Laboratories, Inc., Chatham, New Jersey

A. D. Russell Welsh School of Pharmacy, University of Wales Institute of Science and Technology, Cardiff, Wales

F. Howard Schneider Bioassay Systems Corporation, Woburn, Massachusetts

Karl H. Wallhäusser Hoechst AG, Phama-Qualitatskontrolle, Frankfurt, West Germany

CONTENTS

About the Series   iii

Foreword (Stanley E. Allured)   v

Preface   vii

Contributors   ix

Part I: Basic Considerations for Cosmetic Preservation

1. Cosmetic Preservation: The Problems and the Solutions   3
   Jon J. Kabara

2. Microemulsions in Relation to Cosmetics and Their Preservation   7
   Stig E. Friberg

3. Composition and Structure of Microorganisms   21
   Jon J. Kabara

Part II: Chemical, Physical, and Microbiological Properties of Common Preservatives

4. Chemical Preservatives: Use of Bronopol as a Cosmetic Preservative   31
   Betty Croshaw and Victor R. Holland

5. Esters of para-Hydroxybenzoic Acid   63
   Thomas E. Haag and Donald F. Loncrini

6. Cosmetically Acceptable Phenoxyethanol   79
   Allen L. Hall

7. Phenols as Preservatives for Pharmaceutical and Cosmetic Products   109
   W. B. Hugo

8. Sodium and Zinc Omadine   115
   Gene A. Hyde and John D. Nelson, Jr.

9. Kathon CG: A New Single-Component, Broad-Spectrum Preservative System for Cosmetics and Toiletries   129
   Andrew B. Law, Jack N. Moss, and Edward S. Lashen

10. Dowicil 200 Preservative   143
    S. Ross Marouchoc

11. Glydant and MDMH as Cosmetic Preservatives   165
    Marvin Rosen

12. Germall 115: A Safe and Effective Preservative   191
    William E. Rosen and Philip A. Berke

Part III: Sterilant Gases and Radiation

13. Inactivation of Microorganisms by Lethal Gases   209
    C. Richards, J. R. Furr, and A. D. Russell

14. Gamma-Radiation Decontamination of Cosmetic Raw Materials   223
    G. P. Jacobs

Part IV: Use of Multifunctional Chemicals in Preservative Systems

15. Aroma Preservatives: Essential Oils and Fragrances as Antimicrobial Agents   237
    Jon J. Kabara

16. Medium-Chain Fatty Acids and Esters as Antimicrobial Agents   275
    Jon J. Kabara

17. Lauricidin: The Nonionic Emulsifier with Antimicrobial Properties   305
    Jon J. Kabara

Contents

18. Chelating Agents as Preservative Potentiators   323

       J. Roger Hart

19. Food-Grade Chemicals in a Systems Approach to Cosmetic Preservation   339

       Jon J. Kabara

Part V: Formulation Principles Involved in Cosmetic Preservation

20. Formulated Factors Affecting the Activity of Preservatives   359

       Terence J. McCarthy

21. Design and Assessment of Preservative Systems for Cosmetics   389

       Malcolm S. Parker

22. Evaluation of Preservatives in Cosmetic Products   403

       Donald S. Orth

23. The Test for the Effectiveness of Antimicrobial Preservation of Pharmaceuticals   423

       3rd Joint Report of the Committee of Official Laboratories and Drug Control Services and the Section of Industrial Pharmacists-Federation Internationale Pharmaceutique (FIP)

24. A Preservative Evaluation Program for Dermatological and Cosmetic Preparations   441

       O. J. Lorenzetti

25. Microbial Challenge and In-Use Studies of Periocular and Ocular Preparations   465

       Donald Joseph Reinhardt

Part VI: Safety, Toxicological, and Regulatory Issues

26. Dermal and Ocular Toxicity of Antiseptics: Methods for the Appraisal of the Safety of Antiseptics   483

       F. Lautier

27. Safety Evaluation of Cosmetic Preservatives   503

       Robert L. Bronaugh and Howard I. Maibach

28. Evaluation of Chemical Toxicology of Cosmetics   533

       F. Howard Schneider

29. Cosmetic Product Preservation: Safety and Regulatory Issues   559
   Heinz J. Eiermann

Part VII: Appendixes

Appendix A. Microbiological Methods for Cosmetics   573
   Joseph M. Madden

Appendix B. Antimicrobial Preservatives Used by the Cosmetic Industry   605
   Karl H. Wallhäusser

Index   747

Cosmetic and Drug Preservation

PART I
BASIC CONSIDERATIONS FOR COSMETIC PRESERVATION

1
COSMETIC PRESERVATION
The Problems and the Solutions

JON J. KABARA *Michigan State University, East Lansing, Michigan*

The term *preservation* will be limited to the protection of cosmetics, toiletries, and drugs from the effects of microbiological contamination. The problem seems simple enough. Just pick a germicidal agent which will kill "bugs," and add it to the product. Unfortunately there is more to the story, since in many cases chemicals which are highly active against microbes also have similar effects against mammalian cells. Therefore a balance needs to be established with the preservative of choice between killing organisms which may be in the cosmetic and injuring cells of the sonsumer who uses the product. In addition, there are a host of variables in the finished product which influence the effectiveness of preservatives active in laboratory media. Predictions of preservative strength based on information gained from laboratory growth media cannot be extrapolated to complex cosmetic formulae without rigorous testing.

Microbial contamination of cosmetics, toiletries, and pharmaceuticals is of grave concern to industry, regulating agencies, and the person most affected—the consumer. The problem, while not a new one, has dimensions which only recently have been fully realized [1]. Microbial spoilage can be caused by bacteria, yeast, or fungi, all of which are extremely versatile in their metabolic activities. The metabolic reactions of microorganisms can present a health hazard because the degradation of emulsion products can be toxic, mutagenic, and so on, while uncontrolled microbial growth can produce toxins and other metabolites which exhibit pharmacological activity [2].

Several species of microorganisms produce toxic molecules and may render a product dangerous if they grow under conditions supporting toxin production. Endotoxins, produced by gram-negative bacteria, are not necessarily inactivated by sterilization, because they are heat stable. Certain strains of *Staphylococcus aureus* produce a toxin, but the organism must grow to a density of several million cells per gram before its toxin becomes a problem.

It is well to remember that some end products of microbial growth can have subtle but important effects even when the organism itself is no longer present. For instance, the water supply which is contaminated but filtered before use can still cause product breakdown. While the filtered water is sterile, metabolic products, for example, toxins and enzymes, may still affect the product. In the case where high lipase activity is an end product of previous bacterial growth, product spoilage (hydrolysis) can occur even though the organism no longer exists. Also, filterable toxin or irritant can cause incidents of irritation following the application of the cosmetic. The contaminant may be a foreign protein which evokes an allergic dermatitis.

For pharmaceutical preparations microbial contamination could mean destruction or inactivation of the biologically active molecule. In toiletries the use of biodegradable detergents can have serious repercussions. Their surface-active properties could be lost owing to degradation of the surfactant by contaminating bacteria.

When microorganisms are actually observed in or on a product, there is no doubt that microbial spoilage occurred; in fact, it is probably the most common way in which spoilage is manifest to the consumer. These visible changes can occur in product color because of shifts in pH or redox potential, or other changes caused by metabolic activity of an organism. Gas production is visible by the presence of bubbles, frothing, an increase in pressure, or other manifestations. Another clue to product spoilage is the composition of a homogeneous product becoming visibly heterogeneous. Odoriferous effects may be produced by a variety of aroma-producing bacteria [3]. Other signs of spoilage of cosmetic material are a "peculiar" taste or changes in the texture of the material; for instance, cream may become lumpy or gritty, and changes in the viscosity of liquid preparations may occur.

The kinds of organisms capable of causing spoilage are varied, and each problem tends to be unique. Generalizations about susceptible products are likely to be oversimplified. The greater use of biodegradable ingredients solves an environmental problem, but vastly increases the difficulty of cosmetic preservation [4].

Although hundreds of chemicals can function as germicidal agents, only a handful of substances have made it to the marketplace. The small list is not based so much on a compound's effectiveness as an anitmicrobial agent as it is on the compound's safety and effectiveness when placed in the final product. The issue of safety is most important, since cosmetics are used for aesthetic reasons and should carry a low health risk to the consumer. Because of the paramount question of safety, much effort is now focusing on newer approaches to cosmetic preservation: use of natural products, application of food-grade chemicals, as well as the use of sterilizing gases and/or irradiation. While progress is being made, the ideal preservative has not been found. In fact, the solution to the problem of preservation is not necessarily to find more powerful germidices but, rather, to build into the product an environment hostile to microbial growth. This means that the development of a new formulation should include early involvement of all the team members at the start of the project. The formulator, fragrance person, microbiologist, and marketer should constitute a team effort in the search for the most acceptable solution to the project. This team's *modus operandi* should replace the current "domino" approach where each group works on the problem in a sequential manner and is usually insensitive to the problem of other project members. In addition, a consultant to the project should be introduced to the issues of the problem at the beginning rather than at the end of the project. Having been faced with product failure issues "after the fact," I have been made to feel more like a mortician rather than a consultant.

Of all the people on the team, the microbiologist is the most vulnerable, since he has the least number of ways to solve the preservative problem. The perfumer has literally thousands of chemicals from which to choose, the formulator has hundreds, whereas the microbiologist has only a handful of chemicals useful for cosmetic preservation. It is important, therefore, that fragrances or essential oils be used, when possible, that contribute antimicrobial activity to the finished product. In this same vein, the formulator should be aware of those surfactants and other ingredients which enhance antimicrobial activity or, in the very least, do not inhibit germicidal action [5,6].

A greater understanding of cosmetic preservation as a science needs to be acquired. Cosmetic preservation has progressed from an empirical science to one which is becoming more exact. More important, the cosmetic scientist is urged to carry out even initial studies on model cosmetic systems rather than laboratory culture media. Extrapolation of results from simple single-phase systems to complex cosmetic formulations have led to a number of misleading conclusions and disregard of potentially promising leads. An early example of such misinformation which is still prevalent today is the effect of nonionic surfactants on emulsion preservation [7]. The failure of many preservatives, especially members of the phenolic family, has been attributed to the formation of a complex between preservative and surfactant; the complex impairs the preservative activity of the "phenol." However, a classic but rarely quoted article by Charles and Carter [8] in 1969 emphasized the need to consider the cosmetic formulation as a whole when one is dealing with the problem of preservation. Their results with parabens reinforced the findings of studies which showed that preservatives were more effective in finished formulations than could have been predicted from their performance in simple component testing.

I strongly feel that the preservative scientist needs to understand the influences of and interactions between emulsion systems and the preservative(s), as well as the composition and structure of microorganisms (the villain in our story). The former is addressed directly in Chap. 2, by S. E. Friberg (Microemulsions in Relation to Cosmetics and Their Preservation), and the latter is covered in Chap. 3, by J. J. Kabara (Composition and Structure of Microorgansims). A clearer idea of the role of these and other variables in preservation will be gained as the reader becomes more familiar with other chapters in this book.

REFERENCES

1. C. W. Bruch, Cosmetics: sterility vs. microbial control, *Amer. Perfum. Cosmet.*, *86*:45 (1971).
2. N. J. Butler, The microbial deterioration of cosmetic and pharmaceutical products, in *Biodeterioration of Materials*, Elsevier, Amsterdam, (1968).
3. V. C. Omelianski, Aroma-producing microorganisms. *J. Bacteriol.*, *8*:393 (1923).
4. G. Sykes, Microbial contamination in pharmaceutical preparations for oral and topical use, *Indian J. Pharm.*, *31*:233 (1969).
5. C. Ishizeki, Studies on the antimicrobial activity of nonionic and anionic surfactants, *Bull. Nat. Inst. Hyg. Sci. (Jpn)*, *88:*75 (1970).
6. C. Ishizeki, S. Iwahora, T. Watanabe, F. Ono, M. Watanabe, and Y. Asaka, Studies on preservatives, especially on the inhibitory activity of polysorbate 80 and lecithin, *Bull. Nat. Inst. Hyg. Sci. (Jpn.)*, *91*:82 (1973).

7. I. R. Gucklhorn, Antimicrobials in cosmetics, TGA Cosmet. J., Fall:15 (1969).
8. R. D. Charles, and P. J. Carter, The effect of sorbic acid and other preservatives on organism growth in typical nonionic emulsified commercial cosmetics, *J. Soc. Cosmet. Chem.*, *10*:383 (1965).

2
MICROEMULSIONS IN RELATION TO COSMETICS AND THEIR PRESERVATION

STIG E. FRIBERG *University of Missouri-Rolla, Rolla, Missouri*

I. Abstract 7
II. Introduction 7
    A. Microemulsions and liquid crystals 9
    B. Microemulsions with nonionic surfactants 15
III. Summary 17
    References 17

I. ABSTRACT

Microemulsions are transparent emulsions that form spontaneously when the ingredients are brought into contact. The droplet size varies for different compositions in the range 25–1500 Å.

The microemulsions are usually stabilized by a combination of an ionic surfactant and a medium chain length alcohol such as pentanol. Such combinations are not useful for cosmetic or pharmaceutical purposes because of the strong irritation action of the alcohol.

This article shows the relation between the biological activity and structure of microemulsions, and it also presents the means of preparing microemulsions with biologically less harmful nonionic surfactants in a systematic manner.

II. INTRODUCTION

Under some conditions water and hydrocarbon may spontaneously form emulsions with a sufficiently small droplet size that cannot be observed in a microscope. The emulsions appear transparent to the eye. These kinds of emulsions have been known in industry for half a century [1]; their application in liquid fuels and polishes [2] is well known.

Their scientific evaluation began with Schulman [3], and his school developed a thesis about the importance of a temporary ultralow or even negative interfacial tension as an explanation for the spontaneous formation of the microdroplets [4–10]. This approach is obviously an oversimplication of a rather complex phenomenon [11–15], but it should be emphasized that an extreme lowering of the interfacial tension is certainly the key factor in the thermodynamics of microemulsion stability.

Apart from the theoretical evaluation of the stability of microemulsions, their properties have been the focus of an intense and extensive research interest during the last decade. This research has been prompted by the economic importance of tertiary oil recovery [16], and since an ultralow interfacial tension is necessary for the deformation of oil droplets to permit their transport through local constrictions in the pore network of the soil matrix, numerous treatments of this phenomenon and its cause have been published by the Texas school of Schechter and Wade [17–19].

The research on the properties of microemulsions has employed a series of investigative methods such as light [20–23] and neutron scattering [24], dielectric properties [25,26], electron microscopy [23,27,28], positron annihilation techniques [29,30], and phase equilibria [31–47]. Complete consensus has not been reached about the structure in all parts of the system, but the following description may be acceptable to a majority of researchers in the field.

The water–oil microemulsions begin as monomers or oligomers at low water concentrations and grow in size with the addition of water [27] to reach a level of 0.1–0.2 μm at high water contents. It should be emphasized that there is no sudden increase of the particle size in a narrow concentration range such as is the case in aqueous solutions with their critical micellization concentration.

The oil–water microemulsions [42] form an extension from the aqueous micellar solutions [41]; their exact structure has not yet been determined.

The fact that microemulsions, by definition, have a small particle size has an important bearing on the question of bacterial life in their presence [48]. Since bacteria cannot grow in pure fat or oil, the presence of water droplets is necessary for their viability. Water droplets that are too small do not provide the necessary environment for bacterial growth. In addition to this fact, the structural characteristics of the microemulsion system are harmful to the biological membranes of the bacteria. The conditions in a microemulsion favor the formation of small droplets of an amphiphilic substance. The bimolecular layer of the biomembrane must retain its lamellar organization in order to be functional. The relation between these two kinds of structures is best illustrated with the equilibria in systems stabilized by nonionic surfactants, and the following treatment will be limited to such systems.

These systems have been chosen also with regard to the fact that the temperature-dependent behavior of nonionic surfactants is the basis for modern techniques of low-energy emulsification [49,50]. In addition, it has to be realized that the conventional stabilization of microemulsions by a combination of an ionic surfactant and a medium chain length alcohol [2] leads to products that are entirely unsuited for pharmaceutical and cosmetical purposes because of the irritating action of the pentanol.

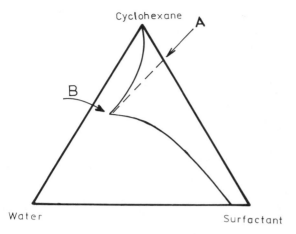

FIGURE 1 The solubility area in the system of water, pentaethylene glycol dodecyl ether (surfactant), and cyclohexane at 30°C. A hydrocarbon–surfactant weight ratio of 4:1 (A) permits addition of water to a final composition: water, 38.7%; surfactant, 12.3%; and hydrocarbon, 49.0%, by weight (B in the diagram).

Considering these factors, the treatment has been organized in the following manner. The first section describes the relationship between microemulsions and liquid crystals for a nonionic system and comparison is made with the structures found in biological membranes. The following section treats the system water–nonionic surfactant–hydrocarbon, showing the relationship between the temperature and phase behavior that gives oil–water and water–oil microemulsions.

A. Microemulsions and Liquid Crystals

The water–oil microemulsions in a system of nonionic surfactants are found by titration of hycarbon–surfactant solutions to appropriate turbidity with water. A typical solubility area is shown in Figure 1. The hydrocarbon, cyclohexane, with no surfactant (top corner) dissolves extremely small amounts of water, but the water solubility is rapidly increased with added surfactant. A hydrocarbon–surfactant solution marked by the arrow (A), gave a maximum water solubility with 38.7% (by weight) water, 12.3% surfactant, and 49.0% hydrocarbon (marked by arrow B in the Fig. 1). Such a composition is typical of water–oil microemulsion; the dissolved water requires approximately one-third its amount in surfactant.

The structure of the aggregates varies in the solubility area [51,52], but close to composition B the small water droplets may be depicted according to Figure 2. The surfactants are organized with their polar parts pointing toward the center of the spherical particle, where the water molecules form a pool that partly penetrates between the polar range of the surfactant chains. The contrast between this structure and those obtained at higher surfactant–hydrocarbon ratios has an important bearing on the interaction with biological microstructures.

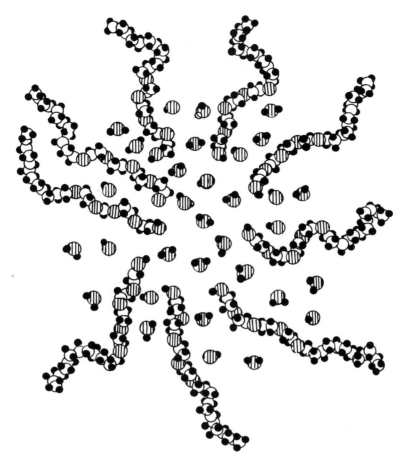

FIGURE 2 In a water—oil microemulsion droplet stabilized by a nonionic surfactant, the polar part of the surfactant points toward the water microdroplet, while the surfactant hydrocarbon chains are directed outward, the hydrocarbon-continuous medium.

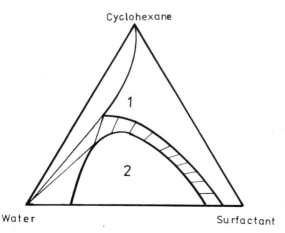

FIGURE 3 The water—oil microemulsion area (1) changes to a lamellar liquid crystalline phase (2) at high hydrocarbon—surfactant ratios.

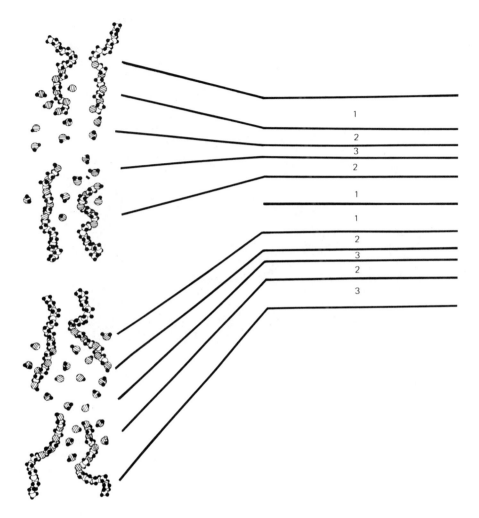

FIGURE 4 In the lamellar liquid crystalline phase (2, Fig. 3) the molecules are organized in "infinite" parallel layers, with the hydrocarbon parts pointing toward each other (layer 1), while the polar parts (layer 2) are separated and penetrated by aqueous layers (layer 3).

This is understood after completion of the phase diagram, as in Figure 3. It shows that a new phase appears when the amount of surfactant is increased. It is called a liquid crystal and its structure is shown in Figure 4. Instead of a liquid with colloidal spherical particles dispersed in it, as in the microemulsion, the colloidal structure is now lamellar, and it occupies the whole space. This means that there is no "continuous phase"; the entire volume is occupied by a network of layers such as those in Figure 4.

Such a structure will be optically anisotropic, a fact used to identify it. The lamellar structure gives the pattern observed in Figure 5. In addition, the lamellae will give a diffraction pattern with x-ray radiation; the reflections show a distance of 1:2:3:4. . . [53].

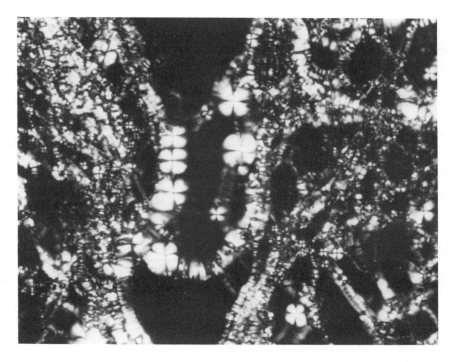

FIGURE 5 The optical pattern in polarized light by the lamellar liquid crystalline phase is typical.

This structure transfer has a decisive bearing on the interaction between microemulsions and living tissue such as microbes. These latter are protected by a membrane like all other cells. The membrane consists of outer layers of biopolymers and a central bimolecular layer of lipids. The main constituents of this bilayer are lecithins. These lecithins are triesters of glycerol with two carbons esterified to carboxylic acids and the third one to a phosphate group, which in turn is also esterified to an alcohol, typically choline (Fig. 6). The lecithin bilayers are organized according to Figure 7, (layer 2), with the hydrophobic part, the hydrocarbons, against each other, with the polar groups pointing toward the hydrophilic environment (layer 1 in Fig. 7).

The lecithin molecule has such a balance between the cross-sectional area of the polar group and the volume of the hydrocarbon part [54,55] that it spontaneously forms bilayers instead of micelles when mixed with water. The structure of the lamellar liquid crystal (Fig. 7) is closely related to the bimolecular layer of the biomembrane (Fig. 7). With some simplification, the

$$\underset{C}{\overset{R_1}{|}} \text{---} \underset{C}{\overset{R_2}{|}} \text{---} C\text{-}O\text{-}\underset{\underset{O}{|}}{\overset{\overset{O}{\|}}{P}}\text{-}O\text{-}C\text{-}C\text{-}N(R)_3$$

FIGURE 6 The lecithin molecule contains two fatty ester residues (R_1 and R_2). The third ester is a phosphate ester with cholin ($R = CH_3$).

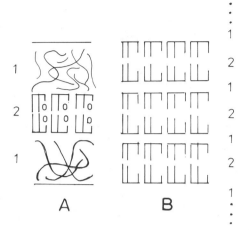

FIGURE 7 The principal structure of the biomembrane (A) consists of a lecithin (□)−chlesterol (0) bilayer (2) with the polar parts pointing toward aqueous layers (1) with biopolymers. This structure compares with the lecithin−water liquid crystalline phase (B) in which lecithin bilayers (2) are separated by aqueous layers (1).

lamellar liquid crystal could be described as the bimolecular layers stacked on each other.

The relation between microemulsions and biological structures is now evident after a comparison of Figures 2 and 7. The microemulsion, by definition, will have a structure that will damage the bimolecular layer of the biomembrane. This is evident from the following discussion. The balance between the area of the polar group and the volume of the hydrocarbon chain of the surfactant [54] in a microemulsion is such that the structure of a small spherical drop is preferred (Fig. 3). This means that the lamellar arrangement (Fig. 7) that is a necessity for the biomembrane cannot persist in the presence of a microemulsion. The lamellar arrangement (Fig. 7) will be broken and its amphiphilic molecules will be part of the microemulsion structure forming small spherical particles (Fig. 2). The incompatibility between the two systems appears evident, and their relations should be investigated more thoroughly in order to clarify an important phenomenon.

It should be emphasized that the incompatibility between microemulsions and microbes is not necessarily a result of the biological aggressivity of the compounds per se. This happens to be the case for the microemulsions with pentanol as a cosurfactant; pentanol in itself is aggressive to biological tissue. But to illustrate the main point, nonionic surfactants such as pentaethylene glycol dodecylether in their native state will be rapidly biodegraded if left without protection [56]. The surfactant per se has no biological activity, at least no harmful biological activity. It is the combination of water, hydrocarbon, and surfactant (Fig. 1) that gives the hydrophilic−lipophilic balance such that the kinds of structures (Fig. 2) formed destroy the lamellar organization (Fig. 7).

These structural relations have so far been little studied, but their importance is obvious. In the following sections some simple rules for microemulsion compositions with biologically harmless substances will be given.

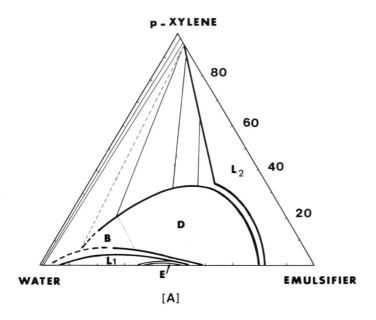

FIGURE 8 A multitude of phases are often observed in a system of water, nonionic surfactant (emulsifier), and hydrocarbon. (From Ref. 66.)

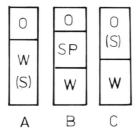

FIGURE 9 The temperature-dependent behavior of the system water–nonionic surfactant (~3%)–hydrocarbon is exemplified by the phases in the three test tubes. At low temperatures (A) two phases are found. The aqueous phase (W) contains the greater part of the surfactant (S), while the oil phase contains only small amounts of it and water. In an intermediate temperature range (B) an isotropic liquid phase, the surfactant phase, is formed, containing most of the surfactant as well as oil and water in an approximate weight ratio of 1. At high temperatures (C) the surfactant is found in the oil phase with solubilized water, while the aqueous phase is water with extremely small amounts dissolved of the two other components.

B. Microemulsions with Nonionic Surfactants

The formulation of microemulsions stabilized by nonionic surfactants can best be understood by a study of the temperature dependence to the total phase diagrams [57-62]. These show a simple and systematic behavior for aliphatic hydrocarbons, and these will primarily be used in the following treatment. The total phase diagrams may show a multitude of phases (Fig. 8) and may at first give a bewildering impression [63]. However, the microemulsions are only a part of the diagrams and the following description facilitates the formulation of these microemulsions.

The key factor from an experimental point of view is to find the hydrophilic—lipophilic balance (HLB) temperature [64]. This is achieved by observation of the temperature dependence of the phases in a water—hydrocarbon mixture with 5% by weight of a nonionic surfactant. The behavior is illustrated in Figure 9. At low temperatures two solutions are in equilibrium with each other. The surfactant is found in the aqueous solutions (Fig. 9a). At high temperatures the surfactant is not soluble in the aqueous phase, since the temperature of the cloud point has been exceeded, and the two solutions are now water with an extremely low concentration of surfactant and the hydrocarbon with the surfactant and some solubilized water. At intermediate temperatures three solutions are found in equilibrium. This level is the HLB temperature, the key to the formulation of microemulsions stabilized by nonionic surfactants. However, before this formulation is described, a word of caution about the experiments is called for. The middle phase in the sample, called the surfactant phase, may have a particle size of $0.1-0.2$ μm and hence be translucent. It cannot be overemphasized that care must be taken to ensure equilibrium with the other two phases. It often happens that a stable, turbid emulsion is collected between the oil and the aqueous phase. This emulsion must not be confused with the surfactant phase; an emulsion is a multiphase system! If a stable emulsion is obtained during the agitation that is necessary to obtain equilibrium, it must be destabilized before continuation. This destabilization may be difficult; in our experience heat and cold shocks are most efficient.

The relation between the conditions at the HLB temperature and the microemulsions can be understood when the variation in phase regions at higher and lower temperatures has been described. The phase regions at the HLB temperature are shown in Figure 10. The surfactant phase occupies a region in the central part of the diagram, being in equilibrium with (1) an aqueous phase with extremely small amounts of surfactant and hydrocarbons dissolved, and (2) a hydrocarbon phase with about 2% surfactant and about 1% water.

A reduction of the temperature (Fig. 10b) leads to a transfer of the surfctant phase region toward the aqueous corner and a coalescence between the two phases. The new combined phase will reach sectorially from the aqueous corner, covering a limited hydrocarbon—surfactant ratio range. *This is an oil—water microemulsion.*

An increase of the temperature (Fig. 10c) leads to a corresponding shift of the surfactant phase region toward the hydrocarbon phase. The result will be a region extending from the hydrocarbon phase, a *water—oil microemulsion region.*

These regions are easy to detect by titrating with water or hydrocarbon and noting the points of clarity and turbidity. The microemulsions formed are thermodynamically stable; however, they suffer from a serious disadvantage: Their temperature stability is totally insufficient for practical application.

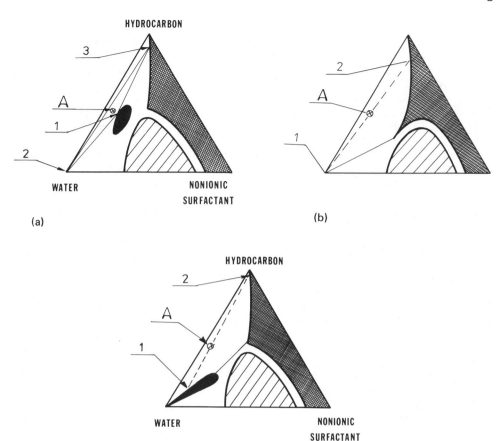

FIGURE 10 At the HLB temperature (a) the sample (A) will be separated into three liquid phases: (1) the surfactant phase, (2) the aqueous phase, and (3) the hydrocarbon phase (compare Fig. 9b). At lower temperatures (b) the surfactant phase region has transferred to the aqueous corner, forming a sectorial solubility region (1). The sample (A) will now separate into two phases: (1) an aqueous phase and (2) a hydrocarbon phase (compare Fig. 9a). At temperatures above the HLB temperature the surfactant phase region has coalesced with the hydrocarbon–surfactant solution (c) (checkered), and the sample (A) will separate into (1) a pure water phase and (2) a hydrocarbon solution with the surfactant and dissolved water.

The temperature range in which a certain formulation remains stable is of the magnitude of 5–10°C.

This problem can be remedied by adding another surfactant. Combinations with ionic surfactants [65] give extremely interesting results, but only preliminary data are available so far. A more extensive study of combinations with lecithin has been made [62], with encouraging results. Oil–water microemulsions were prepared with a temperature range for stability of more than 25°C. It appears highly probable that with a limited amount of development the temperature range could be extended to enhance the hydrocarbon–surfactant ratio.

III. SUMMARY

The structure of microemulsions and their relation to the bimolecular layer of the biomembrane were discussed. Emphasis was given to the fact that substances which are biologically harmless per se may, if their hydrophilic–lipophilic balance is changed by the addition of water and hydrocarbon, form structures that strongly interact with the bimolecular layer of the biomembrane. The preparation of microemulsions stabilized by nonionic surfactants was described.

REFERENCES

1. V. R. Kokatner, U.S. Patent 2,111,100 (1935).
2. L. Prince, ed., *Microemulsions: Theory and Practice*, Academic, New York (1967).
3. T. P. Hoar, J. H. Schulman, Transparent water-in-oil dispersions: the oleopathic hydro-micelle, *Nature*, 152:102 (1943).
4. J. H. Schulman, and D. P. Riley. X-ray investigation of the structure of transparent oil-water disperse systems. I, *J. Colloid Sci.*, 3:383 (1948).
5. J. L. Bowcott, and J. H. Schulman. Emulsions—control of droplet size and phase continuity in transparent oil–water dispersions stabilized with soap and alcohol, *Z. Electrochem.*, 59:283 (1955).
6. D. I. Sears, and J. H. Schulman. Influence of water structure on the surface pressure, surface potential, and area of soap monolayers of lithium, sodium, potassium, and calcium, *J. Phys. Chem.*, 68:3529 (1964).
7. D. O. Shah, and R. M. Hamlin. Structure of water in microemulsions: electrical, birefringence, and nuclear magnetic resonance studies, *Science*, 171:483 (1971).
8. H. L. Rosano, N. Schiff, and J. H. Schulman. Molecular interactions between phospholipids and salts at air and liquid–liquid interfaces, *J. Phys. Chem.*, 66:1928 (1962).
9. L. M. Prince. A theory of aqueous emulsions. I. Negative interfacial tension at the oil/water interface, *J. Colloid Interface Sci.*, 23:165 (1967).
10. H. L. Rosano. Microemulsions, *J. Soc. Cosmet. Chem.*, 25:609 (1974).
11. E. Ruckenstein, and J. C. Chi. Stability of microemulsions, *J. Chem. Soc. Faraday Trans. II*, 71:1690 (1975).
12. H. Reiss. Entropy-induced dispersion of bulk liquids, *J. Colloid Interface Sci.*, 53:61 (1975).
13. E. Ruckenstein. Thermodynamics of microemulsification with ionic surfactants, *J. Dispersion Sci. Technol.*, 2:1 (1981).
14. A. M. Bellocq, D. Bourbon, and B. Lemanceau. Thermodynamic, interfacial and structural properties of alcohol–brine–hydrocarbon systems, *J. Dispersion Sci. Technol.*, 2:27 (1981).
15. J. Biais, P. Bothorel, B. Clin, and P. Lalanne. Theoretical behaviour of microemulsions: geometrical aspects, dilution properties and role of alcohol. Comparison with experimental results, *J. Dispersion Sci Technol.*, 2:67 (1981).

16. D. O. Shah, and R. S. Schechter, eds., *Improved Oil Recovery by Surfactant and Polymer Flooding*, Academic, New York (1967).
17. J. L. Cayias, R. S. Schechter, and W. H. Wade. Modeling crude oils for low interfacial tension, *Soc. Pet. Eng. J.*, *16*:351 (1976).
18. P. H. Doe, W. H. Wade, and R. S. Schechter. Alkyl benzene sulfonates for producing low interfacial tension between hydrocarbons and water, *J. Colloid Interface Sci.*, *59*:525 (1977).
19. P. H. Doe, M. El-Emary, and R. S. Schechter. Surfactants for producing low interfacial tensions: II. Linear alkylbenzenesulfonates with additional alkyl substituents, *J. Am. Oil Chem. Soc.*, *55*:505 (1978).
20. E. Gulari, B. Bedwell, and S. Alkhafaji. Quasi-elastic light-scattering investigation of microemulsions, *J. Colloid Interface Sci.*, *77*:202 (1980).
21. A. M. Bellocq, and G. Fourche. Micellization study of two microemulsion systems by Rayleigh scattering, *J. Colloid Interface Sci.*, *78*:275 (1980).
22. R. Finsy, A. Devriese, and H. Lekkerkerker. Light scattering study of the diffusion of interacting particles, *J. Chem. Soc. Faraday Trans. II*, *76*:767 (1980).
23. E. Sjöblom, and S. E. Friberg. Light-scattering and electron microscopy determinations of association structures in W/O microemulsions, *J. Colloid Interface Sci.*, *67*:16 (1978).
24. R. Ober and C. Taupin. Interactions and aggregation in microemulsions. A small-angle neutron scattering study, *J. Phys. Chem. 84*:2418 (1980).
25. C. Boned, M. Clausse, B. Lagourette, V. E. R. McClean, and Sheppard. Detection of structural transitions in water-in-oil microemulsion-type systems through conductivity and permittivity studies, *J. Phys. Chem.*, *84*:1520 (1980).
26. V. K. Bansal, K. Chinnaswamy, C. Ramachandran, and D. Shah. Structural aspects of microemulsions using dielectric relaxation and spin label techniques, *J. Colloid Interface Sci.*, *72*:524 (1979).
27. Y. Talmon, H. T. Davis, L. E. Scriven, and E. L. Thomas. Cold-stage microscopy system for fast-frozen liquids, *Rev. Sci. Instrum.*, *50*:698 (1979).
28. J. Biais, M. Mercier, P. Lalanne, B. Clin, A. M. Bellocq, and B. Lemanceau. Electron microscopy study of freeze-fractured microemulsions, *C. R. Acad. Sci. Ser. C*, *285*:213 (1977).
29. A. Boussaha, and H. J. Ache. Microemulsion formation in sodium stearate−alcohol−hexadecane−water mixtures: studies by positron annihilation techniques, *J. Colloid Interface Sci.*, *78*:257 (1980).
30. A. Boussaha, B. Djermouni, L. A. Fucugauchi, and H. Ache. Microemulsion systems studied by positron annihilation techniques, *J. Am. Chem. Soc.*, *102*:4654 (1980).
31. E. Ruckenstein, and R. Krishnan. Effect of electrolytes and mixtures of surfactants on the oil−water interfacial tension and their role in formation of microemulsions, *J. Colloid Interface Sci.*, *76*:201 (1980).
32. C. Tondre, and R. Zana. Rate of dissolution of n-alkanes and water in microemulsions of water/sodium dodecyl sulfate (1/3) + 1-pentanol (2/3)/n-dodecane in rapid mixing experiments, *J. Dispersion Sci. Technol.*, *1*:179 (1980).
33. B. Lindman, N. Kamenka, T. Kathopoulis, B. Brun, and P. G. Nilsson, Translational diffusion and solution structure of microemulsions, *J. Phys. Chem.*, *84*:2485 (1980).

34. A. Vrij, E. A. Nieuwenhuis, H. M. Fijnaut, and W. G. M. Agterof. Application of modern concepts in liquid state theory to concentrated particle dispersions, *J. Chem. Soc. Faraday Discuss.*, 65:101 (1978).
35. J. T. G. Overbeek. Microemulsions, a field at the border between lyophobic and lyophilic colloids, *J. Chem. Soc. Faraday Discuss.*, 65:7 (1978).
36. S. Friberg, and I. Burasczewska. Microemulsions in the water—potassium oleate—benzene system, *Prog. Colloid Polym. Sci*, 63:1 (1978).
37. D. G. Rance, and S. Friberg. Micellar solutions versus microemulsions, *J. Colloid Interface Sci.*, 60:207 (1977).
38. J. Biais, P. Botherel, B. Clin, and P. Lalanne. Theoretical behavior of microemulsions: geometrical aspects and dilution properties, *J. Colloid Interface Sci.*, 80:136 (1981).
39. H. F. Eicke, and J. Markovic. Temperature-dependent coalescence in water—oil microemulsions and phase transitions to lyotropic mesophases, *J. Colloid Interface Sci.*, 79:151 (1981).
40. J. Sjöblom, J. Model calculations of aggregation numbers and radii of water/oil microemulsions, *Colloid Polym. Sci.*, 258:1164 (1980).
41. M. Podzimek and S. E. Friberg. O/W microemulsions, *J. Dispersion Sci. Technol.*, 1:341 (1980).
42. H. F. Eicke. Aggregation in surfactant solutions: formation and properties of micelles and microemulsions, *Pure Appl. Chem.*, 52:1349 (1980).
43. V. K. Bnsal, D. D. O. Shah, and J. P. O'Connell. Influence of alkyl chain length compatibility on microemulsion structure and solubilization, *J. Colloid Interface Sci.*, 75:462 (1980).
44. N. Borys, S. Holt, and R. Barden. Detergentless water/oil microemulsions. III. Effect of KOH on phase diagram and effect of solvent composition on base hydrolysis of esters, *J. Colloid Interface Sci.*, 71:526 (1979).
45. H. F. Eicke. On the cosurfactant concept, *J. Colloid Interface Sci.*, 68:440 (1979).
46. Y. Talmon and S. Prager. Statistical mechanics of microemulsions, *Nature*, 267:333 (1977).
47. L. E. Scriven. Equilibrium bicontinuous structures, *Proc. Int. Symp. Solubilization Microemulsions*, 2:877 (1977).
48. Grindsted Products, Technical Memorandum, TM 5-le.
49. T. J. Lin. Low energy emulsification. I. Principles and applications, *J. Soc. Cosmet. Chem.*, 29:117 (1978).
50. T. J. Lin, T. Akabori, S. Tanaka, and K. Shimura. Low energy emulsification. IV. Effect of emulsification temperature, *Cosmet. Toilet.*, 96:31 (1981).
51. H. Christenson, and S. E. Friberg. Spectroscopic investigation of the mutual interactions between nonionic surfactant, hydrocarbon, and water, *J. Colloid Interface Sci.*, 75:276 (1980).
52. H. Christenson, D. W. Larsen, and S. E. Friberg. NMR investigations of aggregation of nonionic surfactants in a hydrocarbon medium, *J. Phys. Chem.*, 84:3633 (1980).
53. K. Fontell. *Liquid Crystals and Plastic Crystals*, Vol. 2 (G. W. Gray and P. A. Winsor, eds.), Ellis Harwood, Chichester (1974), p. 80.
54. J. N. Israelachvili, D. J. Mitchell, and B. W. Ninham. Theory of self-assembly of hydrocarbon amphiphiles into micelles and bilayers, *J. Chem. Soc. Faraday Trans. II*, 72:1525 (1976).

55. B. W. Ninham, and D. J. Mitchell. Micelles, vesicles and microemulsions, *J. Chem. Soc. Faraday Trans. II*, *76*:201 (1980).
56. M. J. Schwuger. Solubilization properties of nonionic surfactants, *Kolloid Z. Z. Polym.*, *250*:703 (1972).
57. K. Shinoda, and T. Ogawa. The solubilization of water in nonaqueous solutions of nonionic surfactants, *J. Colloid Interface Sci.*, *24*:56 (1967).
58. H. Saito, and K. Shinoda. The stability of W/O type emulsions as a function of temperature and of the hydrophilic chain length of the emulsifier, *J. Colloid Interface Sci.*, *32*:647 (1970).
59. K. Kuriyama. Temperature dependence of micellar weight of nonionic surfactant in the presence of various additives. Part I. Addition of n-decane or n-decanol, *Kolloid Z.*, *180*:55 (1962).
60. A. Kitahara, T. Ishikawa, and S. Tanimori. The study on solubilization of water from its vapor pressure over nonionic and ionic surfactant solutions in various nonpolar solvents, *J. Colloid Interface Sci.*, *23*:243 (1967).
61. S. Friberg, and I. Lapczynska. Microemulsions and solubilization by nonionic surfactants, *Prog. Colloid Polym. Sci.*, *56*:16 (1975).
62. S. Friberg, I. Lapczynska, and G. Gillberg. Microemulsions containing nonionic surfactants—the importance of the PIT value, *J. Colloid Interface Sci.*, *56*:19 (1976).
63. S. Friberg, L. Mandell, and K. Fontell. Mesomorphous phases in systems of water—nonionic emulsifier—hydrocarbon, *Acta Chem. Scand.*, *23*:1055 (1969).
64. K. Shinoda, and H. Arai. The correlation between phase inversion temperature in emulsion and cloud point in solution of nonionic emulsifier, *J. Phys. Chem.*, *68*:3485 (1964).
65. G. Gillberg, L. Eriksson, and S. Friberg. New microemulsions using nonionic surfactants and small amounts of ionic surfactants, in *Emulsions, Lattices, Dispersions* (Becher and Yudenfreund, eds.), Marcel Dekker, New York (1978), p. 201.
66. S. Friberg and L. Mandell, Phase equilibrum and their influence on the properties of emulsions, *J. Am. Oil. Chem. Soc.* 47:149 (1970).

3
COMPOSITION AND STRUCTURE OF MICROORGANISMS

JON J. KABARA *Michigan State University, East Lansing, Michigan*

I. Introduction 21
II. Cell Membranes 22
III. Cell Wall 23
    References 26

I. INTRODUCTION

In order to interpret the mechanism of action of cosmetic preservatives, it is well to understand the composition and structure of microorganisms. For the sake of this discussion the focus of primary concern will be on cell walls and membranes—the outer periphery of the organism. It is the cell surface, more than any other single feature, that differentiates cells from each other and is a focus for interpreting effects caused by germicidal agents, particularly lipophilic compounds. The cell envelope becomes a rate-limiting barrier for the selective partition of nutrients, waste products, and other chemicals. These outer envelopes protect the internal milieu of the organism from hostile chemicals. Since the normal function of the cell is so strongly dependent upon the integrity of the outer wall, the barrier structures and compositions of prokaryotic and eukaryotic organisms will be discussed. Only a broad discussion of the factors involved can be given here. The reader is advised to obtain more specific information on the properties and structural basis of membranes from several excellent sources [1-4].

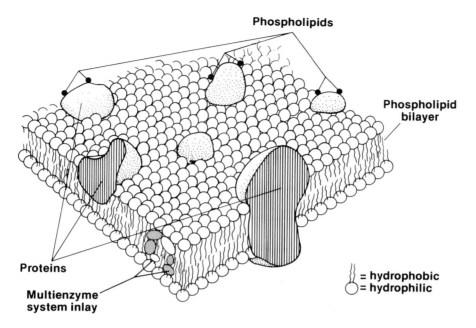

FIGURE 1 Schematic for a biological membrane.

II. CELL MEMBRANES

The plasma or cell membrane is a thin structure that completely surrounds the cell. This structure is a critical barrier vital to the function of the cell. The permeability of the cell is important to growth and survival. If membrane function is altered, both reversible and irreversible changes may lead to the death of the cell.

The main components of the plasma membrane are phospholipids and proteins. The phospholipid molecules disperse themselves in water in such a way that the water-insoluble (hydrophobic) groups associate. This attraction leads to the formation of a double-layered membrane or bimolecular leaflet (Fig. 1). The major proteins of the membrane are hydrophobic and become embedded in the phospholipid matrix. The structure of the plasma membrane is stabilized mainly by hydrogen and hydrophobic bonding. Cations such as Mg^{2+} and Ca^{2+} combine with phospholipids and help stabilize the membrane structure.

The plasma membrane is a barrier to the penetration of materials and is involved in the select transportation or uptake of materials. In general, small nonpolar and lipophilic substances may penetrate cell membranes readily by becoming dissolved in the lipid phase of the membrane. Ionized molecules do not readily bridge the membrane barrier because of charge or size.

The specific effect of fatty acids and fatty acid esters (lipophilic germicides) on cell membranes and cell walls will be discussed in a later chapter dealing with their mechanisms of action.

III. CELL WALL

One of the more important structural features which distinguish prokaryotic from eukaryotic organisms is the cell wall. This difference is important to understanding the effects of various biocidal agents. The prokaryotic gram-negative and gram-positive cells and the eukaryotic fungal and yeast cells differ considerably in cell wall structure. The differnces between bacteria and fungi are listed in Table 1.

The cell walls of gram-negative and gram-positive organisms also have significant structural and chemical compositions which explain the differential susceptibility to lipophilic preservatives (Fig. 2). Gram-positive organisms are simpler in structure than gram-negative organisms. In gram-positive bacteria as much as 90% of the wall consists of peptidoglycan. In gram-negative bacteria the same wall contains only 5–20% peptidoglycan, the rest consisting of lipids, polysaccharides, and proteins, usually in a layer outside the peptidoglycan layer.

The most important distinguishing feature of between gram-positive and gram-negative organisms is the latter's additional lipopolysaccharide and protein layer (Fig. 2). While Table 2 lists the differences between gram-positive and gram-negative bacteria, it is the lipopolysaccharide layer that accounts for gram-negative bacteria being more resistant to the effects of preservatives than gram-positive bacteria. In this regard yeast and fungi are also gram-positive organisms. It is not the chemical constituents alone that determine susceptibility to lipophilic preservatives, but also the physical structure of the wall.

The effect of preservative agents on a bacterial cell varies according to the microorganisms used. This is understandable in terms of the composition of the cell wall, which varies from species to species and even from strain to strain.

TABLE 1 Differences Between Bacteria and Fungi

Property	Bacteria	Fungi
Cell size (μm)	0.5–5	Yeasts, 20–50; molds, $\geqslant$100
Cell wall	Teichoic acids, muramic acid, peptides	Chitin, glucans, mannans, diamino pimelic acid
Cytoplasmic membrane	No sterols	Sterols
Cytoplasm	No mitochondria or endoplasmic reticulum	Mitochondria and endoplasmic reticulum
Nucleus	Prokaryotic (no membrane)	Eukaryotic (defined membrane)

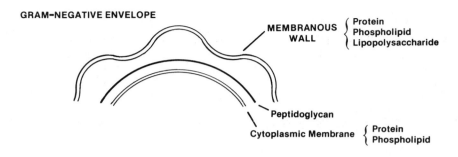

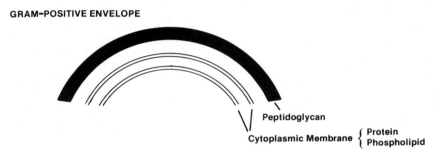

FIGURE 2 Structure of the cell envelopes of gram-negative and gram-positive bacteria.

TABLE 2 Differences Between Gram-Positive and Gram-Negative Bacteria

Property or constituent	Gram-positive bacteria	Gram-negative bacteria
Liposaccharide outer layer	Absent	Present
Polysaccharide	High	Low
Lipids	2–4%	15% or more
Solubility of lipids in fat solvents	Resistant	Less resistant
Amino acids	4–5	Complete range
pH of optimal growth	Relatively high	Relatively low
Nature of toxin	Exotoxin	Endotoxin
Bacteriostatic action of fatty acids and derivatives	Susceptible	Resistant

In addition to structural dissimilarities, there are striking differences between the cell wall compositions of gram-positive and gram-negative organisms [5]. The walls of gram-positive organisms, after removal of surface protein, contain three of four amino acids (aromatic and sulfur-containing amino acids are absent) [6], whereas the walls of gram-negative organisms contain a variety of amino acids comparable to those found in proteins [7]. Many gram-positive organisims possess a higher amino sugar content and, in general, a much lower lipid content than gram-negative organisims [17]. Nucleic acids are not present in the isolated cell wall of either type of organism.

Although it has been claimed that walls of gram-positive organisms contain only small amounts (2–4%) of lipid [8], in contrast to the high (15%) content of gram-negative cells, it must be pointed out that the amount of cell wall lipid is very dependent on the nature of the growth medium [9–12]. Furthermore, a number of gram-positive strains resistant to antibiotics or bacteriostatic agents have been isolated and found to contain large quantities of cell wall lipids [12–14]. Small amounts of cell wall lipids are not, therefore, a diagnostic feature of gram-positive organisms.

The capsular material of gram-negative organisms is chemically more complex than that of gram-positive organisms and includes polysaccharides, protein—polysaccharide—lipid complexes, and lipopolysaccharides. The capsular material of gram-positive organisms is, in general, a simple polysaccharide or a polypeptide (Fig. 2).

The cell walls of yeast and fungi are composed of two of more protein polysaccharide complexes held together by a variety of covalent bonds. The presence of mannan as a cell wall component distinguishes yeasts from other fungi (Fig. 3). The chemical composition of fungal cell walls has been reviewed by Bartnicki [15]. A close correlation was drawn between wall composition and toxonomic grouping based on morphology. A detailed review of the chemical composition of yeast cell walls has appeared [16].

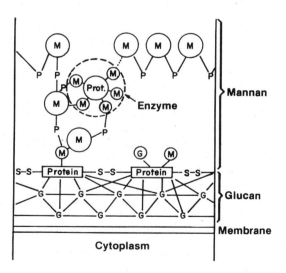

FIGURE 3 Schematic structure of the yeast cell wall: (M) mannan, (G) glucan, (P) phosphates, and (S) sulfur. [From J. O. Lampen, *Antonie von Leewenhoek, J. Microbiol. Serol.*, 34:1 (1968)].

All this information about the overall composition of the cell wall and capsule has been obtained by gross chemical destruction, for example, acid or alkaline hydrolysis, of the capsular isolated cell wall material, followed by analysis and identification of the individual components. The results of such chemical treatments, however, give no clear indication of the orientation or arrangement of the component molecules throughout the cell wall. Furthermore, and possibly more important, the chemical description alone provides no information on the nature or orientation of the groups or components at the surface, that is, the fluidity of the membrane. Such information is essential for the comprehension of the processes associated with biological activity. The cell surface is a barrier whereby the living organism establishes with its environment the level of substrates required for cell metabolism, waste products, and toxins produced by the complex chain reactions occurring within. It is the first and often final site of attack by drugs, antibiotics, and particularly surface-active agents. In the case of biologically active lipophilic agents, the cell surface is the most important site of action. This will be substantiated in later sections dealing with the mechanism of action of various lipophilic germicidal preservatives, as well as the role of ethylenediaminetetraacetic acid.

REFERENCES

1. L. Bolis, A. Katchalsky, R. D. Keynes, W. R. Lowewenstein, and B. A. Pethica, eds., *Permeability and Function of Biological Membranes*, North-Holland, London (1970).
2. A. B. Pardee, The surface membrane as a regulator of animal cell division, *In Vitro*, 7:95 (1971).
3. W. O. Russell, Ninth Annual ASCP Research Symposium on Cell Membrane and Disease, *Am. J. Clin. Pathol.*, 63:618 (1975).
4. B. Deuticke, Properties and structural basis of simple diffusion pathways in the erythrocyte membrane, *Rev. Physiol. Biochem. Parmacol.*, 78:1 (1977).
5. M. R. J. Salton, Studies of the bacterial cell wall. IV. Composition of the cell walls of some gram-positive and gram-negative bacteria, *Biochim. Biophys. Acta*, 10:512 (1953).
6. C. S. Cummins, and H. Harris, Cell wall composition and toxonomy of Actinomycetales and related groups, *J. Gen. Microbiol.*, 14:583 (1958).
7. M. R. J. Salton, *Bacterial Anatomy*, (E. T. C. Spooner and B. A. D Stocker, eds), Cambridge University Press, Cambridge (1956), p. 81.
8. M. R. J. Salton, in *The Bacteria*, Vol. 1, Cambridge (I. C. Gunsalus and R. Y. Stanier, eds.) Academic, New York (1960) p. 97.
9. L. W. Larson, and W. P. Larson, Factors governing the fat content of bacteria and the influence of fat on pellicle formation, *J. Infect. Dis.*, 31:407 (1922).
10. M. T. Kyar, and E. J. Ordal, Electrokinetic studies on bacterial surfaces. (1) Effects of surface-active agents on the electrophoretic mobilities of bacteria, *J. Bacteriol.*, 51:149 (1946).
11. C. Larche, Electrophoresis of *Micrococcus pyogenes*, *Acta Pathol. Microbiol. Scand. Suppl.*, 98:1 (1953).
12. M. J. Hill, A. M. James, and W. R. Maxted, Physical investigations of the behavior of bacterial surfaces. IX. The streptococcal cell wall, *Biochim. Biophys. Acta*, 75:414 (1963).

13. A. Few, Interaction of polymixin E with bacterial and other lipids, *Biochim. Biophys. Acta,* 75:414 (1963).
14. L. Vaczi, and L. Frakas, Association between lipid metabolism and antibiotic sensitivity. (1) Lipid composition of antibiotic sensitive and resistant *Staphylococcus aureus* strains, *Acta Microbiol. Acad. Sci. Hung.,* 8:205 (1961).
15. G. S. Bartnicki, Cell wall chemistry, morphogenesis and taxonomy of fungi, *Ann. Rev. Microbiol.,* 22:87 (1968).
16. G. S. Bartnicki, and I. McMurrough, in *Yeasts,* Vol. 2 (A. H. Rose and J. S. Harrison, eds.), Academic, New York (1971), p. 441.

PART II
CHEMICAL, PHYSICAL, AND MICROBIOLOGICAL PROPERTIES OF COMMON PRESERVATIVES

4
CHEMICAL PRESERVATIVES
Use of Bronopol as a Cosmetic Preservative

BETTY CROSHAW* and VICTOR R. HOLLAND *The Boots Company PLC, Nottingham, England*

I. Names and Synonyms 32
II. Background Information 32
III. Structure and Chemical Properties 33
    A. Synthesis 33
    B. Physical properties 33
    C. Stability 34
    D. Analytical methods 38
IV. Microbiological Activity 41
    A. Antibacterial activity 41
    B. Antifungal activity 48
V. Safety Studies 51
    A. Metabolism 51
    B. Acute toxicity 51
    C. Repeated-dose toxicity 52
    D. Irritancy and contact sensitivity 52
    E. Phototoxicity 54
    F. Reproduction toxicity 54
    G. Mutagenicity 55
    H. Carcinogenicity 55
    I. General pharmacology 55
    J. Acute toxicity of decomposition products and impurities 55
    K. Experience in use 57
VI. Cosmetic Applications 57
    A. Advantages 57
    B. Limitations 58

---

*Now retired in Burton Joyce, Nottingham, England

VII. Other Applications 58

VIII. Regulatory and Independent Review Status 59

References 59

I. NAMES AND SYNONYMS

The chemical name of the subject of this chapter is 2-bromo-2-nitropropane-1,3-diol, and its nonproprietary name is Bronopol. Other commonly used names are BNPD and Bronopol-Boots. Myacide BT, and Myacide AS are industrial grades marketed by The Boots Company PLC, England.

The studies described in this chapter have almost exclusively been carried out on the pharmaceutical grade of Bronopol now marketed by The Boots Company PLC as Bronopol-Boots; for brevity, this material is referred to subsequently as Bronopol. Much of this work has already been reviewed by Bryce et al. [1].

II. BACKGROUND INFORMATION

Work by Hodge et al. [2] and later by Zsolnai [3] indicated that geminal bromonitroalkanes had antifungal activity. The broad-spectrum antibacterial properties of Bronopol, a compound of this type, have been described in a preliminary communication by Croshaw et al. [4], in comparison with other members of a series of antimicrobial aliphatic halogenonitro compounds by Clark et al. [5], and in more detail by Bryce et al. [1].

Bronopol has been used as an effective antibacterial preservative at concentrations of 0.01-0.02% in various cosmetic, toiletry, household, and pharmaceutical formulations over the past 16 years because of its high activity against gram-negative bacteria, especially members of the genus *Pseudomonas*. The most frequently reported contaminants in toiletries and cosmetics in recent years have been gram-negative bacteria belonging to such genera as *Pseudomonas*, *Klebsiella*, *Achromobacter*, and *Alcaligenes* [6-9]. These bacteria are common residents in aqueous environments, and as water is a raw material common to many products susceptible to microbial contamination, it is generally believed to be a prime source of contamination [6]. Nonionic surfactants, which are widely used in toiletries and cosmetics, can also support the growth of gram-negative bacteria and particularly pseudomonads and thus be a source of contamination of the final product [9-12].

Bronopol is an effective antibacterial preservative over a wide pH range. It is stable at acid pH values and is also useful as a labile preservative in alkaline media. Bronopol is more active against bacteria than against fungi and yeasts, although it may be effective against the latter at higher concentrations. In products where fungal contamination and spoilage are likely to be major problems, Bronopol can be used in combination with other preservatives such as 2,4-dichlorobenzyl alcohol (Myacide SP) or the parabens. The high water and low oil solubility of Bronopol together with the fact that, unlike some other preservatives, it is not antagonized by anionic and nonionic surfactants enhances its usefulness.

Animal toxicity studies and human patch tests have demonstrated the safety of Bronopol when used at concentrations between 0.01 and 0.1%, and

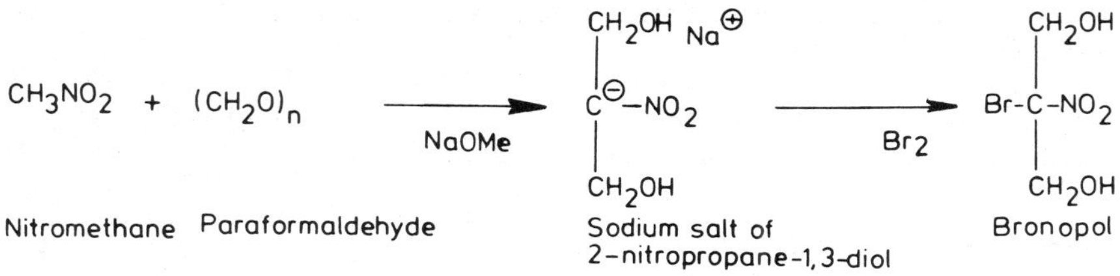

Bronopol

CH_3NO_2 + $(CH_2O)_n$ →[NaOMe] Sodium salt of 2-nitropropane-1,3-diol →[Br_2] Bronopol

FIGURE 1 Synthesis of Bronopol.

there has been no evidence of human skin sensitization, even at levels considerably in excess of these.

III. STRUCTURE AND CHEMICAL PROPERTIES

A. Synthesis

The preparative method for Bronopol is summarized in Figure 1. Paraformaldehyde and nitromethane are mixed under alkaline conditions (sodium methoxide) to give the sodium salt of 2-nitropropane-1,3-diol. This product is treated with bromine to give Bronopol, which can be recovered by crystallization from the concentrated solution.

B. Physical Properties

Bronopol is a white or almost white crystalline powder with a faint characteristic odor. Its molecular formula is $C_3H_6BrNO_4$ (molecular weight, 200) and its melting point is approximately 130°C. Bronopol is readily soluble in water and polar organic solvents. Its approximate solubilities in some common solvents at room temperature are given in Table 1. Bronopol partitions effective-

TABLE 1 Approximate Solubility of Bronopol at Room Temperature (22-25°C) in Some Common Solvents

Solvent	Percent wt/vol
Water	25
Ethanol	50
Isopropanol	25
Chloroform	0.9
Glycerol	1
Propylene glycol	14
Polyethylene glycol 300	11
Diethyl sebacate	10
Isopropyl myristate	<0.5
Arachis oil	<0.5
Liquid paraffin	<0.5
Cottonseed oil	<0.5
Olive oil	<0.5
Castor oil	<0.5

TABLE 2 Partition Coefficients of Bronopol in Some Common Solvent Combinations at 22-24°C

Solvent combination	Partition coefficient
Hexanol–water	0.74
Cyclohexanol–water	1.63
Isooctane–water	<0.01
Liquid paraffin–water	0.043
Arachis oil–water	0.11
Chloroform–water	0.068
Dichloromethane–water	0.045

ly into the aqueous phase in most aqueous–organic solvent combinations. Its partition coefficients in some common solvent combinations at room temperature are given in Table 2.

C. Stability

No evidence of instability has been observed when Bronopol is stored in pure crystalline form for a period of up to 1 year at temperatures up to 45°C. No evidence of photodecomposition was obtained during this period. The results are presented in Table 3, the assays being carried out by gas–liquid chromatography (g.l.c.) of the trimethylsilylated material, using both the internal standard and normalization methods. Samples of Bronopol stored for longer periods (up to 2 years) in the dark at room temperature have also shown no evidence of decomposition (see Table 4). The assays were again carried out by g.l.c. following derivatization.

TABLE 3 Stability of Bronopol During Storage Determined by Gas-Liquid Chromatography

Storage conditions	Internal standard method (Bronopol % wt/vol)	Normalization method (Bronopol % wt/vol)
Initial	100.0	99.6
12 weeks		
At 20–25°C		
At normal RH[a]	99.8	99.5
At 90% RH	100.1	99.6
At 37°C	100.2	99.5
At 45°C	99.6	99.6
In north window	99.8	99.6
52 weeks		
At 20–25°C		
At normal RH	100.2	99.5
At 90% RH	99.2	99.5
At 37°C	100.3	99.5
At 45°C	100.3	99.5
In north window	99.3	99.4

[a]RH, relative humidity.

TABLE 4 Stability of Batches of Bronopol Determined by Gas-Liquid Chromatography After Storage at Room Temperature in the Dark

Batch number	Initial assay (Bronopol % wt/vol)	Time of storage (months)	Assay after storage (Bronopol % wt/vol)
1	99.2	24	98.2
			99.2
2	101.1	22	101.5
			102.2
3	100.2	21	101.4
			102.4
4	99.9	18	99.7
			99.5

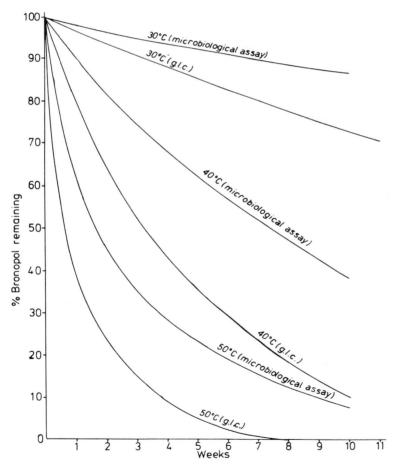

FIGURE 2 Effect of temperature on the stability of aqueous solutions of Bronopol at pH 6 (initially 0.2% wt/vol).

In aqueous solution, Bronopol is stable at low pH. Decomposition is accelerated by increasing the pH or the temperature of the solution. In order to investigate this more thoroughly, aqueous solutions of Bronopol buffered at pH 4, 6, and 8 were stored at various temperatures between 30 and 50°C and then assayed for remaining Bronopol at appropriate time intervals using microbiological, polarographic, g.l.c., and thin-layer chromatographic (TLC) techniques. The results of these experiments are summarized in Figures 2 and 3. It was found that the rate of decomposition at a given pH was accelerated approximately fourfold by each 10° increase in the temperature of the solution. Using these results, it is possible to calculate by extrapolation the times for 50% decomposition at 20°C, as shown:

pH	Time for 50% decomposition
4	Over 5 years
6	1.5 years
8	2 months

When Bronopol decomposes in aqueous solution, a number of complex chemical reactions take place. Among the decomposition products, formaldehyde, bromide ion, nitrite ion, 2-bromo-2-nitroethanol, and 2-hydroxymethyl-2-nitropropane-1,3-diol have been detected. Overall, decomposition results in a lowering of the pH of an unbuffered solution; such solutions therefore tend to be self-stabilizing.

The initial process in the decomposition of Bronopol appears to be a retroaldol reaction with the liberation of formaldehyde:

$$\underset{\text{Bronopol}}{\text{HOCH}_2-\underset{\underset{\text{Br}}{|}}{\overset{\overset{\text{NO}_2}{|}}{\text{C}}}-\text{CH}_2\text{OH}} \rightarrow \underset{\text{Formaldehyde}}{\text{CH}_2\text{O}} + \underset{\text{2-Bromo-2-nitroethanol}}{\text{HOCH}_2-\underset{\underset{\text{Br}}{|}}{\overset{\overset{\text{NO}_2}{|}}{\text{CH}}}}$$

2-Bromo-2-nitroethanol is considerably less stable than Bronopol, and its concentration under the conditions investigated never rose above 0.5% of the initial Bronopol concentrations. The formaldehyde released appears to be consumed in a second reaction involving unchanged Bronopol:

$$\underset{\text{Formaldehyde}}{\text{CH}_2\text{O}} + \underset{\text{Bronopol}}{\text{HOCH}_2-\underset{\underset{\text{Br}}{|}}{\overset{\overset{\text{NO}_2}{|}}{\text{C}}}-\text{CH}_2\text{OH}} \rightarrow \underset{\text{2-Hydroxymethyl-2-nitropropane-1,3-diol}}{\text{HOCH}_2-\underset{\underset{\text{NO}_2}{|}}{\overset{\overset{\text{CH}_2\text{OH}}{|}}{\text{C}}}-\text{CH}_2\text{OH}}$$

The product of this reaction, 2-hydroxymethyl-2-nitropropane-1,3-diol, itself slowly decomposes to release formaldehyde.

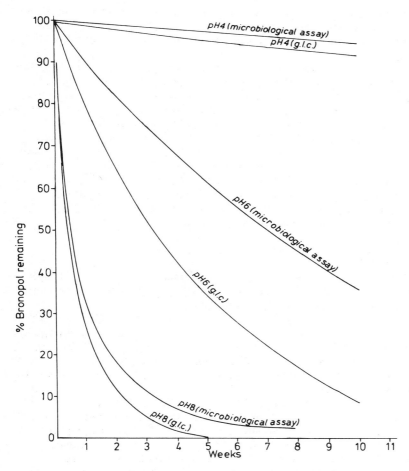

FIGURE 3 Effect of pH on the stability of aqueous solutions of Bronopol at 40°C (initially 0.2% wt/vol).

A number of the identified decomposition products have microbiological activity, and it is this activity which helps to explain the differences in Bronopol's apparent decomposition rates when followed by g.l.c. and microbiological methods (see Figs. 2 and 3). However, an inspection of the microbiological activities of these compounds reveals that they are not great enough to account for the high microbiological activity of Bronopol itself when one considers the levels at which these decomposition products are present. A study designed to examine the release of formaldehyde from shampoo and emulsion bases initially containing 200 ppm Bronopol revealed that at room temperature over 3 months extremely low levels of formaldehyde were present, the highest concentration detected being 6.3 ppm. Even after storage at 45°C for the same period formaldehyde levels of less than 15 ppm were observed.

When exposed to light, especially under alkaline conditions, solutions of Bronopol may become yellow or brown, but there is no evidence of a close correlation between color change and loss of antimicrobial activity. Solutions appear stable in contact with stainless steel and tin, although deactivation, discoloration, and corrosion occur to a limited extent when in contact with iron and aluminium.

An additional mode of decomposition of Bronopol results in the formation of nitrite ion. Nitrate ion, however, has not been detected. The rate of formation of nitrite is slower than the overall rate of decomposition of Bronopol as measured by g.l.c., and the final organic products have not been identified. The formation of nitrite makes theoretically possible its subsequent interaction with secondary and tertiary amines to give rise to nitrosamines which may be carcinogenic. In the opinion of the authors, it is advisable that formulators using Bronopol, or any other material which could contain or give rise to nitrite, take steps to ensure that if nitrosamines are produced, their presence will not represent a health hazard to the user. The complex subject of the formation of nitrosamines in cosmetics and the relevance of this to the use of Bronopol have recently been reviewed by Holland [13].

D. Analytical Methods

The methods described in this section have been used to obtain the results recorded in the preceding sections and to assay Bronopol in the types of formulations in which it is likely to be incorporated.

Pure Bronopol has been assayed by the determination of its bromine content, by the determination of its nitrogen content, and by g.l.c. of the acetylated and of the trimethylsilylated material, the methods using g.l.c. being the most specific. In formulations, Bronopol has been estimated by TLC, a polarographic procedure, a microbiological procedure, and g.l.c.. The procedure by TLC has been applied to ointments (at a concentration of 0.1%), barrier creams (at concentrations of 0.1 and 0.2%) and aerosol concentrates (at a concentration of 0.05%). Gas-liquid chromatography has been applied to aqueous formulations (at concentrations of from 5 to 50 ppm). The polarographic procedure has been applied to ointments, suppositories, creams, and gels (all at a concentration of 0.2%) and has also been used to estimate Bronopol in buffered aqueous solutions and in blood serum. The microbiological procedure has been applied to creams, including barrier creams (at concentrations of 0.1 and 0.2%), liquid shampoos, and buffered aqueous solutions. The polarographic method estimates the alkyl nitro group and therefore, although an acceptable procedure for freshly prepared formulations, is not very specific and is subject to interference from breakdown products. In the absence of interfering substances, the precision of this procedure is about ±2%. The method using TLC is more specific, but the spot-comparison procedure that has been used is liable to relative errors of about 15%; errors of this magnitude may be acceptable, however, at the concentrations in which Bronopol is usually incorporated in formulations. The microbiological method had an error of about ±10% on aqueous solutions and about ±10–20% on creams.

1. Polarographic Assay

The base electrolyte was McIlvaine's buffer solution containing 2% vol/vol of 0.2% vol/vol Triton X-200 as a maximum suppressor. McIlvaine's buffer solution, at pH 4, was prepared by mixing 12.29 parts by volume of 0.1 M citric acid solution and 7.71 parts by volume of 0.2% disodium phosphate (Na_2HPO_4) solution. The test solution was prepared as follows. Aqueous solutions only required dilution with base electrolyte to a Bronopol concentration between

10^{-3} and 10^{-5} M. Bronopol in gels, creams, and other fatty-base formulations could be extracted with base electrolyte by warming gently on a steam bath, after which any insoluble matter in the aqueous phase was removed by centrifugation and the aqueous phase diluted to give appropriate concentrations of Bronopol. Other formulations were more appropriately treated by dissolution in chloroform and extraction with base electrolyte.

The determination was carried out by transferring a portion of the solution in base electrolyte to the cell of a suitable polarograph. A stream of oxygen-free nitrogen was passed through the solution for 10–15 min to remove dissolved oxygen. The height of the mercury reservoir was adjusted to give a constant drop rate appropriate to the apparatus, this drop rate being identical to that used for the preparation of the calibration curve.

The polarogram was recorded over the range 0 to -1 V relative to the quiescent mercury pool using the appropriate recorder or galvanometer sensitivity to give a suitable wave. The diffusion current at -0.8 V relative to the mercury pool was measured, and the concentration of Bronopol read from the calibration curve.

Since the polarographic response was affected by the composition of the test solution, it was necessary to prepare a calibration curve for each formulation examined. Such calibration curves were obtained by adding known amounts of Bronopol to blank formulations and processing in the required manner.

2. Microbiological Assay

Bronopol can be assayed microbiologically by agar diffusion using *Pseudomonas aeruginosa* in agar of the following composition:

Compound	Percent wt/vol
Dextrose	0.1
Lemco beef extract	0.15
Difco yeast extract	0.3
Sodium chloride	0.5
Difco casitone	0.08
Magnesium sulfate ($7H_2O$)	0.004
Oxoid peptone	0.6
Davis agar	1.8
Distilled water	to 100, pH adjusted to 5.3

Alternatively, Difco Assay Agar No. 11 (pH 7.9) with *Bacillus subtilis* NCIB 8054 can be used. The minimum detectable level of Bronopol in water with *B. subtilis* is 0.005%.

Moore and Stretton [14], using *P. aeruginosa* NCIB 8295, found that in buffered agar with unbuffered Bronopol solutions at 250, 500, or 1000 µg/ml the maximum zone sizes occurred at the pH extremes, that is, at pH 5.61 and >7.0, with the smallest zones at pH 6.47. When the pH of the agar was kept constant and buffered Bronopol solutions were used over the range pH 5.1–8, there was little significant difference in the zone sizes.

A rapid diffusion method using *Bacillus stearothermophilus* has been described by Kabay [15].

3. Gas–Liquid Chromatographic Assay

Although Bronopol is a water-soluble compound, it can be extracted from aqueous solution into diethyl ether or ethyl acetate after the addition of sodium chloride. The extract can then be evaporated to dryness, the residue acetylated, and the Bronopol estimated by means of g.l.c. with electron-capture detection. This procedure offers a means of determining Bronopol in aqueous formulations and has been applied to Bronopol concentrations down to 5 ppm.

In aqueous formulations containing concentrations of Bronopol down to 50 ppm, the Bronopol has been determined by a similar procedure, but using n-pentadecane as the internal standard, acetyl chloride in chloroform as an acetylating reagent, carbon disulfide as the final solvent, and flame ionization detection.

The following are examples of the methods which have been used to assay Bronopol by g.l.c..

Method Based on the Acetylated Material: The sample (about 0.15 g accurately weighed) was dissolved in 15 ml of chloroform with the aid of minimum heating, 5 ml of a 2% solution of n-pentadecane (as internal standard) in chloroform was added, and the solution was diluted to 25 ml. To 1 ml of this solution in a vial was added 0.3 ml acetyl chloride, and the vial was sealed and then heated on a steam bath for 3 hr. The mixture (2 µl) was subjected to g.l.c. in a glass column (183 cm × 3 mm) packed with 10% of silicone JXR on Gas Chrom Z (70–80 mesh), operated at 150°C with nitrogen (20 ml/min) as carrier gas, and flame ionization detection. The ratio of the product of the peak height and retention time for the Bronopol diacetate (relative retention time = 1.00) to that for n-pentadecane (relative retention time = 1.54) was calculated and compared with the ratio for a standard containing purified Bronopol which had been similarly treated. The relative standard deviation of the method was found to be 1.5%.

Method Based on the Trimethylsilylated Material: The sample (about 0.15 g, accurately weighed) was dissolved in 15 ml of chloroform with the aid of minimum heating, 4 ml of a 1.4% solution of n-tridecane (as internal standard) in chloroform was added, and the solution was diluted to 25 ml. To 1 ml of this solution in a vial was added 0.1 ml of silylating reagent (prepared by mixing 1 part trifluoroacetic acid and 2 parts hexamethyldisilazane and filtering the mixture rapidly under dry conditions), and the vial was sealed and then heated on a steam bath for 1 hr. The mixture (1 µl) was subjected to g.l.c. in a glass column (152 cm × 3 mm) packed with 10% of silicone JXR on Gas Chrom Q (80–100 mesh), operated at 125°C with nitrogen (40 ml/min) as carrier gas, and flame ionization detection. The ratio of the product of the peak height and retention time for Bronopol di(trimethylsilyl)-ether (relative retention time = 1.92) to that for n-tridecane (relative retention time = 1.00) was calculated and compared with the ratio for a standard containing purified Bronopol which had been similarly treated.

4. Thin-Layer Chromatographic Assay

A 10% Bronopol solution was examined by TLC and bioautography using 0.25 mm Kieselgel G, with chloroform–methanol (4:1) as developing solvent. Two-microliter aliquots of the solution were spotted on the plates. A similar method for ointment formulations has been devised using an initial water:chloroform extraction system to remove excipients, followed by chromatography on Kieselgel GF_{254} using isopropanol as the developing solvent.

5. Determination of Bromide Ion

Bromide ion was determined by potentiometric titration. Bronopol solution (5 ml) was acidified and titrated with 0.02 M silver nitrate solution.

6. Determination of Formaldehyde

Formaldehyde was determined by reaction with chromotropic acid. A 0.2% Bronopol solution (0.5 ml) was diluted to 25 ml with 12 N sulfuric acid. To this solution (1 ml) was added a 5% solution of chromotropic acid in 12 N sulfuric acid (1 ml), and the mixture heated at 100°C for 30 min. Concentrated sulfuric acid (2 ml) was added, and the absorbance at 570 nm measured against the appropriate blank.

7. Determination of Nitrite and Nitrate

Nitrite and nitrate were determined by reaction with 2,6-xylenol before and after decomposition of nitrite with sulfamic acid. This method was not used after the preliminary work, as the results were in good agreement with the polarographic estimation of alkyl nitro groups.

IV. MICROBIOLOGICAL ACTIVITY

Bronopol has broad-spectrum antibacterial activity, and although it has some activity against both pathogenic and spoilage fungi and yeasts, this activity is less than that against bacteria.

A. Antibacterial Activity

1. Bacteriostatic Activity

The spectrum of bacteriostatic activity of Bronopol was determined in vitro by twofold serial dilution of the compound in nutrient agar. Dried plates of these dilutions were surface inoculated using a multipoint inoculator [16] with 0.01 ml of 24-hr broth cultures of the test bacteria. Ox serum and glucose were added to the inocula where necessary for the test organisms in Table 5, and the plates inoculated with *Clostridium perfringens* were incubated anaerobically. The minimum inhibitory concentration (MIC) of Bronopol was recorded after 24 hr at 37°C (48 hr at 37°C for the anaerobes) for the test organisms shown in Table 5, and after 3 days at 32°C for those in Table 6, since the optimum temperature for some of the latter is below 37°C. All bacteria tested are inhibited by 12.5–50 µg/ml Bronopol.

The bacteriostatic activity of Bronopol in liquid medium is shown in Table 7. Twofold serial dilutions were prepared in nutrient broth, and these were inoculated with overnight broth cultures of the test organisms to give about 5×10^5 cells per milliliter. The MIC for each organism was recorded after 24 hr at 37°C; all bacteria tested were inhibited by 3.1–25 µg/ml Bronopol.

The marked activity of Bronopol against pseudomonads is important, since these organisms, and particularly *P. aeruginosa*, show an exceptionally high resistance to chemical antibacterial agents [17,18] in addition to being able to grow on simple substrates under a wide range of environmental conditions.

TABLE 5 Antibacterial Spectrum of Bronopol by Dilution in Nutrient Agar

Test organism	Number of strains tested	Minimum inhibitory concentration after 24 hr at 37°C (μg/ml)
Gram-positive bacteria		
Staphylococcus aureus	30	12.5–50
Staphylococcus epidermidis	2	25–50
Streptococcus faecalis	2	50
Streptococcus pyogenes	2	12.5–50
Corynebacterium pyogenes	4	12.5–50
Bacillus subtilis	2	12.5–25
Clostridium perfringens	5	25–50
Gram-negative bacteria		
Pseudomonas aeruginosa	54	12.5–50
Pseudomonas sp.	9	12.5–25
Klebsiella pneumoniae	3	25
Escherichia coli	15	12.5–50
Salmonella sp.[a]	20	12.5–25
Proteus sp.[b]	37	12.5–50
Shigella sp.[c]	6	25

[a]Salmonella sp. includes Salmonella gallinarum, typhimurium, typhi, paratyphi B, enteritidis, ser. dublin, ser. thompson, ser. heidelberg, and ser. seftenberg.
[b]Proteus sp. includes Proteus vulgaris, mirabilis, morganii, and rettgeri.
[c]Shigella sp. includes Shigella sonnei and flexneri.

The broad spectrum of antibacterial activity of Bronopol and particularly its activity against Pseudomonas sp. has been confirmed by workers in various parts of the world [19–22].

Comparisons of the bacteriostatic activity of Bronopol and other commonly used cosmetic and toiletry preservatives reported by Croshaw [23] and Bryce et al. [1] show that it is one of the most active preservatives available at the present time.

2. Effect of pH on Bacteriostatic Activity

Bronopol is effective over a wide pH range, although the bacteriostatic activity decreases two to eight times as the pH increases from 5.3 to 8 [4] (see Table 8). This was confirmed by Moore and Stretton [14], who demonstrated that over the pH range 5.1–7.9 activity decreased as the pH increased by a factor of 3–8. On the other hand, Saito and Onoda [21] found the activity of Bronopol to be optimal at pH 7 and that MIC values increased when the pH was decreased to pH 6 or increased to pH 8 or 9. Tuttle et al. [24] obtained variable results in their tests; in some cases Bronopol was more active at higher pH values, and in other situations it was less active. As suggested by Moore and Stretton [14], the seemingly contradictory results of various workers may be explained by the type of test and the stability of Bronopol over the pH range considered. This observation is different from that found with other preservatives which are markedly affected by pH because of dissociation, for example, benzoic and sorbic acids.

TABLE 6 Antibacterial Activity of Bronopol by Dilution in Tryptone Soya Agar

Test organism	Strain	Minimum inhibitory concentration after 3 days at 32°C (μg/ml)
Gram-positive bacteria		
Staphylococcus aureus	8452	25
Staphylococcus aureus	NCIB 9518	50
Staphylococcus epidermidis	NCTC 7291	25
Streptococcus faecalis	NCTC 8213	50
Micrococcus lutea		25
Bacillus cereus	NCIB 8012	25
Gram-negative bacteria		
Pseudomonas aeruginosa	NCTC 6750	25
Pseudomonas aeruginosa	NCIB 11338	25
Pseudomonas putida	NCIB 9034	25
Pseudomonas cepacia	NCIB 9085	25
Pseudomonas stutzeri	NCIB 9040	12.5–25
Pseudomonas fluorescens	NCIB 9046	12.5
Proteus vulgaris	NCTC 4635	25
Proteus morganii	NCTC 10041	25
Escherichia coli	NCTC 5934	25
Escherichia coli	NCIB 9517	25
Klebsiella aerogenes	NCTC 418	25
Enterobacter cloacae	Ref. 146	50
Salmonella typhimurium	NCTC 74	25
Serratia marcescens	Ref. 131	25
Flavobacterium meningosepticum	NCTC 10016	25
Alcaligenes sp.		50
Citrobacter freundii	Ref. 139	50

TABLE 7 Antibacterial Activity of Bronopol by Dilution in Nutrient Broth

Test organism	Number of strains tested	Minimum inhibitory concentration after 24 hr at 37°C (μg/ml)
Gram-positive bacteria		
Staphylococcus aureus	8	6.12–12.5
Gram-negative bacteria		
Pseudomonas aeruginosa	25	3.1–25
Proteus vulgaris	6	3.1–6.25
Proteus rettgeri	2	6.25
Proteus morganii	2	6.25–12.5
Proteus mirabilis	2	6.25–12.5
Escherichia coli	3	3.1–6.25
Salmonella typhimurium	1	3.1
Salmonella ser. heidelberg	2	6.25
Shigella sonnei	2	6.25

TABLE 8 Effect of pH on the Bacteriostatic Activity of Bronopol Against *Pseudomonas aeruginosa*

pH	Minimum inhibitory concentration	
	In agar (6 strains) (µg/ml)	In broth (3 strains) (µg/ml)
5.5	6.25	0.78–3.1
6.0	25	3.1–6.25
7.0	25–50	6.25
8.0	25–100	6.25–12.5

3. Effect of Organic Matter on Bacteriostatic Activity

The bacteriostatic activity of Bronopol is not greatly reduced in the presence of low concentrations of organic matter (e.g., 10% serum or milk) or 5% Crotein A, a protein hydrolysate used in shampoos. Oxalated horse blood at a concentration of 50% markedly reduced the activity (see Table 9).

4. Effect of Surfactants and Possible Antagonists

The bacteriostatic activity of Bronopol is not antagonized in the presence of anionic and nonionic surfactants [4,25], although nonionic surface active agents are known to antagonize the action of some preservatives. Brown [19] and Allwood [26] have confirmed this finding; in fact, Allwood has reported synergism between Bronopol and some octyl or nonyl phenol nonionic surfactants. On the other hand, sodium thiosulfate, sodium metabisulfite, and particularly sulfhydryl compounds adversely affect activity [4] (see Table 10). The marked antagonism of thiol-containing compounds to Bronopol has been confirmed by Stretton and Manson [27].

The antagonism of Bronopol by cysteine hydrochloride is also shown by means of turbidimetric measurements of the logarithmic growth rates of *P. aeruginosa* in the presence of the antibacterial agent alone and with the anta-

TABLE 9 Effect of Organic Matter on the Bacteriostatic Activity of Bronopol

Additive	Decrease (fold) in bacteriostatic activity[a]
10% milk	0
10% ox serum	0–2
50% ox serum	4–8
10% human serum	2
50% human serum	4
10% oxalated horse blood	4–8
50% oxalated horse blood	32–64
5% Crotein A (protein hydrolysate)	0

[a]Twofold serial dilution in agar using 27 strains of *P. aeruginosa* as test organisms.

TABLE 10 Effects of Various Surfactants and Possible Antagonists on the Bacteriostatic Activity of Bronopol

Additive	Decrease (fold) in bacteriostatic activity[a]
5% Tween 80 (nonionic)	0
5% Span 80 (nonionic)	0
5% Empicol MD (anionic)	0
5% Empicol ESB 3 (anionic)	2
5% Empigen BB (amphoteric)	2
1% Lubrol W	0
0.1% lecithin	0
0.1% sodium stearate	0-2
0.1% cystine hydrochloride	4
0.1% sodium thioglycollate	8-16
0.1% dimercaprol (BAL)	32
0.1% cysteine hydrochloride	16-64
0.1% sodium thiosulfate	4-16
0.1% sodium metabisulfite	8-16

[a]Twofold serial dilution in agar using 27 strains of *P. aeruginosa* as test organisms.

gonist (see Fig. 4). The reversal of the inhibitory effect of Bronopol by the late addition of cysteine hydrochloride is also demonstrated.

5. Compatibility with Other Antimicrobial Agents

With the filter paper strip technique of King et al. [28] and *Staphylococcus aureus* as the test organism, Bronopol has been shown to be compatible with Cetrimide BP, domiphen bromide BP, benzalkonium chloride, Cetylpyridinium Chloride BP [29], trichlorocarbanilide, and 2,4-dichlorobenzyl alcohol (Myacide SP) [30].

6. Bactericidal Activity

Bronopol is bactericidal in liquid medium after 24 hr at 37°C at concentrations only two to four times higher than the bacteriostatic levels. Subcultures were made from tubes of nutrient broth containing inhibitory concentrations of Bronopol after 24 hr at 37°C onto the surface of nutrient agar containing cysteine hydrochloride as the inactivating agent. Absence of growth on the agar was taken to indicate bactericidal activity after 24 hr at 37°C (see Table 11).

In extinction time tests aqueous solutions of Bronopol were challenged with about 1×10^6 organisms per milliliter of *P. aeruginosa* NCIB 11338, *Escherichia coli* NCTC 86, or *S. aureus* FDA. These were held at 37 or 22°C and, after various contact times up to 2 hr, 0.1-ml aliquots were transferred to broth containing cysteine hydrochloride as inactivating agent. These broths were incubated for 48 hr at 37°C to determine the presence or absence of growth (see Table 12). A 0.2% aqueous solution of Bronopol was bactericidal within 24 hr at 22 or 37°C against the two gram-negative organisms tested, and a 0.1% solution was effective in 2 hr at 37°C but not at 22°C. The bac-

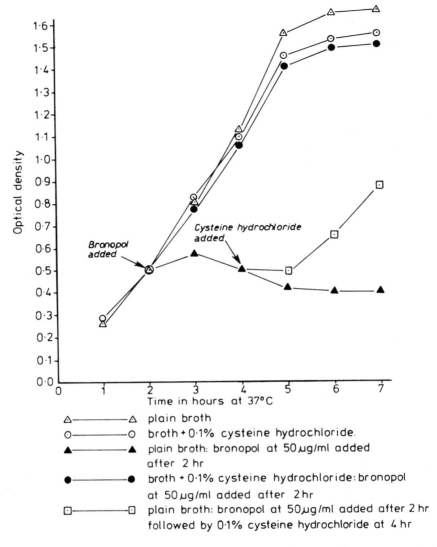

FIGURE 4 Effect of Bronopol and Bronopol with cysteine hydrochloride on the growth rate of *Pseudomonas aeruginosa* NCIB 11338.

TABLE 11 Comparison of the Bacteriostatic and Bactericidal Concentrations of Bronopol in Nutrient Broth

Test organism[a]	Bacteriostatic concentration after 24 hr at 37°C (µg/ml)	Bactericidal concentration after 24 hr at 37°C (µg/ml)
Staphylococcus aureus 8452	12.5	25
Pseudomonas aeruginosa NCTC 6750	6.25	25
Pseudomonas aeruginosa NCIB 11338	12.5	50
Escherichia coli NCTC 86	3.1	12.5

[a]Inoculum about 5×10^5 cells per milliliter.

TABLE 12 Bactericidal Activity of Bronopol

Concentration of aqueous Bronopol solution (% wt/vol)	Temperature of test (°C)	Bactericidal activity[a]											
		Staphylococcus aureus FDA				Pseudomonas aeruginosa NCIB 11338				Escherichia coli NCTC 86			
		0 hr	0.5 hr	1 hr	2 hr	0 hr	0.5 hr	1 hr	2 hr	0 hr	0.5 hr	1 hr	2 hr
0.2%	22	−	−	−	−	−	−	−	+	−	−	−	+
0.1%	22	−	−	−	−	−	−	−	−	−	−	−	−
0.2%	37	−	−	−	−	−	±	±	+	−	±	±	+
0.1%	37	−	−	−	−	−	±	±	+	−	−	±	+

[a] +, bactericidal, that is, no growth in recovery medium; −, not bactericidal, that is, growth in recovery medium; ±, results variable in different tests.

tericidal activity is greater against gram-negative bacteria than against the staphylococcus used in this test, since a 0.2% solution was not bactericidal to this organism in 2 hr at 37°C.

The presence of an inactivating agent (cysteine hydrochloride) in the recovery medium affects the bactericidal activity. Solutions of Bronopol held at 37 or 22°C were challenged with about 1×10^5 organisms per milliliter. At various times 1-ml aliquots were removed and (1) diluted in peptone water and plated in nutrient agar without an inactivating agent and (2) diluted in peptone water and plated in agar containing cysteine hydrochloride. The results are shown in Table 13. When cysteine hydrochloride was incorporated in the recovery medium, the percent reduction in the number of viable organisms was greatly reduced at 22°C, but less so at 37°C, indicating that cell-bound Bronopol, present on damaged cells, cannot be removed by dilution alone.

The bactericidal activity is not markedly affected by changes in pH over the range pH 5-8. Moore [31] and Moore and Stretton [14] investigated the effect of a combination of temperature and pH on the extinction time of Bronopol. The shortest extinction time occurred at a combination of the highest temperature, pH, and Bronopol concentration, while the longest extinction time was at the lowest pH, temperature, and Bronopol concentration. For any one concentration of Bronopol (400, 100 or 25 µg/ml) at a fixed temperature (37, 30, or 25°C) the extinction time against *P. aeruginosa* NCIB 8295 decreased with increase in pH over the pH range 5.7-7.2. The extinction time also decreased with increase in temperature for any one concentration of Bronopol at a fixed pH [31].

Bronopol shows little sporicidal activity.

7. Resistance Studies

Further work [1] has confirmed the original findings of Croshaw et al. [4] that there is no evidence of the development of Bronopol-resistant organisms after passage in the presence of the agent for 20 subcultures. In practice, Bronopol-resistant organisms have not occurred. This has been confirmed by Moore [31].

8. Mode of Action of Bronopol

Since Bronopol is more active against metabolizing cells than against resting cells and since its antibacterial activity is reversed by thiol-containing compounds [4], thiol-containing enzymes would appear to be implicated. Zsolnai [3] suggested that halogenonitro compounds could react with sulfhydryl groups which are widely distributed in essential enzyme systems.

Bronopol forms disulfide bonds from thiol groups, and Stretton and Manson [27] have suggested that these may account for the observed inhibition of dehydrogenase activity by the compound at concentrations approximating the minimum inhibitory value for each organism. They also noted that Bronopol affects the integrity of the cell membrane and suggested that this was a secondary action of the agent. On the other hand, Japanese workers [22] using electron microscopy concluded that Bronopol was a slow-acting, cell-wall-affecting compound.

B. Antifungal Activity

Bronopol has some antifungal activity, although this activity is much lower than the antibacterial activity.

TABLE 13 Effect of an Inactivating Agent (Cysteine Hydrochloride) in the Recovery Medium on the Bactericidal Activity of Bronopol

Concentration of aqueous Bronopol solution (% wt/vol)	Temperature of test (°C)	Recovery medium	Percent reduction in the number of viable organisms			
			Pseudomonas aeruginosa NCIB 11338		Escherichia coli NCTC 86	
			0.5 hr	1 hr	0.5 hr	1 hr
0.2	22	Without inactivator	95	99.5	98	99.9
0.2	22	With inactivator	<50	85	<50	70
0.1	22	Without inactivator	90	99	95	99
0.1	22	With inactivator	<50	<50	<50	<50
0.2	37	Without inactivator	>99.9	>99.9	99	>99.9
0.2	37	With inactivator	99.9	99.9	>99.9	>99.9
0.1	37	Without inactivator	>99.9	>99.9	99	>99.9
0.1	37	With inactivator	99.9	99.9	99	>99.9

1. Fungistatic Activity

The inhibitory activity of Bronopol against pathogenic and spoilage yeasts and fungi is shown in Table 14 using twofold serial dilutions in Sabouraud agar. The plates were inoculated with 0.01 ml of spore suspensions prepared from 7-day cultures of the fungi or 0.01 ml of 48-hr cultures of the yeasts.

2. Compatibility with Myacide SP

Bronopol is compatible with the antifungal agent Myacide SP, and it has been suggested that these two agents together may produce a satisfactory preservative system in certain formulations where bacterial and fungal contamination may occur [30].

TABLE 14 Fungistatic Activity of Bronopol by Dilution in Sabouraud Agar

Test organism	Number of strains	Minimum inhibitory concentration after 5 days at 25°C ($\mu g/ml$)
Pathogenic yeasts		
Pityrosporum ovale[a]	1	200
Candida albicans	4	400–1600
Candida tropicalis	1	3200
Spoilage yeasts		
Saccharomyces cerevisiae	1	3200
Spoilage yeast isolate	1	1600
Pathogenic fungi		
Trichophyton mentagrophytes	2	200
Trichophyton rubrum	2	100–200
Trichophyton tonsurans	1	100
Trichophyton interdigitale	1	100
Trichophyton quinckeanum	1	100
Trichophyton violaceum	1	100
Trichophyton verrucosum var. *discoides*	1	100
Trichophyton schoenleinii	1	50
Microsporum audouinii	1	50
Microsporum canis	1	100
Epidermophyton floccosum	1	100
Aspergillus sp. (ear infection)	1	400–800
Spoilage fungi		
Aspergillus niger	2	3200–6400
Myrothecium verrucaria	1	1600–3200
Chaetomium globosum	1	800
Cladosporium herbarum	1	3200
Penicillium funiculosum	1	1600–3200
Penicillium roqueforti	1	400–800
Penicillium canadense	1	1600–3200

[a]In neopeptone bile salt agar; 5 days at 32°C.

V. SAFETY STUDIES

A. Metabolism

After oral administration of [2-^{14}C]Bronopol to rats and dogs, radioactivity was found to be rapidly absorbed and evenly distributed in the tissues. Excretion of the majority of the dose was complete in 24 hr. No unchanged compound was detected in plasma and urine, indicating rapid and extensive metabolism. The major urinary metabolite (40% of administered radioactivity) was 2-nitropropane-1,3-diol. Other minor metabolites have not been identified. Significant radioactivity was found in expired air, indicating complete metabolism.

When applied in acetone solution to rat skin, a smaller proportion of the dose was absorbed than when administered orally. The use of a small area of skin for dosing may in part account for this. When dosed in the same vehicle to the skin of rabbits, the radioactivity was found to be mainly localized on the epidermis around hair follicles. This suggests that limited percutaneous absorption may occur through the follicles.

The pattern of urinary metabolites was similar when the compound was administered orally or dermally, indicating no difference in metabolism related to the route of administration.

B. Acute Toxicity

Bronopol administered in single doses by the oral or intraperitoneal routes to rodents caused gastrointestinal lesions and peritonitis. The LD_{50} values are shown in Table 15. A small number of rats were injected subcutaneously with Bronopol, and those which died had hemorrhage and edema at the site of the injection, stomach lesions, and lung congestion and edema; the LD_{50} was between 100 and 200 mg/kg. After dermal application of acetone solutions of Bronopol to rats, deaths occurred at 160 mg/kg or more.

Oral administration of single doses of 40 mg/kg to dogs caused gastric irritation. Single oral doses of up to 25 mg/kg did not affect methemoglobin levels in cats over a 24-hr period, whereas the positive control acetanilide elicited a marked rise in concentration.

TABLE 15 LD_{50} Values for Bronopol in Mice and Rats

Species	Sex	Route	LD_{50} (mg/kg)
Mouse	Male	Oral	374
	Female	Oral	327
House	Male	Intraperitoneal	35
	Female	Intraperitoneal	33
Rat	Male	Oral	307
	Female	Oral	342
Rat	Male	Intraperitoneal	22
	Female	Intraperitoneal	30

C. Repeated-Dose Toxicity

In the repeated-dose studies, the observations and laboratory investigations generally included signs of poisoning, body weight, food consumption, hematology, blood biochemistry, ophthalmoscopy, organ weights, macroscopic appearance at autopsy, and histopathology.

Male and female rats were given daily doses by oral intubation of 20 or 80 mg/kg Bronopol for 90 days; 160 mg/kg proved too toxic to another group and was terminated after 9 days. The 20-mg/kg dose was well tolerated, wheras 80 mg/kg caused gastrointestinal lesions, repiratory distress, and some deaths. When Bronopol was given in the drinking water, rats maintained on 160 mg/kg Bronopol per day for 6 weeks drank less and had slightly enlarged kidneys, while among those given the highest dose level of 300 mg/kg per day, a few deaths occurred. In dogs given a maximum daily dose of 20 mg/kg per day by oral intubation for 90 days, the only toxic reaction of note was vomiting.

Aqueous 2.5% methylcellulose solutions containing 0.2 or 0.5% Bronopol were applied once daily at a dosage of 1 ml/kg for 3 weeks to the clipped and abraded dorsal skin of rabbits. The vehicle alone and the 0.2% Bronopol solution elicited local skin erythema and the 0.5% Bronopol solution produced moderate erythema, edema, and scabbing; otherwise, the rabbits showed no ill effects clearly attributable to treatment.

D. Irritancy and Contact Sensitivity

1. Animal Studies

Bronopol was tested for local effects on the mucous membranes of the eyes of rabbits. Neither 0.5% Bronopol in normal saline applied once daily for 4-consecutive days nor a single application of 2% in polyethylene glycol 400 was irritant, though a single application of 5% in polyethylene glycol 400 was irritant.

Skin irritancy was investigated by application of Bronopol in a variety of solvents to the nonabraded, clipped, and shaved backs of rabbits for 6 hr, with or without occlusion. Single applications of 1% Bronopol in acetone or 0.5% in 2.5% aqueous methyl cellulose under occlusive dressing were not irritant, while 2% concentrations were irritant.

Under similar experimental conditions 5% Bronopol in polyethylene glycol 300 was not irritant; therefore the skin reaction elicited by the compound appears to be dependent upon the vehicle employed.

Bronopol was without skin-sensitizing activity in the guinea pig when tested as a 1% solution in acetone by the ear-flank method; the positive control dinitrochlorobenzene was strongly positive.

The sensitization potential of Bronopol in the guinea pig was assayed in a Magnusson and Kligman guinea pig maximization test conducted by Unilever Research. Sensitization was induced in guinea pigs by intradermal injections of both Bronopol (at a concentration of 0.03%) and complete Freund's adjuvant. The induction process was supplemented after 7 days by a 1.5% aqueous solution of Bronopol applied to the shoulder injection sites, under occlusion. The animals were then challenged 14 days later on the flank with a 0.4% solution of Bronopol under occlusion; further challenges were made at weekly intervals, as required.

By this method, sensitization is normally assessed after one challenge, at which stage there was no evidence of sensitization. Three challenges were

TABLE 16 Skin-Sensitizing Potential of Bronopol in the Guinea Pig: Summary of Challenge Results

Flank challenge by occluded patch	Bronopol concentration (% wt/vol)	Positive skin response in guinea pigs after 24 hr	Positive skin response in guinea pigs after 48 hr
Challenge 1	0.4	0/10	0/10
Challenge 2	0.4	0/10	1/10 scattered mild erythema
Challenge 3	0.4	2/10	2/10 moderate diffuse erythema
Challenge 4	0.4	0/9	1/9 scattered mild erythema
	0.2	1/9	2/9 moderate diffuse erythema

necessary before 2 out of 10 animals became sensitized and, hence, the sensitization potential was regarded as less than mild. A summary of the challenge results is given in Table 16.

2. Human Studies

The skin irritant potential of Bronopol was investigated in volunteers and in patients attending a contact dermatitis clinic.

The volunteer study showed that Bronopol was slightly irritant to human skin at 1% in soft paraffin (petroleum), and at o.25% in aqueous buffer at pH 5.5. The study consisted of a closed patch test using 1 cm × 1 cm lint squares backed with Blenderm surgical tape on the forearms of 10 subjects. Concentrations of 0, 0.5, 1, and 2% Bronopol in soft paraffin and of 0, 0.005, 0.1, and 0.25% in aqueous buffer at pH 5.5 were used. Any skin reaction after 24 hr was graded from 0 (normal skin) to 5 (marked erythema with vesicles and induration). The results are shown in Table 17.

The study carried out in patients attending a contact dermatitis clinic showed that Bronopol was a mild irritant when applied in yellow soft paraffin (yellow petrolatum) at 0.25%. No evidence of sensitization was seen in this study, nor was there any suggestion of cross-sensitization with a range of other substances with which it is likely to be formulated. The compound was applied as one of a battery of closed patch tests used in that clinic to screen the patients for a potential allergen. The patches were applied for 48 hr and examined on the second and fourth days after the application. Of the 149 patients studied, 3 showed a slight erythema on the second day that had faded by the fourth day; in addition, 1 had moderate erythema on the second day, but this patient did not return for the second examination.

Marzulli and Maibach [32,33] studied in man the contact sensitization of a number of commonly used biocides and concluded that, under the conditions of a closed patch test, Bronopol in yellow soft paraffin was a potential sensitizer. The challenge concentration in these studies was 2.5%, which, according to these authors, was not irritant. However, the investigations on volunteers and patients attending a contact dermatitis clinic, reported

TABLE 17 Irritancy of Bronopol to Human Skin

Base	Bronopol concentration (% wt/vol)	Positive skin response	Degree of reaction
Soft paraffin	0	0/10	—
	0.5	0/10	—
	1	2/10	Both slight erythema
	2	4/10	All moderate erythema
Aqueous buffer, pH 5.5	0	0/10	—
	0.05	0/10	—
	0.1	0/10	—
	0.25	1/10	Slight erythema

above, are not consistent with this view. The patch tests carried out by Marzulli and Maibach showed a dose–response relation and, since the response decreased very rapidly to zero at an induction concentration of 2% (which is considerably greater than that used in formulations), the authors suggested that Bronopol may be used safely in cosmetic formulations. In a further study, Maibach [34] confirmed that under similar conditions, Bronopol was a direct irritant to human skin at concentrations greater than 1%. A subsequent sensitization test included 93 normal subjects who were induced with 10 applications of 5% Bronopol in yellow soft paraffin under an occlusive dressing over a period of 3 weeks. After a rest period of 2 weeks, the subjects were challenged at a different site with 0.25% Bronopol in yellow soft paraffin. No evidence of contact sensitization was observed.

E. Phototoxicity

Raab [35] dispersed 0.1 ml of a 0.5% aqueous solution of Bronopol on a 1-cm-diameter circle of healthy shaved skin of rats and guinea pigs and subsequently irradiated the areas with ultraviolet light for 3 min. Control sites were treated with water. There were no differences over the observation period of 3–5 days.

In six healthy humans, 2-cm-diameter areas of the forearm were treated with 0.2 ml of a 0.25% aqueous solution of Bronopol and were then irradiated for 3 min; this procedure was repeated after 2 days and 2 weeks. Control areas were treated with water. There were no differences between test and control areas over the 2-week period.

F. Reproduction Toxicity

The effects of Bronopol on reproduction were investigated in rats and rabbits. In rats dosed from day 1 to 20 of pregnancy with 10, 30, or 100 mg/kg daily by oral intubation, the dams had a dose-related retardation in body weight gain and some died from gastric and lung lesions. Aside from a slight delay in calcification of the fetal skeleton at the highest dose level, there were no embryotoxic or teratogenic effects observed.

Daily applications of 0.5 or 2% aqueous solutions of Bronopol thickened with 2.5% methyl cellulose in a dose of 1 ml/kg to the clipped dorsal skins of rats from day 6 to 15 of pregnancy had no adverse effects on the dams or fetuses, aside from causing local skin reactions at the site of application.

Oral administration of Bronopol at 1, 3.3, or 10 mg/kg daily to rabbits from day 8 to 16 of pregnancy did not elicit embryotoxic or teratogenic effects, though the highest dose level suppressed maternal weight gain during the dosing period.

Bronopol, 20 or 40 mg/kg daily, given orally to rats from day 15 of gestation and throughout lactation did not affect parturition, litter size, postnatal survival, or development of the young. Fertility and general reproductive performance of rats were unaffected by these dose levels given to males from 63 days before mating, and females from 14 days before mating up to day 12 of pregnancy or until the litters were weaned 21 days postpartum. In this study, body weight gain of the males that received 40 mg/kg daily was slightly reduced.

G. Mutagenicity

Bronopol was not mutagenic under in vitro or in vivo conditions. It was tested using *Salmonella typhimurium* in an Ames-type assay and in a host-mediated assay in mice; in a dominant lethal assay in mice, the only noteworthy finding was antifertility arising from toxicity rather than dominant lethality.

H. Carcinogenicity

A tumorigenicity study in mice involved the application of 0.3 ml of aqueous acetone solutions containing 0.2 or 0.5% Bronopol to the shaved backs three times weekly for 80 weeks. The concentrations were selected after a preliminary tolerance study showed that 1% or more evoked a local reaction. Bronopol did not alter the spontaneous tumor profile, either locally or systemically.

Furthermore, a 2-year toxicity and tumorigenicity test, in which rats received Bronopol at 10, 40, or 160 mg/kg in the drinking water, provided no evidence to suggest any effects on tumor incidence. There was no indication of toxicity at 10 mg/kg per day, whereas the higher dose levels adversely affected growth, food intake, and survival rate. In addition, renal changes associated with diminished water intake, histological reactions in stomach and gastric lymph nodes (probably due to irritancy from prolonged exposure to Bronopol), and an exacerbation of spontaneous morphological alterations in the salivary gland were also observed at the higher dose levels in a dose-related manner.

I. General Pharmacology

Preliminary studies showed that Bronopol had only nonspecific effects on the cardiovascular system of the anesthetized cat and on the response of isolated smooth muscle to standard stimulants.

J. Acute Toxicity of Decomposition Products and Impurities

In mice, intraperitoneal injection of a 1% aqueous solution of Bronopol at pH 7–7.5, which had previously been maintained at 100°C for 1 hr, was less toxic than a freshly prepared solution.

As part of a preliminary acute toxicity screen, compounds identified as Bronopol impurities or degradation products have been examined in mice. The

TABLE 18 Acute Toxicity of Bronopol Impurities and Degradation Products [General formula: $HOCH_2 \cdot C(NO_2)R_1R_2$]

R_1	R_2	Occurrence[a]	Dose (mg/kg)	Number of deaths per number treated (mice)	Published oral LD_{50} (mg/kg)	Reference
H	CH_2OH	I*	1000	2/2	—	
			250	0/2		
Me	CH_2OH	I	1000	0/2	4000 (mouse)	36
					2000–3000 (rat)	37
					1000–2000 (rabbit)	37
CH_2OH	CH_2OH	D,I	1000	1/2	1900 (mouse)	38
					500–1000 (rat)	37
					250–500 (rabbit)	37
H	Br	D	1000	2/2	—	
			250	1/2		

[a]D, decomposition product; I, impurity; I*, impurity as sodium salt.

results are shown in Table 18, together with published LD_{50} values in various species.

K. Experience in Use

Bronopol has been marketed in the United Kingdom since 1965, and is now widely used in cosmetic, toiletry, and pharmaceutical formulations throughout the world. In a recent survey of customer complaints over a 3-year period about Bronopol-containing products marketed by The Boots Company Ltd., no evidence of complaints ascribable to the introduction of the compound has been observed.

A survey of the medical records of Boots staff involved over 8 years in the manufacture of the compound has produced no evidence of sensitization.

VI. COSMETIC APPLICATIONS

Bronopol is now widely used at a level of 0.01 or 0.02% as a preservative in various shampoos, bath foams, hair conditioners, rinses, creams, lotions, aqueous eye cosmetics, fabric softeners, and detergents; at higher concentrations it has been used as an active agent, for example, in aerosol and cream formulations [1]. It has been reported to show persistent activity on the skin by Marples and Kligman [39], and in vitro it has been shown to have a weak growth-inhibiting effect on cultured human skin cells [40].

The properties of the ideal preservative have been listed by many workers [23,41,42], but no single agent has yet matched up to the ideal. It is probably unrealistic to expect any one compound to possess all the properties attributed to the ideal preservative [43]; safety, compatibility with the ingredients and the packaging of the formulation, and microbiological efficacy are the most important factors. Other aspects to be considered are water solubility, the oil–water partition coefficient, and the effects of pH, temperature, and storage time on the stability of the preservative. Ideally, the preservative should be stable during the manufacture and for the shelf life of the product; it should also be cheap, odorless, and colorless at use concentrations.

A. Advantages

When the various properties of the ideal preservative are considered, it will be seen from the data presented here that Bronopol is safe at use concentrations.

Incompatibility with newer cosmetic ingredients and particularly with nonionic surfactants has accounted for the failure of some of the older preservatives which at one time were considered to be the ideal agents [41]. Unlike these, Bronopol is not adversely affected by nonionic or anionic surfactants [4,19,25,26]; in fact, synergism has been reported by Allwood [26]. The activity of Bronopol is not markedly affected by low concentrations of proteins [1,25] or by protein solutions used in shampoos [1,24,25].

The use of Bronopol in the preservation of shampoos, including those containing protein hydrolysates which are difficult to preserve [24], has been described by several workers [44–46]. For instance, it has been used for many years at a level of 0.01–0.02% in conventionally formulated shampoos based on sodium lauryl ether sulfates and alkanolamine alkyl sulfates with 2% or more of a foam-boosting alkanolamide. When challenged with about 1 ×

10^6 pseudomonads per milliliter, the bacterial count of these shampoos is reduced to less than 10 per milliliter in 24 hr and the inclusion of protein-derived materials, for example, 0.1–0.5% Crotein C (hydrolyzed collagen, Croda Chemicals Ltd.), does not affect this result [1]. Bryce and Smart [44] showed that Bronopol at 0.025% in a 30% aqueous solution of Empicol ESB (27% sodium diethoxylauryl sulfate) and in a 40% solution of Empicol TLP (38% triethanolamine lauryl sulfate with 1% alkylolamide) was lethal to a strain of *P. aeruginosa* within 1 hr at room temperature.

Bronopol is an effective antibacterial preservative in preparations for application to the skin [8,47]. Barnes and Denton [48] found that at a concentration of 0.02% it was one of the most satisfactory preservatives against gram-negative bacteria in a cream, suspension, or solution in their capacity test.

The fact that Bronopol-resistant organisms have not been encountered is also important. Bronopol is highly soluble in water, whereas its solubility in oils is very low. The oil–water partition coefficient is low: for example, 0.043 for liquid paraffin–water and 0.11 for arachis oil–water. Thus Bronopol will be available in the aqueous phase where microbial growth occurs.

Bronopol is an effective preservative over a wide pH range. Although it is more stable under acid than under alkaline conditions, it can be used as a labile preservative to combat contamination during manufacture in products at alkaline pH in combination with another more stable but perhaps less active preservative to combat in use contamination.

Combinations of preservatives can be justified on several grounds, one being increased spectrum of antimicrobial activity and another being synergism. The antibacterial activity of Bronopol is greater than its antifungal activity, but its spectrum can be increased by the addition of Myacide SP [30]. Parker [49], Proserpio [50] and Jacobs et al. [51] have examined combinations of Bronopol with other agents in cosmetics and oil–water emulsions. The latter workers found that this preservative plus parabens was satisfactory in an alkaline anionic lotion, whereas Bronopol alone was unsatisfactory. Croshaw [47] showed that while Bronopol alone was an effective antibacterial preservative in Oily Cream BP, combination with methyl p-hydroxybenzoate increased the antifungal protection.

B. Limitations

The main limitation of the use of Bronopol as a preservative is its instability at alkaline pH; its decomposition is also accelerated by increase in temperature. In the presence of light decomposition may be accompanied by discoloration of the product.

Bronopol is incompatible with some metals, for example, aluminium, so that the type of packaging must be carefully considered. It has been found to be stable as an active agent in a personal hygiene spray in a lacquered aluminium can and in an alcoholic deodorant spray in a lacquered tinplate can, and no corrosion of the cans was observed [1].

VII. OTHER APPLICATIONS

The broad antibacterial spectrum of activity of Bronopol, together with its extensive safety data, makes the compound potentially valuable in many areas where microbiological contamination is a problem. Although first employed in the 1960s primarily as a preservative and active agent in pharmaceuticals and as a preservative for cosmetics and toiletries, interest is now being shown in many other fields where protection from microbiological attack is required. Examples include emulsion paints; textile fiber lubricants and spin finishes; consumer products such as detergents, fabric softeners, and cleaning agents; processing solutions for hides and skins; oil extraction drilling muds and flooding waters; cooling towers and air conditioning humidifiers; and paper manufacturing water. Environmental toxicity studies have been carried out on an industrial grade of Bronopol to support its use in these applications, and no untoward findings have been observed.

VIII. REGULATORY AND INDEPENDENT REVIEW STATUS

In the United States a Drug Master File has been opened with the Food and Drug Administration, and the compound is included in the Toxic Substances Control Act inventory. As part of the Cosmetic Ingredient Review currently being carried out by the Cosmetic, Toiletry, and Fragrance Association, Inc., a safety assessment of Bronopol has recently been completed [52], with the following conclusions: "The evidence at hand indicates 2-bromo-2-nitropropane-1,3-diol to be safe as a cosmetic ingredient at concentrations up to and including 0.1% except under circumstances where its action with amines or amides can result in the formation of nitrosamines or nitrosamides."

In Europe the European Economic Community's provisional list of preservatives for use in cosmetics includes Bronopol for use without restriction at levels up to 0.1%. The final list of definitely permitted preservatives is not expected until 1984. In addition, the Council of Europe has classified antimicrobial preservatives in their booklet entitled *Cosmetic Products and Their Ingredients* [PA/SG (78)1]. Bronopol is included in Group A at a level of 0.1% without restriction. This group "contains ingredients which, in the opinion of the working Party, do not present a hazard to health when incorporated in cosmetic up to the concentration stated and in product types where specified."

In Switzerland Bronopol is provisionally approved for use as a pharmacologically active substance in cosmetics. Levels up to 0.05% are permitted in skin care formulations, and up to 0.1% in rinse-off presentations.

ACKNOWLEDGMENTS

The authors wish to acknowledge the assistance of other members of the staff of The Boots Company PLC in the production of the data included here.

REFERENCES

1. D. M. Bryce, B. Croshaw, J. E. Hall, V. R. Holland, and B. Lessel, The activity and safety of the antimicrobial agent Bronopol (2-bromo-2-nitropropane-1,3-diol), *J. Soc. Cosmet. Chem.*, 29:3–24 (1978).
2. E. B. Hodge, J. R. Dawkins, and E. Kropp, A new series of antifungal compounds, *J. Am. Pharm. Assoc. Sci. Ed.*, 43:501–502 (1954).
3. T. Zsolnai, Versuche zur Entdeckung neuer Fungistatika. II. Nitroverbindungen, *Biochem. Pharmacol.*, 5:287–304 (1961).
4. B. Croshaw, M. J. Groves, and B. Lessel, Some properties of Bronopol, a new antimicrobial agent active against *Pseudomonas aeruginosa*, *J. Pharm. Pharmacol. Suppl*, 16:127T–130T (1964).
5. N. G. Clark, B. Croshaw, B. E. Legetter, and D. F. Spooner, Synthesis

and antimicrobial activity of aliphatic nitro compounds, *J. Med. Chem.*, 17:977-981 (1974).
6. S. A. Malcolm and R. C. S. Woodroffe, The relationship between waterborne bacteria and shampoo spoilage, *J. Soc. Cosmet. Chem.*, 26:277-288 (1975).
7. L. J. Morse and L. E. Schoenbeck, Hand lotions—a potential nosocomial hazard, *N. Engl. J. Med.*, 278:376-378 (1968).
8. G. Sykes and R. Smart, Preservation of preparations for application to the skin, *Am. Perfum. Cosmet.*, 84:45-49 (1969).
9. R. Smart and D. F. Spooner, Microbiological spoilage in pharmaceuticals and cosmetics, *J. Soc. Cosmet. Chem.*, 23:721-737 (1972).
10. S. Tenenbaum, Significance of pseudomonads in cosmetic products, *Am. Perfum. Cosmet.*, 86:33-37 (1971).
11. H. S. Bean, S. M. Heman-Ackah, and J. Thomas, The activity of antibacterials in two-phase systems, *J. Soc. Cosmet. Chem.*, 16:15-30 (1965).
12. Y. Wachi, M. Yanagi, H. Katsura, and S. Ohta, Decomposition of surface-active agents by bacteria isolated from deionized water, *J. Soc. Cosmet. Chem.*, 31:67-84 (1980).
13. V. R. Holland, BNPD and nitrosamine formation, *Cosmet. Technol.*, 3:31-36 (1981).
14. K. E. Moore and R. J. Stretton, The effect of pH, temperature and certain media constituents on the stability and activity of the preservative, Bronopol, *J. Appl. Bacteriol.*, 51:483-494 (1981).
15. A. Kabay, Rapid quantitative microbiological assay of antibiotics and chemical preservatives of a non-antibiotic nature, *Appl. Microbiol.*, 22:752-755 (1971).
16. L. J. Hale and G. W. Inkley, A semiautomatic device for multiple inoculation of agar plates, *Lab. Pract.*, 14:452-453 (1965).
17. M. R. W. Brown and D. A. Norton, The preservation of ophthalmic preparations, *J. Soc. Cosmet. Chem.*, 16:369-393 (1965).
18. A. Mates and P. Zand, Specificity of the protective response induced by the slime layer of *Pseudomonas aeruginosa*, *J. Hyg. Lond.*, 73:75-84 (1974).
19. M. R. W. Brown, Turbidimetric method for the rapid evaluation of antimicrobial agents—inactivation of preservatives by non-ionic agents, *J. Soc. Cosmet. Chem.*, 17:185-195 (1966).
20. K. H. Wallhäusser, Die mikrobielle Kontamination von Kosmetika Rohstoffe—Produktion—Konservierung, *Perfuem. Kosmet.*, 53:305-319 (1972).
21. H. Saito and T. Onoda, Antibacterial action of Bronopol on various bacteria, especially on *Pseudomonas aeruginosa*, *Chemotherapy*, 22:1466-1473 (1974).
22. R Naito, T. Itoh, E. Hasegawa, H. Arimura, Y. Fujita, K. Hasegawa, T. Inaba, Y. Kagitani, S. Komeda, T. Matsumoto, H. Okamoto, K. Okano, Y. Oguro, and T. Ogushi, Bronopol as a substitute for thiomersal, *Dev. Biol. Stand.*, 24:39-48 (1974).
23. B. Croshaw, Preservatives for cosmetics and toiletries, *J. Soc. Cosmet. Chem.*, 28:3-16 (1977).
24. E. Tuttle, C. Phares, and R. F. Chiostri, Preservation of protein solutions with 2-bromo-2-nitro-1,3-propanediol (Bronopol), *Am. Perfum. Cosmet.*, 85:87-89 (1970).
25. Anonymous, Preservative properties of Bronopol, *Cosmet. Toilet.* 92:87-95 (1977).
26. M. C. Allwood, Inhibition of *Staphylococcus aureus* by combinations of

non-ionic surface active agents and antibacterial substances, *Microbios*, 7:209–214 (1973).
27. R. J. Stretton and T. W. Manson, Some aspects of the mode of action of the antibacterial compound bronopol (2-bromo-2-nitropropan-1,3-diol), *J. Appl. Bacteriol.*, 36:61–76 (1973).
28. M. B. King, R. Knox, and R. C. S. Woodroffe, Investigation of antituberculous substances—an agar diffusion method using *Mycobacterium smegmatis*, *Lancet*, 264:573–575 (1953).
29. O. G. Clausen, An examination of the bacteriostatic and bactericidal, fungistatic and fungicidal effects of cetylpyridinium chloride and 2-bromo-2-nitropropan-1,3-diol, separately and in combinations also including benzyl alcohol, *Pharm. Ind.*, 35:726–729 (1973).
30. K. D. Brunt, B. Croshaw, V. R. Holland, B. Marchant, and M. C. Meyer, An answer to fungal contamination, *Cosmet. Technol.*, 3:25–34 (1981).
31. K. E. Moore, Evaluating preservative efficacy in pharmaceutical and cosmetic products, Ph.D. thesis, University of Technology, Loughborough, England (1978).
32. F. N. Marzulli and H. I. Maibach, Antimicrobials: experimental contact sensitization in man, *J. Soc. Cosmet. Chem.*, 24:399–421 (1973).
33. F. N. Marzulli and H. I. Maibach, The use of graded concentrations in studying skin sensitizers: experimental contact sensitization in man, *Food Cosmet. Toxicol.*, 12:219–227 (1974).
34. H. I. Maibach, Dermal sensitization potential of 2-bromo-2-nitropropane-1,3-diol (Bronopol), *Contact Dermatitis*, 3:99 (1977).
35. W. Raab, 2-Nitro-2-bromopropan-1,3-diol (Bronopol) and UV-rays, *Arch. Dermatol. Res.*, 269:211–212 (1980).
36. IMC Chemical Group, Technical Data Sheet, NP Series TDS No. 15.
37. W. Machle, E. W. Scott, and J. Treon, The physiological response of animals to some simple mononitroparaffins and to certain derivatives of these compounds, *J. Ind. Hyg. Toxicol.*, 22:315–332 (1940).
38. IMC Chemical Group, Technical Data Sheet, NP Series TDS No. 5.
39. R. R. Marples and A. M. Kligman, Methods for evaluating topical antibacterial agents on human skin, *Antimicrob. Agents Chemother.*, 5:323–329 (1974).
40. T. Onoda and H. Saito, Influence of a new antibacterial agent, Bronopol, upon the growth of cultured cells, *Chemotherapy Tokyo*, 22:196–197 (1974).
41. D. L. Wedderburn, Preservation of emulsions against microbial attack, *Adv. Pharm. Sci.*, 1:195–268 (1964).
42. D. Coates, Preservatives—microbiological and physical factors, *Manuf. Chem. Aerosol News*, 44:35–37 (1973).
43. M. S. Parker, Requirements of a preservative system, *Soap Perfum. Cosmet.*, 46:291–292 (1973).
44. D. M. Bryce and R. Smart, The preservation of shampoos, *J. Soc. Cosmet. Chem.*, 16:187–201 (1965).
45. G. Schuster, Die Conservierung von Shampoos und Schaumbademitteln, *Seifen Oele Fette Wachse*, 99:489–492 (1973).
46. J. I. Yablonski and C. L. Goldman, Microbiology of shampoos, *Cosmet. Perfum.*, 90:45–52 (1975).
47. B. Croshaw, Bronopol as a preservative for oily cream B.P., *J. Appl. Bacteriol.*, 39:x (1975); paper read at the Summer Conference of the Society for Applied Microbiology, University of Nottingham, Nottingham, July 1975.

48. M. Barnes and G. W. Denton, Capacity tests for the evaluation of preservatives in formulation, *Soap Perfum. Cosmet.*, 42:729–733 (1969).
49. M. S. Parker, Some aspects of the use of preservatives in combination, *Soap Perfum. Cosmet.*, 46:223–224 (1973).
50. G. Proserpio, Protection des cosmétiques par des mélanges synergiques de préservateurs a dosage microbiocide, *Parfum. Cosmet. Savons, 2*: 305–315 (1972).
51. G. Jacobs, S. M. Henry and V. F. Cotty, The influence of pH, emulsifier, and accelerated ageing upon preservative requirements of o/w emulsions, *J. Soc. Cosmet. Chem.*, 26:105–117 (1975).
52. Cosmetic, Toiletry and Fragrance Association, Inc., Final report on the safety assessment for 2-bromo-2-nitropropane-1,3-diol (1980), *J. Environ. Pathol. Toxicol.*, 4:47–61 (1980).

5
ESTERS OF PARA-HYDROXYBENZOIC ACID

THOMAS E. HAAG and DONALD F. LONCRINI *Mallinckrodt, Inc., St. Louis, Missouri*

    I. Introduction 64
   II. Production 64
  III. Physical and Chemical Properties 65
  IV. Antimicrobial Activity 65
   V. Factors Influencing Antimicrobial Activity 69
      A. Surfactants 69
      B. Partition coefficients 70
      C. pH 70
      D. Other inactivating agents 71
  VI. Applications 71
 VII. Methods of Incorporation 71
VIII. Toxicity 72
      A. Acute toxicity 72
      B. Subchronic toxicity 72
      C. Chronic toxicity 72
      D. Biochemical behavior 72
      E. Carcinogenicity 73
      F. Teratogenicity 73
      G. Skin irritation studies 73
      H. Eye irritation studies 73
  IX. Commercial Products 74
   X. Methods of Detection and Determination 74
      References 74

INTRODUCTION

The alkyl esters of p-hydroxybenzoic acid, more commonly known in the United States as the parabens, are the most widely used preservatives in the cosmetic industry today [1]. They were first introduced by Sabalitschka in 1924 [2], when he presented a series of reports on their use as antimicrobial agents. Sabalitschka's primary goal was to find replacements for salicylic and benzoic acids, both of which are effective preservatives only in the highly acid pH range. His research was rewarding when he found that by esterifying p-hydroxybenzoic acid with various alkyl and aryl alcohols, the resulting products were effective over a wide pH range.

Official recognition of the parabens came in 1934 with the adoption of the methyl ester in the fifth edition of the *Swiss Pharmacopeia*. In the United States the methyl ester was first adopted in the *National Formulary VII* in 1942; both the methyl and propyl esters were admitted to the U.S. *Pharmacopeia XIII* in 1947. Methylparaben and propylparaben are also contained on the "generally recognized as safe" list (*Code of Federal Regulations*, Title 21 §184.1490 and 184.1670), with a maximum permissible use quantity of 0.1%.

In many ways the parabens are ideal cosmetic preservatives. They are essentially colorless, odorless, nonvolatile, stable, effective over a wide pH range, and relatively active against a broad spectrum of microorganisms. Their cost is low in relation to their use concentration. Numerous studies have indicated that the parabens have a low order of acute and chronic toxicity [3], although in some isolated cases contact dermatitis has been reported [4]. Therefore, considering the widespread usage of the parabens and the low incidence of adverse reactions, it is generally accepted that the parabens are among the safest of all cosmetic preservatives.

II. PRODUCTION

The esters of p-hydroxybenzoic acid are prepared by reacting the corresponding alcohols with p-hydroxybenzoic acid in the presence of an acid catalyst. The reaction products are formed in high yields, especially if the by-product water is removed azeotropically according to the following equation:

$$\text{HO-C}_6\text{H}_4\text{-COOH} + \text{ROH} \xrightarrow{\text{H}^+} \text{HO-C}_6\text{H}_4\text{-COOR} + \text{H}_2\text{O}$$

Generally, sulfuric acid or p-toluenesulfonic acid is recommended as the catalyst.

p-Hydroxybenzoic acid is produced commercially by treating the dry potassium salt of phenol with carbon dioxide at temperatures above 200°C under 4–7 atm of pressure. The resulting salt is then treated with acid to yield p-hydroxybenzoic acid:

$$\text{KO-C}_6\text{H}_5 \xrightarrow[\text{pressure}]{\text{CO}_2} \text{KO-C}_6\text{H}_4\text{-C(=O)-OK} + \text{C}_6\text{H}_5\text{OH}$$

III. PHYSICAL AND CHEMICAL PROPERTIES

The methyl, ethyl, propyl, and butyl esters of p-hydroxybenzoic acid (shown below) are the most widely used by the cosmetic industry [1].

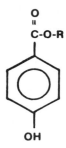

Methyl   R = CH_3
Ethyl   R = C_2H_5
Propyl   R = C_3H_7
Butyl   R = C_4H_9

These esters may be described as small colorless crystals or as white crystalline powders, slightly to very slightly soluble in water, and soluble to freely soluble in nonpolar vehicles. Their water solubility decreases with increasing molecular weight, while their lipid solubility increases with increasing molecular weight (see Table 1). A linear relationship exists between the alkyl chain length and the log of the water solubility, and also between chain length and the log of the partition coefficient [5].

The paraben esters are colorless in solution. The methyl ester has been reported to have a slight characteristic odor, whereas the other esters are practically odorless in their purified form [5]. No particular taste is noticeable in solutions of the esters, but in saturated aqueous solution these esters exhibit a slight tang or weak burning sensation in the mouth [5].

Both methylparaben and propylparaben undergo acid-catalyzed and base-catalyzed ester hydrolysis. Maximum stability for the paraben esters is in the pH range of 4–5 [6]. For example, negligible hydrolysis occurred after several hours of boiling an aqueous solution of paraben esters at neutral or slightly acidic pH levels, while slight hydrolysis occurred after the same length of time at pH 8 or above [5]. At higher temperatures and at higher pH ranges, hydrolysis occurred rapidly. To exemplify this, Raval and Parrott [7] found that after autoclaving (121°C) a methylparaben solution for 30 min at pH 6 and 9, there remained 94.5 and 58.0% of the initial paraben concentrations, respectively.

IV. ANTIMICROBIAL ACTIVITY

Table 2 shows the minimal inhibitory concentrations of methyl-, ethyl-, propyl-, and butylparaben against a variety of molds, yeasts, and bacteria common in cosmetic spoilage. In general, the parabens are more effective against yeast and molds than they are against bacteria, and more effective against gram-positive bacteria than against gram-negative bacteria. Because of their limited water solubility, parabens are not particularly effective

TABLE 1 Properties of the para-Hydroxybenzoic Acid Esters

Property	Methyl	Ethyl	Propyl	Butyl
Molecular weight	152.14	166.17	180.20	194.23
Appearance	White powder	White powder	White powder	White powder
Melting point (°C)	125–127	116–118	96–98	69–71
Solubility (g/100g at 25°C)				
Water				
25°C	0.25	0.11	0.04	0.015
15°C	0.16	0.08	0.023	0.005
80°C	3.2	0.86	0.45	0.15
Methanol	59	115	124	220
Ethanol	52	70	95	210
Ethanol (10%)	0.5	—	0.1	—
Ethanol (50%)	18	—	18	—
Propylene glycol	22	25	26	110
Propylene glycol (10%)	0.3	—	0.06	—
Propylene glycol (50%)	2.7	—	0.9	—
Glycerin	1.7	0.5	0.4	0.3
Peanut oil	0.5	1	1.4	5
Mineral oil	0.01	0.025	0.03	0.1

TABLE 2 Paraben Antimicrobial Activity[a]

Microorganism	Concentration Required to Inhibit Growth (ppm)			
	Methylparaben	Ethylparaben	Propylparaben	Butylparaben
Bacillus subtilis ATCC 6633	2000	1000	500	250
Staphylococcus aureus ATCC 6538	2000	1000	500	125
Staphylococcus epidermidis ATCC 12228	2000	1000	500	250
Escherichia coli ATCC 8739	2000	1000	500	500
Klebsiella pneumoniae ATCC 8308	1000	500	500	250
Salmonella typhosa ATCC 6539	1000	1000	500	500
Proteus vulgaris ATCC 13315	1000	500	250	125
Serratia marcescens ATCC 8100	1000	1000	500	500
Enterobacter cloacae ATCC 23355	1000	1000	500	250
Pseudomonas aeruginosa ATCC 9027	4000	>2000	>1000	>1000
Pseudomonas aeruginosa ATCC 15442	4000	>2000	>1000	>1000
Pseudomonas stutzeri	2000	1000	500	500
Candida albicans ATCC 10231	1000	500	250	125
Saccharomyces cerevisiae	1000	500	125	32
Aspergillus niger ATCC 9642	1000	500	250	125
Penicillium chrysogenum ATCC 9480	500	250	125	63
Trichophyton mentagrophytes	250	125	63	32

[a]A standard agar inhibition test utilizing tryptic soy agar, at pH 7.3, was employed. The inoculum size equaled 10^6–10^7 colony-forming units per milliliter.

against *Pseudomonas aeruginosa,* since more preservative is required for inhibition than would be soluble in water at room temperature.

The antimicrobial activity of the p-hydroxybenzoic acid esters increases with increasing chain length of the alkyl moiety up to and including the butyl ester. The benzyl ester has been reported to have even higher activity [8], but it has not gained widespread acceptance in the cosmetic industry. The limitations of use imposed on the longer-chain esters are due to their poor aqueous solubilities. Thus, in practice, the lower esters (i.e., methyl, ethyl, propyl, butyl) are most commonly used. The problem of effectiveness versus aqueous solubility can be overcome somewhat by combining two or more paraben esters in the product formulation. Aalto et al. [5] found that *Aspergillus niger* was inhibited by 0.05% methylparaben or 0.015% propylparaben. By combining these two, one can reduce the paraben concentration required for complete inhibition of this microorganism to 0.025% methylparaben and 0.01% propylparaben. This is an additive antimicrobial effect. According to Gottfried [9], the main advantage of using paraben mixtures is that the solubilities of the individual esters are independent of each other. Thus a higher total concentration of preservative may be incorporated into the product.

O'Neill et al. [10] compared the preservative efficacy of methyl-, ethyl-, propyl-, and butylparaben in "problem" products. (Problem products were defined as products containing high levels of vegetable oils emulsified with nonionic surfactants and thickened with weak acid polyelectrolytes at pH values above 7.0). Methylparaben was found to be the most effective ester in terms of preservative capabilities. The authors did not show any examples of better preservation with a mixed paraben system than with an equal weight of the methyl ester. Thus they concluded that the least soluble paraben in the product is the most efficient in initial potency and capacity. The general greater efficacy of methylparaben in such systems was attributed in part to its lower solubility in vegetable oils when compared to the longer-chain esters.

The mechanism of action of the p-hydroxybenzoates corresponds in principle to that of phenol. One proposed mechanism involves the destruction of the cell membrane and protein denaturation in the interior of the cell. Another involves competitive reactions with coenzymes [11].

Development of resistance was shown not to occur by Lück and Rickerl [12]. (Resistance is defined here as an increase in the minimum inhibitory concentrations under the influence of subthreshold concentrations of p-hydroxybenzoic acid esters.) However, more recent work by Peterson et al. [13] indicated that *Escherichia coli* has the ability to adapt to paraben solutions. Similarly O'Neill and Mead [14] stated that the rate of adaptation increased markedly with increasing molecular weight; *E. coli* adapted readily to propylparaben and hardly at all to the methyl analog. As a result, O'Neill ranked the parabens in the order methyl > ethyl > propyl > butyl > benzyl in their antimicrobial capacities.

Various agents have been reported to potentiate the antimicrobial activity of the parabens. Prickett et al. [15] demonstrated that 2–5% propylene glycol enhanced the antimicrobial activity of the parabens against vegetative cells, but not against bacterial spores. Phenylethanol was reported by Richards and McBride [16] to enhance paraben antimicrobial activity in ophthalmic preparations. Similarly, ethylenediaminetetraacetate (EDTA) increased the activity of the p-hydroxybenzoic acid esters against gram-negative organisms, especially *P. aeruginosa* [17].

V. FACTORS INFLUENCING ANTIMICROBIAL ACTIVITY

The antimicrobial action of the parabens is inactivated by several known mechanisms. The more notable ones that can significantly influence activity are the type and concentration of surfactant, the oil-to-water ratio, pH, and the presence of certain macromolecules or other potentially binding solids.

A. Surfactants

Paraben inactivation by nonionic surfactants has been well known for several decades. Their inactivating effects are of greater significance than those of anionic and cationic systems because nonionic surfactants have no growth-inhibiting properties. Thus the necessity for adequate preservation of systems containing them is much greater.

Bolle and Mirimanoff [18] were the first to emphasize the reduced effect of methylparaben in the presence of several structurally different nonionic surfactants. Their findings stimulated a widespread interest in the pharmaceutical and cosmetic industries. Investigations by de Navarre [19] and deNavarre and Bailey [20] led to the conclusion that any soluble ethoxylated nonionic surfactant would inactivate the paraben esters at use concentrations in an agar medium. For example, deNavarre found that 0.1% paraben was ineffective in the presence of 1% nonionic surfactant. Similar studies by Blaug and Ahsan [21] dealing with the interaction of the methyl, ethyl, propyl, and butyl esters with several nonionic macromolecules indicated that the higher molecular weight esters were inactivated to a greater extent than those of lower molecular weight. Other studies and reviews in the literature have confirmed the ability of nonionic surfactants to reduce paraben activity [22–25].

The mechanism of paraben inactivation by nonionic surfactants has also received considerable study. Aoki et al. [26] attributed the decrease of paraben activity to the envelopment of the paraben in a micelle formed by the nonionic surfactant. Evans [27] explained inactivation on the basis of solubilization of the preservative in the nonionic micelles. Thus compounds with high micellar solubility would be present in insufficient quantities in the aqueous phase to inhibit microbial growth.

Patel and Kostenbauder [28] employed a semipermeable membrane to evaluate the degree of binding or intermolecular association between paraben esters and Tween 80. Both the methyl and propyl esters were investigated, and a relatively high degree of interaction with Tween 80 was observed for both. These authors concluded that the binding was a function of both the concentration of unbound p-hydroxybenzoate and the concentration of Tween 80. Patel and Kostenbauder also noted that the propyl ester had a far greater affinity for Tween 80 than the methyl ester. At a concentration of 5% Tween 80, 22% of the total methylparaben existed as a free preservative, whereas in the case of the propyl ester, only 4.5% existed as free propylparaben. paraben.

Later studies by Pisano and Kostenbauder [29] demonstrated that the activity of p-hydroxybenzoates in the presence of nonionic surfactants is a function of the unbound preservative. They suggested that the total concentration of preservative required to protect a nonionic formulation could be estimated from predetermined binding data. These investigations suggest that the required concentration of p-hydroxybenzoates can be estimated by multiplying the preservative use level by the ratio of total to free preservative at any given Tween 80 concentration. This approach to emulsion preservation assumes, however, that other factors such as the oil–water ratio or pH do not contribute to the efficacy of the preservative. Since it is known that these factors can exert powerful influences, it would be wrong to rely on this kind of prediction to the exclusion of other considerations.

While most nonionic inactivation studies were conducted in relatively simple aqueous systems, Charles and Carter [30] determined the activity of p-hydroxybenzoates in 18 typical nonionic cosmetic emulsions. They found the preservatives to be more effective in these products than would be anticipated from results obtained with simpler systems. Their results indicated the need for testing the final finished product with all components present.

Propylene glycol has been shown to potentiate the activity of the p-hydroxybenzoates in the presence of nonionic surfactants. According to Poprzan and deNavarre [31], 10% propylene glycol prevented the interaction between methylparaben and Tween 80.

B. Partition Coefficients

Owing to their high oil solubility, p-hydroxybenzoates have relatively unfavorable partition coefficients between the oil and water phases of emulsions. Protection of these emulsions requires that lethal levels of the preservative contact the microorganisms in the aqueous phase. In an emulsion containing 50% or more oil, more preservative may be present in the oil phase than in the aqueous phase after equilibrium is achieved. Thus a considerable proportion of the preservative can become inactivated by migration into the wrong phase. This migration can occur over a considerable period of time, which is one more reason why prolonged microbiological testing of formulated products is required.

Hibbot and Monks [32] determined the oil—water partition coefficient of methylparaben in a number of oils and waxes. They then tested the activity of this preservative in a series of creams prepared from these oils and waxes covering a range of calculated partition coefficients. Results of their tests showed good correlation between activity and partition coefficients, indicating that this property can be used to estimate the performance of a preservative in an emulsion.

Bean et al. [33] determined the parition coefficients of methylparaben in aqueous mixtures and emulsions of mineral oil, safflower oil, isopropyl palmitate, and lanolin. The oil—water coefficient for the preservative in mineral oil was 0.11 because of its limited solubility in this vehicle. The partition coefficients in the other oils, on the other hand, were about 100—150 times greater than in mineral oil.

The partition coefficient of many preservatives can be reduced by the addition to the aqueous phase of a substance in which the preservative is preferentially soluble. Propylene glycol, glycerin, sorbitol, and hexylene glycol are examples which can have advantageous effects in certain systems [34]. Humectants can, of course, provide an added benefit by increasing the osmotic pressure of the aqueous phase and thus increasing the inhibitory effect of the preservative. Propylene glycol, in addition, appears to have toxic effects on bacteria and fungi in its own right if used in the aqueous phase at a concentration of 15%.

C. pH

The parabens are effective over a pH range of 4—8. As the pH rises above 8, the parabens become less effective because of disassociation of the molecule.

At a pH of 8.5, 50% of the compound is disassociated, and considerable loss of antimicrobial activity occurs. At pH 7 the parabens are 60-65% effective [35].

D. Other Inactivating Agents

Binding of parabens by other macromolecular materials has been reported. Interaction between methyl cellulose and the parabens was reported by Tillman and Kuramoto [36]. The methyl p-hydroxybenzoate showed the greatest binding effect, while the butyl ester showed the least. According to Miyawaki et al. [37], methyl- and propylparaben are bound to some extent with polyethylene glycol, polyvinylpyrrolidone, and gelatin. No significant interaction was noted with carboxymethylcellulose or tragacanth. These authors concluded that the concentrations of macromolecular polymers commonly used were not sufficient to interfere seriously with the use of the parabens.

VI. APPLICATIONS

The paraben esters are commonly used at levels of 0.1-0.8% in cosmetic products. Combinations of these esters are generally found to be more effective against bacteria, yeasts, and molds than single esters. These combinations take advantage of the fact that each ester is independently soluble, and as a result, a higher combined concentration can effectively be introduced into a formulation.

In many cosmetic formulations the paraben esters are used in combination with other preservations. Advantages of using preservative combinations include the following aspects: (1) The spectrum of activity can be increased, (2) the toxicological hazard can be reduced by using lower concentrations of component preservatives, (3) the development of resistance by an organism to one preservative alone may be prevented, and (4) additive or synergistic activities may result. The combination of Germall 115 (Sutton Laboratories, Inc.) and the paraben esters has been recommended [38]. Mixtures of the paraben esters with Glydant (Glyco Chemicals, Inc.), Bronopol (The Boots Co., Ltd.), and Dowicil 200 (Dow Chemical Co.), among others, are also common throughout the cosmetic industry.

VII. METHODS OF INCORPORATION

The incorporation of the p-hydroxybenzoic acid esters into various formulations requires a familiarity with their solubility characteristics, particularly in water. Even though their water solubility is limited, they can be dissolved in aqueous products at room temperature, provided that good agitation is maintained. The rate of solution is a function of particle size. With the fine powders that are normally available, aqueous concentrations of 0.20% methyl, 0.10% ethyl, 0.03% propyl, or 0.015% butyl can be achieved in 20 or 30 min. A coarse powder, on the other hand, dissolves only after a long period of continuous stirring. Dissolution can be accomplished more readily by heating the medium to 70 or 75°C and stirring until solution is complete. This temperature should be kept about 20°C below the melting point of the paraben ester used to prevent agglomeration of the paraben particles.

Another method of incorporation involves the preparation of stock solutions of the paraben esters in solvents such as alcohol or propylene glycol.

Such a solution must be added slowly to the aqueous phase of the formulation in order to prevent precipitation of the ester.

VIII. TOXICITY

A. Acute Toxicity

Sokol [3] reported the acute oral LD_{50} in mice to be greater than 8 g/kg of body weight for the methyl, ethyl, and propyl esters, and greater than 5 g/kg for the butyl ester. In a similar study, Matthews and co-workers [39] determined the LD_{50} in mice of orally administered methylparaben to be 8 g/kg of body weight, and the intraperitoneal LD_{50} to be 960 mg/kg of body weight. These authors observed the acute toxicity to increase with an increase in the molecular weight of the ester, and they judged the butyl ester to be approximately three times as toxic as the methyl ester. Symptoms of acute toxicity observed include ataxia, paralysis, hypotension, and general and myocardial depression. The animals that survived the test, however, recovered rapidly.

B. Subchronic Toxicity

Matthews et al. [39] reported that dogs fed 500 and 1000 mg/kg methylparaben per day were unaffected after 1 year on the regimen. Sabalitschka and Neufeld-Crzellitzer [40] reported that a man was unaffected after taking 2000 mg of methylparaben daily for 1 month. Similar studies by Braccesi [41] confirmed these observations.

C. Chronic Toxicity

Weanling Wistar rats, fed 0.9—1.2 g/kg of either methyl- or propylparaben for 96 weeks remained indistinguishable from the controls. Autopsies revealed no pathological changes in kidney, liver, heart, lung, spleen, or pancreas. When the dosage of either compound was increased to approximately four times these amounts, rats showed a slower rate of weight gain than the controls. These authors estimated that the toxic threshold for rats of both methyl- and propylparaben is at least 3000 mg/kg per day [39].

Sokol [3] reported that propylparaben stimulated growth in rats when fed at a dosage of 150 mg/kg per day for 18 months. On the other hand a dosage of 1500 mg/kg per day for the same period resulted in a growth depression without specific pathology.

A survey by the U.S. Department of Commerce [42] revealed that there are no short-term toxicological consequences in the rat, rabbit, cat, dog, or man, and no long-term toxicological consequences in rats when the parabens are consumed in amounts greatly exceeding those currently consumed in the normal diet of the U.S. population.

D. Biochemical Behavior

p-Hydroxybenzoic acid esters are readily and completely absorbed from the gastrointestinal tract. In a study by Matthews et al. [39] in which dogs were orally administered 1.0 g of methyl- or propylparaben per kilogram body weight per day for 1 year, between 96 and 100% of the daily administered dose was accounted for in the urine. In a similar study Jones et al. [43] found no accumulation of methyl- or propylparaben or their metabolites in dogs which had been given 1.0 g/kg per day for 1 year. These authors were able to account for 96% of the daily administered dose in the urine in the form of free or conjugated p-hydroxybenzoic acid. By giving a fasted man 70 mg of methylparaben per kilogram body weight, they concluded that the paraben is handled similarly in human metabolism.

E. Carcinogenicity

A carcinogenic study reported by Homberger [44] which involved several techniques for ascertaining carcinogenicity, that is, subcutaneous injection—secondary host transfer, intravenous injection—observation of lung adenomas, and cocarcinogenesis, showed that methylparaben did not have carcinogenic activity. Boyland et al. [45] dissolved methylparaben in polyethylene glycol and, by introducing it twice weekly into the vaginas of weanling mice for 18 months, found no initiation of carcinomas.

F. Teratogenicity

Methylparaben elicited no teratogenic response in pregnant mice or rats when these were fed up to 550 mg/kg per day for 10 consecutive days, or in pregnant hamsters when these were fed up to 300 mg/kg per day for 5 consecutive days [46].

G. Skin Irritation Studies

Matthews et al. [39] reported Shelanski's [47] results on human patch tests carried out with the paraben esters. The esters were dissolved in propylene glycol at 5, 7, 10, 12, and 15%. The highest percentages not causing irritation were methyl, 5%; ethyl, 7%; propyl, 12%; and butyl, 5%. Repeated insult tests were also carried out on 50 subjects and no sensitization was observed.

The parabens, when used in topical therapeutic agents, have been reported to sensitize approximately 1% of the population with eczematous dermatitis [4]. However, these identical parabens are generally considered "safe" in cosmetic preparations widely used by thousands of persons. The reason for this is that cosmetics are usually applied to normal skin, whereas therapeutic agents are applied to inflamed, eczematous, excoriated, or otherwise damaged skin, which is much more readily sensitized than even the thinnest normal skin.

Simpson [48] reported a rare instance of a patient with allergic hypersensitivity to parabens in several cosmetic preparations. Similarly, Henry et al. [49] reported a case of contact urticaria developed in a patient after topical application of several paraben-containing compounds. While occasional cases of paraben sensitivity occur, the incidence of sensitization is low when the extensive use of the material is considered.

H. Eye Irritation Studies

Methyl- and propylparaben as saturated aqueous solutions have been reported to be moderately irritating to the eye [50]. According to Foster [51], nearly saturated solutions of parabens are too irritating to the eyes for inclusion in ophthalmic preparations.

IX. COMMERCIAL PRODUCTS

The methyl, ethyl, propyl, and butyl esters of p-hydroxybenzoic acid are generally sold directly by domestic producers, although some imported material is handled through distributors and agents. Blends of the methyl and propyl esters are also available to the industry.

A liquid paraben is now commercially available to the cosmetic and toiletry industry. LiquaPar (Mallinckrodt, Inc.), a 50% by weight paraben oil-in-water emulsion, consists of 20% isopropyl-p-hydroxybenzoic acid, 15% isobutyl-p-hydroxybenzoic acid, and 15% n-butyl-p-hydroxybenzoic acid. The remaining 50% is water containing emulsion stabilizers. The principle behind the product is that the three solid paraben esters, when combined in the proper ratio, form a eutectic oil which has a freezing point near 0°C. This oil can then be easily emulsified.

According to the manufacturer, LiquaPar offers the following advantages over conventional crystalline paraben esters: ease of incorporation into formulations, since additional solvents are not necessary; a decreased chance of error in formulation measurements; elimination of paraben particle size problems, thus eliminating the chance of grittiness in creams or lotions; substantial savings in labor and product preparation time; and cost effectiveness in many formulations.

X. METHODS OF DETECTION AND DETERMINATION

The literature contains a number of papers on the detection and determination of the p-hydroxybenzoates in various formulations. A colorimetric method, described by Jones et al. [43], makes use of the reagent 4-aminoantipyrine. Higuchi and co-workers [52] described a method based on preliminary extraction of the paraben from the product, separation of the component esters by partition chromatography, and subsequent determination of the individual constituents by ultraviolet spectrophotometry.

Liem [53] described a quantitative determination of the paraben esters using gas–liquid chromatography. In this procedure, the parabens are extracted from the product with methanol, purified by thin-layer chromatography, and derivatized with N,O-bis-trimethyl-silyl acetamide prior to injection. The concentrations of the esters are calculated from a standard calibration curve.

Methods of detection and determination have also been developed using high-performance liquid chromatography [54].

REFERENCES

1. E. Richardson. Update-frequency of preservative use in cosmetic formulas as disclosed to FDA, *Cosmet. Toiletr.*, 96:91–92 (1981).
2. T. Sabalitschka. Chemische Konstitution und Conservierungsvermogen, *Chem. Ztg.*, 48:703 (1924).
3. H. Sokol. Recent developments in the preservation of pharmaceuticals, *Drug Stand.*, 20:89–106 (1952).
4. A. Fisher. Cosmetic dermatitis part II, reactions to some commonly used preservatives, *Curr. Contact News*, 26:136–148 (1980).
5. T. Aalto, M. Firman, and N. Rigler. p-Hydroxybenzoic acid esters as preservatives I, *J. Am. Pharm. Assoc. Sci. Ed.*, 42:449–457 (1953).
6. K. Connors, G. Amidon, and L. Kennon. *Chemical Stability of Pharmaceuticals*, Wiley, New York, (1979), pp. 260–265, 307–310.

7. N. Raval and E. Parrott. Hydrolysis of methylparaben, *J. Pharm. Sci.*, *56*:274–275 (1967).
8. E. MacDonald. The preservative p-hydroxybenzoic acid esters, *J. Am. Pharm. Assoc. Prac. Pharm. Ed.*, *3*:181 (1942).
9. N. Gottfried. Alkyl p-hydroxybenzoate esters as pharmaceutical preservatives. A review of the parabens, *Am. J. Hosp. Pharm.*, *19*:310–314 (1962).
10. J. O'Neill, P. Peelor, A. Peterson, and C. Strube. Selection of parabens as preservatives for cosmetics and toiletries, *J. Soc. Cosmet. Chem.*, *30*:25–38 (1979).
11. E. Lueck. *Antimicrobial Food Additives*, Springer-Verlag, New York (1980), p. 223.
12. H. Lück, and E. Rickerl. Untersuchungen an *Escherichia coli* über eine Resistenzsteigerung gegen Konservierungsmittel und Antibiotica, *Z. Lebensm. Unters. Forsch.*, *109*:322–329 (1959).
13. A. Peterson, J. O'Neill, and C. Mead. Preservation: from art to science, *Dev. Ind. Microbiol.*, *21*:161–165 (1980).
14. J. O'Neill and C. Mead. The parabens: bacterial adaptation and preservative capacity, presentation made to the Society of Cosmetic Chemists, Washington, D.C. (1981).
15. P. Prickett, H. Murray, and N. Mercer. Potentiation of preservatives (parabens) in pharmaceutical formulations by low concentrations of propylene glycol, *J. Pharm. Sci.*, *50*:316–320 (1961).
16. R. Richards and R. McBride. Phenylethanol enhancement of preservatives used in ophthalmic preparations, *J. Pharm. Pharmacol.*, *23*:141S–146S (1971).
17. Mallinckrodt, Inc., unpublished data, Preservatives Laboratory, Mallinckrodt, Inc., St. Louis, Mo (1981).
18. A. Bolle and A. Mirimanoff. Antagonism between nonionic detergents and antiseptics, *J. Pharm. Pharmacol.*, *2*:685–691 (1950).
19. M. deNavarre. The interference of nonionic emulsifiers with preservatives with special reference to cosmetics, *J. Soc. Cosmet. Chem.*, *8*:371–380 (1957).
20. M. deNavarre, and H. Bailey. The interference of nonionic emulsifiers with preservatives. II, *J. Soc. Cosmet. Chem.*, *7*:427–433 (1956).
21. S. Blaug and S. Ahsan. Interaction of parabens with nonionic molecules, *J. Pharm. Sci.*, *50*:441–443 (1961).
22. A. Beckett and A. Robinson. The inactivation of preservatives by nonionic surface active agents, *Soap Perfum. Cosmet.*, *31*:454–459 (1958).
23. M. deNavarre. The interference of nonionic emulsifiers with preservatives, *Am. Perfum. Aromat.*, 1st Doc. Ed., 99–100 (1960).
24. G. Nowak. The preservation of nonionic emulsions, *Soap Perfum. Cosmet.*, *36*:914–924 (1963).
25. D. Wedderburn. Preservation of emulsions against microbial attack, in *Advances in Pharmaceutical Sciences* (H. Bean, A. Beckett, and J. Carless, eds.), Academic, New York (1964), pp. 195–268.
26. A. Aoki, A. Kameta, I. Yoshioka, and T. Matsuzaki. Application of surface active agents to pharmaceutical preparations. I. Effect of Tween 20 upon the antifungal activities of p-hydroxybenzoic acid esters in solubilized preparations, *J. Pharm. Soc. Jpn.*, *76*:939–943 (1956).

27. W. Evans. The solubilization and inactivation of preservatives by non-ionic detergents, *J. Pharm. Pharmacol.*, *16*:323–331 (1964).
28. N. Patel and H. Kostenbauder. Interaction of preservatives with macromolecules. I, *J. Am. Pharm. Assoc. Sci. Ed.*, *47*:289–293 (1958).
29. F. Pisano and H. Kostenbauder. Interaction of preservatives with macromolecules. II, *J. Am. Pharm. Assoc. Sci. Ed.*, *48*:310–314 (1959).
30. R. Charles and P. Carter. The effect of sorbic acid and other preservatives on organism growth in typical nonionic emulsified commercial cosmetics, *J. Soc. Cosmet. Chem.*, *10*:383–394 (1959).
31. J. Poprzan and M. deNavarre. The interference of nonionic emulsifiers with preservatives. VIII, *J. Soc. Cosmet.*, *10*:81–87 (1959).
32. H. Hibbot and J. Monks. Preservation of emulsions—p-hydroxybenzoic ester partition coefficient, *J. Soc. Cosmet. Chem.*, *12*:2–8 (1961).
33. H. Bean, G. Konning, and J. Thomas. Significance of the partition coefficient of a preservative in cosmetic emulsions, *Am. Perfum. Cosmet.*, *85*:61–65 (1970).
34. D. Wedderburn. Interactions in cosmetic preservation, *Am. Perfum. Cosmet.*, *85*:49–53 (1970).
35. I. R. Gucklhorn. Antimicrobials in cosmetics, Part 3, *Manuf. Chem. Aerosol News*, *40*:71–75 (1969).
36. W. Tillman and R. Kuramoto. A study of the interaction between methylcellulose and preservatives, *J. Am. Pharm. Assoc.*, *46*:211–214 (1957).
37. G. Miyawaki, N. Patel, and H. Kostenbauder. Interaction of preservatives with macromolecules. III. Parahydroxybenzoic acid esters in the presence of hydrophilic polymers, *J. Am. Pharm. Assoc. Sci. Ed.*, *48*:315 (1959).
38. Sutton Laboratories, Germall 115 Technical Bulletin (1973).
39. C. Matthews, J. Davidson, E. Bauer, J. Morrison and A. Richardson. p-Hydroxybenzoic acid esters as preservatives II. Acute and chronic toxicity in dogs, rats, and mice, *J. Am. Pharm. Assoc. Sci. Ed.*, *45*:260–267 (1956).
40. T. Sabalitschka and R. Neufeld-Crzellitzer. Behavior of p-hydroxybenzoic acid in the human body, *Arzneim. Forsch*, *4*:575–579 (1954).
41. M. Braccesi. Antiseptic properties of a derivative of salicylic acid: methyl ester of p-hydroxybenzoic acid, *Boll. Soc. Ital. Biol. Sper.*, *14*:265–266 (1939).
42. Informatics, Inc., GRAS (Generally Recognized as Safe) food ingredients: methyl and propyl paraben, PB-221209, National Technical Information Service, U.S. Department of Commerce, Springfield, Va. (1972).
43. P. Jones, D. Thigpen, J. Morrison, and A. Richardson. p-Hydroxybenzoic acid as a preservative. 3. The physiological disposition of p-hydroxybenzoic acid and its esters, *J. Am. Pharm. Assoc. Sci. Ed.*, *45*:268–273 (1956).
44. F. Homberger. Carcinogenicity of several compounds, PB-183027, National Technical Information Service, U.S. Department of Commerce, Springfield, Va. (1968).
45. E. Boyland, R. Charles, and N. Gowing. The induction of tumours in mice through intravaginal application of chemical compounds, *Br. J. Cancer*, *15*:252–256 (1961).

46. Food and Drug Research Laboratories, Inc., Teratologic evaluation of FDA 71-38 (methyl paraben) in mice, rats, and hamsters, report prepared under DHEW contract No. FDA 71-260, Maspeth, N.Y. (1972).
47. H. Shelanski. Industrial Toxicology Laboratories, Philadelphia, Pa.
48. J. Simpson. Dermatitis due to parabens in cosmetic creams, Contact Dermatitis, 5:311 (1978).
49. J. Henry, E. Tschen, and L. Becker. Contact urticaria to parabens, Arch. Dermatol., 115:1231 (1979).
50. D. Vaughan and S. Riegelman. Management of corneal ulcers, in Ocular Pharmacology and Therapeutics and the Problems of Medical Management (S. Kimura and E. Goodner, eds.), Publisher, City (1963), pp. 129–135.
51. J. Foster. Solution for eye-drops, Pharm. J., 192:429 (1964).
52. T. Higuchi, K. Patel, E. Bonow, and J. Landsman. Chromatographic separation and determination of mixtures of p-hydroxybenzoate esters, J. Am. Pharm. Assoc. Sci. Ed., 41:293–295 (1952).
53. D. Liem. Analysis of antimicrobial compounds in cosmetics, Cosmet. Toiletr., 92:59–72 (1977).
54. Mallinckrodt, Inc., Technical Bulletin (1981).

6
COSMETICALLY ACCEPTABLE PHENOXYETHANOL

ALLEN L. HALL *Emery Industries, Inc., Cincinnati, Ohio*

    I. Introduction 79

        A. Historical perspective 79
        B. Physical and chemical properties 80

   II. Manufacture 83

  III. Microbiology 84

        A. Literature and history 84
        B. Minimum inhibitor concentration procedures and results 84
        C. Mechanism 85
        D. Microbial testing 86

  IV. Formulation effects 100

        A. Formulations 100
        B. Viscosity and stability testing 103

   V. Toxicology 106

  VI. Summary 107

       References 107

I. INTRODUCTION

A. Historical Perspective

Phenoxyethanol is one of those simple materials which has been known and used in the chemical industry for decades. It is synthesized by a simple, well-known process, and has been the object of a steady demand because of its solvent properties, particularly for inks, dyes, and many other materials [1].

The microbiological properties of phenoxyethanol were noted early in its chemical history in an article published in the British journal *Lancet* [2]. In that report it was observed that *Pseudomonas* infections following gunshot injuries could be controlled or eliminated by treatment with a 2% aqueous solution of phenoxyethanol.

Since those early days, the number of published reports of utilization of phenoxyethanol as an antimicrobial has risen and fallen in an almost whimsical fashion. Phenoxyethanol has been "rediscovered" in response to specific needs and then cavalierly discarded as new products of higher technology have been introduced. Now again, phenoxyethanol is being rediscovered as a cosmetic preservative owing to a set of market conditions which find virtue in the old, proven remedies.

B. Physical and Chemical Properties

Because of the age and history of this compound, it has picked up many synonyms, probably a new one every time it was "rediscovered." A short list of these names together with the Chemical Abstracts Service (CAS) Registry Number and the Cosmetic, Toiletry, and Fragrance Association (CTFA) appellation is presented in Table 1.

Some of the common physical property values are listed in Table 2. It should be specifically noted that these are literature values for absolute phenoxyethanol, not what would be seen with a cosmetic-grade material. In all other cases the values reported throughout this chapter are those obtained using Emeressence 1160 Rose Ether phenoxyethanol, a product of Emery Industries, Cincinnati, Ohio. Typically, Rose Ether phenoxyethanol is 92—94% phenoxyethanol, and the balance of the material is the diethoxylate.

The major physical difference between absolute phenoxyethanol and the cosmetically acceptable grade is that in melting point. The presence of the diethoxylate significantly depresses the melting point so that, rather than having a melting point of 14°C, the material tends to form heavy syrup or glass at temperatures below -10°C. This is a significant advantage for the formulator, because it permits colder storage temperatures without subsequent use of melting facilities and energy demands.

TABLE 1 Phenoxyethanol Synonyms

Chemical formula	$C_8H_{10}O_2$
CAS registry number	122-99-6
CTFA label name	Phenoxyethanol
Synonyms	Ethylene glycol phenyl ether
	Glycol monophenyl ether
	2-Phenoxyethanol
	β-Phenoxyethanol
	Phenoxyethyl alcohol
	Phenyl Cellosolve
	Phenoxetol

TABLE 2 Physical Properties of Phenoxyethanol (Absolute)

Molecular weight	138.2
Specific gravity	1.1094 (20/20°C)
Boiling point	244.9°C
Melting point	14°C
Vapor pressure	<0.01 mm (20°C)
Flash point	107°C

Perhaps the best way to describe phenoxyethanol is by comparison with other materials. Based on the melting and boiling points of pure materials, phenoxyethanol falls between n-decyl (C-10) and n-dodecyl (C-12) saturated aliphatic alcohols (Table 3). Given the similar molecular weights and functional groups present, what differences do exist are derived directly from the functional groups. In all three cases the alcohol group (C—OH) is the most reactive group present, and thus the chemistry of phenoxyethanol is that of an alcohol. It is subject to oxidation to yield aldehydes and carboxylic acids, as well as to condensations to form esters and, to a much lesser extent,

TABLE 3 Comparison of Phenoxyethanol to Alkanols

	Alcohol		
Physical property	1-Decanol (C_{10})	Phenoxyethanol (C_8O_2)	1-Dodecanol (C_{12})
Melting point (°C)	6.4	14	24
Boiling point (°C)	232.9	244.9	259
	Functional group present		
Chemical reactivity	Alcohol aliphatic chain	Alcohol aliphatic chain ether aromatic ring	Alcohol aliphatic chain
Oxidation	Yes	Yes	Yes
Stability to acids	Stable	Ether hydrolysis with very strong acid	Stable
Electrophilic substitution	None	Susceptible	None
Microbiological activity	Some	Active	Some

ethers. The ether linkage present can do little more than undergo hydrolysis under strong acid conditions (a difficult reaction at best). The aromatic ring is much less prone to chemical attack and, except for substitution reactions, is really not that much different from ordinary aliphatic chains.

The final comparison to be made is that of microbiological activity: Only phenoxyethanol has activity suitable for use as a preservative.* While the chemical behaviors of these three alcohols are similar at the molecular level of the microbes, the compounds are, in fact, very different. It is then these functional group differences that account for the observed differences in properties and, ultimately, the differences in utility.

A further effect of the functional groups present is manifest in the solubility of phenoxyethanol in various liquids. Related to this, then, is the partition coefficient which describes the distribution of phenoxyethnaol between water and another water-immiscible phase. Table 4 lists both the solubilities and partition coefficients of phenoxyethanol in various solvents and cosmetic oils.

Of particular interest in Table 4 is the indication that phenoxyethanol is a good solvent. This, together with partition coefficients that exhibit a relatively even distribution between organic and aqueous phases, strongly suggests that phenoxyethanol should be well dispersed throughout a cos-

TABLE 4 Solubilities and Partition Coefficients of Phenoxyethanol

Solubility of phenoxyethanol in	
Water	2.6 g/100 ml
Mineral oil	0.7 g/100 ml
Isopropyl palmitate	3.9 g/100 ml
Aqueous alkali	Soluble
Glycerine	Miscible
Propylene glycol	Miscible
Alcohol	Miscible
Ether	Miscible
Benzene	Miscible
Solubility of water in phenoxyethanol	9.3 g/100 ml
Partition coefficients for phenoxyethanol	
Isopropyl palmitate–water	2.9
Peanut oil–water	2.6
Mineral oil–water	0.3

*Editor's Note: Both C_{10} and C_{12} alcohols have been shown to be active, Kabara et al., Antimicrob. Agents Chemother., 2:23 (1972).

Cosmetically Acceptable Phenoxyethanol

metic formulation and adds significantly to its character. This has been amply demonstrated several times and will be discussed later.

II. MANUFACTURE

The usual process for manufacturing phenoxyethanol is the same as that for producing an ethoxylated derivative of any acidic material. Phenol is treated with ethylene oxide in the presence of a basic catalyst under pressure and with heating:

$$\text{C}_6\text{H}_5\text{OH} + \text{(ethylene oxide)} \xrightarrow[\text{Heat/Pressure}]{\text{Base}} \text{C}_6\text{H}_5\text{-O-CH}_2\text{CH}_2\text{-OH}$$

Typically, following completion of the reaction, the catalyst is neutralized with acid, and then some combination of filtration, bleaching, distillation, and so on, is used to "finish off" the industrial-grade material. These phenoxyethanols can range from 70 to >99% of the monoethoxylate, and usually are more than 90% pure, *but* the last few percent causes unpleasant sour or metallic odors, has colors anywhere from clear water white to stray yellow, and has a residual free phenol content of a few tenths of a percent.

Phenoxyethanol, which is truly suitable for cosmetic or fragrance use, is a very different entity. For these applications, a different set of standards are needed. First, the actual mild, green, rose note must not be obscured, and exactly the same odor must be produced each time. Next, the color must be very light, and the free phenol essentially nonexistent.

There are two basic ways this may be accomplished. The most common method is to "clean up" an industrial-grade material. This can frequently involve multiple steps and contribute additional processing costs equal to the initial value of the material. The alternative is to develop a process which produces a uniquely cosmetic/fragrance-grade material. The best method to accomplish this is one that produces a 92—96% pure phenoxyethanol with the remainder (4—8%) consisting almost entirely of the diethoxylate. This process represents a compromise between concentration of active ingredient and free phenol content; that is, the ethoxylation is continued to a point where the desired product is partially ethoxylated in order that the last traces of phenol may be reacted and thus produce the 4—8% of diethoxylate:

$$\text{C}_6\text{H}_5\text{-O-CH}_2\text{CH}_2\text{-OH} + \text{(ethylene oxide)} \xrightarrow{\text{Base}} \text{C}_6\text{H}_5\text{-O-CH}_2\text{CH}_2\text{-O-CH}_2\text{CH}_2\text{-OH}$$

Naturally, there is a limit to the utility of this approach: The diethoxylate is not an active biocide and, therefore, is a diluent of microbial activity. Fortunately, this material has almost no detectable effect on fragrance characteristics.

The final manufacturing concern is the handling and even selection of shipping containers. Care must be exercised through all phases of manufacturing to ensure that accidental contamination does not occur. Contaminants can be potentially harmful in the formulated cosmetic—toiletry item or result in off-odors in the finished fragrance.

III. MICROBIOLOGY

A. Literature and History

As was mentioned earlier, the antimicrobial history of phenoxyethanol (POE) is fairly long. It has been primarily a European product. It was introduced in the 1950s and, subsequently, has been pushed about as newer, more potent preservatives arising from higher technology have been marketed. The material is known in Europe primarily as Phenoxetol or a proprietary blend with parabens known as Phenonip. In fact, much of the early literature uses only these two names [3].

Perhaps the most interesting examples of usage of phenoxyethanol for its antimicrobial properties were those where it was included in major drug formulations. First, it was used to prevent attack on penicillin preparations primarily by gram-negative bacteria [4], and it also prolonged the action of the penicillin [5].

A second use with implications for today was preservation of various vaccines. Specifically mentioned were poliomyelitis (Salk) vaccine, diphtheria and tetanus toxoids [6], smallpox vaccine [7], and measles and rubella viruses [8]. This type of application suggests that an absolute minimum of destructive interaction or reaction occurred between POE and these exotic materials; by extrapolation the simpler ingredients of cosmetics and toiletries ought to be similarly unaffected.

An additional medically oriented use has been to preserve reagents, solutions, and other preparations used in various testing procedures [9]. Primarily, this has been for reagents and samples of blood so that test results truly represented the condition of the patient, not the state of health of a colony of microorganisms living in a tube of human blood.

Currently, in the United States and abroad, there is an active interest in replacing a number of the high-technology preservatives with less controversial materials. One of the preferred means of doing this is to combine phenoxyethanol with other preservatives which will complement the limited antimicrobial activity of the former. Active interest and research continue in this area, and new products are expected to be on the market in the near future.

B. Minimum Inhibitor Concentration Procedures and Results

Phenoxyethanol has variously been listed in the literature with minimum inhibitor concentrations ranging from ~1000 to 10,000 ppm [10]. The variety of methods and organisms account for the variability, but there remains the impression that phenoxyethanol is not a truly high-activity material. In a set of proprietary tests conducted for Emery Industries by Hill Top Research, a similar set of values was obtained (see Table 5). These values were obtained in a slightly unusual manner. Dilutions of phenoxyethanol were made with the proper nutrient broths, and these solutions were then inoculated. The optical density of the inoculated broths was determined before and after incubation. The change in light transmission was plotted against concentration of phenoxyethanol. In this method the greater the measured change, the more the organism had grown. A least-squares regression equation was developed to fit the data, and an extrapolation was made to the concentration corresponding to no change in optical transmission. This value of concentration was taken as the minimum inhibitory concentration. By this method a realistic set of values was obtained which ranged from a low of 3200 ppm

TABLE 5 Minimum Inhibitory Concentrations

	Organism	Concentration of Emeressence 1160 Rose Ether Phenoxyethanol	
Gram-negative bacteria	Pseudomonas aeruginosa ATCC No. 9027	0.32%	(3200 ppm)
	Escherichia coli ATCC No. 8739	0.36%	(3600 ppm)
Gram-positive bacteria	Staphylococcus aureus ATCC No. 6538	0.85%	(8500 ppm)
Yeast	Candida albicans ATCC No. 10231	0.54%	(5400 ppm)
Fungus	Aspergillus niger ATCC No. 16404	0.33%	(3300 ppm)

against *Pseudomonas aeruginosa* to a high of 8500 ppm for *Staphylococcus aureus*.

Even in this rather crude evaluation, the antimicrobial nature of phenoxyethanol is evident: It has a broad range of activity, with highest activity against gram-negative organisms and lesser activity against the gram-positive organisms. These data perhaps as well as any others, suggest the way that a cosmetically acceptable grade of phenoxyethanol may best be utilized as a preservative. The stronger showing against the troublesome gram-negative *P. aeruginosa* indicates that it can be used in combination with a number of common preservatives which have strong activity against gram-positive organisms but fail with gram-negative organisms. Combinations of this sort are currently being researched and can be expected on the market soon.

C. Mechanism

The mechanism of activity of phenoxyethanol has been studied for at least 30 years. An initial report [11] indicated that, as a spermicidal agent, the material was active by virtue of its detergent properties. This conclusion was reached when intracellular proteins were detected in the extracellular medium. In more sophisticated studies [12] the effect on O_2 uptake by suspensions of *Escherichia coli* supplied with specific oxidizable substrates was examined. It was found that low concentrations of phenoxyethanol stimulated O_2 uptake in the presence of sugars and retarded O_2 uptake with lactate, pyruvate, and so on. At higher concentrations O_2 uptake was retarded with all substrates. These data were rationalized by saying that the sugar-oxidizing enzymes may be within the cell while the lactate- and succinate-oxidizing systems are at the cell surface and thus immediately exposed to the preservative. This was even further refined by the suggestion derived from working with cell-free preparations that inhibitory action was not associated with cell respiration, but with other centers involved in cell multiplication [13].

From those early days of intense observation within a poorly defined scientific framework which led to much unsupported conjecture, a better definition and understanding of the mechanism through the work of Gilbert et al. [14] have resulted. In a definitive set of experiments the tricarboxylic acid cycle malate dehydrogenase enzyme was identified as the site of inhibition. This effect was competitive for the oxalacetic acid-limited reaction and noncompetitive for the NADH-limited reaction. A related finding [15] confirmed earlier membrane disruption observations, but also added much more. Leakage of cytoplasmic constituents, increased preservative uptake, and cellular penetration of N-tolyl-α-napthylamine-8-sulfonic acid all occurred at bactericidal concentrations. Furthermore, at levels well below lethal levels potassium efflux from the cells and disorganization of the outer liposaccharide-rich region of the envelope of *E. coli* were noted. Apparently there is a secondary effect at concentrations below lethal levels that resulted in suble disorganization of the cytoplasmic membrane and outer regions of the cell envelope and which is distinct from the bactericidal action.

The final work in this area is yet to be done, and there is still active pursuit in search of better, more definitive knowledge of the mechanism of activity.

D. Microbial Testing

1. Formulations

In building cosmetic formulations, all the skills of the microbiologists, synthetic chemists, chemical engineers, and quality assurance people are without value if a compatible, stable, attractive formulation cannot be created by the formulations chemist. Such creations, however, do not necessarily yield significant data when subjected to microbial testing. They are less a model system and much more a world unto themselves.

In order, then, to generate preservative data to support the usage of a cosmetically acceptable grade of phenoxyethanol, a set of formulations was developed. The formulations chosen for these tests represent simple, noncomplicated creams, lotions, and shampoos. They are basic starting points for a formulation chemist and were designed to show the effects of various types of typical cosmetic components on the microbiology. Included were anionic and nonionic emulsifiers and many other assorted ingredients.

Notable in its absence is any formulation containing protein. Frequently this is an especially difficult material to preserved and would thus represent a serious omission. The literature, however, is replete with examples of proteinaceous materials preserved with phenoxyethanol. This experience ranges from preservation of teleost fish tissue [16], chicken heart or skeletal muscle [17], and zooplankton samples from the tropics [18]. Typically these samples were successfully preserved with phenoxyethanol for subsequent study by biochemical taxonomists or others interested in maintaining the functioning of protein. Even more related to cosmetics was a report where phenoxyethanol in combination with other preservatives was used to preserve collagen preparations [19]. Thus, unlike a great many preservatives, phenoxyethanol will preserve protein with minimum bonding-type interaction with the protein, at least as measured by such sensitive parameters as enzyme function.

The actual formulations constructed for the preservation efficacy testing are presented in Table 6.

TABLE 6 Cosmetic Formulae for Efficacy Testing

	Parts
Shampoo	
Emersol 6400 (sodium lauryl sulfate)	30.0
Rewomid DC-212/S (Cocamide DEA)	5.0
Deionized water	64.0
Preservative	See individual tests

   Preparation: Heat all ingredients to 45–50°C; cool to room temperature

Cream rinse	
A. Stearalkonium chloride (50% solids)	1.5
Sodium chloride	0.1
Deionized water	93.9
B. Emerest 2642 (polyethylene glycol 8 distearate)	1.5
Ethoxyol AC (laneth-10 acetate)	2.0
Preservative	See individual tests

   Preparation: Heat both phases to 70°C; add A to B; cool to room temperature

Water–oil cream	
A. Beeswax	12.0
Paraffin wax	6.0
Cetyl alcohol	2.0
Mineral oil (70 viscosity)	47.0
Preservative	See individual tests
B. Borax	0.9
Deionized water	31.1

   Preparation: Heat both phases to 75°C; add B to A; cool to room temperature

TABLE 6 Continued

	Parts
Oil–water cream	
A. Clearlan (lanolin)	2.5
Stearic acid	3.0
Cetyl alcohol	1.0
Preservative	See individual tests
B. Glycerine	5.0
Triethanolamine	1.0
Deionized water	86.5
Preparation: Heat both phases to 70°C; add A to B; cool to room temperature	

2. Challenge Testing

Fundamentally, two types of challenge testing were done to illustrate the preservation efficacy of phenoxyethanol. These tests were sponsored by Emery Industries, the producer of Emeressence 1160 Rose Ether phenoxyethanol, and conducted by Hill Top Research, Inc. They were intended only as guidelines to teach the usage of phenoxyethanol, and not as guarantees of performance in similar formulations. As is always true, each formulation must be adequately tested by microbial challenges before being presented to the market for use by the consumer.

The organisms chosen for these tests are common ones specified in the U.S. *Pharmacopoeia XX* (see Table 7). Represented are gram-positive and gram-negative bacteria, yeasts, and fungi, all of which are relevant as individuals and representatives of broad classes of organisms.

Pure Culture:

Procedure: Bacterial and yeast inocula were prepared by harvesting trypticase soy agar slants with 10.0 ml of sterile phosphate buffer solution at pH 7.2. The bacterial cultures had been incubated 24 hr at 35°C, and the *Candida albicans* culture was incubated 48 hr at 25°C. Cells were removed from the agar surface by gently agitating the tube, and, if necessary, mechanically removing with a sterile pipet. The cell suspensions were then transferred to a clean, sterile culture tube (18 X 150 mm), and cell populations estimated by determining the percentage transmittance of the suspensions on a spectrophotometer (Coleman Junior II Spectrophotometer) and consulting established growth curves of the specific organism. Suspensions were adjusted to a concentration of between approximately 1.0×10^8 and 1.0×10^9 organisms per milliliter.

The *Aspergillus niger* spore suspension was prepared in the following manner. Approximately 50 ml of Sabouraud dextrose agar were placed into a 1000 ml flask and autoclaved at 15 psi for 15 min. The agar was allowed to solidify, creating maximum surface area. The desired organism was seeded at the center of each flask and incubated for 5–7 days at 25°C. A diluent was prepared which contained 0.85% NaCl and 0.2% Triton X-100 in distilled

TABLE 7 Organisms Chosen for Challenge Testing

	Organism
Gram-negative bacteria	*Pseudomonas aeruginosa* ATCC No. 9027
	Escherichia coli ATCC No. 8739
Gram-positive bacteria	*Staphylococcus aureus* ATCC No. 6538
Yeast	*Candida albicans* ATCC No. 10231
Fungus	*Aspergillus niger* ATCC No. 16404

water, and autoclaved at 15 psi for 20 min. Approximately 100 ml of the sterile diluent were poured over the culture, sterile glass beads were added, and the flask was shaken vigorously to dislodge spores.

This spore suspension was poured into a tissue grinder which had been heat sterilized, and the piston reciprocated several times to break up spore chains. The suspension was filtered through a ½-in. layer of sterile absorbant cotton to remove spore chains and hyphal elements. Conidial suspensions were standardized to contain 5×10^6 conidia per milliliter (determined by spore counts with a hemocytometer). By the addition of sterile diluent, suspensions were adjusted to a concentration of between approximately 1.0×10^8 and 1.0×10^7 organisms per milliliter just prior to use as inoculum.

Within 15 min of the preparation of the test organisms, 0.1-ml amounts of each test organism were added to separate 50-g portions of each test material and stirred vigorously 100 times to ensure an even dispersal of the challenging organisms.

Each inoculated sample was examined to determine the microbial population immediately (~0.2 hr) after inoculation as follows: An initial 1:10 dilution of inoculated sample was prepared by adding 1.0 ml or 1.0 g of inoculated sample to 9.0 ml of dilution water which consisted of a 0.75 M phosphate buffer diluent with 5% Tween 80 and 0.5% triethanolamine as a neutralizer. Additional 10-fold dilutions were prepared if the sample was suspected to contain greater than 3000 organisms per milliliter.

The prepared dilutions were stirred vigorously, and an aliquot of 1.0 ml of each prepared dilution was pipetted into each of two Petri dishes.

To Petri dishes containing sample inoculated with bacteria, 15—20 ml of trypticase soy agar with 5% Tween 80 and 0.5% triethanolamine was added. The sampling procedures were repeated for all test organisms, except that potato dextrose agar with 5% Tween and 0.5% triethanolamine was used as a plating medium for *A. niger*.

Plates containing samples challenged with bacteria were incubated for 48 hr at 35°C. Those challenged with *C. albicans* were incubated for 48 hr at 25°C, and those challenged with *A. niger* were incubated 5—7 days at 25°C. Following incubation, plates yielding 30—300 colony-forming units (CFUs) were counted, and the number of organisms per milliliter of test material was determined by multiplying the number of CFUs obtained by the dilution factor yielding the count.

The inoculated test material was stored at room temperature and resampled at 2, 4, and 24 hr for samples inoculated with bacteria or *C. albicans*, and at 4, 8, and 24 hr for samples inoculated with *A. niger*. Additional samples were taken at 3, 5, 7, and/or 14 days after inoculation if populations were not reduced to less than 10 viable organisms per milliliter by the above time periods.

Following the final assay described above, each sample showing a reduction in microbial populations was rechallenged with test organism suspensions. Following the rechallenge, microbial populations were determined, as described above, at the time at which a reduction to less than 10 viable organisms per milliliter was noted in the previous challenge. If rechallenge organisms were recovered at this time interval, a second assay was conducted 7 days following initiation of the rechallenge. Three additional rechallenges were conducted in the above manner, unless reductions in population were no longer noted.

Results: In the first set of results (see Tables 8–15), for each formulation only the initial time at which total kill was observed is reported. The variable in these cases is the concentration of Rose Ether phenoxyethanol, either with or without parabens. As a reference or control, a formulation containing only parabens was also tested; the parabens used were methyl (0.2%) and propyl (0.1%). Typically, this "control" was not adequately preserved, so Tables 8–15 also include the number of viable organisms remaining per gram of formulation at the time listed. The time chosen to evaluate the paraben control is the same as that where the paraben phenoxyethanol blend showed total kill. By this methodology, a real measure of the effect of phenoxyethanol can be made. Frequently this meant that a formulation which was not adequately preserved by parabens alone was now protected. The remaining three entries (C–E) in Tables 8–11 show the effect exerted using only Rose Ether phenoxyethanol. Most interesting here is the observation that phenoxyethanol at about 1% levels is as good as a blend (70:30) of phenoxyethanol–parabens. At these higher levels such properties of the phenoxyethanol as odor and color and formulation effects become very significant, and, hence, the value of a cosmetically acceptable grade becomes even more apparent.

In Tables 12–15 the results of rechallenging the formulations preserved only with Rose Ether phenoxyethanol are presented. One procedural matter must be noted. In the initial challenge, an exact and fairly accurate time of total kill is determined; however, in the rechallenges, samples were tested *only* at the same time as reported in the initial challenge. If this was not successful, then a 7-day sample was examined. As an example, in Table 12 with 0.7% phenoxyethanol concentration against *P. aeruginosa* the initial challenge showed total kill in 2 hr. The first rechallenge was sampled at 2 hr and total kill was not found. No further samples were examined until 7 days had lapsed, at which time no surviving organism was found. Thus the preservative may have effected total kill in 3 or 24 hr, but it was not resampled soon enough to detect this. While this is a limitation of the procedure, the final result is an *under*statement of the power of Rose Ether phenoxyethanol as a preservative.

In general, phenoxyethanol at higher levels than are perhaps traditionally used for preservatives is a good cosmetic ingredient capable of withstanding repeated challenges and adding to the overall effect of the formulation.

Mixed Culture:

Procedure: Following the same procedure reported earlier, the organisms were prepared for the study. Within 15 min of the preparation of the mixed inoculum suspension, 0.1 ml of the suspension was added to 50

TABLE 8 Water-Oil Cream: Initial Observed Time of Total Kill (Hours)

	Organism				
	P. aeruginosa	E. coli	S. aureus	C. albicans	A. niger
a. Rose Ether phenoxyethanol–parabens (0.7; Me 0.2%; Pr 0.1%)	0.2	0.2	4	4	4
b. Control–parabens (Me 0.2%; Pr 0.1%)	0.2	0.2	4	4	4
Surviving organisms at specified time	120	30,000	75,000	78,000	2,600
Rose Ether phenoxyethanol					
c. 0.7%	2	2	24	24	24
d. 1.0%	0.2	0.2	4	4	8
e. 1.5%	0.2	2	2	2	4

TABLE 9 Oil-Water Cream: Initial Observed Time of Total Kill (Hours)

	Organism				
	P. aeruginosa	E. coli	S. aureus	C. albicans	A. niger
A. Rose Ether phenoxy-ethanol–parabens (0.7%; Me 0.2%; Pr 0.1%)	0.2	2	24	24	120
B. Control–parabens (Me 0.2%; Pr 0.1%)	0.2	2	24	24	120
Surviving organisms at specified time	280	8900	80	100	30
Rose Ether phenoxyethanol					
C. 0.7%	2	>24	120	120	>336
D. 1.0%	0.2	2	120	24	168
E. 1.5%	0.2	2	24	24	168

TABLE 10 Cream Rinse: Initial Observed Time of Total Kill (Hours)

	Organism				
	P. aeruginosa	E. coli	S. aureus	C. albicans	A. niger
A. Rose Ether phenoxyethanol–parabens (0.7%; Me 0.2%; Pr 0.1%)	0.2	0.2	2	4	120
B. Control–parabens (Me 0.2%; Pr 0.1%)	0.2	0.2	2	4	120
Surviving organisms at specified time	<10	20	21,000	20	12,000
Rose Ether phenoxyethanol					
C. 0.7%	2	24	2	2	24
D. 1.0%	2	4	2	2	24
E. 1.5%	0.2	2	0.2	2	24

TABLE 11 Shampoo: Initial Observed Time of Total Kill (Hours)

	Organism				
	P. aeruginosa	E. coli	S. aureus	C. albicans	A. niger
A. Rose Ether phenoxy-ethanol–parabens (0.7%; Me 0.2%; Pr 0.1%)	0.2	2	a	2	168
B. Control–parabens (Me 0.2%; Pr 0.1%)	0.2	2	a	2	168
Surviving organisms at specified time	<10	60	a	<10	<10
Rose Ether phenoxy-ethanol					
C. 0.7%	0.2	4	120	2	24
D. 1.0%	0.2	2	120	4	24
E. 1.5%	0.2	2	24	2	24

[a]No data collected.

TABLE 12 Water-Oil Cream: Observed Time of Total Kill[a]

Concentration of Rose Ether phenoxyethanol	P. aeruginosa	E. coli	S. aureus	C. albicans	A. niger
0.7% initial	2	2	24	24	24
Rechallenge					
1	7 days	7 days	7 days	7 days	24
2	7 days	7 days	(10^4/7 days)[b]	(10/7 days)[b]	7 days
3	7 days	7 days	(10^3/7 days)[b]	7 days	7 days
4	7 days	7 days	—	(10^2/7 days)[b]	—
1.0% initial	0.2	0.2	4	4	8
Rechallenge					
1	7 days	7 days	7 days	7 days	7 days
2	7 days	7 days	7 days	(10^3/7 days)[b]	7 days
3	7 days	7 days	7 days	(10^4/7 days)[b]	7 days
4	7 days	7 days	7 days	—	—
1.5% Initial	0.2	2	2	2	4
Rechallenge					
1	2	2	7 days	7 days	4
2	2	2	7 days	7 days	7 days
3	2	2	7 days	7 days	7 days
4	2	2	7 days	7 days	—
Rose Ether phenoxy-ethanol–parabens (0.7%; 0.2% Me; 0.1% Pr)	0.2	0.2	4	4	4

[a]Time given in hours, except as noted.
[b]Number of surviving organisms per gram of formulation at time given.

TABLE 13 Oil-Water Cream: Observed Time of Total Kill[a]

Concentration of Rose Ether phenoxyethanol	Organism				
	P. aeruginosa	E. coli	S. aureus	C. albicans	A. niger
0.7% initial	2	(10^5/24 hr)[b]	5 days	5 days	(10^2/14 days)[b]
Rechallenge 1	2	7 days	5 days	(10^2/7 days)[b]	(10^2/14 days)[b]
2	2	7 days	5 days	7 days	—
3	2	7 days	5 days	—	—
4	2	7 days	—	7 days	—
1.0% Initial	0.2	2	5 days	24	7 days
Rechallenge 1	2	2	5 days	7 days	(10^4/14 days)[b]
2	2	2	5 days	(10/ 7 days)[b]	(10^3/14 days)[b]
3	2	2	5 days	7 days	—
4	2	2	5 days	(10^5/7 days)[b]	—
1.5% initial	0.2	2	24	24	7 days
Rechallenge 1	2	2	7 days	24	7 days
2	2	2	7 days	24	7 days
3	2	2	7 days	24	—
4	2	2	7 days	(10^6/7 days)[b]	—
Rose Ether phenoxyethanol–para- (0.7%; 0.2% Me, 0.1% Pr)	0.2	2	24	24	120

[a]Time given in hours, except as noted.
[b]Number of surviving organisms per gram of formulation at time given.

TABLE 14 Cream Rinse: Observed Time of Total Kill[a]

Concentration of Rose Ether phenoxyethanol	Organism				
	P. aeruginosa	E. coli	S. aureus	C. albicans	A. niger
0.7% initial					
Rechallenge					
1	2	24	2	2	24
2	2	(10^4/7 days)[b]	2	2	7 days
3	7 days	(10^4/7 days)[b]	2	2	7 days
4	(10^4/7 days)[b]	(10^4/7 days)[b]	2	2	(10^5/7 days)[b]
	—	—	2	2	—
1.0% initial					
Rechallenge					
1	2	4	2	2	24
2	2	24	2	2	14 days
3	2	24	2	2	7 days
4	2	24	2	2	(10^4/7 days)[b]
	2	7 days	2	2	—
1.5% initial					
Rechallenge					
1	0.2	2	0.2	2	24
2	2	2	2	2	7 days
3	2	2	2	2	7 days
4	2	7 days	2	2	7 days
	2	7 days	2	2	—
Rose Ether phenoxyethanol–para- (0.7%; 0.2% Me, 0.1% Pr)	0.2	0.2	2	4	120

[a]Time given in hours, except as noted.
[b]Number of surviving organisms per gram of formulation at time given.

TABLE 15 Shampoo: Observed Time of Total Kill[a]

Concentration of Rose Ether phenoxyethanol	Organism				
	P. aeruginosa	E. coli	S. aureus	C. albicans	A. niger
0.7% initial	0.2	4	5 days	2	24
Rechallenge					
1	2	7 days	5 days	2	24
2	2	7 days	7 days	2	7 days
3	2	7 days	7 days	2	—
4	2	7 days	7 days	2	—
1.0% initial	0.2	2	5 days	4	24
Rechallenge					
1	2	7 days	5 days	4	24
2	2	7 days	5 days	7 days	7 days
3	2	7 days	5 days	7 days	7 days
4	2	7 days	—	(10^5/7 days)[b]	—
1.5% initial	0.2	2	24	2	24
Rechallenge					
1	2	2	7 days	2	24
2	2	7 days	7 days	7 days	7 days
3	2	7 days	7 days	7 days	7 days
4	2	7 days	7 days	(10^5/7 days)[b]	—
Rose Ether phenoxyethanol–parabens (0.7%; 0.2% Me, 0.1% Pr)	0.2	2	—	2	168

[a]Time given in hours, except as noted.
[b]Number of surviving organisms per gram of formulation at time given.

g of the test formulation contained in an 8-oz jar. Immediately following inoculation, the contents of the jar were stirred vigorously 100 times, and 11 ml were transferred into 99 ml of dilution broth containing 0.5% triethanolamine and 5% Tween 80. Additional 10-fold dilutions were prepared if the sample was suspected to contain greater than 3000 organisms milliliter. The initial 1:10 dilution and any subsequently prepared dilutions were shaken 25 times prior to removing an aliquot. From each prepared dilution, 1.0 ml was pipetted into each of four Petri dishes. Two plates were poured trypticase soy agar containing 5% Tween 80 and 0.5% triethanolamine, and two poured with potatoe dextrose agar containing 5% Tween 80 and 0.5% triethanolamine. Trypticase soy agar plates were incubated at 35°C for 48 hr, and potatoe dextrose agar plates were incubated at 25°C for 5 days.

Following incubation, plates yielding 30—300 colony-forming units (CFUs) were counted, and the number of organisms per milligram per milliliter of test material was determined by multiplying the number of CFUs obtained by the dilution factor yielding the count.

The inoculated test material was stored at room temperature and resampled in the above manner at time intervals of 7, 14, 21, and 28 days after inoculation. The time intervals selected were in accordance with those outlined by the *Cosmetic, Toiletry, and Fragrance Association*.

Results: As the results in Table 16 attest in these simple formulations, a blend of Rose Ether phenoxyethanol and mixed parabens can produce a monotonous repetition of complete kill over a range of formulations and organisms. There is little to be interpreted when all the samples at all the time intervals were free of organisms. Obviously the formulations were well protected.

Table 17, however, includes some interesting data deserving interpretation. In the first cream rinse example the level of Rose Ether phenoxy-

TABLE 16 Mixed-Culture Challenge Testing for the Preservative System 0.7% Rose Ether Phenoxyethanol—0.2% Methylparaben—0.1% Propylparaben

Formulation	Time (days)	Surviving organisms[a]		
		Bacteria	Yeast	Fungus
Shampoo	7	<10	<10	<10
	14	<10	<10	<10
	21	<10	<10	<10
	28	<10	<10	<10
Water—oil cream	7	<10	<10	<10
	14	<10	<10	<10
	21	<10	<10	<10
	28	<10	<10	<10
Oil—water cream	7	<10	<10	<10
	14	<10	<10	<10
	21	<10	<10	<10
	28	<10	<10	<10

[a]Inoculum count per gram of product: bacteria, 2.4×10^6; yeast, 4.3×10^5; fungus, 5.2×10^6.

TABLE 17 Effect of Phenoxyethanol Concentration and pH on Cream Rinse Challenge Test

Formulation	pH	Time (days)	Surviving organisms[a]		
			Bacteria	Yeast	Fungus
Cream rinse (0.3% Rose Ether phenoxyethanol– 0.2% methylparaben– 0.1% propylparaben)	3.67	7	120	20	5900
		14	<10	<10	2300
		21	<10	<10	7500
		28	<10	<10	340
Cream rinse (0.7% Rose Ether phenoxyethanol– 0.2% methylparaben– 0.1% propylparaben)	3.66	7	<10	<10	470
		14	<10	<10	320
		21	<10	<10	55
		28	<10	<10	10
Cream rinse (0.7% Rose Ether phenoxyethanol– 0.2% methylparaben– 0.1% propylparaben)	6.55	7	<10	<10	10
		14	<10	<10	<10
		21	<10	<10	
		28	<10	<10	

[a]Inoculum count per gram of product: bacteria, 2.4×10^6; yeast, 4.3×10^5; fungus, 5.2×10^6.

ethanol was set at 0.3%, and the pH was measured to be 3.67. The result of testing must be to acknowledge the formulation as an inadequately preserved cream rinse. The second formulation shows the effect of increasing the Rose Ether phenoxyethanol content to 0.7%. In this case only the fungus survived. In the final case, adjusting the pH back toward neutrality resulted in a much more acceptable performance by the preservative package. Thus the recommendation arising from this testing is to avoid excessively low pH values or, alternatively, to use slightly more Rose Ether phenoxyethanol.

IV. FORMULATION EFFECTS

A. Formulations

A study was designed to determine the effect phenoxyethanol had on the stability and viscosity of cosmetic formulations. The four basic but relatively intricate formulations investigated were a cream rinse, a liquid hand soap, a water–oil cream, and an oil–water lotion. The formulations were prepared with 1% Emeressence 1160 Rose Ether phenoxyethanol and duplicate omitting the phenoxyethanol. These latter formulations, serving as controls, were

TABLE 18 Cream Rinse Formulation[a]

Ingredient	Active	Control
Emerest 2400 (glyceryl stearate)	2.0	2.0
Emerest 2717 (polyethylene glycol 100 stearate)	2.0	2.0
Cetyl alcohol	3.0	3.0
Ammonyx 4 (stearalkonium chloride; 25% active)	2.5	2.5
Emerest 2711 (polyethylene glycol 8 stearate)	1.5	1.5
Propylene glycol	2.5	2.5
Emeressence 1160 Rose Ether (phenoxyethanol)	1.0	—
Deionized water	85.5	86.5

[a]Procedure: Heat and mix to 75°C; cool to room temperature.

very carefully handled to prevent microbial contamination, since they were not adequately preserved. The formula compositions are listed in Tables 18–21.

The amount of phenoxyethanol chosen for inclusion in these formulations represents an arbitrary value not related to the amount needed to actually preserve the formulations. This 1% value was picked to illustrate in an exaggerated fashion the effects which might result from using phenoxyethanol as a preservative.

TABLE 19 Liquid Soap Formulation[a]

Ingredient	Active	Control
Emersal 6400 (sodium lauryl sulfate)	30.0	30.0
Emid 6511 (lauramide DEA)	5.0	5.0
Emery 5430 (cocamidopropyl betaine)	3.0	3.0
Emerest 2350 (glycol stearate)	1.0	1.0
Emeressence 1160 Rose Ether (phenoxyethanol)	1.0	—
Deionized water	60.0	61.0

[a]Procedure: Heat all ingredients to 75–80°C with mixing until a homogenous blend is obtained; cool to 45°C with stirring and then allow to cool to room temperature without mixing.

TABLE 20 Oil-Water Lotion Formulation[a]

Ingredient	Active	Control
A. Lantrol 1673 (lanolin oil)	7.5	7.5
Emerest 2300 (octyl isononanoate)	7.5	7.5
Emsorb 2505 (sorbitan stearate)	4.0	4.0
Emersol 132 (stearic acid)	1.5	1.5
Emerest 2400 (glyceryl stearate)	1.0	1.0
Emerest 2717 (polyethylene glycol 100 (stearate)	0.5	0.5
B. Emery 916 (glycerine)	5.0	5.0
Triethanolamine	0.7	0.7
Emsorb 2728 (Tween 60)	1.0	1.0
Emeressence 1160 Rose Ether (phenoxyethanol)	1.0	—
Deionized water	70.3	71.3

[a]Procedure: Heat both phases to 70°C; add B to A and cool to 40° with mixing, package.

TABLE 21 Water-Oil Cream Formulation[a]

Ingredient	Active	Control
A. A-C polyethylene grade 617 (polyethylene)	3.0	3.0
Beeswax	2.0	2.0
Nimlesterol 1732 [mineral oil (and) lanolin alcohol]	5.0	5.0
Mineral oil, 70 viscosity	8.2	8.2
Silicone 200 fluid (Dimethicone)	1.0	1.0
Emerest 2307 (octyl pelargonate)	10.0	10.0
Emerest 2452 (polyglycerol-2 diisostearate)	5.5	5.5
B. Sorbo (sorbitol)	5.0	5.0
Borax	0.3	0.3
Emeressence 1160 Rose Ether (phenoxyethanol)	1.0	—
Deionized water	59.0	60.0

[a]Procedure: Heat A to 85°C and B to 88°C; add A to B in homomixer; mix until all water is taken up (inversion occurs at approximately 68°C).

Cosmetically Acceptable Phenoxyethanol

B. Viscosity and Stability Testing

The formulations (see above) were evaluated by two different performance-related methodologies, and, correspondingly, the samples were divided into two different sets. One of these sets was maintained at 40°C in an accelerated aging test. The samples were visually inspected for emulsion breakdown and discoloration at irregular intervals over a total period of 40 days. This is a fairly standard method of assessing the effect of a component on emulsion stability.

The second type of testing was the measurement of viscosity at irregular intervals on samples stored at room temperature (20 ± 2°C) using a Brookfield LTV viscometer with a Helipath Stand. To ensure accurate, meaningful measurements, all samples were equilibrated for 24 hr at room temperature after formulation before the initial measurement.

Taken together, these two sets of experiments produced some significant and unusual results. The correlation of the findings together with the microbiological studies will permit optimum utilization of the unique properties of Rose Ether phenoxyethanol as a cosmetic ingredient.

1. Cream Rinse

It is, of course, to be expected that for each change in the composition of a formulation a change ought to be expected in the performance characteristics of the formulation. Thus the viscosity of a cream rinse might well be expected to change with the addition of phenoxyethanol, but an increase of 2.3 is significantly greater than expected (see Table 22). Because of phenoxyethanol's single hydroxyl group, it functions as an emulsifier, which is probably working in combination with the cetyl alcohol to increase viscosity. However, the dramatic increase in viscosity defies explanation by current theory. Perhaps, owing to its aromatic nature, strong affinity for oil, and yet slight water solubility, it is in some way complexing with the stearalkonium chloride to produce a higher viscosity. Any current micellar theories relative to its structure and solubilities are inadequate to explain the observation. This is especially true when comparison is made with another oil-in-water lotion, which is essentially what a cream rinse is. In this case the viscosity was drastically reduced (see Sec. IV.B.3). This is indeed an interesting phenomenon and illustrates the wisdom of considering phenoxyethanol as an active cosmetic ingredient and not just a simple preservative.

2. Liquid Soap

As may be expected, Rose Ether phenoxyethanol caused a decrease in viscosity (see Table 23). However, since finished formulations are typically viscosity modified with sodium chloride or ammonium chloride, the use of phenoxyethanol poses no great problem. The only exception would be when a very viscous product is to be made and the salt required becomes excessive. Most detergent systems can only be modified to a point, and then they begin to thin out again. This kind of problem may be solved by further after-the-fact adjustments of the formulation. It would, perhaps, be easier to avoid the problem. It is therefore essential that the Rose Ether phenoxyethanol be in the formulation from the start, as with any other preservative, rather than being added as an afterthought.

TABLE 22 Comparative Viscosity Data[a] of Cream Rinse Formulations

	Viscosity (cP)			
Sample	Initial	17 days	40 days	Average
1% phenoxyethanol	58,300	52,500	42,900	51,200
Control	21,600	20,800	16,200	19,500
Δ(1% phenoxyethanol−control)	36,700	31,700	26,700	31,700
Relative viscosity (1% phenoxyethanol−control)	2.7	1.66	2.65	2.34

[a] 6 rpm, TE spindle.

TABLE 23 Comparative Viscosity Data[a] of Liquid Soap Formulations

	Viscosity (cP)				
Sample	Initial	6 days	22 days	45 days	Average
1% phenoxyethanol	16,100	15,300	16,600	14,600	15,600
Control	21,900	26,000	26,000	25,000	24,000
Δ (1% phenoxyethanol−control)	-5,800	-11,000	-9,400	-10,000	-9,000
Relative viscosity (1% phenoxyethanol−control)	0.74	0.59	0.64	0.58	0.64

[a] 3 rpm, TD spindle.

TABLE 24 Comparative Viscosity Data[a] of Oil-Water Lotion Formulations

Sample	Viscosity (cP)			
	Initial	15 days	38 days	Average
1% phenoxyethanol	7,000	3,660	2,160	4,270
Control	24,300	17,300	15,000	18,900
Δ (1% phenoxyethanol– control)	-17,300	-13,700	-12,800	-14,600
Relative viscosity (1% phenoxyethanol– control)	0.29	0.21	0.14	0.21

[a] 6 rpm, TD spindle.

3. Oil–Water Lotion

In the oil–water lotion, a resulting effect was obtained which was almost the complete opposite of that obtained with the cream rinse. Both the 1% phenoxyethanol sample and the control showed a significant decrease in viscosity with time (see Table 24), yet both showed no indication of emulsion breakdown and phase separation at either room temperature or at 40°C. The fact that phenoxyethanol was able to thin out a lotion and still maintain stability is an interesting point. This again relates to the emulsifying properties of phenoxyethanol. A rationale to explain why phenoxyethanol thins out an oil–water lotion must be based on the chemical structure:

[Chemical structure: phenyl-O-CH₂CH₂-OH]

In this kind of emulsion the oil phase is encapsulated in the water; that is, the micelles have a hydrophobic interior and a hydrophilic surface. Phenoxyethanol has a short, compact cyclic structure, as opposed to the typical ethoxylated straight-chain alcohol. This would mean that the entire molecule must remain fairly close to the surface of the micelle, possibly filling surface cracks and crevices. The result would be a more stable micelle, and this could then account for a thin viscosity without loss of emulsion stability.

4. Water–Oil Cream

As shown by the data in Table 25, phenoxyethanol may slightly increase the viscosity of a water–oil cream. This increase is to be expected, since this type of emulsion is oil continuous, and phenoxyethanol is primarily an oil-soluble emulsifier, and it should force the system to be thicker. Again, all samples showed excellent stability at room temperature and 40°C.

TABLE 25 Comparative Viscosity Data[a] of Water-Oil Cream Formulations

Sample	Viscosity (cP)				
	Initial	14 days	37 days	43 days	Average
1% phenoxy-ethanol	240,000	236,000	143,000	146,000	191,000
Control	186,000	186,000	166,000	140,000	170,000
Δ (1% phenoxy-ethanol−control)	54,000	50,000	−23,000	6,000	20,000
Relative viscosity (1% phenoxyethanol−control)	1.29	1.27	0.86	1.04	1.12

[a]3 rpm, TF spindle.

5. Conclusions

As shown by the above data, the incorporation of phenoxyethanol had an effect on each formulation. In each system the viscosity was altered by the phenoxyethanol, but, equally important, the stability even in an accelerated test was *not* affected. In a detergent system and an oil−water lotion the viscosity decreased, whereas in a cream rinse and an oil−water cream the viscosity became greater than in the control. It is therefore essential that Rose Ether phenoxyethanol be added to a formulation as a contributing ingredient.

V. TOXICOLOGY

Phenoxyethanol is an interesting and typical example of compounds which came into commercial existence some decades ago. Its antimicrobial activity was established early on, as was its value as an industrial chemical. Since that time a history of exposure and usage has developed that does not to this time give any indication of impending disasters. In fact, the history of phenoxyethanol may well have begun long before its initial synthesis; humans may have been ingesting or inhaling it as a natural product for centuries. There are reports in the literature which demonstrate the presence of phenoxyethanol as a volatile, naturally occurring substance. The first of these reports, published in the *Journal of Agricultural and Food Chemistry* [20], claims that phenoxyethanol is a volatile constituent of one of the highest grades of grean tea, gyokuro (*Camellia sinensis* L. var. Yabukita). The other reported occurrence was as an airborne material found in the vicinity of cotton fields [21]. While neither of these should be taken as absolute indications of safety for human exposure, they do suggest a long history of low-level human inhalation−ingestion without overt problems.

Beyond this, most of the toxicology information on phenoxyethanol may be obtained from a collection of standard toxicology references (see Table 26). The values presented in Table 26 are perhaps most notable in comparison with

TABLE 26 Toxicology of Phenoxyethanol

Acute oral LD_{50} (rats)	1.26–2.33 ml/kg
Acute dermal LD_{50} (rabbits)	2.0 ml/kg
Eye irritation	
Undiluted, 0.1 ml unwashed (rabbits)	Irritant
2.2% in distilled water, 0.1 ml unwashed (rabbits)	Nonirritant

other preservatives. Rarely can a useful preservative exhibit such a lack of toxic effects. It is useful to note that, while undiluted phenoxyethanol is an eye irritant and appropriate precautions should be used during its handling, when it is dissolved in distilled water near the saturation limit, it is not an irritant. This suggests that incorporation into finished products at typical use levels should pose few problems of eye irritation for the consumer and, with reasonable, adequate protection for workers, be of minimal risk for cosmetic manufacturers. Within the fragrance industry there is also some information relative to skin irritation. Using current best evaluation methods at a 10% application level, phenoxyethanol was found to be nonirritating in the sensitization maximization test.

All of this information confirms that phenoxyethanol is cosmetically acceptable and, when used at appropriate levels, can be an effective preservative.

VI. SUMMARY

Examination of the data really yields no surprises; Rose Ether phenoxyethanol remains available to preserve the formulation under repeated challenges because it lacks chemical reactivity. It does not depend on reaction to generate the active antimicrobial chemical species as many preservatives do, nor is it inactivated by a variety of chemical reactions. In other words, what is put into the product to preserve the cosmetic stays in the formulation.

REFERENCES

1. D. R. Nutter, *Res. Discuss.*, *195*:270 (1980); H. J. Santemma and R. S. Fisch, U.S. Patent 3,996,054 (1976); V. Hederich, H. S. Bien, and G. Gehrke, Ger. Offen. 2,531,557 (1977); H. Abel, Ger. Offen. 2,938,607 (1980); R. C. Nahta, U.S. Patent 4,192,648 (1980); T. Shirakawa, T. Ito, Y. Nakamura, and N. Araya, Jpn. Kokai Tokkyo Koho 78,111,178 (1978).
2. H. Berry, *Lancet*, *II*:175–176 (1944); J. Gough, H. Berry, and B. M. Still, *Sem. Med.*, *I*:311 (1945), H. Berry, *Sem. Med.*, *I*:1027 (1946).
3. E. Boehm, British Patent 566,139 (1944); U.S. Patent 2,451,149 (1948); E. Boehm and R. Williams, Ger. Offen. 856,043 (1952); H. Schonenberger, *Pharm. Ztg. Nachr.*, *87*:752 (1951); H. Marx, *Riechst Aromen Koerperpflegem*, *19*:381 (1969).

4. T. Sabalitschka and H. Marx, *Pharm. Ztg. Apoth. Ztg.*, 90:244 (1954).
5. C. Marini and R. Becchiati, *Minerva Med.*, 41:718 (1950).
6. H. Pivnick, J. M. Tracy, A. L. Tosoni, and D. G. Glass, *J. Pharm. Sci.*, 53:899 (1964); J. Cameron, *Dev. Biol. Stand.*, 24:155 (1974).
7. E. Kaiser, K. Megay, and M. Pantlitschko, *Boll. Ist. Sieroter. Milan.*, 29:237 (1950).
8. A. Gray, L. F. Schuchardt, and J. J. Hanson, *Dev. Biol. Stand.*, 24:123 (1974).
9. D. Armstrong, U.S. Patent 3,962,125 (1976); Ger. Offen. 2,709,576 (1977).
10. G. Z. El-Din, *Acta Pharm. Intern.*, 1:327 (1950); H. Benger and K. Megay, *Zentralbl. Bakteriol. Parasitend. Infektionskr. Hyg. Abt. 1: Orig.* 159:105 (1953); K. Kjeldgaard, *Farm. Tid.*, 66:671 (1956); J. C. Maruzzella and E. Bramnick, *Soap Perfum. Cosmet.*, 34:743 (1961); G. Jacobs, S. M. Henry, and V. F. Cotty, *J. Soc. Cosmet. Chem.*, 26:205 (1975).
11. T. Mann, *Proc. Soc. Study Fertil.*, 6:41 (1954).
12. W. B. Hugo and H. E. Street, *J. Gen. Microbiol.*, 6:90 (1952); W. B. Hugo, *Congr. Int. Biochim. Resumes Communs. Ze Congr. Paris, 1952:*447 (1952).
13. W. B. Hugo, *J. Gen. Microbiol.*, 15:315 (1956).
14. P. Gilbert, E. G. Veveridge, and P. B. Crone, *J. Pharm. Pharmacol. Suppl.*, 28:51 (1976); *Microbios*, 79:29 (1977).
15. P. Gilbert, E. G. Beveridge, and P. B. Crone, *Microbios*, 19:17, 125 (1977).
16. P. J. Smith and J. Crossland, *N.J.Z. Mar. Freshwater Res.*, 12:341 (1978).
17. M. Nakanishi, A. C. Wilson, R. A. Nolan, G. C. Gorman, and G. S. Bailey, *Science*, 163:681 (1969).
18. T. Balachandran, *Curr. Sci.*, 43:380 (1974).
19. R. Riemschneider and W. H. Chik, *Kosmetika Zurich*, 5:119 (1977); *Dtsch. Apoth. Ztg.*, 117:1557 (1977).
20. K. Yamaguchi and T. Shibamoto, *J. Agric. Food Chem.*, 29:366 (1981).
21. P. A. Hedin, A. C. Thompson, and R. C. Gueldner, *Phytochemistry*, 14:2088 (1975); P. A. Hedin, *Environ. Entomol.*, 5:1234 (1976).

7
PHENOLS AS PRESERVATIVES FOR PHARMACEUTICAL AND COSMETIC PRODUCTS

W. B. HUGO   *The University of Nottingham, Nottingham, England*

I. Introduction 109
II. Historical Survey 109
III. Phenolic Preservatives in Cosmetic Products 110
   A. Structure–activity relationships 110
   B. Mode of action 110
   C. Phenols used in cosmetic preservation 110

References 112

I. INTRODUCTION

Phenols [1] share with hypochlorites [2-4], iodine [5], and mercury II (mercuric) salts [6] a pride of place in the rational use of antiseptics and disinfectants, and were in use in the United States and Europe in the middle and late nineteenth century.

II. HISTORICAL SURVEY

Phenols of a complex nature occurring in balsams and essential oils were used as preservatives since early recorded historical times, but they were probably only used in a form that could be regarded as truly phenolic since the introduction of pyrolytic distillation, first of wood to yield the wood tars and later of coal to yield the coal tars. Wood and coal tars were used, as obtained from the manfacturer, as preservatives for wood and cordage, especially in maritime practice.

As the art of chemistry progressed and with the rise of an organic chemical and dyestuffs industry based at first on coal tar products, many more phenolic substances became available [1].

Phenol and its homologs will be discussed below.

III. PHENOLIC PRESERVATIVES IN COSMETIC PRODUCTS

A. Structure—Activity Relationships

As more and more phenolic compounds became available and were tested for their antimicrobial activity, it became possible to extract some general relationships between structure and activity. This relationship was reviewed in depth by Suter [7].

In brief, the findings may be summarized as follows:

1. Para substitutions of an alkyl chain up to six carbon atoms in length increases the antibacterial action of phenols, presumably by increasing the surface activity and ability to orient at an interface. Activity falls off after this, owing to decreased water solubility. Again, because of the conferment of polar properties, straight-chain para substituents confer greater activity than branched-chain substituents containing the same number of carbon atoms.
2. Halogenation increases the antibacterial activity of phenol. The combination of alkyl and halogen substitutions that confers the greatest antibacterial activity is that where the alkyl group is ortho to the phenolic group and the halogen para to the phenolic group. Halogenation of the bis-phenols confers additional biocidal activity; in fact, without halogenation, activity is low or nonexistent.

B. Mode of Action

The mechanism behind the lethal action of the phenols has been studied from the early 1900s, and the first papers indicated that they interacted with protein [8] and were dubbed general protoplasmic poisons. Later they were shown to promote leakage of cellular constituents [9,10], which suggested an interaction with the bacterial cytoplasmic membrane.

More recently they have been shown to collapse all or part of the proton motive force, with the metabolic consequences this implies [11,12].

These three effects are elicited at various concentrations, the first being associated with a rapid bactericidal effect, and the others with bacteriostatic effects. The last two are reversible; the first is irreversible.

Phenols are generally inactivated in the presence of nonionic surface-active agents, a point to bear in mind when considering their use in cosmetic formulations.

C. Phenols Used in Cosmetic Preservation

1. General Introduction and Regulatory Constraints

Of the vast array of commercially available phenols, it is difficult to prepare a list of recommended products. In 1973 the U.S. Food and Drug Administration instituted a voluntary cosmetics registration program asking makers to disclose the ingredients present together with their role in the product. In 1977 the findings from those responding to the scheme were published in the form of a frequency of use list, which itself was drawn from the contents of some 18,000 products [13]. The list includes both substances used as antimicrobial agents and antioxidants, as well as a very useful overview of the substances being used and their frequency of usage.

The Comité de Liaison des Syndicats Européens de l'Industrie de la Parfumerie et des Cosmétiques also publishes lists of preservatives according to the acceptability status. The European Economic Community also publishes in this area.

A recent summary of bactericides and preservatives used for cosmetics has been drawn by Davis [14] of the United Kingdom. From these sources it is possible to draw up a list of phenolic preservatives for consideration in this chapter. Their inclusion does not imply that they may be used in every country and in every type of preparation.

Regulations vary from country to country, and a product preserved to acceptable standards in one country may be unacceptable in another. Overnight a preservative may be banned if an unexpected toxic hazard is exposed; similarly, a banned product may be rehabilitated after further investigation.

It is the responsibility of the manufacturer to ensure that the current regulations as to the use of preservatives in cosmetics in the countries where the product is destined for use are compiled with when formulating their products.

In consideration of these restrictions, the following sections will deal with a selection of phenols.

2. Individual Phenols

This group comprises the phenols with a single benzene ring and substituted with alkyl or aryl groups.

Phenol: Phenol, C_6H_5OH, the parent compound of this group, as has been said, was among the first disinfectant compounds to be used. It is accorded a place of distinction in the family of disinfectants, as it was used by Lister to establish his principles of antiseptic surgery [15]. It is also used as a standard in many disinfectant tests.

Phenol is a white crystalline solid with a melting point of 39–40°C. It becomes pink and finally almost black on exposure to air. It is soluble in water 1:14 and in alcohol (ethanol) 1:6.

Phenol is a toxic substance causing corrosion of the skin and mucous membranes and subsequent damage to many essential body functions. Fatalities have been reported from doses as low as 1 g.

Concentrations of 0.5–0.8% are bacteriostatic to most microorganisms, and higher concentrations are bactericidal. It is only slowly effective against bacterial spores and acid-fast organisms. It is active against fungal spores.

It becomes progressively less active in alkaline solution (pK_a = 10), on dilution (concentration exponent 6), and with decreasing temperatures, (temperature coefficient ca. 7.5 for each 10°C rise in temperature over the range 10–40°C) [16].

It is inactivated by organic matter. Phenol itself was reported to have been used in one instance in Richardson's survey [13], and it is probably not the preservative of first choice by any means for cosmetic preparations.

Cresol: Cresol [$C_6H_4(OH)_2$] is a term applied to a mixture of 1-, 2-, and 3-hydroxy phenols or o-, m-, and p-hydroxyphenols. The individual members are often spoken of as o-, m-, or p-cresol and occur in coal tar.

As was stated, the effect of alkylation is to enhance antimicrobial activity; at the same time water solubility is reduced and the compounds are slightly less corrosive. They form the active ingredient of the disinfectant fluid Lysol. The toxicity pattern is similar to that of phenol.

Cresol is not among the cosmetic preservatives listed by Richardson [13], but is included in Davis' list [14] and is unlikely to be a preservative of choice.

Thymol: Thymol is 2-*iso*-propyl-3-methyl phenol. It is a white crystalline solid, with a melting point of 52°C, and is soluble 1:1000 in water; it is much more soluble in organic solvents; for example, 1:0.3 for alcohol (ethanol), and 1:200 for glycerol. It occurs naturally in the volatile oils from species of thyme from which it is largely obtained.

Thymol is a more potent bactericide than phenol, but its low water solubility reduces its usefulness. It occurs in one formulation in the Richardson [13] list.

2-Phenylphenol, or o-Phenylphenol: This aryl phenol ($C_6H_5 \cdot C_6H_4 \cdot OH$) is, according to Richardson [13], the most frequently used of all the phenols in cosmetic preservation, it appears either as the phenol itself or as its sodium salt in 19 instances.

2-Phenylphenol is a white crystalline compound, with a melting point of 57°C. It is soluble 1:1000 in water, and the sodium salt is readily soluble in water. It is used extensively as a preservative in industrial products, and is active against bacteria and fungi.

It is active over the range 1:2000–1:6000 and has low toxicity, skin irritancy, and sensitization. Detailed data on its antimicrobial activity appear in the paper by Harris and Christiansen [17], and in the manufacturers' literature. It is used typically at 0.1–0.5% as a cosmetic preservative. Imperial Chemical Industries, Manchester, England markets the product as Topane and Topane WS, an aqueous solution of the sodium salt, and Dow Chemicals, Michigan, markets it as Dowicide 1 and Dowicide A (aqueous solution of sodium salt).

In addition to its use in cosmetics themselves, 2-phenylphenol is added to the paraffin wax used in the preparation of waxed paper and card liners for bottle and jar caps.

3. Summary and Conclusions

It can be seen that, with the exception of 2-phenylphenol, the phenols as a preservative class so not find extensive application in cosmetic preservation. For further details see Russell et al. [18], Chapters 2 and 10.

REFERENCES

1. W. B. Hugo, *Microbios*, 23:83 (1979).
2. T. Alcock, *Lancet*, 1:643 (1827).
3. G. Lefevre, Lancet, 1:145 (1843).
4. J. P. Semmelweiss, *Die Atiologie der Bergriff und die Prophylaxis des Kindbettfiebers*, Pest, Vienna (1861).
5. J. Davies, *Selections in Pathology and Surgery*, Part II, Orme, Brown, Green, and Longmans, London (1839).
6. R. Koch, *Ueber Mittheilungen aus dem Kaiserlichen Gesundheitsamte*, 1:234 (1881).
7. G. M. Suter, *Chem. Rev.*, 28:269 (1941).
8. W. B. Hugo, *J. Pharm. Pharmacol.*, 9:145 (1957).
9. E. F. Gale and E. S. Taylor, *J. Gen. Microbiol.*, 1:77 (1947).

10. J. Judis, *J. Pharm. Sci.*, *51*:261 (1962).
11. W. B. Hugo and J. G. Bowen, *Microbios*, *8*:189 (1973).
12. W. B. Hugo, *Int. J. Pharm.*, *1*:127 (1978).
13. E. L. Richardson, *Cosmet. Toiletr.*, *92*:85 (1977).
14. J. G. Davis, *Soap Perfum. Cosmet.*, *53*:133 (1980).
15. J. Lister, *Lancet*, *1*:362; *2*:353 (1867).
16. F. W. Tilley, *J. Bacteriol.*, *43*:521 (1942).
17. S. E. Harris and W. G. Christiansen, *J. Am. Pharm. Assoc. Sci. Ed.*, *23*:530 (1934).
18. A. D. Russell, W. B. Hugo and G. A. J. Ayliffe, Eds. Principles Practice of Disinfection Preservation and Sterilization, Blackwells

8
SODIUM AND ZINC OMADINE

GENE A. HYDE and JOHN D. NELSON, JR. *Olin Corportion, New Haven, Connecticut*

- I. Nomenclature 116
- II. Production 116
- III. History 117
- IV. Chemistry 117
    - A. Preparation 117
    - B. Physical properties 118
    - C. Chemical properties 118
    - D. Analytical methods 120
- V. Antimicrobial Action 120
    - A. Spectrum 120
    - B. Mechanism of action 120
    - C. Efficacy in formulations 122
- VI. Formulating Properties 123
    - A. Effect of heat 123
    - B. Effect of pH 123
    - C. Effect of light 123
    - D. Chemical reactivity 123
    - E. Solubility 124
    - F. Specific gravity and particle size 124
    - G. Compatibility 124
- VII. Toxicology 124
- VIII. Biodegradation 125
- IX. Summary 126
    - References 126

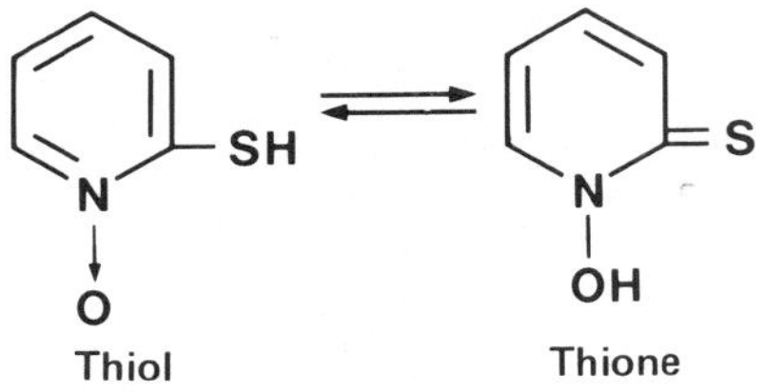

FIGURE 1 Tautomeric forms of pyrithione.

I. NOMENCLATURE

The two compounds discussed in this chapter are known to the cosmetic and toiletry industry as zinc and sodium pyrithione [1]. The parent compound exists in tautomeric form [2] (Fig. 1) and names have been assigned to it based on both structures. These include 2-pyridinethiol-1-oxide (CAS Registry No. 1121-31-9), 1-hydroxypyridine-2-thione (CAS Registry No. 1121-30-8), 2-mercaptopyridine-1-oxide (or N-oxide), 1-hydroxy-2(1H)-pyridinethione (CAS Registry No. 1121-30-8), and pyrithione.

The Chemical Abstracts Service (CAS) Registry names and numbers for sodium pyrithione are 2-pyridinethiol-1-oxide, sodium salt (CAS Registry No. 3811-73-2), and 1-hydroxy-2-(1H)-pyridinethione, sodium salt (CAS Registry No. 15922-78-8). For zinc pyrithione the CAS Registry name and number is bis[1-hydroxy-2(1H)-pyridinethionato-O,S]-(T-4)-zinc (CAS Registry No. 13463-41-7). The structures are shown in Fig. 2.

II. PRODUCTION

Zinc and sodium pyrithione are produced by several companies. By far the largest producer is Olin Corporation, which markets both products throughout the world under their registered Omadine trademark as Zinc Omadine and Sodium Omadine microbiostats. Other producers are Reutgers-Nease Chemical Co. in the United States (Zinc Pyrion), Yoshitomi Pharmaceutical Industries, Ltd., in Japan (Tomicide Z-50), and Pyrion-Chemie, GmbH, in West Germany (Zinc Pyrion).

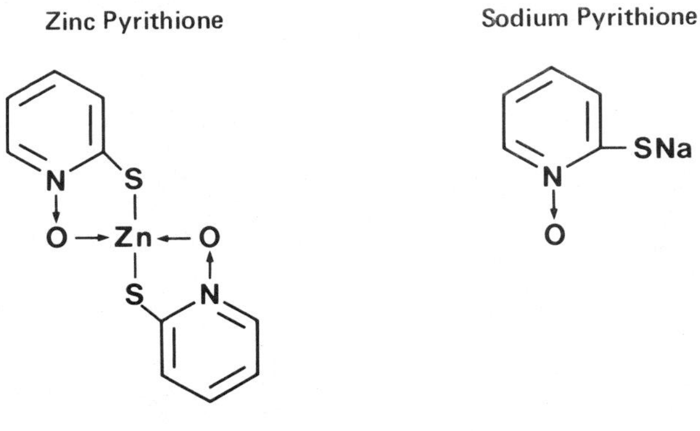

FIGURE 2 Structures of zinc and sodium pyrithione.

III. HISTORY

Pyrithione was first prepared in 1950 by Shaw et al. [3] at the Squibb Institute for Medical Research. At that time Squibb was owned by the Matheson Chemical Co., which was soon to merge with the Olin Chemical Co. to form what eventually became the present Olin Corporation.

In 1954 the first U.S. patent was granted [4] covering the composition and manufacture of pyrithione and its sodium salt. This was followed by a second patent in 1957 [5] which covered the heavy-metal derivatives.

In the early 1960s investigators at the R. T. Vanderbilt Co. and the Procter and Gamble Co. discovered the antidandruff properties of zinc pyrithione. On February 14, 1964, Procter and Gamble's antidandruff shampoo Head and Shoulders was approved for sale by the Food and Drug Administration (FDA) [6] and was launched shortly thereafter. By the time use patents were issued in 1966 and 1968 [7,8], Head and Shoulders dominated the antidandruff shampoo market.

The antimicrobial properties of the pyrithiones were known from the beginning and they were used as cosmetic preservatives by the late 1960s. Several studies reported in the literature during the 1970s included the pyrithiones [9-13], and their use continued to grow slowly. In 1981 the FDA reported [14] that 29 formulations registered with the agency contained zinc pyrithione and 7 contained the sodium salt.

IV. CHEMISTRY

A. Preparation

Pyrithione can be prepared by reacting a 2-halopyridine with a peracid, such as peracetic, to form the 2-halopyridine-N-oxide. The halogen is then replaced with sulfur by reaction with an alkali metal hydrosulfide [3]. If sodium hydrosulfide is used, this sequence produces sodium pyrithione, which, after purification and concentration, is usually sold as a 40% aqueous solution.

Zinc and other heavy-metal complexes of pyrithione are prepared from sodium pyrithione by precipitation using the appropriate heavy-metal salt [4].

TABLE 1 Typical Physical Properties of Technical-Grade Material

	Zinc pyrithione		Sodium pyrithione	
	Powder	48% dispersion	Powder	40% solution
Molecular weight	317.7	–	149.2	–
Specific gravity (25°C)	1.78	–	1.17	1.22
Density (lb/gal)	–	10.0	–	10.2
Bulk density (g/cm^3)	0.35	–	0.50	–
pH (10% in H_2O, average)	7.3	7.5	9.8	8.3
Melting point (°C)	240 (d)	–	250 (d)	–
Particle size (%)	90 < 841 μm	90 < 5 μm	65 < 25 μm	–

B. Physical Properties

The physical properties and solubilities given in Tables 1 and 2 are for technical-grade zinc and sodium pyrithione and are from an Olin Corporation data bulletin [15].

C. Chemical Properties

Pyrithione exists in tautomeric form [2] as shown in Figure 1. At or below pH 3 the prevailing form is the thione. The thiol form appears as the pH is raised, the equivalence point being about 7.6. Between pH 7.6 and about

TABLE 2 Solubilities of Technical-Grade Material (wt/wt % at 25°C)[a]

Solvent	Zinc pyrithione powder	Sodium pyrithione powder
Water (pH 7)	0.0015	53
Ethanol, 40A	0.01	19
Isopropanol	0.008	0.8
Propylene glycol	0.02	13
Polyethylene glycol 400	0.02	12
Poly-Solv EM[b]	0.09	32
Poly-Solv DE[b]	0.01	12
Mineral oil, light	<0.0001	<0.0001
Solulan 98[c]	0.1	1
Olive oil	<0.0003	<0.0005
Castor oil	<0.0003	<0.0005
Isopropyl myristate	<0.0001	<0.0001
Isopropyl palmitate	<0.0001	<0.0001
Tween 40[d]	0.1	~4
Duponol WAQE[e]	—	~24
Sipon LLS[f]	—	~31
BioSoft EA-8[g]	0.07	~9

[a] Approximate solubilities.
[b] Olin Corporation.
[c] Amerchol Unit of CPC International.
[d] ICI United States, Inc.
[e] E.I. Du Pont de Nemours & Co.
[f] Alcolac Inc.
[g] Stepan Chemical.

10.0 the thiol form (or its salts) is relatively stable, but above pH 10 it is rather easily oxidized to the sulfimic acid anion [16]. The thione form is more easily oxidized than the thiol at lower pH values.

Pyrithione is rather easily oxidized by many oxidizing agents. The most common of these found in cosmetics and toiletries are organic peroxides, found as impurities in glycols and polyglycol derivatives. The oxidation of pyrithione produces the disulfide (bispyrithione), which under alkaline conditions undergoes the classic hydrolysis of organic disulfides [17], eventually being converted to the sulfonic acid (Fig. 3).

Pyrithione will also undergo self-oxidation in the absence of oxygen to yield bispyrithione, 2,2'-bisthiopyridine, and, probably, the mixed disulfide. Reduction of pyrithione takes place at the N → O linkage to yield the thiopyridine.

Pyrithione is subject to photolysis by ultraviolet and visible light. Aqueous solutions exposed to daylight will degrade rather rapidly [18]. The photolysis reaction appears to proceed along the same lines as the oxidation-hydrolysis already discussed, with the disulfide intermediate and the sulfonic acid as the end product.

The pyrithione molecule is an excellent chelating agent, forming neutral complexes with almost every known metal under almost any pH condition [19, 20]. The most well known of these, of course, is the zinc complex used in antidandruff shampoos. Most of the metal complexes are only slightly soluble in water, the values ranging from a few thousand parts per million to less than one.

Transchelation will also occur. For example, if zinc pyrithione is added to a solution containing ferric ions, ferric pyrithione will form because the ferric complex is stronger than the zinc complex.

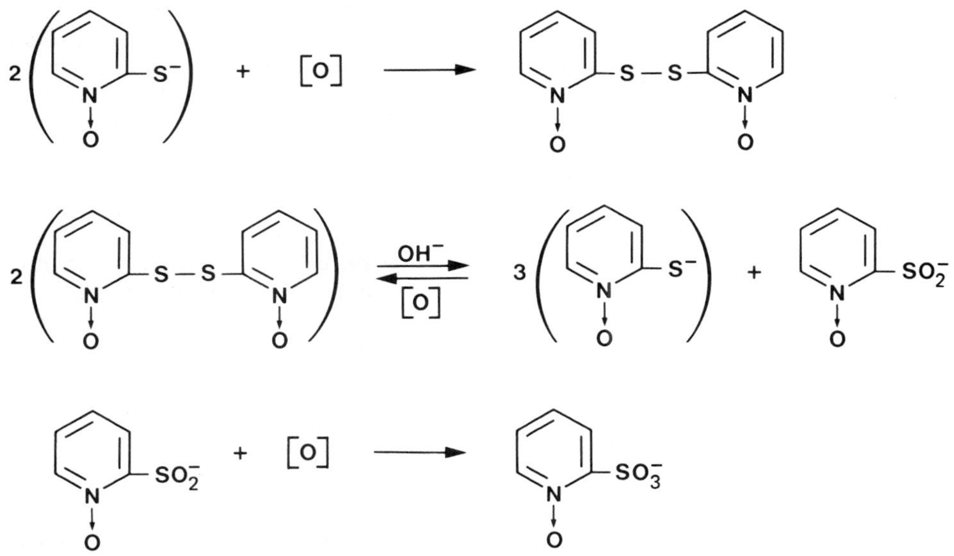

FIGURE 3 Reaction sequence for the oxidation of pyrithione.

D. Analytical Methods

Several analytical methods for pyrithiones have appeared in the literature [21-29]. In general, they involve polarography or other electroanalytical techniques, chromatography, spectrophotometry, or titration procedures. The method of choice is usually dictated by the formulation involved. In the authors' experience, the best general procedure for cosmetics is to extract the sample with caustic and analyze the extract by polarography. In some cases direct potentiometric titration with iodine is also possible.

V. ANTIMICROBIAL ACTION

A. Spectrum

The pyrithiones are broad-spectrum antimicrobial agents, effective against both bacteria and fungi. Table 3 gives minimum inhibitory concentrations (MICs) for the zinc and sodium derivatives as reported by Olin Corporation in their data bulletins [15,30].

Minimum inhibitory concentration values are, of course, only an indication of a preservative's activity, and comparisons should be used with caution. Many factors in a use situation can cause partial deactivation of a preservative so that much more than is indicated by the MIC value is actually needed. For this, and other reasons discussed below, actual use tests should be performed to determine the best preservative and its optimum use level for a given formulation.

B. Mechanism of Action

The authors have discussed the mechanism of action of the pyrithiones in a previous paper [31]:

> The relatively broad spectrum of activity against both primitive (procaryotic) bacterial cells and plant/animal-like (eucaryotic) fungal cells, as well as a low incidence of resistance suggests that the pyrithione molecule affects microorganisms in more than one way. Lott and Shaw hypothesized that the antibacterial activity of the cyclic hydroxamate arrangement of structurally-related 1-hydroxy-2-pyridones was due to the chelating activity of the latter [32]. More recently it was shown that the corresponding cyclic thio-hydroxamate structure is responsible for the antimicrobial activity of 1-hydroxypyridinethiones [33]. The dipole structure of the molecule creates a pseudo-quaternary ammonium group, which also may be the basis of its activity. A third mechanism is suggested by the observation that pyrithione (1-hydroxypyridinethione) inhibits the growth of the mold, *Penicillium*, by interfering with the assimilation of various nutrients [33]. Because the inhibition increases as the pH of the environment decreases, it has been suggested that pyrithione, like other organic acids, disrupts a proton gradient across the cell's membrane and thus inhibits the transport of solutes through the membrane barrier. There is also indirect evidence that pyrithione is an antimetabolite of the pyridine derivative, pyridoxal, a member of the vitamin B_6 complex [33]. If this is correct, the activity of pyrithione should be analogous to the inhibition of microbial folate production by sulfa drugs.

TABLE 3 Antimicrobial Spectrum Minimum Inhibitory Concentrations (ppm)[a]

Organism	ATCC number	Zinc pyrithione[b]	Sodium pyrithione
Gram-positive bacteria			
Bacillus cereus	11778	1	16
Bacillus cereus[c]	14579	4	16
Bacillus subtilis	9524	–	0.5
Micrococcus agilis	9814	–	≤0.25
Sarcina lutea	9341	16	8
Staphylococcus aureus[c]	6538P	8	16
Streptococcus faecalis[c]	19433	16	2
Gram-negative bacteria			
Enterobacter aerogenes	15038	–	128
Escherichia coli	9637	32	16
Escherichia coli[c]	11229	16	8
Flavobacterium flavescens	8315	–	128
Klebsiella pneumoniae	9997	–	256
Proteus vulgaris	9920	1	8
Pseudomonas aeruginosa	9721	512	128
Pseudomonas aeruginosa[c]	10145	512	512
Pseudomonas glycinea	8727	≤0.25	4
Pseudomonas lachrymans	7386	2	16
Pseudomonas phaseolicola	11355	≤0.25	32
Salmonella typhimurium	7823	32	16
Salmonella typhimurium[c]	13311	16	64
Xanthomonas celebensis	19045	8	64
Xanthomonas cucurbitae	23378	2	4
Xanthomonas juglandis	11329	0.5	8
Xanthomonas vesicatoria	11551	32	128
Molds			
Aspergillus niger[c]	9642	16	≤0.25
Aspergillus oryzae	15-9550[d]	–	≤0.25
Aspergillus terreus	10690	0.5	16
Aureobasidium pullulans	SC2599	≤0.25	≤0.25
Cephalosporium acremonium	1014P	–	≤0.25
Cercospora nicotianae	18366	≤0.25	4
Chaetomium globosum	6205	–	≤0.25
Cladosporium resinae	22712	≤0.25	≤0.25
Cylindrocladium scoparium	11614	≤0.25	0.5
Helminthosporium populosum	11616	≤0.25	≤0.25
Mucor abundans	11010	–	4
Penicillium vermiculatum[c]	1124	1	2
Phytophthora infestans	13196	≤0.25	1
Pythium irrigularae	16970	≤0.25	≤0.25
Rhizoctonia fragariae	14691	≤0.25	≤0.25
Trichopyton mentagrophytes	9533	≤0.25	≤0.25
Trichophyton mentagrophytes[c]	9129	≤0.25	0.5

TABLE 3 Continued

Organism	ATCC number	Zinc pyrithione[b]	Sodium pyrithione
Yeasts			
Candida albicans[c]	10231	≤0.25	4
Candida albicans	11651	≤0.25	≤0.25
Rhodotorula glutinis	16726	–	≤0.25
Saccharomyces cerevisiae	7752	≤0.25	≤0.25

[a]Obtained by using an Autotiter IV (Product #7302, Ames Company, Division of Miles Laboratories, Elkhart, Indiana 46514). The Autotiter automatically performs serial dilutions and inoculations in nutrient broth. Bacteria concentrations were approximately 10^5 organisms per milliliter. Fungal concentrations were approximately 10^5 colony-forming units per milliliter.
[b]Because of the low solubility of zinc pyrithione in water, these values were obtained using dimethylsulfoxide as a solvent.
[c]Organisms specified in the American Society for Testing and Materials test method E 640-78 (Preservatives in Water-Containing Cosmetics) [35].
[d]Carolina Biological Supply Co.
Source: Refs. 15 and 30.

Albert et al. [34] have suggested that in order to penetrate the cell wall, the pyrithione must first form a 2:1 complex with iron, but that once inside the cell, the iron complex must be broken down before biocidal action can occur.

C. Efficacy in Formulations

Most of the published work on efficacy of the pyrithiones reports only MIC values or test results obtained on aqueous slurries or solutions. These types of data, while useful as a screening tool, bear little resemblance to conditions in actual use.

One article by Jacobs et al. [11] is an exception to the above. This article reports the results of tests using zinc and sodium pyrithione (and 27 other preservatives) in an anionic and a nonionic lotion both in an acid and alkaline version. The results show that 0.1% of sodium pyrithione satisfactorily protected both versions of the nonionic lotion and the acid version of the anionic lotion. Zinc pyrithione at 0.1% was not a satisfactory preservative in any of the lotions.

A data bulletin published by Olin Corporation [30] gives the results of tests done on cosmetic formulations using the American Society for Testing and Materials (ASTM) test method E 640-78 [35], which is designed to show if a compound has potential as a cosmetic preservative. In these tests Zinc and Sodium Omadine brands of pyrithione were tested at concentrations of 250 to 1000 ppm (active basis) in a variety of cosmetic and toiletry formulations, including an alkaline and an acid oil-water lotion; an alkaline oil-water hand cream; a clear, alkaline protein shampoo; an alkaline water-oil cream; an acid shampoo; and an eye shadow.

The ASTM E 640-78 test method involves an initial challenge of the test samples with the challenge organisms (see Table 3), followed by a rechallenge on the 28th day. Plate counts are made weekly for 4 weeks after each challenge. To show potential as a cosmetic preservative, the test substance must reduce the bacteria and yeast by 99.9% within 1 week after each challenge, and the molds by 90% within 4 weeks after each challenge. In all cases there must be no increase in counts for the remainder of the test period.

On the basis of these tests, the conclusion is drawn that both Zinc and Sodium Omadine show excellent potential as cosmetic preservatives and that, while data should be developed on individual formulations to determine the most effective concentration, a good starting point is 250 ppm on an active basis.

VI. FORMULATING PROPERTIES

The following discussion of those physical and chemical properties of the pyrithiones which have an influence on their formulating properties is based on the authors' experience. The preceding sections on physical and chemical properties should also be consulted.

A. Effect of Heat

Neither zinc nor sodium pyrithione (in powder form) show any significant decomposition when heated at 100°C for periods up to 120 hr. Aqueous solutions of sodium pyrithione do not lose microbiological activity after 3 months at 40°C.

B. Effect of pH

The pyrithiones can be used over the pH range of 4.5-9.5.

C. Effect of Light

Both zinc and sodium pyrithione will degrade if exposed to ultraviolet or visible light. Any of the common ultraviolet screens used in cosmetics can be used to retard the degradation, or suitable screens can be incorporated in the bottle itself [18]. The extent and speed of degradation appears to be somewhat dependent on the other formulation ingredients, as well as the pH, so stability tests should be done on each formulation to determine if a problem exists.

D. Chemical Reactivity

Oxidizing or reducing agents will react with the pyrithiones to form products which are inactive microbiologically.

Heavy-metal ions in the formulation will form metal complexes with pyrithione, some of which are highly colored. Ferric ions, for example, will form blue ferric pyrithione from either sodium or zinc pyrithione. The transchelation reaction of ferric ions with zinc pyrithione can be prevented by adding zinc ions in the form of any soluble zinc salt [36].

E. Solubility

Sodium pyrithione is very soluble in water (see Table 2), and, in theory, can be used in any aqueous formulation. However, it is a salt and as such it can upset some emulsions or have an adverse effect on thickening agents.

The solubility of zinc pyrithione in water is very low, and it is therefore suitable only for cream- or lotion-type formulations. Because of its insolubility and lack of ionic character, it will not normally affect emulsion stability.

F. Specific Gravity and Particle Size

The specific gravity of zinc pyrithione (1.78) is considerably higher than that of other cosmetic ingredients and it will settle out of low-viscosity formulations. It is usually not suitable for formulations with viscosities below 2500 cP.

Zinc pyrithione is available commercially as a powder or an aqueous dispersion. The particle size in the dispersion is much smaller than in the powder (see Table 1) and these smaller particles are much easier to keep suspended. The dispersion, then, is the preferred choice, unless a dry powder is essential, such as when it is used in talcum powder.

G. Compatibility

In general, the pyrithiones appear to be compatible with most cosmetic ingredients. As with any preservative, however, each formulation must be tested for microbial resistance, since one never knows when some new ingredient or combination of ingredients will act as an inhibitor.

The authors have found sodium pyrithione to be compatible with hydrolyzed animal protein from several sources. Nonionic surfactants and emulsifiers, however, tend to increase slightly the MIC values of sodium pyrithione, making it necessary to use a little more of the preservative than would be necessary if the nonionics were not present.

Ethylenediaminetetraacetic acid (EDTA) and its derivatives should not be used with zinc pyrithione, EDTA is a stronger complexing agent toward zinc than is pyrithione, so that the two will react to form a pyrithione salt and zinc EDTA. This will not affect the preservative action, but the EDTA will not be available to do the job it was intended to do.

The use of combinations of preservatives is becoming very common in the cosmetic industry. This technique is used to broaden the spectrum of microorganisms controlled and also to decrease the chances that organism immunity will develop. In some cases true synergism will occur between two preservatives, but more often the effect is simply additive. The latter is the case with the pyrithiones. The authors have tested sodium pyrithione in 1:1 combinations with the preservatives listed in Table 4. In all cases but two, the two preservatives were compatible, showing neither synergism nor antagonism toward each other. The two cases of antagonism were with Lauriciden and Bronopol.

VII. TOXICOLOGY

Because Olin Corporation was for many years the sole producer of zinc and sodium pyrithione, most of the toxicological information in the open literature

TABLE 4 Preservatives Tested in Combination with Sodium Pyrithione

Benzyl Alcohol	Kathon CG
Bronidox L	Lauricidin[a]
Bronopol[a]	Lexgard M
Chlorobutanol	Myacide SP
Dehydroacetic acid	Ottasept Extra
Dowicil 200	Oxadine A
Germall II	Phenoxyethanol
Germall 115	Sodium benzoate
Giv-Gard DXN	Sorbistat K
Glydant	Suttocide A
	Undebenzophene

[a]Not compatible with sodium pyrithione.

was obtained on Zinc and Sodium Omadine brands of pyrithione. Many articles, too numerous to list here, have appeared in the past 20 years. We will only mention two recent reviews [37,38] which include most of the pertinent data. Olin Corporation's application data sheet [30] includes the following statement:

> Zinc Omadine: The acute oral LD_{50} (male rats) is 200 mg/kg. Both the 48% dispersion and the powder are irritating to the skin, and extremely irritating to rabbit eyes. Zinc Omadine is not an allergic sensitizer. Sodium Omadine: The acute oral LD_{50} (rats) is approximately 875 mg/kg. Both the 40% solution and the powder are irritating to the skin. The 40% solution is non-irritating to rabbit eyes. Sodium Omadine is not an allergic sensitizer or a photosensitizer.

We are not aware of any data which indicate that either zinc or sodium pyrithione is irritating to the skin or eyes at the recommended use concentrations.

VIII. BIODEGRADATION

Tests conducted in the authors' laboratories show that sodium pyrithione, when used in the activated sludge test recommended by the Toxic Substance Control Act, exceeds the Environmental Protection Agency's criterion of 90% degradation of 250 ppm within 4 days. No fish toxicity (96-hr bioassay) was found for the effluent from activated sludge digestion of 1 ppm of sodium pyrithione.

Neihof et al. [39] have reported that pyrithione biocides are rapidly degraded to nontoxic products when discharged into sunlit natural waters.

IX. SUMMARY

All of the available data indicate that sodium and zinc pyrithione are effective preservatives which can be easily formulated into a variety of cosmetic and toiletry products. As with any preservative however, efficacy, stability, and toxicology testing should be done on the complete product, as formulation ingredients can and do have an effect on all three parameters.

The manufacturers of the preservatives should also be consulted, as there frequently are products in which an individual preservative should not or cannot be used.

REFERENCES

1. Cosmetic, Toiletry, and Fragrance Association, Inc., *Cosmetic Ingredient Dictionary*, 3rd Ed. (N. F. Estrin, ed.), Cosmetic Toiletry, and Fragrance Association, Washington, D.C. (1982).
2. R. A. Jones and A. R. Katritzky, N-Oxides and related compounds. Part XVII. The tautomerism of mercapto- and acylamino-pyridine-1-oxides, *J. Chem. Soc.*, 2937-2942 (1960).
3. E. Shaw, J. Bernstein, K. Losee, and W. A. Lott, Analogs of aspergillic acid. IV. Substituted 2-bromopyridine-N-oxides and their conversion to cyclic thiohydroxamic acids, *J. Am. Chem. Soc.*, 72:4362-4364 (1950).
4. E. N. Shaw and J. Bernstein, N-Hydroxy-2-pyridinethiones, U.S. Patent 2,686,786, Mathieson Chemical Co., August 17 (1954).
5. J. Bernstein and K. Losee, Heavy metal derivatives of 1-hydroxy-2-pyridinethiones and method of preparing same, U.S. Patent 2,809,971, Olin Mathieson Chemical Corp., October 15 (1957).
6. *Federal Register*, 29:5769 (1964).
7. K. S. Karsten, W. S. Taylor, and J. J. Parran, Method of combatting dandruff with pyridinethione metal salts detergent compositions, U.S. Patent 3,236,733, R. T. Vanderbilt Co. and The Procter and Gamble Co., February 22 (1966).
8. K. S. Karsten and W. S. Taylor, Germicidal detergent compositions, U.S. Patent 3,412,033, R. T. Vanderbilt Co., November 19 (1968).
9. S. Tenenbaum and D. L. Opdyke, Antimicrobial properties of the pyrithione salts. VII. In vitro methods for comparing pyrithiones to standard antimicrobials, *Food. Cosmet. Toxicol.*, 7:223-232 (1969).
10. A. E. Elkhouly, Studies on the antibacterial activity of zinc pyridinethiol oxide, *Pharm. Unserer Zeit*, 29:45-47 (1974).
11. G. Jacobs, S. M. Henry, and V. F. Cotty, The influence of pH, emulsifier, and accelerated aging upon preservative requirements of o/w emulsions, *J. Soc. Cosmet. Chem.*, 26:105-117 (1975).
12. K. H. Wallhaüsser, Is it necessary to preserve cosmetics?, *Sonderdruck Parfuem. Kosmet.*, 56:121-129 (1975).
13. K. H. Wallhaüsser, Antimicrobial contamination of cosmetics. Raw materials—manufacture—preservation, *Sonderdruck Parfuem. Kosmet.*, 53:305-319 (1972).
14. E. L. Richardson, Update—frequency of preservative use in cosmetic formulations as disclosed to FDA, *Cosmet. Toiletr.*, 96:91-92 (1981).
15. Olin Corporation, *Zinc Omadine® and Sodium Omadine®*, Product Data Bulletin No. 735-009R3, Olin Corp., Stamford, Conn. (1979).

16. R. J. Fenn and D. A. Csejka, The stability of 2-pyridinethiol-1-oxide, sodium salt as a function of pH, *J. Soc. Cosmet. Chem.*, *33*:243-248 (1982).
17. J. P. Danehy and K. N. Parameswaran, The alkaline decomposition of organic disulfides. III. Substituent effects among aromatic disulfides, *J. Org. Chem.*, *33*:568 (1968).
18. G. A. Hyde and M. M. Auerback, Formulation techniques for zinc pyrithione antidandruff shampoos, *Cosmet. Toiletr.*, *94*:57-59 (1979).
19. M. Edrissi, A. M. Massoumi, and J. A. W. Dalzeil, Comparative studies of 1-hydroxy-2-pyridinethione and its methyl derivatives as analytical reagents for metal ions, *Microchem. J.*, *16*:538-547 (1971).
20. A. I. Buseu, V. M. Byr'ko, H. K. Yen, and N. N. Novikova, 2-Mercaptopyridine-N-oxide and possibilities of using it in concentrations, *Vestn. Mosk. Univ. Khim.*, *27*:319-322 (1972).
21. A. F. Krivis, E. S. Gazda, G. R. Supp, and M. A. Robinson, Polarographic analysis of a series of metal ion 1-hydroxypyridine-2-thione systems, *Anal. Chem.*, *35*:966-968 (1963).
22. A. F. Krivis and E. S. Gazda, Polarographic behavior of 1-hydroxypyridine-2-thione, *Anal. Chem.*, *41*:212-214 (1969).
23. D. A. Csejka, S. T. Nakos, and E. W. DuBord, Determination of the sodium salt of 2-mercaptopyridine-N-oxide by differential pulse cathodic stripping voltammetry, *Anal. Chem.*, *47*:322-324 (1975).
24. C. H. Wilson, Identification of preservatives in cosmetic products by thin-layer chromatography, *J. Soc. Cosmet. Chem.*, *26*:75-81 (1975).
25. S. Oliveri-Vigh and H. L. Karageozian, Colorimetric determination of sodium 2-pyridinethiol-1-oxide with ferric ammonium sulfate, *Anal. Chem.*, *48*:1001-1003 (1976).
26. B. L. Kabacoff and C. M. Fairchild, Determination of zinc pyrithione by chelate exchange, *J. Soc. Cosmet. Chem.*, *26*:453-459 (1975).
27. M. D. Seymour and D. L. Bailey, Thin-layer chromatography of pyrithiones, *J. Chromatogr.*, *206*:301-310 (1981).
28. M. B. Graber, I. I. Domsky, and M. E. Ginn, A TLC method for identification of germicides in personal care products, *J. Am. Oil Chem. Soc.*, *46*:529-531 (1969).
29. R. T. Brooks and P. D. Sternglanz, Titanometric determination of the N-oxide group in pyridine-N-oxide and related compounds, *Anal. Chem.*, *31*:561-565 (1959).
30. Olin Corporation, *Omadine Antimicrobials for Cosmetic Preservation*, Application Data Bulletin No. 735-026, Olin Corp., Stamford, Conn. (1980).
31. J. D. Nelson and G. A. Hyde, Sodium and Zinc Omadine antimicrobials as cosmetic preservatives, *Cosmet. Toiletr.*, *96*:87-90 (1981).
32. W. A. Lott and E. Shaw, Analogs of aspergillic acid. I. The tautomerism of the hydroxypyridine N-oxides, *J. Am. Chem. Soc.*, *71*:67-70 (1949).
33. C. J. Chandler and I. H. Segel, Mechanism of the antimicrobial action of pyrithione; effects on membrane transport, ATP levels, and protein synthesis, *Antimicrob. Agents Chemother.*, *14*:60-68 (1978).
34. A. Albert, C. W. Rees, and A. J. H. Tomlinson, The influence of chemical constitution on anti-bacterial activity. Part VIII. 2-Mercaptopyridine-N-oxide, and some general observations on metal binding agents, *Br. J. Exp. Pathol.*, *37*:5000-5011 (1956).

35. American Society for Testing and Materials, Standard test method for preservatives in water containing cosmetics, ASTM E 640-78, in *Annual Book of ASTM Standards Part 46*, American Society for Testing and Materials, Philadelphia, Pa. (1981).
36. W. G. Gorman, Zinc salt prevention or removal of discoloration in pyrithione, pyrithione salt, and dipyrithione compositions, U.S. Patent 4,161,562, Sterling Drug, Inc., July 17 (1979).
37. N. P. Luepke and P. Preusser, Antidandruff cosmetics—efficacy and toxicology, *Aerztl. Kosmetol.*, *8*:269-280 (1978).
38. J. G. Black and D. Howes, Toxicity of pyrithiones, *Chem. Toxicol.*, *13*:1-26 (1978).
39. R. A. Neihof, C. A. Bailey, C. Patouillet, and P. J. Hannan, Photodegradation of mercaptopyridine-N-oxide biocides, *Arch. Environ. Contam. Toxicol.*, *8*:355-368 (1979).

9
KATHON CG
A New Single-Component, Broad-Spectrum Preservative System for Cosmetics and Toiletries

ANDREW B. LAW, JACK N. MOSS, and EDWARD S. LASHEN *Rohm and Haas Company, Spring House, Pennsylvania*

I. Introduction 129
II. Chemical Identification 130
III. Composition 130
IV. Physical and Chemical Properties 131
V. Microbiostatic Activity 131
   A. Minimal inhibitory concentration values 131
   B. Efficacy in several typical cosmetic-toiletry formulations 131
   C. Efficacy as a preservative for cosmetic raw materials 132
VI. Toxicological Properties 132
   A. Animal studies 139
   B. Genotoxicity studies 140
   C. Human studies 140
VII. Summary 141

I. INTRODUCTION

Research into the chemistry of isothiazolones in the early 1960s by the Rohm and Haas Company of Philadelphia led to the development of several commercial antimicrobial products currently in use in a variety of industrial applications. These include mildewcides for leather and fabric; preservatives for metal working fluids, emulsion polymers, and latex paints; and antibiofoulants for cooling towers, paper mills, and oil recovery applications. Their most recent development based on isothiazolone chemistry is Kathon CG,* a broad-spectrum preservative agent for cosmetic and toiletry products.

The preservative has been shown to provide control of gram-negative and gram-positive bacteria, as well as fungi and yeasts, at unusually low concentration levels in a wide variety of cosmetic and toiletry formulations. This preservative has good compatibility with common cosmetic ingredients such as protein, emulsifiers, and anionic, cationic, and nonionic surfactants. It is also effective over the pH range normally encountered in cosmetic and toiletry products, and does not impart color or odor to most finished formulations.

This new antimicrobial agent is supplied as a 1.5% active aqueous solution that is readily miscible in water, lower alcohols, glycols, and other hydrophilic organic solvents. It is recommended for use in cosmetic and toiletry formulations at levels ranging from 0.02 to 0.1% as supplied (3-15 ppm active ingredient).

II. CHEMICAL IDENTIFICATION

The active ingredient of Kathon CG consists of two isothiazolones, named by the Chemical Abstracts Services (CAS) as 5-chloro-2-methyl-3(2H)-isothiazolone (CAS Registry No. 26172-55-4) and 2-methyl-3(2H)-isothiazolone (CAS Registry No. 2682-20-4). The Cosmetic, Toiletry, and Fragrance Association (CTFA) adopted names for these two components are methylchloroisothiazolone and methylisothiazolone, respectively, the latter ingredient being a by-product in the manufacturing process. While the chlorinated molecule is the more active of the two, both ingredients are considered biologically functional. The corresponding structural formulas for the two components are shown below.

5-chloro-2-methyl-3(2H)-isothiazolone

2-methyl-3(2H)-isothiazolone

III. COMPOSITION

The composition of the preservative as supplied is given below:

Active ingredients	
5-Chloro-2-methyl-3-(2H)-isothiazolone	1.15%
2-Methyl-3(2H)-isothiazolone	0.35%
	1.50%
Inert ingredients	
Magnesium salts	23.0%
Water	75.5%
	98.5%

---

*A registered trademark of the Rohm and Haas Company.

Kathon CG

IV. PHYSICAL AND CHEMICAL PROPERTIES

Appearance	Clear liquid
Color	Light amber
Odor	Mild
Specific gravity at 20°C	1.19
Density (lb/gal)	9.9
pH (as supplied)	3.5
Active ingredient content (%)	1.5
Viscosity	5.0 cP (± 0.2 cP) at 23°C
Freezing point	-18 to -21.5°C
Miscibility	Miscible with water, lower alcohols, and glycols
Compatibility	Biologically and physically compatible with emulsifiers, proteins, and anionic, nonionic, and cationic surfactants, active ingredient may be inactivated by amines, mercaptans, sulfides, and sulfites
Stability	Stable for at least 1 year at ambient temperature, and for at least 6 months at 50°C

V. MICROBIOSTATIC ACTIVITY

A. Minimal Inhibitory Concentration Values

The results of minimal inhibitory concentration (MIC) tests against a variety of bacterial and fungal organisms are shown in Table 1. Not all of the organisms listed are found in cosmetic and toiletry products, and the values shown do not necessarily reflect the concentrations of Kathon CG required in such products to control these organisms. The data do, however, show Kathon CG to possess broad-spectrum antimicrobial activity at low concentrations, and provide a comparison of its relative activity against specific microbial species.

B. Efficacy in Several Typical Cosmetic-Toiletry Formulations

The results of repeated-insult tests in three typical cosmetic-toiletry formulations are shown in Tables 2-4. In these tests several concentrations of Kathon CG were compared to recommended concentrations of a mixture of methyl and propyl p-hydroxybenzoates for their ability to prevent microbial buildup in the three formulations. The composition of the formulations employed in the studies are given as follows:

Ingredient	Weight percent
Formulation I: Nonionic ointment	
Cetyl alcohol	20.0
Mineral oil	20.0
Sorbitan monooleate	0.5
Polyoxyethylene sorbitan monooleate	4.5
Water	
Formulation II: Anionic hand lotion	
Glycerol monostearate	4.0
Lanolin absorption base	1.0
Stearic acid	1.5
Glycerol	3.0
Sodium lauryl sulfate	1.0
Water	89.5
Formulation III. Anionic protein shampoo	
Triethanolamine lauryl sulfate (40%)	25.0
Lauric diethanolamide	5.0
Coconut dicarboxylic imidazoline	5.0
Polyoxyethylene lanolin (50%)	3.0
Phosphoric acid (85%)	0.2
Cosmetic polypeptides	9.0
Water	52.8

In addition to the types of formulations described above, Kathon CG has also been found to be effective at its recommended use levels for preserving hair shampoos, hair conditioners, body shampoos, baby lotions, liquid skin cleansers, bath foam liquids, and skin creams.

C. Efficacy as a Preservative for Cosmetic Raw Materials

Kathon CG has also been shown to have utility as a preservative for many cosmetic and toiletry components which are themselves susceptible to microbial attack. Examples of its efficacy in several surfactant products are shown in Table 5. In these surfactants the new preservative was effective at a concentration of approximately 0.025% (3.8 ppm active ingredient).

VI. TOXICOLOGICAL PROPERTIES

The studies described below show Kathon CG to be safe at use levels. However, the unformulated product is a corrosive material and may cause eye damage and skin burns, as well as allergenic skin reactions. It is also toxic to fish. Skin and eye contact should be avoided, and users should follow the manufacturers handling and disposal recommendations which appear on the label and in product bulletins.

Toxicological discussions continued on page 139.

TABLE 1 Microbiostatic Concentrations of Kathon CG

Organism	ATCC number	Kathon CG as supplied (ppm)	Active ingredient (ppm)
Gram-positive bacteria			
Bacillus cereus var. *mycoides*	RandH No. L5	150	2.25
Bacillus subtilis	RandH No. B2	150	2.25
Brevibacterium ammoniagenes	6871	150	2.25
Sarcina lutea	9341	300	4.5
Staphylococcus aureus	6538	150	2.25
Staphylococcus epidermidis	155	150	2.25
Streptococcus pyogenes	624	600	9
Gram-negative bacteria			
Achromobacter parvulus	4335	150	2.25
Alcaligenes faecalis	8750	150	2.25
Enterobacter aerogenes	3906	300	4.5
Escherichia coli	11229	300	4.5
Flavobacterium suaveolens	958	600	9
Proteus vulgaris	8427	300	4.5
Pseudomonas aeruginosa	15442	300	4.5
Pseudomonas cepacia	GBL #165	50	0.75
Pseudomonas fluorescens	13525	150	2.25
Pseudomonas oleoverans	8062	300	4.5
Salmonella typhosa	6539	300	4.5
Shigella sonnei	9292	150	2.25
Fungi			
Aspergillus niger	9642	600	9
Aspergillus oryzae	10196	300	4.5
Candida albicans (yeast)	11651	300	4.5
Chaetomium globosum	6205	600	9
Cladosporium resinae	11274	300	4.5
Gliocladium fimbriatum	QM7638	600	9
Mucor rouxii	RandH L5-83	300	4.5
Penicillium funiculosum	9644	300	4.5
Penicillium variabile (glaucum)	USDA	150	2.25
Phoma herbarum (pigmentivora)	12569	150	2.25
Pullularia (Aureobasidium) pullulans	9348	300	4.5
Rhizopus stolonifer	10404	300	4.5
Rhodotorula rubra (yeast)	9449	150	2.25
Saccharomyces cerevisiae (yeast)	2601	150	2.25
Trichophyton mentagrophytes (interdigitale)	9533	300	4.5

TABLE 2 Antimicrobial Activity of Kathon CG in a Nonionic Ointment (Formulation I)[a]

Preservative	Level (%)	Parts per million of active ingredient	Replicate number[b]	Colonies per gram of formulation (days after)										
				First inoculation				Second inoculation					Third inoculation	
				4	7	14	28	4	7	14	28	182	14	
Kathon CG	0.13	19.5	1	<10	<10	<10	<10	20	<10	<10	<10	280	<10	
			2	<10	<10	<10	<10	20	<10	<10	<10	<10	<10	
	0.08	12	1	40	<10	<10	<10	30	<10	<10	<10	<10	<10	
			2	20	<10	<10	<10	20	<10	<10	<10	<10	<10	
	0.04	6	1	150	<10	<10	<10	100	140	<10	<10	<10	<10	
			2	<10	<10	<10	<10	160	<10	<10	<10	<10	<10	
	0.02	3	1	600	<10	4500	4200	2×10^5	36,000	3.4×10^6	$>22 \times 10^6$	7×10^5	$>22 \times 10^6$	
			2	200	<10	$>22 \times 10^6$	3.7×10^6	c	—	—	—	c	—	
	0.01	1	1	85,000	6×10^5	<10	5600	14×10^6	$>22 \times 10^6$	$>22 \times 10^6$	19×10^6	—	—	
			2	160	<10	<10								
70:30 methyl-propyl p-hydroxy-benzoates	0.50	5,000	1	11×10^6	14×10^6	16×10^6	20,000	c	—	—	—	—	—	
			2	8×10^6	13×10^6	13×10^6	4×10^5	c	—	—	—	—	—	
	0.30	3,000	1	16×10^6	13×10^6	15×10^6	88,000	c	—	—	—	—	—	
			2	$>22 \times 10^6$	15×10^6	22×10^5	7×10^5	c	—	—	—	—	—	
	0.10	1,000	1	$>22 \times 10^6$	$>22 \times 10^6$	$>22 \times 10^6$	1.6×10^6	c	—	—	—	—	—	
			2	$>22 \times 10^6$	$>22 \times 10^6$	$>22 \times 10^6$	8×10^5	c	—	—	—	—	—	
None (controls)	—	—	1	94×10^6	27×10^7	22×10^7	6×10^6	25×10^7	25×10^7	10×10^7	66×10^6	c	—	
			2	3×10^6	8×10^6	4×10^6	4×10^6	6×10^6	3×10^6	5×10^6	8×10^6	c	—	
			3	23×10^6	32×10^6	41×10^6	4×10^6	34×10^6	30×10^6	40×10^6	16×10^6	c	90×10^6	
			4	2×10^5	90,000	80,000	93,000	68,000	55,000	2×10^6	8×10^5	c	—	

[a]See Table 5 for composition.
[b]All samples, except Controls 2, 3, and 4, were inoculated with a combined inoculum of yeasts, bacteria, and fungi. Control number 2 was inoculated with only yeasts and gram-positive bacteria, control number 3 with only gram-negative bacteria, and control 4 with only fungi.
[c]Testing was discontinued.

TABLE 3 Antimicrobial Activity of Kathon CG in an Anionic Hand Lotion (Formulation II)[a]

Preservative	Level (%)	Parts per million active ingredient	Replicate number[b]	First inoculation				Second inoculation					Third inoculation
				4	7	14	28	4	7	14	28	182	14
Kathon CG	0.13	19.5	1	<10	<10	<10	<10	<10	<10	<10	<10	<10	<10
			2	<10	<10	<10	<10	>10	<10	<10	<10	<10	<10
	0.08	12	1	<10	<10	<10	<10	<10	<10	<10	<10	<10	<10
			2	<10	<10	<10	<10	<10	<10	<10	<10	<10	<10
	0.04	6	1	<10	<10	<10	<10	<10	<10	<10	<10	<10	<10
			2	<10	<10	<10	<10	<10	<10	<10	<10	<10	<10
	0.02	3	1	150	<10	<10	<10	<10	<10	<10	<10	<10	<10
			2	7,100	<10	2,800	<10	<10	<10	<10	<10	<10	<10
	0.01	1.5	1	<10	<10	<10	<10	2×10^6	1.4×10^6	5×10^5	4×10^5	<10	<10
			2	480	88,000	38,000	31,000	890	1.3×10^5	3×10^5	6×10^5	<10	<10
70:30 methyl-propyl p-hydroxy-benzoates	0.50	5,000	1	<10	<10	<10	<10	<10	<10	<10	<10	<10	<10
			2	<10	<10	<10	<10	<10	<10	<10	<10	<10	<10
	0.30	3,000	1	<10	<10	<10	<10	<10	<10	<10	<10	<10	<10
			2	<10	<10	<10	<10	<10	<10	<10	<10	<10	<10
	0.10	1,500	1	1.0×10^6	1.9×10^6	1.6×10^6	7×10^6	2.3×10^6	4.9×10^6	2.7×10^6	1.3×10^6	c	—
			2	1.5×10^6	1.3×10^6	4×10^6	4×10^5	2.7×10^6	4.1×10^6	1.8×10^6	3×10^5	c	—
None (controls)	—	—	1	3.5×10^6	3.7×10^6	2.7×10^6	1.3×10^6	4.3×10^6	5.8×10^6	2.0×10^6	3.9×10^6	—	6×10^5
			2	2.5×10^5	2×10^5	2×10^6	98,000	4×10^5	5×10^5	3×10^5	2×10^5	c	—
			3	9×10^5	1.1×10^6	1.0×10^6	28,000	7×10^5	4.2×10^6	4.2×10^6	3×10^5	c	—
			4	11,000	2,800	1,500	280	18,000	10,500	7,200	5,600	c	—

[a] See Table 5 for composition.
[b] All samples, except controls 2, 3, and 4, were inoculated with a combined inoculum of yeasts, bacteria, and fungi. Control number 2 was inoculated with only yeasts and gram-positive bacteria, control number 3 with only gram-negative bacteria, and control number 4 with only fungi.
[c] Testing was discontinued.

TABLE 4 Antimicrobial Activity of Kathon CG in an Anionic Protein Shampoo (Formulation III)[a]

Preservative	Level (%)	Parts per million of active ingredient	Replicate number[b]	First inoculation				Second inoculation			
				4	7	14	28	4	7	14	28
Kathon CG	0.13	19.5	1	<10	<10	<10	<10	<10	<10	<10	<10
			2	<10	<10	<10	<10	<10	<10	<10	<10
	0.08	12	1	<10	<10	<10	<10	2.9×10^6	3.4×10^6	2×10^5	c
			2	<10	<10	<10	<10	3.1×10^6	5.2×10^6	2×10^5	c
	0.04	6	1	<10	<10	<10	<10	3.2×10^6	3.3×10^6	98,000	c
			2	<10	<10	<10	<10	2.9×10^6	6.4×10^6	2×10^6	c
	0.02	3	1	2,700	4×10^5	3×10^5	3,700	3.6×10^6	5.0×10^6	19,000	c
			2	1,500	3×10^5	3×10^5	1×10^5	6.2×10^6	2.2×10^6	12,000	c
	0.01	1.5	1	7×10^5	1.8×10^5	2×10^5	66,000	c	—	—	—
			2	3.4×10^6	8.3×10^6	3×10^5	4500	c	—	—	—
70:30 methyl-propyl p-hydroxy-xybenoates	0.50	5,000	1	6,600	11,000	1×10^5	11,000	15,000	5,000	4,000	4,000
			2	8,600	11,000	2×10^5	8,300	13,000	10,000	5,000	1,000
	0.30	3,000	1	5×10^5	1.8×10^5	1.0×10^6	1×10^5	c	—	—	—
			2	4×10^5	1.0×10^6	1.1×10^6	3,600	c	—	—	—
	0.10	1,000	1	2.5×10^6	3.5×10^6	1×10^5	1,500	c	—	—	—
			2	2.7×10^6	2.5×10^6	8×10^5	3,700	c	—	—	—
None (controls)	—	—	1	6.7×10^6	1.5×10^5	2×10^5	23,000	4.9×10^6	5.6×10^6	70,000	36,000
			2	5.6×10^6	1.3×10^5	2×10^5	7,000	3.7×10^6	8.5×10^5	90,000	37,000

[a]See Table 5 for composition.
[b]All samples, except control 2, were inoculated with a combined inoculum of yeasts, bacteria, and fungi. Control number 2 was inoculated with only yeasts and gram-positive bacteria.
[c]Testing was discontinued.

TABLE 5 Antimicrobial Activity of Kathon CG in Cosmetic Raw Material Products[a]

		Number of colony-forming units per milliliter of product in test		
Product type[a]	Kathon CG (ppm)	Week 2	Week 4	Week 6
Olefin sulfonate	None	129,000	324,000	624,000
	230	<10	<10	<10
	460	<10	<10	<10
	920	<10	<10	<10
Sodium ether sulfate (4 mole)	None	960,000	7,700,000	7,700,000
	230	<10	<10	<10
	460	<10	<10	<10
	920	<10	<10	<10
Sodium ether sulfate (7 mole)	None	170	8,700	230,000
	230	<10	<10	80
	460	<10	<10	<10
	920	<10	<10	<10
Sodium lauryl sulfonate	None	50,000	27,900	24,000
	230	<10	<10	<10
	460	<10	<10	<10
	920	<10	<10	<10
Sodium alkylaryl polyether sulfonate	None	372,000	268,000	—
	250	<10	<10	—
	500	<10	<10	—
	1,000	<10	<10	—
Alkylglucoside	None	289,000	400,000	460,000
	250	<10	<10	<10
	500	<10	<10	<10
	1,000	<10	<10	<10

[a]All samples except the alkylarylpolyether sulfonate were inoculated at zero time, 2 weeks, and 4 weeks with a suspension of microorganisms (≥640,000 culture-forming units per milliliter of test sample) representing 30 cultures of bacteria, fungi, and yeasts isolated from naturally contaminated products. The alkylarylpolyether sulfonate was naturally contaminated with bacteria before testing and was further challenged at zero time with an inoculum from three yeast cultures isolated from this product.

TABLE 6 Toxicological Properties

Acute tests		
Oral toxicity (LD_{50}), rats	3350 mg/kg	
Dermal toxicity (LD_{50}), rabbits	>5000 mg/kg	
Inhalation toxicity (LC_{50}), rats	>4.6 mg/liter air[a]	
Eye, rabbits	Corrosive	
Skin, rabbits	Severely irritating	
Subchronic tests		
90-day feeding in diet, rats	448 ppm active ingredient (equivalent to 31 mg/kg per day)[b]	Nontoxic, no pathological findings
90-day feeding in diet, dogs	840 ppm active ingredient (equivalent to 28 mg/kg per day)[b]	Nontoxic, no pathological findings
90-day drinking water and one-generation reproduction, rats	225 ppm active ingredient (equivalent to ~20 mg/kg per day)	Slight gastric irritation, no other pathological findings; no effect on fertility or reproduction
Teratology, rats	1.5, 4.5, and 15 mg/kg per day, active ingredient	Not a teratogen
Teratology, rabbits	1.5, 4.4, and 13.3 mg/kg per day, active ingredient	Not a teratogen
Skin sensitization tests		
Beuhler technique, guinea pigs		Sensitizer, EC50 induction = 90 ppm active ingredient; EC50 elicitation = 429 ppm active ingredient
Magnusson-Kligman test, guinea pigs		No sensitization at 56 ppm active ingredient[c]
Phototoxicity, guinea pigs		Not phototoxic
Photosensitizer, guinea pigs		Not photosensitizing
Bacterial mutagen (Ames) test		Positive response in TA 100 without activation; negative response in TA 100 with activation;

TABLE 6 Continued

Bacterial mutagen (Ames) test (Continued)	negative response in TA 1535, TA 1537, and TA 98 with or without activation
In vitro mouse lymphoma assay	Positive response
In vivo cytogenetics, mouse or rat	No chromosomal damage
In vivo sex-linked recessive lethal, *Drosophila*, oral and by injection	Negative response, no increase in mutation frequency
In vitro mammalian cell transformation	No transformed cells
Chronic toxicity (dermal oncogenic test) 30-Month skin painting, mouse	Slight skin irritation; no other treatment-related pathology, no increase in tumor frequency (400 ppm active ingredient 3 times per week)

[a] Air saturated with solution containing a 10 times greater level of active ingredients than Kathon CG.
[b] Highest concentration tested.
[c] Only concentration tested.

A. Animal Studies

In acute animal tests (see Table 6), Kathon CG as supplied was slightly toxic to rats orally, essentially nontoxic to rabbits dermally, but corrosive to the eyes and severely irritating to the skin of rabbits. In subchronic 3-month feeding tests, doses in food for rats of up to 3.0% as supplied (448 ppm active ingredient) and for dogs of up to 5.6% as supplied (840 ppm active ingredient) were not toxic. In a 3-month drinking water and 1 generation reproduction study in rats, concentrations up to 1.5% Kathon CG as supplied (225 ppm active ingredient) in the drinking water were not toxic and had no adverse effects on fertility, reproduction, fetal survival, or fetal health. No treatment-related toxicity occurred in subacute and subchronic rabbit dermal tests.

Kathon CG, administered by gavage to pregnant rabbits and rats during the period of organogenesis, produced dose-related toxicity to the dams, but was not teratogenic in either species.

Kathon CG produced delayed contact hypersensitivity (skin sensitization) in guinea pigs, but no phototoxicity or photosensitization.

In a test to determine dermal oncogenicity, after 30 months of topical application to mice three times per week with 2.67% Kathon CG as supplied (400 ppm active ingredient), aside from slight skin irritation, no increase in tumor incidence or occurrence of other lesions was observed.

B. Genotoxicity Studies

The active ingredient of Kathon CG gave a positive mutagenic response in an in vitro bacterial mutagen assay (Ames test) only against *Salmonella typhimurium* TA 100 without activation, and in an in vivo mammalian test system utilizing mouse lymphoma cells. It gave negative responses in a hepatocyte unscheduled DNA synthesis *in vitro* assay, in an *in vitro* cytogenetics test with the Chinese hamster cell, and in in vivo cytogenetics tests with mice and rats. When tested in a sex-linked recessive lethal assay in *Drosophila melanogaster*, the active ingredient did not produce any increase in mutation frequency following administration by the oral route or by injection. The active ingredient also gave a negative response in an in vitro mammalian cell transformation test, a prophetic test for carcinogenic potential.

C. Human Studies

A combined repeat occluded patch and arm dip test was conducted on 10 human volunteers who were exposed to an aqueous solution containing 0.37% Kathon CG (56 ppm active ingredient). In the occluded patch test the solution was applied 24 hr per day, 5 days per week, for 4 consecutive weeks (20 exposures). After a rest period of 2 weeks, each volunteer was challenged with the same solution for 24 hr. Arm immersion tests were run simultaneously on the same volunteers. They dipped their arms into the test solution twice daily for 15 min, 5 days per week, for 4 weeks. After the rest period, they immersed their arms once more. No visible changes characteristic of irritation or sensitization were observed in the skin of any individual subjected to the tests.

In another repeat patch test with 18 volunteers, a 0.17% aqueous solution of Kathon CG (25 ppm active ingredient) was applied under occluded conditions 24 hr per day, 3 days per week, for 3 consecutive weeks. After a 2-week rest, each subject was challenged for 1 day by another patch containing the same concentration of the preservative. One subject showed a response suggesting sensitization.

Occluded patch tests were also run with cosmetic formulations (lotions and ointments) to observe sensitization at several levels of Kathon CG. The results are summarized in Table 7. Subjects who manifested sensitization exhibited only localized dermal response.

The safety of Kathon CG as a preservative for shampoos was assessed in a 3-month study conducted on 248 human subjects. Subjects shampooed their hair for 90 consecutive days with a shampoo product containing 0.06% Kathon CG as supplied (9 ppm active ingredient). At the end of the test period each subject was patch tested with an aqueous solution containing 0.084 and 0.18% Kathon CG as supplied (12.5 and 27 ppm active ingredient,

TABLE 7 Patch Tests with Cosmetic Formulations

Test formulation	Kathon CG (ppm active ingredient)	Number sensitized per[a] number tested
Anionic hand lotion	56	4[b]/50
	28	0/10
Nonionic ointment	56	2/10
	28	0/10
Nonionic lotion	28	0/10
Aqueous solution	6	0/105

[a]Localized dermal response only.
[b]Two of the sensitized subjects were rechallenged at 22 ppm active ingredient and showed no response.

respectively). No sensitization was elicited in any of the subjects. Kathon CG was neither phototoxic nor a photosensitizer and did not produce contact dermatitis when tested on human subjects patched with an aqueous dilution of Kathon CG (15 ppm active ingredient) and then challenged with 50 ppm active ingredient.

In summary, contact sensitization occurred in humans at 0.17-0.4% Kathon CG (25-60 ppm active ingredient), but not at use concentrations. Accordingly, Kathon CG is judged to be safe in cosmetic formulations.

VII. SUMMARY

Kathon CG, a recent development of the Rohm and Haas Company of Philadelphia, is a highly effective antimicrobial agent which has shown utility as a preservative for cosmetics and toiletry products and their components. Its broad-spectrum activity, low toxicity at use levels, good compatibility with common cosmetic and toiletry ingredients, and efficacy at extremely low dosage levels make it one of the more practical preservative agents currently available for the type of products for which it was developed.

Editors Note: the positive genotoxicity found for Kathon CG in the Ames test (S. typhimurium TA100) and for mouse lymphoma cells need to be evaluated against the negative findings in other systems.

10
DOWICIL 200 PRESERVATIVE

S. ROSS MAROUCHOC *The Dow Chemical Company, Midland, Michigan*

I. Introduction 144
II. Structure 144
III. Solubility 145
IV. Antimicrobial Properties 146
    A. Agar inhibition 146
    B. Effectiveness in cosmetic formulations 146
V. Formulating with Dowicil 200 147
    A. Method of addition 147
    B. Temperature 148
    C. pH 148
    D. Discoloration 148
    E. Solutions of Dowicil 200 148
    F. Lab samples of Dowicil 200 148
VI. Toxicology (Safety Substantiation) 149
    A. Toxicity by single-dose oral administration 149
    B. Eye irritation 149
    C. Skin irritation 150
    D. Acute toxicity by percutaneous absorption 151
    E. Thirty-day dermal toxicity 152
    F. Ninety-one-day dermal toxicity 152
    G. Guinea pig skin sensitization 153
    H. Human skin phototoxicity and photosensitization study 153
    I. Human skin irritation studies on normal and eczematous skin 154

J. Skin sensitization study on formaldehyde-sensitized subjects 155
K. Mutagenicity 155
L. N-Nitrosamine analysis 156

VII. Polarographic Determination of Methylene Glycol* and Dowicil 200 Preservative in Cosmetic Formulations 156

A. Introduction 156
B. Scope 157
C. Safety precautions 157
D. Principle 157
E. Apparatus 157
F. Reagents 158
G. Methylene glycol 158
H. Total formaldehyde equivalent 159
I. Dowicil 200 preservative 160
J. Precision 160
K. Notes 161
L. Extraction and Filtration Procedure 161

Bibliography 164

I. INTRODUCTION

Dowicil 200 preservative, a highly active, broad-spectrum antimicrobial, is designed specifically for the preservation of cosmetics, cosmetic raw materials, and personal care products. Its major features are the following:

1. It provides effective antimicrobial activity against bacteria, yeasts, and molds at low concentrations — generally 0.02-0.3 wt % in formulations.
2. It is particularly effective against *Pseudomonas*.
3. It is not inactivated by nonionic, anionic, or cationic formulation ingredients.
4. It has high water solubility and is virtually insoluble in oils and organic solvents.
5. It is not inactivated even by relatively high concentrations of proteinaceous matter.
6. It offers an excellent toxicological profile.
7. Its activity is independent of pH.

---

*Trademark of The Dow Chemical Company.

II. STRUCTURE

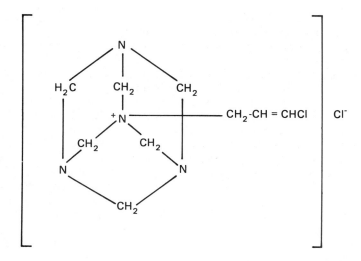

Dowicil 200 is the cis isomer of 1-(3-chloroallyl)-3,5,7-triaza-1-azoniaadamantane chloride ($C_6H_{12}N_4[CH_2CHCHCl]Cl$), with a molecular weight of 251.2.

III. SOLUBILITY

TABLE 1   Solubility of Dowicil 200

Solvent	Solubility (grams per 100 grams of solvent at 25°C)
Dowanol[a] DPM glycol ether	0.2
Dowanol PM glycol ether	1.07
Ethanol (absolute)	2.40
Isopropanol (anhydrous)	<0.1
Methanol (anhydrous)	20.8
Glycerine (99.5%)	12.6
Mineral oil	<0.1
Propylene glycol, USP	18.7
Water	127.2

[a]Trademark of The Dow Chemical Company.

IV. ANTIMICROBIAL PROPERTIES

A. Agar Inhibition

The primary function of a preservative is to prevent the growth of microorganisms. Data obtained using an agar inhibition technique (Table 2) demonstrate that Dowicil 200 is effective against both bacteria and molds.

The cosmetic chemist must determine the activity of any given preservative in the specific system of intended use. While a number of procedures can be used, a multiple challenge study is one of the most demanding and is probably the most valuable type of procedure to follow. A multiple challenge test gives some assurance that the product will maintain control of microorganisms during use — and possible misuse — by the consumer.

For the past 10 years Dow has used a multiple-challenge procedure. The relative effectiveness of various preservatives in specific formulations can be rapidly determined in the laboratory with this test. The data obtained can then be used to calculate the relative cost effectiveness of each preservative, enabling the formulator to adequately preserve each of his products at minimal cost.

B. Effectiveness in Cosmetic Formulations

The high antimicrobial activity of Dowicil 200 is reproducible in many different formulations, which is due largely to its high water solubility. The low solubility in nonaqueous solvents (Table 1) means that little or no Dowicil 200 migrates to the oil phase of a cosmetic product. Since microorganisms must utilize the water phase in order to metabolize, Dowicil 200 is present where it can be most effective.

The multiple-challenge procedure was used to evaluate the performance of Dowicil 200 versus that of parabens in three fairly typical end products: a hand cream, a hand lotion, and a protein hair shampoo.

TABLE 2 Antimicrobial and Antifungal Efficacy: Agar[a] Inhibition Data

Test organism	Concentration required to inhibit growth (ppm)
Bacteria	
Alcaligenes faecalis	50
Bacillus subtilis ATCC 8473	100
Escherichia coli ATCC 11229	100
Klebsiella pneumoniae ATCC 8308	100
Proteus vulgaris ATCC 881	50
Pseudomonas aeruginosa PRD-10, ATCC 15442	400
Staphylococcus aureus ATCC 6538	100
Fungi	
Aspergillus niger ATCC 9642	1200
Penicillium chrysogenum	1200
Trichophyton mentagrophytes	50

[a]Bacterial tests were run with nutrient agar (Difco Laboratories); fungal tests were run on malt agar (Difco) with added yeast extract.

TABLE 3 Level of Preservative Required for Preservation

Preservative	Organism	Formulation (ppm)		
		Hand cream	Hand lotion	Protein shampoo
Dowicil 200	Bacterial pool	250	250	2000
	Aspergillus niger	500	250	2000
	Saccharomyces cerevisiae	500	250	2000
Methyl-propyl-paraben[c] (50-50 blend)	Bacterial pool	3000	Failed[a] at 1400	Failed at 7500[b]
	Aspergillus niger	Failed at 3000	Failed at 1400	Failed at 7500[b]
	Saccharomyces cerevisiae	Failed at 3000	Failed at 1400	Failed at 7500[b]

[a]Note: All "failures" represent the highest level of parabens tested; all lower levels of parabens tested also failed.
[b]Methylparaben only was used in this formulation.
[c]Parabens were added to test formulations at or prior to heating.

Dowicil 200 and parabens were added at various levels to each of the formulations to determine their optimum level in each. Samples of each level of preservative in each formulation were evaluated separately against a mold (*Aspergillus niger*), a yeast (*Saccharomyces cerevisiae*), and a bacterial pool consisting of individually grown pure cultures of primarily gram-negative organisms. *Bacillus subtilis* was also included. Again, the criterion for adequate preservation is that the test samples remain uncontaminated throughout the entire test.

The results seen in Table 3 show that Dowicil 200 outperformed the parabens in all three formulations against all the types of microorganisms tested, and also demonstrate the ability of Dowicil 200 to act as the sole preservative in many types of cosmetic and personal care products.

V. FORMULATING WITH DOWICIL 200

A. Method of Addition

Dow laboratory studies have shown that Dowicil 200 can be incorporated by different methods, all resulting in equal preservative effectiveness:

    The antimicrobial may be added either as a free-flowing powder or as a freshly made water concentrate to the final product during the cooling phase.
    It is often more convenient to add the antimicrobial to the water phase immediately before mixing it with the oil phase during the formulating process.

B. Temperature

Concentrated (>1%) aqueous solutions of Dowicil 200 should not be heated above 50°C, as the effectiveness of the product may be rapidly reduced. However, preliminary results of new Dow studies indicate that temperatures as high as 80°C do not affect the antimicrobial activity of Dowicil 200 at the levels commonly used to preserve cosmetic products (0.02-0.3%). Formulations stored at 50 and 80°C for 1 month have shown no loss in the effectiveness of Dowicil 200.

Since these data are preliminary in nature, but potentially of great significance to the cosmetic formulator, it is suggested that cosmetics containing Dowicil 200 be individually evaluated for effectiveness after exposure to these temperatures. Results of temperature studies on Dowicil 200 will be made available by Dow as soon as testing is completed.

C. pH

Fresh aqueous solutions of Dowicil 200 will have a pH of 4-6 and will subsequently drift alkaline, attaining an equilibrium pH of 7.5-8.5 in a few hours. These solutions, or products containing Dowicil 200, can be pH adjusted with organic acids or bases. The use of strong acids and bases (e.g., HCl and NaOH) should be avoided, since Dowicil 200 becomes unstable in solution below pH 4 and above pH 10.

D. Discoloration

Some acidic formulations containing Dowicil 200 will demonstrate a slight to pronounced yellow discoloration. The use of small quantities (0.05-0.1%) of sodium borate or sodium sulfite has proven, in many instances, effective in preventing this discoloration. Other similar additives may also prove effective; however, the formulator should avoid using strong oxidizing or reducing agents in his or her evaluations, since these additives may adversely affect the antimicrobial efficacy of Dowicil 200.

The propensity for Dowicil 200 to discolor in any particular formulation can be rapidly determined simply by storing the formulation at 45-50°C for 48 hr. Discoloration will not be a factor if none has occurred during this time.

E. Solutions of Dowicil 200

Concentrated (>1%) aqueous solutions of Dowicil 200, such as those described previously (Sec. V.A), should be stored for no longer than 2 weeks. While the dilute levels (0.02-0.3%) of Dowicil 200 used to preserve cosmetics will remain effective for several years, concentrated solutions will gradually diminish in antimicrobial activity.

F. Lab Samples of Dowicil 200

Dowicil 200 is hygroscopic, and the small lab samples supplied to formulators may, as a result, begin to discolor and become caked after a few months. Therefore it is suggested that fresh samples be obtained from the manufacturer every 6 months.

VI. TOXICOLOGY (SAFETY SUBSTANTIATION)

A. Toxicity by Single-Dose Oral Administration

1. Method

The acute oral toxicity of Dowicil 200 was assessed by feeding the material, either undiluted or as an aqueous solution, to laboratory animals by single-dose oral gavage.

All animals, with the exception of rabbits, were fasted overnight prior to administration of the test material. After dosing, all animals were allowed to eat and drink ad libitum. The rats and cavies were caged in groups of two to five, and the rabbits and chickens were caged individually. All animals were weighed immediately prior to administration of the material, the day following dosing, and at weekly intervals thereafter, or until the animals regained any weight loss resulting from acute toxic effects.

2. Results

Results of acute oral toxicity studies are summarized in Table 4.

B. Eye Irritation

1. Study 1

Method: The eyes of three albino rabbits were established as being free of corneal injury by applying a drop of aqueous fluorescein (5.0%) stain into the conjunctival sac, allowing it to remain about 30 sec and washing it out thoroughly with flowing tap water. About 24 hr later, approximately 0.1 g of Dowicil 200 was applied directly into the conjunctival sacs of both eyes of the rabbits. Within 30 sec following instillation, one eye of each animal was washed for 2 min with tepid, flowing tap water. The other eye was left unwashed. Immediately following either the application of the ma-

TABLE 4 Toxicology by Single-Dose Oral Administration

Animal	Preparation fed	LD_{50} (g/kg)	95% confidence interval (g/kg)
Rat (female)	50% aqueous solution	1.55^a	0.91-2.68
Rat (male)	10% aqueous solution	0.94^b	0.61-1.44
Rabbit (adult-mixed)	50% aqueous solution	0.079^a	0.043-0.14
Rabbit (juvenile)	50% aqueous solution	0.047^a	Could not be calculated
Chicks (male)	Undiluted (capsules)	2.80^b	Could not be calculated
Cavies (male)	10% aqueous solution	0.71^b	Could not be calculated

[a] Calculated by the method of D. J. Finney, *Probit Analysis*, 3rd Ed., Cambridge University Press, Cambridge, (1972).
[b] Calculated by the Weil modification of the method of W. R. Thompson and C. S. Weil, *Biometrics*, 8:51-54 (1952).

terial or the washing procedure, the eyes were examined for signs of conjunctival irritation, corneal injury, and internal effects. The behavior of animals was also observed for indication of pain. All eyes were subsequently examined at intervals of 1 hr, 24 hr, 48 hr, and 1 week after instillation. Here again, soluble fluorescein stain was employed to assist in detecting corneal injury.

Results: The results of this study indicate that approximately 0.1 g of powdered Dowicil 200, when applied directly into the conjunctival sacs of laboratory rabbits, produced essentially no reaction in two of the three animals employed and slight conjunctival redness in the third animal. The redness subsided between 1 and 24 hr following instillation of the material. Signs of corneal injury or internal effects were not apparent.

2. *Study 2*

Method: In this probe study, 0.1-ml aliquots of 3, 5, and 10% aqueous solutions of Dowicil 200 were instilled into rabbit eyes three times per day, 5 days a week, for 3 weeks. The solutions were made up prior to the first instillation in amounts sufficient for the 3 weeks of testing. Each day the eyes were examined prior to dosing, and on Fridays a drop of 5% aqueous fluorescein was used to aid in the assessment of possible corneal injury.

Results: No evidence of irritation or injury was seen at any of the readings throughout the test.

C. Skin Irritation

1. *Study 1*

Method: Six albino rabbits were prepared for skin irritation studies by shaving the abdomen free of hair with a straight razor and barber soap. The animals were then rested for several days to permit healing of minute razor abrasions and to be certain that their skin was free of regrowth. The animals were caged individually and allowed to eat and drink ad libitum through the experimental period of 2 weeks. Dowicil 200 was evaluated for skin irritation by applying the undiluted material or preserved cosmetic formulation to intact and abraded rabbit skin (abdomen) in both the moist and dry states, on a repeated, prolonged contact basis. Fresh amounts (2-3 g) were applied under a cloth bandage each day to each exposure site. The bandage was taped in place in a manner to provide intimate skin contact with the test material. In the case of intact skin, the test program represented a continuous exposure for 14 days. The continuous exposure to abraded skin was of 3 days' duration. The animals were weighed at intervals during the test period and were observed frequently to detect any departure from normal, insofar as general appearance and behavior were concerned. The condition of the skin was assessed and the severity of the responses recorded daily (5 days per week during the exposure).

Results: The results of this study indicate that powdered Dowicil 200 is nonirritating to dry rabbit skin as a result of 14 days of continuous, confined contact. When confined to skin in the presence of excess moisture, however, undiluted Dowicil 200 produced slight eschar in two rabbits and moderate necrosis in one rabbit. The undiluted material proved to be moderately irritating to freshly abraded skin in either the dry or moist states. There was no evidence that 2% or less Dowicil 200 increased or influenced the irritancy of the cosmetic formulations.

2. Study 2

Method: Following the skin testing regimen described previously, albino rabbits received applications of aqueous solutions of Dowicil 200. One rabbit received a 1% solution, one received a 5% solution, and one received a 10% solution. In addition, two rabbits received applications of undiluted Dowicil 200; one rabbit had dry skin, while the other had moistened skin.

Results: Applications of the 1 and 5% aqueous solutions of Dowicil 200, as well as Dowicil 200 on dry rabbit skin, resulted in essentially no irritation. Applications of a 10% aqueous solution resulted in slight erythema and exfoliation. Undiluted Dowicil 200 was moderately irritating to moistened rabbit skin, and necrosis resulted from repeated application.

D. Acute Toxicity by Percutaneous Absorption

1. Rabbit Study 1

Method: Acute percutaneous toxicity was evaluated on laboratory rabbits employing the conventional cuff technique. Twenty-four hours prior to applying the material, the entire trunks of 26 albino rabbits were clipped free of hair with electric clippers. The test material was applied in graded doses under a heavy-gauge plastic cuff which was held in place with a cloth bandage taped to the hair. Following application of the material under the cuffs, the animals were placed individually in holding cages and were allowed to eat and drink ad libitum. After a 24-hr exposure period, the cuffs were removed and the exposed skin areas were washed with soap and water, rinsed, and dried with a towel. The animals were observed during and after the exposure period and were weighed at intervals up to 2 weeks after treatment or, when practical, until any weight loss had been regained and the animals appeared healthy.

Results: Four grams per kilogram body weight of undiluted Dowicil 200 in the presence of intact skin produced no deaths in three rabbits as a result of 24-hr exposure under an impervious cuff. The same dose, however, in the presence of abraded skin proved lethal to all three test rabbits. When Dowicil 200 was tested as a freshly prepared aqueous solution (50%), the LD_{50} was 0.565 g/kg body weight (0.227-1.4 g/kg, 95% confidence interval).

2. Rabbit Study 2

Method: Following the method of acute percutaneous absorption testing described previously, two male and two female albino rabbits per dose level with intact skin received undiluted Dowicil 200.

Results: No mortality occurred among rabbits dosed with 0.63 g/kg body weight. A dose of 5 g/kg produced only two deaths in the four animals tested at this dose. Assuming 100% mortality at the 10 g/kg level, the acute percutaneous absorption LD_{50} for rabbits with intact skin was 2.88 g/kg (1.57-6.14 g/kg, 95% confidence interval).

3. Rat Study

Method: Six laboratory rats were prepared for skin absorption studies by clipping their bellies free of hair with electric clippers. Dowicil 200, as a 50% aqueous solution, was applied to the skin at doses of 0.50, 1.0, and 2.0 g/kg body weight (two rats per dose). Immediately following application of this solution, the entire trunk of each rat was wrapped with plastic film to prevent evaporation. Owing to the extreme trauma afforded the animals by the restraining technique, the exposure period was limited to 6.5 hr. At the end of this period the skin was washed free of residual test material and the animals were wiped dry and returned to holding cages, where they were allowed to eat and drink ad libitum.

Results: The results of this test indicate that Dowicil 200 as a concentrated aqueous solution (50%) is not absorbed through the intact skin of rats in acutely toxic amounts. All animals appeared normal during and after 6.5 hr exposure to 0.5, 1.0, and 2.0 g/kg body weight.

E. Thirty-Day Dermal Toxicity

1. Method

Studies were conducted which were designed to evaluate the toxic effects of Dowicil 200 when applied to rabbit skin on a repeated, daily basis. Dowicil 200, as a freshly prepared 20% aqueous solution, was applied daily, 5 days per week, over a 20-day period to three groups of albino rabbits, consisting of seven male rabbits per group. An additional group served as control (these rabbits received applications of water). Amounts representing 25, 50, or 100 mg/kg per day of Dowicil 200 were applied to abraded skin. Topical skin response, body weights, food consumption values, hematology, clinical chemistry, organ weights, and gross and microscopic pathological examinations were evaluated.

2. Results and Conclusions

Repeated, uncovered application to the clipped, abraded backs of rabbits resulted in a dose-related occurrence of chronic inflammation, degeneration, and necrosis of the epidermis and dermis at the site of application. Body weights and food consumption of the rabbits receiving 100 mg/kg per day were significantly depressed throughout the test when compared to controls. No significant differences in clinical chemistry and hematology values occurred in any of the test groups when compared to controls.

Liver weights of rabbits receiving 50 or 100 mg/kg per day were depressed when compared to controls, which corresponds with significantly decreased food consumption at the 100 mg/kg per day level, with a trend for depressed food consumption at the 50 mg/kg per day level. No other organ weight effects were observed among test rabbits.

No other compound-related effects were found.

F. Ninety-One-Day Dermal Toxicity

An additional dermal study was conducted on a prototype cosmetic formulation to obtain data to assist in the evaluation of the safety of Dowicil 200 under use conditions.

A 91-day dermal study was conducted in which Dowicil 200 in a cosmetic formulation was applied to the backs of male and female rabbits at doses of 31.3, 10.5, or 1.04 mg/kg per day. Applications were made daily, 5 days a week, resulting in 62 treatments during the 91-day test period. No compound-related effects were found, as judged by evaluation of the following parameters: mortality, symptoms of toxicity, local skin reaction, body weights, food consumption, hematological and clinical chemistry studies, final organ weights, and gross and microscopic examination of the tissues. It is concluded that comparable doses (31.3 mg/kg per day or below) of Dowicil 200 under use conditions in a cosmetic formulation should not present a hazard due to absorption of the test material through the skin.

G. Guinea Pig Skin Sensitization

1. Method

In a modification of the Maguire method [H. C. Maguire, *J. Soc. Cosmet. Chem.*, 24:151-162 (1973)] groups of 10 male albino guinea pigs received applications of Dowicil 200. Some received a 2% aqueous solution of Dowicil 200 during the insult phase of testing, while others received Dowicil 200 as a 2% suspension in petrolatum. Additional groups of 10 guinea pigs received either an epoxy resin or formaldehyde (known sensitizers) to serve as positive controls. The insult phase consisted of four applications of the selected test material in 9 days. At the time of the third application, 0.2 ml of Freund's Bacto-Adjuvant complete (H37Ra) was injected intradermally at a point adjacent to the application site of each guinea pig. After a 2-week rest period, each guinea pig was challenged on the clipped left flank with the test material it received during the insult phase. In addition, guinea pigs which originally received Dowicil 200 solutions were challenged on the right flank with formaldehyde to determine if, in the event of a positive response, formaldehyde was the sensitizing agent. Topical responses were observed and graded for the extent of hyperemia and edema 24 and 48 hr after application.

2. Results

Neither the 2% aqueous solution of Dowicil 200 nor the 2% Dowicil 200 suspension in petrolatum proved to be positive guinea pig sensitizers, while both positive control samples sensitized all guinea pigs tested. These data indicate that Dowicil 200 at the dilutions tested should not be considered a potential human skin sensitizer.

H. Human Skin Phototoxicity and Photosensitization Study

A human skin photosensitization study was conducted on prototype cosmetic formulations. The samples were applied 25 times to 50 volunteer subjects (25-35 years of age), who subsequently had the test sites exposed for 30 sec to a Fischer Quartz Sunlamp, Model 88. After 18 days the subjects were challenged for observation of a sensitization response. The subjects were also observed during the following summer for reaction to sunlight. Table 5 lists the materials applied to the indicated number of subjects, with the indicated number of responses occurring after the challenge application.

It was concluded by Dr. Morris Shelanski of the Food and Drug Research Laboratories, Inc., who conducted the study, that "Dowicil 200 in 1% concentrations of aqueous formulations, new or aged, was neither a primary irritant, fatiguing agent or sensitizer." He further concluded that "Dowicil 200 in concentrations of 1% or lower in aqueous based formulations may be considered safe to use, if the conditions of exposure do not exceed those of the test procedure."

TABLE 5 Human Skin Phototoxicity and Photosensitization Study

Samples		Male	Female
A.	1% Dowicil 200 in water prepared 24 hr prior to application to allow pH to equilibrate	0/25	0/25
B.	0.25% Dowicil 200 in facial gel cleanser	0/25	0/15
C.	0.75% Dowicil 200 in newly prepared hand lotion	0/25	0/15
D.	0.1% Dowicil 200 in aged hand lotion	0/25	0/15
E.	Facial gel cleanser, 0% Dowicil 200	0/25	0/15
F.	Newly prepared hand cream, 0% Dowicil 200	0/25	0/15
G.	Aged hand cream, 0% Dowicil 200	0/25	0/25
H.	2% TCSA in petrolatum	3/4	5/6
	Control: petrolatum	0/25	0/25
I.	2% TCSA in methanol	4/4	5/6
	Control: methanol (also control for J and K)	0/25	0/25
J.	2% Dowicil 200 in methanol	1/4[a]	2/6[a]
K.	2% G-11 hexachlorophene in methanol	1/4[a]	1/6[a]

[a]The subjects responded only while they were showing an active reaction to the positive control tetrachlorosalicylanilide (TCSA). No resposne occurred from challenges made after the TCSA reaction had subsided, nor did any response ever occur from any cosmetic formulation or the water solution.

I. Human Skin Irritation Studies on Normal and Eczematous Skin

Skin irritation studies designed to evaluate the effect of various concentrations of Dowicil 200 in a hand cream and in a nonirritating ointment* were conducted on test subjects with healthy skin and on patients displaying eczema of various degrees, as far as virulence and affected areas were concerned, but not at the test site.

1. Hand Cream

Method: A total of 32 eczema patients were patch tested with hand cream containing 0.5, 0.2, or 0% Dowicil 200. The patches remained in contact with a 1-cm^2 area of lateral back skin for 24 hr, after which the patches were removed and the condition of the exposed skin observed. A second observation was made 24 hr later.

Results: No skin reactions attributable to the inclusion of Dowicil 200 in the hand cream at either the 0.5 or 0.2% level were observed among any of the 32 eczema subjects.

2. Ointment

Method: Ten subjects with healthy skin and ten eczema patients were patch tested with the ointment containing 2.5, 1.25, or 0.5% Dowicil 200. The

---

*Eucerin-anhydrous (a mixture of 95% refined aliphatic hydrocarbons and 5% lanolin).

patches remained in contact with 1 cm^2 of lateral skin on the back for 24 hr, after which the patches were removed and the condition of the exposed skin observed. A second observation was made 24 hr later.

Results: No skin reactions were observed among any of the subjects with normal skin or among the eczema patients exposed to the ointment containing 2.5, 1.25, or 0.5% Dowicil 200.

J. Skin Sensitization Study on Formaldehyde-Sensitized Subjects

Skin studies were designed to elucidate whether or not Dowicil 200 causes a provocation of the allergic eczema reaction for formaldehyde-sensitized subjects.

1. Method

Six subjects previously established as being sensitized to formaldehyde were patch tested with a 2% formaldehyde solution; a hand cream containing 0.5, 0.2, or 0% Dowicil 200; a cream shampoo containing 0.5, 0.2, or 0% Dowicil 200; and two aqueous solutions containing 5 or 20% Dowicil 200. Each patch remained in contact with a 1-cm^2 area of lateral back skin for 24 hr, after which the patches were removed and the condition of the exposed skin observed. A second observation was made 24 hr later.

2. Results

All six subjects displayed an equivocally positive reaction to the 2% formaldehyde solution, whereas no corresponding allergic reactions were caused by any of the preparations containing Dowicil 200 or the hand cream or cream shampoo preparations containing no Dowicil 200 and used in these studies as vehicle controls. One subject responded with a temporary erythema at the skin area patched with the hand cream containing 0.5, 0.2, and 0% Dowicil 200. The cream shampoo containing 0.5, 0.2, and 0% Dowicil 200 caused nonspecific, nonallergic skin irritation reactions in four of the six subjects.

K. Mutagenicity

1. Ames Test

Dowicil 200 was tested for mutagenicity by the Ames et al. Standard Plate Incorporation Test and was found to be nonmutagenic (see Table 6).

TABLE 6 Results of the Ames Test: Induced Versus Spontaneous Revertants

Salmonella strain	No S-9 mix added	S-9 mix added	Mutagenic
TA 98	4/39	0/46	No
TA 100	36/152	55/136	No
TA 1535	8/32	3/22	No
TA 1537	0/9	0/16	No
TA 1538	0/16	6/18	No

2. Unscheduled DNA Synthesis

The genetic activity of Dowicil 200 was evaluated in the rat hepatocyte unscheduled DNA synthesis assay. Cellular toxicity was observed at aqueous Dowicil 200 concentrations of 2×10^{-2}, 4×10^{-2}, 8×10^{-3}, 4×10^{-3}, and 4×10^{-4} M. At concentrations of 4×10^{-5}–4×10^{-8} M, Dowicil 200 did not elicit statistically significant DNA repair which would result from interaction of Dowicil 200 with the DNA of the hepatocytes. Therefore Dowicil 200 demonstrated no genetic activity in this test system.

L. N-Nitrosamine Analysis

1. N-Nitrosodiethanolamine

Dowicil 200 sample extracts were partitioned between chloroform and water following silica gel chromatography. N-Nitrosodiethanolamine was transferred from the aqueous phase to ethyl acetate by the addition of excess drying agent. Ethyl acetate extracts were then concentrated for high-pressure liquid chromatography-thermal energy analyzer determination. The detection limit for N-nitrosodiethanolamine is 20 ppb. No N-nitrosodiethanolamine was detected in Dowicil 200.

2. Other N-Nitrosamines

Dowicil 200 was dissolved in water-methanol solution and analyzed for N-nitrosamines using a liquid chromatography-pyrolysis-chemiluminescence technique. The purpose of this analysis was to determine the presence of any N-nitrosamines other than N-nitrosodiethanolamine. Although analytical standards exist for less than 10% of the known N-nitrosamines, a total N-nitrosamine analytical sensitivity of 1 ppm can be achieved by this technique. At this sensitivity level no N-nitrosamines were detected in Dowicil 200.

VII. POLAROGRAPHIC DETERMINATION OF METHYLENE GLYCOL* AND DOWICIL 200 PRESERVATIVE IN COSMETIC FORMULATIONS

A. Introduction

Aqueous solutions of Dowicil 200 preservative have been reported, by Dow and others, to generate small amounts of free formaldehyde, because the analytical techniques employed (colorimetric, titrimetric, polarographic, etc.) cannot differentiate between formaldehyde and methylene glycol. Recent Dow experiments using ^{13}C nuclear magnetic resonance (NMR) spectroscopy, which can differentiate between formaldehyde and methylene glycol in aqueous solution, have shown that aqueous solutions of Dowicil 200 contain methylene glycol rather than formaldehyde. However, ^{13}C-NMR techniques are not practical for routine analytical determinations in cosmetic matrices. Polarographic techniques, such as the one described in this method, are preferred for the analysis of Dowicil 200 and methylene glycol in cosmetics. For convenience and continuity with our previous methods, the methylene glycol found in aqueous solutions (including cosmetic products) of Dowicil 200 will be calculated and expressed as its formaldehyde equivalent.

---

*Methylene glycol calculated as its formaldehyde equivalent.

Dowicil 200 Preservative

B. Scope

This method is applicable to the determination of 5-5000 ppm of methylene glycol (expressed as its formaldehyde equivalent) and 5-5000 ppm of Dowicil 200 preservative in cosmetic formulations.

C. Safety Precautions

1. Hydrazine sulfate is a powerful reducing agent. It is a skin sensitizer and a poison capable of causing permanent injury or death resulting from exposure. Avoid all contact.†
2. Glacial acetic acid is corrosive. Wear impervious gloves and chemical workers' goggles to prevent skin and eye contact. If exposure occurs, wash the affected area with copious quantities of water.
3. Mercury is employed as the working electrode. Exposure to mercury vapor may result in damage to the central nervous system and gastrointestinal symptoms. Mercury is poisonous, and repeated or prolonged contacts may cause appreciable systemic injury due to absorption. Care should be taken to avoid contamination of the laboratory and exposure to mercury vapors.
4. Good laboratory practices dictate that each analyst should be thoroughly acquainted with potential hazards of the reagents, products, and solvents before commencing laboratory work. For information on the products being analyzed, consult the current Dow material safety data sheet supplied to Dow customers. Safety information on reagents may be obtained from the supplier. Disposal of reagents, reactants, and solvents must be in compliance with local, state, and federal laws and regulations.

D. Principle

The sample is dissolved in water and an appropriate aliquot is taken for each determination. Methylene glycol is determined by measuring the formaldehyde hydrazone formed in an acidic hydrazine solution at room temperature to minimize hydrolysis of any Dowicil 200 preservative. Dowicil 200 preservative is determined by measuring the formaldehyde hydrazone formed after quantitative hydrolysis in a hot, acidic hydrazine solution. Since this determination includes any methylene glycol originally present in the sample, the difference between the total methylene glycol released and any methylene glycol initially present (calculated as its formaldehyde equivalent) is a measure of the amount of Dowicil 200 preservative present in the sample.

E. Apparatus

1. Polarographic analyzer: PARC Model 174A; available from EG&G Princeton Applied Research, Box 2565, Princeton, New Jersey 08540, or equivalent.
2. Polarographic cell: equipped with a dropping-mercury electrode, a saturated calomel reference electrode (SCE), and a platinum auxiliary electrode; available from EG&G Princeton Applied Research, or equivalent.

---

†Editor's Note: Hydrazine and its salts are carcinogenic in mice and rats: IARC Monographs, 4:127-136 (1974).

3. X-Y Recorder: Houston Omnigraphic; available from Houston Instrument, 8500 Cameron Rd., Austin, Texas 78753, or equivalent.
4. Eppendorf pipets: 100- and 1000-µl capacities; available from Fisher Scientific Company, 711 Forbes Avenue, Pittsburgh, Pennsylvania 15219, or equivalent.
5. Buchner funnel with medium porosity sintered glass disk: available from Fisher Scientific Company.
6. Separatory funnel: 30-ml capacity.
7. Steam bath.
8. Erlenmeyer flask: glass, sidearm.

F. Reagents

1. Formaldehyde solution, ~37% (formalin); glacial acetic acid; sodium acetate trihydrate, crystal; hydrazine sulfate, crystal; all "Baker Analyzed" reagent grade; available from J. T. Baker Chemical Company, Phillipsburg, New Jersey 08865, or equivalent.
2. Prepurified nitrogen; available from Matheson Gas Products, 1275 Valley Brook Avenue, Lyndhurst, New Jersey 07071, or equivalent.
3. Diatomaceous earth (analytical filter aid): available from J. T. Baker Chemical Company.
4. Acetate buffer: 0.1 M at pH 4.5. Add 5.70 ml of glacial acetic acid to a 1-liter volumetric flask. Add 13.6 g of sodium acetate trihydrate and dilute to volume with water.
5. Hydrazine reagent: 2% solution. Add 5.0 g of hydrazine sulfate ($N_2H_4 \cdot H_2SO_4$) to a 250-ml volumetric flask and dilute to volume with water.
6. Formaldehyde standard solution: 110 µg/ml. A 37% aqueous formaldehyde solution is used as the most convenient source of formaldehyde equivalent for determination of both methylene glycol and Dowicil 200 preservative. Weigh approximately 0.3 g of 37% formaldehyde solution to the nearest 0.1 mg into a previously tared weighing bottle containing a small amount of water. Quantitatively transfer the contents of the bottle to a 1-liter volumetric flask and dilute to volume with water. Use the assay percentage on the label of the 37% formaldehyde solution to calculate the exact concentration of the formaldehyde standard solution. This solution is stable for 2 weeks under normal laboratory conditions. Formaldehyde at this dilution level (13:1) exists as methylene glycol in aqueous solution (1).
7. Dow Corning Antifoam DB: available from Dow Corning Corporation, Midland, Michigan 48640. Prepare a 1% solution as follows: Add 1 g of Dow Corning Antifoam DB to a 100-ml volumetric flask and dilute to volume with water.

G. Methylene Glycol

1. Weigh approximately 0.5 g of sample to the nearest 0.1 mg into a 10-ml volumetric flask and dilute to volume with water. Mix well.
2. Transfer 1.0 ml or a suitable portion of sample solution from Section VII.F.1 to a 10-ml volumetric flask. Add 5.0 ml of acetate buffer and dilute to volume with water. Transfer to the polarographic cell and purge the solution for 5 min with prepurified nitrogen (see Sec. VII.K.1). Then add 1.0 ml of hydrazine reagent and purge for precisely 2 min. After deaeration, adjust the position of the nitro-

Dowicil 200 Preservative

gen flow so that a blanket of nitrogen is maintained above the sample, but so no sample stirring is observed. Start the scan immediately.

3. Record a differential pulse polarogram under the following conditions:

Range	-0.6 to -1.6V versus saturated calomel electrode
Current sensitivity	5 µA full scale
Drop time	0.5 sec
Modulation amplitude	50 mV
Scan rate	10 mV/sec

4. Measure the peak height in scale divisions at approximately -1 V versus the saturated calomel electrode (SCE) and record this as A (see Sec. VII.K.2). This measures methylene glycol.

5. Transfer 1.0 ml or a suitable portion of sample solution from Section VII.G.1 to a 10-ml volumetric flask and add exactly 100 µl of formaldehyde standard solution from Section VII.F.6 to the flask. Add 5.0 ml of acetate buffer solution and dilute to volume with water. Transfer the solution to the polarographic cell and deaerate for 5 min with prepurified nitrogen. Then add 1.0 ml of hydrazine reagent and purge for precisely 2 min. Adjust the nitrogen flow as in Section VII.G.2 and continue as in Section VII.G.3 above. Start the scan immediately.

6. Measure the peak height in scale divisions at approximately -1 V versus the SCE and record this as B (see Sec. VII.K.2). This calibrates the system.

7. Transfer 1.0 ml of sample solution from Section VII.G.1 to a 10-ml volumetric flask. Add 5.0 ml of acetate buffer and dilute to volume with water. Transfer the solution to the polarographic cell, add 1.0 ml of water, and deaerate for 7 min with prepurified nitrogen. Continue as in Section VII.G.3. Measure the peak height as before and record this as C. This measures any noncarbonyl interferences.

8. Calculations

Calculate a response factor for formaldehyde as follows:

$$D = \frac{\text{micrograms of formaldehyde added in VII.G.5}}{B - A} = \text{scale division/micrograms}$$

Calculate the ppm methylene glycol as formaldehyde equivalent as follows:

$$\frac{(A - C) \times D}{(1/10) \times \text{grams of sample}} = \text{ppm methylene glycol (formaldehyde equivalent)}$$

H. Total Formaldehyde Equivalent

1. Transfer 100 µl or a suitable aliquot (see Sec. VII.K.5) of the sample solution from Section VII.G.1 to a 10-ml volumetric flask using an Eppendorf pipet. Add 5.0 ml of acetate buffer and 1.0 ml of hydrazine reagent and dilute to volume with water. Mix well.

2. Remove the stopper from the flask and place the flask on a steam bath for 10 min. Remove the flask from the steam bath, replace the stopper, and cool the contents of the flask by holding it under a stream of cold water for several minutes.
3. Transfer the contents of the flask to the polarographic cell and deaerate with prepurified nitrogen for 5 min. Adjust the nitrogen flow as in Section VII.G.2 and continue as in Section VII.G.3.
4. Measure the peak height in scale divisions at approximately -1 V versus the SCE and record this as E (see Sec. VII.K.3). This measures total formaldehyde.
5. Transfer 100 µl or a suitable portion (see Sec. VII.K.5) of sample solution from Section VII.G.1 to a 10-ml volumetric flask. Add exactly 100 µl of formaldehyde standard solution from Section VII.F.6 to the flask using an Eppendorf pipet. Add 5.0 ml of acetate buffer and 1.0 ml of hydrazine reagent and dilute to volume with water. Continue as in Sections VII.H.2 and VII.H.3 above.
6. Measure the peak height in scale divisions at approximately -1 V versus the SCE and record this as F. This calibrates the system.
7. Transfer 100 µl of sample solution or a suitable portion from Section VII.G.1 to a 10-ml volumetric flask. Add 5.0 ml of acetate buffer and dilute to volume with water. Continue as in Sections VII.H.2 and VII.H.3 above.
8. Measure the peak height in scale divisions at approximately -1 V versus the SCE and record this as G. This measures any noncarbonyl interferences.
9. Calculate a response factor as follows:

$$H = \frac{\text{micrograms of formaldehyde added in Section VII.H.5}}{F - E}$$

= scale division micrograms

Calculate the ppm total formaldehyde equivalent as follows:

$$\frac{(E - G) \times H}{(1/100) \times \text{grams of sample}} = \text{ppm total formaldehyde equivalent}$$

(see Sec. VII.K.4)

I. Dowicil 200 Preservative

Calculate the ppm Dowicil 200 preservative as follows:

ppm Dowicil 200 preservative = 1.394 × (total formaldehyde equivalent − methylene glycol as its formaldehyde equivalent)

where 1.394 is the molecular weight of Dowicil 200 preservative divided by 6 times the molecular weight of formaldehyde, 251 is the molecular weight of Dowicil 200 preservative, 180 is 6 times the molecular weight of formaldehyde, and 30 is the molecular weight of formaldehyde.

J. Precision

Data obtained by this procedure indicates a relative standard deviation of 15% for a sample containing about 1000 ppm Dowicil 200 preservative and using

Dowicil 200 Preservative

the extraction and filtration steps. The values obtained may be expected to vary from the average by not more than ±30% relative at the 95% confidence level (see Sec. VII.K.5).

K. Notes

1. For determinations in samples that easily foam, two drops of Dow Corning Antifoam DB 1% solution may be added to the polarographic cell to prevent loss of liquid while purging with nitrogen.
2. The peak potential for formaldehyde hydrazone is -0.90 V versus the SCE in aqueous solution. In the presence of cosmetic additives the peak potential often shifts negatively by as much as 200 mV. The peak potential should remain constant for any given formulation and can be determined from the scans in Sections VII.G.5 and VII.H.5 where formaldehyde standard has been added.
3. If a polarogram with an unacceptable level of noise is obtained, the sample must be extracted and filtered prior to hydrolysis and analysis. A polarogram with an unacceptable level of noise is shown in Figure 1. An acceptable polarogram is shown in Figure 2. See Section 12 for the extraction and filtration procedure.
4. If the amount of total formaldehyde equivalent found is less than 5 µg, that is, $(E - G) \times H < 5$ µg, prepare and analyze a heated reagent blank without the sample present as described in Sections VII.H.1-VII.H.4. Subtract the value obtained for the reagent blank from E to obtain the correct value at low levels of total formaldehyde equivalent.
5. It is recommended that samples be run in duplicate to improve the precision and reliability of the results. If the sample does not give an adequate response, the sample size should be increased. Decrease the sample size if the sample or the formaldehyde-spiked sample is offscale using a current range of 5 µA.

L. Extraction and Filtration Procedure

1. Add 5 ml of acetate buffer, 0.3-1 g of sample, and 3 ml of methylene chloride to a 30-ml separatory funnel, and shake the funnel vigorously. Draw off the methylene chloride and discard. Add 3 ml of fresh methylene chloride. Shake the funnel vigorously, draw off the methylene chloride, and discard it.
2. Make a slurry of diatomaceous earth and prepare a uniform pad about 1/8 in. thick in a Buchner funnel with a medium-porosity sintered glass disk using vacuum. Filter the aqueous (top) layer through the pad of diatomaceous earth and recover the filtrate in a sidearm Erlenmeyer flask.
3. Transfer an appropriate aliquot of the filtrate to a 10-ml volumetric flask. Add 5.0 ml of acetate buffer and 1.0 ml of hydrazine reagent, and dilute to volume with water. Mix well.
4. Continue as in Sections VII.H.2-VII.H.6. Substitute the filtrate for the sample solution in Section VII.G.1 in Section VII.H.5.
5. Calculate the total formaldehyde equivalent using the appropriate factor to correct for dilution.

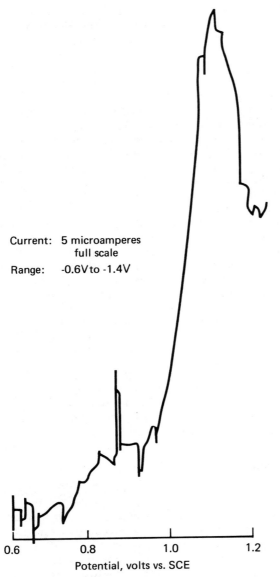

FIGURE 1 Polarogram of unacceptable level of noise.

Dowicil 200 Preservative

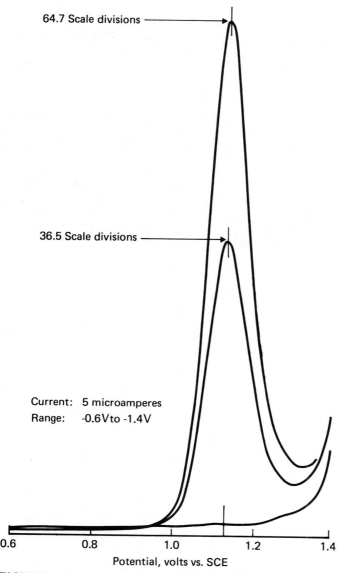

FIGURE 2 Polarogram of acceptable level of noise.

ACKNOWLEDGMENT

The data in this chapter have been made available through the courtesy of The Dow Chemical Company, Midland, Michigan.

BIBLIOGRAPHY

1. J. F. Walker, *Am. Chem. Soc.* Monograph No. 159: Robert E. Krieger Publishing Co., Huntington, N.Y., 1975, 3rd Ed., pp. 52-82.
2. N. Irving Sac, *Dangerous Properties of Industrial Materials*, 5th Ed., Van Nostrand Reinhold, New York (1979).
3. J. D. McLean and John F. Holland, Development of a portable polarographic determination of aldehydes in automotive exhaust and production plant samples, *Environ. Sci. Technol.*, 9:127-131 (1975).
4. Dow Chemical, "Dowicil 200 Preservatives: Formaldehyde Release in Aqueous Systems," Form No. 192-747-80, Dow Chemical, Specialty Chemicals Department, Midland, Mich.

11
GLYDANT AND MDMH AS COSMETIC PRESERVATIVES

MARVIN ROSEN *Glyco Inc., Williamsport, Pennsylvania*

    I. Introduction   166
   II. History and Background   166
  III. Structure and Nomenclature   167
  IV. Physical Properties   168
       A. Stability   169
       B. Corrosion   169
    V. Synthesis   170
   VI. Analysis   170
  VII. Cosmetic Applications   177
VIII. Toxicology: Glydant   183
       A. Acute oral LD_{50} 183
       B. Acute primary skin irritation   183
       C. Eye irritation   184
       D. Acute dermal toxicity   184
       E. Four-day static aquatic toxicity studies   184
       F. Eight-day dietary LC_{50}: Bird wildlife   184
       G. Acute oral toxicity: Mallard ducks   186
       H. Twenty-eight-day subacute dermal toxicity   186
       I. Acute inhalation LC_{50} 186
       J. Human photosensitization   186
       K. Mutagenicity evaluation: Ames *Salmonella*-microsome plate test   187

IX. Toxicology: MDMH   187
   A. Acute oral toxicity   187
   B. Primary skin irritation   187
   C. Eye irritation   187
   D. Acute dermal toxicity   188

X. Other Uses   188

   References   188

I. INTRODUCTION

Formaldehyde can be reacted with nitrogen-based organics to yield microbiologically active substances. For example, the reaction product of 2 mol of formaldehyde and dimethylhydantoin is dimethyloldimethylhydantoin; reaction of 1 mol of formaldehyde yields monomethyloldimethylhydantoin. Both materials are active against microorganisms.

Dimethyloldimethylhydantoin is water soluble, and a 55% solution is marketed under the trade name Glydant. In recent years Glydant has proven itself to be a highly active broad-spectrum preservative for use in cosmetics and toiletries. It is effective at low concentrations in cosmetics, and functional over a wide range of pH and temperature. It is noncorrosive, toxicologically acceptable, and ecologically safe. Glydant is stable, colorless, and essentially odorless. Because of its water solubility, it incorporates itself easily into cosmetic formulations.

While monomethyloldimethylhydantoin has physical, chemical, and antimicrobial properties similar to those of Glydant, Glydant has become the preferred dimethylhydantoin preservative of use because of its superior preservative properties.

Glydant, as referred to herein, is a 55% solution of the active substance dimethyloldimethylhydantoin. Dimethyloldimethylhydantoin, herein, is abbreviated as DMDMH, and monomethyloldimethylhydantoin is identified as MDMH.

II. HISTORY AND BACKGROUND

The first commercial-scale production of dimethylhydantoin and some of its derivatives was carried out in the United States by Du Pont. This technology was licensed to Glyco Inc. in 1956 [1]. Glydant was introduced to the market by Glyco Inc. in 1978, whereas DMDMH and MDMH were available in the late 1960s.

Dimethylhydantoin reacts readily with formalin to form hydroxymethyl products. Controlled reaction conditions with 1 mol of formaldehyde and 1 mol of dimethylhydantoin yields MDMH; similarly, reaction of 2 mol of formaldehyde gives DMDMH. Polymeric resins of dimethylhydantoin and formaldehyde can be formed by varying the mole ratio of these reactants and reaction conditions.

DMDMH and MDMH are crystalline solids and can be manufactured as such. Both are extremely soluble in water, and therefore Glydant is offered commercially as an aqueous solution to facilitate handling and formulation.

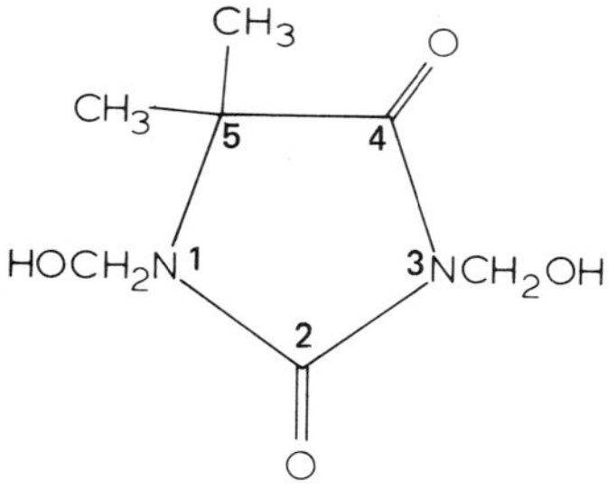

FIGURE 1 Dimethylol-5,5-dimethylhydantoin.

III. STRUCTURE AND NOMENCLATURE

Dimethyloldimethylhydantoin (DMDMH), the active ingredient in Glydant, is represented in Figure 1.

Chemical Abstracts nomenclature for DMDMH include 1,3-bis(hydroxymethyl)-5,5-dimethylhydantoin (8CI) and 1,3-bis(hydroxymethyl)-5,5-dimethyl-2,4-imidazolidinedione (9CI). A number of synonyms are also listed, including dimethylol-5,5-dimethylhydantoin, N,N'-dimethylol-5,5-dimethylhydantoin, DMDMH, DMDM hydantoins, Dantoin DMDMH, 1,3-dimethylol-5,5-

Chemical Abstract Service (CAS) Registry Numbers assigned to this material are 6440-58-0 and 37871-06-0. Only 6440-58-0 is cited in the Initial Inventory, May 1979, Toxic Substances Control Act (TSCA) Chemical Substances Inventory. The Cosmetic, Toiletry, and Fragrance Association (CTFA) designation for this compound is DMDM hydantoin. The monomethylol derivatives of dimethylhydantoin can exist as two isomers (see Figs. 2 and 3). There are literature references for each of these isomers, as well as for structures where the hydroxymethyl cited is unknown.

Chemical Abstracts names for the structure in Figure 2 include 1-(hydroxymethyl)-5,5-dimethylhydantoin (8CI) and 1-(hydroxymethyl)-5,5-dimethyl-2,4-imidazolidinedione (9CI). Synonymous designations include MDM and hydantoin. The latter name is used as the CTFA designation. The CAS Registry Number assigned to this compound is 116-25-6; the substance is listed in the Toxic Substances Control Act Initial Inventory.

It is interesting to note that there are no listings in Chemical Abstracts, 1967-1981, for the corresponding 3-hydroxymethyl analog.

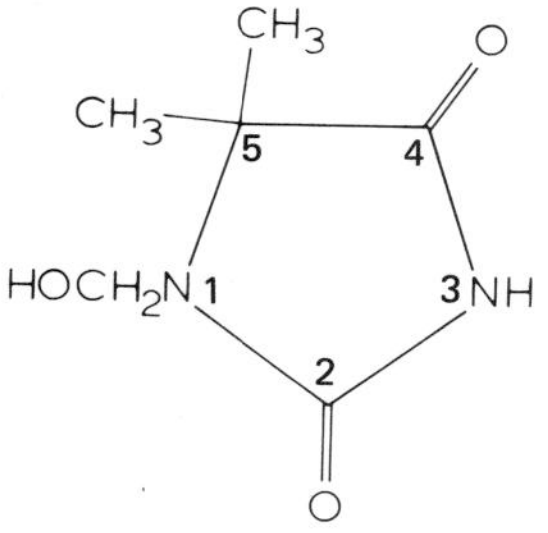

FIGURE 2 1-Hydroxymethyl-5,5-dimethylhydantoin.

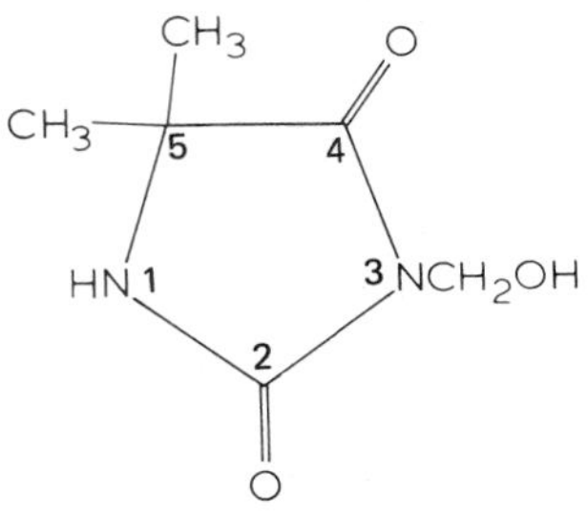

FIGURE 3   3-Hydroxymethyl-5,5-dimethylhydantoin.

Where the position of the hydroxymethyl group is not specified, CAS nomenclature is identical to the 1-(hydroxymethyl) compound, except that the side chain is recorded as "hydroxymethyl." CAS Registry Number 27636-82-4 has been assigned to this product; this compound is not listed in the TSCA Initial Inventory.

IV. PHYSICAL PROPERTIES

The physical properties of Glydant are presented in Table 1. Properties of DMDMH and MDMH are given in Table 2.

TABLE 1   Glydant: Physical Properties

Appearance	Colorless liquid
Color, American Pharmaceutical Association (APHA)	10
Odor	Mild
Freezing point (°C)	-11
Density (25°C)	1.158
pH (25°C)	
As is	6.9
1:10 dilution	6.0
1:20 dilution	5.9
Viscosity (cP)	
15°C	8.4
25°C	5.5
Formaldehyde content	
Combined (%)	17.0-17.6
Free (%)	0.5-2.0

TABLE 2 DMDMH and MDMH: Physical Properties

	DMDMH	MDMH
Molecular formula	$C_7H_{12}N_2O_4$	$C_6H_{10}N_2O_3$
Molecular weight	188.19	158.16
Combined formaldehyde (%)	31.91	18.99
Appearance	White crystal	White crystal
Odor	Very slight	Very slight
Melting point (°C)	102-104	116-121
Boiling point (°C)	Decomposes	Decomposes
Vapor pressure (60°C, mm)	0.5	—
Formaldehyde vapor equilibrium (ppm)[a]	—	20
Solubility (g/100 g of solvent)		
Water (20°C)	177.3	83.3
Water (30°C)	>200	125.4
Methanol	107.5	84.8
Acetone	20.2	23.4
Ethanol	56.4	54.0
Isopropanol	15.3	17.8
Chloroform	1.52	0.84
Methylene chloride	0.93	0.78
Toluene	0.09	0.11
Hexane	0.02	0.11

[a]Concentration of formaldehyde vapor in equilibrium with a 1 M solution of MDMH at 35°C. This compares to 625 ppm for a 1 M formaldehyde solution.

A. Stability

In laboratory tests at -18, 25, and 50°C for as long as 1 year, Glydant showed no change in composition as determined by measurements of free and total formaldehyde. Similarly, there was no change in composition after 9 months of warehouse storage. Heating Glydant at 80°C reduces its free formaldehyde content with no significant change in combined formaldehyde.

No change in free and total formaldehyde was found in Glydant after 32 days of storage at pH levels of 5, 7, and 9.

B. Corrosion

The corrosion rate for Glydant was determined by the procedure given in National Association of Corrosion Engineers Standard TM-01-69. The average rate, based on three samples, was 0.0024 and 0.0001 in./year for SAE 1020 steel and nonclad aluminum 7075-T6, respectively.

EQUATION 1 Synthesis of DMDMH.

V. SYNTHESIS

One reference to the synthesis of DMDMH employs the use of 3-5 mol of formaldehyde, as the 37% by weight aqueous solution, in reaction with 1 mol of dimethylhydantoin at 84°C [2]. Yields were diminished by reaction with less than 3 mol of formaldehyde.

A convenient means for preparing a highly concentrated aqueous solution of DMDMH is reported by Foelsch [3]. Two moles of formaldehyde, again as 37% formalin, are reacted with dimethylhydantoin at 38-50°C in the pH range 8.1-8.3. The exothermal reaction is complete in about 30 min (Equation 1). Glydant is prepared from this reaction mass by adjusting the solids content to 55%.

MDMH is prepared by a similar reaction using 1 mol of formaldehyde per mole of dimethylhydantoin. It is crystallized out of solution, dried, and isolated as a white powder.

VI. ANALYSIS

DMDMH and MDMH can be analyzed readily per se or in aqueous solutions. They provide characteristic infrared, ultraviolet, and nuclear magnetic resonance (NMR) spectra. Gas chromatography can be used for compositional information. Differential scanning calorimetry is an additional tool for characterization.

Infrared spectra for DMDMH and MDMH are given in Figures 4 and 5, respectively. Characteristic group frequencies are given in Tables 3 and 4.

Ultraviolet absorption spectra of DMDMH (Fig. 6) and MDMH (Fig. 7) at concentrations of 2×10^{-4} M in distilled water are very similar. DMDMH shows absorption at 205 and 224 nm with absorptivities of 9.30×10^3 and 6.80×10^3 liter/mol cm, respectively. Similarly, MDMH absorbs at 207 and 220 nm with absorptivities of 8.70×10^3 and 6.11×10^3 liter/mol cm.

Nuclear magnetic resonance spectra of DMDMH (Fig. 8) and MDMH (Fig. 9) were prepared in deuterium oxide on a 60-MHz instrument. As expected, DMDMH displays its methyl protons at the 5-position of the ring at 1.54 ppm δ. The doublet at 4.97 and 5.02 ppm δ is associated with the protons of the methylene portion of the hydroxymethylol groups. MDMH shows its methyl protons on the 5-position at 1.48 ppm δ and its methylene proton at 4.98 ppm δ.

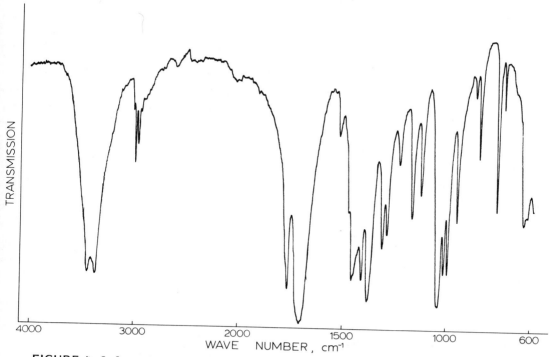

FIGURE 4 Infrared spectrum of DMDMH.

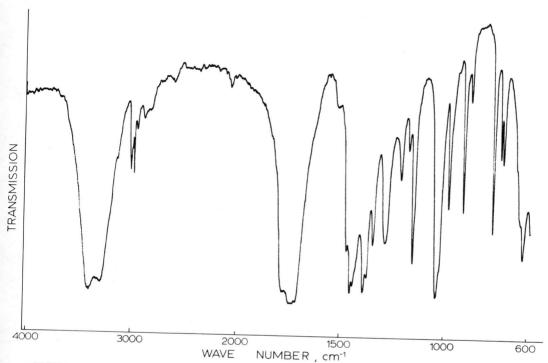

FIGURE 5 Infrared spectrum of MDMH.

TABLE 3 Characteristic Infrared Frequencies for DMDMH

Wave number (cm^{-1})	Group	Characteristic
3552	OH	Stretch
3430	OH	Stretch showing intermolecular hydrogen bonding
2962	CH$_3$	Stretch
1767	C=O	Stretch associated with adjacent imide group
1700	C=O	Stretch associated with adjacent imide group
1380	C−OH	Hydroxyl deformation
1047	C−OH	Stretch

DMDMH shows a major melt endotherm at 96.9°C by differential scanning calorimetry (Fig. 10). The minor endotherm at 104.9°C is unidentified and it is speculated that the series of endotherms at 143-178°C may be due to decomposition. MDMH produces a single melt endotherm at 113.1°C (Fig. 11).

Hydroxymethyldimethylhydantoin can be derivatized by reaction with hexamethyldisilazane and trimethylchlorosilane in pyridine. This derivative, probably the trimethylsilyl, is sufficiently volatile to be gas chromatographed

TABLE 4 Characteristic Infrared Frequencies for MDMH

Wave number (cm^{-1})	Group	Characteristic
3380	OH	Stretch showing intermolecular hydrogen bonding
3280	NH	Stretch
2965	CH$_3$	Stretch
1771	C=O	Stretch associated with adjacent imide group
1720	C=O	Stretch associated with adjacent amide group
1285	$\mathrm{NH-\overset{\overset{\displaystyle O}{\|}}{C}}$	Mixed CH stretch and NH bend associated with secondary amide
1035	C−OH	Stretch

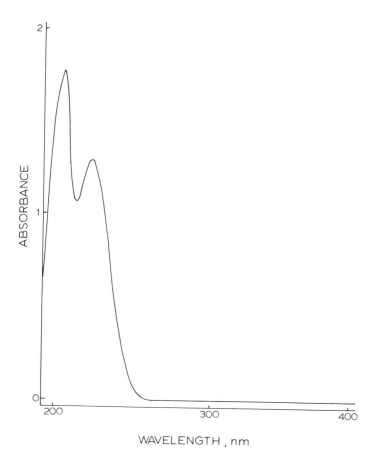

FIGURE 6 Ultraviolet spectrum of 2×10^{-4} M DMDMH.

using a 10% UCW 98 on 100/120 mesh Supelcoport column at 100-300°C. Analysis of both Glydant and DMDMH by gas chromatography shows the following compositional range: 94-98% DMDMH, 2.5-3.0% MDMH, and the balance as other dimethylhydantoin formaldehyde products.

MDMH has been analyzed by gas chromatography and shown to contain 99+% MDMH, 0.4% DMDMH, and 0.3% dimethylhydantoin. The MDMH chromatograph shows two peaks for the MDMH species. They account for 88 and 9% of total MDMH content. It is fair to conclude that these two peaks represent the 1- and 3-hydroxymethylol derivatives.

A variety of analytical techniques have been reported for hydantoin structures. In a drug-related study, Rhodes and Thornton [4], showed that hydantoin, various 5,5-disubstituted hydantoins, and some formaldehyde-derived dimethylhydantoins, namely, MDMH, methylene-bis-dimethylhydantoin, and dimethylhydantoin-formaldehyde resins, respond qualitatively to color and microcrystalline tests.

Thin-layer chromatography has been shown to be a qualitative and sometimes quantitative technique for separating formaldehyde-derived preserva-

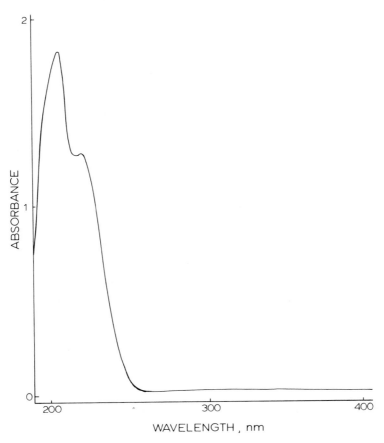

FIGURE 7 Ultraviolet spectrum of 2×10^{-4} M MDMH.

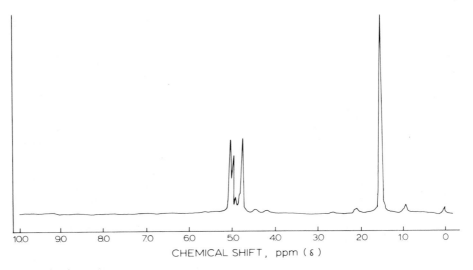

FIGURE 8 NMR spectrum of DMDMH in D_2O.

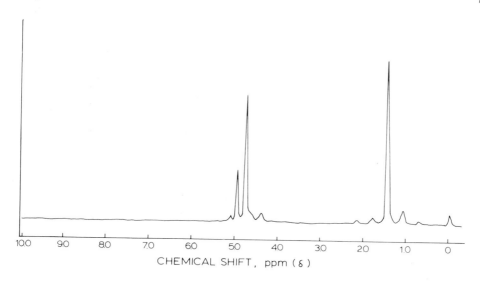

FIGURE 9 NMR spectrum of MDMH in D_2O.

tives from cosmetic products. Liem [5] resolved Bronopol, Dowicil 200, and Germall 115 using silica sheets, a solvent system comprising ethyl acetate-methanol-ammonia, and Hantzsch reagent as the color developer. He reported that MDMH will produce a violet-red color at a level of 0.01% when reacted

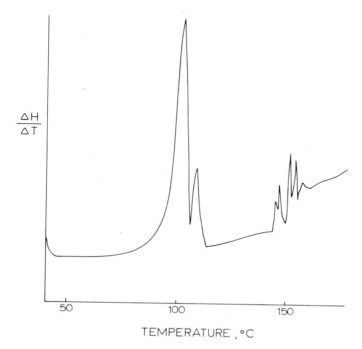

FIGURE 10 Differential scanning calorimetry curve of DMDMH.

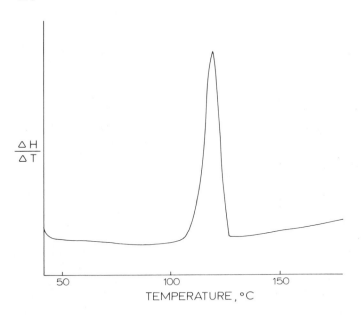

FIGURE 11 Differential scanning calorimetry curve of MDMH.

with 4-amino-3-hydrazino-5-mercapto-1,2,4-triazole. Wisneski [6] separated Dowicil 200 from Bronopol, Germall 115, and MDMH using an alumina-coated plate, a solvent system consisting of hexane-tert-butylamine, and the Hantzsch reagent. In this case, MDMH and Germall 115 did not move in the system and therefore the technique is applicable only to Dowicil 200 and Bronopol. Schmahl and Hieke [7] also reported thin-layer chromatographic separation techniques for MDMH and DMDMH. Column chromatography followed by gas chromatography was used to separate and identify antimicrobials in cosmetic products [8].

A convenient means for analyzing methylol-derived preservatives is to determine the contained formaldehyde and relate this back to the preservative concentration. One has to be especially careful in interpreting these results, since other formaldehyde-containing moities may be present, and also formaldehyde may be used as a preservative for the raw materials employed to prepare the cosmetic product. In addition, it should be realized that proteinaceous material will combine with any free formaldehyde present.

Formaldehyde content can be expressed as total or free formaldehyde. Free formaldehyde defines only the formaldehyde which has been released by hydrolysis of the methylol groups of the parent molecule. In aqueous solutions formaldehyde exists in a hydrated form and is sometimes called methylene glycol. Irrespective of the form or nomenclature, the molecule reacts classically as formaldehyde. Total formaldehyde is the sum total of free formaldehyde, bound formaldehyde as methylol groups in the parent molecule, and formaldehyde not attributable to the methylol-derived preservative, for example, other formaldehyde-based raw materials and formalin used as a preservative for some raw materials such as the lauryl sulfates.

In 1974 Wilson [9] developed a method for determining formaldehyde in cosmetics by forming a fluorescent lutidine derivative and measuring its fluorescence on a spectrophotometer. It was reported that recoveries of formaldehyde added to shampoo and other cosmetic products were good and that formaldehyde was completely released from MDMH, since a quantitative recovery was obtained. It was inferred that MDMH completely hydrolyzes in aqueous cosmetic products. However, this is a misinterpretation, since the lutidine technique measures total formaldehyde, as will be explained below, and not free formaldehyde that is present. The work was expanded [10] and it did indeed confirm that total formaldehyde can be determined in a variety of cosmetic products. Data were presented to demonstrate that it is necessary to vary the reaction conditions for color development from preservative to preservative in order to obtain quantitative results.

Liem [5] used the same fluorometric method plus polarography to determine formaldehyde in cosmetics. It was confirmed that the amount of formaldehyde determined was dependent on the temperature of the color-forming step and the analytical techniques used.

It must be understood in interpreting these data that the release of formaldehyde in the systems studied by Liem, Shepphard, and Wilson is really an expression of the reaction of acetylacetone with the formaldehyde existing as the methylol substituents in the preservative. True formaldehyde-releasing properties of a preservative in water or in a cosmetic property must be attributable only to the water present and not to a reagent causing a reaction between itself and the methylol groups. It is to be noted that DMDMH was not examined in these studies.

A single report of formaldehyde release from DMDMH is given in the literature [11]. The purpose of this work was to clarify the molecular nature of the "active principle" of the antimicrobial action of DMDMH. DMDMH was treated with 0.68 mol eq of sodium oxydeuteride wherein the pH shifted to 10.3. After addition of the alkali, the NMR peaks associated with the two methylol groups disappeared, and a new peak, associated with methylene glycol, appeared. Addition of 0.01-0.34 eq of alkali showed intermediate peak disappearance and appearance. Hence the reaction is pH dependent. Based on this, it was concluded that the principal mechanism of action of DMDMH is through formaldehyde production. The conclusion is not necessarily sound, since DMDMH is used in cosmetics over a range of about 5.0-8.5, not at a pH of 10.3, and it would be necessary to carry out NMR studies at the lower pH values to characterize the stability of the hydroxymethyl moieties in DMDMH.

Work is presently in progress which will offer techniques for determining free and total formaldehyde in cosmetic products, namely, shampoos. A number of hydroxymethyl preservatives are under study. These data will be the first to demonstrate true hydrolysis in cosmetics.

VII. COSMETIC APPLICATIONS

Glydant has been proven to be an effective preservative by the cosmetic industry. It has received wide acceptance in the past 5 years, and its use in a variety of cosmetic products has far exceeded the 15 cosmetic products it was known to preserve in 1977 [12]. It can be used in percentages as low as 0.10-0.15%, depending on the composition of the product and the susceptibility of the formulation to microbial degradation. Glydant can be used in combination with the parabens where microbial control is difficult to attain. It has shown excellent activity against gram-positive and gram-negative bacteria, fungi, and yeast.

Glydant has many of the attributes desired for the perfect shampoo preservative. Since it is extremely water soluble, its sphere of antimicrobial action is in the water phase and at the water-oil interface of a water and oil emulsion — the site of microbial growth. It is compatible with almost all cosmetic products, especially those containing anionic surfactants which cause rapid growth of gram-negative bacteria. The preservative action of Glydant is unaffected by the presence of hydrolyzed protein.

Minimum inhibitory concentrations (MICs) of DMDMH-55 against bacteria and fungi as presented by Maeda et al. [11] is given in Table 5. In an unpublished private study, MIC values for *Pseudomonas aeruginosa* (AATCC 9027), *Staphylococcus aureus* (AATCC 6538), and *Aspergillus niger* (AATCC 16404) were 800, 500, and > 1000 ppm, respectively.

The MIC level of DMDMH was determined to be 0.1% in acidic and alkaline anionic lotions and in an alkaline nonionic lotion. The value was greater than 0.1% in an acidic nonionic lotion. MDMH had a value of 0.2% in the same acidic and alkaline anionic lotions. In the acidic and alkaline nonionic lotions, the MIC of MDMH was greater than 0.5%. *Staphylococcus faecalis, P. aeruginosa, Candida albicans,* and *A. niger* were used as the challenge organisms [13]. In the same study a mixture of DMDMH and Sodium Omadine was an effective preservative in the four lotions; a mixture of MDMH and Sodium Omadine was also effective across the board.

In another study MDMH was shown to have an MIC of 0.8% in solutions containing 10% sodium lauryl sulfate and 1% hydrolyzed protein. The addition of ethoxylated (10 mol) oleyl alcohol, 3%, did not affect this value [14].

An early disclosure of the use of MDMH as an antimicrobial in shampoos is described in U.S. Patent 2,773,834. Excellent preservative action was found in cream, paste, or liquid shampoos using MDMH in the concentration range 0.1-1.0% [15].

The effectiveness of Glydant as a cosmetic preservative was demonstrated by Schanno et al. [16]. Excellent antimicrobial activity was demonstrated in a lotion shampoo (0.20% protein), a liquid hair conditioner (0.30% protein), and a henna cream conditioner (18.1% protein). Tests were conducted using single organisms, namely, *A. niger, C. albicans, Escherichia coli, S. aureus,* and *P. aeruginosa*. DMDMH was effective at 0.15% against bacteria in 7 days in the lotion shampoo, but required 14 days to be completely effective against the mold and yeast. Incorporation of Glydant into shampoo at 80°C produced no change in microbiological activity (completely effective), and complete control was retained on rechallenge after 28 days. Glydant performed perfectly in the liquid hair conditioner at a level of 0.15% at formulation temperatures of 23 and 80°C. It controlled all the organisms studied in the henna cream conditioner at levels of 0.25 and 0.50%; rechallenge against *S. aureus* and *P. aeruginosa* showed no growth. In the products studied, Glydant was not affected by the presence of protein, by high temperatures used in formulating the products, or by varying pH. Since Glydant is highly water soluble, it is not necessary, in practice, to use processing temperatures above ambient conditions.

In an heretofore unpublished work, Glydant was shown to be a fully effective preservative against four organisms in an antiperspirant roll-on lotion (formula I), a cream rinse (formula II), a face cream makeup (formula

TABLE 5 Minimum Inhibitory Concentrations of DMDMH Against Bacteria and Fungi

Strain	Source unknown MIC (ppm)
Salmonella typhi H-901	250
Salmonella paratyphi A 1015	500
Salmonella paratyphi B 8006	250
Salmonella typhimurium	250
Shigella flexneri 2a	250
Shigella flexneri 3a	250
Shigella sonnei	500
Escherichia coli 0-55	500
Escherichia coli Sakai	500
Escherichia coli 0-6	250
Pseudomonas aeruginosa A3	250
Klebsiella pneumoniae	500
Proteus vulgaris OX-19	500
Bacillus subtilis PCI-219	500
Staphylococcus aureus 209-P	500
Staphylococcus aureus Imashige	500
Staphylococcus aureus STP-182	500
Staphylococcus aureus Smith	250
Staphylococcus aureus Neuman	500
Sarcina lutea PCI-1001	360
Trichophyton asteroides	31.2
Microsporum gypseum	31.2
Candida albicans	250
Aspergillus fumigatus	250
Aspergillus niger ATCC 9508	250

Source: Ref. 11.

III), and a night cream (formula IV). The test procedure used was as described in the *U.S. Pharmacopeia XIX*, Revision 1975, *Antimicrobial Agents — Effectiveness*. Microbial data for the 28-day test is given in Table 6.

Glydant has been reported [17] as being used in body lotions, liquid soft soaps, and bubble bath products.

TABLE 6 Twenty-Eight-Day Preservative Test

Test material	Organisms[a]	Microorganism count per gram of product				
		0	7 days	14 days	21 days	28 days
Antiperspirant, control	1	5.9×10^6	2.4×10^4	1.2×10^2	8.0×10^2	6.0×10^2
Antiperspirant, control	2	4.3×10^6	2.0×10^2	2.0×10^2	2.0×10^2	0
Antiperspirant with 0.4% Glydant	1	5.9×10^6	0	0	0	0
Antiperspirant with 0.4% Glydant	2	4.3×10^6	0	0	0	0
Cream rinse, control	1	5.9×10^6	4.1×10^6	2.4×10^6	1.8×10^6	2.7×10^6
Cream rinse, control	2	4.3×10^6	4.8×10^6	5.2×10^6	7.9×10^6	9.9×10^6
Cream rinse with 0.4% Glydant	1	5.9×10^6	0	0	0	0
Cream rinse with 0.4% Glydant	2	4.3×10^6	0	0	0	0
Face cream, control	1	5.9×10^6	6.2×10^6	5.8×10^6	5.9×10^6	6.4×10^6
Face cream, control	2	4.3×10^6	2.5×10^6	2.7×10^6	2.6×10^6	2.8×10^6
Face cream with 0.4% Glydant	1	5.9×10^6	0	0	0	0
Face cream with 0.4% Glydant	2	4.3×10^6	0	0	0	0
Night cream, control	1	5.9×10^6	4.2×10^6	4.6×10^6	4.3×10^6	2.8×10^6
Night cream, control	2	4.3×10^6	6.6×10^5	4.2×10^5	2.4×10^6	4.8×10^7
Night cream with 0.4% Glydant	1	5.9×10^6	0	0	0	0
Night cream with 0.4% Glydant	2	4.3×10^6	0	0	0	0

[a]1, mixed culture of *C. albicans* and *A. niger*, 2, mixed culture of *E. coli*, *P. aeruginosa*, and *S. aureus*.

FORMULA I Antiperspirant Roll-On Lotion[a]

	Weight percent
A. Pegosperse 1750 MS[b]	4.0
Glycosperse 0-20[b]	1.5
Cetyl alcohol	2.5
Mineral oil (70 viscosity)	2.5
Glycerin, USP	2.0
Propylparaben	0.1
B. Veegum HV[c]	1.0
Glydant[b]	0.2-0.4
Water, deionized	49.9-50.1
Methylparaben	0.1
C. Aluminum chlorohydrate (50%)	36.0
D. Perfume	q.s.

[a]Procedure: Disperse the Veegum thoroughly in the water using high-speed agitation. Add the Glydant and paraben and heat to 75°C. Weigh and heat phase A to 75°C. Add phase A to phase B with good mixing. Mix and cool to 40°C. Slowly add phase C, and then phase D. Continue mixing and cool to 30°C.
[b]Glyco Inc., Greenwich, Connecticut.
[c]R. T. Vanderbilt, Norwalk, Connecticut.

FORMULA II Cream Rinse[a]

	Weight percent
A. Pegosperse 400 DS[b]	2.0
Ethosperse CA-2[b]	0.5
Stearalkonium chloride (25%)	7.0
Sodium chloride, USP	0.1
Glydant[b]	0.2-0.4
Water, deionized	89.8-90.0
Methylparaben	0.1
Propylparaben	0.1
B. Perfume	q.s.

[a]Procedure: Weigh and heat phase A to 80°C. Mix until uniform and free from lumps. Cool with mixing to 40°C and add phase B. Continue mixing and cool to 30°C.
[b]Glyco Inc.

FORMULA III Face Cream Makeup[a]

		Weight percent
A.	Veegum[b]	2.8
	CMC-7 LF[c]	0.4
	Water, deionized	41.8-42.0
	Glydant[d]	0.2-0.4
	Methylparaben	0.1
B.	Darvan #1[b]	0.3
	Propylene glycol	5.0
	Water, deionized	12.3
C.	Nytal 400[b]	18.5
	Kaolin	1.3
	Titanium dioxide, USP	3.7
	Iron oxides	1.5
D.	Glycomul L[d]	0.7
	Glycosperse L-20[d]	2.3
	Isopropyl myristate	5.0
	Stearyl alcohol	2.0
	Amerchol L 101[e]	2.0
	Propylparaben	0.1
E.	Perfume	q.s.

[a]Procedure: Dry blend the Veegum and cellulose gum. Sift slowly into the water with high shear agitation until smooth. Add Glydant and paraben. Micropulverize phase C, add to phase B, and mill to a smooth paste. Add the paste phase to phase A and heat to 70°C. Heat phase D separately to 70°C and add to paste blend with mixing. Continue mixing and cool to 40°C. Add phase E. Mix and cool to 30°C.
[b]R. T. Vanderbilt, Norwalk, Connecticut.
[c]Hercules, Wilmington, Delaware.
[d]Glyco Inc.
[e]Amerchol Corp., Edison, New Jersey.

FORMULA IV Night Cream[a]

	Weight percent
A. Aldo MS[b]	5.0
Lanolin	1.5
Lanolin alcohol	0.5
Cocoa butter	7.5
Synthetic spermaceti	3.5
Stearic acid	5.0
Cetyl alcohol	2.5
Silicone 200 fluid[c]	1.0
Mineral oil (70 viscosity)	10.0
Propylparaben	0.1
B. Triethanolamine	1.5
Propylene glycol	4.5
Glydant[b]	0.2-0.4
Water, deionized	57.1-58.9
Methylparaben	0.1
C. Perfume	q.s.

[a]Procedure: Weigh and heat phases A and B separately to 75°C. Add phase A slowly to phase B with good mixing. Avoid excessive aeration by angling mixer blade. Continue mixing and cool to 40°C. Add phase C. Mix and cool to 30°C.
[b]Glyco Inc.
[c]Dow Corning, Midland, Michigan.

VIII. TOXICOLOGY: GLYDANT

A. Acute Oral LD_{50}

Ten each of male and female albino rats (Sprague-Dawley strain) were fed single doses of Glydant via stomach tubing. After a 2-week observation period, the acute oral LD_{50} for males was determined to be 3.5 g/kg of body weight, and 2.00-3.65 g/kg for females [18].

B. Acute Primary Skin Irritation

Glydant, 0.5 ml per area, was applied to intact and abraded areas of six albino rabbits. The treated areas were covered with gauze and readings taken at 24 and 72 hr. No effects were observed in 24 hr; after 72 hr defined erythema and barely perceptible edema occurred in all animals. The skin irritation index was determined to be 1.5 [18].

C. Eye Irritation

Glydant was considered to be nonirritating to the eye, as determined by the following procedure [19]. One eye of each of nine New Zealand white rabbits were instilled with 0.1 ml of a 1.0% wt/vol solution of the test material. The contralateral eye served as the control. The treated eye of three rabbits was rinsed with 30 ml of tap water 15 sec after compound administration. The six remaining eyes received no further treatment. No signs of irritation were noted in any of the unwashed or washed eyes.

In another test, 0.1 ml of Glydant was instilled into one eye of each of six adult albino rabbits. The untreated eyes served as a control. Eye irritation scores recorded after 24, 48, and 72 hr and 7 days were 8.00, 7.00, 5.33, and 0.00, respectively. Glydant per se was judged to be irritating to the eye [18].

D. Acute Dermal Toxicity

Up to 20 g/kg of body weight of the test material were applied under a rubber sleeve fastened about the trunk of adult male rabbits for 24 hr. The animals were then observed for a 2-week period. Eight rabbits were used for the test. The acute dermal LD_{50} was determined to be in excess of 20 g/kg body weight [18].

E. Four-Day Static Aquatic Toxicity Studies

Rainbow trout (*Salmo gairdnerii*) and bluegill sunfish (*Lepomis macrochirus*) were used. Ten of each were tested at each concentration and ten were used in the untreated control group. Water was controlled at a pH of 7.2-7.6, hardness at 40-48 ppm as calcium carbonate, and an alkalinity at 30-35 ppm also as calcium carbonate. Temperature was 10°C for the trout and 18°C for the bluegills; dissolved oxygen was 5 ppm for the trout and 4 ppm for the bluegills. After a 24-hr acclimation period, the fish were fed and observed for 96 hr. A referenced pesticide, p,p'-dichlorodiphenyltrichloroethane (p,p'-DDT) was employed.

Daphnia (*Daphnia magna*) were introduced into water which had been aged for 7 days. Water was controlled at 20°C, pH 8.2-8.6, a hardness of 96-100 ppm as calcium carbonate, and an alkalinity of 40-45 ppm as calcium carbonate. The young of these, 24 hr old, were used for the test. Five animals were studied at each concentration level and in the untreated control group. Observations were made for 96 hr. The p,p'-DDT was again used as the reference pesticide.

The results of all tests are presented in Table 7 [20].

F. Eight-Day Dietary LC_{50}: Bird Wildlife

Mallard ducklings and bobwhite quail were employed using five control groups, five positive control (Dieldrin) groups, and five test groups; each group contained 10 birds and each test group studied a different combination of the Dieldrin or Glydant. Glydant was incorporated into the diet and fed to the ducklings for 5 days at five different levels. The same procedure was used for the positive control. The control groups received a standard diet for the 8-day test period. The test materials were no longer added to the diets of the positive and test groups after the fifth day, and all birds were

TABLE 7 Aquatic Toxicity (ppm)

Species	Glydant			p,p'-DDT		
	4-day TL_{50}	4-day TL_1	4-day TL_{99}	4-day TL_{50}	4-day TL_1	4-day TL_{99}
Rainbow trout	514	835	317	0.025	0.055	0.011
Bluegills	173	371	80	0.0108	0.0332	0.0035
Daphnia	37	301	4	0.0019	0.0055	0.0007

then placed on a plain diet for the balance of the 8-day test period. The dietary median lethal concentration LC_{50} for both species for Glydant was in excess of 5000 ppm. The dietary LC_{50} for mallard ducklings for Dieldrin was 94 ppm, that for bobwhite quail 42.5 ppm [21].

G. Acute Oral Toxicity: Mallard Ducks

Five male and five females were fed by gavage on the first day of the test and then allowed to feed ad libitum on the laboratory diet plus water for the balance of the 21-day test. A control group consisted also of five males and five females. Regurgitation and generalized weakness was observed within 30 min postdosing. All surviving birds appeared normal approximately 24 hr postdosing. Postmortem examination revealed no gross pathological alterations. The acute oral LD_{50} was determined to be in excess of 10.0 g/kg body weight [22].

H. Twenty-Eight-Day Subacute Dermal Toxicity

Eighteen adult albino rabbits (New Zealand strain) were employed in the tests. About 30% of the back area of each animal was clipped and, after a 24-hr recovery period, the clipped site of two males and two females was abraded. Clipping was repeated once a week, abrading repeated twice a week. Test materials, at dose levels of 8 and 800 mg/kg applied as 0.4 and 40% wt/vol solutions, were distributed over 20% of the body surface; dosing was performed 5 days per week for 5 weeks. Control animals were dosed with tap water and treated in every respect as the test animals [23].

The test material was practically nonirritating to the skin of the albino rabbits upon repeated dermal application of the 0.4% solution (8 mg/kg), and mildly to moderately irritating following subacute exposure to the 40% solution (800 mg/kg). None of the rabbits died and no pharmacotoxic symptoms were noted. No treatment-related effects were observed based on body weight examination. No abnormalities were found in hematological and chemical blood chemistry studies, urinalysis, and gross and microscopic pathological studies.

I. Acute Inhalation LC_{50}

Ten control and ten test animals, male albino rats (Sprague-Dawley strain) were exposed for 1 hr at a level of 204 mg/liter. While the animals developed signs of discomfort, they all survived the exposure and a 14-day holding period. Therefore the LC_{50} was established to be in excess of 204 mg/liter of air [18].

J. Human Photosensitization

A total of 25 subjects were evaluated in this study. Prior to testing, a minimal erythema dose (MED) was determined for each subject by irradiating separate skin sites on their backs with a xenon arc system. The MED is the time of irradiation on normal skin required to produce a faintly perceptible erythema 24 hr following irradiation. A 2-week induction period was employed applying triplicate occlusive patches on Monday, Wednesday, and Friday which remained in place for 24 hr. Patches applied on Tuesdays and Thursdays remained in contact with the skin for 30 min. Test sites were exposed to twice the individual MED after patch removal. After the 2-week induction period there was a 14-day rest period with no exposure to natural sunlight.

Triplicate patches were then applied to the above skin sites during the 4-day photosensitization test period. After a 24-hr contact, two patches were removed: One site was irradiated for 0.5 sec less than the observed MED, and the second was irradiated with the delayed erythema dose. The delayed erythema dose is the photoallergic response provoked by irradiating a site with eight times the MED of ultraviolet light.

After 24 hr the last patch (control) was removed and the skin compared to the other two patch sites. Examinations were repeated after 74 hr.

No skin irritation was produced at any time during the study. Glydant was determined to be neither a phototoxic nor a photoallergenic agent [24].

K. Mutagenicity Evaluation: Ames *Salmonella*-Microsome Plate Test

The test compound was examined for mutagenic activity in a series of in vitro microbial assays employing *Salmonella* and *Saccharomyces* indicator organisms. The compound was tested directly and in the presence of liver microsomal enzyme preparations from Aroclor-induced rats.

The compound was tested over a series of concentrations such that there was either quantitative or qualitative evidence of some chemically induced physiological effect at the highest dose level. The low dose in all cases was below a concentration that demonstrated any toxic effect. The dose range employed for the evaluation of this compound was from 0.001 to 5 µl per plate. The test compound exhibited toxicity with all the strains at the 5 µl per plate does level [25].

The results of the tests conducted on the compound (1) in the absence of a metabolic activation system and (2) in the presence of a rat liver activation system were all negative.

Since Glydant did not demonstrate greater activity in these assays, it was considered not mutagenic under these test conditions.

IX. TOXICOLOGY: MDMH

A. Acute Oral Toxicity

The oral LD_{50} of MDMH was determined to be 2.1 g/kg in a test using 10 male adult albino rats (Sprague-Dawley strain). Similarly, the oral LD_{50} for female rats was 1.9 g/kg. Powdered MDMH was administered directly to the stomachs of the rats [18].

B. Primary Skin Irritation

MDMH was demonstrated to be nonirritating to rabbit skin. The material was applied to intact and abraded areas on each of six albino rabbits. The treated areas were covered with gauze and readings were taken at 24 and 72 hr. No observable irritation was noted [18].

C. Eye Irritation

Solid MDMH was instilled in one eye of each of six adult albino rabbits (New Zealand White). The untreated eyes served as a control. Readings were made after 24, 48, and 72 hr and 7 days. While there was no effect or damage to the cornea or iris, redness, chemosis, and discharges were observed at all times. Therefore MDMH is rated as irritating to the eye [18].

D. Acute Dermal Toxicity

In this test the test material, up to 10 g/kg, was applied under a sleeve of rubber fastened about the clipped trunk of the animal for 24 hr. Body weights were then observed for a 2-week period. Eight rabbits were used for the test. No mortalities were observed. The acute dermal LD_{50} was in excess of 10 g/kg [18].

X. OTHER USES

DMDMH, MDMH, and other derivatives of dimethylhydantoin and formaldehyde have been reported to have a wide variety of uses. Hydroxymethyl derivatives are claimed to improve the optical density of pressure-sensitive copying paper [26]. A 2% methylolated dimethylhydantoin mixture is reported to provide improved chalking resistance, firmness, elasticity, pilling resistance, napping resistance, and dry touch to polyacrylonitrile fibers [27]. Water-soluble resins are formed by polymerizing dimethylhydantoin and formaldehyde first in alkali and then in acid [28,29]. Condensation products of formaldehyde and dimethylhydantoin are claimed as monomeric hardening agents for hydrophilic film-forming alkaline-soluble polymers used in color diffusion transfer photography [30].

DMDMH is stated to impart improved nerve and wrinkle resistance to solution-spun and solution-drawn acrylic fibers [31]. As a cotton-crosslinking agent, DMDMH provides resistance to acid treatment [32].

In the microbiological areas, condensation products of dimethylhydantoin and formaldehyde provide photographic color materials, unexposed as well as completely processed, with improved resistance to fungal growth without impairing photographic properties [33]. MDMH in aerosol form is claimed as a treatment for foot rot in cattle and sheep [34]. MDMH has been recommended as a preservative in adhesives and textile sizing agents [1]. MDMH also has been suggested as a preservative for air fresheners, water colors, latex-based paints, and floor waxes. Finally, DMDMH by itself and in combination with chelating agents is claimed to provide antimicrobial action in cutting oils [35,36].

ACKNOWLEDGMENTS

All of the figures and tables, with the exception of Table 5, are reprinted courtesy of Glyco Inc., Williamsport, Pennsylvania.

REFERENCES

1. Anonymous, New chemicals with unusual ring, Chem. Eng. News, 34:5577 (1956).
2. Y. Shimodoi, Dimethylolhydantoin, Japanese Patent 4709 (1966).
3. D. H. Foelsch, Dimethyloldimethylhydantoin solution, U.S. Patent 3,987,184 (1976).
4. E. F. Rhodes and J. I. Thornton, Microchemical tests for hydantoins and barbiturate-hydantoin mixtures, Microchem. J., 17:528-539 (1972).

5. D. H. Liem, Analysis of antimicrobial compounds in cosmetics, *Cosmet. Toiletr.*, *92*:59-92 (1977).
6. H. H. Wisneski, Thin layer chromatographic detection and fluorimetric determination of the preservative cis-1-(3-chloroallyl)-3,5,7-triaza-1-azoniaadamantane chloride (Quaternium-15) in cosmetics, *J. Assoc. Off. Anal. Chem.*, *63*:864-868 (1980).
7. H. J. Schmahl and E. Heike, Separation and identification of various antimicrobials by thin layer chromatography, *Fresenius Z. Anal. Chem.*, *304*:398-404 (1980).
8. R. Matissek, H. J. Schmahl, and E. Hieke, Isolation of antimicrobial substances from surfactant based cosmetics by column chromatography, *Fresenius Z. Anal. Chem.*, *303*:31 (1980).
9. C. H. Wilson, Fluorometric determination of formaldehyde in cosmetic products, *J. Soc. Cosmet. Chem.*, *25*:67-71 (1974).
10. E. P. Sheppard and C. H. Wilson, Fluorometric determination of formaldehyde-releasing cosmetic preservatives, *J. Soc. Cosmet. Chem.*, *25*:655-666 (1974).
11. H. Madea, N. Yamamoto, T. Nagoya, K. Kurosawa, and F. Kobayashi, Release of formaldehyde from dimethyloldimethylhydantoin, a possible antiseptic agent, *Agric. Biol. Chem.*, *40*:1705-1709 (1976).
12. E. L. Richardson, Preservatives: frequency of use in cosmetic formulas as disclosed to FDA, *Cosmet. Toiletr.*, *92*:85-86 (1977).
13. G. Jacobs, S. M. Henry, and V. F. Cotty, The influence of pH, emulsifier, and accelerated aging upon preservative requirements of O/W emulsions, *J. Soc. Cosmet. Chem.*, *26*:105-117 (1975).
14. G. Shuster, Preservation of shampoos and bubble baths, *Seifen Oele Fette Wachse*, *99*:489-492 (1973).
15. H. Henken, Shampoo compositions containing monomethyloldimethylhydantoin, U.S. Patent 2,773,834 (1956).
16. R. J. Schanno, J. R. Westlund, and D. H. Foelsch, Evaluation of 1,3-dimethylol-5,5-dimethylhydantoin as a cosmetic preservative, *J. Soc. Cosmet. Chem.*, *31*:85-96 (1980).
17. Glyco Inc., private communication.
18. Glyco Inc., WARF study for Glyco Inc., (1976).
19. Glyco Inc., Bio/Dynamics Inc. study for Glyco Inc., (1978).
20. Glyco Inc., Industrial BioTest study for Glyco Inc., (1976).
21. Glyco Inc., Industrial BioTest study for Glyco Inc., (1976).
22. Glyco Inc., Industrial BioTest study for Glyco Inc., (1976).
23. Glyco Inc., Industrial BioTest study for Glyco Inc., (1977).
24. Glyco Inc., Industrial BioTest study for Glyco Inc., (1977).
25. Glyco Inc., Litton Bionetics study for Glyco Inc., (1977).
26. H. Shigetoshi and M. Sadao, Pressure-sensitive copying paper, Japanese Patent 34,916 (1980).
27. N. Kobayashi and T. Higuchi, Dimension-stable polyacrylonitrile fibers, Japanese Patent 29,560, (1969).
28. R. A. Jacobson, Resin compositions, U.S. Patent 2,155,863, (1939).
29. A. F. Chadwick, Production of hydantoin formaldehyde resins, U.S. Patent 2,532,278, (1950).
30. T. Yoshida and N. Yamamato, Image-receiving element for color diffusion photography, Ger. Offen. 2,441,644 (1975).
31. H. Yamaguchi and T. Naoki, Acrylonitrile copolymer fibers, Japanese Patent 40,697, (1970).

32. S. Ender and G. Pusch, Crosslinking of cotton in relation to the chemical structure of cyclic methylol compounds, *Text. Prax.*, *21*: 581-586 (1966).
33. E. S. Mackey, Fungus resistant overcoating for color silver halide emulsion layers, U.S. Patent 2,762,708, (1956).
34. A. I. Frew and G. Whitehouse, Veterinary preparations, British Patent 1,037,690, (1965).
35. S. E. Shull and E. O. Bennett, Antimicrobial hydantoin derivative compositions and method of use, U.S. Patent 4,172,140, (1979).
36. E. O. Bennett, Formaldehyde preservatives for cutting fluids, *Int. Biodeterior. Bull.*, *9*:95-100 (1973).

12
GERMALL 115
A Safe and Effective Preservative

WILLIAM E. ROSEN and PHILIP A. BERKE   *Sutton Laboratories, Inc., Chatham, New Jersey*

I. Preservative Name and Synonyms   191
II. History   192
III. Properties and Analysis   193
   A. Properties   193
   B. Analysis   194
IV. Production   196
V. Antimicrobial Action   196
   A. General criteria of action   196
   B. Spectrum of action   200
VI. Toxicology   202
VII. Regulatory Status   203
   References   204

I. PRESERVATIVE NAME AND SYNONYMS

Germall 115 was first described by Berke and Rosen [1] in 1970 as "the first... of a family of substituted imidazolidinyl urea compounds." In a 1971 review of cosmetic preservation and preservative testing methods, Bruch [2] listed Germall 115 as imidazolidinylurea. In 1973 the Cosmetics, Toiletry, and Fragrance Association (CTFA) adopted the name Imidazolidinyl Urea in its *CTFA Cosmetic Ingredient Dictionary* [3].

II. HISTORY

Germall 115 became available to cosmetic chemists [1] at a time when the cosmetic industry urgently needed new cosmetic preservatives. Early attitudes toward cosmetic preservation had been limited in large part to concern over the microbiological condition of cosmetic products when they were shipped from the manufacturer's plant. Thus a total microbial count on a product was often considered sufficient microbiological quality control to justify shipping. In the mid-1960s not all manufacturers routinely incorporated a preservative into their cosmetic products to prevent later microbial contamination or product spoilage. It was a rare cosmetic manufacturer who had a microbiologist on the professional staff. Consideration of cosmetic preservation was often treated as an unwelcome burden rather than as an essential part of cosmetic formulation.

Following a series of observations in Europe and the United States involving gram-negative bacterial contamination of eye ointments [4], a hand cream [5], a baby lotion [2], and other topical products [6], the U.S. Food and Drug Administration (FDA) conducted its own study and found that 20% of the 169 topical drugs and cosmetics surveyed were contaminated with microorganisms [7]. Once the nature and extent of the problems were recognized by government and industry, changes in attitude followed quickly. By 1970 both the FDA and the cosmetic industry had assigned a new and important role to cosmetic preservatives. Cosmetic manufacturers set out to incorporate preservative systems so that their products not only were free of contamination when they left the plant, but also remained uncontaminated indefinitely in the hands of the consumer.

With this change in attitude toward cosmetic preservation, cosmetic manufacturing firms soon established in-house microbiology departments, and industry microbiologists sought ways to better use or to upgrade existing cosmetic preservative systems. Special preservation problems caused by the resistance of gram-negative bacteria to parabens and to certain other preservatives were soon recognized, as described by Goldman [8]. Challenge testing of products preserved with parabens alone showed that such products often were vulnerable to contamination with gram-negative bacteria. Since Germall 115 had been found to be especially effective against gram-negative bacteria and was easily incorporated into new or established products without any problems of incompatibility, Germall 115 was widely added to cosmetic products by manufacturers in order to fill the preservative gap. By 1977 Germall 115 had become the most frequently used preservative in the cosmetic industry after the parabens, as reported by the FDA [9]. An update report on the frequency of preservative use in cosmetic formulas as disclosed to the FDA [10] showed that between 1977 and 1980 the use of methylparaben and propylparaben increased by 19 and 16%, respectively, while the use of Germall 115 by the cosmetic industry increased by 34%.

In 1972 and 1973 numerous American and European investigators discovered and described the advantages in stability, compatibility, safety, and desirability of incorporating Germall 115 into a wide variety of cosmetic products [11-18].

III. PROPERTIES AND ANALYSIS

A. Properties

Germall 115 is a substituted imidazolidinyl urea compound with the formula $C_{11}H_{16}N_8O_8 \cdot H_2O$ (mol. wt. 406.33)[51].

Its Chemical Abstracts Service (CAS) Registry Number is 39236-46-9, and its Chemical Abstracts Index Name is N,N'-methylenebis[N'-[1-(hydroxymethyl)-2,5-dioxo-4-imidazolidinyl]urea]. It exists as a hydrate and is in equilibrium with isomeric species having hydroxymethyl groups on other nitrogens and with cleavage products about the central methylene group.

The composition of Germall 115 is as follows: imidazolidinyl urea, 96% minimum; water, 3% maximum; and sodium ion, 1% maximum. Approximately 10% of the imidazolidinyl urea is in the form of its sodium salt, resulting from the addition of sodium hydroxide for pH control during its manufacture.

Germall 115 is stable under normal conditions of storage. It is a stable, odorless, white, free-flowing fine powder with high water solubility and low oil solubility. It is therefore easily incorporated into water solutions or emulsions. It does not migrate out of the water phase of emulsions, where bacteria grow, into the oil phase. Germall 115 powder decomposes at temperatures over 160°C. It does not absorb ultraviolet light. Specifications for Germall 115 include the following: percentage nitrogen (Kjeldahl method), 26.0-28.0%; and pH (1% aqueous solution), 6.0-7.5. Other quality control tests carried out by the manufacturer of Germall 115 to assure consistent high purity include the following:

- Solubility (30% aqueous solution). Clear and colorless. (Clear is defined as 6.0 NTU maximum as measured on a Hach Nephelometer. Colorless is defined as not darker than APHA 20.)
- Loss on drying: Maximum of 3.0%. Water content is measured by the percentage loss of weight on drying over P_2O_5 in an evacuated dessicator for 48 hr. Users of Germall 115 must be sure to close and reseal the inner polyethylene bag to avoid uptake of water from humid air by the hygroscopic powder.
- Residue on ignition. Maximum of 3.5%. The sodium ion maximum of 1.0% is measured either by direct residue on ignition, for example, 600-700°C/2 hr, maximum of 2%, or as sulfated ash, for example, 500-600°C/2 hr, maximum of 3.5%.
- Heavy metals. Maximum of 10 ppm.

Identification by infrared spectrum is included by some users as a control specification for bulk Germall 115. In a Nujol mull, the ratio of absorbances of the peak at 5.79 μm (1727 cm^{-1}) to that at 6.48 μm (1543 cm^{-1}) is 2.30, with both peaks measured from a base line drawn between minima at about 5.25 μm (1905 cm^{-1}) and 6.70 μm (1493 cm^{-1}). Potassium bromide pellets, pressed at 23,000 lb to give a 1-mm-thick pellet, are conveniently prepared at a concentration of 0.33% (10 mg in 2990 mg of KBr). At this concentration the absorbance of the peak at 13.02 μm (768 cm^{-1}) to that at 13.50 μm (741 cm^{-1}), is approximately 0.146, with the base line drawn from about 12.25 μm (816 cm^{-1}) to 13.5 μm (741 cm^{-1}).

B. Analysis

A qualitative colorimetric assay useful for both bulk material and finished cosmetic formulations is carried out as follows. Mix about 50 mg of Germall 115 with 2.5 ml of a freshly prepared and filtered 1% aqueous solution of 1,3-dihydroxynaphthalene, and add 10 ml of 3 N hydrochloric acid. A white turbidity results. After brief heating either in a steam bath or over a flame, the turbidity remains and the mixture becomes deep red.

For a semiquantitative estimation of Germall 115 in cosmetic products, use of this qualitative identification test in concert with standard products containing various but known amounts of Germall 115 permits the determination of the Germall 115 content quickly and easily. The color test has also been modified to permit quantitative assay of Germall 115 in a cosmetic emulsion, as follows:

Preparation of stock solution. Transfer exactly 500 mg of Germall 115 into a 100-ml volumetric flask, add about 50 ml of water, dissolve, dilute to volume, and mix well. Dilute exactly 1.0 ml of this solution to 100 ml with water to give a standard solution containing 0.05 mg of Germall 115 per milliliter.

Preparation of standard solutions. Into 50-ml volumetric flasks, pipette aliquots of 0 (blank), 2, 4, 6, 8, and 10 ml of the standard solution (0.05 mg/ml Germall 115):

2 ml of Germall 115 standard solution	=	0.1 mg Germall 115
4 ml of Germall 115 standard solution	=	0.2 mg Germall 115
6 ml of Germall 115 standard solution	=	0.3 mg Germall 115
8 ml of Germall 115 standard solution	=	0.4 mg Germall 115
10 ml of Germall 115 standard solution	=	0.5 mg Germall 115

Add sufficient water to bring each to a volume of 10 ml, and add 15 ml of freshly prepared and filtered color reagent solution.* Heat in a steam bath (~100°C) for 15 min or in a constant-temperature water bath at 70°C for 1 hr, cool in an ice bath to room temperature, then make up to 50 ml with absolute methanol. Using methanol for zero absorbance before each reading, measure the absorbances of all solutions at about 520 nm in a photometer. (The position of maximum absorption is actually below 520 nm, and any convenient wavelength from 490 to 529 nm can also be used for this assay.) Subtract

---

*Preparation of color reagent solution: Dissolve 0.25 g of 1,3-dihydroxynaphthalene in 50 ml of water and add a mixture of 30 ml of concentrated hydrochloric acid and 70 ml of water. Filter and use the same day.

the absorbance of the blank from the observed absorbance to get the corrected absorbance.

Extraction and assay of unknown. Weigh out enough sample to contain 4-5 mg of Germall 115 (e.g., for an emulsion that is 0.5% Germall 115, weigh out approximately 1000 mg of sample, weighed to the nearest milligram, to obtain 5 mg of Germall 115) into a separatory funnel, and add 10 ml of chloroform. Extract the mixture with five 10-ml portions of water (for some preparations it may be necessary to add 0.3 ml of concentrated hydrochloric acid to each 10-ml extract to help separate the two layers). Backwash the combined water extraction layers with two 10-ml chloroform extractions. Filter the chloroform-washed aqueous layer by gravity to obtain a clear, colorless filtrate, and dilute to volume with water in a 50-ml volumetric flask. Pipette a 5-ml aliquot of this clear aqueous extract into another 50-ml volumetric flask, and add 5 ml of H_2O. Add 15 ml of color reagent, and heat in a steam bath (~100°C) for 15 min or in a constant temperature water bath at 70°C for 1 hr. Cool to room temperature in an ice bath and dilute to 50 ml with anhydrous methanol. Measure the absorbance at about 520 nm as described above for the standard solutions.

Select the corrected absorbance of the Germall 115 standard solution closest to that of the found corrected absorbance of the unknown test sample, and calculate the percentage concentration of Germall 115 in the emulsion as follows:

Found concentration (%) of Germall 115 in emulsion =

$$\frac{\text{corrected absorbance of test sample}}{\text{corrected absorbance of selected standard}} \times \text{total miligrams in selected standard} \times \frac{1000}{\text{sample weight (mg)}}$$

Correction for "extraction efficiency." Extraction of Germall 115 from some emulsions is not quantitative. To determine the extraction efficiency for a given emulsion, extract a sample of the emulsion which contains a *known* concentration of Germall 115. Determine the found concentration of this "known" emulsion as described above, and divide the found concentration by the known concentration. The ratio of these two concentrations is the extraction efficiency ratio for that emulsion:

$$\text{Extraction efficiency ratio} = \frac{\text{found concentration (\%) of known emulsion}}{\text{known concentration (\%) of known emulsion}}$$

To determine the actual concentration of Germall 115 in an unknown emulsion, divide the found concentration of the unknown emulsion by the extraction efficiency ratio:

$$\text{Actual concentration (\%) of unknown emulsion} = \frac{\text{found concentration (\%) of unknown emulsion}}{\text{extraction efficiency ratio}}$$

Other analytical methods, which have however proven to be of less practical application, include thin-layer chromatography [19-22], high-performance liquid chromatography [23], and destructive fluorometric analysis [24].

Germall 115 is very soluble in water, but it is insoluble in almost all organic solvents. Solubility in hydroxylic solvents is intermediate: ethylene glycol, propylene glycol, and glycerin are good solvents, but methanol and ethanol are poor solvents. If some water is present in these hydroxylic solvents, higher concentrations can be obtained. When Germall 115 is first dissolved in water and the solution is diluted with ethanol at 25°, it is possible to get stable solutions of reasonable concentrations:

Solvent	Solubility[a]	Solvent	Solubility[a]
Water	200	Ethanol, 70%	0.3
Ethylene glycol	150[b]	Ethanol, 60%	4
Propylene glycol	120[b]	Ethanol, 50%	40
Glycerin	100[b]	Isopropanol	<0.05
Methanol	0.05	Sesame oil	<0.05
Ethanol	<0.05	Mineral oil	<0.05
Ethanol, 90%	<0.05		

[a] In grams per 100 g of solvent.
[b] Slow to dissolve; requires heating and stirring.

IV. PRODUCTION

Manufacturing procedures for the production of Germall 115 are complex and proprietary, but it is clear from the previous discussion of properties and analysis that slight changes in the manufacturing procedure of Germall 115 can easily result in a product having different properties and/or different toxicities.

V. ANTIMICROBIAL ACTION

A. General Criteria of Action

A preservative in a cosmetic product is not a germicide, a sterilizer, an antiseptic, or a disinfectant. Like these other agents, a preservative must be antimicrobial, but its function must also be considered specifically in terms of the requirements of cosmetic formulations [25]. A preservative prevents product spoilage, most notably that caused by microbial growth.

Unlike disinfectants and many antiseptics, which must act quickly and powerfully, often against specific organisms, to accomplish their tasks in a short period of contact, preservatives must act steadily and effectively against a wide range of microorganisms over a long period of time. The fast kill of a successful antiseptic may, in fact, be a disadvantage for a cosmetic preservative, because such lethal effects against microorganisms usually coincide with toxic or irritant properties toward all living tissues.

Two essential requirements for a cosmetic preservative are the ability to meet the challenge of microbial insult and stability to storage. Germall 115, and especially the synergistic combination of Germall 115 and parabens [26], meets these two requirements.

1. Meeting the Challenge of Microbial Insult

A cosmetic preservative system must meet and overcome the challenge of various kinds of microbial contamination when in the hands of the consumer. (It should not be called on to counter massive contamination from raw materials, manufacturing or packaging operations, or contaminated containers. These sources of contamination must be eliminated by good housekeeping procedures.)

A typical satisfactory challenge test result is that shown below for a protein shampoo, using the preservative system 0.5% Germall 115, 0.2% methylparaben, and 0.1% propylparaben. Cultures of organisms actively growing in unpreserved cosmetic formulation were inoculated into preserved formulation. The challenged formulation was streaked on agar daily in order to determine microbial levels. After successfully meeting the challenge, the formulation was rechallenged to determine the ability of the preservative system to withstand multiple insult. Challenge tests were carried out using a gram-positive organism, a gram-negative organism, a yeast, and a mold. Levels are given for observed colonies on agar (see Table 1).

Elimination of microbial contamination over several days, or even over a period of a week, is not only satisfactory, but desirable for the reasons given above.

The combination of Germall 115, methylparaben, and propylparaben takes advantage of a synergistic action which exists between Germall 115 and parabens and provides a generally applicable, wide-range, versatile preservative system [27]. The basic system which has been successful in numerous cosmetic products is 0.30% Germall 115, 0.20% methylparaben, and 0.10% propylparaben.

Every cosmetic formulation needs a preservative system which has been hand tailored to meet its specific requirements. In order to hand-tailor the basic Germall 115-paraben preservative system to a specific cosmetic formulation, factors such as the following must be considered:

> The quantity of parabens may be limited by the nature of the formulation. For wholly aqueous systems, for example, the propylparaben content may have to be reduced because of its low solubility in water.

TABLE 1 Challenge Test Results on a Preserved Protein Shampoo[a]

	Gram-positive bacteria	Gram-negative bacteria	Yeast	Mold
Inoculation + 0 day	2+	4+	4+	4+
Inoculation + 1 day	0	1+	1+	0
Inoculation + 2 days	0	0	0	0
Reinoculation + 0 day	4+	4+	4+	4+
Reinoculation + 1 day	0	1+	0	1+
Reinoculation + 2 days	0	0	0	0

[a] Key: 4+, too many colonies to count; 3+, 100-300 colonies; 2+, 50-100 colonies; 1+, 1-50 colonies; 0, no growth.

Germall 115 should always be incorporated into the water phase. Methods for incorporating the parabens may differ, depending on the nature of the formulation and on personal preference.

Where large quantities of nonionic emulsifiers, proteins, or other paraben-deactivating components are present, the Germall 115 content may have to be increased (e.g., to 0.5%) to make up for the diminished contribution of the parabens.

If challenge tests include challenging with an unusually resistant strain, the Germall 115 content may have to be increased (e.g., to 0.5%).

Certain surfactants (e.g., sodium lauryl sulfate) enhance the preservative potency of Germall 115, and may permit a reduction in the amount of Germall 115 needed.

The following systems are typical of preservative combinations presently being used by cosmetic manufacturers:

	Percentage Germall 115	Percentage methylparaben	Percentage propylparaben
Creams	0.25	0.20	0.10
	0.40	0.15	0.10
	0.50	0.10	0.10
	0.20	0.25	0.05
Eyeliners	0.25	0.10	0.10
	0.50	0.10	0.10
Hair conditioners	0.25	0.20	0.10
	0.20	0.10	0.10
Lotions	0.40	0.20	0.10
	0.25	0.15	0.10
	0.50	0.10	0.10
	0.20	0.10	0.10
Mascara	0.25	0.15	0.15
	0.30	0.20	0.10
	0.50	0.30	0.20
	0.25	0.10	0.10
Powders	0.20	0.15	0.15
	0.25	0.15	0.10
	0.20	0.10	0.10
Rouge	0.50	0.10	0.10
Shampoo	0.25	0.20	0.10
	0.20	0.20	0.10
	0.50	0.20	0.10
	0.40	0.10	0.01
Suntan lotion	0.25	0.10	0.05

2. Stability to Storage

A preservative system must be able to protect a cosmetic product throughout the period of its storage and use. A volatile, reactive, or unstable preservative will not meet this requirement. Germall 115 has preserved samples of milk for over 1 year at room temperature, and the preservative system 0.3% Germall 115, 0.2% methylparaben, and 0.1% propylparaben has preserved whole egg for over 1 year at room temperature. The stability of Germall 115 to storage and use conditions has been established in many millions of actual cosmetic units. A typical stability test result for a protein shampoo using the preservative system 0.3% Germall 115, 0.2% methylparaben, and 0.1% propylparaben is as follows.

Standard challenge tests with bacteria (including spore-forming *Bacillus subtilis*) and mold showed satisfactory preservation. When the cosmetic containing the Germall 115-paraben preservative system was held for 13 weeks at 44°C (equivalent to normal storage for 1 year) and the challenge tests were repeated, the cosmetic showed no change in its capacity for preservation against microbial challenge.

3. Reasons for Incorporating Germall 115

Bacterial contamination of cosmetic products is a major concern of cosmetic formulators, microbiologists, and government officials. Unfortunately, even under the best of conditions most preservatives have only weak activity against gram-negative bacteria, and the deactivating effect of certain common cosmetic components, such as emulsifiers and proteins causes such preservatives to be ineffective against the growth of these bacteria. The parabens, for example, give only marginal protection against *Pseudomonas aeruginosa* in lotions and shampoos.

In a paper published in 1980 entitled "Are Cosmetic Emulsions Adequately Preserved Against *Pseudomonas*?" Berke and Rosen [28] described the purchase of 26 commercial cosmetic products, all emulsions, which were preserved solely or primarily with parabens. Each cosmetic product was challenged with two different strains of the gram-negative *P. aeruginosa*, which is a major cause of cosmetic contamination and a potential health hazard. Only 3 of the 26 cosmetic emulsions as purchased killed both *Pseudomonas* types within 24 hr, and only 13 of the 26 cosmetic emulsions killed both *Pseudomonas* types even after 7 days. When these same cosmetic emulsions were fortified by the addition of 0.3% Germall 115 and rechallenged, all but four products killed both *Pseudomonas* types within 24 hr, and *all* the fortified cosmetic emulsions killed both *Pseudomonas* types within 72 hr. It is clear from these results that parabens alone provide inadequate protection against *P. aeruginosa* for many creams and lotions. In all such cases the preservative system was dramatically improved by the addition of Germall 115 to the emulsion.

In a paper published in 1977 [29] Rosen et al. studied a model cosmetic lotion containing 0.2% methylparaben and 0.1% propylparaben previously shown by the CTFA Microbial Preservation Subcommittee to fail against *P. aeruginosa*. Challenge testing confirmed that the test lotion failed to kill *P. aeruginosa*, which continued to grow vigorously over the 28-day test period in spite of the presence of paraben preservatives. After 0.3% Germall 115 had been added, the cosmetic lotion killed *P. aeruginosa* within 2 days, and the lotion continued to show the absence of *P. aeruginosa* throughout the 28-day test period. The lotion containing Germall 115 plus the parabens also met a second and a third challenge with *P. aeruginosa*, successfully eliminating all pseudomonads all three times.

In addition to the Germall 115-paraben preservative system, other systems using Germall 115 in combination with sorbic acid, dehydroacetic acid, quaternary ammonium compounds, phenoxyethanol, Irgasan DP-300, and other preservatives have also been used successfully in cosmetic products. A key property of Germall 115 is its ability to act synergistically with other preservatives, resulting in a preservative system which gives not only a wider range of protection against bacterial insult, but also greater preservative capacity.

B. Spectrum of Action

As mentioned above, Germall 115 has its primary activity against bacteria, with only selective activity against yeast and molds. As such, it is usually used in combination with an antifungal preservative such as the parabens. However, in certain cosmetic formulations, such as some anionic shampoos, where the formulation itself is hostile to yeast or mold growth, Germall 115 alone appears to be as satisfactory a preservative system as the Germall 115-paraben combination. For example, the shampoo formulation shown in Table 2 was tested with no preservative (i.e., shampoo alone), with 0.2% methylparaben and 0.1% propylparaben added, with 0.3% Germall 115 alone, and with a combination of 0.3% Germall 115 plus 0.2% methylparaben plus 0.1% propylparaben. Challenge testing with bacteria, yeast, and mold showed that Germall 115 alone provided satisfactory protection against all challenge [31]. In this case the test results with Germall 115 alone were as good as those with Germall 115 plus parabens (see Table 3). However, since variations exist even among different types of the same microbial species, and since the Germall 115-paraben preservative system has wide-range effectiveness, it is wise even in this case to include the parabens in the preservative system as added insurance against contamination.

It is possible, however, to use Germall 115 alone to preserve certain anionic surfactant raw materials. The rapid growth of gram-negative bacteria in unpreserved anionic surfactants is well documented. Unpublished

TABLE 2 Model Shampoo Formulation (Water Base Detergent System)[a]

Ingredient	Percentage by weight	Amount (g/kg)
Triethanolamine lauryl sulfate (40%)	25.00	250.0
Lauryl diethanolamide	5.00	50.0
Amphoteric-2	5.00	50.0
Polyoxyethylene lanolin (50%)	3.00	30.0
Phosphoric acid	0.20	2.0
Demineralized water	qs to 100%	

[a]Detergent system containing an amphoteric. Add all ingredients to mixing vessel and warm to 150°F. Sweep stir to 90°F.
Source: Ref. 30.

TABLE 3 Challenge of Shampoo Formulation and With Added Preservatives[a]

	Pseudomonas aeruginosa 15442	*Escherichia coli* 10536	*Staphylococcus aureus* 6538	*Candida albicans* 10231	*Aspergillus niger* 9642
Shampoo alone	+	−	−	+	+
Shampoo + 0.2% methylparaben + 0.1% propylparaben	+	−	−	+	+
Shampoo + 0.3% Germall 115	−	−	−	−	−
Shampoo + 0.3% Germall 115 + 0.2% methylparaben + 0.1% propylparaben	−	−	−	−	−

[a]Subcultures after 2 days incubation; +, growth; −, no growth.

studies by the authors with 28-30% solutions of sodium lauryl sulfate and sodium lauryl ether sulfate showed that whereas unpreserved solutions of these surfactants supported vigorous growth of *P. aeruginosa*, solutions containing 0.1% Germall 115 eliminated *P. aeruginosa* challenges of 10^6 microorganisms per milliliter within 24 hr.

Minimum inhibitory concentration testing results often bear little resemblance to the effectiveness of a preservative in a cosmetic formulation because microorganisms are usually under nutritional stress in a cosmetic product. Rarely does a cosmetic formulation supply all the nutrients needed by microorganisms. In minimum inhibitory concentration testing, all required nutrients are carefully provided in the growth media. It is therefore essential that a preservative system be tested in the finished cosmetic product, preferably in the same container in which it will eventually be sold.

Pseudomonas contamination of cosmetic products is a major concern in the cosmetic industry because pseudomonads are so widely distributed in nature, so adaptable, and so resistant to most antimicrobials [32-36]. A series of 11 ATCC-type [37] pseudomonads, representing varieties known to contaminate cosmetics, and 17 "house" pseudomonads, isolated from a variety of contaminated cosmetic products, were found to differ in their vulnerability to Germall 115 alone or in combination with parabens [31]. The few resistant strains that were not killed fast enough by Germall 115 were killed quickly using a Germall 115-paraben combination system.

Nonionic emulsifiers are well known for their ability to inactivate preservatives by extracting and/or binding them, causing them to become unavailable for antimicrobial action [38,39]. Most cosmetic preservatives, including the parabens, are inactivated by nonionic emulsifiers, especially by polyoxyethylene derivatives of fatty acids, esters, and alcohols. In fact, polysorbate 20 and polysorbate 80 alone or in combination with lecithin are classified as preservative inactivators [40,41]. Germall 115 has been shown to retain its antimicrobial activity in the presence of polyoxyethylene nonionic emulsifiers and/or lecithin [42].

Recently a new member of the imidazolidinyl urea family, Germall II, has become commercially available. Its CTFA adopted name is Diazolidinyl Urea, and it resembles Germall 115 in many of its physical, chemical, and toxicological properties. However, Germall II is microbiologically superior because it has a wider spectrum of activity, including greater activity against troublesome "house" microorganisms. Not only is Germall II more active against gram-negative bacteria such as *Pseudomonas*, but it also has increased activity against yeast and mold. It is therefore an excellent preservative for shampoos, either alone or in combination with parabens. Creams and lotions preserved with a Germall II-paraben combination system will retain activity against yeast and mold even when paraben activity has been diminished by interaction with nonionics or proteins or has migrated into the oil phase. The Germall II-paraben combination is the preservative system of choice for creams and lotions.

VI. TOXICOLOGY

Germall 115 has a high degree of safety, is impressively nontoxic, and can be used in essentially all cosmetic products that require preservation.

Germall 115 has been reviewed by the Expert Panel of the Cosmetic Ingredient Review. The final report of the safety assessment of Germall 115 was published in 1980 [43] and covered most toxicological documentation through 1978.

Wide use of Germall 115 for over 10 years has shown that it is safe when incorporated in cosmetic products in amounts similar to those presently marketed. Food and Drug Administration product formulation data reported in the Cosmetic Ingredient Review Report list more than 50 different product types, with eye products, creams and lotions, powders, and shampoos leading the list. The most common concentration used for Germall 115, as reported by FDA, was 0.1-1.0% [43]; in fact, the vast majority of cosmetic products use 0.2-0.5%. At these levels Germall 115 is undoubtedly a safer ingredient than most of the other ingredients used in cosmetic formulations. In fact, Germall 115 has even been reported to have *anti*-irritant properties [44].

Acute toxicity studies have had to be conducted at extremely high concentration levels in order to detect significant toxicological symptoms. Low concentration levels with Germall 115 are nontoxic. The LD_{50} for Germall 115 (oral, rat) has been reported as being as low as 5200 mg/kg and as high as 11,300 mg/kg. The acute dermal LD_{50} in rabbits is greater than 8000 mg/kg. No skin irritation of rabbits was found at concentrations of 1, 2.5, or 5%; at 50% concentration no erythema or edema occurred on intact skin, but some irritation was noted on abraded skin. No visible irritation of the rabbit eye was found at concentration levels of 5, 10, or 20% with single instillations, or with 5% solution instilled for three successive days. The powder itself did produce mild transient conjunctival irritation of the rabbit eye, but there was no effect on the cornea or iris, and the conjunctival irritation cleared by the second or third day. Intraperitoneal injections of 50% Germall 115 solutions caused death of a rat at a dose of 4000 mg/kg, but not at 1000 or 2000 mg/kg. Intravenous injection of 2000 mg/kg as a 50% solution caused discomfort to the rabbit, but the animal recovered fully within 24 hr [1,43].

Subchronic (90 day) oral studies on rats at levels up to 600 mg/kg per day showed Germall 115 to be essentially nontoxic. Subacute (21 day) dermal studies on rabbits using undiluted powder at levels up to 200 mg/kg per day showed no evidence of any effect on growth, hematology, urinalysis, or gross pathology related to treatment, but there was slight to mild inflammation at the highest levels. Germall 115 did not produce sensitization or phototoxicity in the guinea pig. In teratology studies on mice Germall 115 appeared to be slightly fetotoxic but not teratogenic.

In human studies Germall 115 at concentrations up to 10% has been shown to be essentially nonirritating and nonsensitizing. Extensive studies by dermatologists have shown that an occasional individual is sensitive to Germall 115 [45,46], but all the evidence points to the absence of cross-sensitization of such patients to formaldehyde [47-49] or other preservatives. In fact, Germall 115 has been recommended as the preservative of choice for patients who are formaldehyde sensitive [47].

It is important to note that all toxicological information supplied to the Cosmetic Ingredient Review Expert Panel and to government agencies and to European cosmetic trade associations for European Economic Community appproval of imidazolidinyl urea has been for Germall 115, and not for copies of Germall 115. Approval of imidazolidinyl urea by many government and quasi-official agencies has been based exclusively on the proof of safety and efficacy of Germall 115.

As standards have become more demanding for the ability of cosmetic products to withstand microbial insult, so has the need increased for preservatives to be relatively free of potential for irritation, sensitization, or toxicity. In actual product use by the cosmetic industry, Germall 115 has proved that it meets both requirements: defense against microbial attack and freedom from harmful side effects.

VII. REGULATORY STATUS

Germall 115 is being used in cosmetic products all over the world. It is readily available from stock in Australia, England, France, Germany, Italy, Mexico, and Switzerland. Government regulatory bodies have acknowledged either explicitly or implicitly the value and importance of preserving cosmetics with Germall 115.

The European Economic Community is working to develop a positive list of preservatives. Toward that end, they have listed Germall 115 (imidazolidinyl urea) on their "Provisionally Permitted Substances" list, with a maximum authorized concentration of 0.6%. The Council of Europe, which issues recommendations but not legislation, has classified Germall 115 in Group A, the safest, least toxic category of cosmetic ingredients. Some European countries have specifically approved Germall 115 for use in cosmetics*, and various European trade associations have listed Germall 115 on their approved lists.

---

*See, for example, the discussion of cosmetic regulation in Germany and the German 1974 list of approved cosmetics in Ref. 50.

REFERENCES

1. P. A. Berke and W. E. Rosen, *Am. Perfum. Cosmet.*, 85:55 (1970).
2. C. W. Bruch, *Am. Perfum. Cosmet.*, 86:45 (1971).
3. The Cosmetic, Toiletry, and Fragrance Association, *CTFA Cosmetic Ingredient Dictionary*, 1st Ed., (N. F. Estrin, ed.), Cosmetic, Toiletry, and Fragrance Association, Washington, D.C. (1973).
4. L. W. Kallings, O. Ringertz, and L. Silverstolpe, *Acta Pharm. Suec.*, 3:219 (1966).
5. L. J. Morse, H. P. Williams, R. P. Grenn, Jr., and J. R. Rotta, *N. Engl. J. Med.*, 227:472 (1967).
6. J. A. Savin, *Pharm. J.*, 11:285 (1967).
7. A. P. Dunnigan and J. R. Evans, *TGA Cosmet. J.*, 2:39 (1970).
8. C. L. Goldman, *Drug Cosmet. Ind.*, 117:40 (1975).
9. E. L. Richardson, *Cosmet. Toiletries*, 92:85 (1977).
10. E. L. Richardson, *Cosmet. Toiletries*, 96:91 (1981).
11. M. Lanzet, *CTFA Cosmet. J.*, 4:4 (1972).
12. P. Rutkin, *ATI Market Comment*, 15:3 (1973).
13. M. S. Parker, *Soap, Perfum. Cosmet.*, 45:621 (1972).
14. M. S. Parker, *Soap, Perfum. Cosmet.*, 46:223 (1973).
15. J. J. Sciarra, *Aerosol Age*, 17:24 (1972).
16. G. Proserpio, *Parfum. Cosmet. Savons France*, 2:305 (1972).
17. M. Fontana and G. Proserpio, *Rivista Italiana*, 54:272 (1972).
18. G. Schuster, *Seifen, Oele, Fette, Wachse*, 99:489 (1973).
19. D. S. Ryder, *J. Soc. Cosmet. Chem.*, 25:535 (1974).
20. C. H. Wilson, *J. Soc. Cosmet. Chem.*, 26:75 (1975).
21. A. Martelli and G. Proserpio, *Rivista Italiana*, 58:23 (1976).
22. H. Gottschalck and T. Oelschläger, *J. Soc. Cosmet. Chem.*, 28:497 (1977).
23. W. Mitchell and P. Rahn, *Drug Cosmet. Ind.*, 123:56 (1978).
24. E. P. Sheppard and C. H. Wilson, *J. Soc. Cosmet. Chem.*, 25:655 (1974).
25. W. E. Rosen and P. A. Berke, *J. Soc. Cosmet. Chem.*, 24:663 (1973).
26. W. R. Markland, *Norda Briefs*, No. 485 (1978).
27. W. E. Rosen and P. A. Berke, *Cosmet. Toiletries*, 92:88 (1977).
28. P. A. Berke and W. E. Rosen, *J. Soc. Cosmet. Chem.*, 31:37 (1980).
29. W. E. Rosen, P. A. Berke, T. Matzin, and A. F. Peterson, *J. Soc. Cosmet. Chem.*, 28:83 (1977).
30. ASTM Subcommittee E-35.15 (Antibacterial and Antiviral Agents), Proposed Collaborative Test Protocol, June 19, 1974; see also Microbiology Forum, *CTFA Cosmet. J.*, 9:40 (1977).
31. P. A. Berke and W. E. Rosen, *J. Soc. Cosmet. Chem.*, 29:757 (1978).
32. F. N. Marzulli, J. R. Evans, and P. D. Yoder, *J. Soc. Cosmet. Chem.*, 23:89 (1972).
33. R. M. Baird, *J. Soc. Cosmet. Chem.*, 28:17 (1977).
34. C. L. Goldman, *CTFA Cosmet. J.*, 8:31 (1976).
35. L. A. Carson, M. S. Favero, W. W. Bond, and N. J. Peterson, *Appl. Microbiol.*, 25:476 (1973).
36. S. Tenenbaum, *Am. Perfum. Cosmet.*, 86:33 (1971).
37. American Type Culture Collection, 12301 Parklawn Drive, Rockville, Md. 20852.

38. M. G. deNavarre, *The Chemistry and Manufacture of Cosmetics*, Vol. 1, 2nd Ed., D. Van Nostrand, Princeton, N.J. (1962), pp. 270-272.
39. J. Schimmel and M. N. Slotsky, in *Cosmetics Science and Technology*, Vol. 3, 2nd Ed. (M. S. Balsam and E. Sagarin, eds.), Wiley, New York (1974), pp. 408-415.
40. *United States Pharmacopeia XX*, Microbiological Tests, Mack Printing Company, Easton, Pa. (1980), p. 873.
41. Association of Official Analytical Chemists, *Official Methods of Analysis*, 12th Ed., Association of Official Analytical Chemists, Washington, D.C. (1975), p. 57.
42. W. E. Rosen and P. A. Berke, *Cosmet. Toiletries*, 94:47 (1979).
43. Cosmetic Ingredient Review Expert Panel, *J. Environ. Pathol. Toxicol.*, 4:133 (1980).
44. R. L. Goldemberg, *J. Soc. Cosmet. Chem.*, 28:667 (1977).
45. S. H. Mandy, *Arch. Dermatol.*, 110:463 (1974).
46. E. J. Rudner, *Contact Dermatitis*, 3:208 (1977).
47. A. A. Fisher, *Cutis*, 21:588 (1978).
48. A. A. Fisher, *Cutis*, 26:136 (1980).
49. F. J. Storrs, *Abstracts, 5th Annual Symposium on Contact Dermatitis*, Barcelona, Spain (1980), p. 51.
50. G. van Ham, *Cosmet. Perfum.*, 90:34 (1975).
51. Cosmetic, Toiletry, and Fragrance Association, Cosmetic ingredient description for imidazolidinyl urea, *J. Environ. Pathol. Toxicol.*, 4:133 (1980).

PART III
STERILANT GASES AND RADIATION

13
INACTIVATION OF MICROORGANISMS BY LETHAL GASES

C. RICHARDS *Llandough Hospital, Cardiff, Wales*

J. R. FURR and A. D. RUSSELL *Welsh School of Pharmacy, University of Wales Institute of Science and Technology, Cardiff, Wales*

I. Introduction 209
II. Ethylene Oxide 210
    A. Physical properties 210
    B. Antimicrobial activity 210
    C. Mechanism of action 215
    D. Residual ethylene oxide 215
    E. Control procedures 216
    F. Applications 216
III. Formaldehyde 216
    A. Physical properties 216
    B. Antimicrobial activity 217
    C. Control procedures 218
    D. Applications 218
IV. Other Lethal Gases 219
    References 219

I. INTRODUCTION

Chemical agents have long been used in the gas or vapor phase to achieve disinfection or sterilization. An early procedure for fumigating sick rooms was the employment of sulfur dioxide, obtained by burning sulfur, or of chlorine.

It is, however, only comparatively recently that a scientific basis for using lethal gases has been established. The insecticidal activity of ethylene oxide (ETO) was first noted by Cotton and Roark [1], and by 1935 a detailed account of its use as an insect fumigant was published [2]. The first experimental data on the bactericidal properties of ETO appeared in a patent filed by Cross and Dixon [3] in 1933. In this patent, granted in 1937, the authors claimed to have achieved sterilization against 48 different microorganisms imbedded in moist cotton and moist sugar. The patent also specifically mentioned the use of ETO in the presence of moisture and gave the exact exposure time, concentration, and conditions under which sterilization was achieved. Hall [4], James [5], McBride [6], Yesair and Cameron [7], Smith [8], and Yesair and Williams [9] have published reports on various aspects of the use of ETO to sterilize or reduce microbiological contamination of spices.

In 1939 Nordgren [10] described the factors influencing the activity of gaseous formaldehyde, and Phillips and Kaye [11] in 1949 reviewed the earlier work which had taken place with ETO. Subsequently a great deal of research has been published on the antimicrobial activity and uses of these and other gases. Because bacterial spores are more resistant than vegetative bacteria to chemical and physical agents [12,13], it was inevitable that most of this work has been directed toward the destruction of spores by such agents, although one review [14] has dealt specifically with their effects on nonsporing organisms.

Various gases will be considered in this chapter, but most attention will be devoted to ETO and formaldehyde. The physical properties of the antimicrobial gaseous agents are depicted in Table 1, which is based on the excellent review by Bruch and Bruch [15]. Table 1 shows that ETO and methyl bromide are gases at normal temperatures, whereas formaldehyde and β-propiolactone require heating to produce the vapor form. Table 1 also demonstrates that differences also exist in their diffusibility or pentrability of materials. The chemical structures of these compounds are shown in Figure 1.

II. ETHYLENE OXIDE

A. Physical Properties

Ethylene oxide (see Table 1 and Fig. 1) is a colorless gas which is soluble in water, most organic solvents, oils, and rubber. It is inflammable when more than 3% of ETO is present in air, but this hazard can be overcome by mixing it with carbon dioxide or appropriate fluorocarbon compounds. Commercial mixtures are frequently composed of 10% ETO and 90% carbon dioxide, the latter serving as an inert diluent to prevent flammability.

Ethylene oxide gas has an irritant effect on the skin and causes smarting of the eyes, headaches, nausea, vomiting, and skin rashes.

B. Antimicrobial Activity

Since the basis of the modern work by Phillips and Kaye [11,16-18], a number of review articles on the activity of ETO have been published, notably those by Bruch and Bruch [15], Kaye [19], Ernst [20-22], and Christensen and Kristensen [23].

Once the effectiveness of ETO as a sterilization method was established, investigations began to use ETO on a scientific basis and studied the inactivation kinetics of the gas. It was soon established that effective sterilization

TABLE 1 Properties of Some Gaseous Disinfectants

Gaseous disinfectant	Molecular weight	Boiling point (°C)	Solubility in water	Sterilizing concentration (mg/liter)	RH requirement (%)	Penetration of materials	Microbicidal activity[a]	Best application as gaseous disinfectant
Ethylene oxide	44	10.4	Complete	400-1000	Nondesiccated, 30-50; large loads, 60	Moderate	Moderate	Sterilization of plastic medical equipment
Propylene oxide	58	34	Good	800-2000	Nondesiccated, 30-60	Fair	Fair	Decontamination
Formaldehyde	30	90 (formalin)[b]	Good	3-10	75	None (surface sterilant)	Excellent	Surface sterilant for rooms
β-Propiolactone	72	162	Moderate	2-5	>70	None (surface sterilant)	Excellent	Surface sterilant for rooms
Methyl bromide	95	4.6	Slight	3500	30-50	Excellent	Poor	Decontamination

[a]Based on an equimolar comparison.
[b]Formalin contains formaldehyde plus methanol.

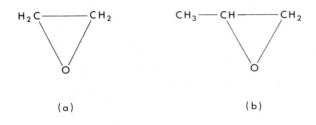

FIGURE 1 Chemical structures of (a) ethylene oxide, (b) propylene oxide, (c) β-propiolactone, (d) formaldehyde, (e) methyl bromide, and (f) glycidaldehyde.

with ETO depended on a number of interrelated factors, the main ones being concentration, temperature, relative humidity (RH), time of exposure, the species and number of organisms present, and the physical and chemical nature of the articles to be sterilized.

1. Concentration and Temperature

In the first paper to describe the kinetics of microorganism inactivation by gaseous ETO, Phillips [16] showed that as the concentration of ETO was increased, the time for sterilization decreased. Also, as the temperature was increased, the time for sterilization likewise decreased. Phillips stated that for the time, concentration, and temperature ranges studied, the death rate k was logarithmic throughout the sterilization period. Phillips calculated the temperature coefficient, or Q_{10} value, for ETO gas and showed that the activity was increased by a factor of 2.74 for each 10° rise in the temperature range studied (5-37°C).

Ernst and Shull [24], however, during their investigations on ETO sterilization at temperatures above 37°C, found that the relationships of concentration, temperature, and time were not as simple as reported by Phillips, and reported that death became zero order with respect to high levels of concentration. They introduced the term *thermochemical death time* to describe the minimal time for complete loss of viability of a microbial population upon exposure to a chemical agent or agents under specified conditions of concentration and temperature. They agreed with Phillips that the death rate was logarithmic and that the Q_{10} of 2.74 described the temperature effect for ETO concentrations below 880 mg/liter at temperatures less than 35°C. However, they found that temperature could be increased until a critical temperature is reached for a particular concentration, after which an increase in concentration has no additional effect on the rate of kill of bacterial spores. The temperature coefficient for their high-temperature segments of all systems was 1.8. This value was characteristic for ETO concentrations of 400 and 880 mg/liter at temperatures above 40.6 and 33.4°C, respectively. Below these critical temperatures the Q_{10} values for the respective systems were 3.2 and 2.3.

In summarizing the work done on the effect of temperature and concentration on ETO sterilization, it is obvious that there are many discrepancies. It has been established that spore inactivation by ETO is logarithmic in nature within the low-temperature ranges (5-32°C), and at concentrations up to 88 mg/liter the work of Phillips [16], Ernst and Shull [24], and El-Bisi et al. [25] proved that the inactivation of microorganisms by ETO gas is basically a first-order chemical reaction with a Q_{10} of approximately 2.74. At higher concentrations, and temperatures above 32°C, Ernst and Shull [24] found that the kinetics became zero order with respect to concentration, with a Q_{10} of 1.9. El-Bisi et al. [25] recorded a Q_{10} of 1.5 in the high-temperature region and thus support the zero-order kinetics found by Ernst and Shull [24]. The zero-order relationship in a chemical reaction exists where the reacting substance is not limiting to the reaction rate.

Another important factor to be taken into consideration is moisture availability, which, in practice, can be the limiting factor. Ernst and Shull [24] and Bruch and Bruch [15], in their review of the literature on reaction kinetics, made a statement that still holds true today: "A theme that runs through most of these kinetic investigations on ETO inactivation is that moisture appears to be the most critical variable for this process after a minimal ETO concentration and temperature are established."

2. Effect of Water Vapor

Griffith and Hall [26] described a vacuum chamber for using ethylene oxide to sterilize spices and pancreation and colloid materials. The role of water vapor in gaseous sterilization with ETO was first established by Kaye and Phillips [18]. There are, however, conflicting values given to the optimum relative humidity (RH) required in the ETO sterilization process. Kaye and Phillips [18], Phillips [27] and Gilbert et al. [28], recommended an RH between 30 and 40%, whereas Ernst and Shull [24], Ernst [20-22], Perkins and

Lloyd [29], Mayr [30], and Kereluk et al. [31] all presented data to show that sterilizing efficiency increased with increasing RH. However, examination of the test conditions used by these different groups showed diverse methods in test procedures [20-22].

The low-level humidity recommendations were based on work where microorganisms and their carrier materials were allowed to equilibrate with the RH of the test environment. It was shown that efficiency dropped off sharply below 20% RH and more gradually when humidity increased beyond 40%. Phillips [27] also gave an optimum RH of 33% at 25°C. The group whose work indicated higher RH levels were more interested in the practical industrial applications of ETO sterilization, and the microbiological test pieces were below equilibrium for the moisture content of the load against the RH of the sterilizing environment.

A model theory was proposed [20-22] to explain why a low RH is indicated under experimental conditions whereas a high RH is required in practice. In this model, spores are characterized with respect to their immediate environment and relative moisture content as compared with the gross environment surrounding them. The basis of this theory is that water molecules carry ETO to reactive sites. Thus, in an environment with a relatively low moisture level with respect to the site, the dynamic exchange is directed outward *from* the spore. This flow of water impedes the movement of ETO gas and also renders the macromolecules of the cell less amenable to alkylation. When the moisture content of the immediate environment increases, the equilibrium condition arises which is intermediate in effectiveness. As the water content of the environment increases further, the dynamic movement of water vapor is *toward* the spore. This represents the most ideal situation in practice.

Another situation must also be considered, namely, that in which the spore and its immediate environment are in equilibrium with the surroundings with regard to moisture exchange. In the case of a relatively dry spore and low-RH environment there is little exchange of moisture in and out of the spore. In practice, this situation is very limiting for sterilization. As the water content of the spore and the environment increases, a relatively wet spore and high RH are obtained, and it is suggested that this would result in a zone of high moisture which would have a diluting effect on the ETO gas, reducing its availability to the cell. This would be the situation in Phillips' [27] experiments when the RH was above 40%. An intermediate RH would thus represent the optimal RH designated by Phillips [27].

3. *Type of Organism*

Bacterial spores are generally only some 2-10 times as resistant as nonsporing bacteria to ETO gas [16]. Spores of the thermophilic *Bacillus stearothermophilus* and of certain other organisms may, in fact, be less resistant to ETO than some vegetative bacteria, such as *Staphylococcus aureus*, *Streptococcus faecalis*, and *Micrococcus radiodurans*.

All data up to now show that ETO inactivates all microorganisms against which it has been tested, with no development of resistance. The virucidal activity of gaseous ETO has been reported against 14 animal viruses [32], foot-and-mouth disease virus [33], and Columbia S. K. *Encephalomyelitis* virus [34]. Hoff-Jorgensan and Lund [35] selected four different virus strains to cover a wide range of stabilities toward chemical and physical treatments and demonstrated that ETO was an effective virucidal agent in the four cases.

C. Mechanism of Action

The first theory to account for the microbiocidal activity of ETO was postulated by Cross and Dixon [3], who stated that ETO sterilization required the presence of sufficient moisture to enable ethylene glycol to be formed. Phillips [16] suggested that the microbiocidal action was due to an alkylation reaction between ETO and protein matter in the microbial cell. Phillips examined a number of compounds related in chemical structure to ETO and found that as their alkylation ability increased, so did their antimicrobial properties. Phillips [16] proposed that alkylation by ETO of free carboxyl, amino, hydroxyl, or sulfhydryl groups on bacterial proteins served to disrupt metabolic processes, eventually causing death of the organism. Phillips [36] suggested that the slightly greater increase in resistance of bacterial spores over vegetative cells could be explained by a folding of protein molecules in spores to provide greater protection of sulfhydryl groups in key enzyme systems. Bruch [37] showed that for the five primary gaseous disinfectants (ETO, propylene oxide, formaldehyde, methyl bromide, and β-propiolactone) the common chemical feature was their alkylation effectiveness, which could be related to the microbiocidal potency. These compounds react most readily with organic and inorganic anions, amino groups, and sulfide groups. The common characteristic of these groups is that they are electron rich. Ethylene oxide reacts extensively with amino groups, but not with carboxyl groups of amino acids. In attacking amino acids and proteins, the reaction of the epoxides involves the hydroxyalkylation of an atom with one or more lone pairs of electrons, either nitrogen or sulfur. Bruch [37] stated that the moisture requirements for ETO steriiization could be explained on the basis that water facilitates proton reactions with tertiary nitrogen compounds. He discounted the hypothesis that enzyme inhibition is the critical cellular reaction by gaseous sterilants, because the concentration of alkylating agents required for enzyme inactivation does not appear to be achieved in vivo when cytotoxic effects have been demonstrated.

The site of action of the gaseous alkylating agents may be the nucleoproteins. Several investigations [38-40] have reported that the reaction of ETO with the purine and pyrimidine bases in nucleic acids, nucleosides, and nucleotides indicated that ETO exerts its principal biological effect by alkylation of the nucleophilic groups of nucleic acids. First the ring nitrogen atoms of the purine and pyrimidine bases are hydroxyethylated, then the phosphate groups are esterified. This was confirmed by three later reports (see Ref. 14 for a fuller discussion).

D. Residual Ethylene Oxide

A comprehensive discussion of the toxic effects arising from residual ETO present in products is outside the scope of this chapter, but full details have bee provided elsewhere [23,41]. Although ETO gas diffuses rapidly in open air, porous materials absorb the gas during the sterilizing cycle in various amounts and then require various periods of time for aerating after sterilization for desorption of the residual gas to take place.

Polyvinyl chloride absorbs more ETO than polyethylene, which in turn absorbs more ETO than rubber. Polyvinyl chloride can absorb very high levels of ETO, and more than 25,000 ppm is not unusual [42].

Ethylene oxide residuals in plastics and rubber are desorbed at a rate which is dependent upon the ratio of the partial pressure of ETO within the item compared to the partial pressure of ETO outside the item. Moving air can therefore facilitate desorption. However, the factors involved in absorption and desorption are complex and can be affected by many variables. In all cases authors have given values for aeration times from experiments under a given set of conditions. It is impossible to take into account all the possible variables which will be met under the various operating conditions at the point of use, such as the effect of the individual packaging materials, methods of storage, conditions of storage (temperature, air velocity), and cycle conditions, as well as the main factors of type of material and thickness, and so on. Ethylene chlorhydrin* may remain with a product after removal of ETO.

In the United States there are three major agencies directly involved in the control of ETO. They are the Environmental Protection Agency, the Occupational Health and Safety Association, and the Food and Drug Administration. In 1971 the Occupational Health and Safety Association adopted the, present threshold limit value (TLV) of 50 ppm as a time-weighted average over an 8-hr period. This is the present standard in the United Kingdom, but it is the intention that the threshold limit value be lowered to 10 ppm, as in the United States.

E. Control Procedures

Control of ETO sterilization procedures must ensure that there are no unacceptable hazards for personnel. Microbiologically, the initial contamination level (bioburden) should be low, and the efficiency of the process can only be measured efficiently by the use of biological indicators [23].

The measurement of physicochemical parameters ensures that the sterilization procedure is within specified limits (gas concentration, temperature, RH, and sterilizing period; see Ref. 23).

F. Applications

A detailed description of the applications of ETO as a sterilizing or disinfecting agent have been provided by Russell [14]. These include its use as a decontaminating agent and as a sterilizing agent for ophthalmic and anesthetic equipment and for crude drugs and powders.

III. FORMALDEHYDE

A. Physical Properties

A summary of the physical properties of formaldehyde (methanal) is provided in Table 1. Formaldehyde gas may be generated by means of either the following: (1) evaporation of commercial formaldehyde solution (formalin) which consists of a 40% solution of formaldehyde in water, plus 10% methanol to prevent polymerization; (2) addition of formalin to potassium permanganate; and (3) volatilization of paraformaldehyde.

The ability of formaldehyde to polymerize at temperatures below 80°C is a problem [23]. Another problem associated with its use concerns its possible toxic effects to the olfactory senses and allergic responses.

---

*Editor's Note: More toxic by skin than orally. Lethal even for rats in air: 32 ppm.

B. Antimicrobial Activity

Vegetative bacteria and bacterial spores are fairly readily killed by formaldehyde gas, and the degree of resistance of spores is only some 2-3 [43] or 2-15 times [36] that of nonsporing bacterial. There is a linear relation between the concentration of formaldehyde and the killing rate [43], but little effect on disinfection rate from variation in temperature over the range 0-30°C has been observed [43]. However, Nordgren [10] observed that the rate of disinfection of spores exposed to formaldehyde vapor increased as the temperature was increased from 10 to 70°C. Organic matter, in the form of blood, sputum, or soil, reduces the rate of bacterial inactivation [10], and cocci suspended in serum are more difficult to kill than those suspended in gelatin [43].

As with other vapor-phase disinfectants, the antibacterial activity of formaldehyde is dependent on RH. Various RH levels for optimum activity of formaldehyde have been proposed: An optimum RH of 80-90% but with no great increase in disinfection rate upon increasing the RH above 58% has been mentioned [43]. Nordgren [10] reached a somewhat similar conclusion in that he observed an increase in the rate of bacterial kill as the RH was raised to 50%, but little increase as the RH was increased from 50 to 90%. Spaulding et al. [44], however, are of the opinion that there is no bactericidal effect unless the RH is 70% or above (see also Ref. 23).

Although paraformaldehyde has been stated as being of little practical use because of its slow volatilization [10], this statement no longer holds true. Paraformaldehyde, a polymer of formaldehyde with the formula $HO(CH_2O)_n \cdot H$, where n = 8-100, is a flake or a fine or coarse powder and is produced by evaporation of aqueous solutions of formaldehyde. When heated, paraformaldehyde depolymerizes rapidly to give formaldehyde, and this is the basis of the process used by Kaitz, cited by Taylor et al. [45], for disseminating formaldehyde gas. This process is now widely used.

Tulis [46] draws attention to the fact that certain organic resins and polymers, when exposed to elevated temperatures, release potentially sterilizing amounts of gaseous formaldehyde. This evolution of formaldehyde is such that the rate of release is a function of time and temperature. Examples of such products are (1) melamine formaldehyde (Fig. 2a), which is formed from formaldehyde and melamine under alkaline conditions; (2) urea formaldehyde products (Fig. 2b), a mixture of monomethylol urea and dimethylol urea; and (3) paraformaldehyde. Tulis [46] considers paraformaldehyde to be an excellent source of monomeric formaldehyde gas because it can be produced in a temperature-controlled reaction, and no contaminating residues (methanol and formic acid) are produced during the evaporation of formalin solutions.

Paraformaldehyde-produced formaldehyde gas is lethal to spores and various nonsporing gram-positive and gram-negative bacteria [45,46], and is considerably more effective as a disinfecting and sterilizing agent than the formaldehyde-releasing resins. The inactivation process is strictly a function of the available formaldehyde gas, and at various temperatures the percentage loss for formaldehyde is much greater from paraformaldehyde than from the resin [46].

Christensen and Kristensen [23] point out that, in principle, the conditions for formaldehyde sterilization are the same as those for ETO. Thus it is essential that the formaldehyde and water vapor gain unhindered access to contaminants.

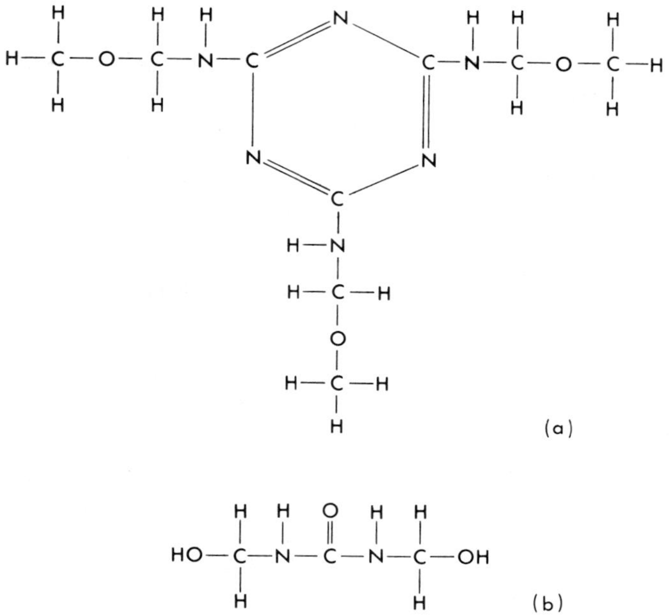

FIGURE 2 Chemical structures of (a) melamine formaldehyde and (b) urea formaldehyde.

Formaldehyde is a highly reactive chemical agent, and appears to kill microorganisms by virtue of its alkylating properties [14].

C. Control Procedures

As with ETO sterilization, various parameters must be controlled to ensure efficient formaldehyde action. These rely partly on physicochemical monitoring and partly on the use of biological indicators. Likewise, adequate precautions should be taken to prevent toxicity to personnel.*
These are considered in more detail by Christensen and Kristensen [23].

D. Applications

Formaldehyde is sometimes employed as a gaseous disinfecting or sterilizing agent [14,23]. It has limited use in the terminal disinfection of premises, but has been suggested for decontaminating microbiological laboratories. Fumigation by formaldehyde has found considerable use in poultry science. Alder and Simpson [47] discuss the use of low-temperature (70-90°C) steam with formaldehyde vapor for disinfecting heat-sensitive material.

---

*Editor's Note: Implications of formaldehyde in genotoxicity and carcinogenicity has recently been reviewed, J. A. Swenberg et al., Carcinogenicity, 4:945-952 (1983).

Manowitz and Sharpell [48] discuss the preservation of cosmetics by formaldehyde solutions and by formaldehyde-releasing agents [49]. Other uses are considered by Hoffman [50].

IV. OTHER LETHAL GASES

Apart from ethylene oxide and formaldehyde, other toxic gases exist (β-propiolactone, methyl bromide, propylene oxide, ozone, nitric oxide, carbon dioxide, glycidaldehyde, vapor-phase glutaraldehyde). These have been utilized for special sterilization purposes [14,50,51], but in general are little used as routine methods [23,52]. The physical properties of some of these agents are presented in Table 1.

β-Propiolactone is a highly active antibacterial agent, but its possible carcinogenicity has limited its applications. Its activity is a direct function of its concentration and the temperature and RH at which it is used. Glutaraldehyde as a vapor-phase agent has been comparatively little studied, although Bovallius and Anas [53] describe surface-decontaminating action in the gas-aerosol phase. Information on the antibacterial activity of other lethal gases is provided by Russell [14] and Kereluk [51]. Glycidaldehyde vapor inactivates sporing and nonsporing bacteria, the inactivation rate depending on the treatment temperature, gas concentration, and RH [54].

The mechanism of action of these various gases has been described by Russell [14]. Propylene oxide and β-propiolactone are known to act by alkylation; methyl bromide is believed to act in a similar manner, although there is scant evidence in support of this contention. In solution, glutaraldehyde acts as a cross-linking agent, but there is a dearth of information on its action as a vapor-phase agent.

REFERENCES

1. R. T. Cotton and R. C. Roark, Ethylene oxide as a fumigant, *Ind. Eng. Chem.*, 20:805 (1928).
2. H. D. Young and R. L. Busby, References to the use of ethylene oxide for pest control, United States Department of Agriculture (Bureau of Entomology and Plant Quarantine) (1935).
3. P. M. Cross and L. F. Dixon, Method of sterilizing, U.S. Patent 2,075,845 (1937).
4. I. A. Hall, Sterilized spices: new factor in food quality control, *Food Ind.*, 10:424-425, 464-467 (1938).
5. L. H. James, Reducing the microbial content of spices, *Food Ind.*, 10:428-429 (1938).
6. R. S. McBride, Automatic equipment for spice fumigation, *Food Ind.*, 10:430 (1938).
7. J. Yesair and E. J. Cameron, Present status of the sterilization of spices, *Canner*, 86:108-110 (1938).
8. H. W. Smith, Treated spices reduce spoilage, *Food Ind.*, 12:50-72 (1940).
9. J. Yesair and O. B. Williams, Spice contamination and its control, *Food Res.*, 7:118-126 (1942).
10. G. Nordgren, Investigations on the sterilizing efficacy of gaseous formaldehyde, *Acta Pathol. Microbiol. Scand. Suppl.*, 40:1-165 (1939).

11. C. R. Phillips and S. Kaye, The sterilizing action of ethylene oxide. I. Review, *Am. J. Hyg.*, *50*:270-279 (1949).
12. T. A. Roberts and A. D. Hitchins, Resistance of spores, in *The Bacterial Spore* (G. W. Gould and A. Hurst, eds.), Academic, London (1969), pp. 611-670.
13. A. D. Russell, The destruction of bacterial spores, in *Inhibition and Destruction of the Microbial Cell* (W. B. Hugo, ed.), Academic, London (1971), pp. 451-612.
14. A. D. Russell, Inactivation of non-sporing bacteria by gases, in *The Inactivation of Vegetative Microbes*, Fifth Symposium of the Society for Applied Bacteriology (F. A. Skinner and W. B. Hugo, eds.), Academic, London (1976), pp. 61-88.
15. C. W. Bruch and M. K. Bruch, Gaseous disinfection, in *Disinfection* (M. A. Benarde, ed.), Marcel Dekker, New York (1970).
16. C. R. Phillips, The sterilizing action of gaseous ethylene oxide. II. Sterilization of contaminated objects with ethylene oxide and related compounds: time, concentration and temperature relationships, *Am. J. Hyg.*, *50*:280-289 (1949).
17. S. Kaye, The sterilizing action of gaseous ethylene oxide. III. The effect of ethylene oxide and related compounds upon bacterial aerosols, *Am. J. Hyg.*, *50*:289-295 (1949).
18. S. Kaye and C. R. Phillips, The sterilizing action of ethylene oxide. IV. The effect of moisture, *Am. J. Hyg.*, *50*:296-306 (1949).
19. S. Kaye, Use of ethylene oxide for the sterilization of hospital equipment, *J. Lab. Clin. Med.*, *50*:289-295 (1950).
20. R. R. Ernst, Ethylene oxide. I. Sterilization, *Int. Anesthiol. Clin.*, *10*:85-100 (1972).
21. R. R. Ernst, *Dev. Biol. Stand.*, *23*:40-50 (1974).
22. R. R. Ernst, Sterilization by means of ethylene oxide, *Acta Pharm. Suec.*, *12*:44-64 (1975).
23. E. A. Christensen and H. Kristensen, Gaseous sterilization, in *Principles and Practice of Disinfection, Preservation and Sterilization* (A. D. Russell, W. B. Hugo and G. A. J. Ayliffe, eds.), Blackwell Scientific, Oxford (1982), pp. 548-568.
24. R. R. Ernst and J. J. Shull, Ethylene oxide gaseous sterilization, *Appl. Microbiol.*, *10*:337-344 (1962).
25. H. M. El-Bisi, R. M. Vondell, and W. B. Esselen, Kinetics of bacterial activity of ethylene oxide in the vapor phase. III. Effect of sterilant temperature and pressure, Bacteriological Proceedings (American Society of Microbiology, Abstracts) (1963), p. 13.
26. C. L. Griffith and L. A. Hall, Sterilization process, U.S. Patent 2,189,947; U.S. Patent 2,189,948; U.S. Patent 2,189,949 (1940).
27. C. R. Phillips, The sterilizing properties of ethylene oxide, in *Recent Developments in the Sterilization of Surgical Materials*, Symposium, The Pharmaceutical Society of Great Britain, Pharmaceutical Press, London (1961), pp. 59-75.
28. G. L. Gilbert, V. M. Gambill, D. R. Spiner, R. K. Hoffman, and C. R. Phillips, Effect of moisture on ethylene oxide sterilization, *Appl. Microbiol.*, *12*:496-503 (1964).
29. J. J. Perkins and R. S. Lloyd, Applications and equipment for ethylene oxide sterilization, in *Recent Developments in the Sterilization of Surgical Materials*, Symposium, The Pharmaceutical Society of Great Britain, Pharmaceutical Press, London (1961), pp. 76-90.

30. G. Mayr, Equipment for ethylene oxide sterilization, in *Recent Developments in the Sterilization of Surgical Materials*, Symposium, The Pharmaceutical Society of Great Britain, Pharmaceutical Press, London (1961), pp. 90-97.
31. K. Kereluk, R. A. Gammon, and R. S. Lloyd, Microbiological aspects of ethylene oxide sterilization, *Appl. Microbiol.*, 19:146-165 (1970).
32. J. Matthews and M. S. Hofstad, The inactivation of certain animal viruses by ethylene oxide, *Cornell Veterinarian*, 43:452-461 (1953).
33. M. Savan, The sterilizing action of gaseous ethylene oxide on foot and mouth disease virus, *American Journal of Veterinary Research*, 16:158-159 (1955).
34. A. Kalrenbreek and H. A. E. Van Tongerin, Virucidal action of ethylene oxide gas, *Journal of Hygiene*, 52:525-528 (1954).
35. R. Hoff-Jorgensan and E. Lund, Studies on the inactivation of viruses by ethylene oxide, *Acta Veterinaria Scandinavica*, 13:520-527 (1972).
36. C. R. Phillips, Symposium on the biology of spores. IX. Relative resistances of bacterial spores and vegetative bacteria to disinfectants, *Bacteriological Reviews*, 16:135 (1952).
37. C. W. Bruch, Gaseous sterilization, *Annual Review of Microbiology*, 15:245-262 (1961).
38. R. K. O'Leary and W. L. Guess, Toxicological studies on certain medical grade plastics sterilized by ethylene oxide, *Journal of Pharmaceutical Sciences*, 57:12-17 (1968).
39. R. K. O'Leary, W. D. Watkins, and W. L. Guess, Comparative chemical and toxicological evaluation of residual ethylene oxide in sterilized plastics, *Journal of Pharmaceutical Sciences*, 58: No. 8, 1007-1010 (1969).
40. D. A. Gunther, Absorption and desorption of ethylene oxide, *American Journal of Hospital Pharmacy*, 26:45-49 (1969).
41. C. Richards, Process control of ethylene oxide sterilization, *M. Pharm. Thesis*, University of Wales (1982).
42. J. B. Stetson, J. E. Whitbourne, and C. Eastman, Ethylene oxide degassing of rubber and plastic materials, *Anesthiology*, 44:2 (1972).
43. Anonymous, Disinfection of fabrics with gaseous formaldehyde, Committee on formaldehyde disinfection, *Journal of Hygiene*, 56:488-515 (1958).
44. E. H. Spaulding, K. R. Cundy, and F. J. Turner, Chemical disinfection of medical and surgical materials, in *Disinfection, Sterilization, and Preservation*, S. S. Block, ed., 2nd Edition, Lea and Febiger, Philadelphia (1977), pp. 654-684.
45. L. A. Taylor, M. S. Barbeito, and G. G. Gremillion, Paraformaldehyde for surface sterilization and detoxification, *Applied Microbiology*, 17:614-618 (1969).
46. J. J. Tulis, Formaldehyde gas as a sterilant, in *Industrial Sterilization: International Symposium, Amsterdam (1972)*, G. B. Phillips and W. S. Miller (eds.), Duke University Press, Durham, North Carolina, U.S.A. (1973).
47. V. G. Alder and R. Simpson, Sterilization and disinfection by heat methods, in *Principles and Practice of Disinfection, Preservation, and Sterilization*, A. D. Russell, W. B. Hugo, and G. A. J. Ayliffe (eds.), Blackwell Scientific Publications, Oxford (1982), pp. 433-453.

48. M. Manowitz and F. Sharpell, Preservation of cosmetics, in *Disinfection, Sterilization, and Preservation*, S. S. Block (ed.), 2nd Edition, Lea and Febiger, Philadelphia (1977), pp. 768-787.
49. W. B. Hugo and A. D. Russell, Types of antimicrobial agents, in *Principles and Practice of Disinfection, Preservation, and Sterilization*, A. D. Russell, W. B. Hugo, and G. A. J. Ayliffe (eds.), Blackwell Scientific Publications, Oxford (1982), p. 8-106.
50. R. K. Hoffman, Toxic gases, in *Inhibition and Destruction of the Microbial Cell*, W. B. Hugo (ed.), Academic Press, London (1971).
51. K. Kereluk, Gaseous sterilization: methyl bromide, propylene oxide and ozone, in *Progress in Industrial Microbiology*, Vol. 10, D. J. D. Hockenhull (ed.), Churchill Livingstone, Edinburgh (1971).
52. C. R. Phillips, Gaseous sterilization, in *Disinfection, Sterilization, and Preservation*, S. S. Block (ed.), Lea and Febiger, Philadelphia (1977), pp. 592-610.
53. A. Bovallius and P. Anas, Surface-decontaminating action of glutaraldehyde in the gas-aerosol phase, *Applied and Environmental Microbiology*, 34:129-134 (1977).
54. F. W. Dawson, Glycidaldehyde vapor as a disinfectant, *American Journal of Hygiene*, 76:209-215 (1962).

14
GAMMA-RADIATION DECONTAMINATION OF COSMETIC RAW MATERIALS

G. P. JACOBS *School of Pharmacy, Hebrew University of Jerusalem, Jerusalem, Israel*

I. Introduction 224
II. Ionizing Radiations 225
    A. Irradiation sources 225
    B. Radiation units and dose 225
III. Applications 226
    A. Solvents 226
    B. Thickening agents 226
    C. Powders 227
    D. Waxes 227
    E. Vegetable oils 227
    F. Animal oils and fats 227
    G. Mineral oils and fats 228
    H. Synthetic oils and fats 228
    I. Essential oils 228
    J. Fatty acids, alcohols, and esters 228
    K. Emulsifiers 229
    L. Antimicrobial preservatives 229
    M. Antioxidants 230
    N. Vitamins 230
IV. Conclusions 230
    References 231

---

*Reprinted with modifications from *Cosmetics and Toiletries*, 96:51 (1981). (1981).

I. INTRODUCTION

There has been an increasing awareness in recent years of the problem of microbiological contamination of pharmaceutical and cosmetic preparations. While such products, with the exception of parenteral and ophthalmic preparations, do not generally have to be "sterile," there is a growing need for reducing their initial microbial contamination, particularly in light of the more vigorous microbiological safety standards that are being introduced in both the pharmaceutical and cosmetic industries.

Progress in the technology of radiation sterilization, including the development of large radiation sources, makes this method of decontamination most feasible. In the case of pharmaceutical and cosmetic creams and ointments, and other cosmetic preparations, reducing the microbial load of individual highly contaminated components may be practicable by use of gamma radiation. Some indication of the severity of such contamination can be gained by examination of the results of Hangay [1] presented in Table 1.

TABLE 1 Microbiological Contamination of Ointment-Basic Materials

Base materials	Number of batches tested	Contamination of samples (%)	Total bacterial count per gram			
			0	10^1-10^2	10^2-10^3	$>10^3$
Emulsifiers						
Tween 60	13	85	2	9	2	–
Span 60	12	33	10	2	–	–
Arlacel C	10	40	7	3	–	–
Tegin	13	30	11	2	–	–
Natural products						
Stearyl alcohol	19	74	8	9	2	–
Cetyl alcohol	17	41	11	3	3	–
Beeswax	32	57	19	6	5	2
Lanolin	35	66	13	17	5	–
Cocoa butter	3	100	–	–	3	–
Synthetic products						
Isopropyl myristate	10	50	6	3	–	1
Miglyol	19	28	16	3	–	–
Propylene glycols	30	55	14	13	3	–
Polyethylene glycols	3	66	1	1	1	–
Glycerin	30	67	11	16	3	–
Others						
Petroleum oil	30	66	12	11	7	–
Petroleum jelly	27	50	13	12	1	1
Borax	11	82	2	3	6	–
Nipagin	17	23	13	4	–	–

It would be useful to recall some pertinent properties of high-energy radiation prior to considering its effects on the various components of topical pharmaceuticals and cosmetics.

II. IONIZING RADIATIONS

Ionizing radiations, which include high-energy electrons and electromagnetic gamma radiations, are lethal to microorganisms. Their use for sterilization purposes has some clear advantages over other methods in that they have high penetrability and cause a minimal temperature rise in the irradiated product. Sterilization can therefore be carried out, if desired, on the finally packaged product and is applicable to heat-sensitive and ethylene oxide-incompatible materials. The lethal action of ionizing radiation is achieved either by "direct" action of the radiation on a specific cellular target, such as a DNA molecule, or by "indirect" action. In the latter, cellular inactivation is a result of free-radical production in the medium, usually a radiolysis product of water, such as $\cdot H$, $\cdot OH$, e^-_{aq}, which initiates a secondary reaction in a target molecule. Whatever detailed mechanisms are involved, a vast amount of knowledge has been accumulated on the radiation inactivation of microbial populations, where lethality is generally measured by the loss of the colony-forming ability of the cells when placed in suitable growth media.

A. Irradiation Sources

There are principally two types of radiation sterilization equipment: electron accelerators, and gamma-radiation sources. Electrons are usually generated by Van de Graaff machines (1-3 MeV electrons) or linear accelerators (3-15 MeV electrons). The dose rate is so high that sterilization is over in a fraction of a second.

Gamma rays are emitted by a number of radioactive sources, the principal ones with industrial application being ^{60}Co and ^{137}Cs. Cobalt-60, with a half-life of 5.3 years, is produced by neutron bombardment of the inactive ^{59}Co. Each disintegrating ^{60}Co atom invokes the emission of a beta particle of energy up to 0.3 MeV and two gamma photons of 1.17 and 1.33 MeV, leaving an atom of stable ^{60}Ni. The less common ^{137}Cs, with a half-life of 30 years, is produced as a fission product of uranium. It emits a single gamma photon of 0.66 MeV. It must be emphasized that the photon energy of both isotopes is well below the threshold for the production of radioactivity in elements of the material being irradiated, and therefore no hazards exist in handling products, no matter how high the dose applied.

B. Radiation Units and Dose

The unit of radiation dose is the rad (*radiation absorbed dose*), which is equivalent to an energy absorption of 100 ergs per gram of material. In practical terms it is more useful to use the unit Mrad (10^6 rads). More recently the Gray (Gy) has been adopted to conform with the changeover to SI units, where 1 Gy is equivalent to 100 rads.

The choice of a *sterilization* dose is dependent on the presterilization microbial load and the specific microbial species present; nevertheless, a generally accepted dose is 2.5 Mrads, or 25 kGy. Naturally, for decontamination purposes (compare sterilization) much lower doses may be used, possibly on the order of 0.5-1 Mrad and less.

III. APPLICATIONS

In the ensuing part of this article, the effects of gamma rays on various components of ointments and creams and certain other cosmetic raw materials have been considered. The classification adopted is somewhat arbitrary and several materials may fall into more than one group. In light of the envisaged readership of this review, it has not been considered necessary to define the cosmeticological purpose of each component.

A. Solvents

A 2.5 Mrad dose caused no significant color or pH change in propylene glycol [1,2] and, furthermore, no toxic products were formed [3]. However, infrared spectra did reveal minimal decomposition [2].

Glycerin is reported to undergo no pH change at doses of up to 2.5 Mrads [1].

Irradiation of aqueous aerated and deaerated solutions of ethanol produces negligible change, although H_2, hydrogen peroxide, acetaldehyde, and, in aerated solutions, glycol are produced in quantities not exceeding 0.01% at radiation doses of up to 1 Mrad [4,5]. Ethanol itself at this radiation dose in the presence or absence of oxygen will yield traces of methane, ethane, propane, and carbon monoxide in addition to the above [6].

Radiolysis of acetone yields small quantities (less than 0.01%) of methane, ethane, and carbon monoxide [7].

Although it is not so usual to decontaminate water per se as a constituent of a cosmetic or pharmaceutical formulation, for completeness mention must be made of reports from a number of laboratories on this possibility [8-11].

B. Thickening Agents

Reduction in the viscosity of sodium carboxymethyl cellulose solutions by 50-60% accompanies exposure to 5.7 krads, and after 250 krads the viscosity of a 0.5% solution was reduced from 4500 cP to less than 500 cP. An 85 krad dose on a 2% solution caused a change in viscosmetric behavior from pseudoplastic to Newtonian in character, indicating a breakdown in the gel structure [12]. Similar findings have been reported by Bor [13].

Measurements on solutions of irradiated (2.5 Mrads) acacia powder showed an 11% reduction in viscosity [14]. However, irradiation of the dried powder caused no change in the acid number [15]. Viscosities of sodium alginate solutions in 0.1 N NaCl showed a 70% decrease in intrinsic viscosity, indicating changes in the molecular structure corresponding to degradation to 30% of the original degree of polymerization [14].

In another study [16] it is reported that deaerated gels of 2% Manucol DM (calcium alginate) are rather sensitive to radiation, the viscosity decreasing until a fluid syrup is produced after a 1 Mrad dose. Oxygen tends to increase the sensitivity [16]. However, the presence of propylene glycol, benzoates, or salicylate greatly reduces the radiation damage. Sodium benzoate, at a 10^{-2} M concentration, gives good protection against a dose of 2.5 Mrads, the gel retaining most of its rigidity [16].

Radiation (2 Mrads) reduces the specific viscosity of gelatin [17] by 15-35% due to a decrease in chain length, although the gel strength remained within British Pharmacopoeia limits.

Irradiation of cellulose causes dehydrogenation and decomposition, with the release of hydrogen, carbon monoxide, and carbon dioxide. Moisture and oxygen affect the radiation products [16].

In a study on the effect of gamma irradiation on the microbial contamination and rheology of tragacanth, it was reported [18] that at 1 Mrad the decrease in the coefficient of viscosity was only 7%.

In another report [19] gels based on tragancanth, methyl celluloses, and carbopols lost their gel structure. However, the addition of 5-10% ethanol prevented the degradation of the carbopols [19].

C. Powders

An increase in the acid-soluble matter of talc has been reported at 2.5 Mrads [20], although no other change was noticed. The color and British Pharmaceutical Codex test for alkalinity were unaffected [20]. Another report confirms these findings [21].

Gamma irradiation of corn, maize, and tapioca starch increases acidity, solubility, and reducing sugar content, but decreases the swelling capacity, gelation temperature, and viscosity [22]. The main degradation products are formaldehyde, malonaldehyde, 1,4-pyrones, formic acid, and glycolaldehyde, although with doses of up to 2.5 Mrads these changes are not appreciable. A decrease in the viscosity of wheat starch gels has been reported [13].

A 2.3 Mrad radiation dose from a 10 MeV electron beam caused a 10% increase in the saponification number of magnesium stearate [15].

D. Waxes

Examination of a straight-chain paraffin wax with no branching or unsaturation, and an average formula of $C_{54}H_{70}$ (mol. wt. 471), shows that at doses of up to 2.5 Mrads, a small number of cross-links take place, resulting in an increased molecular weight of 516 [23]. In a separate study [1], there was no reported pH change at similar doses in either paraffin wax, beeswax, polyethylene glycol, or carbopol 940 [1]. However, as stated above, gels based on carbopols lost their structure [19].

Other investigators have reported little change in irradiated beeswax and carnauba wax [15].

E. Vegetable Oils

Gamma irradiation of sunflower oil (1 Mrad) produced small quantities of H_2O_2 and epoxy products [24], although only the carotenes and tocopherols are destroyed, while the sterols and linoleic acids are unaffected [25]. Similar products were formed on irradiation of groundnut, rape, and neutral oils [26]. No information is available for these products irradiated at low doses (<1 Mrad). Irradiation of olive oil at 5 Mrads also produces significant quantities of organic peroxides [27]. Irradiation produced no change in castor oil [15].

F. Animal Oils and Fats

No pH change has been reported for lanolin at 2.5 Mrads [1]. Stearin is also unaffected at this dose level [15].

G. Mineral Oils and Fats

Liquid paraffin undergoes no pH change following an irradiation dose of 5 Mrads [1], although it has been reported to be discolored [20]. Bubbles of gas were apparent after irradiation. There was an increase in the iodine value from 2 to 5 when carried out by the British Pharmacopoeia ICl method [20]. Other tests indicated no significant change in liquid paraffin [2]. Tests in this laboratory showed that both irradiated (up to 2.5 Mrads) and unirradiated liquid paraffin behaved as typical Newtonian liquids. Values for the coefficient of viscosity at 26°C were 124 cP for the unirradiated material and 120 cP for that irradiated at either the 1 or 2.5 Mrad dose level [28].

White petrolatum was reported to be slightly discolored with a slight decrease in viscosity at 2.5 Mrads [2,20,21,29,30]. When in tubes, some distention occurred. Similarly, when in glass pots, some leakage was apparent. The parameters of the irradiated and nonirradiated samples equalized in the course of storage [2]. A minimal change in pH value was also observed [1]. The water number remained constant.

In a study carried out in this laboratory, it was found that when white soft paraffin was irradiated in tubes, 90% was unaffected following a 2.5-Mrad dose (10% either distended or developed leaks). Peculiarly, 85% was unaffected following a 1 Mrad dose [28]. No significant differences in viscosity were observed between irradiated (2.5 Mrads) and unirradiated batches of white soft paraffin when determined by comparison of rheological profiles and comparison of shear stress values at different shear rates using a Haake Rotovisco RV3 Viscometer with an appropriate plate and cone sensor system [28]. Similarly, the melting point of white soft paraffin remained unchanged following a 2.5 Mrad dose and complied with the pharmacopoeial stipulation for the absence of light-absorbing impurities [28].

H. Synthetic Oils and Fats

Testing for pH change in a number of synthetic materials showed that there was no change in silicone oil, polyethylene glycols, and carbopol 940 [1].

I. Essential Oils

Gas-liquid chromatographic examination of oil of thyme indicated no difference between the irradiated (3 Mrads) and the unirradiated oil [31]. Similarly, no changes have been reported in gamma-irradiated limonene [32].

J. Fatty Acids, Alcohols, and Esters

Cetyl alcohol remains unchanged at 2.5 Mrads as indicated by the results of a wide variety of tests [2].

Stearic acid undergoes minimal degradation at low doses [33], although there is a sharp drop in pH [1]. Both Miglyol 812 (triglycerides of saturated fatty acids) and isopropyl myristate undergo no pH change at doses of up to 2.5 Mrads [1].

The index of refraction of triethenolamine remained unchanged following a 2.3 Mrad dose [15].

TABLE 2 Changes in pH of Emulsifying Agents After Gamma Irradiation

Raw material	Radiation dose (Mrads)				
	0	0.5	1.0	1.5	2.5
Polysorbate 60	4.5	4.4	4.5	4.6	4.7
Polysorbate 80	6.7	6.6	6.8	6.8	6.9
Arlaton 983	5.1	4.9	4.5	4.4	3.8
Tegin	8.0	–	7.8	7.9	7.8
Texapon K12	7.7	7.5	7.2	7.1	7.0

K. Emulsifiers

Using color change as a criterion for chemical decomposition of Tween 60 (polyoxyethylene 20 monostearate), values were within a ±2% change in color brightness at doses of up to 1.5 Mrad [1]. Values of pH for a number of emulsifiers tested by Hangay [1] are presented in Table 2. It is apparent that at doses of up to 2.5 Mrads, there is a significant decomposition of the commercial preparations Arlaton and Texapon, although the change at very low doses (0.5 Mrad) is only slight. The other materials are hardly affected [1].

On the basis of ultraviolet spectrophotometric examination, polyoxyethylene 1500 was unaffected at a 2.5 Mrad radiation dose [1].

L. Antimicrobial Preservatives

Extensive degradation has been reported for aqueous solutions of chlorbutanol, alkyl dimethyl benzalkonium chloride, and the methyl- and propylparabens following radiation doses of up to 2.5 Mrads [34]. Methylparaben is, however, less sensitive than the propyl ester. Similarly, aqueous solutions of phenylmercuric acetate and chloride undergo radiolysis to give phenol and inorganic mercury salts [16]. Irradiation in the solid state gives the same products [16]. In a spectrophotometric study [34] solid phenylmercuric nitrate was reported to be unchanged at 2.5 Mrads, although there was a slight color change; however, this study did indicate extensive degradation when irradiation was carried out in aqueous solution [34]. Nipagin M (methyl ester of p-hydroxybenzoic acid) undergoes only a slight pH change following low radiation doses [1]. The radiolysis products of chlorbutanol were found to be acetone, hydrogen peroxide, and an unidentified yellow precipitate. A 37 krad dose caused the pH of a 0.5% solution to fall from 5.8 to 2.4. It was found that concentrated solutions were more stable to gamma irradiation than dilute ones [12].

In an extensive study on the effect of gamma irradiation on selected aqueous preservative *solutions*, McCarthy reported that few remained unaffected following this treatment [35].

M. Antioxidants

The effect of gamma rays on three antioxidants has been studied by Chipault and Mizuno [36], who reported that solutions of propyl gallate (0.01%), butylated hydroxyanisole, and α-tocopherol, irradiated to 2 Mrads in methyl myristate in vacuo, resulted in 16, 8, and 90% breakdowns, respectively. In O_2 the destruction was 89, 90, and 100%, respectively.

Octyl gallate has been irradiated as a 0.5% solution in cod liver oil. After 2 Mrads there was a reduction in the active oxygen number, both in the solution and in a sample of the oil. On storage the active oxygen number increased in all cases, suggesting that protective antioxidant action had been destroyed. Change in acid values have also been reported [20].

N. Vitamins

Irradiation of pyridoxine (vitamin B_6) in aqueous solution leads to its partial or total destruction, depending on the dose; however, irradiation in the dry state caused no appreciable change at 2 Mrads [37].

Similarly, an earlier report on electron irradiation [38] of two multivitamin preparations comprising thiamine, riboflavine, pyridoxine, nicotinamide, and calcium pantothenate indicates that the irradiated preparations are as stable as the unirradiated ones.

IV. CONCLUSIONS

Although many of the above compounds undergo degradation at the commonly employed *sterilization* dose of 2.5 Mrads, much smaller radiation doses may be employed if we are simply considering reducing the microbial load of the product. Provided that the degradation pathways are known, considerable protection may be achieved by suitable elimination or conversion of the species responsible for breakdown by the addition of suitable radical scavengers. However, consideration has to be given to whether such radical scavengers may not be simultaneously diminishing the efficiency of the decontamination or sterilization process.

Generally, components irradiated in the dry state are much more radiation stable than those irradiated in aqueous solution, a factor which is particularly relevant when considering the decontamination of specific highly contaminated components.

While many of the investigations cited in this article have been concerned with a somewhat superficial examination of irradiated products, these data are useful insights into their radiation stability and indicate whether more extensive and thorough testing of a product is worthwhile. Naturally, even if only traces of radiolysis products are formed, it has to be conclusively established by appropriate toxicological screening that such products are without any adverse action at the concentration found.

It is certainly encouraging that a recent report of the World Health Organization [39] has recommended the acceptability of food irradiated up to an average dose of 1 Mrad and has stated that up to this dose level there is no toxicological hazard.

Although a great deal is now known about the radiation stability or otherwise of many chemical compounds or moieties, it is still necessary to examine each individual compound in order to assess the feasibility of its irradiation. In considering complete products, it must be emphasized that the stability of an individual component may be quite different when irradiated alone or as part of a product.

No doubt the higher microbiological standards demanded for cosmetics will make gamma irradiation a very useful tool for reducing the microbial loads of contaminated cosmetic raw materials.

REFERENCES

1. G. Hangay, On the theoretical and practical aspects of the use of radiation sterilization and radiation pasteurization in the pharmaceutical and cosmetic industry, in *Sterilization by Ionizing Radiation*, Vol. 2, (E. R. L. Gaughran and A. J. Goudie, eds.), Multiscience, Montreal (1978), pp. 247-263.
2. G. Hangay, Sterilization of hydrocortisone eye ointment by gamma-irradiation. I. Physical and chemical aspects, in *Radiosterilization of Medical Products*, International Atomic Energy Agency, Vienna (1967), pp. 55-62.
3. J. R. Hickman, Acute toxicity of radiation-sterilized propylene glycol, *J. Pharm. Pharmacol.*, 17:255-256 (1965).
4. G. G. Jayson, G. Scholes, and J. Weiss, Chemical action of ionizing radiations in solution. Part XX. Actions of x-rays (200 kV) on ethanol in aqueous solution, *J. Chem. Soc.*, 1957:1358-1368 (1957).
5. J. T. Allan, E. M. Hayon, and J. Weiss, The chemical action of ionizing radiations in solution. Part XXIII. The action of ^{60}Co-gamma-rays on aqueous solutions of ethanol in the liquid and frozen state, *J. Chem. Soc.*, 1959:3913-3919 (1959).
6. E. Hayon and J. J. Weiss, The chemical action of ionizing radiations on simple aliphatic alcohols. Part I. Irradiation of ethanol in the liquid and in the solid state and in the presence of solutes, *J. Chem. Soc.*, 1961:3962-3970 (1961).
7. P. Ausloos and J. F. Paulson, Radiolysis of simple ketones, *J. Am. Chem. Soc.*, 80:5117-5121 (1958).
8. N. Hilmy and S. Sadjirun, Polyethylene plastics as containers for water for injection and a material for disposable medical devices sterilized by radiation, in *Radiation Sterilization of Medical Products 1974*, International Atomic Energy Agency, Vienna (1975), pp. 145-157.
9. T. A. du Plessis, Radiation sterilization of pyrogen-free water in polyethylene packets, *Report PER-17 Atomic Energy Board, South Africa* (1977).
10. G. P. Jacobs, M. Donbrow, E. Eisenberg, and M. Lapidot, The use of gamma-irradiation for the sterilization of water for injections and normal saline solution for injection, *Acta Pharm. Suec.*, 14:287-292 (1977).
11. G. P. Jacobs and E. Eisenberg, The reconstruction of powders for injection with gamma-irradiated water, *Int. J. Appl. Radiat. Isotopes*, 32:180-181 (1981).
12. L. J. Rasero and D. N. Skauen, Effect of gamma radiation on selected pharmaceuticals, *J. Pharm. Sci.*, 56:724-728 (1967).
13. C. Bor, Gamma radiation sterilization of pharmaceutical basic materials, adjuvants and packaging materials, *Zf I - Mitt.*, 43b:445-458 (1981).
14. D. W. Hartman, R. U. Nesbitt, F. M. Smith, and N. O. Nuessle, Viscosities of acacia and sodium alginate after sterilization by cobalt-60, *J. Pharm. Sci.*, 64:802-805 (1975).

15. T. Achmatowicz-Szmajke, T. Bryl-Sandelewska, and M. Galazka, Radiation sterilization of some cosmetic raw materials and preparations, *Radiochem. Radioanal. Lett.*, *38*:5-14 (1979).
16. R. Blackburn, B. Iddon, J. S. Moore, G. O. Phillips, D. M. Power, and T. W. Woodward, Radiation sterilization of pharmaceuticals and biomedical products, in *Radiation Sterilization of Medical Products, 1974*, International Atomic Energy Agency, Vienna (1975), pp. 351-363.
17. K. M. Patel, M. Tantry, G. Sharma, and N. G. S. Gopal, Effect of dry heat, ethylene oxide and gamma radiation on gelatin and gelatin capsules, *Indian J. Pharm. Sci.*, *41*:209-213 (1979).
18. G. P. Jacobs and R. Simes, The gamma irradiation of tragacanth: Effect on microbial contamination and rheology, *J. Pharm. Pharmacol.*, *31*:333-334 (1979).
19. I. Adams, S. S. Davis, and R. Kenhaw, Formulation of a sterile surgical lubricant, *J. Pharm. Pharmacol.*, *24*:178P (1972).
20. Association of the British Pharmaceutical Industry, *Use of Gamma Radiation Sources for the Sterilization of Pharmaceutical Products*, Association of the British Pharmaceutical Industry, London (1960).
21. Report of the Bhaba Atomic Research Centre, No. 737, Bhaba Atomic Research Centre, India (1974).
22. N. G. S. Gopal, Radiation sterilization of pharmaceuticals and polymers, *Radiat. Phys. Chem.*, *12*:35-50 (1978).
23. M. F. Mayahi and M. M. Mousa, Effect of gamma radiation on pure paraffin wax and on wax acetone system, *J. Indian Chem. Soc.*, *49*: 981-984 (1972).
24. S. A. Ivanov and S. D. Stamatov, *Seifen, Oele, Fette, Wachse*, *101*: 589 (1975), (quoted in Ref. 20).
25. S. A. Ivanov and S. D. Stamatov, *Seifen, Oele, Fette, Wachse*, *102*: 145 (1976), (quoted in Ref. 20).
26. R. Huettenrauch and I. Keiser, Use of high energy radiation in pharmaceutical technology. Part 10. Effect of high energy radiation on oils and oil-like substances, *Pharmazie*, *31*:110-114 (1976).
27. M. D. Astudilo, E. Fures, and F. Sanz, Effectos de la radiacion sobre el aceite de olivia, *An. R. Soc. Esp. Fis. Quim. Ser. B.*, *64*:787-789 (1968).
28. G. P. Jacobs, unpublished data (1976).
29. G. Hortobagyi, G. Hangay, B. Lukats, G. Muranyi, and A. Zarandy, Data on the radiosterilization of some basic pharmaceutical materials, in *Radiosterilization of Medical Products*, International Atomic Energy Agency, Vienna (1967), pp. 25-32.
30. M. Haraszti, K. Czeh, and G. Hangay, Sterilization of hydrocortisone eye-ointment by gamma-irradiation, II. Bacteriological aspects, in *Radiosterilization of Medical Products*, International Atomic Energy Agency, Vienna (1967), pp. 63-68.
31. N. A. Diding, G. Redmalm, G. Samuelsson, L. Frigren, B. M. Yman, and I. Van Katjwik, Studies on the effect of gamma irradiation of crude drugs, *Sven. Farm. Tidskr.*, *77*:622-630 (1973).
32. V. J. Buchi and N. Iconomou, Untersuchungen ueber die Beeinflussung von Arzneistoffen durch radioaktive Strahlen, *Pharm. Acta Helv.*, *40*:421-431 (1965).
33. G. S. Wu and D. R. Howton, Gamma-Radiolysis of stearic acid: studies of nongaseous products, *Radiat. Res.*, *61*:374-392 (1975).

34. N. G. S. Gopal, S. Rajagopalan, and G. Sharma, Feasibility studies on radiation sterilization of some pharmaceutical products, in *Radiation Sterilization of Medical Products, 1974*, International Atomic Energy Agency, Vienna (1975), pp. 387-402.
35. T. J. McCarthy, The effect of gamma-irradiation on selected aqueous preservative solutions, *Pharm. Weekbl.*, 78:698-700 (1978).
36. J. R. Chipault and G. R. Mizuno, Effect of high energy radiations on antioxidants and on the stability of fats, *Abstr. Pap. Am. Chem. Soc.*, 150:19A-20A (1965).
37. I. Galatzeanu and F. Antoni, Effets des rayons gamma (cobalt-60) sur la vitamine B_6 et l'acide folique, in *Radiosterilization of Medical Products*, International Atomic Energy Agency, Vienna (1967), pp. 33-48.
38. G. C. Colovos and B. W. Churchill, The electron sterilization of certain pharmaceutical preparations, *J. Am. Pharm. Assoc. Sci. Ed.*, 46:580-583 (1957).
39. World Health Organization, Wholesomeness of irradiation food, Report of a Joint FAO/IAEA/WHO Expert Committee, *Technical Report Series, 659*, World Health Organization, Geneva (1981).

PART IV
USE OF MULTIFUNCTIONAL CHEMICALS IN PRESERVATIVE SYSTEMS

15
AROMA PRESERVATIVES
Essential Oils and Fragrances as Antimicrobial Agents

JON J. KABARA *Michigan State University, East Lansing, Michigan*

I. Early History of Aroma Chemical 237
II. Antimicrobial Activity of Essential Oils and Fragrances 240
III. Chemical Constituents of "Active" Principles in Aroma Preparations 265
    A. Compounds containing carbon and hydrogen 265
    B. Compounds containing carbon, hydrogen, and oxygen 265
IV. Toxicology of Aroma Chemicals 270

References 271

I. EARLY HISTORY OF AROMA CHEMICAL

Since the beginning of time man has been preoccupied with the preservation of "things." A concern to keep the body whole and free from deterioration, before and after death, occupied a high priority with ancient man. Since early Neanderthal man, plants have been used for their healing and preservative powers. Modern man has disdained plant products as too old-fashioned and without scientific bases. Despite this disdain, plants have been, nevertheless, the basis for almost all of our modern drugs. Ephedrine (ma-huang shrub), aspirin (spirea plant), quinine (cinchona bark), digitalis (foxglove), curare (vines of the Menispermaceae in particular from the genera *Stychnos* and *Chendodedron*) come quickly to mind as medicinals based on herbal medicine [1]. For a full history of herbals in medicine the reader is urged to read the detailed review by Garland [2]. What follows is a résumé of that history.

Spices and essential oils which have the characteristic or flavor of the plant from which they are obtained have long been used as preservatives. Their chemistry is complex, but generally most active extracts are comprised of one or more of the following: alcohols, phenols, esters, acids, aldehydes, and/or terpenes. It is believed that the evaporation of essences from plant surfaces act as a defense mechanism against infection by bacteria, fungi, and pests. Although it is impossible to point to a date when plants and plant products were used by man for some medicinal reason, there are papyri recording the medicinal use of plants dating back to the reign of Khufu, who built the Great Pyramid around 2800 B.C. Even before this time Shen Nung, a Chinese emperor of nearly 5000 years ago, was responsible for the first known herbals.

It was widely recorded that Egyptians used "cedarwood oil" in the process of mummification. They used either the pure essence extracted by pressing the wood or a primitive form of distillation; if this is true, it means distillation was known in Egypt at least 2000 years before the Arabs were supposed to have invented it. A Persian called Avicenna (born 980 A.D.) wrote a standard text (*Canon*), and in this "book" he described the means of obtaining the volatile oil of herbs and flowers by distillation.

Historically aromatic substances were used in one form or another to combat disease. During the Great Plague of the Middle Ages, all types of fumigantes were used to destroy the "aura" or poison of the disease. Perfumed candles were used in sickrooms and hospitals. They contained red roses, cloves, storax, frankincense, lemon peels, juniper berries, musk, and ambergris. The mixture was formed into candles with gum tragacanth which had been dissolved in rose water. Aromatics were the best antiseptics available at the time. Up until the nineteenth century medical practitioners carried a little cassolette filled with aromatics on the top of their walking sticks. This acted as a personal antiseptic and would be held up to the nose when visiting any contagious cases.

The long and universal association between herbal medicine and religion continued throughout recorded history until comparatively recently. The emperor Charlemagne (812 A.D.) paid special note to the medicinal power of herbs and for this reason gave his protection to the Benedictine monastery of St. Gall in Switzerland. The monastery became the model for all imperial farms. During the next 600 years many similar gardens were established throughout Europe and Britain. These botanical gardens were attached to universities and medical schools and used for both medical and botanical study.

The herbalists applied their aromatic materials to certain ailments by intuition and empirical experience. Out of early alchemy, essences and essential oils of plants were incorporated into medicine, food, and cosmetics for their flavoring and therapeutic qualities. Urdang in his "Pharmacy in Ancient Greece and Rome" [3] states that reference to essential oils is scarce and vague in early publications. The Greek historian Herodotus (484-425 B.C.) and the Roman naturalist Pliny (23-79 A.D.) mention oil of turpentine. Plants that elaborate essential oils are mentioned in the Bible. There is historical reference to the trade in spices, odoriferous oils, and fatty extracts which was carried on in the Orient and in Greek and Roman commerce. A Catalan physician, Arnold de Villanova (1235-1311), is believed to be the first to have described, in print, the actual distillation of essential oils. His praise of the remedial qualities of distilled aromatic waters resulted in an awareness of the value of essential oils.

Loncier in the *Krauterbuch* (1550) stressed the medicinal value of "many marvelous and efficient oils of spices and seeds." In the second half of the sixteenth century, Reiff, a Strasburg physician, published his book *Neu Gross Destillirbuch* (1556) in which he referred to clove, mace, nutmeg, anise, spice, and cinnamon. The second official Nuremburg edition of the *Dispensatorium Valerii Cordi*, issued in 1592, lists 61 distilled essential oils.

Urdang [3], further reviewing the history of essential oils, writes, "In the 17th and 18th centuries it was chiefly the pharmacists who improved methods of distillation and made valuable investigations into the nature of essential oils." Of special importance was the work of the French apothecaries, Charas (1618-1698), Lemery (1645-1715), Geoffroy (1685-1752), Rouelle (1703-1770), Demachy (1728-1803), and Baume (1728-1804); their German colleagues, Newmann (1683-1737), Wiegleb (1732-1800) and Green (1760-1798); and the German-Russian pharmacist Bindheim (1750-1825).

In the nineteenth century there was considerable research on the nature of essential oils and aromatic compounds. Physical and chemical procedures were established for the analysis of aromatics. Standards were prepared for widespread use and reference. Of considerable importance in the further development of the chemistry of volatile oils were the investigations of the French chemist Berthelot (1827-1907), devoted primarily to the hydrocarbons contained in these oils. Dumas (1800-1884) made a scientific investigation of stearoptens. About 1866 the word *terpene* was mentioned in a textbook written by Kekulé (1829-1896).

In the United States, the production of essential oils was started from three idigenous American plants: sassafras, American wormseed, and wintergreen. These oils, along with oil of turpentine, were introduced into the first *U.S. Pharmacopoeia* published in 1820. The current *U.S. Pharmacopoeia XVIII* lists no essential oils, while the *National Formulary* (13th Ed.) shows only eight. At one time between 1920 and 1940, the *U.S. Pharmacopoeia XI* listed 24 essential oils.

Until the nineteenth century, an exact herbal prescription had never been possible because of the widely varying strength of the active principle possessed by each individual plant. That the general climate, local weather conditions, habitat, soil, time of gathering, and methods of preserving all have an effect upon the properties of the plant was appreciated by most practitioners. The customs and superstitions surrounding the growing and harvesting of herbs, although on the face of it bizarre, in fact often concealed a sensible method for preserving these properties as well as possible. Indeed, Dioscorides, writing about herbals nearly 2000 years ago, noted that "it is proper to use care...in the gathering of herbs each at its due season, for it is according to this that medicines either do their work or become quite ineffectual." Chemistry made it possible to extract a measured quantity of the active principle from a plant and to make it up into exact doses. Chemistry also provided the means for synthetically reproducing the isolated principles which has led to the modern development of thousands of potent synthetic drugs. Although many of the synthetic drugs are of great value, there has been little time to discover the long-term action of such complex drugs, and only too often the benefits are countered by unpleasant side effects. These problems have led, during the present century, to a revival of interest in medical botany and our quest for safer agents. The literature dealing with the use and isolation of natural active substances is reviewed in the following sections. It is of special note that in this age of "high technology" many have "returned to nature" in the belief that Mother Nature, and not the organic chemist, has a better product for our needs.

II. ANTIMICROBIAL ACTIVITY OF ESSENTIAL OILS AND FRAGRANCES

Early attention was drawn to the curative powers of herbal antiseptics used in folk medicine. These remedies suffered in that many of the plant parts (roots, bark, fruit, leaves, etc.) were called for in recipes, but not necessarily described in detail. Adding to the confusion were the source and strength of the herbal extract. Historically, to obviate this problem many apothecaries and physicians grew herbs in their own gardens.

Attempts to apply some science to discovering the principle of a herbal remedy are fraught with problems. In an early study (1969) only 50% of the plants tested showed antibacterial activity [4]. It could be that inactive components were included in a remedy either for ritualistic or cosmetic purposes.

Another explanation for negative findings may be due to the method of screening for antimicrobial effects. Most studies used the conventional agar streak or disk method. These methods have serious limitations in that the strength of the extract, the migration of the active ingredient in agar, and the use of mixed microbial cultures influenced the results. Only those principles which migrated from the paper or agar-well onto the agar and formed an area of no growth could be labeled active. Thus hydrophilic agents automatically measure as being more "active" than similarly active lipophilic agents since the latter do not migrate or migrate poorly.

Much of the early studies dealing with the antimicrobial activity of aromatic substances claimed comparisons to phenol. Martindale [5] published his results on the phenol coefficient of a number of essential oils. The following are a few of his results:

	Phenol coefficient
Oil of clove	9
Oil of cinnamon	9
Oil of eucalyptus	4
Oil of rose	6
Oil of lavender	5
Oil of lemon	4
Oil of thyme	15
Oil of rosemary	6
Oil of sandalwood	1.5

Macht and Kunkel [6] in 1920 and Dyche-Teague [7] in 1924 describe the antimicrobial effects of volatile oils or their vapors. The latter reported on the phenol coefficient of alcoholic perfumes on a mixed culture of nasal bacteria, *Micrococcus catarrhalis*. In the same year Bryant [8] also reported on the activity of alcoholic perfumes. His test organism was a pure culture of *Escherichia coli communis*.

Rideal and co-workers [9] published an extensive list of the Rideal-Walker coefficients of essential oils about the time that Dyson [10] reviewed the physiological aspects of a long list of essential oils.

One must not overlook the important contributions of Gattefossé on the antimicrobial activity of essential oils, especially lavender, that continued over a period of many years, part of which were summarized in his book *Aromatherapie* [11] and in another French publication [12] covering the years 1680-1933.

TABLE 1 Inhibitory Activity of Oils

Volatile oils	Zone of inhibition (mm)[a]					
	S. typhose	M. citreus	P. morgani	B. brevis	M. albus	Sum[b]
Amber, rectified	0[c]	2	0	2	2	6
Anise	0	3	0	4	2	9
Bay	0	7	3	3	3	16
Bergarmot	4	7	0	6	3	20
Cajuput	0	10	0	11	5	26
Caraway	0	10	0	5	3	18
Cedar wood	0	0	0	0	0	0
Chenopodium	0	5	3	3	3	14
Cinnamon	0	11	8	12	10	41
Clove	0	3	0	3	2	8
Cubeb	0	0	0	2	0	2
Dwarf pine needle	3	3	5	3	3	17
Eucalyptus	5	10	10	12	8	45
Geranium Algerian	2	12	2	5	2	23
Hemlock	1	0	2	0	0	3
Juniper berries, rectified Dutch	5	2	9	2	3	21
Juniper tar	0	10	0	2	3	15
Lemon	0	0	0	13	0	13
Myrrh	0	0	0	0	0	0
Neroli	5	3	4	10	3	25
Niaouli	0	10	6	10	3	29
Origanum red	0	15	0	15	10	40
Peppermint	5	0	0	5	2	12
Petigrain, Paraguay	3	0	3	4	2	12
Pimento	0	10	0	4	3	17
Pinus sylvestris	8	5	0	6	5	24
Rectified tar	0	10	0	0	6	16
Rosemary	5	7	0	5	6	23
Sandalwood	0	3	0	2	2	7
Sassafras	0	3	0	3	2	8
Spike lavender	0	5	0	5	3	13
Sweet birch	0	0	0	2	0	2
Sweet orange	0	0	0	2	0	2
Turpentine	2	3	8	4	3	20
Wintergreen	0	5	0	3	0	8

[a]Measurement from disk edge to zone edge.
[b]Sum of zones of inhibition for each oil.
[c]Zone of inhibition absent.
Source: Ref. 18.

TABLE 2 Inhibitory Activity of Combinations of Oils

Oil combinations	Zone of inhibition (mm)[a]					
	S. typhose	M. citreus	P. morgani	B. brevis	M. albus	Sum[b]
Volatile (1:1)						
Anise-clove	0[c]	3	0	4	2	9
Anise-hemlock	0	3	0	4	2	9
Chenopodium-cedar wood	0	10	0	14	7	31
Cinnamon-clove	0	10	0	10	10	30
Cinnamon-cubeb	0	10	0	13	10	33
Cinnamon-dwarf pine needle	0	15	0	15	5	35
Cinnamon-juniper berries	0	10	0	15	5	30
Cinnamon-niaouli	0	10	0	11	10	31
Cinnamon-*Pinus sylvestris*	0	10	0	12	3	25
Cinnamon-sandalwood	0	10	0	12	5	27
Cinnamon-sweet birch	0	7	0	10	3	20
Cinnamon-turpentine	0	10	0	15	14	39
Cubeb-cedar wood	0	0	0	1	0	1
Cubeb-dwarf pine needle	3	2	4	3	2	14
Cubeb-juniper berries	3	2	5	2	3	15
Cubeb-*Pinus sylvestris*	2	3	0	2	1	8
Dwarf pine needle-cedar wood	3	2	4	2	2	13
Eucalyptus-anise	0	5	0	5	5	15
Eucalyptus-cajuput	0	4	0	12	5	21
Eucalyptus-cedar wood	3	5	0	5	5	18
Eucalyptus-chenopodium	0	9	0	8	6	23
Eucalyptus-cinnamon	0	12	0	15	15	42
Eucalyptus-clove	0	5	0	5	5	15
Eucalyptus-cubeb	5	5	4	5	6	25
Eucalyptus-dwarf pine needle	8	10	0	8	6	32
Eucalyptus-juniper berries	8	9	0	10	5	32
Eucalyptus-myrrh	3	5	0	5	3	16
Eucalyptus-niaouli	0	8	0	12	5	25
Eucalyptus-*Pinus sylvestris*	7	10	0	15	3	35

TABLE 2 Continued

Oil combinations	Zone of inhibition (mm)[a]					
	S. typhose	M. citreus	P. morgani	B. brevis	M. albus	Sum[b]
Volatile (1:1)						
Eucalyptus-rosemary	9	8	0	10	9	36
Eucalyptus-sandalwood	3	5	5	4	5	22
Eucalyptus-sweet orange	7	5	0	6	4	22
Eucalyptus-wintergreen	0	5	0	5	5	15
Juniper berries-cedar wood	2	0	3	2	2	9
Juniper berries-dwarf pine needle	2	2	7	2	2	15
Niaouli-clove	0	5	0	4	3	12
Niaouli-cubeb	0	2	3	3	1	9
Niaouli-*Pinus sylvestris*	0	8	0	12	7	27
Niaouli-rectified tar	0	6	0	6	4	16
Niaouli-sweet birch	0	4	0	4	3	11
Niaouli-wintergreen	0	3	0	4	3	10
Peppermint-pimento	0	4	0	3	3	10
Turpentine-clove	3	5	0	4	4	16
Turpentine-hemlock	0	0	4	2	3	9
Turpentine-juniper berries	3	2	9	2	3	19
Turpentine-origanum red	0	10	0	0	8	18
Turpentine-wintergreen	0	5	0	4	3	12
Volatile (1:1:1)						
Cedar wood-cubeb-bay	0	3	2	3	2	10
Eucalyptus-chenopodium-anise	0	5	0	3	4	12
Eucalyptus-chenopodium-cedar wood	4	6	0	4	5	19
Eucalyptus-chenopodium-hemlock	5	6	0	5	4	20
Eucalyptus-chenopodium-peppermint	3	5	0	4	3	15

TABLE 2 Continued

Oil combinations	Zone of inhibition (mm)[a]					
	S. typhose	M. citreus	P. morgani	B. brevis	M. albus	Sum[b]
Volatile (1:1:1)						
Eucalyptus-cinnamon-clove	0	4	3	4	3	14
Eucalyptus-cinnamon-dwarf pine needle	8	11	0	15	18	51
Eucalyptus-cinnamon-juniper berries	10	12	0	17	19	58
Eucalyptus-cinnamon-niaouli	7	14	0	15	15	51
Eucalyptus-cinnamon-rosemary	0	5	3	6	4	18
Sassafras-origanum red-chenopodium	0	5	3	3	3	14
Turpentine-Algerian geranium-rosemary	0	5	3	6	5	19
Turpentine-neroli-Algerian geranium	0	2	2	3	5	12
Wintergreen-sweet birch-lemon	0	2	0	2	0	4
Wintergreen-sweet orange-sassafras	0	4	0	3	0	7
Volatile and fixed (1:1)						
Castor-cedar wood	0	0	0	0	0	0
Castor-eucalyptus	0	5	0	4	5	14
Castor-origanum red	0	2	0	3	0	5
Castor-turpentine	0	0	0	1	0	1
Cod liver-cedar wood	0	0	0	0	0	0
Cod liver-eucalyptus	0	5	0	4	7	16
Cod liver-origanum red	0	3	0	4	3	10
Cod liver-turpentine	0	0	0	2	2	4
Olive-rectified amber	0	0	0	2	0	2
Oive-bay	0	2	0	3	3	8
Olive-bergamot	0	1	0	2	2	5

TABLE 2 Continued

	Zone of inhibition (mm)[a]					
Oil combinations	S. typhose	M. citreus	P. morgani	B. brevis	M. albus	Sum[b]
Volatile and fixed (1:1)						
Olive-cajuput	0	5	0	4	5	14
Olive-caraway	0	0	0	2	0	2
Olive-cedar wood	0	0	0	0	0	0
Olive-chenopodium	0	2	0	3	0	5
Olive-clove	0	3	0	2	2	7
Olive-eucalyptus	0	5	0	4	8	17
Olive-juniper tar	0	2	0	2	2	6
Olive-origanum red	0	4	0	5	2	11
Olive-peppermint	0	0	2	2	0	4
Olive-sassafras	0	0	0	1	0	1
Olive-turpentine	0	2	4	1	0	7
Olive-wintergreen	0	0	0	0	0	0
Vitamin K_1-eucalyptus	0	7	10	4	6	27
White mineral-cedar wood	0	0	0	0	0	0
White mineral-eucalyptus	0	7	0	5	7	19
White mineral-origanum red	0	7	0	8	0	15
White mineral-turpentine	0	2	6	2	3	13
Volatile and infused (1:1)						
Asafetida-eucalyptus	0	3	3	3	5	14
Burdock-eucalyptus	0	3	5	2	5	15
Henbane-eucalyptus	0	6	5	5	5	21
Lobelia-eucalyptus	0	5	7	5	5	22
Mullein-eucalyptus	0	6	3	3	5	17

[a]Measurement from disk edge to zone edge.
[b]Sum of zones of inhibition for each oil.
[c]Zone onf inhibition absent.
Source: Ref. 18.

    Naves [13] reported that the U.S. Food and Drug Administration phenol coefficient test method was better than that of the French Hygienic Laboratory when applied to essential oils.
    Potter [14] found spice oils often to be more effective antiseptics than some well-known aromatic compounds. McLachlen [15] made reference to

essential oils as antiseptics. Rolle and Mayr [16] reported that conifer oils were useful as disinfectants against *Mycobacterium tuberculosis*, *Escherichia coli*, and *Staphylococcus aureus*.

The most extensive studies were reported by Maruzzella et al. [17-23]. This group of workers published some 18 papers over the period of several years on the broad subject of essential oils and fragrances as antimicrobial agents. Maruzzella and Henry [18] presented data on 35 volatile oils, 5 fixed oils, 5 infused oils, and 95 combinations. The filter paper disk method was used. The five organisms used were *Salmonella typhosa*, *Micrococcus citreus*, *Proteus morganii*, *Bacillus brevis*, *Micrococcus pyogenes*, *Var albus*. Of the 35 oils tested, 33 were found to have some antibacterial activity on at least one of the organisms. Details on the inhibitory activity of oils singly and in combination are given in Tables 1 and 2.

In 1960 Maruzzella also studied the action of essential oil vapors on bacterial and fungal growth. He reported the results as follows:

> Early studies and more recent observations have clearly demonstrated that certain essential oils possess markedly antibacterial vapors. When one considers the wide use of essential oils in perfumes, cosmetics and pharmaceutical preparations, generally added to enhance the ornamental value of the preparation or to mask unpleasant odors, little attention is given to their germicidal properties. In many instances the selection of an essential oil with the desired scent and demonstrated antimicrobial activity would certainly give the product some germicidal properties. Since essential oils are generally associated with the vaporous state, studies of the antibacterial-antifungal properties of the vapors would assist in the selection of oils with germicidal properties.

The essential oil vapors were tested in the following manner;* approximately 15 ml of Sabouraud's maltose agar were poured into Petri dishes and allowed to harden. The surface of the agar was then streaked with 0.5 ml of a 5-day-old broth culture of the test organism. Aluminum caps (20 mm in diameter and 5 mm deep) were placed on the inner surface of the Petri dish top. Each cap contained 0.5 ml of the essential oil. When the dishes were inverted and incubated, the surface of the oil in the cap was about 5 mm from the surface growth of the test organism. The essential oil vapors were allowed to emanate for the 5-day incubation period at normal room temperature. After incubation the presence of a definite clear zone of inhibition of growth on the surface of the agar was taken as an indication that the vapor possessed antibacterial-antifungal activity. The diameters of the zones of inhibition were measured with a metric ruler and with the aid of an illuminated Quebec colony counter. All tests were run in triplicate, with one cap in each dish. When all growth was inhibited, the zone of inhibition was recorded as 90 mm (the inside diameter of the Petri dish). Of the 136 essential oil vapors tested, 114 were found to have some antifungal activity against at least one of the five test organisms. Of the active 114 oil vapors, 84 produced zones of inhibition against all the test organisms [20].

In 1972 Munzing and Schels added their test results for the potential replacement of preservatives in cosmetics by essential oils [24]. However, they pointed out that the inhibitory activity of the oils was on the order of

---

*Editor's Note: This method is no longer considered useful. Its inclusion is from a historical reference point only.

TABLE 3 Effectiveness of Aroma Chemicals in a Cosmetic Oil-Water Lotion[a]

	Staphylococcus aureus		Escherichia coli		Proteus vulgaris		Pseudomonas aeruginosa	
	1:500	1:1000	1:500	1:1000	1:500	1:1000	1:500	1:1000
Citral	0	<30	0	0	0	0	R	R
Orange oil	<30	>300	R	R	>300	R	R	R
Lemon oil	0	>300	R	R	R	R	R	R
Bergamot oil	>300	>300	R	R	>300	R	R	R
Linalool	0	204	0	0	R	R	R	R
Spiköl	0	R	0	R	R	R	R	R
Geraniol	0	0	0	0	R	R	R	R
Nerol	0	0	0	0	R	R	R	R
Phenylethyl alcohol	R	R	R	R	R	R	R	R
Eugenol	0	<30	0	R	R	R	>300	R
Zimtaldehyd	0	0	0	0	0	0	0	R
Vetiveröl Bourbon	<30	<30	0	0	0	0	0	0
Sandelholzöl ostind.	<30	<30	R	R	R	R	R	R
Cedrol	0	0	R	R	R	R	R	R
Abs. M. d. Chêne Daksa	0	0	R	R	R	R	R	R
Benzyl acetate	194	R	0	R	0	R	R	R
α-Amylzimtaldehyd	220	>300	R	R	R	R	>300	R
Hydroxycitronellal	>300	R	R	R	>300	R	R	R
Linallyformiat	<30	>300	R	R	R	R	R	R
Linalyl acetate	R	R	R	R	R	R	R	R
Aldehyde C 12 MNA	0	0	R	R	R	R	R	R
Decylalcohol	0	0	R	R	R	R	R	R
Dimethylphthalat	R	R	R	R	R	R	R	R
Tonalide	R	R	R	R	R	R	R	R
Cumarin	178	R	50	R	R	R	R	R
Heliotropin	0	<30	>300	R	R	R	R	R
Benzyl benzoate	R	R	R	R	R	R	R	R

TABLE 3 Continued

	Staphylococcus aureus		Escherichia coli		Proteus vulgaris		Pseudomonas aeruginosa	
	1:500	1:1000	1:500	1:1000	1:500	1:1000	1:500	1:1000
Lilial	<30	<30	R	R	R	R	R	R
Terpineol	0	<30	0	0	0	R	R	R
Thymol	0	0	0	0	0	0	30-300	R
Eucalyptus oil	36	R	R	R	R	R	R	R
Peppermint oil	0	<30	0	R	0	0	R	R
Rosmarinöl	>300	R	R	R	R	R	R	R
Salbeiöl dalmat.	<30	>300	R	R	R	R	R	R
Galbanumöl	52	R	R	R	R	R	R	R
Methylsalicylate	>300	>300	0	0	0	0	R	R
Phenol	200	—	>300	—	>300	—	200	—

[a] 0 = no growth, <30 = <30 cal, >300 = >300 cal, R = complete overgrowth as in control.
Source: Ref. 24.

magnitude of grams per liter rather than milligrams per liter (see table 3).

Richards and McBride reported, in 1973, that 3-phenylpropan-1-ol, 2-phenylethanol, and benzyl alcohol were investigated for their inhibitory action against *Pseudomonas aeruginosa* [25]. 3-Phenylpropan-1-ol was the most effective and benzyl alcohol was the least effective as shown by (1) growth rate studies using subinhibitory concentrations of the alcohols, (2) determination of minimum inhibitory concentrations, and (3) determination of sterilization times. The three compounds enhance the bactericidal action of benzalkonium chloride against *P. aeruginosa* in the same ranking order. It was suggested that 3-phenylpropan-1-ol may be a suitable preservative for oral suspensions and mixtures [24]. Aliphatic alcohols also were shown by several authors to be effective antimicrobial agents [26-28].

One of the more ambitious programs to screen aroma chemicals was carried out in 1979 by Morris [29]. The study was designed to provide quantitative data. In this paper the results from a three-stage study was provided. First materials were screened in a standardized Petri plate procedure. Second, active compounds found by the first test were tested for minimum inhibitory concentration (MIC) values in a liquid culture. Third, the most active fragrance material was incorporated into a soap and then tested in handwashing panel studies.

Table 4 gives the full results of their ambitious study while in Table 5 the microbiological activity of the 521 fragrance raw materials is summarized. The screen identified 309 materials with activity against at least one of four organisms used in the screen; 212 were found to be active against a diphtheroid organism isolated from human skin. Again, it must be emphasized that the size of the zone of inhibition on the Petri plate is a poor indicator of the relative antimicrobial activity of the test material. The size of the cleared zone by this method is dependent on the solubility and rate of diffusion of the sample in the aqueous medium.

The above data held some surprises, since previous work had indicated, contrary to these findings, that the gram-negative organisms were less sensitive than gram-positive ones [18,30]. Also the results against the diphtheroid organisms was probably affected by the inclusion of Tween 80 in the growth medium. Tween 80 is known in the state of the art to inhibit the effect of lipophilic germicides.

The inclusion of the more active ingredients into a soap useful for skin degerming was also investigated [29]. The composition of the soap fragrances and their MIC values are presented in Table 6. The test as carried out had some technical limitations and the reader is encouraged to refer to the original paper [29]. Because of these problems, no skin-degerming effect was measured. The authors suggested that perhaps a better test for the usefulness of aroma chemicals and essential oils as active ingredients would be in cosmetic products which stay on the skin or where the active antimicrobial materials are used as preservatives rather than "active material."

One aspect of aroma chemicals that has received much attention is the use of fragrances which not only have odorizing properties but which are also antimicrobial. An example of this application is given by a provocative article by Sturm [31]. Deosafe fragrances were formulated by the deliberate combination of aromatic chemicals and antimicrobial essential oils. Sturm used as examples the inhibition of *Staphylococcus* by decyl alcohol, citral, eugenol, and cinnamaldehyde as well as by thyme, clove and spike oil.

TABLE 4 Fragrance Materials with Antimicrobial Activity

Chemicals[a]	S. aureus Zone diameter (mm)	S. aureus MIC (ppm)	E. coli Zone diameter (mm)	E. coli MIC (ppm)	Candida albicans Zone diameter (mm)	Candida albicans MIC (ppm)	Diphtheroid Zone diameter (mm)	Diphtheroid MIC (ppm)
Acetanilide	0[b]	NT[c]	13	NT	11	NT	NT	NT
N-Acetyl methyl anthranilate	0	>1000	0	>1000	0	>1000	0	>1000
Aldehyde, C$_8$	0	500	12	500	12	500	12p[d]	1000
Aldehyde, C$_9$	0	NT	0	NT	13	NT	NT	NT
Aldehyde, C$_{11}$ Undecylenic	0	50	0	>1000	22	50	17	500
Aldehyde, C$_{16}$	0	NT	15	NT	0	NT	NT	NT
Aldehyde, C$_{18}$	0	>1000	15	1000	20	500	15	1000
Allyl amyl glycolate	0	NT	0	NT	11P	NT	NT	NT
Amaryllide	14	>1000	13	1000	19	1000	17	1000
Ambrarome, absolute	13	NT	0	NT	0	NT	NT	NT
Amyl cinnamic aldehyde coeur	0	>1000	0	>1000	0	>1000	0	500
Amyl cinnamyl alcohol	0	50	0	>1000	0	500	12	500
Amyl salicylate	0	>1000	0	>1000	0	>1000	0	500
Amyris oil	13	NT	0	NT	0	NT	NT	NT
Anethol, USP	0	500	0	500	0	100	0	500
Anisyl acetate	0	>1000	13	>1000	12P	1000	0	1000
Armoise essence	0	1000	0	>1000	0	1000	0	>1000
Arras aldehyde, 50%	20	500	14	1000	17	500	26	500
Aubepine	0	>1000	18	>1000	11	1000	0	1000
Auralva (Schiff base)	11	NT	12	NT	12	NT	NT	NT
Balsam copaiba, USP	0	>1000	0	>1000	0	>1000	0	>1000
Balsam Peru oil	18	>1000	15	>1000	11	>1000	14	500

Aroma Preservatives

Basil oil	0	500	0	500	0	500	0	1000
Bay oil	18	500	13	1000	20	500	14	1000
Benzaldehyde	0	1000	0	1000	0	1000	0	>1000
Benzoic acid	21	1000	28	1000	21	1000	28	1000
Benzoin coeur	18	>1000	16	1000	10	>1000	12	1000
Benzophenone	0	NT	0	NT	12	NP	NT	NT
Benzyl acetate	0	>1000	0	>1000	0	1000	0	1000
Benzyl alcohol	0	>1000	12P	>1000	0	>1000	0	500
Benzyl benzoate	0	>1000	0	>1000	0	>1000	0	500
Benzyl propionate	0	>1000	0	1000	0	500	0	1000
Benzyl salicylate	0	>1000	0	>1000	0	>1000	0	1000
Bergamot MPF	0	1000	0	>1000	0	500	0	1000
Beta gamma hexenyl formate	13	NT	17	NT	0	NT	NT	NT
Beta naphthyl anthranilate	16	1000	20	1000	22	500	15	1000
Beta pinene coeur	0	500	0	>1000	0	1000	0	>1000
Birch tar, rectified	14	500	14	>1000	17	1000	14	500
Boise de rose filtered	0	>1000	17	1000	0	500	0	1000
Borneol	0	1000	0	1000	0	500	12	500
Camphene 46	0	1000	0	>1000	0	1000	0	>1000
Camphor oil white	12	500	*e	>1000	0	500	0	>1000
Caraway oil	0	500	0	>1000	0	500	12	500
Cardamom oil, Guatemala	0	>1000	0	>1000	0	500	0	500
Carvone, dextro	10	1000	11	1000	0	500	0	1000
Carvone, laevo	0	>1000	11	1000	11	500	0	1000
Cashmeran	0	500	0	1000	0	>1000	14	500
Castoreum, absolute, 50%	11	NT	0	NT	0	NT	NT	NT
Cedarleaf oil	0	1000	0	1000	0	500	0	>1000
Cedarwood, white	0	1000	0	>1000	0	>1000	0	500
Cedrone S.	12	NT	0	NT	0	NT	NT	NT
Cedrus Atlantica coeur	11	500	0	>1000	0	>1000	11	1000
Celery seed oil	0	NT	0	NT	12	NT	NT	NT

TABLE 4 Continued

Chemicals[a]	S. aureus		E. coli		Candida albicans		Diphtheroid	
	Zone diameter (mm)	MIC (ppm)	Zone diameter (mm)	MIC (ppm)	Zone diameter (mm)	MIC (ppm)	Zone diameter (mm)	MIC (ppm)
Chamomile oil	0	1000	0	>1000	0	500	0	>1000
Cinnamalva	0	>1000	19	1000	13	500	0	1000
Cinnamic alcohol	14	>1000	19	>1000	27	500	16	1000
Cinnamon leaf oil, Ceylon	18	500	17	1000	14	500	12	500
Ciste, colorless	12	NT	0	NT	0	NT	NT	NT
Citral, dimethyl acetal	0	500	0	>1000	33	500	14	500
Citral, refined	15	500	14	500	46	500	16	500
Citronama (Schiff base)	0	NT	0	NT	11	NT	NT	NT
Citronella, Formosa, Java	11	NT	0	NT	17	NT	NT	NT
Citronellal	0	500	0	>1000	12	500	0	500
Citronellol coeur	12	1000	10	1000	48	100	18	500
Citronellyl acetate	0	>1000	0	>1000	0	>1000	0	500
Citronellyl ethyl ether	11	NT	0	NT	0	NT	NT	NT
Citronellyl isobutyrate	0	>1000	0	>1000	0	>1000	0	>1000
Citrus oil, distilled	0	1000	0	>1000	0	500	0	>1000
Clove leaf oil	14	500	19	1000	19	500	15	500
Clove bud oil	16	500	16	1000	18	500	14	500
Cocal	15	500	14	>1000	11	1000	15	500
Coniferan	0	1000	0	>1000	0	>1000	0	500
Coriander oil	0	1000	11	1000	0	500	0	1000
Corn mint oil	10	NT	0	NT	0	NT	NT	NT
Coronal, beta	12	NT	0	NT	0	NT	NT	NT

Aroma Preservatives

Cortex aldehyde, 50%	17	1000	21	500	16	>1000	500
Coumarin	13	>1000	15	>1000	16	1000	1000
Cumin oil	0	500	0	1000	0	500	1000
Cuminyl alcohol	15P	1000	14	500	23	500	1000
Cuminyl acetate	0	>1000	*	>1000	0	100	500
Cyclamal extra	0	NT	0	NT	23P	NT	NT
2-Cyclohexyl cyclohexanone	0	NT	0	NT	10	NT	NT
Cyclosia base	14	>1000	16	>1000	14	>1000	1000
Cymene coeur	0	1000	0	>1000	0	500	>1000
Cypress oil, French	0	1000	0	>1000	0	1000	>1000
Decalactone	15	>1000	0	>1000	12	1000	1000
n-Decanol	20	NT	0	NT	13	NT	NT
Dibenzyl ether	0	>1000	0	>1000	0	>1000	500
Dibutyl sulfide, 10%	0	NT	16P	NT	0	NT	NT
Diethyl phthalate	0	>1000	0	>1000	14P	500	1000
Dihydro cuminyl alcohol	14	1000	16	500	23	500	500
Dimethyl anthranilate	0	1000	0	1000	14P	500	500
Dimethyl benzyl carbinol	0	>1000	16P	>1000	10	1000	>1000
Dimethyl octanol	0	NT	0	NT	12	NT	NT
Dimethyl phenyl acetaldehyde	14	NT	0	NT	0	NT	NT
Dimethyl phenyl ethyl carbinol	0	>1000	14	1000	21	1000	1000
Dimethyl phthalate	0	NT	11	NT	12	NT	NT
Dimethyl sulfide	0	1000	0	>1000	0	500	1000
Diphenyl oxide	0	>1000	0	>1000	0	500	500
Dipropylene glycol	0	>1000	0	>1000	0	>1000	>1000
p-Ethyl acetophenone	0	NT	0	NT	13P	NT	NT
Ethyl benzaldehyde	0	500	11	500	11	500	500
Ethyl benzoate	0	1000	0	1000	0	500	1000
Ethyl-3-hydroxy-3-phenyl propionate	0	NT	13	NT	12	NT	NT

TABLE 4 Continued

Chemicals[a]	S. aureus Zone diameter (mm)	MIC (ppm)	E. coli Zone diameter (mm)	MIC (ppm)	Candida albicans Zone diameter (mm)	MIC (ppm)	Diphtheroid Zone diameter (mm)	MIC (ppm)
Ethyl linalool	18	500	14P	1000	11	500	20	500
Ethyl methacrylate	0	>1000	0	>1000	0	>1000	0	>1000
Ethyl phenyl glycidate	0	NT	0	NT	11	NT	NT	NT
Ethyl vanillin	14	>1000	18	1000	19	1000	15	1000
Eucalyptus oil, 70-75%	0	>1000	0	>1000	0	1000	0	>1000
Eugenol, USP	16	500	21	500	22	500	15	500
Fennel oil, sweet	0	500	0	500	0	500	0	500
Fir balsam, absolute	12	500	15P	1000	0	>1000	0	1000
Fraistone	0	NT	0	NT	26P	NT	NT	NT
Furfural	12	>1000	11	>1000	0	1000	0	>1000
Galaxolide	0	1000	0	>1000	0	>1000	0	500
Galbanum coeur	0	1000	0	>1000	0	>1000	0	500
Geraniol coeur	14	1000	16	500	38	500	20	500
Geraniolene, light	13	NT	0	NT	11P	NT	NT	NT
Geranium, African	12	NT	0	NT	19	NT	NT	NT
Geranoxy acetaldehyde, 50%	0	NT	0	NT	13	NT	NT	NT
Geranyl benzoate	0	>1000	0	>1000	0	>1000	0	>1000
Geranyl methyl tiglate	0	NT	0	NT	14	NT	NT	NT
Geranyl propionate	0	>1000	0	>1000	0	>1000	0	500
Grapefruit oil	0	500	0	>1000	0	500	0	>1000
Guaiene	0	500	0	>1000	0	>1000	12	500
Guaiewood oil	13	500	0	>1000	0	>1000	0	500
Hay, absolute	12	NT	0	NT	0	NT	NT	NT
Hedione	0	>1000	0	>1000	0	1000	12	500

Aroma Preservatives

Helional	0	500	0	1000	14	500	16	500
Heliotropyl acetate	0	NT	14	NT	12P	NT	NT	NT
n-Hexanol	0	NT	13P	NT	0	NT	NT	NT
Hexyl cinnamic aldehyde	0	>1000	11	>1000	0	>1000	0	500
Hydratropal acetone	18	NT	0	NT	21	NT	NT	NT
Hydratropic alcohol, white	12	>1000	14	1000	13	1000	11	>1000
Hydroxy citronellal dimethyl acetal	0	NT	0	NT	11	NT	NT	NT
Hydroxy citronellal	20	>1000	16	>1000	13	1000	14	>1000
Hyssop oil	11	NT	0	NT	0	NT	NT	NT
Indisan	12	500	0	>1000	0	>1000	10	500
Indole	18	1000	23	500	17	500	22	500
Iralia	11	NT	0	NT	0	NT	NT	NT
Iritone	10	NT	0	NT	0	NT	NT	NT
Isoamyl pentenoate	12	NT	0	NT	0	NT	NT	NT
Iso beta gamma hexenyl acetate	18	>1000	18	1000	12	1000	12	>1000
Isoborneol	0	>1000	0	500	0	500	0	1000
Isobutyl benzyl carbinol	11	500	12	1000	15	500	14	500
Isobutyl cinnamate	11	NT	0	NT	0	NT	NT	NT
Isobutyl furyl propionate	0	NT	0	NT	10	NT	NT	NT
Isobutyl quinoline	29	100	13	>1000	26P	50	17	100
Isocitral	19	500	11	1000	20	500	18	100
Isoeugenol	23	500	18	500	14	500	14	500
Isoeugenyl benzoate	0	>1000	0	>1000	0	>1000	0	>1000
Isojasmone	12	NT	0	NT	12	NT	NT	NT
Isomuguet aldehyde, 50%	25	NT	0	NT	11	NT	NT	NT
Isopropyl cyclohexyl propanol	15	NT	0	NT	18	NT	NT	NT
Isopropyl quinoline	29	500	13	500	16	100	20	500
Isopulegol M Extra	10	1000	12	1000	0	1000	0	1000
Jasmonate	0	NT	0	NT	11	NT	NT	NT

TABLE 4 Continued

Chemicals[a]	S. aureus		E. coli		Candida albicans		Diphtheroid	
	Zone diameter (mm)	MIC (ppm)	Zone diameter (mm)	MIC (ppm)	Zone diameter (mm)	MIC (ppm)	Zone diameter (mm)	MIC (ppm)
Jasmone, cis	19	1000	13	1000	21	500	15	500
Jasmutone	11	500	12	>1000	13	500	14	500
Labdanax	0	NT	13	NT	12	NT	NT	NT
Labdanol	11	NT	0	NT	0	NT	NT	NT
Labdanum resin, absolute	14	500	0	>1000	0	>1000	12	1000
Lactone HB	17	1000	12	500	16	500	14	500
Lauryl alcohol	0	1000	0	>1000	0	>1000	10	500
Laurine extra	17	>1000	12	>1000	13	1000	0	1000
Lavandin abrialis	0	NT	12	NT	0	NT	NT	NT
Lavandulol	11	1000	12	1000	13	500	0	500
Lavender, absolute, Camilli	0	>1000	0	>1000	0	1000	0	1000
Lemma (Schiff base)	18	100	11	>1000	12	500	12	500
Lemon oil, Calif.	0	500	0	>1000	0	500	0	>1000
Lemongrass	13	500	14	500	15	500	17	500
Lime oil, washed	0	1000	0	>1000	0	500	0	1000
Limonene	0	>1000	0	>1000	0	500	0	>1000
Linalool oxide	0	NT	10	NT	0	NT	NT	NT
Linalool	0	1000	18	1000	0	500	0	500
Linalyl acetate	0	>1000	0	>1000	0	>1000	0	500
Linalyl cinnamate	0	>1000	0	>1000	0	>1000	0	>1000
Lovage oil	14	NT	0	NT	14	NT	NT	NT
LRG No. 182 (ethoxycyclohexanone)	11	NT	0	NT	0	NT	NT	NT
LRG No. 1181 (neo-,isomenthones)	13	NT	12	NT	0	NT	NT	NT

Aroma Preservatives

Lyral	16	1000	13	>1000	12	1000	14	1000
Mace, whole extract	12	>1000	0	>1000	0	>1000	0	500
Maltol	0	>1000	16	>1000	16	1000	14	>1000
Mandarin oil	0	1000	0	>1000	0	500	0	>1000
Menthol, USP	11	500	11	>1000	10	500	0	500
Methallyl pentenoate	14	NT	0	NT	0	NT	NT	NT
p-Methoxy hydrotropic aldehyde	18	1000	26	1000	15	500	16	500
Methyl anthranilate	12	>1000	14	>1000	16	1000	0	1000
p-Methoxy phenoxy acetaldehyde	18	1000	20	>1000	16	1000	19	500
Methyl benzoate	0	>1000	0	>1000	0	1000	0	1000
p-Methoxy phenoxy acetaldehyde dimethyl acetal	0	>1000	0	>1000	11	1000	0	1000
Methyl cinnamate	11	1000	12	>1000	13	500	0	500
α-Methyl cinnamic aldehyde	20	1000	17	500	18	500	0	500
Methyl cyclocitrone	10	NT	0	NT	0	NT	NT	NT
Methyl eugenol	12	1000	11	1000	16	500	12	500
Methyl heptenol	12	NT	13	NT	0	NT	NT	NT
Methyl hexyl acetaldehyde	34	>1000	22	>1000	15	1000	16	500
Methyl isoeugenol	0	>1000	0	>1000	0	>1000	0	100
Methyl lavender ketone	13	100	13	1000	18	500	15	500
Methyl β-naphthyl ketone	0	NT	0	NT	15	NT	NT	NT
Methyl octin carbonate	0	>1000	0	>1000	0	500	0	500
Methyl octyl acetaldehyde	18	NT	0	NT	0	NT	NT	NT
2-Methyl-2-pentenoic acid	19P	1000	35	1000	18	1000	25	1000
Methyl-p-cresol	0	>1000	0	1000	0	500	0	>1000
Methyl phenyl ethyl alcohol	13	>1000	19	>1000	16	1000	0	>1000

TABLE 4 Continued

Chemicals[a]	S. aureus		E. coli		Candida albicans		Diphtheroid	
	Zone diameter (mm)	MIC (ppm)	Zone diameter (mm)	MIC (ppm)	Zone diameter (mm)	MIC (ppm)	Zone diameter (mm)	MIC (ppm)
Methyl p-toluate	0	1000	0	1000	0	500	0	1000
Miel Blac, Delaire	11	NT	17P	NT	0	NT	NT	NT
Mousse, absolute verte								
Maroc	21	50	15	1000	18	500	18	500
Muguet aldehyde	40	NT	0	NT	14	NT	NT	NT
Muscagene	12	NT	0	NT	0	NT	NT	NT
Musk ambrette	0	>1000	0	>1000	0	>1000	13P	500
Musk ketone	0	>1000	0	>1000	0	>1000	12P	100
Musk xylol	0	>1000	0	>1000	0	>1000	11P	50
Myrac aldehyde	17	NT	0	NT	33	NT	NT	NT
Myrrh coeur	13	1000	0	>1000	0	>1000	12	1000
Myrtenal	13	NT	10	NT	0	NT	NT	NT
Myrtle oil, Charabot	12	NT	11	NT	0	NT	NT	NT
Naame (Schiff base)	12	NT	0	NT	13	NT	NT	NT
Narcisse ketone	0	NT	0	NT	13	NT	NT	NT
Narcitol	14	NT	18	NT	41	NT	NT	NT
Nerol	14	500	13	1000	0	500	12	500
Nerolidol	15	NT	0	NT	0	NT	NT	NT
Neroly blanc	13	NT	10	NT	17	>1000	NT	NT
Nortonkalactone	29	>1000	13	>1000	0	500	0	>1000
Nutmeg oil	0	500	0	>1000	12	1000	0	1000
Oakmoss essence	22	50	0	>1000	13	500	14	500
Ocimene	22	500	14	1000	13	50	16	500
Ocmea (Schiff base)	13	50	10	500	0	NT	13	100
Opoponax oil	10	NT	0	NT			NT	NT

Aroma Preservatives

Orange oil, Fla.	0	500	20	>1000	0	500	500
Orange terpeneless, absolute	11	NT	0	NT	0	NT	NT
Orange terpenes	12	NT	16	NT	0	NT	NT
Orenyle	14	NT	0	NT	0	NT	NT
Origanum oil, Spanish	33	500	24	500	13	500	500
Orivone	11	NT	0	NT	39	NT	NT
Oxyphenylon	0	>1000	12	>1000	13	>1000	>1000
para-Cresol	20	>1000	29	1000	16	1000	1000
para-Cresyl acetate coeur	0	>1000	0	>1000	12P	1000	>1000
para-Isopropyl hydratropic aldehyde	13	NT	0	NT	12	NT	NT
para-Methyl benzyl acetate	11	NT	0	NT	0	NT	NT
para-Methyl dimethyl benzyl carbinol	14	>1000	14	1000	13	1000	1000
para-tert-Butyl cycohexanone	11	1000	0	>1000	27P	500	500
para-tert-Butyl-meta-cresol	85	50	26	500	85P	50	100
para-Tolyl alcohol	12	>1000	16	>1000	17	>1000	>1000
Patchouli oil, dark	12	100	0	>1000	0	>1000	500
Patchouli oil, light	0	500	0	>1000	0	>1000	500
Peach aldehyde coeur	12	NT	0	NT	14	NT	NT
Pepper oil, black	0	1000	0	>1000	0	>1000	>1000
Peppermint	0	NT	0	NT	10	NT	NT
Persicol (γ-undecalactone)	0	NT	0	NT	11	NT	NT
Petinerol	0	1000	12	1000	14	500	1000
Petitgrain S.A.	0	>1000	0	>1000	0	500	1000
Petitgrain terpenes	14	500	11	>1000	10	500	>1000
Phellandrene	18	NT	18	NT	0	NT	NT

TABLE 4 Continued

Chemicals[a]	S. aureus		E. coli		Candida albicans		Diphtheroid	
	Zone diameter (mm)	MIC (ppm)	Zone diameter (mm)	MIC (ppm)	Zone diameter (mm)	MIC (ppm)	Zone diameter (mm)	MIC (ppm)
Phenoxyl ethyl propionate	0	>1000	0	>1000	12	500	0	1000
Phenyl acetaldehyde	21	100	40	1000	33	500	18	100
Phenyl ethyl acetate	0	NT	25P	NT	0	NT	NT	NT
Phenyl ethyl alcohol	0	>1000	16	>1000	0	>1000	0	>1000
Phenyl ethyl cinnamate	0	>1000	0	>1000	0	>1000	0	>1000
Phenyl ethyl phenyl acetate	13	>1000	0	>1000	0	>1000	0	500
Phenyl propyl alcohol	12	>1000	18	>1000	16	1000	14	1000
Phenyl propyl aldehyde	12	500	27	500	14	1000	0	500
Phixia	16	>1000	15	>1000	12	1000	14	1000
Piconia	13	NT	0	NT	0	NT	NT	NT
Pimento berry oil	16	500	17	1000	18	500	14	500
Pine needle oil, Siberian	0	500	0	>1000	0	1000	0	1000
Pine oil	12	1000	14	>1000	11	500	12P	1000
Propylene glycol, USP	0	>1000	0	>1000	0	>1000	0	>1000
Rosacene	14	1000	12	1000	29	500	0	1000
Rosalva	16	1000	0	>1000	16	100	14	500
Rosemary oil, Spanish Tunisia	0	1000	0	>1000	0	1000	0	>1000
Rosetone	NT	>1000	0	>1000	0	>1000	0	>1000
Rosin gum	12	NT	0	NT	0	NT	NT	NT
Sandalwood	11	50	0	>1000	*	>1000	11	500
Santalol	13	500	0	>1000	0	>1000	13	500

Aroma Preservatives

Sauge sclaree, absolute	12	500	0	>1000	0	11	500
Sesquiterpenes PC	13	500	12P	>1000	0	12	500
Spearmint oil	0	1000	0	>1000	0	0	1000
Spruce oil	0	500	0	>1000	0	0	1000
St. Guaiol	13	500	0	>1000	0	12	500
St. John's bread, conc. 10%	0	NT	15P	NT	0	NT	NT
Styrax alva essence	21	NT	0	NT	0	NT	NT
Styrax clarified, extra	11	>1000	0	>1000	0	11	>1000
Surfleurs hay	11	NT	0	NT	0	NT	NT
Tabac absolute	10	NT	0	NT	0	NT	NT
Tangerine oil, Fla.	0	500	0	>1000	0	0	>1000
Terpineol	12	1000	19	1000	20P	0	1000
Thuja oil	0	500	0	>1000	0	0	1000
Thyme, white	25	500	27	500	14	16	500
Tiglyl piperidide	0	NT	13	NT	11	NT	NT
Tolu resin, absolute, 50%	17	>1000	17	>1000	11	13	1000
Tonalid	0	>1000	0	>1000	0	0	500
Tonka, absolute	11P	>1000	14	>1000	14P	0	>1000
trans-Decahydro beta naphthol	15	1000	15	1000	22P	16	500
trans-3-Pentenyl acetone	0	NT	18	NT	0	NT	NT
Treemoss, absolute, French, 50%	18	100	0	>1000	0	14	500
Trimethyl cyclohexanol	0	NT	11	NT	0	NT	>1000
Trimethyl cyclohexenone	0	>1000	12	1000	10	0	>1000
Trimethyl cyclohexenol	15	NT	13	NT	0	NT	NT
Undecylenic acid	21	NT	0	NT	13	NT	NT
Vanillin	16	>1000	22	>1000	19	12P	>1000
Veltol plus	0	>1000	20	>1000	12	0	>1000
Veramoss	0	500	0	>1000	12	12	100
Verdural extra	13	NT	13	NT	0	NT	NT
Violettone A, colorless	16	NT	12	NT	0	NT	NT
Wintergreen oil	0	>1000	*	>1000	0	0	1000

TABLE 4 Continued

Chemicals[a]	S. aureus		E. coli		Candida albicans		Diphtheroid	
	Zone diameter (mm)	MIC (ppm)	Zone diameter (mm)	MIC (ppm)	Zone diameter (mm)	MIC (ppm)	Zone diameter (mm)	MIC (ppm)
Yaracetal	0	NT	13	NT	0	NT	NT	NT
Ylang concrete	0	>1000	0	>1000	0	1000	0	1000
Zingerone	11	>1000	12	>1000	11	>1000	0	>1000
Controls								
Hexachlorophene	13	0.10	12	50	0	50	0	100

Source: Ref. 29.

TABLE 5 Antimicrobial Activity of Fragrance Materials in Petri Plate Cultures

Organism	Number tested	Number positive	Percentage positive
S. aureus (gram positive)	521	162	31.1
E. coli (gram negative)	521	136	26.1
C. albicans (yeast)	521	149	28.6
Diphtheroid	212	101	47.6
All four organisms	212[a]	64	30.2
Three organisms	212	89	42.0
Two organisms	212	118	55.7
One organism	212	144	67.9
Negative	212	68	32.1

[a]Only those materials tested against all four organisms are included in this total.
Source: Ref. 29.

TABLE 6 Composition of Soap Fragrance for Hand-Degerming Test

Ingredient	Percentage in fragrance	MIC (ppm)
Aldehyde, C_{11}, undecylenic	0.4	50 (S. aureus)[a]
Amyl cinnamyl alcohol	18.0	50 (S. aureus)
Anethol	10.0	100 (C. albicans)
p-tert-Butyl-m-cresol	0.5	50 (S. aureus)
Citronellol coeur	18.0	100 (C. albicans)
Cuminyl acetate	5.0	100 (C. albicans)
Isobutyl quinoline	0.1	50 (C. albicans)
Isocitral	0.5	100 (diphtheroid)
Lemma (Schiff base)	5.0	100 (S. aureus)
Methyl isoeugenol	5.0	100 (diphtheroid)
Methyl lavender ketone	7.0	100 (S. aureus)
Mousse, absolute ess.	4.0	100 (S. aureus)
Musk ketone	5.0	100 (diphtheroid)
Musk xylol	5.0	50 (diphtheroid)
Ocmea (Schiff base)	8.0	50 (S. aureus)
Patchouli oil, dark	3.0	100 (S. aureus)
Rosalva	2.0	100 (C. albicans)
Sandalwood	0.5	50 (S. aureus)
Veramoss	3.0	100 (diphtheroid)

[a]Indicates organism for which the MIC was determined. In some cases this MIC applies to more than one organism.
Source: Ref. 29.

Sturm found that the method developed by Heiss for the determination of the contact growth index (CGI) provided exact and reproducible results. Using this CGI method, perfumery raw material which gave a CGI value of zero at a concentration of ≤ 3 mg/16cm^2 were, for example, citronellol, geraniol, para-cresol, para-tolyl aldehyde undecyl alcohol, undecylenic alcohol, cinnamic aldehyde, iso-eugenol, clove oil, clove bud oil, pimento oil, rose oil, cossia oil, and sandalwood oil [31].

As example of this approach further informative results were supplied by a sniff test in which the deodorizing properties, with respect to intensity and duration of effect, of a "normal" fragrance, a deosafe fragrance, and Irgasan DP 300 were compared (Fig. 1). These deosafe fragrance compositions can be formulated to yield safe and effective deodorizing agents.

Earlier it was reported by Isacoff that two fragrance mixtures were formulated based on the antimicrobial quality of the fragrance ingredient [32]. In this instance the aroma chemicals were based to aid preservation and not incorporated as an "active" ingredient. The cosmetic products examined had no need for the usual cosmetic preservatives, because after addition of the perfumes the products were rendered free of microbial growth within 96 hr of the challenge.

In view of such data on antimicrobial fragrances, it seems reasonable to suggest that the selection of a fragrance for a cosmetic product be based not only upon the desired scent but also upon some of its inherent germicidal properties.

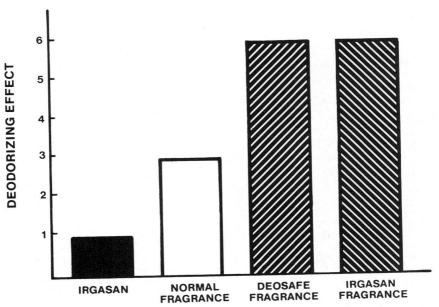

FIGURE 1 A sniff test in which the deodorizing properties of formulation were compared. (From Ref. 31.)

III. CHEMICAL CONSTITUENTS OF "ACTIVE" PRINCIPLES IN AROMA PREPARATIONS

Most essential oils are composed of simple and complex terpenes and oxygenated derivatives. In addition to this group are acidic compounds (phenols and fatty acids), lactones, and coumarin. Most of the compounds contain carbon, hydrogen, and oxygen. This classification excludes most of the alkaloid, amine, and nitro compounds, which do appear, but which are quantitatively fewer in number but not necessarily in importance. The principal essential oil constituents can be divided into two broad classes of organic compounds: those that contain only carbon and hydrogen and those which contain carbon, hydrogen, and oxygen.

A. Compounds Containing Carbon and Hydrogen

The hydrocarbons which have been identified in essential oils include paraffins, olefins, aromatic hydrocarbons terpenes (olefinic as well as mono- and bicyclic) a few lower terpene homologs, sesqui terpenes (aliphatic as well as mono-, bio-, and tricyclic, diterpenes, and azulenes. Of these, the terpenes and the sesquiterpenes are most characteristic of essential oils. Oil of turpentine, for example, is composed exclusively of the terpene hydrocarbons alpha- and beta-pinene oil; oil of orange contains about 90% of the terpene d-limone.

The fundamental building block of the terpene series is the isoprene unit C_5H_8, sometimes called a hemiterpene:

$$CH_2 = CH - C \begin{smallmatrix} \\ \diagup\!\!\diagup CH_2 \\ \diagdown CH_3 \end{smallmatrix}$$

Isoprene Unit

Two such units give rise to the terpene $C_{10}H_{16}$, three units to the sesquiterpenes $C_{15}H_{24}$, and four units to the diterpenes $C_{20}H_{32}$. As such and without functional groups, hydrocarbons tend to be weakly active.

B. Compounds Containing Carbon, Hydrogen, and Oxygen

These are the constituents responsible for the great difference of order observed in various essential oils. They usually represent the more soluble portion of the oil. In this broad class are included the important alcohols, aldehydes, ketones, and esters. As in the case of hydrocarbons, many types are to be found: saturated and unsaturated aliphatic as well as aromatic compounds, and also terpenic and sesquiterpenic compounds (aliphatic, monocyclic, bicyclic, and in the case of the sesquiterpene compounds even tricyclic). Other important constituents are phenols and phenolic ethers, acids (usually esterified), lactones, oxides, acetals, amines, and nitro compounds (see Table 7 and Ref. 33).

Certain essential oils consist largely of one chemical compound; for example, methyl salicylate comprises about 98% of sweet birch oil, anethole about 90% of anise oil, linalool about 85% of boise de rose oil, and safrole about 80% of sassafras oil. This fact frequently makes possible the commercial use of an oil for the isolation of a constituent useful per se or as an intermediate for synthesis. Citral (comprising 75-80% of lemongrass oil) is used in lemon

TABLE 7 Aromatic Chemicals

Atoms in functional groups				Part of systematic name referring to functional group	Aroma chemicals containing corresponding functional groups		
Arrangement	Formula as usually written	Type of molecule in which group is found	Class of compound		Type examples	Stability	Reactivity
$-O-$	$-O-$	$R-O-R^a$	Ethers	Ether	Diphenyl oxide	Very stable	Unreactive, but solvent ethers are highly inflammable
$-O-H$	$-OH$	$R-OH$	Alcohols	$-ol$	Linalool, alcohols C_7 to C_{12}	Most are stable; some sensitive to light	Forms esters with acids; many attack aluminum
$-O-H$	$-OH$	$R-OH$	Phenols	$-ol$	Eugenol	Sensitive to light	Readily oxidized; forms colored compounds with iron
$\overset{O}{\underset{}{\overset{\|\|}{C}}}-H$	$-CHO$	$R-CHO$	Aldehydes	$-al$	Citral, aldehydes C_7 to C_{12}	Higher fatty aldehydes polymerize if not diluted	Highly reactive, e.g., with oxygen to form acids
$\overset{O}{\underset{}{\overset{\|\|}{C}}}-O-H$	$-COOH$	$R-COOH$	Carboxylic acids	$-ic$ acid or $-oic$ acid	Phenylacetic acid	Stable	Reactive, e.g., with alcohols to form esters

Aroma Preservatives

Structure	Formula	Class	Suffix	Example	Light stability	Reactivity
$\begin{matrix} O \\ \parallel \\ -C-O- \end{matrix}$	$-COO-$ (or $-OOC-$)	Esters	-ate	Benzyl acetate	Stable if perfectly dry	Split by water, especially in the presence of an acid or alkali to form alcohols and carboxylic acids
$\begin{matrix} O \\ \parallel \\ -C-O- \end{matrix}$	$O-CO$	Lactones	lactone or -ide	Aldehyde C_{18} (so called)	Stable	Relatively unreactive
$\begin{matrix} O \\ \parallel \\ -C- \end{matrix}$	$-CO-$	Ketones	-one	Ionones, methyl ionones	Many are sensitive to light	Readily oxidized by oxygen
$\begin{matrix} H \\ \mid \\ -N \\ \mid \\ H \end{matrix}$	$-NH_2$	Amines	amine or amino-	Methyl anthranilate (also an ester)	Darkens on exposure to light	Forms colored compounds with aldehydes
$\begin{matrix} O^b \\ \uparrow \\ -N \\ \downarrow \\ O \end{matrix}$	$-NO_2$	Nitro compounds	nitro-	Musk ambrette	Most darken on exposure to light	Not highly reactive in perfumery use
$-C\equiv N$	$-CN$	Nitriles	nitrile	Geranonitrile	Stable	May be sensitive to acid or alkali, depending on pH and temperature

[a] R and R' may be identical or different radicals.
[b] The arrow signifies a bond which behaves as an ordinary single bond in compound.
Source: Ref. 33.

flavors; it is also the basic raw material for the manufacture of the ionones, important for perfumery.

Many of the constituents of essential oils can be synthesized economically from raw materials from other sources, notably coal-tar and petroleum derivatives [34,35].

The comparison of antimicrobial activity and structure is only feasible when isolated and purified components are tested. Although it is difficult to correlate antimicrobial activity with chemical structure, there are some generalizations which are useful. In 1953 Villanova and Casanovas [36] found two types of phenolic configurations with activity represented by eugenol and isoeugenol:

OH
⬡—O·CH₃
CH₂-CH=CH
eugenol

OH
⬡—O·CH₃
CH=CH-CH₃
isoeugenol

Nadal et al. (1973) showed that the o-methoxy group was not necessary for activity, since the structure without it, chavicol, was more active [37]. The change of the phenolic group to a methoxy or acetyl group lowers biological activity.

The oxidation of isoeugenol's hydrocarbon chain to an aldehyde retains activity of the parent phenolic derivative, vanillin.

The other type of compound, hexavalent aliphatic terpene (doubly unsaturated, of myrcene structure) is represented by citral:

$$H_3C\!\!-\!\!C(H_3C)\!=\!CH-CH_2-CH_2-C(CH_3)\!=\!CH\cdot CHO$$

citral

Geraniol and neral are similar in constitution but the former is cis whereas the latter is trans. Generally the cis configuration is more active than the trans [38]. As with similar conclusions made by Kabara for fatty acids, the singly unsaturated derivative, α-citronellol, is less active than geraniol which is polyunsaturated [38].

One can generalize about certain functional groups being important to germicidal activity: From the group of hydrocarbons, p-cymene has been reported to be active; aliphatic terpene alcohols (linalool, geraniol, neral); phenolic derivatives (thymol carvacrol, eugenol, safrae); aromatic aldehydes and ketones (benzaldehyde, cumene aldehyde, cinnamic aldehyde, pulegon, carvone, thujone).

For an aliphatic series the alcohol group is expected to be more active than the aldehyde (ketones) and acid groups [26]. The esterification of an aliphatic group with a monohydric alcohol yields an inactive functional group, while esterification with a polyhydric alcohol gives derivatives more active than the fatty acid [38].

Aroma Preservatives

Chain length is important to antimicrobial activity and determines not only the activity of the parent compound, but also whether it is active against gram-positive or gram-negative organisms. A chain length of C_{12} (laurel derivatives) is most effective against gram-positive organisms. Derivatives with carbon lengths slightly lower (C_8-C_{10}) tend to be more active against fungi, while chain lengths lower than C_8 affect gram-negative organisms. For unsaturated species higher chain lengths are more important; $C_{16:1}$, $C_{18:1}$, and $C_{18:2}$, whether acids and alcohols, have proven to be effective antimicrobial agents [39].

Similar principles for aromatic derivatives are not easily obtained since the effect of each group singly and collectively is more complex. Katayama and Nagai [40] showed that, contrary to what was found for aliphatic compounds, there was no difference in the antibacterial activity according to the number of double bonds in a cyclohexane ring:

d-limonene   α-phellandrene   p-cymene

The addition of a ketone group (ortho position) gave rise to carvone, but there was no increase in antibacterial activity. However, the addition of a hydroxyl group to either a cyclohexane or benzene ring also resulted in increased activity:

thymol   carvacrol

The above examples represent generalities in terms of structure-function relationships. They can only partially explain the action of plant extracts, since aroma chemicals are usually mixtures of two or more distinct substances. The overall effect of the mixture may be additive or synergistic. Some of the increased biological effects noted by other authors, however, were merely keeping the usual volatile components "trap" in a solvent system [32,41]. On the other hand, Miller (1931) found that emulsified mixtures of two or more oils increased their efficiency as compared with each principle used separately [42]. This synergy depends on the right combination of essential oils. Of the 96 combinations studied by Maruzzella and Henry [18], only 3 combinations were found to be more active than the individual ingredients. When fixed or infused oils were added to volatile oils, the antibacterial activity of the latter was markedly diminished. The specificity for synergy was confirmed by Kar and Jain [43].

Further synergy between acid, salt, and essential oils was recently (1982) found by Japanese workers [44]. Their results strongly suggest that acetic acid (0.1%), 3% NaCl, and certain essential oils (citric oil, etc.) at 0.5 mm when combined together were effective and synthetic preservatives were not needed.

It is important to emphasize that while many of the essential oils and fragrances are not the most powerful germicides available to the formulator, they can however, in combination, be part of a preservative system which will protect the cosmetic product from spoilage.

IV. TOXICOLOGY OF AROMA CHEMICALS

While the average consumer equates a "natural" substance with a lack of or low toxicity, it is obvious that natural products are not inherently safe. Digitalis, one of the most toxic drugs used in medicine, comes from the foxglove plant; cobra venom is also a natural product. However, it must be conceded that products that have been used for eons must at least have only subtle rather than acute toxic effects.

Many of the spice extracts and essential oils have been used with greater diligence in food as compared to cosmetics. Research into the toxicity of a number of essential oils has begun by the FAO/WHO Expert Committee on Food Additives. Regulations restricting the use of aromatic substances were made in 1955 [45].

One of the few papers dealing directly with the toxicity of essential oils was written by Chopra et al. [46]. Two examples were presented. The LD_{50} of the oil of *Alpinia galanga willed* for guinea pigs was 0.68 ml/kg, while that of *Acorus calamus* was 0.25 ml/kg. The oils, when given to guinea pigs in sublethal doses (<1/3 of the LD_{50}), produced no pathological changes which could be attributed to the oils.

Hamburger and Boger (1968) have reviewed the carcinogenicity of essential oils, flavors, and spices [47]. Fortunately, relatively few essential oils have been shown to cause cancer in animals. This paper summarizes knowledge on the carcinogenic activity of safrole, citrus oil (d-limone), turpentine oil (1-pinene), eucalyptus oil (phellandrene), bergamot oil, and red pepper.

The results of testing are not always clear or simple, and the reader interested in this problem should read the full review. In addition to governmental agencies interested in this problem, the fragrance industry has stepped forward in taking measures to assure the safety of chemicals. In 1966 principal suppliers of aroma and fragrances established the Research Institute for Fragrance Materials (RIFM). The sole purpose of this council is to assure the safety of perfumery raw materials. At present the RIFM is supported by over 70 companies representing nearly 97% of the industry in the United States and Europe.

In addition to the RIFM, another national group was established in 1973: The International Fragrance Association (IFRA) has the task of drawing up a set of rules of procedure on the production, handling, and safety of aroma chemicals and related materials. These are two nongovernmental sources which can provide the interested formulator or microbiologist information on the safety of various aroma chemicals. In establishing such self-policing efforts, the cosmetic industry deserves much credit.

REFERENCES

1. F. K. Oldham, F. E. Kelsey, and E. M. K. Heiling, *Essentials in Pharmacology*, Lippincott, Philadelphia (1951).
2. S. Garland, *The Complete Book of Herbs and Spices*, Viking Press, New York (1979).
3. G. Urdang, *The Essential Oils*, Vol. 1 (E. Guenther, ed.), Van Nostrand, New York (1948), p. 3.
4. S. A. Malcolm and E. A. Sofowora, Antimicrobial activity of selected nigerian folk remedies and their constituents plants, *Lloydia*, 32:512-517 (1969).
5. W. H. Martindale, Antiseptic powers of essential oils, *Perfum. Essent. Oil Rec.*, 1:266 (1910).
6. D. I. Macht and W. M. Kunkel, Concerning the antiseptic action of some aromatic fumes, *Proc. Soc. Exp. Biol. Med.*, 18:68 (1920).
7. F. C. Dyche-Teague, Bactericidal value of some commercial perfumes, *Perfum. Essent. Oil Rec.*, 15:6-8, 40-43, 81-83 (1924).
8. J. J. Bryant, Detergent and antiseptic value of perfume material, i.e. essential oils and synthetics, in toilet soap, *Perfum. Essent. Oil Rec.*, 15:426-431 (1924).
9. S. Rideal, E. Rideal, and A. Sciver, Investigation into the germicidal powers and capillary activities of certain essential oils. *Perfum. Essent. Oil Rec.*, 19:285-305 (1928).
10. G. M. Dyson, Physiological aspects of the essential oils, *Perfum. Essent. Oil Rec.*, 21:287-314 (1930).
11. R. M. Gattefossé, *Aromatherapie*, Girardot, Paris (1937).
12. R. M. Gattefossé, Essential oils as vegetable hormones, *Parfum. Mod.*, 31:57, 59 (1937).
13. Y. Naves, Evaluation of the bactericidal power of essential oils by determination of their phenol coefficients, *Parfums France*, 13:273-284 (1935).
14. R. S. Potter, Antiseptic value of flavors, *Flavours*, 2:7-12 (1939).
15. T. McLachlen, Microorganisms in cosmetics, *Soap Perfum. Cosmet.*, 19:666-672, 674 (1946).
16. M. Rolle and A. Mayr, Conifer oils as disinfectants against tuberculosis, *Monatsh. Tierheilkd.*, 4:421-432 (1952).
17. J. C. Maruzzella and M. B. Lichtenstein, The in vitro antibacterial activity of oils, *J. Am. Pharm. Assoc. Sci. Ed.*, 45:378-381 (1956).
18. J. C. Maruzzella and P. A. Henry, The in vitro antibacterial activity of essential oils and oil combinations, *J. Am. Pharm. Assoc.*, 47:294-296 (1958).
19. J. C. Maruzzella, J. Balter, and A. Katz, The action of perfume-oil vapors on fungi, *Am. Perfum. Aromat.*, 74:21-22 (1959).
20. J. C. Maruzzella, The antifungal properties of essential oil vapours, *Soap Perfum. Cosmet.*, 33:835-837 (1960).
21. J. C. Maruzzella and E. Bramnick, The antibacterial properties of perfumery chemicals, *Soap Perfum. Cosmet.*, 35:743-745 (1961).
22. J. C. Maruzzella, The germicidal properties of perfume oils and perfumery chemicals, *Am. Perfum.*, 77:67-70 (1962).
23. J. C. Maruzzella, An investigation of the antimicrobial properties of absolutes, *Am. Perfum. Cosmet.*, 78:19-20 (1963).
24. H. Munzing and H. Schels, Potential replacement of preservatives in cosmetics by essential oils, *J. Soc. Cosmet. Chem.*, 23:841 (1972).

25. R. Richards and R. McBride, Enhancement of benzalkonium chloride by aromatic alcohols, *J. Pharm. Sci.*, 62:2035-2037 (1973).
26. J. J. Kabara, D. M. Swieczkowski, A. J. Conley, and J. P. Truant, Fatty acid derivatives as antimicrobial agents, *Antimicrob. Agents Chemother.*, 2:23-28 (1972).
27. R. D. Fletcher, J. R. Gilbertson, A. C. Albers, and S. D. White, Inartination of myroplasma by long-chain alcohols, *Antimicrob. Agents Chemother.*, 19:917-921 (1981).
28. R. J. Crout, J. R. Gilbertson, J. D. Gilbertson, D. Platt, H. H. Langkamp, and R. H. Connamacher, Effect of linolenyl alcohol on the in-vitro growth of the oral bacterium *Streptococcus mutans*, *Arch. Oral Biol.*, 27:1033-1037 (1982).
29. J. A. Morris, A. Khettry, and E. W. Seitz, Antimicrobial activity of aroma chemicals and essential oils, *J. Am. Oil Chem. Soc.*, 56:595-605 (1979).
30. J. C. Maruzzella and N. A. Sicurella, Antibacterial activity of essential oil vapors, *J. Am. Pharm. Assoc. Sci. Ed.*, 49:692 (1960).
31. W. Sturm, Deosafe fragrances: fragrances with deodorizing properties, *J. Cosmet. Toiletries*, 94:35-48 (1979).
32. H. Isacoff, Preservation with essential oils and aroma chemicals, in *The Chemistry and Manufacture of Cosmetics* (M. G. DeNavarre, ed.), Continental Press (1975).
33. D. G. Williams, Aroma chemicals, *Cosmet. World News*, May, p. 20 (1976).
34. E. E. Langenau, *Encyclopedia of Chemical Technology*, Vol. 9, pp. 569-591 (1952).
35. E. J. Parry, *Chemistry of Essential Oils and Artificial Perfumes*, 4th Ed., Vols. 1 and 2 (1922).
36. X. Villanova and M. Casanovas, The fungistatic and fungicidal activity of vegetable essential oils and their components, *J. Invest. Dermatol.*, 20:447 (1953).
37. N. G. M. Nadal, A. E. Montalvo, and M. Seda, Antimicrobial properties of bay and other phenolic essential oils, *Cosmet. Perfum.*, 88(10):37-38 (1973).
38. J. Kabara, Fatty acids and derivatives as antimicrobial agents – a review, in *Pharmacological Effects of Lipids*, American Oil Chemists' Society, Champaign, Ill. (1978), pp. 1-14.
39. J. A. Sands, D. D. Auperin, P. D. Landin, A. Reinhardt, and S. P. Codden, Antiviral effects of fatty acids and derivatives, in *Pharmacological Effects of Lipids*, American Oil Chemists' Society, Champaign, Ill. (1978), pp. 75-95.
40. T. Katayama and I. Nagai, Chemical significance of the volatile components of spices from the food preservative viewpoint. IV. Structure and antibacterial activity of some terpenes, *Nippon Suisan Gakkaishi*, 26:29-32 (1960).
41. J. Risler, Immediate and prolonged antiseptic action of essential oils, *Acad. Sci. Paris*, 203:517-519 (1936).
42. R. E. Miller, Bactericidal efficiency of essential oils, *Am. J. Pharm.*, 103:324-328 (1931).
43. A. Kar and R. Jain, Antibacterial evaluation of some indigenous medicinal volatile oils, *Qual. Plant. Mat. Veg.*, 20:231-237 (1971).
44. N. Kurita and S. Koike, Synergistic antimicrobial effect of acetic acid, sodium chloride and essential oil components, *Agric. Biol. Chem.*, 46:1655-1660 (1982).

45. Anonymous, Flavoring substances, *Br. Food J.*, Nov./Dec., (1973), pp. 175-183.
46. I. C. Chopra, N. Khajuria, and C. L. Chopra, Antibacterial properties of volatile principles from *Alpinia galanga* and *Acorus calamus*, *Antibiot. Chemother.*, 7:378-383 (1957).
47. F. Hamburger and E. Boger, The carcinogenicity of essential oils, flavors and spices: a review, *Cancer Res.*, 28:2372-2374 (1968).

16
MEDIUM-CHAIN FATTY ACIDS AND ESTERS AS ANTIMICROBIAL AGENTS

JON J. KABARA *Michigan State University, East Lansing, Michigan*

I. Background Information 275
II. Lipophilic Antimicrobial Agents 276
    A. Fatty acids as antimicrobial agents 276
    B. Monoesters as antimicrobial agents 282
III. Mechanism of Action of Lipophilic Compounds 286
    A. General considerations 286
    B. Specific considerations 287
VI. Toxicology of Lipophilic Derivatives 290
    A. Toxicology of fatty acids 290
    B. Toxicology and safety of glycerides 293
    C. Toxicology and safety of polyol esters 295
V. Summary 296
    References 297

I. BACKGROUND INFORMATION

Soaps (salts of fatty acids) have long been used as cleaning and disinfecting agents. On a 4000-year-old clay tablet uncovered at Tello, Mesopotamia, soap was mentioned. Soap was well known to "barbarous" people rather than the discovery of "civilized" groups. It was used by the Gauls and Germans before the Christian era. Soap came hence to the Romans prior to the year 79 A.D., when Vesuvius overwhelmed Pompeii. From the Romans the Greeks learned the value of soap. The pharmacological use of soaps followed its detergent uses, while its germicidal value was probably recognized doubtless much earlier than meager records show. The use of soap as an insecticide was recommended by Goeze in 1787. Much of the literature on soap as an insecticide can be found in a marvelous old monograph [1]. In more recent history, the earliest mention (1899) of soap as an antifungal agent can be found in a paper by Clark [2]. Another early worker, Reichenback (1908), studied the action of soaps on *Escherichia coli* and found that potassium myristate, palmitate, and stearate were effective against this organism, while oleate, erucate ($\Delta_9 C_{22:1}$), and linoleate were inactive [3].

Much of the work carried out on the germicidal activity of soaps can be found in literature between 1920 and 1940 [4-7]. Before 1930 the lack of success in the search for active antibacterial agents other than fatty acids evolved a pessimism regarding whether an active agent would ever be found. Two important discoveries changed the course of thinking on the subject: First was the serendipitous finding of Alexander Fleming [8]. In 1928 he observed the lysis of *Staphylococcus* colonies in an area surrounding the growth of a *Penicillum* mold, and thus antibiotics were added to the small arsenal of compounds lethal to microorganisms [8]. Second, the discovery of antimicrobial sulfonilamides, first observed by Trefouel, Nitti, and Boret in Fourneau's laboratory (Pasteur Institute) in 1935, gave hope for finding new and more effective synthetic biocidal agents [9]. Thus these two major discoveries in less than a decade gave tremendous impetus to seeking still more active compounds. The work being carried out with fatty acids was to suffer, since the biocidal activity of the isolated antibiotics and synthesized germicides was so much greater than fatty acids. In proportion to the number of new and more powerful drugs that were discovered, interest in the action of fatty acid biocides waned. The frantic research activities in the 1950s and 1960s produced hundreds of antibiotics and synthetic products useful to the pharmaceutical and agricultural field, but of little use to the food and cosmetic industries. Interestingly, the cosmetic industry kept distant from these new discoveries for the most part and continued to use old standbys. Not without reason was this to be the case.

Several explanations can be given for the "nonparticipation" of the cosmetic microbiologist: First, antibiotics and synthetic germicides were not without health risks in terms of toxicity and the possible development of resistant strains. Second, the cost of these newer agents also precluded their use in quantities normally required in the cosmetic industry. Third, the use of old standbys was usually sufficient for the standards of the day [10]: These included acetic acid from ancient Egypt, sorbic acid isolated (1859) from rowanberry oil by Hoffman [11], benzoic acid discovered (1875) by Fleck [12], and propionic acid recommended (1906) for the preservation of bakery goods by Watkins [13]. From these few references it becomes obvious that fatty acids have played an important historical role in preservation. It is noteworthy, however, that other fatty acids, although more active, have not been examined as potential preservatives or that detailed studies dealing with structure-function relationships been carried out. In order to collate and evaluate the literature dealing with this problem, the following review of antimicrobial fatty acids and their corresponding esters is presented.

II. LIPOPHILIC ANTIMICROBIAL AGENTS

A. Fatty Acids as Antimicrobial Agents

Walker showed that sodium and potassium soaps of the same acid did not vary greatly in their germicidal action [14-16]. The lower members of this series,

C_4-C_{10}, had little or no germicidal effect. He found the pneumococcus to be very susceptible to laurates, oleates, linoleates, and linolenates. Streptococci were killed much like pneumococci, but at higher concentrations of soap.

Hettche [17] found that the unsaturated soaps, such as oleate, linoleate, and linolenate, were germicidal toward the staphylococcus but not toward the colon bacillus.

Bayliss [18] studied the structure-function relationship of soaps on their germicidal properties. A comparison of the pneumococcidal properties of the saturated soaps, laurate, myristate, palmitate, and stearate, showed that there was a maximum in this property with the soap containing 14 carbon atoms. The presence of unsaturated groups in the molecule in some cases altered tremendously the effect of the soap upon pneumococci. The sodium oleate was more than 100 times as effective as either sodium stearate or phenol in destroying the reproductive ability of the pneumococci. The addition of one, two, or three more double bonds, as in the oleic, linoleate, or linolenate soaps, respectively, caused no change in this property. However, the two soaps, sodium α- and β-elaeostearates, although isomeric with sodium linolenate, were relatively ineffective as pneumococcides. The three unsaturated soaps showed pneumococcidal properties considerably less than those of sodium oleate, and so on. The ricinelaidate, a geometric isomer of sodium ricinoleate which has physical properties much like the stearate and palmitate, was somewhat less effective than the ricinoleate. The ricinstearoleate, containing a triple bond instead of a double bond and with physical properties very similar to those of the ricinoleate, is even less active against bacteria.

Sodium undecylenate, containing 11 carbon atoms and a double bond at the end of the hydrocarbon chain, required a fairly high concentration in order to destroy the pneumococcus.

The effect of the soaps upon the streptococci, which cause souring milk and which are among the most resistant to disinfection of the streptococci, did not parallel the effect of the same soaps on the pneumococci. In no case, however, did it require a lower concentration to kill streptococci than to kill pneumococci. Of the saturated soaps, only the myristate and laurate exhibited killing action on *Streptococcus lactis*. The addition of one double bond as in sodium oleate gave the soap the ability to kill this organism in concentrations equal to that required by sodium laurate and myristate. The addition of a second double bond as in linoleate enhanced this killing property considerably. Further addition of a third double bond, however, conferred no greater ability to kill streptococci.

Kodicek [19] tested a number of fatty acids on *Lactobacillus casei* and found that in the saturated series lauric acid depressed growth, but the effect was not reversible by cholesterol. The inhibition of other fatty acids, including the more active and branched chain α-methyl lauric acid, could be reversed by addition of the sterol. As found by Bayliss [18], Kodicek confirmed that, of the unsaturated series, activity was limited to the cis form [19]. Similar bacteriostatic effects were observed by Hettche [17] with *Staphylococcus aureus*, by Chaix and Baud [20] with *Glaucoma piriformis*, and by Berstom et al. [21] and Dubos and Davis [22,23] on the tubercle bacilli. The evidence of the action of unsaturated fatty acids on bacteria in vitro accounted for the general biocidal activity of some biological materials. Humfield [24] found an antibiotic fatty acid in wheat bran, and McKee et al. [25] have found antibacterial lipids from *Tetrahymena geleii*.

By the early 1950s it was well established that unsaturated fatty acids exert an antibacterial influence on gram-positive microorganisms. The subject was reviewed by Nieman [6], who concluded that the inhibitory effects of unsaturated fatty acids increased as the number of double bonds in the molecule increased. These findings [26] supported the conclusion in that the inhibitory effect of linoleic acid was far greater than that of oleic acid. However, the minimum inhibitory concentration of arachidonic acid was about the same as that of linoleic acid, whereas the minimum inhibitory concentration of linolenic acid was somewhat greater than that of linoleic acid. Many of the results on chain length versus antimicrobial activity were confirmed in our laboratory (see Table 1).

The antifungal activity of fatty acids has been recognized for many years [27,28]. The preceding authors and others [29-33] have demonstrated that the fungitoxicity of these compounds is dependent on chain length and the concentration and pH of the medium. Several materials which act as protective agents against the antimicrobial action of the fatty acids include serum albumin, starch, cholesterol, lecithin, saponin, and charcoal [6]. Although a completely satisfactory explanation of the mode of action of these compounds has not yet been presented, the evidence [30] seems to indicate that growth inhibition is due to alteration in cell permeability [33].

As systemic antifungal agents, the fatty acids, in spite of their low order of toxicity, have not been effective. This may be due to their being readily metabolized by the host through the usual fatty acid pathways. In general, they may be esterified to form glycerides and/or may be degraded to small fragments by beta-oxidation. Asami et al. [34] have recently demonstrated that 11-iodo-10-undecenoic acid was esterified, in part, in the rat to a glyceride. Although the effect of fluorine in the 2-position of fatty acids on esterification has not yet been reported, these fatty acids are believed not to undergo beta-oxidation [35]. Thus the 2-fluoro fatty acids possess at least one potential advantage over the nonfluorinated analogs which would be useful for systemic antifungal activity [36].

The antifungal activity of 15 (2-fluoro) fatty acids to a chain length of 20 carbon atoms was determined in parallel with the corresponding nonfluorinated analogs against four fungi [36]: *Aspergillus niger*, *Trichoderma viride*, *Myrothecium verrucaria*, and *Trichophyton mentagrophytes*. Both series of compounds were about equally active, except that the nonfluorinated fatty acids showed maximal activity at chain lengths of 4-10 carbons, whereas the 2-fluoro fatty acids were most active at chain lengths of 8-14 carbons. Since the pK_a values of the 2-fluoro fatty acids are generally two units lower than those of the nonfluorinated fatty acids, it appears that the pK_a of the fatty acid does not play an important role in antifungal activity.

The current interest in dematomycoses has suggested further investigation of the reported activity of fatty acids and related compounds as antimycotic agents. This problem was first studied by Clark [2] in 1899, who reported the effect of acids, including fatty acids, on the germination of fungi. The systematic investigation of Kiesel [27] at the Pasteur Institute in 1913 revealed many of the facts known today about the characteristics of the antimycotic action of fatty acids. He found that (1) the activity of saturated fatty acids increased as the number of carbon atoms in the fatty acid chain increased up to 11 carbon atoms, (2) the branched chain fatty acids were less active than those with straight chains and an equal number of carbon atoms, and (3) substitution of hydrogen by hydroxyl decreased activity.

TABLE 1 Minimal Inhibitory Concentrations of Saturated and Unsaturated Fatty Acids[a]

Fatty acid	Pneumococci	Streptococcus group A	Streptococcus, beta-hemolytic non-A	Candida albicans	Staphylococcus aureus
Caproic	NI	NI	NI	NI	NI
Caprylic	NI	NI	NI	NI	NI
Capric	1.45	1.45	2.9	2.9	2.9
Lauric	0.062	0.124	0.249	2.49	2.49
Myristic	0.218	0.547	2.18	4.37	4.37
Myristoleic	0.110	0.110	0.110	0.552	0.441
Palmitic	0.48	3.9	3.9	NI	NI
Palmitoleic	0.024	0.098	0.049	0.491	0.983
Stearic	NI	NI	NI	NI	NI
Oleic	NI	1.77	NI	NI	NI
Elaidic	NI	NI	NI	NI	NI
Linoleic	0.044	0.089	0.089	0.455	NI
Linolenic	0.179	0.35	0.35	NI	1.79
Linolelaidic	NI	NI	NI	NI	NI
Arachidonic	NI	NI	NI	NI	NI

[a]Results are millimolar; NI = not inhibitory at the concentrations tested (1.0 mg/ml, or 3-6.0 mM).
Source: Ref. 70.

Japanese workers [29-31] in 1931 and 1933 essentially confirmed Kiesel's work, but found that with the saprophytic wood-rotting fungi the optimum activity was with the 12-carbon acid. Unsaturated acids were found to be more active than the corresponding saturated acids, and dicarboxylic acids were almost without activity. Kirby et al. [37], Peck et al. [38,39], Baechler [40], Hoffman et al. [41,42], and Rigler and Greathouse [43] confirmed the findings of these earlier workers and extended them to other conditions and other organisms. Cowles [44] showed that at low pH values, fatty acids have a bactericidal action and their activity increases with chain length. Keeny et al. [45,46] reported fungicidal activity for the long-chain acids, but much less or no killing action by the short-chain members of the fatty acid series. Data showed that the optimum chain length was apparently determined by the resistance of the organisms and the solubility of the fatty acids in question. Thus for *A. niger* the optimum chain length was 11-carbon atoms, but for the more sensitive *T. interdigitale*, the 13-carbon acids were most active. *T. purpureum* is even less resistant and was inhibited by the 14-carbon acids, suggesting that the longer-chain compounds failed to show antimycotic effects because, owing to lack of solubility, a static concentration cannot be obtained. The data showed that the fatty acids increase in activity with decreasing pH, provided that the low pH values do not make the compound so insoluble that a static concentration for the organism under test cannot be obtained. The change in activity with hydrogen ion concentration was much greater for the short-chain acids, suggesting that the ion of the long-chain compound — or some aggregate or micellar form of it which may exist in solution — exerts additional action.

Eggerth [47-50] found that the α-bromo fatty acids were usually more germicidal than the unsubstituted soaps. The effect of a hydroxyl in the α-position was to increase the selective germicidal action of saturated soaps and to diminish that of unsaturated.

Larson [51] found that pneumococci and streptococci would not grow in the presence of even small amounts of sodium ricinoleate. Larson and Nelson [52] reported that pneumococci instantly lost their pathogenicity on treatment with castor oil soap at a final dilution of 0.1%. Scarlet fever streptococci lost their power to grow after 5 min in 0.5% sodium ricinoleate. Miller and Castles [53] found the same fatty acid to inhibit the growth of gonococci on artificial media in dilutions of 1:20,000. Violle [54] studied the effect of a 1:1000 solution of fatty acids on many kinds of bacteria. The common and pathogenic bacteria of the intestinal tract were unaffected. Streptococci were killed, but staphylococci were not. Barnes and Clark [55], from their experiments, determined that 0.004% sodium ricinoleate and 0.0004% sodium oleate were approximately the minimal pneumococcidal concentrations of the soaps against three types of pneumococci. In other words, the oleate was much more effective by a factor of 10.

Kolmer et al. [56] reported that a 20% solution of sodium ricinoleate was completely bactericidal for *S. aureus* in an exposure of 5 min, yet a 10% solution was not completely bactericidal in exposures as long as 1 hr when tested according to the Reddish method.

The hydroxylated salts, gluconate and trihydroxy stearate, seem to lack any ability to kill the pneumococci under the conditions employed [4]. The stearate is peculiar in that the addition of a fourth hydroxyl group restored the power to destroy this organism in fairly low concentrations. The hydroxyl group, in the case of the ricinoleate, enhanced the bactericidal activity against the streptococcus, whereas there was a decrease against the pneumococcus. The replacement of the ethylenic linkage by an acetylenic linkage again caused a marked decrease in germicidal activity [4].

Sodium abietate, which is an alicyclic soap, has a destructive action on the pneumococcus in 1/10 the concentration necessary for phenol. The same is true of chalmaugrate, which has been used as a chemotherapeutic agent in the treatment of leprosy [4].

The introduction of a phenyl group for 6 of the carbon atoms of an 11-carbon acid decreases the activity markedly, regardless of the degree of unsaturation in the fatty acid chain. Substitutions in the phenyl ring with an amino or hydroxyl group further depress the activity. The presence of a cyclohexyl group on the fatty acid chain results in activity comparable to the corresponding straight-chain acids. No variation which has been made in the arrangement of the carbon atoms has brought about greater activity than that displayed by the straight-chain acids [4].

Since early times sulfur has been considered efficacious in the treatment of fungous diseases. Since a sulfur atom is considered equivalent to $-CH=CH-$, n-heptylmercaptoacetic acid may be regarded as the isostere of an unsaturated 11-carbon fatty acid. Neither this compound nor a variety of other substituted mercaptoacetic acids showed activity of the same magnitude as the fatty acids [4].

With the exception of 11-thiohendecanoic acid, none of the other thiohendecanoic acids approached undecylenic acid in activity. The latter fatty acid was found by Rothman et al. [57] to be responsible for the spontaneous cure of ringworm of the scalp in puberty.

Branching of the carbon chain decreases the activity if one considers the total number of carbon atoms in the acid. The effect of branching is less, however, when one considers only those carbon atoms in the longest chain. The aldehyde, acetate, ethyl ester, amide, and substituted amide have considerable activity, but are less active than the corresponding acid. Alcohols have high activity, but owing to limited solubility, the effectiveness of long-chain alcohols can be demonstrated only on the more sensitive organisms [7].

The antibacterial effects of the fatty acids have been discussed by many authors and the consensus of opinion is that their action is due to the undissociated molecule, and not the anion [58-62]. If this is so, the activity of soaps would be profoundly affected by pH, since hydrogen ion concentration determines the degree of dissociation of the acid. The dissociation effect was shown by more rapid killing at lower pH. Lower pH alone was not lethal [63, 64]. The minimum inhibitory concentrations of each fatty acid are given in Table 2. An increase from pH 6.5 to pH 7.5 increased the minimum inhibitory concentrations of the short-chain acid (caproic, caprylic, capric), but decreased the minimum inhibitory concentrations of the two medium-chain fatty acids (lauric, myristic). The minimum inhibitory concentrations of the unsaturated fatty acids were unaffected by a change in pH value in the medium. Of the fatty acids tested, those with the lowest minimum inhibitory concentrations were the polyunsaturated acids, such as linoleic, linolenic, and arachidonic acids. However, the minimum inhibitory concentration (MIC) of linoleic acid, with two double bonds, was somewhat less than that of linolenic acid, with three double bonds.

In our own attempt to bring some order into this vast amount of literature of structure-function activity, we carried out studies with fatty acids and derivatives which were very pure (99%). The results in Table 1 and in several published papers [7,65-68] are the basis for a number of generalizations which help explain most of the data in the literature:

TABLE 2 Minimum Inhibitory Concentrations Against *Clostridium welchii*[a]

Fatty acid	Minimal inhibitory concentration (mg/100 ml)	
	pH 6.5	pH 7.5
Caproic (6:0)	1160.00	5810.00
Caprylic (8:0)	721.00	3610.00
Capric (10:0)	172.00	862.00
Lauric (12:0)	1000.00	200.00
Myristic (14:0)	2280.00	457.00
Palmitic (16:0)	2560.00	2560.00
Stearic (18:0)	2850.00	2850.00
Oleic (18:1)	283.00	283.00
Erucic (22:1)	339.00	339.00
Linoleic (18:2)	5.61	5.61
Linolenic (18:3)	27.90	27.90
Arachidonic (20:4)	6.09	6.09
Methyl linoleate	146.00	733.00
Methyl linolenate	293.00	1460.00

[a]Note: Pure fatty acids and methol esters in 0.1 m phosphate buffer (pH 6.5 or 7.5) containing 0.1% (wt/vol) Tween 80 [31].

1. Except for short-chain fatty acids (less than C_8), lipids do not affect gram-negative organisms.
2. The most active chain length for saturated fatty acids is C_{12}; the most active monounsaturated fatty acid is $C_{16:1}$; $C_{18:1}$ is the most active polyunsaturated fatty acid.
3. The position and number of double bonds is more important to long- ($>C_{12}$) than short- ($<C_{12}$) chain fatty acids.
4. The cis form is active, while the trans isomer is inactive.
5. Acetylenic derivatives as compared to ethylenic fatty acids are more active against fungi.
6. Yeasts are affected by fatty acids with short-chain lengths (C_6-C_{12}), that is, slightly shorter than those affecting gram-positive organisms.
7. Fatty acids esterified to monohydric alcohol became inactive.

B. Monoesters as Antimicrobial Agents

Esterification of fatty acids with a monohydric alcohol (methanol, ethanol, etc.) resulted in an ester which was inactive [7]. This was not true when an α-hydroxy fatty acid was esterified [7]. From these initial results it was concluded that some hydrophilic group was necessary for biological activity and that the esterification of a polyhydric alcohol would yield active derivatives [66].

From these and other experiments the need for a free single or multiple hydroxy group before biological activity of the ester was made obvious. One of the more common polyhydric alcohols, glycerol, was esterified and found to be more active than the corresponding fatty acids (see Table 3). Details

TABLE 3 Minimum Inhibitory Concentrations (g/ml) for Fatty Acids and Their Corresponding Monoglycerides

Organism[a]	Unde-canoic acid	10-Unde-cenoic acid	10-Undecenoyl monoglyceride	10-Undecynoyl monoglyceride	11-Dode-cenoic acid	Dode-canoic acid	Dodecanoyl monoglyceride	12-Tride-cenoic acid	Tride-canoic acid	Tridecanoyl monoglyceride
Streptococcus faecalis	NI[b]	NI	500	500	NI	500	NI	1000	NI	NI
Streptococcus pyogenes	125	1000	125	125	250	62	8	125	1000	62
Staphylococcus aureus	1000	1000	500	500	NI	500	250	1000	NI	NI
Corynebacterium sp.	31	31	62	62	125	31	16	31	NI	NI
Norcardia asteroides	62	125	125	62	62	62	16	125	1000	125
Candida albicans	1000	1000	250	100	1000	1000	500	1000	NI	NI
Saccharomyces cerevisiae	500	500	250	100	500	1000	250	500	1000	NI

[a]Escherichia coli and Pseudomonas aeruginosa were not affected.
[b]NI = MIC > 1000 g/ml.
Source: Ref. 67.

of these findings have been published [66-68] and in all cases the data have been confirmed by others [69-72]. There seems to be some controversy or confusion on which fatty acid ester is the most biocidal. As a general statement, the fatty acid used to esterify the polyol determines the potency of the ester. The structure-function relationship for saturated and unsaturated esters follows the activity of their respective fatty acids as reviewed in an earlier section. This means that lauric acid (C_{12}) and palmitoleic acid ($C_{16:1}$) form the most active saturated and unsaturated derivatives, respectively [66,70].

The above statement concerning structure-function relationships is true except where specific organisms and cosmetic products are concerned. Because the monoglycerides can be neutralized by a variety of polymers, including starch and protein, and because binding seems to follow biological activity, several Japanese papers have reported that the volatile fatty acids esters were more active than the higher homologs [73,74]. These reports, negating the generalizations found later for fatty acids, must be viewed as exceptions to be found under particular circumstances. More recent papers by Beuchat [72] and Shibasaki [75] have supported the finding that the lauric acid derivative is the most active monoglyceride even in the presence of other complexing stuffs. Beuchat [72] compared the effects of glycerides, sucrose esters, benzoate, sorbic acid, and potassium sorbate against Vibrio parahaemolyticus. His results indicated that the C_{12} monoglyceride was more active than lower (C_8, C_{10}) or higher (C_{14}) chain length derivatives. Also, the low MIC value for monolaurin (5 g/ml or less) indicated it to be more effective than sodium benzoate (300 g/ml) or sorbic acid (70 g/ml).

Because these esters were regarded not to be active against gram-negative organisms [76], the inhibition of V. parahaemolyticus growth was somewhat surprising, even though Kato and Shibasaki [77,78] reported effects on gram-negative bacteria. In these latter cases the Japanese workers demonstrated that chelating acids (citric and polyphosphoric acid) were necessary to have an enhancing effect on the biocidal action of monoglycerides against several gram-negative organisms. Their interpretation was that acids which are chelators release a significant amount of cellular lipopolysaccharide from bacterial walls and cause the organism to behave like a gram-positive bacteria. Apparently the presence of chalators is not always a prerequisite for demonstrating inhibitory action of monoglycerides against all gram-negative bacteria, since V. parahaemolyticus is affected in a manner similar to gram-positive organisms [72].

Antimicrobial activity was also found for esters of more complex polyhydric alcohols [66,79]. Kato and Arima [79] reported that the sucrose ester of lauric acid was active against a gram-negative organism, whereas Conley and Kabara [66] and Shibasaki and Kato [71] indicate that these and other esters are primarily active against gram-positive bacteria and fungal organisms. In contrast to data generated for glycerides, the diester of sucrose rather than the monoester was more active. Sucrose dicaprylate possessed the highest activity of the sucrose esters, but was still less active than monolaurin [72,77,78].

The monoesters of glycerol and the diesters of sucrose not only have higher antimicrobial activity than their corresponding free fatty acids, but also compare favorably in activity with commonly used antiseptics such as parabens, sorbic acid, and dehydroacetic acid (see Tables 4 and 5).

Another series of food-grade esters was derived from a group of compounds called polyglycerol esters. These linear polymers of glycerol, dis-

TABLE 4 Comparison of the Antifungal Activities of Fatty Acid Esters and Some Commonly Used Preservatives

Food additive	Minimum inhibitory concentration (g/ml)[a]		
	Aspergillus niger	Candida utilis	Saccharomyces cerevisiae
Monocaprin	123	123	123
Monolaurin	137	69	137
Butyl-p-hydroxybenzoate	200	200	200
Sodium lauryl sulfate	100	400	100
Sorbic acid	1000	1000	1000
Dehydroacetic acid	100	200	200

[a]By the agar dilution method.
Source: Ref. 77.

covered and advocated for food use by Babayan et al. [80], were supplied for biocidal testing. Similar to results with sucrose esters, the medium-chain fatty acids appeared to be the most active [66]. It was of interest that, regardless of whether the polyol was tri-, hexa-, or decapolyglycerol, the fatty acid moiety seemed to determine overall biocidal activity. Generally, as the polyol became bulkier, the spectrum of biocidal activity became narrower. Other polyol esters have been tested and, except for special applications, none have proven more useful than monolaurin (see Table 6).

TABLE 5 Comparison of the Antibacterial Activities of Fatty Acid Esters and Some Commonly Used Preservatives

Food additive	Minimum inhibitory concentration (g/ml)[a]		
	Bacillus subtilis	Bacillus cereus	Staphylococcus aureus
Sucrose dicaprylin	74	74	148
Monocaprin	123	123	123
Monolaurin	17	17	17
Butyl-p-hydroxybenzoate	400	200	200
Sodium lauryl sulfate	100	100	50
Sorbic acid	4000	4000	4000

[a]By the agar dilution method.
Source: Ref. 77.

TABLE 6 Minimum Inhibitory Concentrations of Various Polyglycerol Esters (mM) Against *Corynebacterium* sp.

Fatty acid moiety	Polyglycerol esters		
	Triglycerol	Hexaglycerol	Decaglycerol
Acetate	>3.55[a]	>1.98[a]	>1.25[a]
Butyrate	>3.55[a]	>1.98[a]	>1.21[a]
Caproate	2.96	>1.79[a]	>1.17[a]
Caprylate	1.37	0.85	1.13
Caprate	0.32	0.39	1.10
Laurate	0.24	0.38	0.53

[a]No inhibition, maximum concentration tested.
Source: Ref. 66.

III. MECHANISM OF ACTION OF LIPOPHILIC COMPOUNDS

A. General Considerations

The primary, most general function of biological membranes is compartmentation. Membranes act as barriers separating different microenvironments. This is a very important and well-recognized means for metabolic regulation. The flux of substrates and cofactors from one compartment to another is limited or prevented by the presence of the membrane [81]. General passive permeation of a solute is hindered in proportion to the water solubility of the molecule; lipids must therefore be the membranous constituents making up the barrier being crossed. Water-soluble molecules or ions may be actively transported across membranes; in many instances it has been found that lipids are required for active transport [82,83].

In order to better understand or postulate a mechanism of action for lipids in membrane systems, a number of possible roles of lipids are outlined. The reader is urged to consult Lenaz's excellent review on the subject for more details [84]. Lipophilic preservatives may interfere with the following processes:

1. Molecularization and membrane formation. Reassembly of disaggregated subunits gives rise to membranes, vesicles, and functional aggregates.
2. Membrane binding. Lipids are required for membranous enzymes. Indeed, lipids have been shown to have a more general function; not only are they required for activation of several enzymes, but they also modify kinematic parameters associated with the enzyme in the absence of lipids [85,86].
3. Latency and compartmentation. The importance of compartmentation is observed at many levels of cellular organization and function [81]. Activation of lysosomal enzymes is an example of compartmentation. Treatments leading to "activation" of these enzymes include mechanical breakage, detergents, phospholipids, and proteolytic enzymes [87.

4. Transport phenomena. Transport phenomena of small molecules or ions across membranes also involve lipids. The net charge of the membrane (pH dependent), the phospholipid-sterol ratio, and the nature of the component fatty acids have large effects on ion transport [88].
5. Lipids as nonpolar media. One reason why lipids are required for certain membrane functions may be to provide a suitable microenvironment for processes requiring nonpolar media. Also, the involvement of lipids in mitochondrial oxidative phosphorylation could be the general requirement for a low dielectric constant medium.
6. Conformational changes. Macromolecules that depend on hydrophobic-hydrophilic interactions for their structure will be affected by lipids. This is particularly true for cell surfaces.

B. Specific Considerations

Some specific examples are provided in order to build a working hypothesis for understanding the mechanism of fatty acids and their esters as antimicrobial agents.

Long-chain fatty acids were shown to stimulate the oxygen uptake of *Bacillus megaterium* and *Micrococcus lysodeikticus* at pH 7.4 with concentrations which are bactericidal. These concentrations approximated those producing protoplast lysis with the fatty acids of greater chain lengths than capric acid. Higher concentrations of the fatty acids produced complete inhibition of oxygen uptake. The order of activity of the fatty acids was similar to that published for other antibacterial effects by Galbraith et al. [89] and Galbraith and Miller [90,91]. The greater effect of the cis isomer of oleic acid compared to its trans isomer, elaidic acid, was clearly demonstrated. Protoplasts and whole cells of *M. lysodeikticus* were more resistant to inhibition by the fatty acids than those of *B. megaterium*, and this reflects the relative behavior of the two organisms in the bactericidal and lytic situation [89-91]. Similarly, the limited response of the oxygen uptake mechanisms of *Pseudomonas phaseolicola* (and *E. coli*) reflected the resistance of these organisms to the bactericidal activity of the fatty acids. Partial removal of the cell wall, however, resulted in a marked sensitivity to inhibition of oxygen uptake, and this strongly suggests that the lipid in the cell wall may be protecting the cytoplasmic membrane from the fatty acids following absorption by the cells [90]. Hamilton [92] has also observed that the formation of spheroplasts from gram-negative bacteria rendered the organisms more susceptible to metabolic inhibition by lipid-soluble antibacterial agents.

These effects on the aerobic respiratory activity of the bacteria provide further evidence that the membrane is a site of action of the fatty acids, since the enzymes involved in oxygen uptake are membrane-bound [93].

In contrast to the irreversible effects associated with the inhibition of respiratory activity, the inhibition of the membrane-located transport of amino acids into cells of *B. megaterium* and *M. lysodeikticus* was observed at levels which reversibly inhibited the uptake [94]. Lower concentrations of the fatty acids stimulated uptake of amino acids into the cells. Gale and Folkes [95] and Gale and Llewellen [96] have observed this effect and also the marked inhibition of amino acid uptake in the presence of higher concentrations of oleic and linoleic acids.

The bacteriostatic nature of the inhibition of amino acid uptake in addition to the subbactericidal concentrations used was evident. The order of activity of the fatty acids was similar to that reported previously for bactericidal effects, protoplast lysis, and inhibition of oxygen uptake, with the unsaturated fatty acids more effective than the most effective saturated fatty acids, lauric and myristic acids. The inhibition of amino acid uptake into cells of *M. lysodeikticus* again required higher concentrations than were necessary to similarly affect *B. megaterium* [94].

In general, the uptake of glutamic acid was more sensitive to inhibition than that of lysine. With *S. aureus*, Hamilton [97] described the greater sensitivity of glutamic acid uptake to inhibition by tetrachlorosalicylanilide compared to that of lysine as being due to the total dependence of glutamic acid alone on energy for transport. The uptake of lysine by the bacteria studied here may also be incompletely energy-dependent and hence less sensitive to inhibition due to the postulated energy-inhibiting action of the fatty acids.

It is of interest that the fatty acids markedly inhibited the uptake of lysine and glutamic acid by *Clostridium welchii* (more effectively than either DNP or sodium azide). Harold and Baarda [98] have observed that uncouplers of oxidative phosphorylation inhibited the energy-dependent transport of amino acids anaerobically with glycolyzing cells of *Streptococcus faecalis*. They proposed that these compounds interfered with the utilization of the energy required for membrane transport, since the generation of cellular ATP and its participation in biosynthesis was not affected.

Fatty acids at subbactericidal levels also induced the leakage of ^{14}C-labeled amino acids from preloaded cells of *B. megaterium*. Sodium azide and DNP also promoted the leakage of amino acids, and it is probable that, as observed by Hamilton [97] with tetrachlorosalicylanilide, the release from the amino acid pool occurred in response to the inhibition of the energy-uncoupling systems necessary to maintain their presence in the pool. As indicated previously by Galbraith and Miller [91], the fatty acids did not appear to induce nonspecific leakage from the cells.

Sheu et al. [99] showed that in both whole cells and isolated membrane vesicles short-chain fatty acids inhibit noncompetitively the concentration uptake of amino acids. The amino acid transport system is energized by the cytochrome-linked electron transport system [100,101]. In contrast, the uptake of α-methylglucoside or fructose, which is affected by the phosphoenolpyruvate transference system [102,103], is only minutely inhibited. Long-chain fatty acids seem to have the same mechanism of action [104].

It is evident that the mode of action and the type of inhibition produced by the fatty acids depend on the concentration used. It was apparent that at high concentrations the effects were irreversible and bactericidal. Under these conditions (at pH 7.4) the fatty acids in general induced lysis of protoplasts and stimulated uptake of oxygen. In this respect their activity resembled that found with mitochondria. Scholefield [105] has reviewed the effects found with mammalian mitochondria, and these include stimulation and, at higher concentrations, inhibition of oxygen uptake, uncoupling of oxidation from phosphorylation, and the interference with inorganic phosphate transfer from ATP. As with bacteria, lauric and myristic acids were the most active saturated fatty acids, but were less effective than the C^{18} unsaturated acids. It is probable that the stimulation of oxygen at low bactericidal concentration was induced by the uncoupling of oxidation from phosphorylation, as with mitochondria, and that this effect occurred at concentrations which produced disaggregation of the membrane. That normal respiratory activity is dependent on the maintenance of membrane integrity may be deduced from the fact that stimulation of oxygen uptake occurs at the concentrations which produce lysis of protoplasts. Similarly, with mitochondria, Green and Perdue [106] have emphasized the importance of steric configuration in the functioning of the energy system, and Scholefield [105] has reported that the uncoupling activity of venom lecithinase was due to the disruption of the lipids necessary to preserve mitochondrial structure. The fatty acids could exert their bactericidal activity by producing irreversible distortions of bonding as a result of surfactant effects on the bacterial cell membrane and hence dislocate the components of the mitochondrial energy system and inhibit the synthesis of ATP. Hardesty and Mitchell [107] have demonstrated, by the interaction of fatty acids with mammalian cytochrome c, that fatty acids interact with components of the mitochondrial energy system. Hamilton [97] and Hugo and Bloomfield [108], have also demonstrated, with Fentichlor, the activity of these compounds against amino acid transport in bacterial cells and have implicated the inhibition as being due to interference with the coupling of energy to transport. These workers, however, observed that these antibacterial agents produced both stimulation of oxygen uptake and inhibition of energy-dependent amino acid transport at bacteriostatic concentrations.

The fatty acids specifically exert their activity on uncoupling of energy systems, and it does appear that the fatty acids have many properties in common with the membrane-active antibacterial agents such as tetrachlorosalicylanilide [92], trichlorophenol [109], and Fentichlor [108].

Another mechanism of action can be deduced from the fact that fatty acids are known to decrease glycolysis and to stimulate gluconeogenesis. The growth inhibition of *Bacillus subtilis* in nutrient sporulation by fatty acids was reduced by glycolytic compounds, especially glucose and fructose, but only slightly or not at all by compounds in the citric acid cycle [110]. The growth inhibition by fatty acids was less pronounced when the cells were grown in a synthetic media where they grow more slowly. Presumably, the intracellular production or uptake of the fatty acid was inhibited. Under the conditions the uptake of fatty acids are growth rate limiting in the rich media as compared to the synthetic media.

Because membranes of all organisms and cells could be affected by lipophilic substances, a comparison of growth, transport, and nerve function of human cells with that of microorganisms may reveal which preservatives are least harmful to man. The growth inhibition of bacteria has been attributed to the inhibitors of amino and keto acid transport, whereas that of human cells has remained unexplained. The labors of Sheu and collaborators have shed some light on this problem [110]. In human fibroblasts or HeLa cells neither amino acids uptake nor ATP synthesis is inhibited by fatty acids at concentrations which completely arrest growth. Therefore lipophilic drugs have different modes of action on bacterial and mammalian cells. This difference may account for their lack of toxicity in mammals. The reader is referred to the earlier discussion on cell membranes and cell walls as barriers.

The above review and discussion was based on available literature. The emphasis was primarily on lipophilic fatty acids, rather than their esters, since only one paper could be found on the possible biological mechanisms of monoesters (monoglycerides) [111]. The action of fatty acids and monoglycerides on a $NADH_2$ oxidase system was similar. The activity of the system was depressed by 50% using 0.64 nmol of fatty acids or 0.14 nmol of monoglycerides per gram of protein, respectively. The effect of these two lipid classes on cellular respiration was also studied. The results of these investigations, together with those of the testing of the inhibitory effects of the monoglycerides in various enzyme systems, showed that the monoglycerides act only on the oxygen side of the flavin of the $NADH_2$ dehydrogenase. The fatty acids, however, are less specific inhibitors acting on several sites.

It is anticipated that the above review will be helpful in postulating a mechanism of action for lipophilic preservatives in general and monoglycerides in particular. Up to this time, few such studies have been carried out.

IV. TOXICOLOGY OF LIPOPHILIC DERIVATIVES
A. Toxicology of Fatty Acids

Fatty acids as normal products found in animals and plants are not orally toxic to mammals in the usual sense of the word, except at heroic dose levels. The most complete and recent report on this subject is a report prepared by Briggs et al. [112]. They reported on the toxicity of materials C_8 to C_{18}, both saturated and unsaturated. The acids used in the study were a composite of materials obtained from 12 member companies of the Fatty Acid Producers Council. The upper level for the saturated fatty acids (C_{10} to C_{18}) was 10.0 g/kg, while octadecenoic ($C_{18:1}$) and octadecadienoic ($C_{18:2}$) were given at 21.5 ml/kg. Caprylic acids were given at the higher dosage. Some of the "toxic" effects noted for these substances at the upper levels will be described.

1. Caprylic Acid

While no deaths occurred at the 10.0 ml/kg dose, all rats given 21.5 ml/kg of this fatty acid died within 2 hr after ingestion. The rats exhibited marked depression, coma, absent righting reflex, labored respiration, and excessive salivation prior to death. The acute oral LD_{50} for the male albino rat is 14.7 ml/kg.

2. Capric Acid

Since death did not occur at the highest level tested, the oral LD_{50} for capric acid is greater than 10.0 g/kg.

While death did not occur at this high level, at 4.64 and 10.0 mg/kg rats showed excessive salivation and diarrhea. Rats at the highest dose level appeared to be depressed and had lower righting and placement reflexes and unkempt fur.

After 4 days and all succeeding days rats exhibited normal behavior and appearance.

Gross necropsies revealed no significant gross pathology.

3. Lauric Acid

The oral LD_{50} for lauric acid is greater than 10.0 g/kg, since none of the animals died. The toxicity signs at higher levels and for a period of 3 days were similar to those detailed for capric acid. The exception was the oily unkempt fur and slight emaciation which were noted for the whole observation period.

The average body weight gains were within normal limits for rats of the age, sex and strain used in the study.

4. Myristic Acid

No mortalities occurred at any dosage level tested; therefore the acute oral LD_{50} is greater than 10.0 g/kg.

Gross signs of toxicity could be seen on the first day. The unkempt fur was stained with diarrhea and a perosanguineous discharge from the nose and eyes. On the third post dosage day and for the remainder of the 2 weeks the rats appeared grossly normal.

5. Other Acids

The pattern of response of the other saturated fatty acids followed similar effects. The levels for oleic acid and linoleic acid (soya fatty acids) were similar, except the highest dose tested was 21.5 ml/kg instead of 10 g/kg. As with the high dosage for saturated fats, toxic signs persist for 2-3 days, after which the rats appeared normal.

While the oral toxicity is very low (>10 g/kg), the intravenous toxicity can be substantial. It is mainly the higher fatty acids which are toxic (LD_{50}, 23 mg/kg). With chain lengths which are lower, the toxicity decreases and is least for caproic acid (LD_{50}, 1724 mg/kg) [113].

6. Primary Skin Irritation

Caprylic Acid: The fatty acid produced necrotic tissue at each intact and abraded site at the 24-hr reading. Very slight to moderate edema was observed. All intact and abraded sites were coriaceous (leathery) at the 72-hr reading.

The primary irritation index was calculated to be 5.46.

Capric Acid: Following application of capric acid to intact and abraded skin sites of albino rabbits, the material produced necrosis or blanching in five of six intact sites and all of the abraded sites at the 24-hr reading. Very slight edema was noted at one intact and two abraded skin sites at this reading. At the 72-hr reading coriaceous tissue was observed at all intact and abraded sites, and slight edema at three intact and abraded sites.

The primary irritation index was calculated to be 4.60.

Lauric Acid: This material produced very slight erythema in five of six intact sites and three of six abraded sites, no irritation in one intact and one abraded site, and blanching in one abraded skin site at the 24-hr reading. No edema was present in either intact or abraded skin sites at this reading. No erythema was noted at any intact or four abraded sites at the 72-hr reading. However, coriaceous tissue was observed at two abraded sites at this reading. Very slight or slight edema was observed at two abraded sites only at the 72-hr reading.

The primary irritation index was calculated to be 1.12.

The other saturated fatty acids (C_{14}, C_{16}, C_{18}) all had primary irritation index of 0.0. No signs of irritation or corrosivity were observed.

In contrast to the lack of irritation in the long-chain (C_{12}) saturated fatty acids, oleic and soya fatty acids (53% linoleic acid) were irritating.

Oleic Acid: This acid produced very slight erythema in six intact and six abraded sites at the 24-hr reading only. No edema was observed at either reading.

The primary irritation index was calculated to be 0.50.

Soya Fatty Acids: This material produced spotty blanching at one intact and one abraded site, and very slight or well-defined erythema at four intact and four abraded sites at the 24-hr reading. Very slight edema was noted at one intact and one abraded site at this reading. At the 72-hr reading coriaceous tissue and slight edema were noted at one intact and one abraded site.

The primary irritation index was calculated to be 1.64.

7. Patch Test for Corrosivity

The results following patch applications of caprylic acid 1922-62, capric acid 1922-63, lauric acid 1922-64, myristic acid 1922-65, palmitic acid 1922-66, stearic acid 1922-67, stearic acid (eutectic) 1922-68, oleic acid 1922-69, and soya fatty acids 1922-70 to the skin of albino rabbits are shown in various tables of Ref. 112.

No corrosive effects were noted in rabbits tested with lauric acid, myristic acid, palmitic acid, stearic acid, oleic acid, and soya fatty acids at any time or site during the study.

Caprylic acid produced necrosis or spotted and/or entire blanching of each site at the 4-hr reading. At the 24- and 48-hr readings, entire or spotted coriaceousness was noted at each site.

Capric acid produced blanching of one site and necrosis of one site at the 4-hr reading. The remaining sites exhibited no corrosive effects at the 4-hr reading. At the 24- and 48-hr readings six sites exhibited coriaceousness, while the remaining six sites exhibited no corrosive effects.

8. Acute Eye Application

Caprylic Acid: This material produced corneal opacity and moderate or marked conjuctivitis in all rabbits, and iritis in three rabbits. In addition, blanching of the conjunctival tissues was noted in one rabbit. Irritative signs did not subside appreciably during the 72-hr observation period.

Capric Acid: The fatty acid produced corneal opacity and moderate conjunctivitis in each rabbit, and iritis in four of six rabbits. There was no decrease in irritation during the observation period.

Lauric Acid: This lipid produced corneal opacity and moderate conjunctivitis in all rabbits, and iritis in five of six rabbits. Irritative signs did not subside appreciably during the 72-hr observation period.

Myristic Acid: The results following application of myristic acid to the eyes of albino rabbits were confined to mild conjunctival erythema in three of six rabbits.

Palmitic Acid: The material produced no signs of eye irritation in any rabbit.

Stearic Acid: No signs of eye irritation were observed at any time during the study.

Oleic Acid: The material produced mild conjunctivitis in five of six rabbits. No other irritative signs were observed and all except one rabbit showed no irritative signs at the 72-hr reading.

Soya Fatty Acids: This fatty acid mixture produced mild conjunctival erythema in only four of the six eyes. No other irritative signs were observed and all eyes were clear at the 72-hr reading.

9. Summary on Fatty Acid Toxicology

The acute oral toxicity and the primary skin and acute eye irritative potentials of the test materials were evaluated in accordance with the techniques specified in the Regulations for the Enforcement of the Federal Hazardous Substances Act (Revised, *Federal Register*, September 17, 1964).

In addition, the corrosive potential of the fatty acids were evaluated in accordance with the procedure described in Section 173.240 under Title 49 of Code of Federal Regulations (*Federal Register*, February 12, 1973).

Caprylic acid and capric acid produced blanching, necrosis, and coriaceousness.

No corrosive effects were noted in any animals which were tested with lauric acid, myristic acid, palmitic acid, stearic acid, oleic acid, and soya fatty acids.

Based on these results, caprylic acid and capric acid are classified as corrosive as defined in the above-cited regulations, while the other fatty acids are not classified as such.

Capric, caprylic, and lauric acid as furnished were determined to be eye irritants. Also, two materials (caprylic acid and capric acid) proved to be corrosive under both the Federal Hazardous Substances Act and D.O.T. definitions.

A point to be made which was not discussed in the original work [112] is that the unsaturated acids were not tested for the possible presence of peroxides. Consequently, the toxicity and irritability found for oleic and linoleic acids may be due to the presence of their peroxides, which are known to be toxic. A discussion of the toxic effects of polyunsaturated fats has recently been published [114].

B. Toxicology and Safety of Glycerides

The safety for the long continued use of mono-, di-, and triglycerides was based earlier on (1) the normal occurrence of such materials in food fats and oils and (2) similarity in chemical structure and metabolism to compounds found in man. Feeding studies carried out since 1941 tended to substantiate the safety assessment of these compounds [115]. The subject of glyceride safety has been reviewed recently, with continued affirmation of their safety [116]. The oral LD_{50} of the monooleate ester has been measured as 50 g/kg in rats and greater than 25 g/kg in mice [117-119].

Hine et al. [120] reported that no measurable irritation was produced by the application of glycerin to the skin and eyes of rabbits. Subcutaneous injections of mono- and diacetin were noted by Li et al. [121] to occasionally cause local irritation in mice and rats. In the rabbit eye, 50% monoacetin caused only a slight degree of irritation, while diacetin and triacetin in similar concentrations caused marked congestion and moderate edema [121]. The daily application, for 45 days, of a 30% acetoglyceride emulsion to the skin of albino guinea pigs was reported by Ambrose and Robbins [122] to result in no local irritation of systemic reactions.

In a study in which weanling male and female Holtzman rats were fed diets with or without 50% saturated, partially acetylated monoglycerides, Hertig and Crain [123] noted the appearance of a foreign body-type reaction in the body fat occurring within 8 weeks of initiation of the test.

In the only reported feeding study using chicks, ten 1-day old chicks were fed diets containing diacetyl tartaryl glycerol monostearate, glycerol lactopalmitate, or succinylated monoglyceride at dosage levels of 570 mg/kg, 2.85 g/kg, and 8.55 g/kg for a period of 90 days [124]. Diacetyl tartaryl glycerol monostearate caused growth depression at all levels at 7 weeks, slight hyperemia of the duodenum and ileum at the lower levels, and moderate to severe hyperemia at the highest level. Gizzard erosion was observed at the high level, with only slight changes at the intermediate level. Glycerol lactopalmitate caused growth depression at 90 days. Liver weight was slightly increased and there were instances of very slight hyperemia and gizzard erosion at the intermediate level at 90 days. Succinylated monoglyceride caused very slight growth retardation at 90 days. Cecal size was increased at each level, and slight hyperemia and one instance of gizzard erosion occurred at the intermediate level.

Wretlind [125] found that intravenously administered emulsions of triacetin (LD_{50} = 1600 mg/kg) into the tail veins of mice produced almost immediate convulsions, failure of the righting reflexes, and respiratory arrest. In some animals respiration did not return and death occurred in 1-3 min; other animals started to breathe spontaneously within a few minutes and survived.

Where triglycerides of different chain lengths (C_2-C_{11}) were evaluated, there was a great difference in the toxicity of the various triglycerides. The most toxic value (82 ± 4 mg/kg) was noted for triisovalerin (C_5). With longer-chain (C_9-C_{11}) fats, the LD_{50} was more than 10 g/kg and could not be determined. The LD_{50} was lowest for triisovalerin and increased when the length of the fatty acid chain was less or greater than C_5.

For a period of 2 years, Fitzhugh et al. [126] maintained 24 rats on a basal diet with a 25% supplement of Myverol (glyceryl monostearate). Compared to a similar number of control animals, growth and longevity of the test rats was normal and detailed microscopic pathological examinations of all major organs and tissues revealed only a single change — an increase in the number of calcified renal tubular casts — attributable to the treatment.

In a study by Mattson et al. [127] in which 12 groups of 10 weanling male Sprague-Dawley rats were fed diets containing various pure mono-, di-, and triglycerides at a level of 25% for a period of 10 weeks, growth of all groups was normal and autopsies revealed no peculiarities. Ames et al. [128] reported that the feeding of monoglycerides (derived from the fatty acids of cottonseed oil) to rats for three generations disclosed no untoward effects attributable to the ingestion of the compounds. The only deleterious effect noted by Braun and Shrewsburg [115] to result from the feeding of 8-24% monostearin or monolinolein in the diets of rats for 8 weeks was a somewhat slower growth of the monostearin-treated rats. On the basis of a feeding study in which rats were fed mono-, di-, and triglycerides (prepared from a mixture of partially hydrogenated soybean and cottonseed oils) at levels of 15 or 25% of the diet for 70 days, Harris and Sherman [129] stated that these compounds exhibit no differences in caloric efficiency, nor do they produce any differences in body weight gain.

Groups of 5 male weanling albino rats were raised by Ambrose and Robbins [122] on diets containing 0, 0.25, 0.5, 1, 2, and 4% acetostearin for a period of 57 weeks. The authors suggested that the testicular hypoplasia and suppression of spermatogenesis, which was observed to various degrees in all rats, may have been the results of insufficient vitamin E in the diet. For periods from 400 to 709 days, Ambrose et al. [118] maintained groups of 10 male and 10 female rats on diets containing acetostearin AG-194, Ag-26, or Ag-31 or aceto-olein Ag-21 or Ag-3 at dietary levels of 5, 10, and 20%. No significant differences in growth were noted between the control animals and those fed acetoglycerides, except in the case of male rats on dietary levels of 20% acetostearins AG-194 and AG-31, and female rats receiving 20% AG-31 [118]. Testes of rats receiving the three acetostearin diets were significantly smaller than those of the control rats or rats fed the two aceto-oleins. The livers of male rats fed 20% acetostearin AG-194 and of both male and female rats fed 20% acetostearin AG-31 were significantly larger than those of the control rats. Livers of female rats fed 20% of either aceto-olein were smaller than those of the control rats. Kidneys of female rats fed 20% acetostearin AG-31 were significantly larger than those of the controls. Acetostearins produced a foreign body reaction in the fatty tissue of a number of the organs, presumably owing to crystalline deposits of a high melting point fat. Acetostearins AG-194, AG-26, and AG-31 produced some scarring of kidney tissue, presumably due to deposited calcium.

In a study by Mattson et al. [130] in which rats were maintained for 8-12 weeks on diets containing 15-50% of various acetin fats, it was found that acetin fats derived from the usual edible triglycerides are nutritious materials.

Growth, organ weights, and mortality were normal for all groups of 8 young male rats which were maintained for 608 days on diets containing 10% glyceryl lactopalmitate with low monoglyceride content, 10% glyceryl lacto-oleate with high or low monoglyceride content, 10% polyglycerol lacto-oleate with high monoglyceride content, or 10% acetylated tartaric acid ester of glycerol monostearate [131].

Low growth rate, attributed to poor palatability of the test diets, was observed for groups of weanling rats which were given, for a period of 7 days, a basal diet containing 20% of either monoglyceride citrate or glyceryl lactopalmitate [132].

In a study in which groups of 50 male and 50 female weanling albino rats were given diets containing 0, 0.5, 5, or 15% oxystearin for 2 years, the only apparent effect was an unusual number of Leydig cell adenomas in the testes of the rats in the 15% oxystearin diet [133].

In a paired feeding study in which the effects of feeding a diet containing 20% triacetin for a period of 7 months were examined, McManus et al. [134] observed no discrepancies between the treated animals and the controls.

In a study by Orten and Dajani [135] in which male weanling hamsters were maintained for 28 weeks on diets containing 5-15% glyceryl monostearate, the animals of the 15% groups showed a slight weight loss, while the 5% hamsters exhibited a higher weight gain than the controls; no consistent pathological changes were observed.

C. Toxicology and Safety of Polyol Esters

While the metabolism of fatty acids and triglycerides are well known, the biotransformation of ester of polyhydric alcohol is less known. What evidence is available continues to support their safety. In general, these esters are cleared from the body in the same manner as triglycerides, so that one needs only to follow the effects of the fatty acid and polyhydric alcohol. This simple concept is supported by the following data.

The chronic toxicity of 2% polysorbate 80 (oleic acid ester) was followed in rats for three generations [119]. The emulsifier was found harmless.

Polysorbate 60 (sterate ester) was studied for 2 years in rat-feeding experiments where the levels in the feed were 2, 5, 10, and 25%. While some tissue changes were noted in the liver, other microscopic changes were not seen in any other tissue [136].

In other studies [137-139] involving several polyoxyethylene sorbitan emulsifiers individually and as a mixture, it was determined that there was no cumulative toxicity over a 2-year period. There was no progressively changing response to the emulsifier over this time (four generations). No evidence was shown to indicate that the emulsifiers had any carcinogenic potential.

The polyglycerol ester, decaglyceral deca-oleate, was fed to rats at levels of 2.5, 5.0, and 10% for 90 days [140]. No adverse effects were found upon survival, organ weight, or hematological values. There were no significant microscopic tissue changes which could be attributed to dietary treatment.

There was a decreased utilization of feed by males fed polyglycerol ester at the 10% level. These data show that absorption of polyglycerol ester was not complete.

Studies dealing with the metabolization of tri- (G_3) and decaglycerol (G_{10}) and their corresponding oleic and eicosanoic acids were presented by Mitchell and Coots [141]. The data showed that the ester bonds were hydrolyzed to a large extent prior to absorption. Oleic and eicosanoic acids were absorbed via the thoracic duct. The free or partially esterified polyglycerols were not well absorbed as the free fatty acids.

In vitro hydrolysis experiments indicated that the oleic ester bond in the G_3 and G_{10} esters was cleaved as readily as is the same bond in triglycerides. Unlike glycerol, the polyglycerols were not retained appreciably in the carcass (25%).

The study by Mitchell and Coots [141] supports the earlier conclusion of Babayan et al. [142] that the fatty acid moiety of the polyglycerol ester was absorbed and utilized, as well as those of natural fats. The polyglycerol portion was not utilized but excreted nearly quantitatively [143].

The two examples given above essentially establish the principle of nontoxicity for esters of polyols. It should be expected that the split products, fatty acid and sucrose, would not create problems. This is also true for sucrose esters, but with an interesting point to be made. While the ester itself is nontoxic, it was first made by using dimethylformamide as a solvent. Under this circumstance, sucrose esters made by the dimethylformamide process are not permitted to be added to foods on account of the toxicity caused by the dimethylformamide solvent. Consequently the development of nontoxic manufacturing processes are a necessary part of the total picture and cannot be ignored. Indeed, the filing of Generally Recognized as Safe petitions in the United States includes specific directions on the manufacturing process in order to obviate this problem.

Finally, as a note of caution, esters of polyols may have biological effects which have nothing to do with toxicity or irritation per se, but may nevertheless cause clinical problems. The paper by Fisherman and Cohen [144] is a case in point. Oral challenge with Tween 80 gave positive results in 21 non-aspirin-sensitive patients with intrinsic chronic rhinitis, nasal polyps, and asthma. These studies suggest that the incidence of Tween 80 intolerance in patients with respiratory disease is about one-half the incidence of aspirin intolerance and twice the incidence of iodide intolerance. Thus these nontoxic compounds can cause or precipitate clinical problems in patients with drug idiosyncrasies.

V. SUMMARY

One concludes that fatty acids and derivatives can cause effects on microorganisms by affecting their lipid membranes (envelopes). These effects lead to perturbations of the lipid phase and subsequent changes in the organisms' permeability. Since cellular membranes and surfaces help distinguish one population from another and since the cell surface offers the first line of attack by a drug, agents designed to take advantage of a cell's individualistic characteristics may have unusual specificity. Besides their probable specificity, fatty acids and derivatives tend to be the least toxic chemical known to man. Not only are these agents nontoxic to man, but they are also actual foods and in the case of unsaturated fatty acids are essential to man for growth, development, and health.

The use of antimicrobial agents which are nontoxic and noninjurious has both legal and moral obligations. An ethically oriented manufacturer is well aware of this responsibility and acts accordingly. The concept of "zero risk," however, for any additive is impossible. In the case of additives in drug preparations, the health benefit may justifiy an assumption of some risk. Cosmetics, unlike drugs, are not essential for health. The assumption of any significant or even insignificant danger in the use of toxic preservatives would be more difficult to defend.

Today the federal Food and Drug Administration is taking a continued close look at all industrial additives, including Generally Recognized as Safe items. Those agents industry has comfortably used for 10, 20, or more years are now undergoing careful scrutiny. There is a definite move toward more detailed regulations concerning additives in our own country, as well as the rest of the world.

Where does this put the chemist, pharmacologist, and toxicologist, whose job it is to maximize effectiveness and minimize the dangers in the use of any antimicrobial preservative? Part of the answer lies in taking another look at chemicals (fatty acids and monoesters) discarded previously because of low activity. The principal attribute for these germicides today is their lack of toxicity rather than high antimicrobial activity. With our present concern of environmental disturbance by chemicals, food-grade germicidal agents offer a bright and hopeful future of safety combined with effectiveness.

REFERENCES

1. H. H. Shepard, *The Chemistry and Toxicology of Insecticides*, Burgess, Minneapolis (1939).
2. J. R. Clark, On the toxic effect of deleterious agents on the germination and development of certain filamentous fungi, *Bot. Gaz.*, *28*:289 (1899).
3. H. Reichenback, The effect of soaps on *E. coli*, *Z. Hyg. Infektionskr.*, *59*:296 (1908).
4. M. Bayliss, Effect of the chemical constitution of soaps upon their germicidal properties, *J. Bacteriol.*, *31*:489 (1936).
5. E. Kodicek, The effect of unsaturated fatty acids on gram-positive bacteria, *Soc. Exp. Biol. Symp.*, *3*:217 (1949).
6. C. Nieman, Influence of trace amounts of fatty acids on the growth of microorganisms, *Bacteriol. Rev.*, *18*:147 (1954).
7. J. J. Kabara, D. M. Swieczkowski, A. J. Conley, and J. P. Truant, Fatty acids and derivatives as antimicrobial agents, *Antimicrob. Agents Chemother.*, *2*:23 (1972).
8. A. Fleming, On the antibacterial action of cultures of a *Penicillum* with special references to their use in the isolation of *B. influenza*, *Br. J. Exp. Pathol.*, *10*:226 (1929).

9. R. R. Mellon, P. Gross, and F. B. Cooper, *Sulfanilimide Therapy of Bacterial Infections*, C. C. Thomas, Baltimore (1938).
10. E. Lueck, *Antimicrobial Food Additives*, Springer-Verlag, New York (1980).
11. A. W. Hoffman, Neue fluchtige saure der Vogelbeeren, *Ann. Chem. Pharmacol.*, 34:129 (1859).
12. H. Fleck, (1875), quoted in Strahlmann, B., Entdeckungsgeschichte antimikrobieller konservierungsstaffe fur lebensmittel, *Mitt. Geb. Lebensmittelunters. Hyg.*, 65:96 (1974).
13. E. J. Watkins, Ropiness in flour and bread and its detection and prevention, *J. Soc. Chem. Ind.*, 25:350 (1906).
14. J. E. Walker, The germicidal properties of chemically pure soaps, *J. Infect. Dis.*, 35:557 (1924).
15. J. E. Walker, The germicidal properties of soap, *J. Infect. Dis.*, 37:181 (1925).
16. J. E. Walker, The germicidal properties of soap, *J. Infect. Dis.*, 38:127 (1926).
17. H. O. Hettche, The nature of bactericidal and hemolytic constituents of lipid from *Pyocyaneus*, *Z. Immunitat.*, 83:506 (1934).
18. M. Bayliss, Effect of the chemical constitution of soaps upon their germicidal properties, *J. Bacteriol.*, 31:489 (1936).
19. E. Kodicek, The effect of unsaturated fatty acids on gram-positive bacteria, *Soc. Exp. Biol. Symp.*, 3:217 (1949).
20. P. Chaix and C. Baud, Lysing of *Glaucoma piriformis* by linoleic acid and several other fatty acids, *Arch. Sci. Physiol.*, 1:3 (1947).
21. S. Berstom, H. Theorell, and H. Davide, Effect of some fatty acids on the oxygen uptake of *Mycobact. Tubercl. Hum.* in relation to their bactericidal action, *Nature*, 157:306 (1946).
22. R. J. Dubos and B. D. Davis, Factors affecting the growth of tubercle bacilli in liquid media, *J. Exp. Med.*, 83:409 (1946).
23. R. J. Dubos and B. D. Davis, The effect of lipids and serum albumin on bacterial growth, *J. Exp. Med.*, 85:9 (1947).
24. H. Humfield, Antibiotic activity of the fatty acid-like constituents of wheat bran, *J. Bacteriol.*, 54:513 (1947).
25. C. M. McKee, J. D. Dutcher, V. Groope, and M. Moore, Antibacterial lipids from *Tetrahymena geleii*, *Proc. Soc. Exp. Biol. Med.*, 65:326 (1947).
26. R. Fuller and J. H. Moore, The inhibition of the growth of *Clostridium welchii* by lipids isolated from the contents of the small intestine of the pig, *J. Gen. Microbiol.*, 46:23 (1967).
27. A. Kiesel, The action of different acids and acid salts upon the development of *Aspergillus niger*, *Ann. Inst. Pasteur*, 27:391 (1913).
28. O. Wyss, B. J. Ludwig, and R. R. Joiner, Fungistatic and fungicidal action of fatty acids and related compounds, *Arch. Biochem.*, 7:415 (1945).
29. K. Kitajima and J. Kawamura, Antiseptic action of higher fatty acids against wood-attacking fungi, *Bull. Imp. For. Exp. Sta. Jpn.*, 31:108 (1931); *Chem. Abstr.*, 26:4693 (1932).
30. S. Tetsumoto, Sterilizing action of acids. II. Sterilizing action of saturated monobasic fatty acids, *J. Agric. Chem. Soc. Jpn.*, 9:388 (1933).

31. S. Tetsumoto, Sterilizing action of acids. III. Sterilizing action of saturated monobasic fatty acids, *J. Agric. Chem. Soc. Jpn.*, 9:563 (1933).
32. R. H. Thornton, Antifungal activity of fatty acids to *Pithomyces chartatum* (Beck and Curt) M. B. Ellis and other fungi, *N.Z.J. Agric. Res.*, 6:469 (1963).
33. E. Kodicek and A. N. Worden, Effect of unsaturated fatty acids on *Lactobacillus helveticus* and other gram-positive microorganisms, *Biochem. J.*, 39:78 (1945).
34. Y. Asami, A. Kusakabe, K. Eriguchi, M. Amemiga, A. Itabe, G. Ueno, S. Saito, Y. Sakai, and Y. G. Tanaka, Biochemical studies on antifungal agents containing iodine, *Rikagaku Kenkyusho Hokoku*, 41:259 (1965); *Chem. Abstr.*, 64:18271 (1966).
35. F. L. M. Pattison, R. L. Buchanan, and F. H. Dean, The synthesis of monofluoroalkanoic acids, *Can. J. Chem.*, 43:1700 (1965).
36. H. Gershon and R. Parmegiani, Organic fluorine compounds. II. Synthesis and antifungal properties of 2-fluoro fatty acids, *J. Med. Chem.*, 10:186 (1967).
37. G. W. Kirby, L. Atkin, and C. N. Frey, Growth of bread molds is influenced by acidity, *Cereal Chem.*, 14:865 (1937).
38. S. M. Peck and H. Rosenfeld, Effects of H-ion concentration, fatty acids, and vitamin C on the growth of fungi, *Invest. Dermatol.*, 1:237 (1938).
39. S. M. Peck, H. Rosenfeld, W. Leifer, and W. Bierman, Sweat as a fungicide — use of constituents of sweat in the therapy of fungous infections, *Arch. Dermatol. Syphilol.*, 39:126 (1939).
40. R. H. Baechler, Toxicity of normal aliphatic alcohols, acids, and sodium salts, *Proc. Am. Wood Preserv. Assoc.*, 364 (1939).
41. C. Hoffman, T. R. Schweitzer, and G. Dalby, Fungistatic properties of the fatty acids and possible biochemical significance, *Food Res.*, 4:539 (1939).
42. C. Hoffman, T. R. Schweitzer, and G. Dalby, A comparison of the fungistatic properties of enanthaldehyde and enonthic acid as related to their possible anticarcinogenic effect, *Am. J. Cancer*, 38:569 (1940).
43. N. E. Rigler and G. A. Greathouse, The chemistry of resistance of plants to *Phymatotri chum* root rot. VI. Fungicidal properties of fatty acids, *Am. J. Bot.*, 27:701 (1940).
44. C. Cowles, The germicidal action of the hydrogen ion and of the lower fatty acids, *J. Biol. Med.*, 13:571 (1941).
45. E. L. Keeney, Fungistatic and fungicidal effects of sodium propionate on common pathogens, *Bull. Johns Hopkins Hosp.*, 73:379 (1943).
46. E. L. Keeney, New preparations for the treatment of fungous infections. In vitro and in vivo experiments with fatty acid salts, penicillin and sodium sulfathiayole, *Clin. Invest.*, 23:929 (1944).
47. A. H. Eggerth, The effect of the pH on the germicidal action of soap, *J. Gen. Physiol.*, 10:147 (1926).
48. A. H. Eggerth, Effect of serum upon the germicidal action of soaps, *J. Exp. Med.*, 46:671 (1927).
49. A. H. Eggerth, Germicidal and hemolytic action of bromo soaps, *J. Exp. Med.*, 49, 50, 53:299 (1929).
50. A. H. Eggerth, Germicidal action of mercapto and disulfo soaps, *J. Exp. Med.*, 53:27 (1931).

51. W. P. Larson, The influence of the surface tension of the culture medium of bacterial growth, *Proc. Soc. Exp. Biol. Med.*, *19*:62 (1921).
52. W. P. Larson and E. Nelson, The antigenic properties of pneumococci and streptococci treated with sodium ricinoleate, *Proc. Soc. Exp. Biol. Med.*, *22*:357 (1925).
53. C. P. Miller and R. Castles, The effect of sodium ricinoleate on the gonococcus, *J. Bacteriol.*, *22*:339 (1931).
54. H. Violle, Bactericidal power of sodium ricinoleate, *Rend. Acad. Sci.*, *197*:714 (1933).
55. L. A. Barnes and C. M. Clarke, Pneumococcidal powers of sodium oleate and sodium ricinoleate, *J. Bacteriol.*, *27*:107 (1934).
56. J. A. Kolmer, A. M. Rule, and B. Madden, Chemotherapeutic studies with sodium ricinoleate, *J. Lab. Clin. Med.*, *19*:972 (1934).
57. S. Rothman, M. Smiljanic, and A. L. Shapiro, Fungistatic action of hair fat of *Microsporon audouini*, *Proc. Soc. Exp. Biol. N.Y.*, *60*:394 (1945).
58. C. E. A. Winslow and E. E. Lochridge, The toxic effect of certain acids upon typhoid and colon bacilli in relation to the degree of their dissociation, *J. Infect. Dis.*, *3*:547 (1906).
59. J. D. Reid, The disinfectant action of certain organic acids, *Am. J. Hyg.*, *16*:510 (1932).
60. C. Hoffman, T. R. Schweitzer, and G. Dalby, Fungistatic properties of the fatty acids and possible biochemical significance, *Food Res.*, *6*:539 (1939).
61. A. S. Levine and C. R. Fellers, Action of acetic acid on food-spoilage microorganisms, *J. Bacteriol.*, *39*:499 (1940).
62. A. Albert, *Selective Toxicity*, Methuen, London (1960).
63. G. G. Meynell, Some factors affecting the resistance of mice to oral infection by *Salmonella*, *Proc. R. Soc. Med.*, *48*:916 (1955).
64. H. N. Prince, Effect of pH on the antifungal activity of undecylenic acid and its calcium salt, *J. Bacteriol.*, *78*:788 (1959).
65. J. J. Kabara, A. J. Conley, D. J. Swieczkowski, I. A. Ismail, M. Lie Ken Jie, and F. D. Gunstone, Unsaturation in fatty acids as a factor for antimicrobial action, *J. Med. Chem.*, *16*:1 (1972).
66. A. J. Conley and J. J. Kabara, Antimicrobial action of esters of polyhydric alcohols, *Antimicrob. Agents Chemother.*, *4*:501 (1973).
67. J. J. Kabara, R. Vrable, and M. Lie Ken Jie, Antimicrobial lipids: Natural and synthetic fatty acids and monoglycerides, *Lipids*, *9*:753 (1977).
68. J. J. Kabara, Food-grade chemicals for use in designing food preservative systems, *J. Food Prot.*, *44*:633 (1981).
69. N. Kato and I. Shibasaki, Comparison of antimicrobial activities of fatty acids and their esters, *J. Ferment. Technol.*, *53*:793 (1975).
70. J. A. Sands, D. A. Auperin, P. D. Landin, A. Reinhardt, and S. P. Cadden, Antiviral effects of fatty acids and derivatives, in *Pharmacological Effect of Lipids* (J. J. Kabara, ed.), American Oil Chemists' Society, Champaign, Ill. (1979), p. 75.
71. I. Shibasaki and N. Kato, Combined effects on antibacterial activity of fatty acids and their esters against gram-negative bacteria, in *Pharmacological Effect of Lipids* (J. J. Kabara, ed.), American Oil Chemists' Society, Champaign, Ill. (1979), p. 15.

72. L. R. Beuchat, Comparison of anti-cibrio activities of potassium sorbate, sodium benzoate and glycerol and sucrose, esters of fatty acids, *Appl. Environ. Microbiol.*, *39*:1178 (1980).
73. S. Ueda and H. Tokunaga, Antiseptic effect of capric monoglyceride on pellicle-forming yeast, *Seasoning Sci.*, *13*:1 (1966).
74. T. Koga and T. Watanabe, Antiseptic effect of volatile fatty acid monoglycerides, *J. Food Sci. Technol.*, *15*:310 (1968).
75. I. Shibasaki, Recent trends in the development of food preservatives, *J. Food Safety*, *4*:35 (1982).
76. J. J. Kabara, Fatty acids and derivatives as antimicrobial agents — a review, in *Pharmacological Effects of Lipids* (J. J. Kabara, ed.), American Oil Chemists' Society, Champaign, Ill. (1979), p. 1.
77. A. Kato and I. Shibasaki, Combined effect of different drugs on the antibacterial activity of fatty acids and their esters, *J. Antibacterial Antifung. Agents*, *8*:355 (1975).
78. N. Kato and I. Shibasaki, Combined effect of citric and polyphosphoric acid on the anti-bacterial activity of monoglycerides, *J. Antibacterial Antifung. Agents*, *4*:254 (1976).
79. A. Kato and K. Arima, Inhibitory effect of sucrose ester of lauric acid on the growth of E. coli, *Biochim. Biophys. Res. Commun.*, *42*:596 (1971).
80. V. K. Babayan, T. G. Kaufman, H. Lehjman, and R. J. Tkaczuk, Some uses and applications of poly-glycerol esters in cosmetics and pharmaceutical preparations, *J. Soc. Cosmet. Chem.*, *15*:473 (1964).
81. D. E. Green and R. Goldberger, in *Molecular Insights in the Living Process*, Academic, New York (1967).
82. L. L. M. Van Deenen, in *Progress in Chemistry of Fats and Other Lipids*, Vol. 3, Pergamon, Oxford (1965), p. 1.
83. G. Lenaz, in *The Role of Membrane Structure in Biological Energy Transduction*, Plenum, New York (1972), p. 445.
84. G. Lenaz, The role of lipids in regulation of membrane-associated activities, *Acta Vitamin Enzymol.*, *27*:62 (1973).
85. D. G. Davis, G. Inesi, and T. Gulik-Krzywicki, Lipid molecular motion and enzyme activity in sarcoplasmic reticulum membrane, *Biochemistry*, *15*:1271 (1976).
86. H. K. Kimelberg, Protein-liposome interactions and their relevance to the structure and function of cell membranes, *Mol. Cell. Biochem.*, *10*:171 (1976).
87. J. A. Lucy, in *Lysosomes in Biology and Pathology* (J. T. Dingle and H. B. Fell, eds.), North Holland, Amsterdam (1969), p. 313.
88. K. J. Rothschild and H. E. Stanley, Models of ionic transport in biological membranes, *Am. J. Clin. Pathol.*, *63*:695 (1975).
89. H. Galbraith, T. B. Miller, A. M. Patton, and J. K. Thompson, Antibacterial activity of long-chain fatty acids and the reversal with calcium, magnesium, ergocalciferol and cholesterol, *J. Appl. Bacteriol*, *34*:803 (1971).
90. H. Galbraith and T. B. Miller, Effect of metal cations and pH on the antibacterial activity and uptake of long chain fatty acids, *J. Appl. Bacteriol.*, *36*:635 (1973).
91. H. Galbraith and T. B. Miller, Physicochemical effects of long-chain fatty acids on bacterial cells and their protoplasts, *J. Appl. Bacteriol.*, *36*:647 (1973).

92. W. A. Hamilton, The mode of action of membrane-active antibacterials, *FEBS Symp.*, 29:71 (1970).
93. R. Storek and J. T. Wachsman, Enzyme localization in *Bacillus megaterium*, *J. Bacteriol.*, 73:784 (1957).
94. H. Galbraith and T. B. Miller, Effect on long-chain fatty acids on bacterial respiration and amino acid uptake, *J. Appl. Bacteriol.*, 36:659 (1973).
95. E. F. Gale and E. S. Folkes, The effect of lipids on the accumulation of certain amino acids by *Staphylococcus aureus*, *Biochim. Biophys. Acta*, 144:161 (1967).
96. E. F. Gale and J. J. Llewellen, The role of hydrogen and potassium ions in the transport of acidic amino acids in *Staphylococcus aureus*, *Biochim. Biophys. Acta*, 266:182 (1971).
97. W. A. Hamilton, The mechanism of the bacteriostatic action of tetrachlorosalicylanilide, a membrane-active antibacterial compound, *J. Gen. Microbiol.*, 50:441 (1968).
98. F. M. Harold and J. R. Baarda, Inhibition of membrane transport in *Staphylococcus faecalis* by uncouplers of oxidative phosphorylation and its relationship to protein conduction, *J. Bacteriol.*, 96:2025 (1968).
99. C. W. Sheu, W. N. Konings, and E. Freese, Effects of acetate and other short-chain fatty acids on sugar and amino acid uptake of *Bacillus subtilis*, *J. Bacteriol.*, 111:525 (1972).
100. H. R. Kaback and L. S. Milner, The relationship of a membrane-bound D(-) lactic dehydrogenase to amino acid transport in isolated bacterial membrane preparations, *Proc. Nat. Acad. Sci. U.S.A.*, 66:1008 (1970).
101. W. N. Konings and E. Freese, L-serine transport in membrane vesicles of *Bacillus subtilis* energized by NADH or reduced phenazine methosulfate, *FEBS Lett.*, 14:65 (1971).
102. H. R. Kaback, Transport, *Annu. Rev. Biochem.*, 39:561 (1970).
103. W. Kundig and S. Roseman, Sugar transport. I. Isolation of a phosphotransferase system from *Escherichia coli*, *J. Biol. Chem.*, 246:1393 (1971).
104. C. W. Sheu and E. Freese, Lipopolysaccharide layer protection of gram-negative bacteria against inhibition by long-chain fatty acids, *J. Bacteriol.*, 115:869 (1973).
105. P. G. Scholefield, Fatty acids and their analogues, in *Metabolic Inhibitors*, Vol. 1 (R. M. Hochster and J. H. Quastel, eds.), Academic, New York (1963).
106. D. E. Green and J. F. Perdue, Membranes as expressions of repeating units, *Proc. Nat. Acad. Sci. U.S.A.*, 55:1295 (1966).
107. B. A. Hardesty and H. K. Mitchell, The interaction of fatty acids with mammalian cytochrome c., *Arch. Biochem. Biophys.*, 100:1 (1963).
108. W. B. Hugo and S. F. Bloomfield, Studies of the mode of action of the phenolic antibacterial agent Fentichlor against *Staphylococcus aureus* and *Escherichia coli*. III. The effect of Fentichlor on the metabolic activities of *Staphylococcus aureus* and *Escherichia coli*, *J. Appl. Bacteriol.*, 34:579 (1971).
109. P. A. Wolf and M. M. Schaeffer, The role of cell membrane permeability in determining the antimicrobial activity of 2, 4, 6-trichlorophenol at pH 6 and pH 8, in *The Biodeterioration of Materials*

(A. H. Walters and J. J. Elphick, eds.), Elsevier, London (1968).
110. C. W. Sheu, D. Salomon, J. L. Simmons, T. Sreevalsan, and E. Freese, Inhibitory effects of lipophilic acids and related compounds on bacteria and mammalian cells, *Antimicrob. Agents Chemother.*, 7:349 (1975).
111. C. Coutelle and T. Schewe, Fettsauren and monoglyzeride aus homogenaten von froscheiern und kaulquappen als hemmstoffe der afmungs kette, *Acta Biol. Med. Ger.*, 25:47 (1970).
112. G. B. Briggs, R. L. Doyld, and J. A. Young, Safety studies on a series of fatty acids, *Am. Ind. Hyg. Assoc. J.*, April, 251 (1976).
113. L. Oro and A. Wretlind, Pharmacological effects of fatty acids, triolein, and cottonseed oil, *Acta Pharmacol. Toxicol.*, 8:141 (1961).
114. H. Kaunitz, Toxic effects of polyunsaturated vegetable oils, in *Pharmacological Effects of Lipids* (J. Kabara, ed.), American Oil Chemists' Society, Champaign, Ill. (1979), p. 203.
115. W. Q. Braun and C. L. Shrewsburg, Nutritive properties of monoglycerides, *Oil Soap*, 18:249 (1941).
116. Anonymous, GRAS Food Ingredients — Glycerine and Glycerides, Ordering Number PB-221 227, National Technical Information Service, U.S. Department of Commerce, Springfield, Va. (1973).
117. E. Eagle and C. E. Poling, Oral toxicity and pathology of polyoxyethylene derivatives in rats and hamsters, *Food Res.*, 21:348 (1956).
118. A. M. Ambrose, D. J. Robbins, and A. J. Cox, Jr., Studies on the long-term feeding of acetoglycerides to rats, *Food Res.*, 23:536 (1958).
119. P. H. Elworthy and J. F. Treon, in *Nonionic Surfactants*, Vol. 1 (M. J. Schick, ed.), Marcel Dekker, New York (1967), p. 923.
120. C. H. Hine, H. H. Anderson, H. D. Moon, M. K. Dunlap, and M. S. Morse, Comparative toxicities of synthetic and natural glycerol, *Arch. Ind. Hyg. Occup. Med.*, 7:282 (1953).
121. R. C. Li, P. P. T. Suh, and H. H. Anderson, Acute toxicity of monoacetin, diacetin, and triacetin, *Proc. Soc. Exp. Biol. Med.*, 46:26 (1941).
122. A. M. Ambrose and D. J. Robbins, Toxicity of acetoglycerides, *J. Am. Pharm. Assoc. Sci. Ed.*, 45:282 (1956).
123. D. C. Hertig and R. C. Crain, Foreign-body type reaction in fat cells, *Proc. Soc. Exp. Biol. Med.*, 98:347 (1957).
124. Anonymous, Food Additive Petition, Number 771.
125. A. Wretlind, The toxicity of low molecular triglycerides, *Acta Physiol. Scand.*, 40:338 (1957).
126. O. G. Fitzhugh, A. R. Bourke, A. A. Nelson, and J. P. Frawley, Chronic oral toxicities of four stearic acid emulsifiers, *J. Toxicol. Appl. Pharmacol.*, 1:315 (1959).
127. F. H. Mattson, F. J. Baur, and L. W. Beck, The comparative nutritive values of mono-, di-, and triglycerides, *J. Am. Oil Chem. Soc.*, 28:386 (1951).
128. S. R. Ames, M. P. O'Grady, N. D. Embree, and P. O. Harris, Molecularly distilled monoglycerides. III. Nutritional studies on monoglyceride derived from cottonseed oil, *J. Am. Oil Chem. Soc.*, 28:31 (1951).
129. R. S. Harris and H. Sherman, Comparison of the nutritive values of mono-, di-, and triglycerides by a modified pair-feeding technique, *Food Res.*, 19:257 (1954).

130. F. H. Mattson, J. C. Alexander, F. J. Baur, and H. H. Reller, Short-term feeding studies on acetin fats, *J. Nutr.*, *59*:277 (1956).
131. Anonymous, Food Additive Petition Number 884.
132. Anonymous, Food Additive Petition Number 31.
133. Anonymous, Food Additive Petition Number 35.
134. T. B. McManus, C. B. Bender, and O. F. Ganett, A comparison of acetic acid, fed in the form of triacetin, with glucose as a nutrient in feeds, *J. Dairy Sci.*, *26*:13 (1943).
135. J. M. Orten and R. N. Dajani, A study of the effects of certain food emulsifiers in hamsters, *Food Res.*, *22*:529 (1957).
136. B. F. Chow, J. M. Burnett, C. T. Ling, and L. Barrows, Effects of basal diets on the response of rats to certain dietary non-ionic surface active agents, *J. Nutr.*, *49*:563 (1953).
137. B. L. Oser and M. Oser, Nutritional studies on rats on diets containing high levels of partial ester emulsifiers. III. Clinical and and lactation, *J. Nutr.*, *60*:489 (1956).
138. B. L. Oser and M. Oser, Nutritional studies on rats on diets containing high levels of partial ester emulsifiers. II. Clinical and metabolic observations, *J. Nutr.*, *61*:149 (1957).
139. B. L. Oser and M. Oser, Nutritional studies on rats on diets containing high levels of partial ester emulsifiers. IV. Mortality and post-mortem pathology; general conclusions, *J. Nutr.*, *61*:235 (1957).
140. W. R. King, W. R. Michael, and R. H. Coots, Subacute oral toxicity of polyglycerol esters, *Toxicol. Appl. Pharmacol.*, *20*:327 (1971).
141. W. R. Michael and R. H. Coots, Metabolism of polyglycerol and polyglycerol esters, *Toxicol. Appl. Pharmacol.*, *20*:334 (1971).
142. V. K. Babayan, H. Kaunitz, and C. A. Slanetz, Nutritional studies of polyglycerol esters, *J. Am. Oil Chem. Soc.*, *41*:434 (1964).
143. V. K. Babayan, T. G. Kaufman, H. Lehman, and R. J. Traczuk, Some uses and applications of polyglycerol esters in cosmetic preparations, *J. Soc. Cosmet. Chem.*, *15*:473 (1964).
144. E. W. Fisherman and G. N. Cohen, Aspirin and other cross-reacting small chemicals in known aspirin intolerant patients, *Ann. Allergy*, *32*:307 (1974).

17
LAURICIDIN
The Nonionic Emulsifier with Antimicrobial Properties

JON J. KABARA *Michigan State University, East Lansing, Michigan*

    I.   Background Information 305

   II.   Name and Synonyms 306

 III.   Chemical and Physical Properties 306

       A. Synthesis of monoglycerides 306
       B. Crystalline formation 306
       C. Analysis of Lauricidin 307

 IV.   Microbiological Data 308

  V.   Toxicology and Safety of Monoglycerides 312

       A. Skin irritation study 313
       B. Eye irritation study 316
       C. Delayed dermal sensitization 316
       D. Acute studies 317
       E. Chronic studies 317

 VI.   Application of Lauricidin in Cosmetics 317

VII.   Regulatory Status of Lauricidin 318

       References 320

I. BACKGROUND INFORMATION

In studies which began in 1965 the biological effects of lipids on simple cell systems were initiated. One objective of this research was to determine the growth effect of simple lipids. At this same time the problem of cell-cell communication was a residing factor in our neurochemical research. Both programs were proceeding in parallel when common denominators began to appear based on structure-function relationships. These studies have been reviewed for the cosmetic [1] and food industries [2], while our neurochemical findings have only recently been published [3,4]. It was from these diverse backgrounds that the antimicrobial lipid Lauricidin was discovered.

II. NAME AND SYNONYMS

Lauricidin is a registered trademark for glyceryl monolaurate. The Chemical Abstract Service (CAS) Registry Number is 27215-38-9. Monolaurin has the empirical formula $C_{15}H_{30}O_4$, which gives it a molecular weight of 274.4.

Lauricidin as a monester of lauric acid and glycerine exists in two isomeric forms:

$$\begin{array}{cc}
\begin{array}{l}
\text{H} \quad\ \ \text{O} \\
\ |\quad\ \ \ \ || \\
\text{H}-\text{C}-\text{O}-\text{C}(\text{CH}_2)_{10}\text{CH}_3 \\
\ | \\
\text{H}-\text{C}-\text{OH} \\
\ | \\
\text{H}-\text{C}-\text{OH} \\
\ | \\
\text{H} \\
\\
\alpha\ \text{form}
\end{array}
&
\begin{array}{l}
\text{H} \\
\ | \\
\text{H}-\text{C}-\text{OH} \\
\ | \\
\text{H}-\text{C}-\text{O}-\text{C}(\text{CH}_2)_{10}\text{CH}_3 \\
\ | \\
\text{H}-\text{C}-\text{OH} \\
\ | \\
\text{H} \\
\\
\beta\ \text{form}
\end{array}
\end{array}$$

III. CHEMICAL AND PHYSICAL PROPERTIES

A. Synthesis of Monoglycerides

Industrial monoglyceride can be prepared in several ways. The direct esterification of glycerol with a fatty acid yields mixtures of mono-, di-, and triglycerides, depending on the molar ratio of the reactants. A common commercial method of preparation for monoglycerides is the glycerolysis of fats and oils, a transesterification reaction. Depending on the reaction conditions, the products are usually mixtures of 40 or 60% monoglycerides, the rest being di-, and triglycerides [5]. The usual commercial monolaurin fits these specifications and is biologically inactive. It is because of this low monoester content that molecular distillation is necessary to obtain a monoester level of 90% or greater. When pure monoglycerides are desired, special methods are required for their preparation [6-8].

The lauric acid which is used for esterification is derived generally from oil of species of palm, such as coconut and babassu. The fats from these sources are distinguished from other oils by their high (40-50%) lauric acid content. Fortunately, the supply of these palm oils are from trees which are perennial and offer a stable source of renewable supply.

B. Crystalline Formation

Because lauric acid is saturated, the monoglyceride tends to be very stable under usual storage conditions. The monoglyceride, however, exists in two isomeric forms, the alpha (α) and the beta (β) forms. The relative amount of the two isomers is dependent upon temperature and age. Since the distilling temperature in the preparation of monoglyceride is rather high, the α-mono content (80-85%) will be lower at first, even though the monoester content is 94-96%. With time the α-form becomes the more stable form and reaches values of 90-95%. Fortunately, both isomeric forms have the same biological activity [9].

One of the chief characteristics of monoglycerides is to form lyotropic mesophases. This property of the monoglyceride is important to recognize in the application of monoesters in water systems [10]. This can best be appreciated by the phase diagram of a saturated monoglyceride in an aqueous system (Fig. 1).

Lauricidin is a distilled monoglyceride which is an off-white powder. The lipid is very slightly soluble in water and soluble in alcohols, chlorinated hydrocarbons, and vegetable oils. Solubility data for Lauricidin in some common solvents are presented in Table 1.

C. Analysis of Lauricidin

Lauricidin can easily be assayed by a number of classic procedures used in lipid biochemistry. The following constituents in the usual commercial product are assayed by the following methods:

1. Free glycerol (maximum 4%). The method is based on the second edition of the *Food Chemical Codex* (FCC) (1972, p. 904). Glycerol is determined from the amount of periodic acid consumed in the oxidation of adjacent hydroxyl groups. The results are calculated as the percentage of glycerol in the product by weight.
2. Free fatty acid (maximum 1%). The amount of free fatty acid is based on the American Oil Chemists' Society Ca 5_a-40 method. Too high a value suggests that the quality of the product is poor.

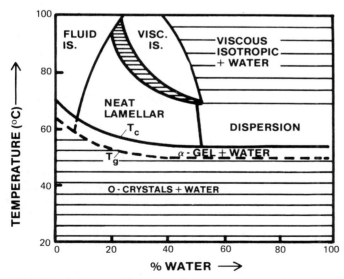

FIGURE 1 Phase diagram of an aqueous system of a saturated distilled monoglyceride (fatty acid composition as fully hydrogenated lard). (From Ref. 10.)

TABLE 1 Solubility of Lauricidin at 25°C

Solvent	Solute dissolved (g/100 ml)
Methanol	250
Ethanol (absolute)	80
Isopropyl alcohol	60
Propylene glycol	4.5
Glycerine	0.2
Mineral oil	0.2
Water	<0.01

3. Melting point (m.p. 55-56°C). The sample is solidified in an open-end capillary tube. The procedure is designated as the softening point by the American Oil Chemists' Society.
4. Iodine value (maximum 0.5). The iodine value is a measure of the unsaturation found in the product and is expressed in the percentage of iodine absorbed. The method used is based on DGF Einheitsmethoden C-V 11d.
5. Saponification value (200-206). The value is expressed as the number of milligrams potassium hydroxide required to saponify 1.0 g of sample. The method is based on the second edition of the FCC (1972, p. 916).
6. Monoester content (90%). The method (American Oil Chemists' Society Cd 11.57, 1971) determines alpha (or 1-mono) from the periodic acid consumed in the oxidation of the adjacent hydroxyl groups. A preferable method would be to use classic thin-layer [11] column [12] or gas-liquid chromatography [13,14].
7. Fatty acid composition. The monoglyceride is saponified and methyl ester formed simultaneously. The methyl esters of the fatty acids are separated and determined by gas chromatography. The fatty acid profile of Lauricidin is represented by C_{10} (maximum 10%), C_{12} (maximum 90%), and C_{14} (maximum 8%).

IV. MICROBIOLOGICAL DATA

The discovery of Lauricidin as a potential antimicrobial agent was made possible by the screening of various related lipid derivatives [9,15-18]. It was only after testing some 600 derivatives that it was decided to focus attention on this particular monoglyceride. This decision was based on the biological activity of this particular derivative and, more important, its lack of toxicity.

The optimum activity for monoglycerides occurred at chain length of C_{12} when tested against gram-positive bacteria (Fig. 2). Optimum activity against yeast, fungi, and molds occurred for monoglyceride derivatives which were shorter in chain length (C_{10}, C_{11}). Mycobacteria and mycoplasma organisms were also found to be affected at very low concentrations by Lauricidin (C_{12}, unpublished information). One of the more exciting effects found for Lauricidin is its effect on envelope viruses [19 and references therein].

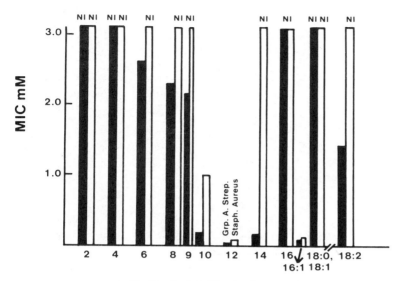

FIGURE 2 Effect of monoglycerol esters of various chain lengths on *Staphylococcus aureus* and *Streptococcus pyogenes* (NI, no inhibition; dark bars, S. pyogenes; light bars, S. aureus).

Monoglycerides had been previously postulated to be one of the active nonprotein fractions in mothers' milk which affected viruses and bacteria [20]. The antiviral effects of Lauricidin were more closely examined by Hierholzer and Kabara [21]. In these studies Lauricidin was tested in cell culture against 16 human RNA- and DNA-enveloped viruses. At relatively low concentrations (1.0%) all viruses were reduced in infectivity by 99.9% in 1 hr or less at 23°C.

Despite its wide spectrum of activity against diverse genera, Lauricidin used alone is generally inactive against gram-negative bacteria. This will be discussed later.

A natural question to be raised for a new preservative is, How does it compare with other cosmetic preservatives which are commonly used? This question was answered nicely by my Japanese friend Professor Shibasaki [22]. In Table 2 the activity of monolaurin was favorably compared against butylparaben, sodium lauryl sulfate, sorbic acid, and dehydroacetic acid. For the three fungal organisms used in the screen, monolaurin was as active or more active than the other cosmetic preservatives. The preservatives had the same relative activity when tested against gram-negative organisms (see Table 3).

As stated earlier, commercial monolaurin represents mixtures of mono-, di-, and triglycerides of lauric acid. It was of interest to prepare a highly purified monoester from less pure lauric acid, that is, fatty acids obtained from the lower cuts of coconut oil, C_8, and C_{10} fatty acids. Three such preparations were synthesized and tested. The results of these monoglycerides against various fungi are present in Table 4. The data shows that monoglycerides can be designed to affect certain organisms better than

TABLE 2 Comparison of Antifungal Activities of Fatty Acid Esters and Some Commonly Used Preservatives

Food additive	Minimum inhibitory concentration (μg/ml)[a]		
	Aspergillus niger	Candida utilis	Saccharomyces cerevisiae
Monocaprin	123	123	123
Monolaurin	137	69	137
Butyl-p-hydroxybenzoate	200	200	200
Sodium lauryl sulfate	100	400	100
Sorbic acid	1000	1000	1000
Dehydroacetic acid	100	200	200

[a]By the agar dilution method.
Source: Ref. 23.

others, depending upon the exact composition of the monoglyceride. The Lauricidin which is presently commercially available has been designed to take advantage of this fact.

As with any preservative, Lauricidin has limitations. This is particularly true for Lauricidin, since it lacks activity against gram-negative organisms. In general, the latter group, because of their lipopolysaccharide envelope, tends to be more resistant to chemical agents. The lack of gram-negative effectiveness can be overcome by the use of Lauricidin and some

TABLE 3 Comparison of Antibacterial Activities of Fatty Acid Esters and Some Commonly Used Preservatives

Food additive	Minimum inhibitory concentration (μg/ml)[a]		
	Bacillus subtilis	Bacillus cereus	Staphylococcus aureus
Sucrose dicaprylin	74	74	148
Monocaprin	123	123	123
Monolaurin	17	17	17
Butyl-p-hydroxbenzoate	400	200	200
Sodium lauryl sulfate	100	100	50
Sorbic acid	4000	4000	4000

[a]By the agar dilution method.
Source: Ref. 23.

TABLE 4 Antifungal Activity of Lauricidin and Derivatives (at a concentration of 0.5%)[a]

Fungus	Lauricidin	Lauricidin 812	Lauricidin 802	Lauricidin 112
Pythium elongatum	32.6	100	100	88.0
Phytophthora citrophthora	70.5	100	100	89.1
Mucor circinelloides	56.4	55.3	64.2	82.7
Rhizopus stolonifer	29.8	42.7	69.9	79.4
Aspergillus flavus	73.6	75.0	65.8	64.5
Aspergillus ochraceus	53.4	100	100	70.3
Aspergillus niger	71.7	100	89.6	82.7
Penicillium digitatum	76.1	100	100	100
Penicillium patulum	34.3	100	38.3	40.1
Fusarium graminearum	57.5	100	100	66.1
Saccharomyces cerevisiae	56.5[b]	51.7[b]	92.7[b]	100[b]
Candida albicans	0[b]	0[b]	35.5[b]	100[b]
Rhodotorula mucilaginosa	9.2[b]	50.0[b]	100[b]	100[b]
Sporobolomyces sp.	79.9[b]	91.2[b]	90.4[b]	100[b]
Sclerotium rolfsii	65.6	93.4	98.7	99.1
Rhizoctonia solani	61.7	85.8	89.7	89.3

[a]Results are expressed as the percentage of mycelial growth inhibition.
[b]Compound concentration was 0.05%.
Source: Ref. 24.

type of chelating agent. The use of the monoglyceride with ethylenediaminetetraacetic acid (EDTA) was found to be very effective, particularly when used to inhibit or kill Pseudomonas aeruginosa [1].

Most recently the application of Lauricidin and EDTA solved a recurring problem involving the resistance of an in-house organism to usual cosmetic germicides [23]. The gram-negative organism (Enterobacter cloacae) was shown to be resistant to 10% benzalkonium chloride, 25% sodium dodecyl sulfate, 25% Triton X-100, and 25% Tween 20, 40, 60, or 80. Lauricidin did not affect this organism at a level of 5.0%, while EDTA inhibited growth at 0.20% or greater. A combination of 0.5% Lauricidin and 0.001% EDTA was inhibitory. Thus true synergism was seen for this combination against a "house-problem organism."

Apparently the presence of a chelating agent is not always necessary for gram-negative activity, since Beuchat found that Lauricidin demonstrated inhibitory action against Vibrio parahaemolyticus [24]. Vibrio bacteria reacts negatively to the Gram stain. It may be that in certain instances, as in this case, there is no correlation between the Gram stain and lipopolysaccharide membrane. In Beuchat's paper the C_{12} monoglyceride was more active than lower (C_8, C_{10}) or higher (C_{14}) chain length derivatives. Also the low minimum inhibitory concentration for monolaurin (5 µg/ml or less) showed it to be more effective than sorbic acid (70 µg/ml) or sodium benzoate (300 µg/ml).

In a number of studies Lauricidin acts additively or synergistically with sorbic acid [1]. This conclusion was substantiated by Lisker and Paster [25], but denied by Robach et al. [26]. Because the latter group used optical density to follow growth, and did not correct for pH differences when comparing the anionic sorbic acid with the nonionic monoglyceride, it is difficult to accept their negative findings. Also, their results indicate that 62.5 ppm monolaurin was as effective in inhibiting "growth" of *Salmonella typhimurium* 13311 as 500 ppm of sorbic acid. Consequently, even their own data show that a combination of the two preservatives would be additive.

V. TOXICOLOGY AND SAFETY OF MONOGLYCERIDES

While only a small portion of existing data has direct reference to the toxicology of Lauricidin, a great deal of knowledge has been gathered summarizing the toxicology and safety of monoglycerides in general. Since there is little to suggest that glycerides of different fatty acids are metabolized in specific manners, except for chains of less or greater than C_{12}, the available data forwarded in this chapter reflect the oral toxicology of monoglycerides in general and topical biological effects of monolaurin more specifically.

The chemistry of mono-, di-, and triglycerides was worked out many years ago, and it was long theorized that triglycerides were broken down to mono- and diglycerides in the body during the process of fat digestion. During the last 30 years it has been repeatedly demonstrated that this is what actually happens. Enzymes, mostly pancreatic enzymes, hydrolyze triglyceride to monoglyceride and free fatty acid [27,28]. These are absorbed into the intestinal wall and are used there to resynthesize triglycerides, which are transported on into the body through the lymphatic system [29-31]. The monoglyceride that passes through the intestinal wall may be formed in the digestive tract by hydrolysis of triglyceride, or it may have been directly provided by the diet; it makes no difference, for the same compound is involved internally in either case.

This is clearly indicated where rats were given either triglyceride or mixtures of triglycerides with mono- and diglyceride. The material in their digestive tracts was analyzed 3 hr later. The same proportions of 1-monoglyceride, 2-monoglyceride, diglyceride, triglyceride, and free fatty acid were found, regardless of which material was fed, except when very high dosage levels of 2-monoglyceride had temporarily overwhelmed the capacity of the system to interconvert the different materials [32].

Nutrition experiments showed that monoglycerides serve very efficiently as energy sources. They provide just as much energy, after a slight correction is made for the fact that partial glycerides contain a larger proportion of oxygen than triglycerides [33,34].

A key point in the safety assessment of the partial glycerides is the conviction that it must be safe to consume small, incremental amounts of materials that are formed in the body naturally, continuously, and abundantly. This seems very reassuring, but a few more questions can be raised. One of these has to do with the possibility that monoglyceride in foods will influence the digestion of other nutrients by lowering the surface tension in the digestive system. This was tested by Dasher [35], who showed that although monoglyceride added to a cottonseed oil-buffer system does indeed lower the interfacial tension, this effect becomes insignificant when the interfacial tension has already been greatly reduced by the bile salts that are normally present in the digestive tract. Later work showed that monoglycerides help solubilize triglycerides into a solution of bile acid conjugates and thus aid in the digestion of fat [36].

Since feeding monoglycerides may cause some effect that would not be anticipated from all that we know about their chemical and biological properties, feeding tests need to be conducted. A number of feeding studies have been carried out with mono- and diglycerides dating back as far as 1941 [37].

Several mono- and diglycerides were fed to rats for 10 weeks as 15 or 25% of the diet (see Table 5). Growth and feed efficiencies varied with the fatty acid compositions of the glycerides, but not with the glyceride type. The composition of depot fat laid down by the animals reflected fatty acid composition, but not glyceride type. Necropsy and histology revealed no adverse effects that could be attributed to the partial glycerides [33].

In other experiments three successive generations of rats were given diets that contained either 15 or 25% monoglycerides as the sole dietary fat. Growth, reproduction, lactation, and lipid absorption were no different from control animals fed cottonseed oil [38].

A few adverse effects have been reported when high concentrations of partial glycerides containing saturated fatty acids were fed to animals. These effects, which are related to poor absorption of saturated fatty acids, were much the same as those seen whenever high concentrations of any saturated lipid are fed [37].

In a study by Mattson et al. [39], 12 groups of 10 weanling male Sprague-Dawley rats were fed diets containing various pure mono-, di-, or triglycerides at a level of 25% for a period of 10 weeks. The mono-, di-, and triglycerides of corresponding fatty acid composition were found to be of equivalent caloric efficiency. The caloric efficiencies of the glycerides of pure lauric or stearic acid were found to be low, possibly as a result of poor absorption. The histological examination of the liver, kidney, spleen, lung, heart, stomach, and small and large intestines, however, revealed no abnormalities. The body fat of the animals was the same regardless of the type of glyceride structure fed; however, the type of body fat deposited reflected, in part, the fatty acid component of the dietary glyceride.

The entire subject of glyceride safety recently has been reevaluated with very reassuring results [40]. The safety of these compounds is beyond question.

A. Skin Irritation Study

The test material, Lauricidin, was prepared as a 20% emulsion by the addition of 10 parts by weight of powdered samples to a mixture of 37.5 parts sterile distilled water at 68°C and 2.5 parts of 0.1 M sodium hydroxide solution. The preparation was kept at a temperature of 65°C for 15 min with slow agitation, and then cooled to ambient room temperature prior to application.

The skin irritation patch test was carried out as described in *The Federal Register*, Vol. 38, No. 187, Section 1500:41, 1973.

The sample (0.5 ml, 20% emulsion) was applied to the abraded and nonabraded skin of six albino rabbits. The sites were examined after 24 and 72 hr for erythema and edema.

TABLE 5 Average Cumulative Values and Standard Errors of the Mean[a] for Gain in Body Weight, Food Consumption, and Caloric Efficiency

	Week 3			Week 6			Week 10		
	Body weight gain	Food consumed	Caloric efficiency[b]	Body weight gain	Food consumed	Caloric efficiency[b]	Body weight gain	Food consumed	Caloric efficiency[b]
Soybean oil/cottonseed oil (partially hydrogenated)									
Triglyceride	132.9 ±2.7	225.7 ±4.7	11.1 ±0.15	240.0 ±12.6	538.1 ±12.6	8.51 ±0.11	305.0 ±7.7	938.1 ±25.0	6.20 ±0.11
Mono + di + tri	133.3 ±3.4	227.0 ±5.4	11.3 ±0.12	242.0 ±6.7	542.8 ±12.4	8.65 ±0.09	314.3 ±8.8	957.3 ±21.7	6.37 ±0.08
Diglyceride	129.3 ±5.9	223.3 ±7.71	11.1 ±0.19	236.2 ±7.4	534.7 ±14.7	8.49 ±0.05	311.8 ±14.8	930.8 ±31.9	6.41 ±0.12
Monoglyceride	122.7 ±3.1	212.8 ±3.2	11.5 ±0.16	207.2 ±9.4	493.6 ±12.6	8.34 ±0.20	275.8 ±9.1	866.5 ±20.3	6.30 ±0.08
Olein									
Triglyceride	115.9 ±2.7	200.4 ±5.0	10.9 ±0.15	200.1 ±3.9	491.8 ±8.0	8.52 ±0.09	287.9 ±6.1	867.2 ±15.4	6.32 ±0.07
Monoglyceride	112.5 ±2.8	198.1 ±3.6	11.1 ±0.20	210.6 ±5.0	475.7 ±8.4	8.77 ±0.13	270.0 ±4.9	850.0 ±11.7	6.29 ±0.08

Sterin								
Triglyceride	108.0 ±1.5	319.4 ±2.0	0.40 ±0.10	200.0 ±1.2	773.7 ±7.3	5.02 ±0.09	274.0 ±4.4	1417.0 ±15.2	3.64 ±0.05
Monoglyceride	95.5 ±1.0	281.0 ±3.0	0.58 ±0.10	202.6 ±2.8	716.6 ±10.0	6.50 ±0.00	271.0 ±3.1	1314.0 ±14.8	4.01 ±0.06
Laurin									
Triglyceride	113.2 ±3.3	225.3 ±4.7	9.75 ±0.18	203.9 ±7.5	518.3 ±10.2	7.60 ±0.17	276.6 ±10.1	931.7 ±19.3	5.77 ±0.90
Monoglyceride	75.0 ±2.0	171.1 ±5.4	8.99 ±0.26	155.4 ±3.3	418.7 ±10.5	7.61 ±0.12	214.8 ±5.0	779.9 ±18.4	5.65 ±0.06
Coconut oi									
Triglyceride	140.8 ±2.2	239.1 4.7	11.4 ±0.11	265.5 ±4.4	605.4 ±10.6	8.90 ±0.14	333.8 ±6.4	991.9 ±18.6	6.60 ±0.08
Mono + di + tri	130.0 ±3.0	222.5 ±4.4	11.6 ±0.11	237.1 ±6.1	624.8 ±6.8	8.83 ±0.07	304.9 ±7.3	923.9 ±12.9	6.54 ±0.11

[a]Standard error of the mean calculated by the formula $\sum(x - x')^2/N(n - 1)$ where x is the individual value, x' is the group mean, and n is the number of observations.
[b]Body weight gain divided by the number of calories consumed × 100 equals the caloric efficiency.
Source: Ref. 35.

The test results were the following:

Sample	Number of animals	Skin irritation Erythema	Edema	Primary irritation score
Lauricidin, 20% concentration	6	7.8	7.8	3.9

The scores given for erythema and edema are the average of the scores obtained for the six rabbits.

A 20% aqueous concentration of Lauricidin produced a primary irritation score of 3.9 and would be classified as a moderate irritant to the skin of the albino rabbit according to the U.S. Federal Register 1973 Skin Test.

Under these testing conditions it is not known how much the 2.5 parts of 0.1 M NaOH contributed to the irritation score.

The evidence suggests that Lauricidin acts as a sensitizer in the guinea pig and, according to Magnusson's method of interpretation of results, would be classified as a mild (grade 11) sensitizer, having a sensitization rate of 9-28%.

The test material was injected at a 10% dilution in ethanol; topical application was a 20% dilution in ethanol. A 10% dilution was used at a challenge phase without causing irritation. The challenge phase used 5 and 10% Lauricidin in petrolatum.

B. Eye Irritation Study

An eye irritancy test was carried out as described in *The Federal Register*, Vol. 38, No. 187, Section 1500:42, 1973.

A total of 0.1 ml of the sample was instilled into the right eye of each of six albino rabbits. The eyes of the rabbits were examined and the grade of ocular reaction recorded 24, 48, and 72 hr after instillation of the sample. The average irritation scores for each group were the following:

Sample	Number of animals	Corneal opacity	Iris inflammation	Conjunctival irritation	Number with positive response
Lauricidin, 20% concentration	6	0.0	0.0	3.7	1

A 20% aqueous concentration of Lauricidin produced positive reactions in only one of the six rabbits in the group and would be classified as a negative eye irritant according to the U.S. Federal Register 1973 Eye Test.

C. Delayed Dermal Sensitization

A delayed dermal sensitization study using the guinea pig maximization test first described by Magnusson and Kligman (*J. Invest. Dermatol.*, 52:268-276, 1969) and revised in 1970 [Identification of Contact Allergens, in *Allergic Contact Dermatitis in the Guinea Pig* (B. Magnusson and A. M. Kligman, eds.), 1970, pp. 102-123].

Ten guinea pigs were treated by intradermal injection in the shoulder region to induce sensitization, and 7 days later the sensitization was boosted by an occluded patch placed over the injection site. Fourteen days later the animals were challenged on the flank by occluded patch for 24 hr. The reaction sites were examined 24 and 48 hr after removal of the occluded patch. Four additional guinea pigs were treated in a similar manner to the test animals, except that the test sample was only applied at the challenge phase.

Two animals in the test group challenged with a 10% dilution of the test material exhibited positive reactions. No visible reaction was exhibited by any animal in the control group challenged with a 10% dilution of the test material or either the test or control group challenged with a 5% dilution.

A second challenge was carried out 7 days after the first challenge. Five animals in the test group and two animals in the control group challenged with a 10% dilution of the test material exhibited positive reactions. Since positive reactions occurred in both the test and control groups, it seems likely that irritation and not sensitization was responsible. Four animals in the test group challenged with a 5% dilution of the test material exhibited positive reactions. No visible reaction was observed in the control group.

D. Acute Studies

The entire subject of glyceride safety recently has been reviewed and its Generally Regarded as Safe status reaffirmed. These compounds, including monoglycerides, have no toxicity in the usual sense of the word. The oral LD_{50} of the mono-oleate has been measured as 50 g/kg in rats and greater than 25 g/kg in mice. In general, all glycerides of natural oils would exhibit similar metabolic and pharmacological fates.

E. Chronic Studies

Groups of rats were fed diets containing monolaurin at a level of 25% for a period of 10 weeks. The histological examination of the liver, kidney, spleen, lung, heart, stomach, and small and large intestines revealed no abnormalities.

VI. APPLICATION OF LAURICIDIN IN COSMETICS

The discovery of a new preservative is always greeted with suspicion and pessimism. Both reactions are human, since of the great number of chemical agents which are toxic to microorganisms, very few make it to the marketplace. Of the few that do appear, none are without some limitation. Today governmental agencies and the court system of the world require strict control of preservative substances. Legislation adds to the problems associated with the introduction of a new agent. Not only is preservative effectiveness a criteria, but toxicology to man and environment is an important legal consideration. The discovery that a food-grade chemical could have a role in cosmetic preservation is very exciting. Here for the first time we have a chemical which is hostile to microorganisms but a food for man.

As we previously described, Lauricidin demonstrated activity against gram-positive yeast, fungi, and mold organisms. In exceptional cases it is active against gram-negative organisms. However, small additions of a chelating agent (citric and lactic acid or EDTA) allows the Lauricidin to be effective against gram-negative genera such as *P. aeruginosa* or *Escherichia coli*.

Acidulants (malic and polyphosphoric acid) were described by Kato and Shibasaki [41] to increase the antimicrobial spectrum and activity of Lauricidin to gram-negative organisms.

Lauricidin needs to be part of a "preservative system" (see Chapter 19) in order to achieve wide-spectrum activity. It can and has been formulated as the only preservative in which a chelating agent has been used to potentiate the effect of the monoglyceride (see Table 6). In Table 6 an oil-water lotion was formulated where the Lauricidin was included as a primary emulsifier for the cosmetic system. Under these conditions the challenge organisms, *P. aeruginosa* and *E. coli*, placed in the cosmetic were reduced to sterile products after 1 week—even at the lowest level (0.1%) of EDTA (Na_2) used. Under these same conditions methylparaben (0.1-0.3%) was not effective except at the highest dose and the longest exposure time (1 week). The paraben was active against *E. coli* but not *P. aeruginosa*. Combinations of methylparaben and EDTA were no more effective in the Lauricidin oil-water emulsions than EDTA alone.

The other uses of Lauricidin are myriad, especially where "medicated" action of the product is desirable but not necessarily claimed. The use of Lauricidin in this manner makes it an ideal "active substance" because of its efficacy and safety. Thus the monolaurin can be placed in oral products as an anticarie and antiplaque agent [42,43] or used in pharmaceutical preparations [44].

VII. REGULATORY STATUS OF LAURICIDIN

Lauricidin is a monoglyceride which is approved for food use as an emulsifier by the Food and Drug Administration, (21 CFR GRAS 182.4505) and has the status of a Generally Regarded as Safe (GRAS) material. Although lauric acid was originally excluded as a component of mono- and diglycerides in the list submitted to the House Interstate Committee in 1958, it has been listed in the December 2, 1964, listing of *The Federal Register*. Its earlier exclusion did not suggest toxicity but, rather, prior to 1958 lauric acid monoglycerides were not used, while subsequent to 1958 additional information was submitted and accepted. Although GRAS food ingredients are approved only for those uses that have been specified, Lauricidin can be used in cosmetics without fear of toxicity and in food as a potentiator of preservative action which lowers or reduces the need for other classic preservatives.

Other than the general listing as a monoglyceride, at least six specific laurate esters are listed as GRAS additives.

Lauric acid and coconut fatty acid derivatives use in cosmetic formulations are so widespread that their lack of toxicity is taken for granted. While this may be true for oral toxicity, their topical use should be applied with prudence. In preservative levels usually found in cosmetics, 0.5% Lauricidin or less should not cause problems. In fact, there is some reason to believe that Lauricidin can act as a counterirritant.

The toxicological data for Lauricidin have been summarized in the preceding sections.

TABLE 6 Preservative Effect of Food-Grade Chemicals in a Lauricidin Lotion Oil-Water System[a]

Preservative	At 24 hr		At 48 hr		At 1 wk	
	Pseudomonas aeruginosa	Escherichia coli	Pseudomonas aeruginosa	Escherichia coli	Pseudomonas aeruginosa	Escherichia coli
EDTA						
0.1%	+	+	+	+	−	−
0.2%	−	±	−	±	−	−
0.3%	−	±	−	±	−	−
Methylparaben						
0.1%	+	+	+	+	+	+
0.2%	+	+	+	+	+	+
0.3%	+	+	+	+	+	−
EDTA + methylparaben						
0.1% + 0.1%	+	+	±	+	−	−
0.2% + 0.2%	−	+	−	+	−	−
0.3% + 0.3%	−	±	−	±	−	−

[a]+, growth; ±, slight growth; −, no growth.
Source: Ref. 44.

REFERENCES

1. J. J. Kabara, GRAS antimicrobial agents for cosmetic products, *J. Soc. Cosmet. Chem.*, *31*:1-10 (1980).
2. J. J. Kabara, Food-grade chemicals for use in designing food preservative systems, *J. Food Prot.*, *44*:633-647 (1981).
3. C. D. Tweedle and J. J. Kabara, Evidence for a lipophilic nerve sprouting factor(s), in *Pharmacological Effect of Lipids* (J. J. Kabara, ed.), The American Oil Chemists' Society, Champaign, Ill. (1979), pp. 169-178.
4. J. J. Kabara and C. D. Tweedle, Changes in lipid levels of three skeletal muscles following denervation, *Neurochem. Res.*, *6*:619-632 (1981).
5. N. O. V. Sonntag, Structure and Composition of Fats and Oils in *Baileys Industrial Oil and Fat Products* (D. Stern, ed.), Wiley, New York (1979), pp. 1-98.
6. J. B. Martin, Preparation of saturated and unsaturated symmetrical monoglycerides, *J. Am. Oil Chem. Soc.*, *75*:5482-5483 (1953).
7. L. Hartman, Advances in the synthesis of glycerides of fatty acids, *Chem. Rev.*, *58*:845-867 (1958).
8. F. H. Mattson and R. A. Volpenhein, Synthesis and properties of glycerides, *J. Lipid Res.*, *3*:281-296 (1962).
9. A. J. Conley and J. J. Kabara, Antimicrobial action of esters of polyhydric alcohols, *Antimicrob. Agents Chemother.*, *4*:501-506 (1973).
10. J. B. Lauridsen, Food emulsifiers: surface activity edibility, manufacture, composition and application, *J. Am. Oil Chem. Soc.*, *53*:400-407 (1976).
11. L. I. Emdur, C. Lyle, and J. J. Kabara, Quantitation of lipid classes following thin-layer chromatography, *Anal. Lett.*, *10*:21-27 (1977).
12. T. Riisom and L. Hoffmeyer, High performance liquid chromatography analyses of emulsifiers: quantitative determination of mono- and diocyl glycerols of saturated fatty acids, *J. Am. Oil Chem. Soc.*, *55*:649-652 (1978).
13. H. Halvarson and O. Qvist, A method to determine the monoglyceride content in fats and oils, *J. Am. Chem. Soc.*, *51*:162-165 (1974).
14. K. Payne-Wahl, G. F. Spencer, R. D. Plattner, and R. O. Butterfield, High-performance liquid chromatographic method for quantitation of free acids, mono-, di- and triglycerides using an infrared detector, *J. Chromatogr.*, *209*:61-66 (1981).
15. J. J. Kabara, D. M. Swieczkowski, A. J. Conley, and J. P. Truant, Fatty acids and derivatives as antimicrobial agents, *Antimicrob. Agents Chemother.*, *2*:23-28 (1972).
16. J. J. Kabara, A. J. Conley, D. M. Swieczkowski, I. A. Ismail, M. S. F. Lie Ken Jie, and F. D. Gunstone, Antimicrobial action of isomeric fatty acids on group A *Streptococcus*, *J. Med. Chem.*, *16*:1060-1063 (1973).
17. J. J. Kabara and A. J. Conley, A non-caloric role for MCT and other lipids, in *Mittelkettige Triglyceride (MCT) in der Diät zur Zeitschrift für Ernahrungswissenschaft supplementa No. 17* (H. Kaunitz, K. Lang, and W. Fekl, eds.) (1974), pp. 17-26.
18. J. J. Kabara, R. Vrable, and M. S. F. Lie Ken Jie, Antimicrobial lipids: natural and synthetic fatty acids and monoglycerides, *Lipids*, *12*:753-759 (1977).

19. J. K. Welsh, M. Arsenakis, and J. T. May, Effect of Sewliki Forest virus and cocksackievirus B_4 of lipids common to human milk, *J. Food Safety*, 3:99-107 (1981).
20. J. J. Kabara, Lipids as host resistance factors of human milk, *Nutr. Rev.*, 38:65-73 (1980).
21. J. C. Hierholzer and J. J. Kabara, in vitro effects of monolaurin compounds on enveloped RNA and DNA viruses, *J. Food Prot.*, 4:1-12 (1982).
22. N. Kato and I. Shibasaki, Combined effect of different drugs on the antibacterial activity of fatty acids and their esters, *J. Antibacteriol. Antifung. Agents Jpn.*, 4:254-261 (1975).
23. K. W. Nickerson, V. C. Kramer, and J. J. Kabara, The effectiveness of Lauricidin preservative systems against detergent-resistant *Enterobacter cloacae*, Soap/Cosm/Chem Specialties, *February*: 50-58 (1982).
24. L. R. Beuchat, Comparison of anti-*Vibrio* activity of potassium sorbate, sodium benzoate and glycerol and sucrose esters of fatty acids, *Appl. Environ. Microbiol.*, 39:1178-1182 (1980).
25. N. Lisker and N. Paster, Antifungal activity of monolaurin and related compounds, *J. Food Safety*, 4:27-34 (1982).
26. M. C. Robach, C. S. Hickey, and E. C. To, Comparison of antimicrobial action of monolaurin and sorbic acid, *J. Food Safety*, 3:89-98 (1981).
27. A. C. Frazer and H. G. Sammons, Formation of mono- and diglycerides during the hydrolysis of triglyceride by pancreatic lipase, *Biochem. J.*, 39:122 (1945).
28. P. Desnuelle, M. Naudet, and M. J. Constantin, A new type of lipolysis in vitro. A lipolysis generator of glycerol, *Biochim. Biophys. Acta*, 7:251 (1951).
29. V. P. Skipsi, M. G. Motehouse, and H. J. Deuel, Jr., Absorption in the rat of a 1,3-dioleoyl-2-deuteriostearoyl glyceride-C^{14} and a 1-monodeuteriostearoyl glyceride-C^{14}, *Arch. Biochem. Biophys.*, 81:93 (1959).
30. B. Clark and G. Hubscher, Direct esterification of monoglycerides in rat intestinal mucosa, *Biochem. J.*, 80:12P (1961).
31. F. H. Mattson and R. A. Volpenhein, The digestion and absorption of triglycerides, *J. Biol. Chem.*, 239:2772 (1964).
32. F. H. Mattson, J. H. Venedict, and L. W. Beck, Composition of intestinal lumen lipides following the feeding of triglycerides, partial glycerides or free fatty acids, *J. Nutr.*, 52:575 (1954).
33. F. J. Mattson, F. J. Baur, and L. W. Beck, The comparative nutritive values of mono-, di-, and triglycerides, *J. Am. Oil Chem. Soc.*, 28:386 (1951).
34. R. S. Harris and H. Sherman, Comparison of the nutritive values of mono-, di-, and triglycerides by a modified pair-feeding technique, *Food Res.*, 19:257 (1954).
35. G. F. Dasher, Surface activity of naturally occurring emulsifiers, *Science*, 116:660 (1952).
36. A. F. Hoffman, Micellar solubilization of fatty acids and monoglycerides by bile salt solutions, *Nature*, 190:1106 (1961).
37. W. Q. Braun and C. L. Shrewsbury, Nutritive properties of monoglycerides, *Oil Soap*, 18:249 (1941).
38. S. R. Ames, M. P. O'Grady, N. D. Embree, and P. L. Harris, Molecularly distilled monoglycerides. III. Nutritional studies on monogly-

ceride derived from cottonseed oil, *J. Am. Oil Chem. Soc.*, 28:31 (1951).
39. F. H. Mattson, J. C. Alexander, F. J. Baur, and H. H. Reller, Short-term feeding studies on acetin fats, *J. Nutr.*, 59:277-285 (1956).
40. Anonymous, GRAS (Generally Recognized As Safe) food ingredients—glycerine and glycerides, Ordering Number PB-221227, National Technical Information Service, U.S. Department of Commerce, Springfield, Va. (1973).
41. N. Kato and I. Shibasaki, Combined effect of citric and polyphosphoric acid on the antibacterial activity of monoglycerides, *J. Antibacteriol. Antifung. Agents Jpn.*, 4:254-261 (1966).
42. P. Lynch, J. J. Kabara, and R. Schemmel, *Streptococcus mutans* colony forming units and severity of dental caries in rats fed three types of diet with and without Lauricidin, *Microbios. Lett.*, 12:7-13 (1979).
43. K. A. Williams, B. R. Schemehorn, J. L. McDonald, Jr., G. K. Stookey and S. Katz, Influence of selected fatty acids upon plaque formation and caries in the rat, *Arch. Oral Biol.*, 27:1027-1031 (1982).
44. J. J. Kabara and C. M. Wernette, Cosmetic formulae preserved with food-grade chemicals—part II, *Cosmet. Toiletries*, 97:77-84 (1982).

18
CHELATING AGENTS AS PRESERVATIVE POTENTIATORS

J. ROGER HART *W. R. Grace & Co., Lexington, Massachusetts*

   I. Introduction 323

      A. Chemistry of chelating agents 324
      B. Preservative functions of chelating agents 325

  II. Biological Activity of EDTA 326

      A. Antimicrobial activity of EDTA 326
      B. Sensitizing and potentiating effects of EDTA 326

III. Mechanism of Action 331

 IV. Toxicology of EDTA 332

  V. Analysis of EDTA in Cosmetic Products 332

      A. Background 332
      B. Titrimetric estimation of EDTA 333

 VI. General Conclusions 333

      References 335

I. INTRODUCTION

Amino acid-chelating agents such as ethylenediaminetetraacetic acid (EDTA) have been well known and used since the late 1930s. Chelating agents of the EDTA type, originally developed to counteract the deleterious effects of water hardness and heavy metal ions on dyestuffs used in textile manufacture, soon found application in such wide-ranging applications from pharmaceuticals to the cleaning of high-pressure steam boilers.

Because EDTA and its related derivatives were safe, inexpensive, and effective, they were soon being added to cosmetic and toiletry formulations, often as a form of "insurance" against spurious metal ion contamination. Since the quantity and cost of the chelating agent were small in comparison to the overall formulation cost, EDTA was often added for rather ill-defined reasons.

There are good and legitimate benefits obtained from the addition of chelating agents to cosmetic and toiletry formulations. These typical uses of EDTA have been extensively reviewed [1]. For example, chelating agents will maintain clarity, protect fragrance components, stabilize polymeric thickeners, prevent rancidity and off-odors, stabilize color additives, prevent discoloration in products containing thioglycolate or p-aminobenzoic acid, retard sulfite decomposition, and, most significantly, increase the preservative effectiveness against gram-negative bacteria, especially *Pseudomonas aeruginosa*.

A. Chemistry of Chelating Agents

In order to understand the nature of chelating agents, it is necessary to examine the effect of metal ions on various systems. Alkaline-earth (mainly calcium and magnesium) and transition metal (iron, copper, manganese, etc.) cations, when present, can react with anionic constituents to form insoluble compounds. Many cosmetic colorants are susceptible to shade changes or even precipitation by transition metal ions. The oxidative degradation of labile fragrance components, lipid emollients, proteins, and other cosmetic ingredients is often accelerated by traces of iron, copper, and manganese. It is readily apparent that these deleterious effects should be eliminated or at least greatly diminished to provide the maximum product protection and performance.

Two approaches are possible. The first consists of total and complete removal of interfering metal ion contamination, which is simply not possible in a practical sense. The minimization of contamination is a worthwhile aim, and the critical choice of pure raw materials, water, and packaging is essential to good manufacturing. Inadvertent contamination with low levels of metal ion is still a reality, however, and can only be countered with the second approach: inactivation of the metal ion with a chelating agent.

Chelating agents currently in wide use in cosmetics and toiletry formulations are ethylenediaminetetraacetic acid (EDTA), hydroxyethyl ethylenediaminetriacetic acid (HEEDTA), and diethylenetriaminepentaacetic acid (DTPA) and their salts (see Fig. 1).

Chelating agents such as EDTA are anionic in nature (EDTA, for instance, forms a tetranegative anion, $EDTA^{4-}$) and are strongly attracted to alkaline-earth and transition metal ions (M^{2+}) to form metal-EDTA chelates ($MEDTA^{2-}$).

In reacting with EDTA, the metal ion is converted to an anionic form and no longer exists in significant quantities as the cationic metal ion, but as part of the anionic metal-EDTA complex ($MEDTA^{2-}$). This anionic nature prevents the metal ion from entering into the harmful reactions described above. Thus hazes and precipitates may be avoided, fragrances protected from oxidation, rancidity prevented in emulsions, and so on.

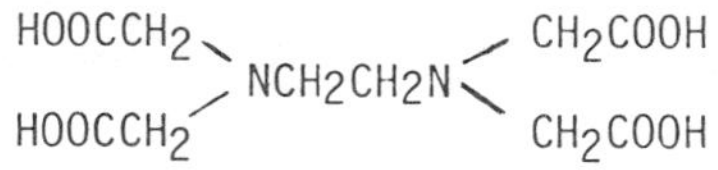

EDTA
Ethylenediaminetetraacetic Acid

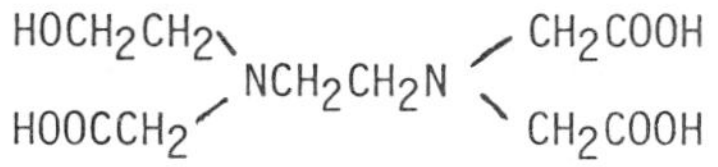

$$\text{(HOOCCH}_2\text{)}_2\text{NCH}_2\text{CH}_2\text{N(CH}_2\text{COOH)}_2$$

HEEDTA
Hydroxyethylethylenediaminetriacetic Acid

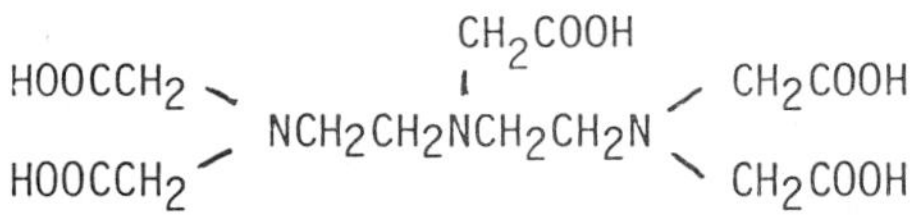

$$\text{HOCH}_2\text{CH}_2\text{-N(CH}_2\text{COOH)-CH}_2\text{CH}_2\text{-N(CH}_2\text{COOH)}_2$$
(with HOOCCH$_2$ on first N)

DTPA
Diethylenetriaminepentaacetic Acid

$$\text{(HOOCCH}_2\text{)}_2\text{NCH}_2\text{CH}_2\text{N(CH}_2\text{COOH)CH}_2\text{CH}_2\text{N(CH}_2\text{COOH)}_2$$

FIGURE 1 Chelating agents commonly used in cosmetics and toiletries.

B. Preservative Functions of Chelating Agents

Two separate preservative functions in cosmetic products are fulfilled by EDTA and related chelating agents: First, EDTA effectively prevents rancidity in emulsified products like hand lotions and skin creams. EDTA ties up trace amounts of transition metal ions (copper, manganese, and iron, for example) which catalyze the oxidation of fats. The ability of EDTA to form anionic metal complexes greatly alters the oxidation-reduction potential of the metal ion and also enhances the partitioning of the metal to the aqueous phase where it is less harmful. This antioxidant function of EDTA prevents product spoilage due to rancidity and has established EDTA as a commonly used preservative in food products like salad dressings and cosmetic products like hand lotions; second, EDTA alone is capable of functioning as an antimicrobial material and, more importantly, as a potentiator or "booster" in combination with other commonly used cosmetic preservative materials. The potentiation of preservative action of other antimicrobials accounts for the widest use of EDTA in cosmetic and toiletry formulations. Especially significant, EDTA's primary activity is against gram-negative bacteria, particularly *Pseudomonas*.

The seriousness of gram-negative bacterial contamination is well documented [2-4]. Yablonski and Goldman [2] state that these organisms account "for approximately 98% of all contaminants found in shampoo products and their associated susceptible raw ingredients. Leading the list of gram-negative offenders is the genus *Pseudomonas*."

The intrinsic sensitivity of *P. aeruginosa* and related species to EDTA is therefore of great importance and has been extensively reviewed in the medical literature as a result.

Russell [5] describes EDTA as having

> considerable microbiological importance for the following reasons: (a) it causes an increase in bacterial permeability to drugs, although it is obvious that more attention must be devoted to determining whether the presence of EDTA during treatment of bacteria with a drug gives a synergistic response or merely an additive one; (b) it is useful in studies on cell wall chemistry of Gram-negative bacteria; (c) EDTA is useful in classifying Pseudomonads; (d) it can be employed for releasing surface-bound enzymes.

Wilkinson [6] offers an extensive review of the antibacterial activity of EDTA against gram-negative bacteria, particularly in regard to its effect on the structure of the cell envelope.

II. BIOLOGICAL ACTIVITY OF EDTA

A. Antimicrobial Activity of EDTA

The antibacterial activity of EDTA was reported in the late 1950s by Repaske [7] and MacGregor and Elliker [8]. Results of these workers and others demonstrates that EDTA exerts a lytic action on the outer layers of the cell walls of gram-negative bacteria. The EDTA-induced lysis releases intracellular solutes and partially solubilizes the cell envelope of sensitive bacteria.

The bactericidal action of EDTA on *P. aeruginosa* is quite significant when used alone, but incomplete. For instance, 250 ppm of EDTA alone gave 99.999% kill of *P. aeruginosa*, while 100 ppm of EDTA killed between 57 and 99.5% at pH 7.4 [8]. Therefore total sterility should not be expected from EDTA alone.

While the minimum inhibitory concentration value for EDTA against *P. aeruginosa* depends somewhat on the strain and growth history of the bacterial culture, values up to about 30 mM for EDTA at pH 7 have been obtained [9,10].

Though sensitivity to EDTA is a general characteristic of pseudomonads, it is not restricted to this species (see Table 1). Sensitivity to EDTA is primarily exhibited by gram-negative bacteria, yeasts [11], and fungi [12]. Gram-positive bacteria do not exhibit a significant sensitivity to EDTA.

The inherent lack of broad-spectrum bacterial activity and the incompleteness of total bacterial kill strongly limits the use of EDTA as the sole preservative in cosmetic systems. The use of EDTA in combination with other antimicrobial agents offers the widest degree of utility to the cosmetic formulator.

B. Sensitizing and Potentiating Effects of EDTA

While the direct toxicity of EDTA is limited to primarily gram-negative species, the ability of EDTA to increase cell wall permeability and thus increase the bacteria's sensitivity to a wide variety of preservatives is quite general.

TABLE 1 Bacterial Species Highly Sensitive to EDTA Alone

Achromobacter sp.	*Spirillum itersonii*
Alcaligenes faecalis	*Vibrio cholerae*
Azobacter vinelandii	*Vibrio eltor*
Escherichia coli	*Vibrio fetus*
Pseudomonas sp.	*Vibrio succinogenes*
Salmonella sp.	

This potentiating effect extends to antibacterial compounds which represent widely differing chemical structures and modes of action.

Voss [13] suggests that the greater antimicrobial resistance of gram-negative species may be due to the greater complexity of their cell wall structure which acts to exclude the antibacterial agent from the interior of the cell. It is believed that EDTA's disruption of the cell's permeability barrier plays the dominant role in the potentiation of preservatives against gram-negative organisms.

While EDTA will kill *P. aeruginosa*, the principal utility of EDTA lies in its ability to increase the permeability of the cell wall structure of gram-negative bacteria. This holds especially true for the troublesome coliform bacteria. Free, uncomplexed EDTA must be present in the system for this permeability change to occur [14]. This action of EDTA to increase cell wall permeability is responsible for the almost universal potentiation of cosmetic preservatives toward gram-negative bacteria when combined with the chelating agent.

Of all gram-negative bacteria encountered in cosmetic products, *P. aeruginosa* generally causes the most concern. *Pseudomonas aeruginosa* is able to survive contact with concentrations of antibacterials which are many times that required to kill other species. It also readily mutates to more resistant forms. Gould [15] describes it thus: "Biologically active to an extreme degree, it can live on simple substrates, survive for long periods under a wide range of environmental conditions apart from severe dehydration, and adapt to the presence of high concentrations of most...antibacterial substances."

Pseudomonas aeruginosa is a ubiquitous organism which has very simple growth requirements. Certain strains have been shown to be capable of growth even in filter-sterilized distilled water [16].

The cosmetic industry is constantly vigilant in its efforts to eliminate *Pseudomonas* contamination. Upon entrance to the corneal epithelium, this organism can be totally destructive to the eye and will almost certainly cause rapid, severe, and permanent damage. Even minute abrasions on the corneal surface can lead to corneal destruction when infected with this organism. Indeed, it has been shown [17] that the corneal damage to a rabbit's eye caused by a commercial shampoo increases the likelihood of infection by this organism.

Shampoos and skin cleansers have been found to be especially vulnerable to contamination by pseudomonads. While the potential for serious eye injury is of greatest importance to the cosmetic formulator, the emulsion breakdown and spoilage due to lipases and other enzymes released by killed *Pseudomonas* cells is a serious problem in itself.

The use of EDTA to increase the sensitivity of *P. aeruginosa* to other preservatives allows the formulator to use less of the primary and usually more expensive preservative. Lower preservative levels also contribute to fewer irritation and sensitization problems.

1. Cationic Preservatives

Quaternary Ammonium Salts: Early in the use of quaternary ammonium germicides, Lawrence [18] observed that the calcium and magnesium ions in hard water reduced the antimicrobial activity. It quickly became accepted practice to add EDTA to quaternary ammonium germicides to improve performance under actual use conditions. Klimek [19], however, found that EDTA increased the antimicrobial activity of cationic germicides even in distilled water. Clearly EDTA was causing some effect on the bacterial cells to render them more sensitive to the germicidal material. MacGregor and Elliker [8] found that *P. aeruginosa* readily forms strains which are resistant to quaternary ammonium compounds. Indeed, this highly adaptive organism has been found growing in solutions of benzalkonium chloride and cetrimonium chloride [20,21]. This acquired resistance of *P. aeruginosa* to quaternary ammonium chlorides is eliminated by EDTA [8].

Quaternary ammonium compounds were reported as being used as preservatives in a total of 116 cosmetic formulations manufactured in the United States during 1980 [22]. As pseudomonads easily acquire resistance to these preservatives, the use of EDTA is strongly recommended as a potentiating additive to quaternary ammonium salts.

Other Cationic Compounds: Chlorhexidine salts are also used as preservatives in cosmetic formulations. During 1980, 38 instances of chlorhexidine usage were reported for cosmetic products in the United States [22]. This cationic compound also suffers from the fact that some bacteria, especially *P. aeruginosa*, become resistant. In fact, chlorhexidine-cetrimonium bromide stock solutions have been found contaminated with *P. aeruginosa* [23]. The addition of EDTA eliminates resistance and lowers the minimum inhibitory concentration for chlorhexidine [24,25].

Quaternium 15 [1-(3-chloroallyl)-3,5,7-triaza-1-azoniaadamantane chloride] is widely used because of its broad-spectrum antimicrobial activity. This popular preservative is strongly potentiated against gram-negative bacteria by EDTA [26]. This enables the formulator to utilize a lower concentration of quaternium 15 when formulated with EDTA. A considerable cost savings results from this approach.

2. Anionic Preservatives

Parabens: EDTA is frequently used in combination with esters of p-hydroxybenzoic acid (parabens) to provide broad-spectrum preservative activity. Very little appears in the literature about EDTA-paraben mixtures, however. This is surprising, since so much evidence exists for the synergism of phenolic preservatives and EDTA. This particular group of phenolics — the parabens — appear to have received very little experimental investigation in regard to the potentiating effect of EDTA and would be a worthy topic for further study. Unpublished work by Haag [27] clearly demonstrates the potentiating effect of EDTA on methylparaben (see Table 2).

Greater control of gram-negative bacteria, especially *P. aeruginosa*, is obtained when EDTA is used in conjunction with the paraben preservatives.

TABLE 2 Effect of EDTA on Antimicrobial Activity of Methylparaben Against P. aeruginosa

Preservative	Time[a]		
	4 hr	24 hr	48 hr
None	+	+	+
Methylparaben, 0.2%	+	+	+
EDTA, 0.2%	+	+	+
MP, 0.2% + EDTA, 0.2%	−	−	−

[a]+, heavy growth; −, no growth.

This combination is a popular and economical system for broad-spectrum antimicrobial protection in cosmetics.

Other Phenolic Compounds: An early observation of the potentiating effect of EDTA on P. aeruginosa with a phenolic germicide was made by Stothart and Beecroft [28]. They found a 250-fold increase in activity of an o-benzyl-p-chlorophenol-based formulation with the addition of the chelating agent. This phenomenon takes on added importance, since P. aeruginosa has been found growing in phenolic [29] and chloroxylenol [30] disinfectant solutions.

The activity of chloroxylenol, mixed phenols, and o-benzyl-p-chlorophenol against P. aeruginosa is strongly potentiated by EDTA [31-34]. These compounds are also found to have increased activity against Escherichia coli and Staphylococcus aureus when EDTA is present.

Triclosan also is strongly potentiated against gram-negative bacteria by EDTA [3]. Increased activity in combination with EDTA is found against P. aeruginosa, E. coli, and Klebsiella aerogenes [35].

Miscellaneous Anionic Preservatives: The addition of EDTA to potassium sorbate was found to potentiate the activity against S. aureus 12600 strain, but not the S 6 strain [36]. The combination of EDTA and sorbate salts has been proposed as a bacterial preservative for food products.

A conventional-type anionic surfactant, sodium myristyl sulfate, was found to have little activity toward Salmonella typhi unless EDTA was present [37]. The LD_{50} for sodium myristyl sulfate against this organism was 10,000 μg. With sensitization by EDTA, the LD_{50} dropped to 230 μg, an increase in bactericidal action of 40 times.

3. Nonionic Preservatives

After paraben esters, imidazolidinyl urea is the most frequently used cosmetic preservative in the United States [22]. In a preservative challenge test with 10^6 pseudomonad microorganisms per milliliter, neither 0.3% imidazolidinyl urea nor 0.3% EDTA alone could effect a complete kill within 3 days. A combination of 0.15% imidazolidinyl urea and 0.15% EDTA did result in complete kill [38].

1,3-Dimethylol-5,5-dimethyl hydantoin (DMDM) hydantoin has become a popular preservative. Record of use has grown from 15 in 1977 to 79 in 1980. Shull and Bennett [39] have recently found that EDTA greatly enhances the antimicrobial activity of DMDM hydantoin. They claim that 1 part of DMDM hydantoin and 1 part of EDTA is more active than 2 parts of DMDM hydantoin alone.

2-Bromo-2-nitropropane-1,3-diol has found wide use because of its high activity and broad preservative spectrum. This material can be made more effective against *P. aeruginosa* through potentiation with EDTA [40]. Lower preservative levels are possible.

A wide variety of lipid materials have been examined for antimicrobial activity [41]. Of these numerous compounds, the monoglyceride of lauric acid (monolaurin) has been found to be most active. This compound is principally active against gram-positive bacteria, fungi, mold, and yeast organisms. The principal limitation to this nontoxic preservative is its lack of gram-negative activity. It has been found, however, that EDTA strongly potentiates the activity of monolaurin against gram-negative bacteria [42-44]. Table 3 demonstrates the potentiation of a monolaurin-preserved hand lotion with EDTA. The presence of methylparaben did not protect this system from *P. aeruginosa* challenge, whereas the EDTA did [45]. It has also been demonstrated that EDTA lowers the minimum inhibitory concentration of monolaurin against yeasts. Shibasaki and Kato [44] found that EDTA tripled the percentage of uptake of monolaurin into the cells of *E. coli*.

Broad-spectrum preservation is claimed for mixtures of monolaurin, a phenolic such as a paraben ester or tert-butyl hydroxyanisole and EDTA [42]. Richards and McBride [46] have concluded from their work that phenethyl alcohol is potentiated by EDTA in its antibacterial action against *P. aeruginosa*.

4. Preservatives Unaffected by EDTA

While the generalized action of EDTA on the bacterial cell wall would suggest that EDTA will potentiate the action of virtually any antimicrobial agent, there are reports to the contrary. Experimental conditions, of course, could

TABLE 3 Effect of Preservative on Oil-Water Hand Lotion Preserved with Monolaurin Against *Pseudomonas aeruginosa* Inoculation

Preservative	Time[a]		
	1 day	2 days	7 days
Control with monolaurin only	+	+	+
EDTA, 0.1%	+	+	−
EDTA, 0.2%	−	−	−
EDTA, 0.3%	−	−	−
Methylparaben, 0.1%	+	+	+
Methylparaben, 0.2%	+	+	+
Methylparaben, 0.3%	+	+	+
Methylparaben, 0.1%, + EDTA, 0.1%	+	±	−
Methylparaben, 0.2%, + EDTA, 0.2%	−	−	−
Methylparaben, 0.3%, + EDTA, 0.3%	−	−	−

[a]+, growth; ±, slight growth; −, no growth.
Source: Ref. 45.

create negative results from EDTA. If sufficient calcium and magnesium ions are present in the media, water, or preservative, then EDTA will not aid in the antimicrobial action. Free, uncomplexed EDTA is essential to potentiation; the calcium and magnesium chelates of EDTA are inactive [5].

When the EDTA content falls to or below the level sufficient to complex the calcium and magnesium present in the formulation, it will often appear that this low level of EDTA causes an increase in bacterial growth. In reality, however, it is simply the loss of the free, uncomplexed EDTA which is required for the potentiation of the primary antimicrobial against gram-negative organisms.

Reybrouck and van de Voorde [32] report that EDTA had no effect on the antimicrobial activity of an iodophor, hydrogen peroxide, formaldehyde, or glutaraldehyde. Combinations of EDTA and chlorobutanol appear to have no advantage over chlorobutanol alone [46].

III. MECHANISM OF ACTION

A lytic action on the cell wall structure of gram-negative bacteria, especially *P. aeruginosa*, is exerted by EDTA. This gross lysis of cells may be measured by the decrease in turbidity of cell suspensions along with a corresponding increase in intracellular solutes. Eagon and Carson [47] found that a loss of viability of 99% correspond to about a 50% decrease in turbidity. The extraction of metal cations, especially calcium and magnesium, by EDTA is rapidly followed by partial solubilization of the cell membrane [6].

The lipopolysaccharide core polysaccharide of *P. aeruginosa* is highly substituted with polyphosphate residues. These polyphosphate residues possess metal cation-binding capacity, and it is believed that this polyphosphate-rich core, located either on or in the inside surface of the outer layer of the outer cell wall membrane, binds to other membrane components (e.g., proteins and phospholipids) through cation bridges; EDTA removes these cations which, in turn, releases a complex of protein and lipopolysaccharide.

The effect of EDTA on isolated cell wall membranes of *P. aeruginosa* has been studied [48]: EDTA lowers the density of both the outer and inner membranes and causes lethal damage to the outer membrane, resulting in membrane disintegration.

Part of the evidence for EDTA's mechanism of action comes from studies of the activity of other aminopoly carboxylic acid chelating agents on *P. aeruginosa*. Table 4 shows those chelating agents which have been found active against *P. aeruginosa* [25,49-51]. They are shown in decreasing order of antimicrobial activity. This order is correlated quite well with chelating ability with magnesium.

TABLE 4 Chelating Agents Active Against *P. aeruginosa*

Cyclohexanediaminetetraacetic acid (CDTA)

Ethylenediaminetetraacetic acid (EDTA)

Diethylenetriaminepentaacetic acid (DTPA)

Hydroxyethylethylenediaminetriacetic acid (HEEDTA)

IV. TOXICOLOGY OF EDTA

EDTA and its salts exhibit a low order of animal toxicity and have been used for many years as a direct food additive and pharmaceutical without incident. The LD_{50} of disodium EDTA when administered to rabbits is 47 mg/kg (intravenous) and 2.3 g/kg (oral) [52].

Published medical records show that when 81 human patients were given 2000 infusions of disodium EDTA over a 2-year period, there were no serious side effects [53]. Slightly more than half of the subjects received 1-20 infusions each, and the others received up to 120 infusions each with no difficulties when the compound was administered intravenously in 3.0-g doses as a 0.5% solution over a period of 2.5-3 hr.

Feeding levels as high as 5% (50,000 ppm) of disodium EDTA have been administered to rats over a 2-year period. No effects were found on gross or histopathological examination of organs or tissues. Toxic symptoms were limited to diarrhea at the highest dosage level and a reduced weight gain in some of the treated groups [54].

EDTA has also been examined for possible carcinogenicity. No compound-related signs of clinical toxicity were noted in a 2-year study [55].

Repeated-insult skin patch tests on guinea pigs have shown that EDTA is not a skin sensitizer [56] and that the small quantities of EDTA in cosmetic products do not represent any appreciable risk of human skin sensitization.

V. ANALYSIS OF EDTA IN COSMETIC PRODUCTS

A. Background

The analysis of small amounts (0.1-0.5%) of EDTA in cosmetic and toiletry products is complicated by the variety of raw materials present in shampoos, skin cleansers, and the like. Several photometric methods have been examined with no success [57]. Techniques included the zirconium-xylenol orange color-bleaching method and the chromium-EDTA photometric estimation.

The technique of Ross and Frant [58], using copper as a titrant and a specific ion electrode for end-point detection, has been examined and modified for the estimation of EDTA in cosmetic and toiletry products [57]. Several commercial shampoo products were analyzed before and after a standard addition of EDTA. Recovery compared quite well with the actual spiked

TABLE 5 Recovery of Added EDTA (0.1%) from Shampoo Samples by Copper-Ion Electrode Titration

Shampoo	EDTA analyzed (%)	Percent recovery
A	0.09	90
B	0.10	100
C	0.08	80

quantities (see Table 5). One cautionary note: Amphoteric surfactants such as Amphoteric 2 produced from fatty acid, aminoethylethanolamine, and chloroacetic acid may contain significant quantities of hydroxyethylethylenediaminetriacetic acid (HEEDTA) salts as a by-product from the reaction of the latter two ingredients. The presence of this chelating agent will interfere with the analysis of EDTA.

B. Titrimetric Estimation of EDTA

1. Principle of Method

The following is the potentiometric titration of EDTA with copper titrant using a copper ion-specific electrode to detect the equivalence point:

> 0.001 M copper nitrate solution. Prepare fresh daily by accurate dilution of standardized 0.100 M cupric nitrate solution.
> Buffer solution. Dilute 5.7 g of ammonium hydroxide and 8.0 g of ammonium nitrate to 100 ml.
> Cupric specific ion electrode (Orion model 94-29A).
> Calomel reference electrode.
> Specific ion meter.
> Weight accurately about 5 g of sample into a 250-ml beaker. Add 100 ml of water and sufficient buffer solution to pH 8.0 Titrate with 0.001 M cupric nitrate solution. Record millivolt readings after each 0.2 ml of titrant as the end point is approached. Plot millivolts versus milliliters of titrant to determine the equivalence point:

$$\text{Percentage of EDTA compound} = \frac{\text{ml of titrant} \times \text{molarity of titrant} \times \text{mol. wt. of EDTA compound}}{\text{wt. of sample} \times 10}$$

VI. GENERAL CONCLUSIONS

EDTA is known to increase the activity of a great many cosmetic preservatives toward gram-negative bacteria. Included are the parabens, imidazolidinyl urea, tert-butyl hydroxyanisole and tert-butyl hydroxytoluene. This effect is especially pronounced in the case of *P. aeruginosa*. The preservatives potentiated by EDTA are shown in Table 6. Compounds within each general category are ranked in decreasing order of frequency of usage by the cosmetic industry in the United States during 1980.

In 1980 there were 21,557 reported instances of use for preservatives other than chelating agents in cosmetic formulations in the United States [22]. Of this total, 18,266 instances involved the use of preservatives which are known to be potentiated by EDTA. Thus over 80% of the preservative usage in cosmetics can be formulated with EDTA to provide increased effectiveness. This broad applicability of EDTA in cosmetic preservative system suggests that EDTA is being underutilized and deserves much greater consideration by the cosmetic chemist.

The low cost and high degree of safety of EDTA offer the formulator a simple and effective technique to improve the control of gram-negative bacterial contamination.

TABLE 6 Preservatives Known to be Enhanced by EDTA

Cationic preservatives
  Quaternary ammonium salts
    Benzalkonium chloride
    Benzethonium chloride
    Myristalkonium chloride
    Cetylpyridinium chloride
    Cetrimonium bromide
    Lauryl pyridinium chloride
    Methylbenzethonium chloride
    Lauryl isoquinolinium bromide

  Other cationic compounds
    Quaternium 15
    Chlorhexidine digluconate
    Chlorhexidine dihydrochloride

Anionic preservatives
  Parabens
    Methylparaben
    Propylparaben
    Butylparaben
    Ethylparaben
    Benzylparaben
    Sodium methyl paraben

  Other phenolic compounds
    tert-Butyl hydroxyanisole
    Di-tert-butyl methylparaben
    Triclosan
    Chloroxylenol
    Sodium o-phenyl phenate
    Salicylic acid
    Resorcinol
    Phenol

  Miscellaneous anionic preservatives
    Sorbic acid
    Potassium sorbate

Nonionic preservatives
  Imidazolidinyl urea
  2-Bromo-2-nitropropane-1,3-diol
  DMDM hydantoin
  Phenethyl alcohol
  Monolaurin

REFERENCES

1. J. R. Hart, EDTA-type chelating agents in personal care products, *Cosmet. Toiletries*, 98:54-58 (1983).
2. J. I. Yablonski and C. L. Goldman, Microbiology of shampoos, *Cosmet. Toiletries*, 90:45-52 (1975).
3. R. A. Cowen and B. Steiger, Why a preservative must be tailored to be a specific product, *Cosmet. Toiletries*, 92:15-20 (1977).
4. K. H. Wallhauser, The problem of preserving cosmetics, *Cosmet. Toiletries*, 91:45-51 (1976).
5. A. D. Russell, Ethylenediaminetetraacetic acid, in *Inhibition and Destruction of the Microbial Cell* (W. B. Hugo, ed.), Academic, London (1971).
6. S. G. Wilkinson, Sensitivity to ethylenediaminetetraacetic acid, in *Resistance of Pseudomonas aeruginosa* (M. R. W. Brown, ed.), Wiley, London (1975).
7. R. Repaske, Lysis of gram-negative organisms and the role of Versene, *Biochim. Biophys. Acta*, 30:225-232 (1958).
8. D. R. MacGregor and P. R. Elliker, A comparison of some properties of strains of *Pseudomonas aeruginosa* sensitive and resistant to quaternary ammonium compounds, *Can. J. Microbiol.*, 4:499-503 (1958).
9. R. Weiser, A. W. Asscher, and J. Wimpenny, In vitro reversal of antibiotic resistance by ethylenediamine tetraacetic acid, *Nature*, 219:1365-1366 (1968).
10. F. W. Adair, S. G. Geftic, and J. Gelzer, Resistance of *Pseudomonas* to quaternary ammonium compounds. II. Cross-resistance characteristics of a mutant *Pseudomonas aeruginosa*, *Appl. Microbiol.*, 21:1058-1063 (1969).
11. K. G. Indge, The effects of various anions and cations on the lysis of yeast protoplasts by osmotic shock, *J. Gen. Microbiol.*, 41:425-432 (1968).
12. C. Y. Li and J. J. Kabara, Effects of Lauricidin on *Fomes annosus* and *Phellinus weirii*, in *The Pharmacological Effect of Lipids* (J. J. Kabara, Ed.), American Oil Chemists' Society, Champaign, Ill. (1979).
13. J. G. Voss, Effect of inorganic cations on bactericidal activity of anionic surfactants, *J. Bacteriol.*, 86:207 (1963).
14. L. Leive, Studies on the permeability change produced in coliform bacteria by ethylenediaminetetraacetate, *J. Biol. Chem.*, 243:2373-2380 (1968).
15. J. C. Gould, *Pseudomonas pyocyanae* infections in hospital, in *Infection in Hospitals: Epidemology and Control*, Blackwell Scientific, Oxford (1963).
16. M. S. Favero, W. W. Bond, N. J. Peterson, and L. A. Carsen, Bacteriological Proceedings, Abstract No. 75, p. 11 (1970).
17. F. N. Marzulli, J. R. Evans, and P. D. Yoder, Induced *Pseudomonas keratitis* as related to cosmetics, *J. Soc. Cosmet. Chem.*, 23:89-97 (1972).
18. C. A. Lawrence, *Surface-Active Quaternary Ammonium Germicides*, Academic, New York (1950).
19. J. W. Klimek, *The Quaternary Interests*, Chemical Specialty Manufacturer's Association, Proceedings, December 1950, pp. 109-115.
20. F. W. Adair, S. G. Geftic, and J. Gelzer, Resistance of *Pseudomonas* to quaternary ammonium compounds. I. Growth in benzalkonium chloride solutions, *Appl. Microbiol.*, 18:299-302 (1969).

21. D. C. Bassett, K. J. Stokes, and W. R. G. Thomas, Wound infection with *Pseudomonas multivorans*; a water-borne contaminant of disinfectant solutions, *Lancet*, 1:1188-1191 (1970).
22. E. L. Richardson, Update-frequency of preservative use in cosmetic formulations as disclosed to FDA, *Cosmet. Toiletries*, 96:91-92 (1981).
23. D. W. Burdon and J. L. Whitby, Contamination of hospital disinfectants with *Pseudomonas* species, *Br. Med. J.*, 2:153-155 (1967).
24. M. R. W. Brown and R. M. E. Richards, Effect of ethylenediaminetetraacetate on the resistance of *Pseudomonas aeruginosa* to antibacterial agents, *Nature*, 207:1391-1393 (1965).
25. H. Haque and A. D. Russell, Effect of chelating agents on the susceptibility of some strains of gram-negative bacteria to some antibacterial agents, *Antimicrob. Agents Chemother.*, 6:200-206 (1974).
26. I. N. Izzat and E. O. Bennett, The potentiation of the antimicrobial activities of cutting fluid preservatives by EDTA, *Lubr. Eng.*, 35:153-159 (1979).
27. T. R. Haag, Mallinkrodt, Inc., unpublished data.
28. S. N. H. Stothart and G. C. Beecroft, Antiseptics and disinfectants, British Patent 858,030 (1961).
29. C. H. Jellard and G. M. Churcher, An outbreak of *Pseudomonas aeruginosa*...in the newborn, *J. Hyg. Cambridge*, 65:219-228 (1967).
30. G. A. J. Ayliffe, E. J. L. Lowbury, J. G. Hamilton, J. M. Small, E. A. Ashshov, and M. T. Parker, Hospital infection with *Pseudomonas aeruginosa* in neurosurgery, *Lancet*, 12:365-368 (1965).
31. G. W. Gray and S. G. Wilkinson, The action of ethylenediaminetetraacetic acid on *Pseudomonas aeruginosa*, *J. Appl. Bacteriol.*, 28:153-164 (1965).
32. G. Reybrouck and H. van de Voorde, Effect of ethylenediaminetetraacetate germicidal action of disinfectants against *Pseudomonas aeruginosa*, *Acta Clin. Belg.*, 24:32-41 (1969).
33. G. Smith, Ethylenediaminetetraacetic acid and the bactericidal efficiency of some phenolic disinfectants against *Pseudomonas aeruginosa*, *J. Med. Lab. Technol.*, 27:203-206 (1970).
34. A. D. Russell and J. R. Furr, The antibacterial activity of a new chloroxylenol preparation containing ethylenediaminetetraacetic acid, *J. Appl. Bacteriol.*, 43:253-260 (1977).
35. A. K. Panezai, Ger. Offen. 2,351,386 (to Beecham Group, Ltd), April 25, 1974.
36. M. C. Robach and C. L. Stateler, Inhibition of *Staphylococcus aureus* by potassium sorbate in combination with sodium chloride, tertiary butylhydroquinone, butylated hydroxanisole or ethylenediamine tetraacetic acid, *J. Food Prot.*, 43:208-211 (1980).
37. L. H. Muschel and L. Gustafson, Antibiotic, detergent, and complement sensitivity of *Salmonella typhi* after ethylenediaminetetraacetic acid treatment, *J. Bacteriol.*, 95:2010-2013 (1968).
38. W. E. Rosen, Sutton Laboratories, unpublished data.
39. S. E. Shull and E. O. Bennett, Antimicrobial hydantoin derivative compositions and method of use, U.S. Patent 4,172,140 (to Glyco Chemicals, Inc.), October 23, 1979.
40. S. E. Lentsch, Control of mastitis and compositions therefore, U.S. Patent 4,199,602 (to Economics Laboratory, Inc.), April 22, 1980.
41. J. J. Kabara, Fatty acids and derivatives as antimicrobial agents -- a review, in *The Pharmacological Effect of Lipids* (J. J. Kabara, ed.), American Oil Chemists' Society, Champaign, Ill. (1979).

42. J. J. Kabara, Synergistic microbiocidal composition and method, U.S. Patent 4,067,997 (to Med-Chem Laboratories) January 10, 1978.
43. J. J. Kabara, GRAS antimicrobial agents for cosmetic products, *J. Soc. Cosmet. Chem.*, *31*:1-10 (1980).
44. I. Shibasaki and N. Kato, Combined effects on antibacterial activity of fatty acids and their esters against gram-negative bacteria, in *The Pharmacological Effect of Lipids* (J. J. Kabara, ed.), American Oil Chemists' Society, Champaign, Ill. (1979).
45. J. J. Kabara and C. M. Wernette, Cosmetic formulas preserved with food-grade chemicals, *Cosmet. Toiletries*, *97*:77-84 (1982).
46. R. M. E. Richards and R. J. McBride, The preservation of ophthalmic solutions with antibacterial combinations, *J. Pharm. Pharmacol.*, *24*: 145-148 (1972).
47. R. G. Eagon and K. J. Carson, Lysis of cell walls and intact cells of *Pseudomonas aeruginosa* by ethylenediamine tetraacetic acid and by lysozyme, *Can. J. Microbiol.*, *11*:193-201 (1965).
48. K. Matsushita, O. Adachi, E. Shinagawa, and M. Ameyama, Isolation and characterization of outer and inner membranes from *Pseudomonas aeruginosa* and effect of EDTA on the membranes, *J. Biochem.*, *83*: 171-181 (1978).
49. N. A. Roberts, G. W. Gray, and S. G. Wilkinson, The bactericidal action of ethylenediaminetetra-acetic acid on *Pseudomonas aeruginosa*, *Microbios.*, *2*:189-208 (1970).
50. A. B. Spicer and D. F. Spooner, The inhibition of growth of *Escherichia coli* spheroplasts by antibacterial agents, *J. Gen. Microbiol.*, *80*:37-50 (1974).
51. H. Haque and A. D. Russell, Cell envelopes of gram negative bacteria: composition, response to chelating agents and susceptibility of whole cells to antibacterial agents, *J. Appl. Bacteriol.*, *40*:89-99 (1976).
52. S. Shibata, Toxicological studies of EDTA salt (disodium ethylenediaminetetraacetate), *Nippon Yakurigaku Zasshi*, *52*:113-119 (1956).
53. L. E. Meltzer, J. R. Kitchell, and F. Palmon, The long-term use, side effects, and toxicity of disodium ethylenediaminetetraacetic acid (EDTA), *Am. J. Med.*, *242*:11-17 (1961).
54. S. S. Yang and M. S. Chan, Summaries of toxicological data: toxicology of EDTA, *Food Cosmet. Toxicol.*, *2*:763-767 (1964).
55. Bioassay of trisodium ethylenediaminetetraacetate trihydrate (EDTA) for possible carcinogenicity, National Cancer Institute, Technical Report Series No. 11, U.S. Dept. of Health, Education, and Welfare, Washington, D.C., National Institutes of Health (1977).
56. J. W. Henck, D. D. Lockwood, and K. J. Olsen, Skin sensitization potential of trisodium ethylenediaminetetraacetate, *Drug Chem. Toxicol.*, *3*:99-103 (1980).
57. M. T. DeGeorge and R. J. Lemery, Organic Chemicals Division, W. R. Grace & Co., unpublished data.
58. J. W. Ross and M. S. Frant, Chelometric indicator titrations with the solid-state cupric ion-selective electrode, *Anal. Chem.*, *41*:1900 (1969).

19
FOOD-GRADE CHEMICALS IN A SYSTEMS APPROACH TO COSMETIC PRESERVATION

JON J. KABARA *Michigan State University, East Lansing, Michigan*

I. Introduction 339

II. Food-Grade Chemicals 342

    A. Monolaurin (Lauricidin) as a preservative 342
    B. Phenolic lipophilic food additives 345
    C. Chelating agents as preservative potentiators 346
    D. Other food ingredients as preservatives 347

III. Application to Model Cosmetic Systems 347

    A. General solution to the preservative system approach 347

    References 354

I. INTRODUCTION

The science of cosmetic preservation is relatively new. It has only been in the last 50 years that the cosmetic scientist has dealt with the problem in a "scientific" manner. Too often in the early days of cosmetic preservation, germicidal agents were used in a haphazard manner. Little thought went into its inclusion in the finished product. Today (1982) a great deal of thought is given to the problem. Responsibility to solve this problem is thrust upon the cosmetic microbiologist because of the concern of product irritation and/or toxicity by governmental agencies as well as the consumer. To find the right germicide that meets all the criteria for the industry is a tall order and one not easily filled by the use of a single compound:

1. The ideal preservative would have to have a wide or correct spectrum of antimicrobial activity and be effective over a wide pH range during the shelf life of the product.
2. It should be effective against ATCC genera as well as house organisms.
3. The distribution between oil and water would have to be appropriate for a given emulsion system.
4. The preservative should be compatible and not interact with ingredients in the formula or package.
5. It should not interfere with the color or fragrance of the finished product.
6. The preservative should be nontoxic and nonirritating.
7. It should have an acceptable cost at its level of insertion.

The probability of any one preservative meeting all of these requisites is low. Consequently the microbiologist is faced with compromise in picking out a preservative. Since it is a "given" that no one chemical can reach the ideal, it becomes necessary to use two or more preservatives which together form an acceptable preservative system. The primary purpose in constructing a preservative system is to choose those chemicals for the system which will eliminate any microorganism that could cause breakdown of the product or nosocomial infection. Along with preservative capacity, the system should be free of toxicity and/or irritation. Toxicity is a likely problem because the metabolism of all cells is essentially similar. A compound which shows marked toxicity to microbial cells is likely to show toxicity to human cells. There is, however, one feature which makes each cell unique: its outer membrane or envelope. It is this characteristic of a cell which can be used as a focal point for attack. It is to this end that I have attempted to look for those agents which produce effects primarily on membrane structures. Since toxicology was considered a primary rather than a secondary goal, I chose to look for such potential candidates in a group of chemicals currently being used in the food industry. It was felt that if the material was a food or food additive, the question of toxicity would be a moot one. Of course, a nontoxic mixture (system) which had no preservative power would not serve the requirements for cosmetic preservation. As explained in the Introduction, I was able to formulate a preservative system from chemicals which had multifunctional properties. Our research evolved the concept of producing a hostile environment—a self-preserving cosmetic composition. The main feature of this approach was to look at each component in the system as a potential antagonist to microorganism growth or viability. This holistic approach was at variance to the simple and traditional use of preservative chemicals. By taking advantage of agents which have multiple function properties, it has been possible to design new, nontoxic, and effective preservative systems. As a corollary to this approach, it means that the cosmetic preparation itself becomes a "hostile environment" in which microorganisms cannot grow or survive. It is important to give the cosmetic product preparation those qualities which make the system microbiocidal rather than microstatic. Moreover, preservative activity should be sufficiently rapid to ensure that any contamination by the consumer is eliminated in the period between uses.

In designing such a nontoxic preservative system a number of chemical classes were examined for their suitability. It was decided that a monoglyceride, food-grade phenolics, and a chelating agent could be used to build the preservative system. The monoglyceride chosen was Lauricidin, a Generally Regarded as Safe food emulsifier. Food-grade phenolics were methyl-

TABLE 1 Antimicrobial Activity of Fatty Acids and Monoglycerides[a]

	Capric acid	1-Monocaprin	Sucrose caprate	Lauric acid	1-Monolaurin	(Tri)glycerol monolaurate
Microsporum audouini 9079	12.5	50	200	100	12.5	100
Microsporum canis 10241	12.5	50	400	100	6.25	100
Tricophyton mentagrophytes (*gypsium*) 9129	25	25	400	100	12.5	100
Tricophyton mentagrophytes (*interdigitale*) 9972	12.5	25	400	100	6.25	100
Tricophyton mentagrophytes (*asteroides*) 8757	25	25	400	100	12.5	50
Candida albicans 10231	100	200	>400	>400	>400	>400

[a]Minimum inhibitory concentration is measured in micrograms per milliliter. The minimum inhibitory concentration is defined as the lowest concentration which will inhibit macroscopic visible growth in 24 hr.
Source: Ref. 2.

and propylparabens and antioxidants; such as, tert-butylhydroxyanisole (BHA) and tert-butylhydroxytoluene (BHT). Ethylenediaminetetraacetic acid (EDTA) represented a chelating agent. These food-grade chemicals have been used to design our food-grade preservative systems. Details on their individual as well as combined activities and functions are detailed. While much of this work has been carried out in foods, the principles involved are applicable to the cosmetic industry. Examples pertinent from both industries and useful to the cosmetic formulator and microbiologist will be presented.

II. FOOD-GRADE CHEMICALS

A. Monolaurin (Lauricidin) as a Preservative

It was found that fatty acids esterified to monohydric alcohols were inactive [1], while esterification to polyhydric alcohols produced derivatives which generally were more active than the free fatty acids (see Table 1). These data have been confirmed by others [3-6]. While much of the data generated to date have been in food systems, the principles and conclusions reached are applicable to problems in the cosmetic industry. Also, it is uncommon that discoveries or progress made in one field of endeavor are so quickly transferred to another research area. I hope the following discussion helps bridge this usual gap in technology that exists between the two industries.

Shibasaki [7] stated, "Of all lipids evaluated lauric acid (esters) appear to have the greatest overall antimicrobial activity. Of these derivatives monolaurin would appear to have the highest potential for use in foods and cosmetics." While this may be true for studies carried out in laboratory media, monolaurin, as is true with other lipophilic preservatives, is influenced by the presence of starchy and proteinaceous material [8]. To avoid this problem the lower molecular weight monoglyceride (MC_8) was used early in Japanese-style food [see Ref. 7 for references]. It is the observation of Shibasaki that the use of MC_8 is not warranted when the MC_{12} isomer is properly used, that is, with homogeneous mixing, proper pH range, and effective acidulants (citric and lactic acid). The purity of the monoglyceride is very important. I have emphasized the use of Lauricidin (94-96% mono content) in my own studies, since commercial glycerylmonolaurate is only 40-50% monoester. Failures to successfully apply our discovery to preservative problems were always traced to the use of monolaurin of low mono content.

In the application of monolaurin (Lauricidin) to complex systems, Kato and Shibasaki [9] demonstrated that lauric acid, monocaprin (MC_{10}), and monolaurin (MC_{12}) had significantly enhanced the thermal death of *Escherichia coli* and *Pseudomonas aeruginosa*. In a later paper the thermodynamic characteristics of the bactericidal action of monolaurin and heat were studied [10]. These authors found that the bactericidal effect of increasing the concentration of monolaurin would be more effective in low temperature-long time treatment than high temperature-short time treatment.

The usefulness of monolaurin in the heat inactivation of spores was substantiated by Kimsey et al. [11]. Inactivation of *Bacillus stearothermophilus* 1518 was increased when heated in the presence of 18 m§m monolaurin. The decimal reduction time (D value, the time required to reduce the number of viable spores by 90%) was reduced by 49.8%; at 120°C the D_{120} for *Bacillus subtilis* was reduced by 47.9%. The effectiveness of various saturated monoglycerides on *B. stearothermophilus* spores is given in Table 2. Monocaprin and mono-

TABLE 2 Effectiveness of Various Saturated Monoglycerides on *B. stearothermophilus* Spores at 120°C

Spores heated in 18 mM	Number of carbon atoms in chain	Average D_{120} (sec)	D_{120} range (sec)
Control	–	265	236-294
Monocaprylin	8	195	180-120
Monocaprin	10	88	84-91
Monolaurin	12	119	117-120
Monomyristin	14	221	210-228
Monopalmitin	16	244	234-248

Source: Ref. 11.

laurin were the most active of the series C_8 to C_{16}. Monolaurin was also tested in 10% nonfat dry milk and in the fluid portion of cream-style corn. In these foods the D_{120} time was lowered by approximately 50%. It was concluded that monolaurin could be used as an additive (emulsifier) to lower the heat treatment required to achieve commercial sterility of foods.

Although Ueda and Tokunaga [12] and Koga and Watanabe [13] reported on the preferred action of monocaprylin and monocaprin in soy sauce, Kato [14] made a direct comparison of fatty acids, monoglycerides, and sugar esters. He found that capric acid and monolaurin were the most active preservatives in soy sauce that contained an osmophilic yeast. In a series of sugar esters, sucrose monolaurate was more active than sucrose monocaprate. The activity of monolaurin compares with sodium lauryl sulfate and is superior to butyl-p-hydroxybenzoate, a paraben, which is approved as an antimicrobial additive for soy sauce.

Smith and Palumbo investigated the inhibition of the aerobic and anaerobic growth of *Staphylococcus aureus* in a model sausage system. They compared potassium sorbate, sorbic acid, and monolaurin [15]. Whereas anaerobic growth of *S. aureus* was inhibited by 2500 ppm of sorbate and monolaurin, twice as much was required to inhibit aerobic growth. For greater preservation efficiency it was necessary to acidulate the system with 20 mEq of lactic acid. Under these same acidulating conditions sorbic acid and Lauribic (a proprietary mixture of sorbic acid and monolaurin) inhibited both aerobic and anaerobic growth equally as well at 500 and 750 ppm, respectively.

In the above experiments the amount of lactic acid added and not pH was critical. The addition of 10-20 mEq of lactic acid resulted in a small change in pH value (0.4 pH) and no change in microbial titer. The addition of 25 and 30 mEq of acid resulted in a lowering of only 0.6 pH units (pH 5.7), but a drop in the microbial population of 4 to 5 log units from control levels. Thus, where sorbic acid is added to such a test system, one is not only adding a preservative but also an acidulating agent. When comparisons between sorbic acid and monolaurin are made, this important factor should be recognized. In studies carried out to date, the acidulating effect of added sorbic acid has not been considered.

A new monolaurin product, Lauricidin 812, was tested in the same meat system and was found to have greater activity than Lauricidin or K-sorbate.

TABLE 3 Antifungal Activity of Lauricidin and Derivatives[a] as Compared with Sorbic and Propionic Acids[b]

Fungus	Lauricidin	Sorbic acid	Lauribic	Propionic acid
Pythium elongatum	10	100	100	100
Phytophthora citrophthora	41.2	100	100	100
Mucor circinelloides	30.5	94.3	100	78.7
Rhizopus stolonifer	10.1	100	100	100
Aspergillus flavus	28.6	83.5	100	70.9
Aspergillus ochraceus	21.1	73.5	100	84.1
Aspergillus niger	40.3	100	100	97.6
Penicillium digitatum	39.3	100	100	100
Penicillium patulum	18.9	75.2	100	65.7
Fusarium graminearum	20.1	100	100	100
Saccharomyces cerevisiae	26.5	100	100	23.2
Candida albicans	17.5	100	100	11.7
Rhodotorula mucilaginosa	98	100	100	100
Sporobolomyces sp.	100	100	100	100
Sclerotium rolfsii	63.3	100	100	100
Rhizoctonia solani	54.9	98.2	100	100

[a]At a concentration of 0.1%.
[b]Results are expressed as percentage of mycelial growth inhibition.
Source: Ref. 17.

This new monolaurin product was five times more effective against both anaerobic and aerobic growth than the other two preservatives (unpublished data).

A wider application of monolaurin in meat emulsions was reported by the Cornell group [16]. Baker et al. added Lauricidin to deboned chicken meat, minced fish, and chicken sausage. Spoilage for these unpreserved meat systems was evident in 5, 8, and 9 days, respectively. The addition of citric acid (0.2%) and Lauricidin (250 ppm) had a pronounced retarding effect on storage spoilage. The shelf life of the deboned chicken meat was increased by 140%, that of minced fish by 63%, and that of chicken sausage by 89%.

Lisker and Paster have carried out extensive studies with monolaurin and related food-grade agents as potential antifungal agents for citrus products [17]. The lipophilic lipids were tested against 16 fungi belonging to different groups and having different cell wall compositions. Results are given in Table 3. The combination of Lauricidin and sorbic acid (Lauribic) proved to be more active than either compound alone (see Table 3).

Chipley et al. [18] were interested in the antifungal and antiaflatoxin effects of monolaurin. In their study monolaurin, Lauribic and sorbic acid were tested for aflatoxin inhibition using *Aspergillus flavus* and *Aspergillus parasiticus* as model organisms. While sorbic acid levels of 1000 ppm were required to inhibit aflatoxin formation, monolaurin and Lauribic were active at 750 ppm. The reduction in mycelial growth was also significant. While control values reached an average of 952 mg per flask, values for sorbic acid, Lauricidin, and Lauribic were 540, 582, and 480 mg per flask, respectively. The preservatives were used at 500 ppm. Since the molecular weight of sorbic acid is 112.1 and the molecular weight of Lauricidin is 274.4, this suggests that Lauricidin is approximately 2.4 times more active than sorbic acid on a mole basis.

Moustafa and Agin first reported the commercial insertion of monolaurin as an emulsifier to help solve a critical antifungal problem in diet margarine (personal communication, 1978). Since the concentration of salt is lower and the amount of water is higher in diet margarine as compared to regular margarine, the use of higher preservation levels of sorbate alone was limited by taste and flavor considerations. The introduction of low levels of monolaurin helped produce an acceptable and marketable product. In the 4 years in which the Lauricidin preservative system has been employed, no product failure has resulted. In the single instance in which a monoglyceride was inadvertently omitted, fungal growth resulted and the batch was discarded (personal communication).

The above are some representative examples wherein monolaurin, a Generally Regarded as Safe emulsifier, has been used to create a hostile environment in foods susceptible to spoilage. Since the monoglyceride has multiple functions, it cannot be considered a chemical preservative in the usual sense of the word. The concept of monolaurin and other food additives being part of a systems approach to preservatives will be amplified in a later section.

B. Phenolic Lipophilic Food Additives

A second class of compounds which are useful to the preservative systems concept are edible phenols. These include the esters of p-hydroxyl benzoic acid, methyl- and propylparaben [19], as well as the antioxidants tert-butyl hydroxyanisol (BHA) and tert-butyl hydroxytoluene (BHT). The latter phenolic antioxidants have only recently been implicated as potential preservatives. The role of these antioxidants as preservatives for foods and cosmetics has recently been reviewed [20-22].

It should not be surprising to note that phenols would have a role in food preservation, since their use as germicidal agents resides in early reports. Kuchenmeister [23] was the first (1860) to use pure phenol as a dressing for wounds, although the use of phenol in medicine is usually associated with the name of Lister [24]. Coal tar, the source of early phenol derivatives, had been used as an antiseptic and disinfectant as early as 1815. A recent review reports the history of phenols and their developments as antimicrobial agents [25].

The antimicrobial nature of the parabens was first described in 1924 [26] and adopted for food use in 1932. Much of their history of use can be found in reviews by Neidig and Burnell [27], Aalto et al. [28], Matthew et al. [29], and Chichester and Tanner [19]. Early data on parabens reflected the use of the shorter alkyl derivatives, C_1 to C_4. Branched-chain esters were less active. The heptyl and octyl esters were proposed as preservatives. Maximum activity for paraben derivatives was found in longer (C_{11} and C_{12}) chain esters [30]. These chain lengths are similar to optimal size

TABLE 4 Minimum Inhibitory Concentration of Antioxidants (pH 7.0) in Liquid Culture Medium

Microorganism	BHT (μg/ml)	BHA (μg/ml)
Escherichia coli	>5000	2000
Pseudomonas aeruginosa	>5000	>5000
Streptococcus mutans	>5000	125
Streptococcus agalactiae	>5000	125
Staphylococcus aureus	>5000	250
Corynebacterium sp.	500	125
Norcardia asteroides	>5000	250
Saccharomyces cerevisiae	>5000	125
Candida albicans	>5000	250

Source: Ref. 22.

for the fatty acid alone and confirm the generalization made by Kabara [31] concerning structure-function relationships. Only methyl- and propylparaben are Generally Regarded as Safe when used as direct food additives and when the total addition is less than 0.1%.

A group of food-grade phenolics which have been overlooked as potential antimicrobials are the phenolic antioxidants [32]. Screenings of BHT, BHA, and other phenolic antioxidants all indicate antimicrobial activity (see Tables 4 and 5). While the accepted use of these phenolic derivatives is the prevention of rancidity due to oxidation [33], their antimicrobial character needs to be exploited as part of a systems approach to cosmetic preservation [22].

A word of caution is advised on being too optimistic in equating the antimicrobial property of an agent under laboratory conditions and its role as a preservative in a complex medium. This is particularly true in the cases of BHA and BHT [20].

C. Chelating Agents as Preservative Potentiators

While the adverse effects of chelators on biological systems is not new [33], there is general ignorance of their use as antimicrobial agents in the food and cosmetic industry. This is rather remarkable in view of their frequent inclusion in products used for sanitation. Chelators were initially added to eliminate the prooxidant effects of metals rather than for their effect on microorganisms. Many chelators occur naturally: polycarboxylic acids (oxalic, succinic), hydroxy fatty acids (lactic, citric, malic, tartaric, polyphosphoric) amino acids (glycine, leucine, cysteine), and various macromolecules (peptides, proteins). In food the most popular chelators are citrates, pyrophosphates, and ethylenediaminetetraacetic acid (EDTA).

Chelators per se are generally not considered as preservatives. In conjunction with other preservatives, chelators can potentiate antimicrobial activity [21]. This potentiation is more noticeable against the usually resistant

TABLE 5 Effect of Structural Changes of Some tert-Butyl Phenol Derivatives on the Minimum Inhibitory Concentration Against *Streptococcus mutans*

2,6-Di-tert-butyl derivatives	Minimum inhibitory concentration (μg/ml)
-phenol	250
-4-methyl phenol (BHT)	>1000
-4-ethyl phenol	62
-4-butyl phenol	31
-4-hexyl phenol	125
-4-octyl phenol	500
-4-decyl phenol	>1000
-4-hydroxymethyl phenol	125
-4-methoxy phenol	>1000

Source: Ref. 22.

gram-negative and fungal organisms as contrasted with the more susceptible gram-positive organisms. Effects on the latter are generally not noticeable. Much of the understanding of chelation effects on organisms is the result of work by Leive [34,35] and Salton [36] and has recently been reviewed [21,37]. Hart discusses the role of EDTA in Chapter 18 of this volume.

D. Other Food Ingredients as Preservatives

Besides monoglycerides, phenolic additives, and chelators, a number of other food-grade materials are available to help formulate a "preservative system." To the above list can be added amino acids, peptides, and enzymes and simple and complex alcohols, spices, and other plant constituents. The functions of these various food-grade substances have been reviewed by Shibasaki [7]. It can be concluded that there is a whole spectrum of foodstuffs which can aid in producing a hostile environment to microorganisms and thereby increase shelf life without adding toxicity to the product.

III. APPLICATION TO MODEL COSMETIC SYSTEMS

A. General Solution to the Preservative System Approach

With more than 250 chemicals available as potential germicidal agents, the problem of finding a suitable cosmetic preservative would not seem to be a great problem. The reality of the situation is that only a handful of biocides are used by the cosmetic industry [38]. Parabens can be found in almost three-quarters of the cosmetics used. The next most popular preservative represents less than 10% of the market; the rest of the preservative agents is divided by the 200 or more germicidal agents available for such uses.

Although the parabens are the most popular preservative group used, by themselves they fail to meet the preservation needs of most cosmetic formulas [39]. In a "model lotion" suggested by the Cosmetic, Toiletry, and Fragrance Association Microbial Preservation Subcommittee [40], the paraben-preserved lotion failed to kill *P. aeruginosa*, which remained at high levels throughout the 28-day study period. Thus there has been an established need to augment or replace parabens with other preservatives [41]. The number of options generally available to the formulator is restricted, particularly with the problems associated with formaldehyde, formaldehyde donors, and other toxic and nonbiodegradable chemicals. Recently some new chemials based upon advances in food technology have surfaced. These materials have been shown to be useful preservatives on their own, but combined, their effectiveness is even more dramatic [42,43]. Much of the basic research carried out by Kabara et al. was conducted without specific application to the cosmetic industry. Because of the wide disparity between laboratory and finished product results, there has been a need to test the early promise of these materials in specific formulations. A technical report on their use of Lauricidin and Lauricidin systems has recently been published [44,45].

In the above report, five cosmetics were used as model emulsion systems: (1) a protein hair treatment, (2) a shampoo for dry hair, (3) a hand and body oil-water lotion, (4) a multipurpose water-oil skin cream, and (5) a blow waving and hair-styling lotion. See Table 6 for exact compositions. Because the details of each formula are considered proprietary, the percent compositions are not given. The preservative systems used to protect the cosmetic products were generally composed of a monoglyceride (Lauricidin 812), a food-grade phenolic (methyl- or propylparaben or both), and a chelator. Where the monoglyceride was used to replace either glyceryl or glycol stearate, Lauricidin was used as an emulsifier. Eight variations of the preservative systems were used in the five cosmetic formulas. These variations in the preservative system are given in Table 7. In these initial experiments no attempt was made to optimize the level of preservative use.

The first formula, which is a protein hair treatment product, was preserved using two different preservative systems. In the first case (1a in Table 7), a combination of Lauricidin 812, parabens, and EDTA was applied; in the second instance (1b in Table 7) Lauricidin was used to replace glyceryl stearate and PEG 100 stearate in the formula, and EDTA was added as a potentiator for Lauricidin. Parabens were not added. With preservative system 1a, counts for all organisms were reduced to essentially zero counts after 48 hr. The second combination was only less effective in that counts were markedly reduced after 96 hr.

Formula 2, a shampoo for dry hair, was preserved using Lauricidin 812, methylparaben, and EDTA or only Lauricidin and EDTA. There were small differences between the two preservative systems. While *P. aeruginosa* colonies were reduced from over 10^6 colony-forming units to less than 100 colony-forming units in 96 hr, *E. coli* and *Candida albicans* were more resistant. A full 7 days or more were needed for complete reduction of the latter two organisms.

The third formula, a typical oil-water emulsion, was individually challenged with the five organisms. Preservative systems similar to 1a and 1b (see Table 7) were used for the hand and body lotion formula (3 in Table 7). This formula was easily preserved, as in the first case (protein hair treatment). Colony-forming unit numbers in the product were generally zero after 96 hr. Again, *E. coli* and *C. albicans* organisms indicated slightly more resistance. However, samples taken between 7 and 14 days were zero.

TABLE 6 Compositions of Test Cosmetic Formulations

Formulation 1: Protein hair treatment	Formulation 2: Shampoo for dry hair	Formulation 3: Hand and body oil-water lotion	Formulation 4: Multi-purpose water-oil skin cream	Formulation 5: Blow waving and hair-styling lotion
Deionized water	Deionized water	Deionized water	Deionized water	Deionized water
Hydrolyzed animal protein	Sodium myreth sulfate	Mineral oil	Mineral oil	SD alcohol 40
Cetyl alcohol	Amphoteric-12	Propylene glycol	Sesame oil	PVP/VA copolymer
Glyceryl stearate (and) PEG-100 stearate	Disodium monococamido MIPA-sulfosuccinate	Sorbitol	White beeswax	Quaternium-23
Quaternium-31	Sodium lauryl sulfate	Glyceryl stearate (and) PEG-100 stearate	Lanolin	Hydrolyzed animal protein
Hydrolyzed milk protein	Hydrolyzed animal protein	Cetyl alcohol	Sunflower oil	Polysorbate 20
Oat flour	Glycol stearate	Stearic acid	Paraffin wax	Fragrance
Cherry kernel oil	Lauramide DEA	Isopropyl palmitate	Oleth-3	Dimethicone copolyol
Sunflower oil	Linoleamid DEA	Triethanolamine	Synthetic spermaceti	Phosphoric acid
Quaternium-41	Glucose glutamate	Soya sterol	Borax	Panthenol
PEG-5 soya sterol	Nonowynol-10	Carbomer-934	Fragrance	Benzophenone-4
Glycerin	Fragrance	Fragrance		Simethicone
Guar hydroxypropyl trimonium chloride	Lecithin	Panthenol		
Titanium bioxide	Autolyzed yeast	Wheat germ glycerides		
Wheat germ glycerides	Rice bran oil	Tocopherol		
Fragrance	Cherry kernel oil	Benzophenone-4		
Autolyzed yeast	Safflower oil	Dimethicone		
PEG-75 lanolin				
Panthenol				

TABLE 7 Compositions of Preservative Systems Used in Various Cosmetic Formulations

Preservative system[a]	Level (%)	Constituents
1a	0.30	Lauricidin 812
	0.30	Methylparaben
	0.10	Propylparaben
	0.20	Disodium EDTA
1b	4.80	Lauricidin (to replace glyceryl stearate and PEG 100 stearate in formulation 1)
	0.30	Disodium EDTA
2a	0.30	Lauricidin 812
	0.25	Methylparaben
	0.25	Disodium EDTA
2b	2.00	Lauricidin (to replace glycol stearate)
	0.30	Disodium EDTA
3a	0.30	Lauricidin 812
	0.30	Methylparaben
	0.10	Propylparaben
	0.25	Disodium EDTA
3b	3.20	Lauricidin (to replace other glycerides)
	0.30	Disodium EDTA
4	0.30	Lauricidin 812
	0.25	Propylparaben
	0.15	Methylparaben
	0.10	BHA
	0.20	Disodium EDTA
5	0.30	Lauricidin 812
	0.25	Methylparaben
	0.25	Disodium EDTA

[a]The number refers to the cosmetic formula in Table 6, while the small letter represents alternative systems used for preservation.

A generally more resistant water-oil emulsion (4 in Table 7, a multipurpose skin cream) was also tested and found to be affected by a preservative system which included not only Lauricidin 812, parabens, and EDTA, but also tert-butyl hydroxyanisole (BHA). This water-oil emulsion was readily preserved. Except for *C. albicans*, samples taken at 96 hr and later were all negative. The more resistant yeast was reduced to zero colony-forming units between 1 and 2 weeks.

Since no attempt was made in the above report to optimize the composition of the preservative system, subsequent work coming from our laboratory attempted to answer this question. Using the three classes of chemicals (monoglycerides, parabens, EDTA) as the preservative system, experiments

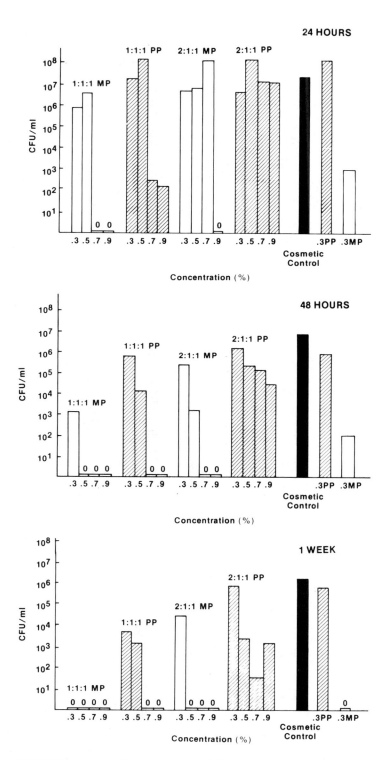

FIGURE 1 The effect of Lauricidin:parabens:EDTA (Na_2) combinations against *P. aeruginosa* in a water-oil emulsion.

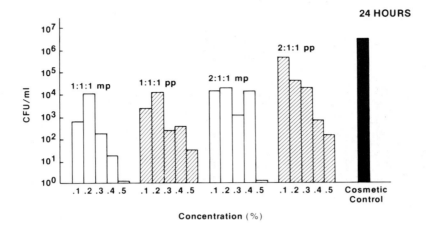
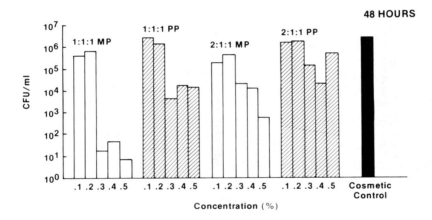
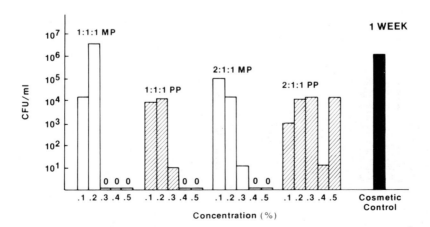

FIGURE 2 The effect of Lauricidin:parabens:EDTA (Na_2) combinations against *P. aeruginosa* in an oil-water emulsion.

TABLE 8 Preservative Effect of Food-Grade Chemicals in a Lauricidin Lotion Oil-Water System[a]

Preservative	At 24 hr		At 48 hr		At 1 week	
	Pseudomonas aeruginosa	*Escherichia coli*	*Pseudomonas aeruginosa*	*Escherichia coli*	*Pseudomonas aeruginosa*	*Escherichia coli*
EDTA						
0.1%	+	+	+	+	−	−
0.2%	−	±	−	±	−	−
0.3%	−	±	−	±	−	−
Methylparaben						
0.1%	+	+	+	+	+	+
0.2%	+	+	+	+	+	+
0.3%	+	+	+	±	+	−
EDTA + methylparaben						
0.1% + 0.1%	+	+	±	+	−	−
0.2% + 0.2%	−	+	−	+	−	−
0.3% + 0.3%	−	±	−	±	−	−

[a]+, growth; ±, slight growth; −, no growth.
Source: Ref. 45.

were conducted in an oil-water (Inolex Formula #7134C) and a water-oil (Inolex Formula #7149A) emulsion system (unpublished data). To the two types of emulsions a mixture of Lauricidin:methyl-, propylparaben:EDTA (Na_2) were added in two different ratios. One ratio represented a 1:1:1 ratio of the three classes of compound; the other ratio was 2:1:1. The challenge organism was either *P. aeruginosa* or *E. coli*. Details and results of these experiments have been published [45]. The water-oil emulsion was more difficult to preserve than the oil-water emulsion (see Figs. 1 and 2). In either case, however, the 1:1:1 ratio of food-grade chemical was more effective than the 2:1:1 ratio. Also, under the conditions of these experiments the methylparaben was more effective than the propyl derivative. The preservative system developed, Lauricidin:methylparaben:EDTA (1:1:1) is effective at a concentration as low as 0.3%. Thus not only the safety but the economics of the system make it an attractive alternative to the use of parabens and other high-priced additional germicidal agents.

Another exciting use of Lauricidin found in our laboratory is its use as a primary emulsifier in a cosmetic. Used in this manner and with the addition of low levels (<0.3%) of EDTA, most cosmetic systems need no other preservative (see Table 8). This is evidenced by showing that methylparaben in the Lauricidin cosmetic emulsion, even at 0.3%, was not an effective preservative against two gram-negative bacteria as compared to EDTA alone.

Thus food-grade preservative systems can be formulated with a special emulsifier, Lauricidin, to create a hostile environment within the product and to minimize the level and number of preservatives needed to protect the formulation. Such preservative systems not only have good efficacy and low toxicity, but also are cost effective to use because of low insertion levels (<0.3%).

REFERENCES

1. A. J. Conley and J. J. Kabara, Antimicrobial action of esters of polyhydric alcohols, *Antimicrob. Agents Chemother.*, 4:501 (1973).
2. J. J. Kabara, R. Vrable, and M. Lie Ken Jie, Antimicrobial lipids: natural and synthetic fatty acids and monoglycerides, *Lipids*, 9:753 (1977).
3. N. Kato and I. Shibasaki, Comparison of antimicrobial activities of fatty acids and their esters, *J. Ferment. Technol.*, 53:793 (1975).
4. J. A. Sands, D. A. Auperin, A. Landin, A. Reinhardt, and S. P. Cadden, Antiviral effects of fatty acids and derivatives, in *Pharmacological Effect of Lipids* (J. J. Kabara, ed.), American Oil Chemists' Society, Champaign, Illinois (1979), p. 75.
5. I. Shibasaki and N. Kato, Combined effects on antibacterial activity of fatty acids and their esters against gram-negative bacteria, in *Pharmacological Effect of Lipids* (J. J. Kabara, ed.), American Oil Chemists' Society, Champaign, Ill. (1979), p. 15.
6. L. R. Beuchat, Comparison of anti-*Vibrio* activities of potassium sorbate, sodium benzoate and glycerol and sucrose, esters of fatty acids, *Appl. Environ. Microbiol.*, 39:1178 (1980).
7. I. Shibasaki, Recent trends in the development of food preservatives, *J. Food Safety*, 4:35 (1982).
8. A. Kato and I. Shibasaki, Combined effect of citric and polyphosphoric acid on the anti-bacterial activity of monoglycerides, *J. Antibacteriol. Antifung. Agents Jpn.*, 4:254 (1976).

9. N. Kato and I. Shibasaki, Enhancing effect of fatty acids and their esters on the thermal destruction of *E. coli* and *Pseudomonas aeruginosa*, *J. Ferment. Technol.*, 53:802 (1975).
10. T. Tsuchido, T. Saeki, and I. Shibasaki, Death kinetics of *Escherichia coli* in a combined treatment of heat and monolaurin, *J. Food Safety*, 3:57 (1982).
11. H. R. Kimsey, J. Adams, and J. J. Kabara, Increased inactivation of bacterial spores in the presence of monolauryl-glyceride, *J. Food Safety*, 3:69 (1981).
12. S. Ueda and H. Tokunaga, Antiseptic effect of capric monoglyceride on pellicle-forming yeast, *Seasoning Sci.*, 13:1 (1966).
13. T. Koga and T. Watanabe, Antiseptic effect of volatile fatty acid monoglycerides, *J. Food Sci. Technol. Jpn.*, 15:310 (1968).
14. N. Kato, Antimicrobial activity of fatty acids and their esters against film-forming yeast in soy sauce, *J. Food Safety*, 3:121 (1982).
15. J. L. Smith and S. A. Palumbo, Inhibition of aerobic and anaerobic growth of *Staphylococcus aureus* in a model sausage system, *J. Food Safety*, 2:221 (1980).
16. R. C. Baker, D. Kline, W. Poon, and D. H. Vadehra, Antimicrobial properties of Lauricidin in mechanically deboned chicken and minced fish, *J. Food Safety*, 4:177 (1982).
17. N. Lisker and N. Paster, Antifungal activity of Lauricidin and related compounds, *J. Food Safety*, 4:27 (1982).
18. J. R. Chipley, L. D. Story, P. T. Todd, and J. J. Kabara, Inhibition of *Aspergillus* growth and aflatoxin release by sorbic acid and derivatives of fatty acides, *J. Food Safety*, 4:27 (1982).
19. D. F. Chichester and F. W. Tanner, Antimicrobial food additives, in *Handbook of Food Additives*, 2nd Ed. (T. E. Furia, ed.), CRC Press, Cleveland, Ohio (1972), pp. 122-129.
20. A. L. Branen, P. M. Davidson, and B. Katz, Antimicrobial properties of phenolic antioxidants and lipids, *Food Technol.*, May: 42 (1980).
21. J. J. Kabara, Food grade chemicals for use in designing food preservative systems, *J. Food Prot.*, 44:633 (1981).
22. J. J. Kabara, GRAS antimicrobial agents for cosmetic products, *J. Soc. Cosmet. Chem.*, 31:1 (1980).
23. F. Kuchenmeister, Ueber Desinfectionsmittel in allgemeinen das Spirol und seine therapeutische Verwendung im besondern. *Dsch. Klin. Eingange Zwanzigster Jahrhunderts*, 23:123 (1860).
24. J. Lister, On a new method of treating compound fractures, abscesses, etc., *Lancet*, 1:326 (1867).
25. W. B. Hugo, Phenols: a review of their history and development as antimicrobial agents, *Microbios*, 23:83 (1978).
26. T. Sabilitschka, Conservation, sterilization and maintenance of sterility by means of Nipasin and Nipasal, *Pharm. Presse Wiss, Prakt Hefte*, 1:173 (1931).
27. C. P. Neidig and H. Burnell, Esters of p-hydroxy benzoic acid as preservatives, *Drug Cosmet. Ind.*, 54:408 (1944).
28. T. R. Aalto, M. C. Firman, and N. E. Rigler, p-Hydroxy benzoic acid esters as preservatives. I. Uses, antibacterial and antifungal studies, properties and determination, *J. Am. Pharm. Assoc.*, 42:449 (1953).

29. C. Matthew, J. Davidson, E. Bauer, J. L. Morrison, and A. P. Richardson, p-Hydroxy benzoic acid esters as preservatives. II. Acute and chronic toxicity in dogs, rats, and mice, *J. Am. Pharm. Assoc. Sci. Ed.*, 45:260 (1956).
30. M. Dymicky and C. N. Huktanen, Inhibition of *Clostridium botulinum* by p-hydroxybenzoic acid n-alkyl esters, *Antimicrob. Agents Chemother.*, 15:798-801 (1979).
31. J. J. Kabara, Structure-function relationship of surfactants, *J. Cosmet. Chem.*, 29:733 (1978).
32. B. N. Stuckey, Antioxidants as food stabilizers, in *Handbook of Food Additives*, 2nd Ed. (T. E. Furia, ed.), CRC Press, Cleveland, Ohio (1972), p. 185.
33. A. E. Martell and M. Calvin, *Chemistry of the Metal Chelate Compounds*, Prentice-Hall, New York (1952).
34. I. Leive, Actinomycin sensitivity in *Escherichia coli* produced by EDTA, *Biochem. Biophys. Res. Commun.*, 18:13 (1965).
35. I. Leive, A nonspecific increase in permeability in *Escherichia coli* produced by FETA, *Proc. Nat. Acad. Sci. U.S.A.*, 53:745 (1965).
36. M. R. J. Salton, Structure and function of bacterial cell membrances, *Annu. Rev. Microbiol.*, 21:417 (1967).
37. J. R. Hart, Chelating agents in cosmetics and toiletry products, *Cosmet. Toiletries*, 93:23 (1978).
38. E. L. Richardson, Update — frequency of preservative use in cosmetic formulas as disclosed to FDA, *Cosmet. Toiletries*, 96:91 (1981).
39. P. A. Berke and W. E. Rosen, Are cosmetic emulsions adequately preserved against *Pseudomonas*?, *J. Soc. Cosmet. Chem.*, 31:37 (1980).
40. J. A. Rogers, G. Jacobs, E. M. Owens, D. C. Pease, Jr., F. Peterson, W. H. Riley, A. G. Schenkel, J. L. Smith, C. E. Stegmann, and J. A. Ramp, *CTFA Cosmet. J.*, 5:2 (1973).
41. W. E. Rosen, P. H. Benke, T. Matzin, and A. F. Peterson, Preservation of cosmetic lotions with imidazolidinyl urea plus parabens, *J. Soc. Cosmet. Chem.*, 28:83 (1977).
42. J. J. Kabara, Lipids as safe and effective antimicrobial agents for cosmetics and pharmaceuticals, *Cosmet. Perfum.*, 90:21 (1975).
43. J. J. Kabara, Monolaurin as an antimicrobial agent, U.S. Patent 4,022,775, Med-Chem Laboratories, January 1977.
44. D. Kenney, Cosmetic formulas preserved with Lauricidin, *Cosmet. Toiletries*, 97:71-76 (1982).
45. J. J. Kabara and C. M. Wernette, Cosmetic formulas preserved with food-grade chemicals, *Cosmet. Toiletries*, 97:77-84 (1982).

PART V
FORMULATION PRINCIPLES INVOLVED IN COSMETIC PRESERVATION

20
FORMULATED FACTORS AFFECTING THE ACTIVITY OF PRESERVATIVES

TERENCE J. McCARTHY *School of Pharmacy, University of Port Elizabeth, Port Elizabeth, Republic of South Africa*

I. Introduction 359

II. The Problem 360

    A. Contamination of products 361
    B. Occurrence of microorganisms 361

III. The Remedy 363

IV. Factors Adversely Influencing the Antimicrobial Action of Preservatives 363

    A. The effect of pH 364
    B. Adsorption-complexation 370
    C. Solubilization-complexation 376
    D. Phase distribution 377
    E. Packaging 378
    F. Deterioration 381

V. Summary 383

    References 385

I. INTRODUCTION

The task of choosing a preservative system for the complex formulations to be found in modern cosmetic and pharmaceutical preparations is an extremely difficult one. In a thought-provoking paper, Cowen and Steiger [1] have shown why a preservative system needs to be tailored to a specific product and not simply selected haphazardly or on grounds of apparent economy. The opening paragraph in the paper concerned highlights many of the factors which will be dealt with more fully in this chapter, and, in consequence, is summarized here.

A preservative has to have the correct antimicrobial spectrum, be compatible with each of the ingredients in the formulation and packaging, and have the appropriate oil-water partition coefficient. It has to be effective at the pH of the product and, in addition, maintain its activity over any pH change that the product may undergo during its shelf life. It must not interfere with color, taste, or odor, depending on the product type. The preservation system must be free from toxic or irritant properties. Finally, it must have an acceptable cost profile.

Needless to say, satisfying all of these requisites is an exceptionally difficult if not impossible task. Many microbiologists, indeed, believe that the ideal preservative will never be found. Consequently, compromises must be made and certain less-than-ideal features of a preservative system must often be accepted.

While some guidance can be given toward a suitable range of preservative systems for testing in actual products, such are the complexities involved that no guarantee of success can be made. Certainly, no preservative should ever be chosen on theoretical grounds alone. Practical testing is essential. With somewhere between 150 and 250 different preservative systems to choose from, selecting the ideal preservative obviously poses problems.

II. THE PROBLEM

There is no doubt that a problem exists with regard to microbial contamination of cosmetic and other topical products. Over the past 20 years a variety of adjuvants have appeared on the market that either form an excellent substrate for microbial growth or inactivate preservatives, and sometimes both. The many investigations that have been launched have revealed that the ubiquitous microorganism can survive in a number of media previously thought unmetabolizable, for example, liquid paraffin, vegetable oils, and even aircraft fuel. One becomes even more wary when one reads the paper by Collins and Deverill on the survival of bacteria in antiseptic skin preparations [2].

A number of adverse effects of the growth of microorganisms in products can be listed. There include discoloration, the formation of odors and/or gas that can cause unsightly bulging of the container, changes in the rheological properties of products, and cracking of emulsions. The visual evidence of surface growth of pigmented fungi can be most disconcerting to a customer who has just purchased and opened a jar of expensive face cream. As Dunnigan [3] has pointed out, decomposition may not reveal itself for many months if conditions are such that growth can only take place after the organisms have adapted themselves to the environment. Chemical changes during storage, such as gradual alteration of pH to a bio-optimal level, may be important for biological adaptation in delayed spoilage problems. Perhaps the product which shows no visible evidence of microbial deterioration is more to be feared, since it is this product which may cause infection, particularly if used on burns or on otherwise damaged epithelium.

Consequently there are two main reasons for controlling the numbers and activity of microorganisms in topical products, whether of cosmetic or pharmaceutical manufacture. The first is to protect the user from any health hazard that may result from the use of microbe-containing preparations, while the second is to protect the product itself — and no less important, the reputation of the firm manufacturing that product.

A. Contamination of Products

This can occur during manufacture or through unhygienic handling by the end user, the consumer. Contamination during manufacture can occur in a number of ways and at various sites in the factory. Without going into tremendous detail, one can think of packaging materials as a source of contamination in the receiving area; of dust, waste matter, and insects as known sources of contamination; of poorly designed manufacturing areas where steam condensates on walls and ceilings invite contamination, particularly of fungal origin; of water supplies, drainage traps, and poorly cleaned pans and flow lines; of final containers of issue, where residues of packaging materials or faulty cleaning processes will be the most probable source of contamination; and of the ingredients themselves and the aerial environment in which the various processes are carried out. Contamination arising from such problem areas would be minimized by Good Manufacturing Procedures (GMP).

The consumer is not easy to regulate. Certain cosmetics, like pharmaceuticals, will end up in the steamy bathroom, where conditions for microbial growth or invasion are optimal. Even those cosmetics which remain on the dressing table will be subject to frequent microbial assault from the aerial environment, fingers, or mascara brushes. Furthermore, some products are more vulnerable to particular types of spoilage organisms than others [1]. Liquid shampoos, for instance, are susceptible to contamination by gram-negative bacterial species, especially *Pseudomonas*, *Aerobacter*, *Klebsiella*, *Achromobacter*, and *Alcaligenes*. Gram-negative bacteria, especially pseudomonads, have also been found in baby lotion. *Serratia* species have been isolated from the same source. Toilet soap, on the other hand, has been colonized by *Pullularia pullulans*, *Stachybotrys atra*, *Scopulariopsis brevicaulis*, and *Trichoderma viride*. Fungal colonies have also been found on mascara, eyeliner, and eye shadow. Talcum powder has, in the past, contained species of *Clostridium tetani* and *Clostridium sporogenes*.

Dunningan [3] has classified the following genera as health hazards: *Pseudomonas*, *Proteus*, *Staphylococcus*, *Streptococcus*, *Serratia*, *Penicillium*, *Aspergillus*, *Candida*, and *Monilia*. This list has been extended by Bruch [4], who has classified several organisms as either objectionable or usually objectionable. For damaged epithelium Bruch rates the following species as objectionable: *Pseudomonas aeruginosa*, *Serratia marcescens*, *Klebsiella* sp., *Pseudomonas multivorans* and *Pseudomonas putida*, *Staphylococcus aureus*, *Clostridium perfringens*, *Clostridium tetani*, and *Clostridium novyi*. For eye products, *Pseudomonas aeruginosa* is listed as always objectionable, while other *Pseudomonas* sp. and *Staphylococcus aureus* are classified as usually objectionable.

B. Occurrence of Microorganisms

There is no doubt that microorganisms are widely spread in manufactured products, and when one considers the number of microorganisms abounding in the air, in raw materials — whether of natural origin or not, in water in particular — and in association with personnel, it is not surprising that in a 3-year period the Food and Drug Administration (FDA) isolated 3400 isolates from production environments for *sterile* drugs. Between 1968 and 1971 the FDA also isolated 1550 gram-negative isolates from raw ingredients and finished topical products [5]. Of these, *Pseudomonas* species accounted for 50%.

The extent to which contamination has taken place during production as compared to in-use contamination is not always easy to assess. The environment in which the product is used may also play a role. According to Wargo [6], it was estimated that in 1970, of the 30 million patients entering U.S. hospitals, 10%, or 3 million, would be exposed to an infected hospital environment. Blyth [7] has reported that in Britain — between the months of June and September — every cubic meter of outdoor air will contain approximately 12,000-15,000 fungal spores. In agricultural areas where hay is being unbaled, and where proliferation in self-heating fodders will have occurred, the astonishing figure of 3000 million aerial spores per cubic meter has been reported. Furthermore, some 68 fungal organisms have been identified in ocular fungal infection [8]. Among these are the accepted human pathogens or saprophytes rarely causing disease in man other than in the eye.

That one does not need to be in an agricultural area to incur risk of ocular contamination is obvious from a comprehensive survey by Myers and Pasutto [9] which reveals that of 327 samples of eye cosmetics examined (mascara, eyeliner, and eyeshadow), 61% showed microbial contamination, with 46% contaminated with fungi [8]. Selecting just a few of the many references cited in this work, one finds mention of hair-conditioning cream containing 1 million gram-negative bacilli per milliliter and of "antiseptic" dusting and talcum powder contaminated with spores of *Clostridium tetani* and *Clostridium sporogenes*, besides other pyogenic bacteria. There are many reports of contamination of such products as hand creams and lotions, hair-conditioning lotions, liquid shampoos, and soaps. However, there are also several reports to show that used cosmetics show higher contamination rates than unused cosmetics. Before manufacturers can sit back contentedly, happy that they are employing good manufacturing procedures, the following question must be posed: On whom does the onus of suitably preserving an in-use product rest — on the consumer, or on the producer? Let us consider this thorny problem later. To return to the results of Myers and Pasutto, a comparison of the incidence of bacterial contamination for all types of cosmetics tested in their study showed contamination to be significantly higher for used cosmetics (49% of 222 samples) than for unused cosmetics (12% of 165 samples). Their findings regarding cosmetic applicators are significant and worthy of note: Although 27.5% of unused cosmetic applicators were found to be contaminated with bacteria (6.9% with *Staphylococcus*), by comparison 100% of used applicators were found to be heavily contaminated with bacteria (84% with *Staphylococcus*). The authors state that it is obvious that the cosmetic applicator must serve as a continuing source of infection or reinfection for the user, and a continuing source of contamination for the remaining cosmetic. They suggest that single-use disposable applicators would break this sequence of events.

In contrast to these findings for cosmetics, Wargo [6] has reported on the microbial contamination of topical ointments used in a hospital patient care area. Six commercial ointment brands were examined and contamination was found in only three of the six product brands. Of 180 previously unopened ointment tubes examined, 11% were found to be contaminated before use, the most common contaminants being gram-positive cocci and diplococci. However, of the 80 tubes partially used in a patient care area, 93% were contaminated. Once again, gram-positive cocci and diplococci were the most common contaminants, but gram-negative bacilli were found in all contaminated tubes.

III. THE REMEDY

The solution to contamination lies in the prevention of contamination, rather than in the final treatment of contaminated products. Final sterilization of a grossly contaminated product would be prohibited on legal grounds so that the answer lies in limitation of microbial contamination to a minimum by Good Manufacturing Procedures. This does not preclude the inclusion in the formulation of an antimicrobial preservative to control limited and acceptable numbers of microbes which could have gained access either before or after reaching the consumer.

Apart from close attention in hygienic procedures from the moment of receipt of raw materials through the various stages of manufacture to the packaging line, the value of microbiological testing of ingredients and the final product should not be underestimated.

If the final product contains unacceptable numbers of microorganisms, it should be within the manufacturer's capability to know the potential areas of contamination within the production process so that such weaknesses can be eliminated. Certain raw materials, particularly those of natural origin, are likely sources of contamination, as are the water supply and parts of equipment not easily cleaned and therefore harboring residues which provide suitable substrates for growth. The hygiene of personnel can be an important factor, and well-planned sampling along the production line can often highlight times of peak contamination (e.g., after canteen or toilet breaks) or areas (a drafty, dusty door or aperture). It is also important to consider carefully the intrinsic ability of each formulation to act as a substrate for growth. Certain products (e.g., those with a large proportion of nonaqueous petroleum-based ingredients) will be far less prone to contamination than those containing amino acids and proteins in an aqueous base. The shortening of time allocated to the compounding and packing of the latter can reduce the risk of contamination considerably, providing that scrupulous attention is paid to hygiene. Furthermore, the inclusion of a preservative in the basic formulation will profoundly influence the control of microorganisms in the product, provided that the preservative(s) selected for inclusion was selected with due consideration of the effect, possibly deleterious, which the other components of the formulation could have on such a system, and not merely on published in vitro reports of the activity of the preservative system in question in aqueous or broth media.

IV. FACTORS ADVERSELY INFLUENCING THE ANTIMICROBIAL ACTION OF PRESERVATIVES

With so many preservatives in use or available on the market and with the large number of adjuvants in use in modern formulations, one can only generalize when discussing the factors causing preservative inactivation. If we look primarily at the types of reactions that may occur, then we may find that in a complex formulation like an emulsion, there may well be a number of factors combining to nullify the activity of any added preservative. Such factors are listed in Table 1. However, these may further be influenced by the basic factors involved in the dynamics of disinfection, namely, time, temperature, and concentration. It is most important to realize that a 50% decrease in the concentration of a preservative having a high concentration exponent (such as a phenolic) will seriously affect the killing time, whereas this killing time is merely doubled if the concentration exponent of the preservative is unity.

TABLE 1 Factors Causing Preservative Inactivation

Type of reaction	Substances implicated
pH	Acids, alkalines (including glass), salts of weak acid-strong base or strong acid-weak base
Adsorption and/or complexation	Powders, gels (natural or synthetic), proteins, filter media
Solubilization and/or complexation	Polyoxyethylene derivatives, soaps, etc.
Phase distribution	Oils, fats, waxes, plastics (certain ones), rubber
Miscellaneous	
Incompatibility	Chlorhexidine (cork)
Light sensitivity	Several (quaternary ammonium compounds, phenolics, mercurials)
Oxygen lability	Sorbic acid
Volatility	Phenolics, chlorbutol

Source: From T. J. McCarthy, Cosmet. Toiletries, 95:23-27 (1980).

Other important factors which need to be considered in the selection of a preservative system are highlighted in a comprehensive document which was issued over a decade ago by the Council of the Society of Cosmetic Chemists of Great Britain [10], but which are still valid today. They cover the multifaceted aspects of container, storage temperature, partitioning, compatability, and so on. They stress that preservative agents cannot be chosen on a theoretical basis alone. Teamwork between the formulation chemist, packaging technologist, physical chemist, and microbiologist is essential, and the choice and testing of the preservative(s) cannot be left until formulation work is thought to be complete.

The following aspects of preservation should be borne in mind during product development:

1. The preservative should have a broad spectrum of activity in low concentration and at probable storage temperatures.
2. Both pH and partition characteristics must be considered.
3. The stability, toxicity, and sensitizing ability of the compound must be examined.
4. Product ingredients may carry a significant level of contaminants, while the formulation itself may contain ingredients capable of maintaining microbial growth.
5. The packaging should be designed to prevent access of contaminants and also inactivation of the preservative agent due to sorption or complexation.

A. The Effect of pH

Much has been written about the effect of pH on the activity of preservatives, yet it is quite remarkable how many manufacturers attempt to "preserve" alkaline products, such as milk of magnesia, using organic acids such as benzoic acid. They are not the only offenders. Later it will be shown that a number of so-called insoluble powders (magnesium trisilicate, veegum) impart marked alkalinity to the medium in which they are used, and unwary cosmetic formulators must beware of such pitfalls.

The organic acids, such as benzoic, sorbic, or dehydroacetic acid, are most affected by the pH of the final formulation. Negatively charged microorganisms repel negatively charged preservative molecules, while charged ions are not readily lipid soluble. If we bear in mind that when the pH of the product equals the pK_a of the preservative the latter is 50% dissociated, then we can understand the inactivity of the above organic acids as neutrality is approached. Thus at pH 5 the undissociated (and therefore active) percentage of preservative with respect to benzoic, sorbic, and dehydroacetic acids is only 13, 37, and 65%, respectively. At pH 7 they are all equally ineffective. Chlorbutanol is unstable in neutral or alkaline media, while thiomersal (pK_a 4) is only bacteriostatic at pH 7. Note that while the parabens have pK_a values around 8, formulations at that pH require double the amount normally used at pH 4, namely, 0.2% wt/vol at the higher pH.

The review article of Cowen and Steiger [1] cited earlier contains an excellent table summarizing, among other important items such as compatability and solubility, the optimal pH activity of some 25 preservatives.

As can be anticipated from their chemical structure, the quaternary ammonium compounds are more active in alkaline medium, activity decreasing to very low levels between pH 3 and 4, as is also the case with phenolics. The optimum range for chlorhexidine is pH 5-8, above which precipitation as phosphate or sulfate salts can occur in the presence of these ions. By contrast, Bronopol is very stable at pH 4, remains active for more than 1 year at pH 6, but is stable for only a few months at pH 7.

Certain preservatives actually rely on their instability for their activity, since they form formaldehyde on decomposition. In the case of hexamethylenetetramine the pH must be below 7 for formaldehyde to be released, but certain formaldehyde donors are pH independent, for example, Dowicil 200, Grotan, and the adamantane derivatives. The high reactivity for formaldehyde unfortunately leads to a reduction in residual active levels, and the use of increased amounts must be balanced against both cost and the possibility of consumer irritation.

As mentioned previously, the effect of pH from apparently insoluble powders in a suspension can adversely affect the activity of pH-sensitive preservatives. Using Bronopol, we [11] found little loss by adsorption of this preservative onto a number of commonly used powders such as aerosil, kaolin, and magnesium trisilicate, but since certain powders definitely influence the pH of the medium, we decided to determine the influence of pH on both the minimum inhibitory concentration (MIC) and contact killing time of Bronopol against *Escherichia coli*. At a pH of 5 the MIC lay between 0.001 and 0.002% wt/vol Bronopol. At pH 7 the figures were 0.025-0.030%, while at pH 9 the MIC had increased to between 0.055 and 0.060%. In like manner the contact killing times for a 0.1% Bronopol solution (aqueous) increased from below 5 min at pH 5 to between 17 and 18 hr at both pH 7 and 9.

Dehydroacetic acid (DHA) has been reported by several workers to be only mildly influenced by nonionics present in the formulation, while certain workers claimed that it retained its activity even under alkaline conditions. Since at pH 6.5 DHA is more than 90% dissociated, these latter claims seemed to conflict with accepted theory. Consequently this writer [12] investigated

TABLE 2 Activity of Dehydroacetic Acid (DHA) at Different pH Values

Solution	pH	Assay wavelength (nm)	MIC (*E. coli*) % Growth	% Inhibition
Aqueous DHA, 0.08%	3.87	304	0.055	0.06
Buffered DHA, 0.08%	5.5	296	0.06	0.07
Buffered DHA, 0.08%	6.5	292	0.25	0.30
Buffered DHA, 0.08%	8.2	292	>0.60	

Source: Ref. 12.

both the stability and MIC values of DHA 0.08% wt/vol (aqueous) against *E. coli*.

Since solubility of DHA was a factor to be considered, it was decided to use 0.08% DHA in aqueous solution, and the same strength in buffer (citric acid, 0.1 M; disodium hydrogen phosphate, 0.2 M) to give final pH values of 5.5, 6.5, and 8.2, while the unbuffered aqueous solution had a pH of 3.87. It was determined that the buffer solutions did not absorb spectrophotometrically at the maxima for the three DNA solutions containing buffer when diluted to assay concentration.

The pH of the broth solutions was checked, and it was found that the buffer capacity of the DHA solutions had maintained the correct pH, indicating that this had not influenced the results.

From Table 2 it is evident that pH profoundly influenced the MIC of the preservative against the test organism *E. coli*, with effectiveness dropping sharply with increasing pH. From Table 3 it can be seen that aqueous solutions of DHA are not stable, stability being greater for less acidic solutions, but with concomitant and overriding loss in effectiveness. The MIC required increases with decrease in concentration as determined spectrophotometrically, but not directly-proportionally.

Storage studies on preservative solutions can yield valuable information, not only for shelf-life predictions (Table 4), but also because results, which at first glance seem anomolous, can lead to investigations which

TABLE 3 Percentage of DHA Remaining After Storage at 30°C

Solution pH	Time in days								Color
	0	8	23	37	56	79	122	223	
3.87	100.0	95.8	92.8	88.7	81.5	75.2	65.6	47.0	Light yellow
5.5	100.0	95.9	92.9	88.9	79.0	73.0	61.7	41.2	Deep yellow
6.5	100.0	99.2	100.0	98.3	94.5	93.2	87.5	81.5	Medium yellow

Source: Ref. 12.

TABLE 4 MIC Values of DHA Solutions After Storage at 30°C

Time (days)	Solution pH	MIC (DHA%)			Concentration (%) (spectrophotometric)
		0.05	0.06	0.07	
0	3.87	+	−	−	100.0
	5.5	+	+	−	100.0
122	3.87	+	+	−	65.6
	5.5	+	+	−	61.7
223	3.87	+	+	+	47.0
	5.5	+	+	+	41.2

Source: Ref. 12.

provide valuable information about the antimicrobial activity of preservatives presumed to be decayed. This writer's experience with sorbic acid is a case in point [13]. During storage studies on a number of aqueous preservatives packed in polyvinyl chloride, polythene, and brown glass [14] and in polypropylene and brown glass [15], this writer observed that sorbic acid (0.1% wt/vol) decayed spectrophotometrically at an alarming rate in all containers except polythene. These findings are reflected in Table 5.

The relatively small loss in polythene was attributed to the presence of antioxidants in this type of plastic, and to confirm this the addition of 0.1% of the antioxidant sodium metabisulfite was made to one-half of a freshly prepared aqueous solution of sorbic acid (0.1% wt/vol). After 12 weeks of storage in the dark at ambient temperatures the metabisulfite-containing solution assayed (spectrophotometrically) at 70.7% of the original amount, compared to only 30.0% for the sorbic acid solution used as control.

TABLE 5 Percentage of Sorbic Acid (0.1% wt/vol) Remaining After Storage in Various Container Types

Container	Temperature (°C)	Time (weeks)	Percent remaining
Polyvinyl chloride	20	12	55.4
Polythene (medium)	20	12	92.2
Glass (brown)	ambient (15-25)	12	44.1
Polypropylene	30	12	26.7
Glass (brown)	ambient (18-29)	12	30.0
Glass (brown)	~5	4	100.0

Source: Ref. 13.

TABLE 6 Minimum Inhibitory Concentration of Sorbic Acid 0.1% wt/vol Solutions After Storage at Various Temperatures for 13 Weeks[a]

Storage temperature (°C)	75	50	30	10	Freshly prepared
Assay concentration (%)	50	45	83	94	100
Theoretical broth concentration (%)					
0.01	+	+	+	+	+
0.02	+	+	+	+	+
0.03	+	+	+	−	−
0.04	+	+	−	−	−
0.05	−	−	−	−	−
0.06	−	−	−	−	−
0.07	−	−	−	−	−

[a] +, growth; −, inhibition using *E. coli*.
Source: Ref. 13.

Working on the assumption (erroneous, as it turned out) that decayed solutions would have less effectiveness and consequently would take longer to kill, stored solutions which spectrophotometrically assayed at about 30% of original concentrations (0.1%) and a freshly prepared 0.1% solution were inoculated with *E. coli* and aliquots were transferred to nutrient broth at hourly intervals. The rather surprising result was that whereas the subcultures from the freshly prepared (0.1%) solution showed growth for up to 23 hr, the approximately 0.03% solutions were both sterile after 5 hr. The pH of the decayed solution stored in polypropylene was found to be 3.2, compared to a pH of 4.2 for the freshly prepared solution, so that either the additional acidity, toxic degradation products, or both resulted in the theoretically deteriorated solution being more active as a contact killer than the original.

From Table 6 it would seem that pH was the important factor, because several decayed solutions were serially diluted according to their label strength and inoculated with *E. coli*, whereupon the freshly prepared solution exhibited the best MIC value. However, perhaps the lesson to be learned is not to assume that analytical results will necessarily reflect microbiological results unless all factors are considered.

However, subsequent work [16] using broths buffered at pH 4 and 6 showed that a freshly prepared 0.1% sorbic acid solution and one reading approximately 30% of this value inhibited *E. coli* at 0.02% of theoretical broth concentration, while at pH 6 the spectrophotometrically decayed solution was actually more effective. Since the unsaturated nature of sorbic acid (2,4-hexadienoic acid) suggests that oxidation would result in small aldehydic moieties, these could well be cytotoxic to microorganisms.

Finally, mention may be made of the relationship of available oxygen (in air) and sorbic acid degradation [16]. These experiments were performed on autoclaved ampuls and clearly show that air volume is directly related to sorbic acid breakdown. Results are presented in Table 7. It is this writer's experience, however (obtained from screening certain formulations over the past few years), that sorbic acid losses in nonaqueous products or even creams appears

TABLE 7 Percentage Content of Autoclaved Sorbic Acid (0.1%) Ampuls Stored at 25°C Under Various Conditions[a]

	Week 0		Week 2½		Week 5		Week 9½	
	Air filled	N₂ filled	Air filled	N₂ filled	Air filled	N₂ filled	Air filled	N₂ filled
Light stored								
2 ml	97.3% (10.65)	99.4% (0)	60.5% (8.80)	102.0% (0)	30.1% (8.7)	101.1% (0)	18.9% (8.7)	100.6% (0)
4 ml	99.9% (6.9)	99.1% (0)	57.3% (6.75)	101.2% (0)	34.5% (6.9)	100.6% (0)	27.0% (6.7)	100.8% (0)
6 ml	94.4% (4.75)	99.5% (0)	53.8% (4.80)	101.1% (0)	50.1% (4.8)	—	39.3% (4.7)	100.8% (0)
Dark stored								
2 ml	—	—	48.1% (8.85)	103.6% (0)	40.7% (8.6)	—	14.8% (8.5)	—
4 ml	—	—	69.2% (4.85)	101.4% (0)	28.0% (6.7)	101.9% (0)	18.3% (6.6)	—

[a]Volume of sorbic acid per ampul. Figures in parentheses denote the air volume per ampul.
Source: Ref. 16.

to be minimal when assayed chemically. Few companies, however, appear to perform regular microbiological challenges after storage periods.

B. Adsorption-Complexation

1. Adsorption onto Particulate Matter

In Section IV.A mention was made that the pH of a medium can be influenced by the powder suspended in it. Furthermore, since preserved particulate suspension systems are to be found both in the cosmetic and the pharmaceutical industries, any adverse interaction between the preservative and the adsorbent needs to be known. However, as Wedderburn [17] stated in 1970, there was little in the literature at that period to indicate specific interaction between preservatives and suspended solid particles. She stated further that in certain circumstances there was every likelihood that adsorption and consequent loss of activity would occur. She reasoned that the extent of adsorption would depend upon the nature of the suspended solids, the surface area they present, the electrical charge at the surface, the polarity of the preservative, and, not least, the pH of the system. In 1967, Bean and Dempsey [18] had shown that the overall effect of adding carbon to a phenol solution was to decrease bactericidal activity due to adsorption of the solute from the aqueous phase.

Since carbon is not widely used in topical cosmetic or pharmaceutical formulations, this writer [19] chose 15 insoluble powders frequently used either internally or topically and at 5% wt/vol aqueous suspension reacted these with a number of preservatives, whereupon the supernatants were examined spectrophotometrically for remaining preservative. These results appear in Table 8 and show that phenolic preservatives are only minimally adsorbed, whereas long-chain preservatives (as can be anticipated by their surface-active properties) were significantly adsorbed. The latter will be discussed in further detail later in this section. However, although sorbic acid showed little adsorption, the pH of only the heavy kaolin, talc, aerosil, and starch suspensions would have prevented dissociation and concomitant loss of activity, so that adsorption losses must be considered simultaneously with the pH of the medium.

This work was extended [20] to cover 10 further preservative systems (see Table 9). From these figures it will be seen how difficult it is to generalize regarding adsorption. The two quaternary ammonium compounds (cetylpyridinium chloride and cetalkonium chloride) showed virtually no adsorption onto both calcium carbonate and zinc oxide, which would suggest that these compounds have a positive — hence repulsive — surface charge. Only aerosil showed no adsorption of 8-hydroxyquinoline sulfate, but, rather surprisingly, aerosil and starch adsorbed significant amounts of phenylmercuric nitrate. Once again the effect of pH must be borne in mind, particularly for dehydroacetic and benzoic acid, which in certain cases appeared to show no adsorption. However, only in the case of talc was the pH above 7.0 for those showing minimal adsorption. In all cases the magnesium compounds caused a pH of 9 or higher, even in the presence of benzoic acid.

Since heat may well be used in the manufacture of a product, it was decided to investigate the effect of the severest moist heat stress (autoclaving) on selected quaternary ammonium compounds (QACs) and organomercurials in the presence of the above-named powders [21]. With few exceptions, autoclaving appeared to have minimal effect on adsorption as compared to controls stored at 25°C for 24 hr. Four QACs showed very similar losses when

TABLE 8 Percentage Decrease in Preservative Concentration (10 ml/0.5 g)[a]

	Chlorocresol, 0.1%	Benzyl alcohol, 2%	Phenoxetol, 1%	Propylene phenoxetol, 1%	Chlorophenoxetol, 0.25%	Sorbic acid, 0.1%	Chlorhexidine diacetate (Hibitane), 0.05%	Benzalkonium chloride (Zephiran), 0.1%
Bismuth carbonate	14	4	0	0	0	72	12	20[b]
Calamine	9	4	0	0	0	9	74	33[b]
Calcium carbonate	10	4	0	2	0	5	47	8[b]
Kaolin, light	5	2	0	2	0	3	66	55[b]
Kaolin, heavy	8	1	0	4	0	2	52[b]	40[b]
Magnesium carbonate, light	11	6	0	1	0	2	94[b]	?[b]
Magnesium carbonate, heavy	9	0	0	4	0	5	97[b]	7[b]
Magnesium trisilicate	11	3	0	4	0	1	99[b]	90[b]
Magnesium oxide	14	4	0	0	0	2	99[b]	11[b]
Zinc oxide	8	2	0	4	0	5	30[b]	8[b]
Titanium dioxide	9	0	0	0	0	5	53	20
Talc	13	0	0	3	0	7	27	29
Starch	17	2	3	3	5	10	97	25
Kieselguhr	7	0	0	0	0	5	22	20
Aerosil (0.10 g)	7	—	0	3	0	6	29	35

[a]After 48 hr of storage at 20 ± 2°C.
[b]After 72 hr of storage.
Source: Ref. 19.

TABLE 9 Percentage Decrease in Preservative Concentration in Contact with Powder (5% wt/vol suspensions) After 2 Weeks Storage at 25 ± 2°C

	8-Hydroxy-quinoline sulfate, 0.1%	Phenylmercuric nitrate, 0.002%	Methyl-paraben, 0.1%	Benzoic acid, 0.1%	o-Chlorobenzoic acid, 0.1%	Dehydro-acetic acid, 0.05%	Phenol, 0.5%	Cetylpyridium chloride, 0.1%	Cetalkonium chloride, 0.1%	Cetalkonium chloride, 0.1%, in 5% propylene glycol
Aerosil[a]	0	87	1	0	2	8	0	60	69	63
Bismuth carbonate	67	6	26	28	36	41	2	81	83	62
Calamine	99	13	1	65	33	7	0	95	98	93
Calcium carbonate precipitate	66	0	0	30	30	0	1	6	2	0
Kaolin, light	56	66	0	12	16	19	1	100	100	100
Kaolin, heavy	37	32	0	0	1	18	0	100	98	98
Kieselguhr	24	0	0	8	32	0	1	15	28	20
Magnesium carbonate, light	99	0	11	26	24	4	0	25	17	36
Magnesium carbonate, heavy	96	0	12	27	31	4	0	22	16	8
Magnesium oxide, light	100	12	19	30	29	23	0	65	47	32
Magnesium trisilicate	100	49	12	29	32	0	0	100	100	100
Starch	10	88	10	11	7	8	1	89	70	61
Talc	99	25	4	27	27	1	1	30	28	28
Titanium dioxide	31	76	1	0	2	0	0	92	100	89
Zinc oxide	99	53	2	25	11	72	1	5	1	2

[a]At 1% wt/vol owing to bulkiness.
Source: Ref. 20.

stored at 25°C, but at 0.01% concentration CTAB showed significantly higher adsorption onto certain powders (CTAB is synonymous with Cetrimide).

However, Bean and Dempsey [22] had showed that adsorption of a preservative did not necessarily mean inactivation of that preservative. Suspensions containing benzalkonium chloride possessed a greater activity than the corresponding supernatant solutions removed from contact with the test powder (kaolin), owing to some of the bactericide adsorbed on the kaolin becoming available to the bacteria. It is well known that bacteria are adsorbed onto certain powders, placing them into areas of high preservative concentration. It was decided to examine benzalkonium chloride BZC (0.01% wt/vol) against the 15 powders used before, using mean death times [21]. Using a threefold stronger solution than Bean and Dempsey, our results differed slightly inasmuch as bactericidal activity decreased in the order suspension > aqueous solution > supernatant liquid. Thus only when 100% adsorption of BZC occurred was the system unprotected. Recently we examined the adsorption of some of the more modern preservatives [23]. These results are shown in Table 10.

We have seen that in systems where powders sedimentate, the supernatant may well be left unprotected, being depleted of preservative, whereas the sediment itself is actually highly bactericidal. The question may then be put: What happens if the powder remains in suspension? This question will be examined in the next section, where gelling agents are discussed.

Before leaving this section on particulate matter, mention should be made of a paper by Yousef et al. [24], where the effect of 11 different (pharmaceutical) materials on the activity of six commonly used preservatives has been investigated using a viable count technique. Possibly the contact time of 30 min was rather short, but the results showed that veegum, magnesium trisilicate, and polysorbate 80 (in descending order) were most antagonistic to the preservatives, while the bactericidal activities of chlorhexidine and methylparaben were antagonized the most and chlorocresol the least.

2. The Effect of Gels

The term *gels* can be used to cover a number of classes of hydrocolloids [25]. These may be either synthetic polymers or of natural origin. In the latter case it should be borne in mind that natural products often abound in microorganisms, and Wedderburn [17] has suggested that the frequently inefficient preservation of formulas containing gum tragacanth could be due to bacterial and fungal contamination of the gum itself, leading to "swamping" of the preservative, rather than to loss of preservative to the gum. This could explain apparent anomalies in the literature with regard to the inactivation of preservatives by tragacanth.

The potential for interaction between preservatives and gelling agent was realized some decades ago, possibly because natural products such as tragacanth, agar, alginates, and starch were so widely used. In 1957 Eisman and colleagues [26] showed that gum tragacanth (at 3% concentration) exerted a strong neutralizing effect upon the bactericidal activity of chlorbutanol, the para-hydroxybenzoates, and benzethonium chloride. Less inhibition of phenol, phenylmercuric acetate, and merthiolate occurred. Furthermore, such jellies were found to lose antibacterial activity during storage, so that tests conducted immediately after manufacture would not give accurate predictions of shelf-life behavior. In 1958 Taub et al. [27] studied the effect of 2% gum tragacanth dispersed with 5% wt/wt propylene glycol at various pH values. Parabens (methyl plus propyl) were shown to be much more

TABLE 10 Adsorption of Preservatives from Solution onto Various Powders After 18 Hr of Storage at 22°C

Powder	Aqueous Bronopol, 0.1%		Aqueous Hyamine 10X, 0.1%		Aqueous Kathon CG, 0.1%		Ethanolic Irgasan DP300, 0.002%		Average pH of aqueous controls
	Average percentage loss	Average pH of test solutions	Average percentage loss	Average pH of test solutions	Average percentage loss	Average pH of test solutions	Average percentage loss	Average pH of test solutions	
Aerosil*	0	5.7	35	4.8	8	6.0	0	6.7	6.3
Bismuth carbonate	5	6.1	44	8.3	3	7.6	0	7.5	7.8
Calamine	0	7.6	71	7.8	4	8.2	0	8.5	7.9
Calcium carbonate	3	7.9	0	8.9	0	8.5	0	7.4	8.0
Heavy kaolin	0	6.3	69	6.6	0	5.6	10	7.5	7.1
Light kaolin	7	7.0	90	6.9	26	7.0	14	7.4	7.4
Magnesium carbonate, light	5	9.1	19	9.7	4	9.7	0	8.9	9.5
Magnesium carbonate, heavy	4	9.0	26	9.7	0	9.8	0	8.9	9.5
Magnesium oxide, heavy	5	9.8	19	9.2	0	9.8	0	8.6	9.7
Magnesium trisilicate	5	8.9	93	9.1	0	9.1	0	8.8	9.3
Talc	0	6.0	36	6.6	3	7.5	40	7.2	6.8
Titanium dioxide	0	6.0	54	6.6	0	6.2	0	6.8	6.5
Zinc oxide	0	6.4	14	6.8	2	6.6	27	6.2	6.5

[1]A = average percentage loss.
[2]B = average pH of test solutions.
*1% wt/vol suspension, all other systems were 5% wt/vol.
Source: Ref. 32.

effective than either chlorbutanol or benzoic acid. Tillman and Kuramoto [28] investigated methylcellulose, using paraben derivatives, and showing by spectrophotometric means that some complexation had occurred.

Although it is generally accepted that methylcellulose derivatives normally do not complex preservatives significantly (although certain workers on contact lens solutions would challenge this claim), sodium carboxymethylcellulose, being anionic, is incompatible with cationics such as acriflavins and quaternary ammonium compounds. The latter also react adversely with the anionic alginates. De Blaey et al. [29] have shown that one molecule of acriflavine is bound by two carboxyl groups in sodium carboxymethylcellulose.

Some of the reactions between polyvinylpyrrolidone (PVP) and preservatives are reviewed by McCarthy [30], who found little reaction between a 2% aqueous polyvinylpyrrolidone solution and 14 commonly used preservatives. Significant loss of phenolic preservatives into the rubber caps and cardboard wadding of closures was found, however, in solutions stored at 30°C for up to 12 weeks, which highlights the importance of choosing suitable packaging materials.

Reverting to tragacanth, it was decided [31] to investigate interaction between eight commonly used preservatives and a 2% wt/vol (aqueous) tragacanth gel because there had been literature reports of contamination of catheter gels. Dialysis studies showed significant loss (about 30%) of chlorhexidine acetate (0.01%) by binding to the gel, but the remaining preservatives were unaffected (benzalkonium chloride, 0.1%; chlorocresol, 0.1%; phenol, 0.5%; methylparaben, 0.1%; benzyl alcohol, 1.0%; benzoic acid, 0.1%; sorbic acid, 0.05% wt/vol). Instead of using stock cultures for the microbiological assessment of the systems, use was made of the inherent microbial flora of the tragacanth under test. This procedure was followed because the use of "pampered" stock cultures has been criticized in the literature, and several workers have found that isolates from contaminated products are considerably more resistant to selected preservatives than the type cultures recommended as test organisms. All systems were sterile within 30 hr, which bears out the dialysis results. Even chlorhexidine acetate at 0.01% (which showed 33% binding) killed within 3 hr, which perhaps emphasizes the point that binding studies must have relevance to the MIC of the preservative under test, and even 90% inactivation is of no significance if the remaining 10% of preservative is still potently inhibitory.

Before concluding this section on gelling agents, it is perhaps relevant to refer back to Section IV.B.1, which dealt with particulate matter. There we saw that in an aqueous-powder system, the sedimenting particles were often highly antibacterial owing to adsorbed preservative, whereas the supernatant was relatively depleted of preservative and hence capable of allowing the growth of surface-growing organisms. This posed the question of what would happen if the particles remained in suspension, as would occur in many preparations (creams, ointments, and gels). Consequently it was decided to suspend selected powders (calamine and zinc oxide) in various gels (bentonite, sodium alginate, veegum) for the preservatives Bronopol (0.1%) and methylparaben (0.1%), while hypromellose was the suspending agent for calamine when testing the QAC preservatives Hyamine 10X (0.02%) and cetylpyridinium chloride (0.02%) [32].

It was found that whereas 0.1% aqueous Bronopol killed $E.$ $coli$ ($\sim 10^5$ per milliliter) within 10 min, the presence of 2.5% bentonite delayed the killing time to 30-60 min, and the further addition of 5% zinc oxide extended the killing time to over 6 hr. A similar pattern was observed for Bronopol, sodium alginate (2.5%), and calamine (5%), while 2.5% veegum without powders adversely affected Bronopol, with growth at 6 hr. The addition of gels brought the pH to around 8, which in the long term will also adversely affect Bronopol.

Methylparaben at 0.1% showed a reaction very similar to that of Bronopol in that each of the gelling agents, in turn, extended the killing time to greater than 6 hr. Conversely, the hypromellose-QAC systems, with or without suspended powders, showed no growth after 10 min of contact.

C. Solubilization-Complexation

There are few cosmetic preparations which are merely simple solutions of the active ingredient(s) so that the preservative can act without competition from interacting ingredients which can nullify its usefulness. Thus a preservative in a modern emulsion can be lost from the aqueous phase (where its activity is required) by solution in the oil phase (if it is oil soluble), by solubilization into micelles formed, for example, by surfactants, or by chemical bonding onto complex (nonionic) emulgents such as the polyoxyethylene condensates, among others. Add to this formulation an insoluble powder, a gel to give good "feel" to the product, and an unfavorable pH, and the preservative system really has not got much chance of success. Pack this product in a container which reacts adversely with the preservative and the comedy of errors becomes a disaster. We see that when molecules such as surfactant molecules reach a certain concentration in solution, they form micelles, which may be spherical or laminar aggregates of some 200 or more molecules. These remain in the solution as tiny droplets that attract preservatives to them from the aqueous phase, thereby seriously depleting the very phase which requires protection of its preservatives. Thus the required inhibition concentrations of QACs, such as cetylpyridinium chloride or benzalkonium chloride, increases about 50-fold in the presence of 0.5% Tween (polysorbate) 80 and increase 1000-fold in the presence of 3% Tween 80 — hence the widespread use of polysorbate as an inactivator of preservatives in sterility testing.

This type of reaction has been known since the 1950s and has given rise to numerous publications. Both solubilization and complexation have been assessed by several dialysis studies using a variety of membranes and by microbiological means, while the types of binding mechanisms have been investigated by a number of sophisticated chemical methods [33-39]. These methods have recently been reviewed by Rieger [40].

Complexation can occur in semisolid and solid formulations where up to 5% or more polyoxyethylene derivatives may be present. In 1957 Barr and Tice [41] screened over 48 preservatives in the presence of various nonionics and found sorbic acid at 0.1-0.2% to be the most satisfactory for sorbitan and polyoxyethylene esters. Thoma and co-workers [42] have shown that the amount of phenolic preservative required to preserve 5% solutions of polyoxyethylene stearates as compared to aqueous solutions had to be multiplied by the following factors: phenol, ~2; cresol, ~4; chlorocresol, ~8; thymol, ~30; chloroxylenol, ~38; and hexachlorophene, ~800. Both hydrogen bonding and solubilization occurred.

Furthermore, chlorocresol has been found to complex with cetomacrogol with time, so that a preparation that initially can suppress organisms may lose this biostasis with time [43]. However, Crooks and Brown [44,45] have shown that when methylparaben plus chloroxylenol was used with cetomacrogol, there was a greater percentage of free preservatives in the system than when each was used singly, owing to competitive displacement from the binding sites.

More recently, Jacobs et al. [46] published interesting data on the effect of pH and emulsifier on the preservative requirements of oil-water emulsions. A total of 29 individual preservatives and 16 combinations of two or more preservatives were tested in anionic and in a nonionic oil-water lotion, each of which was formulated at an acid pH and at an alkaline pH. Using four representative challenge microorganisms, MIC values were determined for the preservatives and/or preservative combinations. The results were revealing. Less than 35% of the preservatives or systems tested in lotions were effective. Anionic emulsions were somewhat easier to preserve than formulas made with nonionic emulsifiers.

All these studies highlight the potential adverse effect of solubilization-complexation on a preservative or preservatives. Too much detailed work is available to be specific, but both Barr and Tice [41] and Patel [47] have shown that both sorbic acid and dehydroacetic acid are not as adversely affected as many other preservatives. Unfortunately they are adversely affected by pH values above 6 and, as has been mentioned previously (and as will appear under plastics later), sorbic acid is affected by oxygen, while DHA is unstable in certain plastics.

D. Phase Distribution

Emulsions are generally two-phase systems consisting of immiscible liquids, one of which (the internal phase) is dispersed in small droplets through the other phase, called the external or continuous phase. In simple systems one phase is predominantly aqueous (hydrophilic) and the other lipid or fatty (hydrophobic). A simple emulsion can thus be either water-in-oil or oil-in-water, with the external phase usually predominating in volume. However, both phases usually contain a number of different substances, either in solution or suspension, and these can influence the activity of any preservative added to the system. Since the two phases (aqueous and oily) are separated by definite boundary surfaces, the film which is adsorbed at the boundary can be regarded as a separate (third) phase.

There is ample evidence that microorganisms are adsorbed at the interface between the oil-water phases, and, in addition, can be found free in the aqueous phase, particularly if motile. Since microorganisms depend on the presence of water for growth, it follows that substances which concentrate in the aqueous phase (humectants, hydrocolloids, preservatives) will influence the growth or survival of these organisms. We know that a large number of materials used in the preparation of emulsions are biodegradable. At low concentrations glycerin and sorbitol can be broken down, as can a number of surfactants, particularly those containing esters and methyl groups. The nutritive value of vegetable gums, proteins, amino acids, carbohydrates, and so on, found as additives in cosmetics is well known, and these form a potential growth medium for microorganisms. These ubiquitous organisms can even metabolize certain preservatives, notably the phenols and para-hydroxybenzoates.

Since the growth of microorganisms in an emulsion can lead to a number of deleterious effects (emulsion cracking, discoloration, rancidity, gas, and/or odor formation), the need to add an antimicrobial preservative becomes paramount. We have seen, however, that the activity of such a preservative (which will reside primarily in the aqueous phase, under ideal conditions) can be adversely affected by the presence of substances also present in the aqueous phase (gums, particulate matter, solubilizers) or, depending upon its chemical nature, by pH. Another very important consideration is the solubility of the preservative in the hydrophobic phase, where its activity is so limited as to be meaningless. Formulators are often not aware of the fact that certain preservatives have high intrinsic oil solubility and even though these preservatives are water soluble, will partition into the oily phase (particularly if this phase is in fine subdivision and hence presents a large surface area) to the detriment of the aqueous phase, which needs protection, since it is here that microorganisms will flourish.

The term *oily phase* includes natural oils (peanut, sunflower, soya, etc.), hydrocarbon compounds such as mineral oil (liquid paraffin), esters such as isopropyl myristate, fats, and waxes. The preservation of purely oleaginous media is extremely difficult. We have found that a number of preservatives in arachis oil, when heated at 150°C for 40-70 min, could not kill *Bacillus subtilis* from soil. The preservatives were methylparaben, 0.2%; benzyl alcohol, 1 and 5%; chlorocresol, 0.2%; chlorbutol, 0.5%; Irgasan DP300, 0.1%; camphor, 10%; chloroform, 5%; and iodine, 1%.

In emulsions, however, the presence of moisture greatly assists in killing microorganisms, but the loss of preservative into the oily phase can be a serious problem. It is now well known that the phase volume ratio and oil-water partition coefficient profoundly affect the distribution of a preservative between oleaginous and aqueous phases.

The formula of Bean and Heman-Ackah correlates the above factors [48], and one soon observes that in a 20% arachis oil-in-water system without an emulgent, the distribution of 0.1% chlorocresol will be

$$C_w = C \frac{\phi + 1}{K_w^o \phi + 1}$$

where C_w is the percentage of preservative left in water, C is the overall percentage of preservative, ϕ is the phase volume ratio, and K_w^o is the partition coefficient (K_w^o chlorocresol = 116.7 for arachis oil);

$$C_w = 0.1 \frac{0.2 + 1}{(116.7 \times 0.2) + 1} = \frac{0.1 \times 1.2}{24.34} = 0.005\% \text{ (approx.) chlorocresol}$$

A quantity as low as 0.005% is patently below an inhibitory value for many organisms. The solubility of chlorocresol in arachis oil is extremely high. Few figures exist for partition coefficients, but some are given in Table 11.

E. Packaging

The importance of selecting a suitable container for the final product cannot be overemphasized, since failure to do so can negate months of preformulation experimentation. The cap lining is frequently a source of contamination, while poorly fitting closures can allow egress of volatile preservatives and essential oils, or ingress of contaminants, particularly in humid storage conditions. Aluminum will inactivate Bronopol and certain mercurials. Parabens are known to bind to nylon containers, so that surface complexation of surface-active agents with container walls must be considered as a potential problem area. Probably the most commonly reported reaction is sorption, with the preservative being lost from the aqueous phase into the container cap or walls.

TABLE 11 Partition Coefficients of Certain Preservatives

Preservative	K_w^0	Oil
Sorbic acid	3.30	Almond oil
	0.21	Mineral oil
	3.00	Soya soil
Methylparaben	7.80	Almond oil
	0.03	Mineral oil
	5.80	Soya oil
Ethylparaben	26.00	Soya oil
Propylparaben	87.00	Soya oil
Butylparaben	280.00	Soya oil
Benzoic acid	6.10	Soya oil
Phenol	5.60	Arachis oil
	0.07	Mineral oil
Thymol	447.00	Arachis oil
Chlorocresol	117.00	Arachis oil
	1.53	Mineral oil
Bronopol	0.04	Mineral oil
	0.11	Arachis oil
Dowicil 200	<1.0	Mineral oil
	<1.0	Corn oil
	<1.0	Isopropyl palmitate

Rubber and certain lipophilic plastics can be regarded as solid oil phases into which preservatives can partition if lipophilic. Losses can occur both by absorption and adsorption, while both rubber and plastics contain ingredients which are known to leach out into the enclosed solution. Some of these are toxic materials, and others have the potential for complexation with preservatives, particularly organomercurials and quaternary ammonium compounds. However, plastic materials can vary from batch to batch, and this emphasizes the importance of running quality control checks and storage tests on containers, even if these are obtained from the same manufacturer.

Table 12 summarizes a number of storage studies performed by this writer [49]. As these could not all be performed at the same time, it will be noted that in the case of polyvinyl chloride and medium-density polyethylene, temperatures of 20 and 25°C, respectively, were used. These approximate to temperate room temperatures, while the 30°C used for polypropylene can be regarded as an accelerated study.

Only sorbic acid and DHA were unstable in (brown) glass containers, while sorbic acid DHA, the phenolics, and chlorinated phenols are worthy of scrutiny, since they show losses in various containers.

TABLE 12 Comparison of Stability of Aqueous Preservatives in Various Containers[a]

Week storage	Brown glass container, week 12, ambient temperature	Polyvinyl chloride container, week 12		Polyethylene container, week 12		Polypropylene container		
		20°C	25°C	20°C	25°C	Week 12, 30°C	Week 32, 30°C	Week 90, 30°C
Chlorocresol, 0.1%	95.8	91.7	—	42.2	—	91.5	88.4	77.1
Benzyl alcohol, 2%	99.8	98.7	—	84.5	—	93.2	87.7	85.9
Phenoxetol, 1%	97.9	97.9	—	90.6	—	96.8	98.0	96.2
Propylene phenoxetol, 1%	96.1	95.0	—	77.7	—	94.7	94.5	88.7
Chlorophenoxetol, 0.25%	98.8	98.5	—	78.3	—	—	—	—
Sorbic acid, 0.1%	44.1	55.4	—	92.2	—	26.7	11.7	6.1
Chlorhexidine acetate, 0.05%	98.7	100.0	—	97.8	—	100.0	105.4	108.0
Benzalkonium chloride, 0.1%	99.3	100.2	—	103.3[b]	—	100.0	107.8	117.1[b]
Methylparaben, 0.1%	98.2	—	101.7	—	100.0	98.4	104.5	113.3
Cetylpyridium chloride, 0.1%	99.2	—	103.2	—	100.8	100.0	108.5	115.1
Phenol, 0.5%	100.9	—	99.8	—	91.1	96.0	98.4	100.4
Benzoic acid, 0.1%	101.2	—	104.2	—	86.6	96.1	97.3	90.4
o-Chlorobenzoic acid, 0.1%	101.3	—	101.0	—	95.9	—	—	—
Dehydroacetic acid, 0.05%	89.0[c]	—	79.2	—	37.3	—	—	—
Phenylmercuric nitrate, 0.05%	100.1	—	100.0	—	98.0	98.5	102.0	111.4[b]
8-Hydroxyquinoline sulfate, 0.1%	99.4	—	96.4	—	88.6	—	—	—
Dichlorophenol, 0.05%	96.7[d]	—	—	—	16.2[b]	41.8	20.0	4.9
o-Phenylphenol, 0.05%	100.0[d]	—	—	—	25.7[b]	47.9	31.3	11.6
p-Chlorometaxyline 0.05% (in 4% ethanol)	96.2[e]	—	—	—	59.4[b]	83.1	78.4	66.6

[a]Results are of the percentage of preservative remaining.
[b]Particles.
[c]After only 6 weeks of storage.
[d]After only 3 weeks of storage.
[e]After only 7 weeks of storage.
Source: Ref. 49.

F. Deterioration

1. Light

A number of organic and inorganic compounds undergo decomposition if exposed to light, and preservatives are no exception. It is possible to eliminate ultraviolet rays by the choice of suitable packaging materials, but it must be stressed that opacity (nontransparency) of a material such as low-density polythene does not ensure light protection. The simplest procedure is to cut a small piece of the container under test and to place this in a cuvette of a spectrophotometer, whereupon it can be scanned from 250 to 400 nm to see if light is transmitted. A number of preservatives appearing in *Martindale's Extra Pharmacopoeia* [50] are recommended to be stored protected from light. Whether the same necessarily applies to dilute solutions of these compounds can be determined experimentally. These compounds include the quaternary ammonium salts, benzyl alcohol, chlorbutol, chloramine, chlorhexidine salts, chlorocresol (and other chlorinated phenolic compounds), domiphen bromide, hexachlorophane, phenylethanol, potassium sorbate and sorbic acid, the mercurials, and zinc pyrithione, to mention but some.

2. Heat

Many of the compounds used as preservatives are thermostable to the extent that they can be autoclaved at 120°C for 30 min. Chlorbutanol is an exception and decomposes rapidly on heating at neutral to weakly alkaline pH levels and is unstable when stored in aqueous solution, particularly in that pH range. Like the phenolics and chlorinated phenolics, it is volatile in warm solutions (60-70°C), and losses can occur during the manufacture of cosmetics, where heat fusion is often involved. It should be noted that the volatility of the chlorinated phenols does not mean that they are thermolabile. They can be autoclaved in sealed containers such as ampuls without decomposition. Bronopol L is stated to be stable at pH 5-7 below 40°C (which possibly implies instability above that temperature). Dowicil 200 is unstable above 60°C, while formaldehyde will volatilize from warm solutions. However, such solutions should not cool below 15°C or deposits of paraformaldehyde may form. Glydant (DMDM hydantoin) can be heated to 90°C without degradation, even though it is a formaldehyde donor. Potassium sorbate is stated to be unstable above 38°C, and this writer was surprised to read that chlorhexidine gluconate was unstable at high temperatures [53], seeing that Martindale states that such solutions can be sterilized by autoclaving. Both Martindale [50] and the comprehensive article by Henry and Jacobs [51] should be consulted for formulators. Where data cannot be obtained, the onus rests on the formulator to subject the preservative (preferably in the final product) to accelerated (high-temperature) stress, whereupon it can be monitored for degradation either chemically or microbiologically.

3. Chemical Reaction

It is not a simple task to pinpoint chemical reactions which will cause deterioration of preservatives. The effect of pH has been highlighted, while the relative insolubility of certain salts of compounds such as chlorhexidine and the phenylmercuric radical is of chemical origin. Physicochemical principles are involved in bonding preservatives onto macromolecules, while the efficacy with which inactivators work in sterility testing procedures is often associated with their ability to complex preservatives in direct competition with microbial cell receptors. With the increasing usage of radiation sterilization in the cosmetic industry it is perhaps timeous to consider this form of chemical reaction also.

With ethylene oxide sterilization under question, due to possible carcinogenic and mutagenic side effects, the increase in the use of radiation sterilization seems undoubted. Some 30 countries already use cobalt-60 isotopic irradiators and the usage in the United States doubled in 1977-1978 and it is predicted to double in the next 5 years [52].

4. Radiation Sterilization

One of the advantages of radiation sterilization lies in the fact that in the field of packaging materials, stronger and thicker polymer films and metal foils can be used than in other sterilization methods. Furthermore, it can be automated, is reproducible and reliable, and is not influenced by variables of, for example, temperature, pressure, humidity, or packaging as are the autoclaving and ethylene oxide sterilization processes.

5. Application of Radiation Sterilization to the Cosmetic Industry

With increasing legislative requirements from authorities both in the United States and Europe in regard to microbiological safety standards, interest in radiation treatment of cosmetics has increased. The most serious problem likely to occur in irradiating cosmetics is the radiation-induced chemical changes and the associated changes in physical properties that some cosmetics may undergo. It would be extremely difficult to predict what changes would occur in a cosmetic containing inorganic and organic molecules (and frequently water), but certain guidelines can be given.

Dry inorganic compounds of mineral origin (e.g., kaolin, talc, bentonite) will undergo very little chemical or physical change when subjected to irradiation, although a slight color change may occur. A host of changes can occur in mixtures of organic compounds and one compound may act as a sensitizer and another as a stabilizer against radiation. Certain polymers tend to degrade when irradiated, with an associated drop in molecular mass. In the case of thickeners, this leads to a dcrease in viscosity. This problem can, however, be overcome by starting with a higher viscosity than normal to compensate for the radiation degradation, but this introduces a cost factor. Unsaturated hydrocarbons have a lower radiation stability than saturated hydrocarbons, while the former may undergo polymerization. As will be seen later, simple aromatic compounds show a remarkable radiation stability, but the nature of substituent groups can play a very important role in this respect. In the case of halogenated compounds, changes in pH may follow irradiation. In this respect the use of polyvinyl chloride as a packaging material should be avoided, since this polymer yields hydrochloric acid on irradiation.

Since the irradiation of water leads to the formation of chemically reactive radicals which can react with the solute, it may be necessary to irradiate potentially contaminated ingredients (mineral powders, natural products) separately, prior to their incorporation in the final product.

Although relatively little radiation sterilization and pasteurization of cosmetics is currently done on an industrial scale overseas, experience has shown that most cosmetics are little affected by radiation treatment. In many other cases the product can be irradiated after a slight change in formulation [50].

Jacobs [53] has recently published a comprehensive article on the gamma-irradiation decontamination of certain cosmetic raw materials.

6. Preservative Losses Due to Irradiation

Thus, with increasing reports in the literature with respect to the irradiation of cosmetics, it is perhaps timeous to look at the effect of gamma irradiation on preservatives. Apart from a reported study on chlorbutanol [50], this writer could find nothing in the literature concerning the breakdown of preservatives following irradiation, and consequently decided to investigate the effect of gamma irradiation on a number of commonly used preservatives. Thus 17 preservatives (in aqueous solution) were selected covering the QAC, organic acids, phenolics, aromatic alcohols, and a small group of miscellaneous structures [54].

As could be anticipated from a dose of 2.5 MR, few of the preservatives withstood sterilizing irradiation, phenol, and chlorocresol being notable exceptions, as reflected in Table 13. No attempt was made to identify degradation products, nor were facilities available for toxicity studies. Whether these same preservatives would fare better in nonaqueous preparations requires investigation, but apart from totally oleaginous preparations, of which a few occur, there will always be some moisture present, so that irradiation of the final product could pose serious problems. As it is, gelling agents such as acacia and sodium carboxymethylcellulose have been shown to lose viscosity on irradiation, while liquid paraffin showed an increase in iodine value [50].

V. SUMMARY

In selecting a preservative system for a cosmetic formulation, there are certain fundamental considerations which, if taken into account, can result in a satisfactory product with minimal adaptations of the original formula. Unlike a pharmaceutical product, cosmetics are seldom labeled "store in a cool place" or "store in a refrigerator," which means that the cosmetic may well end up in an environment suitable for the growth of microorganisms at temperatures between 15 and 30°C at certain times of the year. This puts full stress on the preservative system, which may be a uni- or multicomponent system.

Primarily, where there is a large aqueous phase, conditions will be far more advantageous to microorganisms than for a nonaqueous product. Where the latter combines both anaerobic and moisture-free conditions, a large number of bacteria and fungi will be automatically eliminated, but it should be cautioned that microorganisms have been isolated from oils after 1 year and longer. Furthermore, preservatives seldom work well, if at all, in oleaginous bases, so that if that environment is unhelpful to most microbes, it is equally unhelpful to most preservatives.

Furthermore, the more complex the formulation with regard to different *types* of ingredients, the more difficult it is to predict how the preservative will react. To clarify, an emulsion may contain five lipophilic ingredients such as a vegetable oil, waxes, and paraffins; yet these are one type of class of compound, and the partition of the preservative between this conjoint oily phase and the aqueous phase (with the emulgent absent) can easily be determined using dialysis or simple partitioning to obtain a measure of the distribution of the preservative in the aqueous phase, where it will be active. The

TABLE 13 Effect of Gamma Irradiation on Preservative Solutions in Clear Glass Ampuls

Preservative (% wt/vol)	Solution age	Appearance after irradiation	Percentage remaining
Methylbenzethonium chloride, 0.1%	1 year	Milky	Not analyzed (NA)
	1 week	Milky	NA
Cetylpyridinium chloride, 0.1%	1 week	Pale yellow	52[a]
Benzalkonium chloride, 0.25%	5 mo	Clear	a,b
	5½ mo	Clear	a,b
Benzoic acid, 0.1%	14 mo	White precipitate	NA
	1 week	White precipitate	NA
	3 weeks	White precipitate	NA
Sorbic acid, 0.1%	15 mo	Clear	0
	1 week	Clear	0
Dehydroacetic acid, 0.1%	1 week	Clear	24[a]
	3 weeks	Clear	25
Phenol, 0.2%	1 week	Clear	100
Phenol, 0.5%	1 day	Clear	100
Chlorocresol, 0.1%	1 week	Clear	95
	3 weeks	Clear	97
Dichlorophenol, 0.05%	2 days	Pale yellow-brown	b
	2 weeks	Pale yellow-brown	79
o-Phenylphenol, 0.05%	2 days	White precipitate	NA
	2 weeks	White precipitate	NA
Benzyl alcohol, 1%	1 week	Faint white precipitate	b
Phenylethanol, 1%	15 mo	Milky and oily deposit	NA
	1 week	Milky and oily deposit	NA
Methylparaben, 0.1%	1 week	Clear	76
	1 day	Clear	69
Vanillin, 0.2%	16 mo	Pale yellow	86
	1 week	Pale yellow faint precipitate	75
Bronopol, 0.1%	10 mo	Clear	0
	1 week	Clear	0
Chlorhexidine acetate, 0.02%	>1 year	Clear	41
	1 week	Flocculent brown precipitate	NA
8-Hydroxyquinoline sulfate, 0.1%	14 mo	Deep yellow brown	84[a]
	1 week	Dark brown	NA

[a]Definite spectral change.
[b]Absorbance readings after irradiation exceeded 100% and decomposition could not be assessed.
Source: Ref. 54.

addition of an alcohol such as ethanol or glycerol (up to 10%) can often make a preservative's distribution into the aqueous phase more favorable to the system.

The choice of pH for the system is of prime importance where pH-dependent entities are being used, particularly the organic acids. In the case of the latter, which are not too adversely affected by nonionic emulgents and are consequently often selected for this reason, a pH above 5 usually inactivates the acid to an extent that renders it useless, so that the final pH required for the type of product being marketed is the all-important criterion in deciding whether organic acids such as sorbic acid should be selected.

To return to the question of the other types or classes of ingredients which could appear in cosmetics, emulgents are almost certain to appear in emulsion-type lotions and creams. Since the nonionic emulgents are known to bind with and hence inactivate a number of preservatives, use of such polymers or large-molecule condensates calls for particular caution. Where the amount of binding is known, the use of a compensatory excess of that preservative is a possibility, due consideration being given to cost and sensitization risk. More popular is the use of di- and tripreservative systems, but these have also been known to fail. Where possible, the quantity of nonionic emulgent used must be kept to a minimum, possibly by the use of suitable thickening or gelling agents, but once again binding of certain preservatives by such agents must be borne in mind, and they themselves form another type or class of compound, as do insoluble powders, which have a far more limited use, but which could have powerful adsorptive (inactivating) properties where preservatives are concerned.

The acid test will, of course, be that of the final formulated product being challenged with suitable hardy strains of microorganisms after storage in its all-important container of issue. If the container is of plain glass with no protective cardboard carton, then storage in a light-free incubator may well give a false result, a factor sometimes overlooked where light-sensitive preservatives are being used. It should also be borne in mind that preservatives may be unstable in solution (benzyl alcohol) or in the presence of ingredients such as macrogols (phenolics) after periods exceeding 6-12 months, so that a careful manufacturer will challenge marketed formulations on a regular basis as a part of "Good Manufacturing Procedure." Since the time taken from the conception of a new product in the board room to its final testing (and possible rejection) in the analytical department can be many months, it behooves the formulator to regard the product as a complex combination, and not a mixture of simple entities. Knowing the pitfalls that can occur is more than halfway to eliminating them.

REFERENCES

1. R. A. Cowen and B. Steiger, Why a preservative system must be tailored to a specific product, *Cosmet. Toiletries, 92*:16-20 (1977).
2. B. J. Collins and C. E. A. Deverill, Survival of bacteria in antiseptic skin preparations, *Pharm. J., 207*:369 (1971).
3. A. P. Dunnigan, Microbiological control of cosmetic products, Proceedings of the Joint Conference Cosmetic Science, Washington, D.C., April 21-23, 1968.
4. C. W. Bruch, Objectionable micro-organisms in non-sterile drugs and cosmetics, *Drug Cosmet. Ind., 110*:32, 116 (1972).

5. J. R. Evans, M. M. Gilden, and C. W. Bruch, Methods for isolating and identifying objectionable gram-negative bacteria and endotoxins from topical products, *J. Soc. Cosmet. Chem.*, 23:549-564 (1972).
6. E. J. Wargo, Microbial contamination of topical ointments, *Am. J. Hosp. Pharm.*, 30:332-335 (1973).
7. W. Blyth, Fungi in relation to allergy and infection, *Pharm. J.*, 206:599-600 (1970).
8. A. Baker, Fungal infections of the eye, *J. Hosp. Pharm.*, 30:45-48 (1972).
9. G. E. Myers and F. M. Pasutto, Microbial contamination of cosmetics and toiletries, *Can. J. Pharm. Sci.*, 8:19-23 (1973).
10. N. J. Van Abbé, H. Dixon, O. Hughes, and R. C. S. Woodroffe, The hygienic manufacture and preservation of toiletries and cosmetics, *J. Soc. Cosmet. Chem.*, 21:719-800 (1970).
11. J. A. Myburgh and T. J. McCarthy, Effect of certain formulation factors on the activity of Bronopol, *Cosmet. Toiletries*, 93:47-48 (1978).
12. T. J. McCarthy, Effect of pH on stored dehydroacetic acid solutions, *Cosmet. Perfum.*, 89:45-47 (1974).
13. T. J. McCarthy, C. R. Clarke, and J. A. Myburgh, Antibacterial effectiveness of stored sorbic acid solutions, *Cosmet. Perfum.*, 88:43-45 (1973).
14. T. J. McCarthy, Interaction between aqueous preservative solutions and their plastic containers. I. *Pharm. Weekbl.*, 105:557-563 (1970).
15. T. J. McCarthy, Interaction between aqueous preservative solutions and their plastic containers. III. *Pharm. Weekbl.*, 107:1-7 (1972).
16. T. J. McCarthy and P. F. K. Eagles, Further studies on glass-stored sorbic acid solutions, *Cosmet. Toiletries*, 91:33-35 (1976).
17. D. L. Wedderburn, Interactions in cosmetic preservation, *Am. Perfum. Cosmet.*, 85:49-53 (1970).
18. H. S. Bean and G. Dempsey, The bactericidal activity of phenol in a solid-liquid dispersion, *J. Pharm. Pharmacol.*, 19:197S-202S (1967).
19. T. J. McCarthy, The influence of insoluble powders on preservatives in solution, *J. Mond. Pharm.*, 4:321-329 (1969).
20. N. R. Horn, T. J. McCarthy, and C. H. Price, Interactions between preservatives and suspension systems, *Am. Perfum. Cosmet.*, 86:37-40 (1971).
21. N. R. Horn, T. J. McCarthy, and E. Ramstad, Interactions between powder suspensions, and selected quaternary ammonium and organomercurial preservatives, *Cosmet. Toiletries*, 95:69-73 (1980).
22. H. S. Bean and G. Dempsey, The effect of suspensions on the bactericidal activity of m-cresol and benzalkonium chloride, *J. Pharm. Pharmacol.*, 23:609-704 (1971).
23. J. A. Myburgh and T. J. McCarthy, Inactivation of preservatives in the presence of particulate solids, *Pharm. Weekbl.*, 115:1405-1410 (1980).
24. R. T. Yousef, M. A. El Nakeeb, and S. Salama, Effect of some pharmaceutical materials on the bactericidal activities of preservatives, *Can. J. Pharm. Sci.*, 8:54-56 (1973).
25. D. Coates, Preservative/colloid interaction, *Manuf. Chem. Aerosol News*, 44:34-37 (1973).
26. P. C. Eisman, J. Cooper, and D. Jaconia, Influence of gum tragacanth on the bactericidal activity of preservatives, *J. Am. Pharm. Assoc. Sci. Ed.*, 46:144-147 (1957).

27. A. Taub, W. A. Meer, and L. W. Clausen, Conditions for the preservation of gum tragacanth jellies, *J. Am. Pharm. Assoc. Sci. Ed.*, 47:235-239 (1958).
28. W. J. Tillman and R. Kuramoto, A study of the interaction between methylcellulose and preservatives, *J. Am. Pharm. Assoc. Sci. Ed.*, 46:211-214 (1957).
29. C. J. De Blaey, O. W. Mulder, and H. J. de Jong, The interaction of acriflavine with sodium carboxymethylcellulose, *Pharm. Weekbl.*, 3:473-479 (1976).
30. T. J. McCarthy, Preservative interaction with PVP, *Pharm. Weekbl.*, 108:1-4 (1973).
31. T. J. McCarthy and J. A. Myburgh, The effect of tragacanth gel on preservative activity, *Pharm. Weekbl.*, 109:265-268 (1974).
32. J. A. Myburgh and T. J. McCarthy, The influence of suspending agents on preservative activity of aqueous solid/liquid dispersions, *Pharm. Weekbl.*, 115:1405-1416 (1980).
33. N. K. Patel and H. B. Kostenbauder, Interaction of preservatives with macromolecules. I. *J. Am. Pharm. Assoc. Sci. Ed.*, 47:289-293 (1958).
34. F. D. Pisano and H. B. Kostenbauder, Interaction of preservatives with macromolecules. II. *J. Am. Pharm. Assoc. Sci. Ed.*, 48:310-314 (1959).
35. P. P. DeLuca and H. B. Kostenbauder, Interaction of preservatives with macromolecules. IV. *J. Am. Pharm. Assoc. Sci. Ed.*, 49:430-436 (1960).
36. C. K. Bahal and H. B. Kostenbauder, Interaction of preservatives with macromolecules. V. *J. Am. Pharm. Assoc. Sci. Ed.*, 53:1027-1029 (1964).
37. W. P. Evans, The solubilization and inactivation of preservatives by non-ionic detergents, *J. Pharm. Pharmacol.*, 16:323-331 (1964).
38. R. A. Anderson and K. J. Morgan, Effect of solubilization on the antibacterial activity of hexachlorophane, *J. Pharm. Pharmacol.*, 18:449-456 (1966).
39. M. J. Crooks and K. F. Brown, Binding of preservatives to sodium lauryl sulfate and polyethylene glycol 1000, *Aust. J. Pharm. Sci.*, NS3:93-94 (1974).
40. M. M. Rieger, The inactivation of phenolic preservatives in emulsions, *Cosmet. Toiletries*, 96:39-43 (1981).
41. M. Barr and L. F. Tice, The preservation of aqueous preparations containing nonionic surfactants. I. *J. Am. Pharm. Assoc. Sci. Ed.*, 46:442-451 (1957).
42. K. Thoma, E. Ullmann, and O. Fickel, Die antibakterielle Wirkung von Phenolen in Anwesenheit von Polyöthylenglykolsteoraten en Polyöthylenglykolen, *Ark. Pharm.*, 303:289-296 (1970).
43. T. J. McCarthy, Dissolution of chlorocresol from various pharmaceutical formulations, *Pharm. Weekbl.*, 110:101-106 (1975).
44. M. J. Crooks and K. F. Brown, A note on the solubilization of preservative mixtures by cetomacrogol, *J. Pharm. Pharmacol.*, 25:281-284 (1973).
45. M. F. Crooks and K. F. Brown, Competitive interaction of preservative mixtures with cetomacrogol, *J. Pharm. Pharmacol.*, 26:235-242 (1974).

46. G. Jacobs, S. M. Henry, and V. F. Cotty, The influence of pH, emulsifier, and accelerated aging upon preservative requirements in o/w emulsions, *J. Soc. Cosmet. Chem.*, 26:105-117 (1975).
47. N. K. Patel, Interaction of some pharmaceuticals with macromolecules. III. *Can. J. Pharm. Sci.*, 2:77-80 (1967).
48. H. S. Bean and S. M. Heman-Ackah, Influence of oil:water ratio on the activity of some bactericides against *Escherichia coli* in liquid paraffin and water dispersions, *J. Pharm. Pharmacol.*, 16:58T-67T (1964).
49. T. J. McCarthy, Storage studies of preservative solutions in commonly used plastic containers, *Cosmet. Perfum.*, 88:41-42 (1973).
50. *Martindale's Extra Pharmacopocia*, 27th Ed. (A. Wade and J. E. F. Reynolds, eds.), Pharmaceutical Press, London, 1979.
51. S. M. Henry and Gene Jacobs, Cosmetic preservatives — 1981, *Cosmet. Toiletries*, 96:29-37 (1981).
52. T. A. du Plessis, Radiation sterilization and its potential role in the field of cosmetic products, *S. Afr. Pharm. Cosmet. Rev.*, 6:27-31 (1979).
53. G. P. Jacobs, Gamma radiation decontamination of cosmetic raw materials, *Cosmet. Toiletries*, 96:51-55 (1981).
54. T. J. McCarthy, The effect of gamma irradiation on selected aqueous preservative solutions, *Pharm. Weekbl.*, 113:698-700 (1978).

21
DESIGN AND ASSESSMENT OF PRESERVATIVE SYSTEMS FOR COSMETICS

MALCOLM S. PARKER *Brighton Polytechnic, Brighton, England*

I. Introduction 390
II. Nutritive Status of Cosmetics for Microorganisms 390
    A. Water availability 390
    B. Nutrient content 391
    C. pH 391
III. Interaction of Microorganisms in Cosmetics 391
IV. Preservation of Cosmetics Against Microbial Spoilage 392
    A. Classification of preservatives 392
    B. The role of formulation 393
    C. Biodeterioration of preservatives 395
    D. The use of combinations of preservatives 395
V. Evaluation of Preservatives 396
    A. Preliminary screening procedures 396
    B. In-use tests 396
    C. Rapid evaluation techniques 397
VI. Conclusion 399
    References 400

I. INTRODUCTION

During the past two decades the literature concerned with the use of preservatives in cosmetics has grown substantially. The reasons for this include the increasing sophistication of cosmetic formulations such that traditional preservation techniques no longer suffice, a growing appreciation of the problems associated with microbial contamination, and a worldwide interest in establishing suitable microbiological standards for cosmetics.

Any complacency that might have been associated with products which had been continuously developed and used for some 5000 years was severely jolted by Kallings and his colleagues [1]. Although this investigation dealt with the microbiological contamination of medical preparations, the incidence of contamination of eye ointments, hand creams, baby creams, and powders had obvious relevance for manufacturers of cosmetics. Increased scrutiny of cosmetic preparations has yielded evidence of microbial contamination, but not of serious infection of users [2]; nevertheless, a potential hazard does exist.

Against this background a three-pronged strategy has emerged for the production of microbiologically acceptable cosmetics. This comprises good manufacturing practice, the use of formulation-integrated preservative systems, and the development of microbiological data and guidelines. It must be appreciated that none of these factors are isolated from each other, for clean manufacture determines the realism of any standards applied and both relate closely to preservative regimes.

II. NUTRITIVE STATUS OF COSMETICS FOR MICROORGANISMS

The general requirements for microbial function and growth, namely, water, energy, nitrogen source together with various minerals and vitamins in an environment with suitable levels of oxygen, pH, and temperature, are readily provided by the majority of cosmetic formulations. To appreciate this, it is necessary to briefly consider the relevant parameters of cosmetic formulations.

A. Water Availability

This is essentially a measure of the water available to microorganisms within a substrate and has been described in terms of availability from aqueous solutions [3]:

$$A_w = \frac{\text{water vapor pressure of solution (substrate)}}{\text{vapor pressure of pure water at the same temperature}}$$

Thus water availability will vary with concentration, and solid substrates of low water content present a special case.

Cosmetics, unless rich in sugars or other substances likely to create high osmotic tension, or in solid form, will not present any problem of water availability to spoilage bacteria, requiring, as they do, levels in excess of 0.95. The exceptions may present A_w values as low as 0.65 for sugary formulations, and lower for powders or "caked" products; under these conditions specialized forms such as osmophilic yeasts or xerophilic molds would be the only types able to establish themselves (see Table 1).

TABLE 1 Utilization of Cosmetic Contents by Microorganisms

Constituent utilized	Microorganisms involved	A_w requirement	Optional pH
Sugars	Osmophilic yeasts	0.60-0.65	4-6
Fats			
Cetostearyl alcohol	Pseudomonads	0.95-1.0	6-8
Polythylene glycol	Staphylococci	0.90-0.95	5-7
Glycerol	Coliforms	0.95-1.0	6-8
Waxes	Aspergilli	0.80-0.87	5-7
Proteins			
Protein residues	Pseudomonads	0.95-1.0	6-8

B. Nutrient Content

The complexity of modern cosmetic formulations is such that they present a wide spectrum of nutrients, from the obvious enriched products containing egg, sugar, beer, or vitamins to a whole range of waxes, fats, esters, and emulgents, all, to some degree, susceptible to biodeterioration (see Table 1).

Those microorganisms which become established in a given formulation will, according to the component they utilize, produce characteristic evidence of their activities. Thus carbohydrates will be broken down to produce acids and possibly gas, which will appear as surface bubbles or even be sufficient to distort containers. Fats and related constituents will yield fatty acids and ketone residues, producing rancidity, whereas proteins and protein residues are metabolized to a wide range of residues with many odors and characteristics, including slime production [4].

C. pH

The microbial flora implicated in cosmetic spoilage are largely tolerant of pH, with the bacteria thriving at pH 5.5-8.0 and a few yeasts and molds preferring more acidic conditions approaching pH values as low as 3.5. Modern cosmetics generally satisfy these pH requirements, except for acidophilic molds and yeasts, but it is of historical interest to note that, prior to the introduction of the nonionic emulgents and stabilizers, creams and emulsions, being soap stabilized, had an ambient pH so alkaline as to be inhibitory to microbial development.

It will be evident from Section II.B that if microbial utilization of a cosmetic is initiated, then the pH may rapidly shift, influencing not only the fate of any indigenous microorganisms, but possibly the effectivity of the preservative system by a series of interactions to be discussed later.

III. INTERACTION OF MICROORGANISMS IN COSMETICS

The fate of the contaminating microorganisms which find their way into a cosmetic will be determined by their ability to survive and multiply in the environment provided by the formulation. Those organisms, if any, which do initially become established may be termed the pioneers, and these, by their metabolic activity, may change the environment such that conditions favorable to other organisms are created. This type of spoilage progress in which one species creates the conditions for others has been termed autogenic succession [5]. An example of such a succession might be the invasion of a formulation with a high osmotic tension due to sugar content by an osmophilic yeast which, as it metabolized the sugars, lowered the overall osmotic pressure, allowing the multiplication of less specialized forms [6]. In this case succession would be by way of acidophilic forms due to the acid production accompanying carbohydrate utilization, and proteolytic spoilage forms, which are inhibited at low pH, would not thrive until the overall pH was about neutral. A more common route of spoilage for cosmetics is via initial attack upon the fatty ingredients. A large number of common contaminants are able to break down fats, including the low-demand gram-negative types such as *Alcaligenes*, *Serratia*, and *Pseudomonas*, gram-negative pathogens including *Staphylococcus* and many molds, for example, *Aspergillus, Mucor*, and *Penicillium*. As with sugar metabolism, lipolytic activities create an acidic, protein-sparing environment in a spoiled product. In addition to this, cells growing in a fatty environment comprising such components as soft paraffin, cetostearyl alcohol, polyethylene glycols, and glycerols are able to accumulate extracellular lipids [7]. Cells which do utilize lipids and incorporate them into their surface layers may show increased resistance to phenolic preservatives [8] and to esters of para-hydroxybenzoic acid [9].

The changes in pH level occurring during spoilage activities will influence the activity of anion-active preservatives such as chlorbutanol, those effective in the undissociated form, like benzoic acid, and the cation-active group, including quaternary ammonium compounds.

The spoilage process is complex, particularly when considered in the context of the many and varied ingredients of cosmetics. There is likely to be a series of microorganisms implicated from pioneer forms to secondary and tertiary invaders, with the original contaminants often creating the conditions for their own demise. In this situation there may be little virtue in concentrating efforts upon eliminating the organism prevalent in a spoiled preparation; the more logical approach would be to seek the source of the pioneer forms which initiated the process.

IV. PRESERVATION OF COSMETICS AGAINST MICROBIAL SPOILAGE

There is no shortage of antimicrobial agents which have a potential role as preservatives for cosmetics [10], although the list diminishes when due consideration is given to formulation effects such as oil-water partitioning, solubilization, adsorption, and pH effects [11]. Despite a plethora of published work commending a variety of new compounds, there remains a major tendency to use the traditional parabens esters, although Germall and Bronopol of the "new generation of preservatives" [12] have achieved general usage.

A. Classification of Preservatives

As the mode of action of antimicrobial agents becomes more fully understood, there is a logical case for grouping them according to their site of activity against the microbial cell [13]. This approach (see Table 2) has found less appeal among cosmetic chemists than the old alternative of classifying according to chemical structure. Perhaps the major reason for this is that most of the problems encountered in preserving cosmetics result from inactivation of preservatives by formulation components. Thus it is essential to know

TABLE 2 Sites of Activity of Antimicrobial Agents

Membrane active	Quaternary ammonium compounds
	Chlorhexidine
	Phenoxetol
	Alcohols and phenylethyl alcohol
	Phenols
SH enzymes	Mercurials and Bronopol
COOH enzymes	Formaldehyde and formaldehyde donors
NH_2 enzymes	Formaldehyde and formaldehyde donors
Nucleic acids	Acridines
Protein denaturation	Phenols
	Formaldehyde

the oil-water partition, the solubilization characteristics, the binding to solids, and the effect of pH if a preservative is to be used efficiently (see Table 3).

B. The Role of Formulation

The complexity of modern cosmetic formulations allows for many mechanisms to render preservatives less effective and sometimes enhance their activity, and it is important that such interactions be appreciated. During the past two decades there have been efforts on the part of pharmaceutical scientists to predict, on a quantitative basis, this loss of preservatives.

The simplest model of an oil-water system as described by Bean et al. [14] will provide a prediction of the preservative concentration available to the aqueous phase (C_w) from a knowledge of the total concentration of preservative in the system (C), the oil-water partition coefficient of the preservative (K_w^o), and the ratio ϕ of oil to water:

$$C_w = C \frac{\phi + 1}{K_w^o \phi + 1}$$

When this equation is considered together with the type of information given in Table 3, the formulator is able to appreciate the problems created by, for instance, the use of vegetable oils in conjunction with esters of para-hydroxybenzoic acid. A further factor in applying this type of equation is the effect of temperature upon the partition coefficient, which, in most cases, changes in favor of the oil as the temperature rises.

Extending this simple treatment to the more realistic situation where a third component is involved, usually an emulgent, Bean and his colleagues [15] extended their equation to allow for partition of preservative into the micellized emulgent or complex formation between these two components:

$$C_w = C \frac{\phi + 1}{K_w^o \phi + R}$$

TABLE 3

Preservative[a]	Physicochemical characteristics			Use concentration (% wt/vol)
	Water solubility	R factor[b]	Optimum pH	
Phenolic/alcoholic				
Chlorocresol	Low	High	5-8	0.1
Myacide	Low	High	3-9	0.05-0.1
Phenoxetol	Medium	Low	3-9	0.5-2.0
Benzyl alcohol	Medium	Low	Above 5	1.0-3.0
Paraben esters (methyl)	Low (decrease with chain length)	High	5-8	0.1-0.2
Phenonip	Low	Medium	5-8	0.5-1.0
Benzoic acid (and salts)	Soluble	Medium	Below 5	0.1-0.2
Quaternary ammonium compounds				
Cetrimide	Soluble	High	Above 5	0.01-0.1
Benzalkonium chloride	Soluble	High	Above 5	0.01-0.3
Mercurials				
Phenylmercuric nitrate	Low	Low	3-7	0.001-0.002
Thiomersal	Soluble	Low	3-7	0.01-0.02
Chlorhexidine gluconate	Soluble	Low	5-8	0.01-0.1
Formaldehyde donors				
Bronopol	Soluble	Low	Below 6	0.01-0.1
Dowicil	Soluble	Low	4-10	0.02-0.3
Germall	Soluble	Low	4-10	0.1-0.5

[a]Bronopol, 2-nitro-2 bromo-1,3-propanediol (Boots Pure Drug Co. Ltd., Nottingham, England); Dowicil, 1-(3chloroallyl)-3,5,7-triazo-1-azo-nia adamantane chloride (Dow Chemical Co., London); Germall, N,N'-methylene-bis-[N'-(1-(hydroxymethyl)-2,5-dioxo-4-imidazaolidinyl)-urea] (Sutton Laboratories, Inc., Rosella, N.J.); Phenonip, a combination of esters of p-hydroxybenzoic acid and phenoxyethanol (Nipa Laboratories, Ltd., Llantwit Fardre, Pontypridd, England); Myacide, 2,-4-dichlorobenzyl alcohol (Boots Pure Drug Co. Ltd., Nottingham, England).
[b]See Section IV.B for the R factor.

where R is the ratio of the total preservative in the aqueous phase to the free preservative in that phase.

The use of mathematical models to calculate the amount of preservative(s) required to protect multicomponent systems has been developed to a high degree of sophistication by workers such as Garrett [16] and Mitchell and Kazmi [17], who have produced equations for preservative interaction in systems containing mixtures of preservatives and several types of macromolecules.

The remaining important factor in preservative activity, as indicated earlier, is the effect of pH, which will influence ionized preservatives. The best-documented examples are among the acid preservatives such as benzoic and sorbic acids, which are principally active in the unionized form, and for a given pH level and a knowledge of the pK_a of the preservative the following equation can be applied:

$$\text{Fraction of undissociated preservative} = \frac{1}{1 + \text{antilog}(pH - pK_a)}$$

This will allow an overall percentage of a preservative present in a formulation to be converted to that proportion which is biologically active.

Another effect that pH can have is upon chemical stability with preservatives, such as thiomersal (sodium ethylmercurithiosalicylate), degrading in acid conditions, and Bronopol (2-bromo-2-nitropropane-1,3-diol), unstable at alkaline pH.

It will be apparent that situations can arise in which a multiplicity of interactions take place in a formulation. Thus, in addition to oils and emulgents, there may be a variety of solids which will adsorb preservatives in association with hydrocolloids acting as suspending agents and thickeners. A further dimension will be added by natural ingredients such as eggs, milk, and protein hydrolysates, all of which can interact with preservatives. An extensive literature has been built up dealing with the inactivation of preservatives by ingredients of cosmetics and pharmaceuticals and the containers, particularly plastic, used to pack them [11], and the only safe procedure is to assess preservative availability in each individual formulation using dialysis cells [18] or dissolution measurements [19,20].

C. Biodeterioration of Preservatives

In addition to the utilization of cosmetic ingredients by invading microflora (see Table 1), the preservatives themselves, particularly if they are aromatic, are prone to microbial attack. The breakdown of parabens esters by *pseudomonads* [21] and by *Cladosporium resinae* [22] has been reported, and the possible mechanisms whereby aromatic molecules are disrupted have been reviewed by Dagley [23]. Beveridge [24] has stressed that, although much of the investigative work was done with low concentrations of antimicrobial agents, their breakdown at normal use concentrations does occur and, indeed, they can be shown to serve as sole carbon substrates for microorganisms.

D. The Use of Combinations of Preservatives

Against the background of the factors described, which can interfere with the protective function of the preservative, it is not surprising that the use of a single antimicrobial compound is sometimes insufficient. With this said, the use of mixtures of preservatives must be rationalized on the grounds that each bestows an advantage in terms of antimicrobial spectrum and in overcoming inactivation by formulation ingredients. Examples of useful combinations cited [11] are parabens esters and Germall (an imidazolidine urea compound), in which the latter complements the deficiency of the traditional esters in terms of eliminating low-demand gram-negative forms, has good water solubility, and is compatible with all ionic species and proteins. Parabens esters may also be assisted by the addition of phenoxetol, as in Phenonip [25], Bronopol, or mercurials [12], while quaternary ammonium compounds such as cetrimide and benzalkonium chloride are used in combination with chlorhexidine gluconate to broaden the antimicrobial spectrum and reduce incompatibility with emulgents.

As well as using mixtures of preservatives, it is possible to render a formulation hostile to microorganisms by the addition of substances which will potentiate the preservative system. Potentiation may be effected in several ways, for example, propylene glycol, butylene glycol, hexylene glycol, and glycerine can decrease the loss of lipophilic preservatives to micellized emulgents, while phenylethyl alcohol and ethylenediaminetetraacetate interfere with membrane permeability and enhance antimicrobial effects.

V. EVALUATION OF PRESERVATIVES

The evaluation of preservatives is concerned with determining the basic parameters of the compound, namely, its antimicrobial spectrum of activity under appropriate conditions of temperature and pH and its toxicity, and if these are satisfactory, it is extended to assess activity in the presence of typical formulation components. The ultimate evaluation procedure should be to assess the protection provided to the final cosmetic preparation.

A. Preliminary Screening Procedures

The simplest tests available for assessing potential preservatives are the broth dilution methods and agar plate diffusion techniques as described in standard bacteriology texts. In these methods the antimicrobial compound is brought into contact with microorganisms in a cultural situation such that the efficacy of the compound can be determined by its ability to prevent visible growth. The technique can be adapted to screen a range of concentrations of preservative against different microorganisms using broth dilutions or to determine the sensitivity of a number of organisms to a single preservative by means of a plate test. In every case it is important to appreciate that what is being determined is an inhibitory rather than a lethal end point under limited, artificial conditions, and that even these simple evaluations are meaningless without standardizations of the age and inoculum size of the test organism, culture medium, pH, and temperature of incubation.

B. In-Use Tests

Attempts to simulate in-use conditions and obtain lethal end points have been developed for testing disinfectants. In these methods the test organism is placed in contact with the antimicrobial agent, perhaps in the presence of organic matter, and after a specified period of time samples are taken, suitably neutralized to prevent carryover of agent, and the survivors estimated by subculture into recovery medium or, more accurately, by counting of colony-forming units. Preservatives can be assessed by methods of this type by mixing them with typical cosmetic ingredients and adding an inoculum of a spoilage organism. Samples taken from such a system into neutralizing recovery media will indicate how efficient the preservative is in a given formulation. The test may be made more demanding by repeating the addition of test organisms to find the capacity of the antimicrobial system to deal with challenge [26]. The ultimate development of this testing procedure is the testing of the final preserved product by the type of challenge test described in the U.S. *Pharmacopoeia* [27] and the *British Pharmacopoeia* [28] and discussed critically by Cowen and Steiger [29], Moore [30], and Parker [11].

C. Rapid Evaluation Techniques

In the general methods so far described both for screening preservatives and assessing final products, there is a considerable time factor, not only in setting up the various broth dilutions or recovery media, but also in waiting for the results. The latter depends essentially upon the time taken for visible microbial growth to be evident, either as turbidity or as colony-forming units, which may range from 24 hr for bacteria to as long as 3 weeks for some molds. A further complication is that inhibitory effects will usually delay development, and if readings are taken too early, a false impression of preservative efficacy is obtained.

In recent years techniques have been developed which quickly detect the onset of microbial growth and its inhibition. This is achieved by selecting some index of cell development which occurs early in the growth cycle and is susceptible to antimicrobial agents. Such a development, although not apparent to the eye, can often be measured accurately by a variety of instruments within a time period of hours rather than the days needed for conventional evaluation methods.

1. Turbidimetric Assays

The first of these rapid techniques to be used was the turbidimetric assay, which involves the use of a spectrophotometer to detect microbial growth in cultures considerably in advance of visible evidence. Brown and Richards [31] used this method to investigate growth rates of *Pseudomonas aeruginosa*, and Brown [32] extended the technique to evaluate preservative efficacy in the presence of nonionic surfactants. The method has also been described by Parker [33] for measuring the effect of preservative against yeasts and molds, using *Saccharomyces rouxii* and *Trichoderma viride*. First indication of cell development and, hence, of any inhibition is detectable within 1 hr for bacteria and 3-6 hr for yeasts or molds; delay is minimized by using prewarmed nutrient medium and incubating in a shaking water bath. A wide range of spoilage organisms, preservatives, and formulation components may be studied, but systems in which there is precipitation, cell clumping, or rapid lysis are not suitable for turbidimetric investigation. Bacterial spores represent a particularly troublesome spoilage vector for cosmetics, and the spectrophotometric technique can provide useful insight into the action of preservatives against them. The application of the method is based upon the consequences of the germination process in which the quiescent spore undergoes dramatic physical and metabolic changes in entering upon the vegetative state. During this transformation the spore swells and loses refractility, with an attendant fall in the optical density of spore suspensions amounting to some 60% of the initial value. Following these early changes, the spore coat opens and vegetative cell outgrowth occurs followed by cell division, so that optical density rises again with normal vegetative development. Using optical density profiles obtained in this way, it is possible to broadly categorize preservatives [34] into (1) those that prevent the initial fall in optical density, that is, germination inhibitors such as phenols, organic acids, hydroxybenzoates, and alcohols and (2) those which suppress the rise in optical density which accompanies the outgrowth of vegetative cells and their division, such as quaternary ammonium compounds, Dowicil, and organic mercurials.

2. Measurement of Spore Swelling

A primary index of the development of both mold and bacterial spores is a swelling process. This comprises two phases: an initial one which is insensitive to preservatives and a second phase which is suppressed by preservatives [35]. The initial phase may be a simple hydration or osmotic phenomenon. The suppression of the swelling of mold spores by antimicrobial agents was used by Mandels and Darby [36] to evaluate antifungal compounds, and they used hematocrit tubes to measure spore volume. Microscopic measurements of spore size are more accurate, but are time-consuming [37], and it is easier to use one of the modern electronic particle size analyzers such as the Coulter Counter (Coulter Electronics, Ltd.). The technique can be used not only to rapidly screen preservatives, but also to investigate the influence of formulation ingredients such as Tween 80 (see Table 4). Preservatives evaluated by this method include parabens, Phenonip, Dowicil, Bronopol, and benzoic acid [33]. In each case the suppression of spore swelling, and hence efficacy of the preservatives, is detected within 4 hr, and the degree to which swelling is retarded compared with that of control spores provides

TABLE 4 Methods for the Rapid Screening of Preservatives

Technique	Index of growth measured	Time involved	Organisms	Applications
Spectrophotometer	Cell mass (turbidity)	1-3 hr	Bacteria, yeasts, molds	Effect of formulation components, provided no precipitation
	Refractility (optical density)	1-4 hr	Bacterial spores	
Coulter Counter (or similar)	Spore volume	1-4 hr	Mold and bacterial spores	Effects of formulation components measured
Radiometry	Carbon metabolism (scintillation counter)	10-30 min	Bacteria, yeasts, molds	Inactivation of preservatives by formulation components detected
Particle microelectrophoresis	Charge and mobility on cell surface	10 min	Bacteria	Antibiotic assay
Flow microcalorimetry	Heat production	1 hr	Bacteria	Phenol coefficients measured effects of "dirty conditions" assessed

a sensitive assay of antimicrobial activity. Similarly, the inactivating or potentiating effects of cosmetic ingredients are rapidly detected.

3. Radiometric Methods

The metabolism by microorganisms of growth medium containing radioactive materials results in the release of, for example $^{14}CO_2$, which can be detected to provide a sensitive index of cell development. Modification of this approach by Judis [38] enabled him to monitor antimicrobial activity by measuring the leakage of cell contents caused by the action of phenolic agents upon ^{14}C-labeled cells. He used this technique to measure the activity of chloroxylenol and its inactivation by Tween 80. Radiometric methods are finding increasing application for estimating microbial levels in foods, biological fluids, and water [39] and have an obvious potential in the field of cosmetics and pharmaceuticals for this role.

4. Particle Microelectrophoresis

Many antibacterial agents are known to affect the charge and mobility on the bacterial surface, and Furr and co-workers [40] have adapted this finding to measure the effect of chlorhexidine upon *Escherichia coli*. The electrophoretic mobility of the cells is measured before and after contact with the preservative, and a result is obtained within 10 min. Within a given concentration range the relationship between chlorhexidine concentration and electrophoretic mobility can be used as a basis for assay. The technique is particularly suited to antibiotic assay but, nevertheless, shows promise for preservative studies.

5. Flow Microcalorimetry

All organisms liberate heat during metabolism, and in recent years microcalorimetry has been used to measure this and provide some insight into the energetics of microbial processes. Beezer and his colleagues [41] have applied flow calorimetry to the measurement of phenol coefficients, taking less than an hour for the determination. The procedures can be readily adapted to simulate "dirty" conditions and automated to give printed readout results. Rapid assessment of use-dilution ranges for preservatives could be carried out and the effects of formulation ingredients determined.

VI. CONCLUSION

The evaluation of preservatives, whether by the tradiational methods or by the novel rapid estimation techniques, must be considered in the context of what is to be achieved, namely, the selection of a preservative or preservatives for a particular cosmetic preparation. In this situation there is no standard procedure for all situations, and the microbiologist must build up from experience of the product history a realistic testing procedure for the preservative system. A real difficulty, however, in this approach is that of simulating in-use conditions without making the test procedure cumbersome. For instance, a challenge test procedure for creams and lotions suggested by the Society of Cosmetic Scientists, England [11], lists some 17 different microorganisms to be included in the challenge, in contrast to the 5 named in the *U.S. Pharmacopoeia* test for pharmaceutical products. Although it can be argued that such a wide range of organisms truly represents typical contaminants isolated, it is doubtful whether their use in laboratory testing provides any more information than would a much smaller selection of marker organisms, such as a mold, a yeast, and a gram-positive and a gram-negative form. The microbial ecology of complex contamination situations is, as we have seen, one of continuous flux and it is completely unrealistic to present a product with a diverse group of microorganisms in a single challenge, when in practice there will be a succession of spoilage organisms involved. What is required of the preservative system is that it eliminate the pioneer spoilage forms which create the conditions for general breakdown of the cosmetic, and these should feature prominently in any challenge test devised, rather than choosing examples of the final organisms overrunning a spoiled product.

REFERENCES

1. L. O. Kallings, O. Ringertz, L. Silverstone, and F. Ernerfeldt, Microbial contamination of medical preparations, *Acta Pharm. Suec.*, 3:219-228 (1966).
2. D. F. Spooner, The microbiological control of pharmaceutical and toiletry raw materials, Proceedings of Microbiological Quality Assurance Conference, University of Sussex, England, Scientific Symposia Ltd., London (1983), pp. 15-19.
3. W. J. Scott, Water relations of food spoilage microorganisms, *Adv. Food Res.*, 7:83-127 (1957).
4. M. S. Parker, The microbiology or cosmetics — an ecological viewpoint, *Soap Perfum. Cosmet.*, 51:90-93 (1982).
5. M. Alexander, *Microbial Ecology*, Wiley, New York (1971).
6. M. S. Parker, Ecology and preserved systems, *Cosmet. Toiletries*, 93:49-51 (1978).
7. T. J. Bradley, R. S. Holdom, and N. H. Khan, Factors in the production and composition of extracellular lipids of *Pseudomonas aeruginosa* NCTC 2000, *Microbios*, 14:121-134 (1975).
8. W. B. Hugo and I. Franklin, Cellular lipid and antistaphylococcal activity of phenols, *J. Gen. Microbiol.*, 52:365-372 (1968).
9. D. I. McRobbie and M. S. Parker, Some aspects of the antifungal activity of esters of p-hydroxybenzoic acid, *Int. Biodeterior. Bull.*, 10:109-112 (1974).
10. E. L. Richardson, Update — frequency of preservative use in cosmetic formulas as disclosed to FDA, *Cosmet. Toiletries*, 96:91-98 (1981).
11. M. S. Parker, Preservation of pharmaceutical and cosmetic products, in *Principles and Practice of Disinfection, Preservation and Sterilization* (A. D. Russell, G. A. J. Aycliffe, and W. B. Hugo, eds.), Blackwell Scientific, Oxford (1982).
12. M. S. Parker, A new generation of preservatives, *Soap Perfum. Cosmet.*, 45:621-624 (1972).
13. W. B. Hugo, Mode of action of non-antibiotic antibacterial agents, in *Pharmaceutical Microbiology* (W. B. Hugo and A. D. Russell, eds.), Blackwell Scientific, Oxford (1977), pp. 202-207.
14. H. S. Bean, S. M. Heman-Ackah, and J. Thomas, The activity of antibacterials in two-phase systems, *J. Soc. Cosmet. Chem.*, 16:15-27 (1965).
15. H. S. Bean, G. H. Konning, and S. A. Malcolm, A model for the influence of emulsion formulation on the activity of phenolic preservatives, *J. Pharm. Pharmacol.*, 21:1735-1805 (1969).

16. E. R. Garrett, A basic model for the evaluation and prediction of preservative action, *J. Pharm. Pharmacol.*, *18*:589-601 (1966).
17. A. G. Mitchell and S. J. A. Kazmi, Chemical preservation of emulsified and solubilized disperse systems, *Cosmet. Toiletries*, *92*:33-43 (1977).
18. S. J. A. Kazmi and A. G. Mitchell, Preservation of solubilized and emulsified systems I & II, *J. Pharm. Sci.*, *67*:1260-1271 (1978).
19. T. J. McCarthy, Determination of preservative availability from creams and emulsions, *Pharm. Weekbl.*, *109*:85-91 (1974).
20. M. S. Parker, The preservation of cosmetic and pharmaceutical creams, *J. Appl. Bacteriol.*, *44*:Sxxix-Sxxxiv (1978).
21. E. G. Beveridge and A. Hart, Metabolism of p-hydroxybenzoate esters by bacteria, *Pharm. J.*, *202*:730 (1969).
22. W. T. Sokolski, C. G. Chidester, and G. E. Honeywell, The hydrolysis of methyl-p-hydroxybenzoate by *Cladosporium resinae*, *Dev. Ind. Microbiol.*, *3*:119-187 (1962).
23. S. Dagley, Catabolism of aromatic compounds, by microorganisms, in *Advances in Microbial Physiology*, Vol. 6 (A. H. Rose and J. F. Wilkinson, Eds.), Academic, London (1971), pp. 1-46.
24. E. G. Beveridge, The microbial spoilage of pharmaceutical products, in *Microbial Aspects of the Deterioration of Materials* (R. J. Gilbert and D. W. Lovelock, Eds.), Academic, London (1975), pp. 213-235.
25. M. S. Parker, M. McCafferty, and S. McBride, Phenonip, a broad spectrum preservative, *Soap Perfum. Cosmet.*, *41*:647-650 (1968).
26. J. C. Kelsey and I. M. Maurer, An improved (1974) Kelsey-Sykes test for disinfectants, *Pharm. J.*, *213*:528-530 (1974).
27. *United States Pharmacopoeia*, 20th rev., United States Pharmacopoeia Convention, Inc. (1980).
28. *British Pharmacopoeia*, Her Majesty's Stationery Office, London (1980).
29. R. A. Cowen and B. Steiger, Antimicrobial activity — a critical review of test methods of preservative efficiency, *J. Soc. Cosmet. Chem.*, *27*:467-481 (1976).
30. K. E. Moore, Evaluating preservative efficacy by challenge testing during the development stage of pharmaceutical products, *J. Appl. Bacteriol.*, *44*:SxIiii-Siv (1978).
31. M. R. W. Brown and R. M. E. Richards, The effect of polysorbate (Tween) 80 on the growth rate of *Pseudomonas aeruginosa*, *J. Pharm. Pharmacol.*, *16*:41T-45T (1964).
32. M. R. W. Brown, Turbidimetric method for the rapid evaluation of antimicrobial agents, *J. Soc. Cosmet. Chem.*, *17*:185-192 (1966).
33. M. S. Parker, The rapid screening of preservatives for pharmaceutical and cosmetic preparations, *Int. Biodeterior. Bull.*, *7*:47-53 (1971).
34. M. S. Parker, Some effects of preservatives on the development of bacterial spores, *J. Appl. Bacteriol.*, *32*:322-328 (1969).
35. M. Barnes and M. S. Parker, Use of the Coulter Counter to measure osmotic effects on the swelling of mold spores during germination, *J. Gen. Microbiol.*, *49*:287-292 (1967).
36. G. R. Mandels and R. T. Darby, A rapid cell volume assay for fungi toxicity using fungal spores, *J. Bacteriol.*, *65*:16-26 (1953).
37. M. Barnes and M. S. Parker, The increase in size of mold spores during germination, *Trans. Br. Mycol. Soc.*, *49*:487-494 (1966).
38. J. Judis, Studies in the mechanisms of action of phenolic disinfectants: release of radioactivity from C-14 labeled *E. coli*, *J. Pharm. Sci.*, *51*: 3, 261-265 (1962).

39. *Second International Symposium on Rapid Methods and Automation in Microbiology*, Learned Information (Europe), Ltd., Oxford (1976).
40. J. R. Furr, G. Davies, A. R. A. Adebyi, and A. D. Russell, A novel method for assaying antimicrobial agents by particle microelectrophoresis, *J. Appl. Bacteriol.*, *50*:173-176 (1981).
41. A. E. Beezer, W. H. Hunter, and D. E. Storey, The measurement of phenol coefficients by flow microcalorimetry, *J. Pharm. Pharmacol.*, *33*:65-68 (1981).

22
EVALUATION OF PRESERVATIVES IN COSMETIC PRODUCTS

DONALD S. ORTH *The Andrew Jergens Company, Cincinnati, Ohio*

I. Introduction 404
II. Preservative Efficacy Testing Methods 405
    A. USP method 405
    B. CTFA method 405
    C. Linear regression method 406
    D. Graphical determination of D values 406
    E. Mathematical determination of D values 406
    F. Reliability of the linear regression method 408
III. Preservative Death Time Curve 408
IV. Selection of the Preservative System 411
    A. The ideal preservative 411
    B. Commonly used preservatives 411
    C. Single- versus combined-preservative systems 411
V. Selection of the Recovery System 412
    A. Diluents and plating media 412
    B. Reliability of the recovery system 412
    C. Stressed microorganisms 413
    D. Recovery experiments 413

This chapter is a modification of a preservatives documentary article by D. S. Orth, entitled "Principles of Preservative Efficacy Testing," which appeared in the March 1981 issue of *Cosmetics and Toiletries* and appears here through the courtesy of this journal.

VI. Selection of Test Organisms 414
   A. Use of pure- versus mixed-culture inoculation 415
   B. Use of repeated-challenge testing 415
   C. Use of adapted organisms 417
VII. Evaluation of Preservative Efficacy Test Results 418
VIII. Acceptance Criteria 419
IX. Summary 419
   References 421

I. INTRODUCTION

All cosmetic products are subject to contamination with microorganisms. The growth of yeasts, molds, and bacteria in products depends upon a number of chemical and physical factors, including the availability of water, the composition of the product which will provide nutrients for organisms, the temperature of storage, and the presence of antimicrobial chemicals. When sufficient water is present, aqueous cosmetics will support the growth of microorganisms unless appropriate preservatives are present. Anhydrous materials, such as mineral oil, eye shadows, and powders, are subject to contamination on repeated use but will not support the growth of organisms unless water is present. Preservatives are included in cosmetic formulations to reduce the likelihood of growth in aqueous products and to reduce the chance of microbial persistence in anhydrous preparations.

Most cosmetics are not considered to be sterile products. They are compounded with raw materials that are not sterile; they may not be sterilized during manufacture; and they are often in multiple-use containers that allow the entry of microorganisms (through repeated use by application on the skin or through use in showers, where water may be splashed into the container, etc.).

Contaminating organisms may be either pathogens or nonpathogens. If contamination is due to pathogenic organisms, such as *Pseudomonas aeruginosa* or *Staphylococcus aureus*, there is a possibility of injury to the consumer during subsequent use of the product. Although contamination by nonpathogens may not be a health hazard, the development of these organisms may produce serious quality defects. Thus growth of yeasts, molds, and bacteria may alter product attributes so that the product becomes unacceptable to the consumer.

Preservative efficacy testing is an essential part of substantiating the safety of cosmetic products. This testing is done to determine the type of preservative to use in a product and the concentration required to satisfactorily preserve the product while in trade channels and in the hands of consumers. Preservative efficacy testing is important to consumer acceptance of a product for the following reasons: (1) Use of too little preservative may result in microbial growth, which will alter product attributes (i.e., color, odor, viscosity) and which may render the product injurious to the user; (2) use of too much preservative may cause skin irritation (i.e., rash, burning, redness, itching), which causes consumer dissatisfaction; and (3) use of too much preservative increases the cost of the product, which is passed on to the consumer. It is apparent that preservative efficacy testing is instrumental in determining the concentration of preservative needed to protect the product and make it safe on the one hand, and to reduce the likelihood of skin irritation and unneeded costs on the other hand [1].

II. PRESERVATIVE EFFICACY TESTING METHODS

Methods for preservative efficacy testing usually follow the guidelines published by the U.S. Pharmacopeia (USP) and the Cosmetic, Toiletry, and Fragrance Association (CTFA) [2,3]. In addition, a linear regression method for the rapid determination of cosmetic preservative efficacy was introduced recently [4]. These methods involve challenging the test sample with different microorganisms and performing aerobic plate counts (APCs) at various times thereafter to determine the number of viable organisms remaining in the test samples.

A. USP Method

The USP method states that the test sample must be inoculated with a ratio of 0.1 ml of inoculum to 20 ml of test sample, to give 10^5-10^6 organisms per milliliter in the test sample. The inoculated test samples are incubated at 20-25°C (or other specified temperature) and APCs are performed at 7, 14, 21, and 28 days subsequent to inoculation. The percentage change in the concentration of each organism during the test is calculated by use of the theoretical concentrations of microorganisms present at the start of the test. The preservative system is judged to be effective in the product when

1. The concentrations of viable bacteria are reduced to not more than 0.1% of the initial concentrations by the 14th day.
2. The concentrations of viable yeasts and molds remain at or below the initial levels during the first 14 days.
3. The concentration of each test organism remains at or below these levels during the remainder of the 28-day test period.

B. CTFA Method

The CTFA method recommends use of an inoculum containing no less than 10^6 organisms per milliliter of broth culture or surface growth on solid media. The amount of test product should be no less than 20 ml (g) of product for each challenge organism. After inoculation and mixing, the test samples are incubated at room temperature (or at the optimum temperature for the test organism). This method recommends that APCs be performed at 0, 1-2, 7, 14, and 28 days; however, the frequency of sampling varies, depending on knowledge of the formulation. The CTFA guideline recommends that the testing be carried out for a minimum of 28 days. A preservative system is judged to be inadequate if a relatively high APC of the challenge organism is obtained after the seventh day. The CTFA guideline for preservation testing of aqueous liquid and semiliquid eye cosmetics offers the following criteria for preservative efficacy assessment:

1. There should be greater than 99.9% reduction of vegetative bacteria within 7 days following each challenge and continued reduction for the duration of the test.

2. There should be greater than 90% reduction of yeasts and molds within 7 days following each challenge and continued reduction for the duration of the test period.
3. There should be bacteriostatic activity against spores through the entire test [5].

C. Linear Regression Method

The linear regression method employs the inoculation of 0.1 ml of a saline suspension of each test organism into approximately 50 ml of test sample in screw-capped bottles. After inoculation the bottles are shaken and the APCs are determined immediately and at various times thereafter — typically at 2, 4, and 24 hr for bacteria, and at 4, 8, and 24 hr for molds. Additional samples are taken and APCs performed at 3, 5, or 7 days after inoculation, unless the previous APC was less than 10 organisms per milliliter.

The rate of inactivation of each challenge organism in the test samples is given by the decimal reduction time (D value). The D value is the time required for 90% inactivation of the population of organisms (that is, 1-log reduction). The rationale for the use of D values is that every organism has a characteristic rate of death when subjected to any lethal treatment [4]. Thus the D value provides a quantitative expression of the rate of death of a specific organism in the test sample.

D. Graphical Determination of D Values

The D value for a test organism in a given test sample is determined from a survivor curve, which is prepared by plotting the number of viable organisms recovered at various times after exposure to the preservative system. For clarification, let us assume that *S. aureus* was inoculated into a lotion and that APCs of 10^6, 4×10^5, 10^5, and less than 10 organisms per milliliter were obtained at 0, 2, 4, and 24 hr, respectively. The survivor curve for *S. aureus* in this product is illustrated in Figure 1. Seven-cycle semilog graph paper is used in our laboratory because it allows one to plot APC values ranging from 10^7 to less than 10 organisms per milliliter. The survivor curve is constructed by drawing a straight line through the points and down to the x axis.

The x intercept represents the time for complete inactivation of *S. aureus* in the test sample. In this example the time is 24 hr. The D value is obtained from the slope of this curve. As indicated in Figure 1, the D value is 4 hr. It is desirable for a cosmetic product to have a D value of ≤4 hr for pathogens, such as *S. aureus*, because this means that a contamination producing 10^6 organisms per milliliter would be inactivated in 24 hr. More will be said on this later.

E. Mathematical Determination of D Values

The D values for yeasts, molds, and bacteria in cosmetic products may be determined by use of a calculator that is capable of least-squares linear regression analysis. In this case one does not need to plot the data on graph paper — the APC values at the time at which they were determined are entered into the calculator, and the linear regression is produced by the calculator. By use of a fairly inexpensive calculator, one is able to calculate the y intercept (the theoretical APC at the time of inoculation), the slope of the survivor

Evaluation of Preservatives

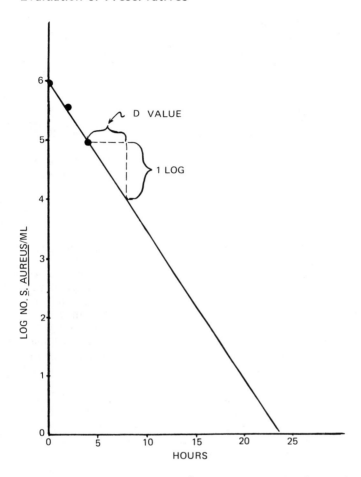

FIGURE 1 Survivor curve for *S. aureus* showing a decrease in the population from 10^6 to less than 10 organisms per milliliter after 24 hr. The D value in this example is 4 hr and is shown to be the time required for a 1-log reduction in the aerobic plate count.

curve, and the correlation coefficient (r). By entering the data used in Figure 1 into the calculator and calculating the linear regression, the following results were obtained:

y Intercept = 1.1×10^6 organisms per milliliter
Slope = -0.2500
D Value = 4 hr (note that this is the negative reciprocal of the slope)
r = -0.9931
x Intercept = 24.1 hr (predicted from the linear regression)

These results indicate that the predicted x intercept was slightly greater than 24 hr. The reason for this is that the data do not fit the linear regression perfectly because r = -0.9931. (Note that a value of r = -1.0000 indicates a perfect fit of the data to the regression — and this did not happen in this example.)

The mathematical determination of D values is preferred over the graphical determination for several reasons. In the first place it is much faster to enter data points in a calculator than it is to construct a graph. The use of linear regression analysis enables one to determine the proper positioning of the curve, should one desire to construct a graph. Also, it enables the analyst to determine the precision of the analysis by use of the r value. There has been an excellent agreement of the time predicted for the APC to be less than 10 organisms per milliliter and the actual time observed by performing an APC [4]. Another advantage of using a calculator is that the APC at any time may be predicted from the linear regression stored in the calculator's memory. In this manner it is possible to predict the APC at any time after challenge — from 5 min to 5 days (if the organisms persist that long).

F. Reliability of the Linear Regression Method

The reliability of the linear regression method was studied using lotion and liquid soap samples, challenged in triplicate on three successive days with *P. aeruginosa, Escherichia coli, S. aureus,* or *Aspergillus niger* [6]. The D values obtained for each set of triplicates were within 0.5 hr for each test organism, and the mean D values for each set of triplicate samples examined on different days were within 1.1 hr for each test organism. An analysis of variance revealed that there were no significant differences ($p < 0.01$) in the triplicate means for *E. coli* and *A. niger* in lotion and for *P. aeruginosa* in liquid soap, whereas the analysis of variance showed that the mean D values for the triplicate analyses for *S. aureus* were significantly different in both cosmetic products.

Although the standard deviation values about the triplicate D value means for *S. aureus* were no larger than were the standard deviation values for means for the other test organisms, it was noted that *S. aureus* triplicate means exhibited more day-to-day variation than did those of the other test organisms. It was suggested that the clumping nature (i.e., the tendency for the cells to stick together) was responsible for the observed variation. Thus clumping makes the cells difficult to disperse in saline and may influence the rate of interaction of the preservative systems with the test organism. Orth and Brueggen [6] suggested that special care be taken to disperse cultures that are known to clump before performing preservative efficacy tests.

Except for the significant difference observed for the mean D values of triplicate analyses obtained using *S. aureus*, which was attributed to clumping, the authors concluded that the linear regression method provides a reliable means of performing preservative efficacy tests when using standard procedures.

III. PRESERVATIVE DEATH TIME CURVE

Use of the linear regression method is ideally suited to selecting the proper concentration of preservative(s) to protect a product adequately. First, samples of the test product containing different concentrations of the preservative are prepared. These samples are challenged with a test organism, and APCs are determined at various times thereafter. Then the D values obtained with the different preservative concentrations are determined. As would be expected, the rates of die-off of the test organism increase with increasing preservative concentration (if it is an effective preservative).

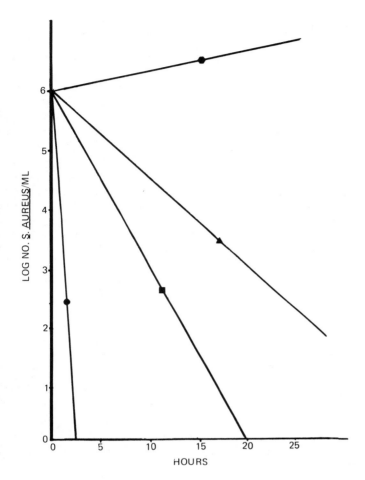

FIGURE 2 Series of survivor curves for *S. aureus* in lotion. D Values of greater than 30, 10, 5, and 0.5 hr were obtained in test samples containing 0% (————— —————), 0.1% (————— ▲ —————), 0.2% (————— ■ —————), and 0.3% (————— ● —————) methylparaben, respectively.

This produces a series of survivor curves, as is shown in Figure 2. For this example, let us assume that 0, 0.1, 0.2, and 0.3% methylparaben were used in a lotion challenged with *S. aureus*, and that D values of more than 30, 10, 5, and 0.5 hr, respectively, were obtained.

The preservative death time curve is constructed by plotting the D values as a function of the concentrations of preservative used to obtain these D values [1]. The preservative death time curve for *S. aureus* in this example is illustrated in Figure 3. This curve is used to select the concentration of preservative needed to inactivate the test organism at the desired rate. *Staphylococcus aureus* is a pathogen; consequently, it is prudent to select the concentration of methylparaben that will give a D value of ≤4 hr. From

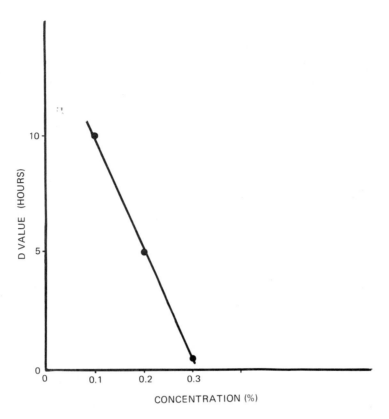

FIGURE 3 Preservative death time for *S. aureus* in lotion test samples used in Figure 2. The D value obtained using 0% methylparaben (i.e., greater than 30 hr) was not used in constructing this preservative death time curve because the bacteria were not being inactivated.

Figure 3, it is seen that about 0.25% methylparaben will be required to obtain a D value of 4 hr.

The preservative death time curve may be determined mathematically. To accomplish this, the D values and the concentration of methylparaben used to obtain these D values are entered into the calculator. Then the linear regression is calculated, and the concentration of preservative required to obtain a D value of 4 hr is predicted. In this case the concentration of methylparaben required is 0.23%. As in the calculation of D values from a survivor curve, use of linear regression is preferable to construction of the preservative death time curve by graphical means. Calculation of this curve enables one to determine precisely the concentration of preservative required to provide a given D value, and calculation of the r value enables one to determine how closely the data fit the linear regression.

IV. SELECTION OF THE PRESERVATIVE SYSTEM

A. The Ideal Preservative

The objective of preservative efficacy testing is to select the most appropriate preservative system for the product being tested. The requirements of an ideal preservative were discussed by Croshaw [7] and are as follows: It must have a broad spectrum of activity; it must be safe to use; it must have a suitable oil-water partition coefficient to be available in an effective concentration in the aqueous phase of the product; it should be stable over the range of pH values encountered in cosmetics; it should be compatible with other ingredients in the formulation; and it should not affect the physical properties of the product (i.e., odor, color). Another desirable feature of a preservative is that it be inexpensive (cost effective) to use.

B. Commonly Used Preservatives

Needless to say, no single preservative meets all of the above requirements for an ideal preservative in all formulations. The frequency of use of various cosmetic preservatives was reported recently by Richardson [8]. Although his data were not all inclusive because they were taken from formulations submitted to the Food and Drug Administration by companies participating in the voluntary cosmetic registration program, these formulations are considered to be representative of industry-wide use of preservatives. The top five preservatives were methylparaben (35%), propylparaben (32%), imidazolidinyl urea (9%), quaternium-15 (5%), and formaldehyde solution (4%). It is apparent that the parabens are used much more often than other preservatives. This provides testimony to their safety and efficacy and leads one to conclude that the parabens come closer to the ideal preservative than any other cosmetic preservative in use today [9].

C. Single- Versus Combined-Preservative Systems

Although a single preservative may provide adequate protection in most instances, there is always the possibility of something going wrong -- such as the occurrence of an adapted organism, or contamination by an organism that is resistant to this preservative. Thus it is usually a good idea to use more than one preservative in a formulation. This provides a backup, just in case the challenge organisms or product contaminants are able to metabolize, inactivate, or otherwise detoxify one preservative. This approach is analogous to combined drug therapy, in which a doctor administers more than one chemotherapeutic agent or antibiotic to a sick person to minimize the spread of infection by resistant organisms.

Generally speaking, each preservative has a specific mechanism of action. It would be expected that increasing the concentration would produce an additive increase in antimicrobial activity, as could be predicted from a dose-response curve. On the other hand, use of two preservatives — each with a different mechanism of action — would be expected to produce a synergistic antimicrobial action. I am unaware of reports in which the quantitative rates of inactivation of test organisms were determined in cosmetic products containing single- and mixed-preservative systems. From the discussion above, it is apparent that the linear regression method is ideally suited to provide such data. It is expected that reports containing quantitative data on the rates of inactivation of microorganisms in cosmetics containing single- and mixed-preservative systems will be published in the near future.

V. SELECTION OF THE RECOVERY SYSTEM

A. Diluents and Plating Media

The recovery system includes the diluent and the plating medium. Use of the proper recovery system is very important in preservative efficacy testing because the recovery system influences the accuracy and sensitivity of the data obtained by use of plating techniques. Diluents used in cosmetic microbiology include distilled water, physiological saline, peptone water, Letheen broth, and Letheen broth with Triton X-100 [2,3]. Commonly used plating media include standard methods agar, tryptic soy agar, and tryptic soy agar with lecithin and Tween 80 (TSALT) [2,3].

Recovery of all organisms that may be in a product by use of one recovery system is nearly impossible because different organisms have different growth requirements. Most organisms of interest to the cosmetic microbiologist are aerobic mesophiles (i.e., they grow best in the presence of oxygen and have optimal growth temperatures in the 25-45°C range). Thus most recovery methods used in examining cosmetic products employ aerobic incubation at 30-37°C for bacteria and 25-30°C for molds [2,3]. Anaerobic organisms, such as members of the genus *Clostridium*, will not grow under these conditions. Similarly, thermophilic organisms may have growth temperature requirements above 45°C, and they may not grow at the incubation temperatures generally used. Thus all recovery procedures are selective to some degree.

Nevertheless, the recovery systems commonly in use are appropriate because they provide the best conditions for recovering the organisms that are encountered most frequently in cosmetics or which are pathogens when they are present in cosmetic products. The ideal recovery system would possess the following characteristics:

1. The diluent should inactivate the preservative.
2. The diluent should be capable of dispersing materials in emulsions.
3. The diluent should stabilize injured cells and prevent them from dying before introduction into the plating medium.
4. The plating medium should inactivate the preservative.
5. The plating medium should facilitate the repair and growth of stressed cells.
6. The plating medium should allow good growth so that colonies are readily observable after an incubation of 48 hr for bacteria and 7 days for yeasts and molds.

It is essential that the recovery system inactivate the preservative used in preventing growth of organisms in cosmetics because "carryover" of the preservative into the plating medium may prevent growth of the viable organisms present. Thus the diluent and plating medium must inactivate preservatives, and they must provide the nutrients required by stressed cells so that they can grow and form colonies. The performance of diluents and plating media is determined by performing reliability studies.

B. Reliability of the Recovery System

It is desirable to know the reliability of any test method before applying it. Reliability depends on the precision, sensitivity, and accuracy of the method. In microbiological testing the precision generally depends on the skill and care used by the analyst, whereas the sensitivity and accuracy reflect the ability of the method to recover all viable organisms. Thus the reliability of the method used in preservative efficacy testing is dependent on the ability of the recovery system to prevent carryover of preservatives in the test sample into the plating medium and to facilitate repair and growth of stressed organisms.

C. Stressed Microorganisms

Stressed microorganisms are injured. Yeasts, molds, and bacteria may be stressed by a number of agents, including heat, sterilizing radiation, and antimicrobial chemicals. Several workers have studied the effects of sublethal stress on microorganisms and their recovery; however, I am unaware of reports in the cosmetic literature that discuss the recovery of stressed organisms from cosmetic products.

Sublethal stress is characterized by a loss of selectivity of the semipermeable membrane of the cell, which results in leakage of some intracellular components into the surrounding medium [10]. In addition, stressed cells have impaired ability to synthesize deoxyribonucleic acid, ribonucleic acid, and proteins — the macromolecules that are needed for repair and growth [10]. Ordal and co-workers reported that the repair of bacteria subjected to sublethal heat involves repair of the RNA-synthesizing mechanism, followed by protein synthesis and regaining normal membrane function [10,11]. The repair process typically requires several hours.

The function of preservatives in cosmetic products is to prevent microbial growth; therefore it seems reasonable to expect that organisms begin to die shortly after being introduced into a product during preservative efficacy testing. As with any population exposed to a lethal agent, not all members of the population die off instantly; sufficient contact time is required. Thus at any time after challenge (and until death of the entire population) some cells present are injured. This may be compared to the portion of the human population that is sick — either at home or in hospitals. These injured microorganisms may be metabolically crippled and may not be capable of growing on plating media that does not enable the cells to repair.

This is why it is especially important that the recovery system contain materials that prevent the carryover of preservatives into the plating medium. Materials that inactivate preservatives should be used in both the diluent and the plating medium. The review by Croshaw indicates that many cosmetic components, such as proteins and some surfactants, are incompatible with many preservatives [7]. Thus Tween 80, lecithin, and proteins are incorporated into diluents and plating media to prevent preservative carryover.

D. Recovery Experiments

Use of the most reliable method possible is necessary because of the importance of preservative efficacy testing. Each microbiological lab should perform recovery experiments on products challenged with appropriate test organisms to determine which recovery system provides the highest APC. The recovery system that yields the highest APC is the most accurate and sensitive, because it allows the most viable organisms to grow. As stated above, this may be due to inactivation of preservatives that may have carried over into the plating medium, or it may be due to facilitation of repair and growth of stressed microorganisms.

TABLE 1 Recovery of *S. aureus* from Lotion Using Standard Methods Agar, Baird-Parker Agar, and Tryptic Soy Agar with 0.07% Lecithin and 0.5% Tween 80 (TSALT)

Recovery medium	APC at 0 hr	APC at 3 hr
Standard methods agar	2.4×10^5/ml	5.0×10^3/ml[a]
Baird-Parker agar	2.4×10^5/ml	1.4×10^4/ml[a]
TSALT	2.7×10^5/ml	5.2×10^4/ml[a]

[a]APC significantly different (at 95% confidence limit) from other APCs at 3 hr
Source: Ref. 12.

An example of the effect of different plating media on the recovery of *S. aureus* from a lotion preserved with parabens and quaternium-15 was reported recently [12] and is illustrated in Table 1. In this study the diluent used for all samples was 0.1% peptone water [13], and the plating media used were standard methods agar, Baird-Parker agar, and TSALT. All plating media gave similar APCs immediately after inoculation, because the *S. aureus* population was not stressed by the brief exposure to the lotion. Significant differences were observed in the APCs obtained with the plating media after 3 hr, and it is believed that these differences reflect the ability of these media to inactivate the preservatives or facilitate repair of the stressed *S. aureus*. This example shows that TSALT gave higher recoveries of *S. aureus* from the test lotion than the other plating media. Thus TSALT gave the most reliable indication of the number of viable organisms present in the test sample.

VI. SELECTION OF TEST ORGANISMS

Both USP and CTFA methods of preservative efficacy testing recommend the use of standard test organisms. The USP method recommends use of *Candida albicans* (ATCC 12031), *A. niger* (ATCC 16404), *E. coli* (ATCC 8739), *P. aeruginosa* (ATCC 9027), and *S. aureus* (ATCC 6538). The CTFA method recommends use of *S. aureus* (ATCC 6538), *P. aeruginosa* (ATCC 15442 or 13388), *A. niger* (ATCC 9642), *Penicillium luteum* (ATCC 9644), *E. coli*, and *C. albicans* (strains not specified). The use of spore-forming bacteria is considered to be optional; however, the CTFA indicates that *Bacillus cereus* or *Bacillus subtilis* var. *globii* be employed when tests on spore formers are performed. The CTFA method also states that formulation isolates should be included in preservative efficacy testing where research has shown them to be resistant to the preservative system of a product.

Use of the above standard test organisms provides a range of morphological and metabolic types of bacteria, yeasts, and molds that are widespread in nature and which have been associated with microbiological problems in cosmetic products. Use of standard test organisms allows comparison of the preservative systems in different products. This enables us to standardize preservative system potency criteria and compare the preservative systems in similar products made by different companies.

The use of product isolates may be advantageous because it is possible that specific products may be more susceptible to these organisms than to the standard test organisms. Also, the company microbiologist may have experience with product isolates in preservative efficacy testing, so that the use of in-house isolates provides a better indicator of susceptibility to contamination than use of the standard test organisms.

Familiarity with testing and knowing how various organisms perform in products enables the cosmetic microbiologist to choose the best test organisms to use in preservative efficacy testing. Thus a profile may be developed for each product so that the constant relationships between different organisms are known. With this information one can select the organisms that are inactivated most slowly (that is, those that have the largest D values) in the product for use in preservative efficacy tests with that product. The selection of test organisms for use in specific products is easy when quantitative data on the rates of death of these organisms have been determined.

A. Use of Pure- Versus Mixed-Culture Inoculation

There are advantages to conducting preservative efficacy tests by use of either pure cultures (i.e., one organism) or mixed cultures (i.e., two or more types of organisms pooled together). Mixed culture studies are performed by inoculating like classes of organisms into the test samples, incubating, and determining the APCs at specified times afterward. Typically, several different bacteria are pooled into one suspension containing 10^5-10^7 organisms per milliliter of each type of bacterium, and this suspension is used to challenge the test samples. Similarly, suspensions of yeasts and molds are pooled and used to challenge the test samples. The primary advantage of using mixed cultures is convenience: It is easier to perform a couple of inoculations with mixed cultures of bacteria and fungi than it is to perform several individual inoculations using each organism in pure culture.

The principal advantage of using pure cultures in preservative efficacy testing is that the rates of die-off of each test organism may be determined. This enables the microbiologist to map the resistance profile for the product. Although pure-culture studies require more time, materials, and effort than parallel mixed-culture studies, the value of the quantitative information obtained by use of pure-culture studies is well worth the extra time and expense. Thus D values may be determined for each different organism in each test sample.

B. Use of Repeated-Challenge Testing

Several publications refer to the use of repeated inoculations in preservative efficacy testing [5,14]. The rationale for the use of repeated inoculations is to determine the number of challenges with a particular test organism that a product can withstand before the preservative system fails. The value of this type of testing was questioned recently when it was demonstrated that the rates of inactivation of different size populations of a specific test organism in a given product were similar [4]. This is illustrated in Figure 4. Note that the rate of inactivation of $S.$ $aureus$ is independent of the initial population, ranging from 1.5×10^3 to 1.8×10^6 $S.$ $aureus$ per milliliter in these test samples.

Orth and Brueggen [6] investigated the need to perform rechallenge testing. They compared the D values obtained using 10 challenges of 0.1 ml on 10 different days versus one challenge of 1.0 ml of suspensions of $S.$ $aureus$ or $P.$ $aeruginosa$. Although the sample receiving 1.0 ml of the suspension

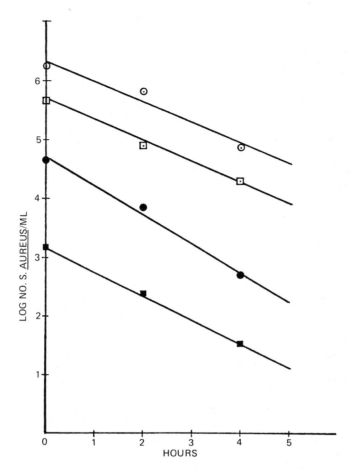

FIGURE 4 Survivor curves showing the effects of initial concentration of *S. aureus* on the rate of death in lotion (⊙ ——— ⊙ , initial concentration of 1.8×10^6/ml; ⊡ ——— ⊡ , initial concentration of 4.6×10^5/ml; ● ——— ●, initial concentration of 4.5×10^4/ml; and ■ ——— ■ , initial concentration of 1.5×10^3/ml). (From Ref. 4.)

of test organisms had 10 times more viable organisms recovered on TSALT at the 0-, 2-, and 4-hr samplings than did the sample receiving the 10th inoculation of 0.1 ml of the suspension, the D values of *S. aureus* were identical in both samples. Similar findings were obtained with *P. aeruginosa*. This study corroborated the findings of the earlier study [4] and demonstrated that little, if any, useful information may be obtained from repeated-challenge testing — up to the point where the preservative system is overloaded or where the sample is diluted as a result of the repeated challenges.

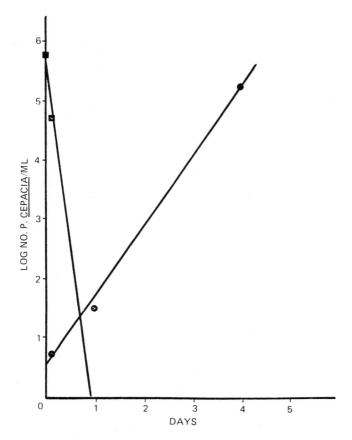

FIGURE 5 Comparison of the fate of adapted and unadapted *P. cepacia* in lotion (●―――●, culture adapted by growth for 24 hr at 37°C in 10% lotion-90% water; ■―――■, culture grown for 24 hr at 37°C on standard methods agar − unadapted). (From Ref. 12.)

C. Use of Adapted Organisms

Microorganisms are capable of adapting to most substrates available in their environment. Many of the materials used in cosmetics become suitable substrates for microorganisms when conditions of water availability, pH, and temperature are favorable for growth. Bacteria, yeasts, and molds may survive, adapt, and grow in adequately preserved cosmetics. This can be illustrated by growing the test organisms in dilute product (that is, a 1 or 10% aqueous dilution of the product) and using this as the inoculum for testing. Often it is found that pure cultures of the adapted organisms are able to survive and grow in the test samples, whereas the same organisms grown on culture media are inactivated readily when introduced into the test samples [12]. The reason for this is that microorganisms must be grown in the presence of a substrate (or substrate analog) to adapt to it, and culture media provide no substrates to adapt test organisms to preservative systems. This is illustrated in Figure 5, in which *Pseudomonas cepacia* was grown on a standard methods agar slant or in 10% lotion-90% water before introduction into lotion test samples. As is apparent in Figure 5, the unadapted organisms grown on the standard methods agar slant died within 24 hr. In contrast, the adapted *P. cepacia* population grew from a low initial count to 1.8×10^5 organisms per milliliter after 4 days. This type of experiment demonstrates that adapted organisms are able to survive and grow in adequately preserved products.

The use of adapted organisms is not recommended for routine preservative efficacy testing, because it is believed that most organisms of importance to the cosmetic industry may be adapted to survive and grow in adequately preserved products. The only instance in which the use of adapted organisms may be informative is in demonstrating that an isolate is capable of growing in a specific product. Again, this may require a period of adaptation by growth on dilute product before introduction into the test samples, and it should be done with the use of the isolate grown on culture medium as a control.

VII. EVALUATION OF PRESERVATIVE EFFICACY TEST RESULTS

The APCs of adequately preserved cosmetics decrease with time. A situation that requires explanation is the appearance of colonies on the plating medium after a negative result (that is, <10 per gram) is obtained. This finding suggests that the organisms died off and then came back to life (i.e., the Phoenix phenomenon). This, of course, cannot happen, and there are several possible explanations for the appearance of colonies on the plating medium after a negative APC is obtained. For example, the sampling may not have been representative of the test sample and may have missed the organisms that were present. This may be due to the use of too small a sample in the test or inadequate mixing of the test material before sampling. Perhaps the recovery system did not prevent carryover of the preservative into the plating medium. Another explanation could be that the preservative system is incompatible with the formulation, and that low levels of the challenge organisms started to grow after the preservative system was inactivated. Also, it is possible that the rate of death was slow enough that the organisms adapted to the test samples. If the organisms adapt, there would be a continually increasing APC until a level of 10^4-10^6 organisms per milliliter was reached. Finally, the colonies on the plating medium may be due to contamination.

When colonies are encountered on the plating medium after the previous APC was less than 10 per gram, the microbiologist should verify that the identity of the organisms in these colonies is the same as the challenge organism. This involves picking the colony and performing a Gram strain to determine if the organism corresponds morphologically to the challenge organism. If the Gram strain and colony appearance indicate that the organism is the same as the challenge organism, then the conservative position is to assume that it is the challenge organism and that the previous APC of less than 10 per gram was in error (for the reasons explained above). If the Gram strain or colony appearance reveals that the organism is not morphologically similar to the challenge organism, then it is correct to conclude that these colonies are due to contamination.

VIII. ACCEPTANCE CRITERIA

The objective of preservative efficacy testing is to determine the type of preservative to use in a product and the concentration required to satisfactorily preserve the product. The CTFA and USP methods do not provide quantitative measures of preservative potency, whereas the linear regression method allows determination of D values for specific organisms in specific products. Thus it is possible to set meaningful guidelines for accepting or rejecting a product based on the rates of die-off (i.e., D values) of the test organisms in the test samples. In an earlier publication it was proposed that the preservative system for each product meet minimum requirements for inactivation of three classes of organisms: pathogenic microorganisms, bacteria not known to be pathogens, and yeasts and molds [4]. It is believed that pathogens should be inactivated completely within 24 hr in cosmetics intended for use in or around the eye and for baby products. Thus eye and baby products should have D values of $\leqslant 4$ hr for pathogens such as *S. aureus* and *P. aeruginosa*, so that these products are able to self-sterilize in 24 hr when they are contaminated with 10^6 organisms per gram. Although specific guidelines need to be set for each product, it is believed that D values of $\leqslant 28$ hr are satisfactory for nearly all nonpathogenic microorganisms. This will result in self-sterilization of 10^6 organisms per gram in 7 days.

The usefulness of preservative efficacy testing is realized only when the test satisfies the reasons for performing the test. Use of the linear regression method to determine D values and construction of a preservative death time curve enable the microbiologist to determine precisely the concentration of preservative required to meet the desired acceptance criteria. Besides conducting tests on lab and pilot batches of product, production samples should be examined. In addition, tests are necessary to establish that the preservative system has adequate storage stability and tolerance to abuse.

IX. SUMMARY

The usefulness of any method of preservative efficacy testing depends on the use of appropriate test conditions, challenge organisms, and acceptance criteria. The objective of this chapter was to provide basic information on preservative efficacy testing for microbiologists and managers to enable them to evaluate preservative systems in cosmetic products in a rational manner. The "principles of preservative efficacy testing" were presented elsewhere [12] and are summarized below:

1. Preservative efficacy testing is performed to determine the type and concentration of preservative to use in a product. All cosmetic products are subject to microbial contamination; consequently there is a need to demonstrate that products are protected from spoilage and are safe to use.
2. The USP, CTFA, or linear regression method may be used for preservative efficacy testing. Although all of these methods may be used for this testing, the use of the linear regression method is recommended because it requires less time to perform than the other methods, it provides quantitative data on the rate of death of specific organisms in test products, and its reliability has been demonstrated.

3. The concentration of preservative needed to provide the desired preservative potency for any test organism is determined by the use of a preservative death time curve.
4. The recovery system that provides the highest APC of the test organisms in test products is the most reliable system. The diluent and plating medium should prevent preservative carryover and should facilitate repair and growth of stressed microorganisms. Letheen broth with Triton X-100 and TSALT have been the most reliable recovery system tested in our laboratories.
5. The use of standard test conditions, such as aerobic incubation at 30-35°C for bacteria and 25-30°C for yeasts and molds, is recommended for routine testing.
6. Testing should be performed using standard test organisms that have been grown on laboratory media. Saline suspensions of the challenge organisms are recommended. The use of standard test organisms allows comparison of the preservative systems in different products and allows development of standardized acceptance criteria. While the use of product isolates may be informative in some situations, the use of adapted organisms is not recommended for routine testing.
7. Rechallenge studies do not provide any useful information to properly designed preservative efficacy testing experiments. It is recommended that these time-consuming studies be replaced by experiments designed to determine the D value for specific organisms in specific products, the concentration of test organisms required to overload the preservative system, or the effects of abuse.
8. There is generally no difficulty in interpreting the results of preservative efficacy tests, because the APCs decrease with time. When successive APCs of a test product reveal that the APC does not decrease with time, or that the APC is greater than 10 per gram when the previous APC was less than 10 per gram, the reason for this must be determined.
9. The quantitative data provided by the linear regression method allow the establishment of quantitative acceptance criteria. It is believed that quantitative acceptance criteria are easier to implement than are qualitative criteria. Adequately preserved cosmetic products should have D values of ≤ 4 hr for pathogens and ≤ 28 hr for nonpathogenic bacteria, yeasts, and molds.
10. Preservative efficacy testing should not be restricted to production samples. The testing should include samples that have been subjected to different storage conditions and samples that have been subjected to conditions of abuse.

The purpose of conducting preservative efficacy tests is achieved only when the test satisfies the reasons for performing the test. The information in this chapter discussed several test methods and test conditions to provide a basis for consideration and selection of the test method that is most suitable for the work to be performed.

REFERENCES

1. D. S. Orth, Establishing cosmetic preservative efficacy by use of D values, *J. Soc. Cosmet. Chem.*, 31:165-172 (1980).
2. Anonymous, Microbiological tests, antimicrobial preservatives — effectiveness, *United States Pharmacopeia XIX*, United States Pharmacopeial Convention, Rockford, Md. (1975), pp. 587-592.
3. Preservation Subcommittee of the CTFA Microbiological Committee, A guideline for the determination of adequacy of preservation of cosmetics and toiletry formulations, *TGA Cosmet. J.*, 2:20-23 (1970).
4. D. S. Orth, Linear regression method for rapid determination of cosmetic preservative efficacy, *J. Soc. Cosmet. Chem.*, 30:321-332 (1979).
5. Preservation Subcommittee of the CTFA Microbiological Committee, A guideline for preservation testing of aqueous liquid and semi-liquid eye cosmetics, Cosmetic, Toiletry, and Fragrance Association, Washington, D.C. (1975).
6. D. S. Orth and L. R. Brueggen, Preservative efficacy testing of cosmetic products, Rechallenge testing and reliability of the linear regression method, *Cosmet. Toiletries*, 97:61-65 (1982).
7. B. Croshaw, Preservatives for cosmetics and toiletries, *J. Soc. Cosmet. Chem.*, 28:3-16 (1976).
8. E. L. Richardson, Update — frequency of preservative use in cosmetic formulas as disclosed to FDA, *Cosmet. Toiletries*, 96:91-92 (1981).
9. D. S. Orth, Use of parabens as cosmetic preservatives, *Int. Soc. Trop. Dermatol.*, 11:505-506 (1980).
10. R. S. Flowers, S. E. Martin, D. G. Brewer, and Z. J. Ordal, Catalase and enumeration of stressed *Staphylococcus aureus* cells, *Appl. Environ. Microbiol.*, 33:1112-1117 (1977).
11. D. R. McCoy and Z. J. Ordal, Thermal stress of *Pseudomonas fluorescens* in complex media, *Appl. Environ. Microbiol.*, 37:443-448 (1979).
12. D. S. Orth, Principles of preservative efficacy testing, *Cosmet. Toiletries*, 96:43-52 (1981).
13. J. M. Madden, Microbiological methods for cosmetics, in *Bacteriological Analytical Manual*, 5th Ed., Association of Official Analytical Chemists, Washington, D.C. (1978).
14. W. E. Rosen and P. A. Berke, Modern concepts of cosmetic preservation, *J. Soc. Cosmet. Chem.*, 24:663-675 (1973).

23
THE TEST FOR THE EFFECTIVENESS OF ANTIMICROBIAL PRESERVATION OF PHARMACEUTICALS

3RD JOINT REPORT OF THE COMMITTEE OF OFFICIAL LABORATORIES AND DRUG CONTROL SERVICES AND THE SECTION OF INDUSTRIAL PHARMACISTS—FEDERATION INTERNATIONALE PHARMACEUTIQUE (FIP)

I. Introduction 424

II. General Principles of Testing 424

III. Parameters of the Basic Test 426

    A. Test organisms 426
    B. Mode of contamination 427
    C. Inoculum 428
    D. The preparation in the test 429
    E. Other test conditions 429
    F. Detection of germs 430
    G. Results and interpretation 431

IV. Method of the Basic Test 431

    A. Strains and inoculum 431
    B. Test with liquid preparations 432
    C. Test with semisolid preparations 432
    D. Count of microorganisms 432

Translated from the french original (*Pharmaceutica Acta Helvetiae*, 12/1979) by M. Devleeschouwer and N. Diding. The following participated in the compilation of this report: for the Committee of Laboratories, J. Dony, M. Devleeschouwer, N. Diding, H. Dressel, G. Estève, O. Flink, L. Møhl, and W. Wozniak; and for the Section of Industrial Pharmacists, J. C. Blanchard, I. de Carneri, M. Gay, H. Lagodsky, H. R. Gravestein, J. Lingnau, E. Olivar, H. P. Riniker, and K. H. Wallhäusser. Reprinted, with permission from *Pharm. Acta Helv.*, 55, No. 2, 40-49 (1980).

E. Interpretation 433
F. Culture media and solutions 433

References 436

I. INTRODUCTION

In the previous FIP Reports 1972 [30] and 1975 [31] emphasis is placed on the importance of effective antimicrobial preservation to ensure the stability and safety of noncompulsory sterile pharmaceutical preparations with regard to microbiological purity. Among these preparations (categories 2 and 3 in Table 1 of the 1975 report [31]), liquid and semisolid forms with aqueous bases are particularly important. The aqueous-based formulation also constitute the majority of sterile injectable preparations, ophthalmics, and others (categories 1a and 1b of Table 1 of the 1975 report [31]) in which the principal aim of the preservation is intended to maintain the initial sterility when this is jeopardized by repeated withdrawals from multidose containers.

Manufacturers as well as health authorities recognize the necessity of demonstrating the efficacy of antimicrobial preservation. This effectiveness cannot easily be tested by a simple chemical or biological analytical test method which complies with a specification. This is why most pharmacopoeias have not included such a test: The *U.S. Pharmacopoeia XVIII* already described a microbiological method [74], and the *British Pharmacopoeia* 1973 stipulated a requirement for antimicrobial effectiveness in one specific case [13]. This fact as well as the large variety of test systems found in the literature and used in various laboratories induced the FIP joint working group to elaborate such a method. The method described below, based on well-known principles and the practical experience of the group members, may be a useful guide for manufacturers and for the official control authorities or independent institutions.

II. GENERAL PRINCIPLES OF TESTING

The effectiveness of antimicrobial preservation in a pharmaceutical preparation depends on many factors:

Components
Composition
Antimicrobial preservative
Container (type, material, closure)
Initial degree of microbiological purity required
Application form
Shelf life

Even with rigid requirements for manufacturing hygiene, good manufacturing procedures, quality control, and use of raw materials with only permissible number of germs, we still must take into account the possibility of secondary contaminations. Owing to the water content of the preparation, some germs may also develop by using some components as a carbon source [7,34,56,86]. On the other hand, other components or systems, without being genuine antimicrobial preservatives, might prevent microbial proliferation. They might even kill the germs or increase the effectiveness of preservatives [8,9,10,29, 57,61,81,84]. If an antimicrobial agent is included in a preparation, it might cause, in addition to its own activity, compatibility problems, such as loss of potency by interaction with components or distribution between different phases [11,22,44,49,57,59,72,85]. Absorption by the container or its closure [28,43,44,45,66,70] must also be taken into account. Finally, we have to consider the resistance of some microorganisms and their ability to decompose the antimicrobial agent [12,38]. The test method should take into account all these different factors.

Study of the real antimicrobial activity of preservatives, either alone or in combination, and their mode of action with regard to their chemical and physicochemical characteristics requires the use of specific methods and assay procedures [14,18,36,37,63,64,65,82,87]. The above-mentioned studies give data for the choice of one or more agents suitable for a given preparation. Nevertheless, however accurate they are, they still represent only fragmentary results which do not provide a guarantee for more complicated systems. For this reason almost all authors agree that *each pharmaceutical preparation must be considered separately and be subjected to a comprehensive test in regard to preservation efficacy.*

In addition to the factors mentioned above, the test conditions and their interpretation must always take into account the initial degree of purity, which depends on the requirements for the form of application and the contamination risk during use. One could infer that each case needs its own assay. But it is an illusion to believe that one can reproduce all conceivable situations. Also, the use of many various test conditions would complicate the comparison of the results and the establishment of interpretation rules. Furthermore, the efforts to increase the parameters applied in order to cover the largest possible number of preparations make the test too expensive [51], and the most comprehensive test that takes into account all possible conditions will still never be more than an artificial model which only approaches reality. Finally, one needs rather a simplified model for routine use, standardized to give quantitative reproducibility, with the possibility for comparative interpretation and correlation with the practical use of the preparations [21].

The interactions mentioned above must also be judged in the light of the shelf life of the preparations. This is based on stability studies, and the antimicrobial preservation test must form part of this study with its parameters of time and temperature. At this stage the test is connected essentially with the developmental phase of the preparations. It can also demonstrate the intrinsic antibacterial effectiveness of a system and show whether there is any need for a specific preservative [81]. Moreover, the test, on the whole, is not intended for the analytical control of each batch of a production, even if some authors make it somewhat easier for that purpose [41]. An analytical method for the microbiological assay of a preservative [27], like that used for antibiotics, is not described here and does not fall within the scope of this report. For the control authorities or the independent laboratories, the test proposed here provides the possibility for verification of the antimicrobial effectiveness of the final product.

Various types of test are described in the literature. They are more or less adapted to certain conditions discussed above. The determination of the minimal inhibitory concentration in a nutritive medium is thus sometimes applied to the preservatives themselves [49]. However, it cannot suitably be applied to galenical preparations, because the germ-killing activity, important for preservation, cannot be so determined. Neither is the agar diffusion test suitable, although it can be performed with the preparation itself [53].

The latter technique involves the liberation and diffusion of the preservative, which can both be influenced by other components of the preparation. The method with inoculated paper strips immersed in ointments [17,53] requires complicated and indirect techniques to detect the surviving germs.

Ever since an artificial contamination test with introduction of germs into the preparation was first proposed [1], this type of test has been most frequently used, with many variations. In some methods only a few bacteria are used and total destruction, as an "end point", is required [1,40]. Other methods recording the presence or absence of visible microbial growth [9,60] become very uncertain in cases where there is insignificant microbial development or the preparation is opaque.

An artificial contamination test using only a single strain and one very short contact time could be valuable for testing eyedrops [2], but cannot be recommended in general. It is not easy to standardize methods using separate artificial inoculations, each with mixtures of several microbial strains of asporogenic bacteria, bacterial spores, yeasts, or fungal spores [42,48,61]. This is also true when a single mixture including pure strains and mixed microbial populations obtained from spoiled products is used [69], and the evolution followed by a total count of viable organisms.

The most comprehensive test system [62] involves — for injectable, ophthalmic, topical, and oral preparations — artificial contamination with separate or mixed strains. The inoculum is standardized, the contact times fixed, the bacteria counted by plate count, and the requirements outlined. Many authors [5,6,23,42,48,59,76], including, for instance, U.S. members of the Pharmaceutical Manufacturers' Association, have worked in this way. A standardized method for injectable and ophthalmic preparations appeared as a result of a collaborative study organized by this association [59], and this method was largely taken into account by the *USP XVIII*. A working group of microbiologists of the Swiss Pharmaceutical Industry also laid down in principles and methods for assessing the microbiological purity of drugs [15] a test method for the determination of the antimicrobial effectiveness intended for all categories of preparations. This system was also proposed for the Swiss Pharmacopoeia [33].

Confirmed by wide experience, this system provides the elements of a basic test, the parameters of which will be discussed in detail in Section III and described in Section IV. *The basic test* is intended to be used as minimal common procedure by research workers and control personnel. As such, it must necessarily be completed, refined, or repeated to give the best possible picture of the effectiveness of the preservation of a preparation in terms of behavior and safety in practice.

III. PARAMETERS OF THE BASIC TEST

A. Test Organisms

The minimal range of microorganisms must contain representatives of the following principal groups: gram-positive and gram-negative bacteria, yeasts, and molds. The organisms chosen must be of significance for the microbial purity of the preparation. It is also desirable, however, that at least part of the organisms be of importance for all categories of preparations. As representatives of pathogenic contaminants of relevance in local infections, *Staphylococcus aureus*, *Pseudomonas aeruginosa*, and *Candida albicans* are adequate for the testing of injectables, ophthalmics, and vulneraries. Although in adequately formulated oral preparations there should be no danger of toxin formation, as in food products, where this could result from an extensive proliferation of, for example, *Staphylococcus aureus* or some molds, it might in any case be justified to choose such organisms for testing preparations of this category.

Among the nonfastidious bacteria able to proliferate in water and adaptable to many various substrates, and also often resistant to antimicrobial agents, *P. aeruginosa* is the most important. Besides its pathogenicity, it is also of great importance for the manufacturing hygiene. Likewise, *Escherichia coli* is relevant as representative of the human and animal intestinal flora. *Pseudomonas aeruginosa* and a mold like *Aspergillus niger* are also of interest as representatives of microbial flora with potentials of degradation.

As regards the origin of the strains used, it is obvious that the best strains are those isolated from a decomposed product of the same kind as that which is to be tested. The most interesting strains of microorganisms of hygienic signification in the manufacturing of pharmaceutical products are those isolated from water [48], the manufacturing environment [51,71], and products decomposed after storage or broken packages incompletely utilized [9, 51]. Experience indicates that some strains isolated from healthy or ill individuals are relatively resistant, whereas those corresponding to the collection strains commonly used might be as resistant or somewhat more sensitive [52]. But the differences may be expected to disappear by in vitro maintenance through several passages. On the other hand, the use of germs from a decomposed or heavily spoiled product may cause practical problems in keeping the initial properties intact. Whatever the maintenance procedure may be — germs in the product itself, isolated germs which are maintained on culture medium or on medium supplemented with the product -- there are always standardization problems. This is the reason why most authors (see, for example, Refs. 15,33,59, and 62) prefer collection strains, with the recommendation that the test, depending on the circumstances, be extended with strains isolated from the product or with strains possessing particular properties, such as osmophilic yeasts [5,26,33] or bacterial spores [26,76], although the latter may not be expected to be killed under ordinary conditions [42,59,71,72].

B. Mode of Contamination

Artificial contamination by mixtures of various strains is used by a number of authors [23,42,47,48,62,72,76], obviously in the hope of increasing the value of the test without increasing the work load to the same extent. Their aim was probably also to obtain a "more realistic test." This is, however, questionable if the microbial count must be done on several selective media [76] or if, as some experience shows, the results are less reliable than those obtained with isolated strains [51,85]. Therefore we admit that this contamination method can be useful in those special circumstances where it has been supported by personal experience, but we do not recommend it for a basic test. Nor will we describe the technique incorporating several strains with soil [42,62,82], a method which has been experienced in particular cases by members of the group. Like the majority of the authors, we will instead choose the contamination by single strains.

Results from repeated inoculations [51,85] are supposed to indicate the actual capacity of the antimicrobial system, bearing in mind the risk of repeated contaminations on successive withdrawals from preparations in multidose containers. However, the value of such models is doubtful, and interpretation of the results is especially difficult. This technique is not practicable for a basic test.

C. Inoculum

The influence of the density of the inoculum on the test results was studied in some experiments. As expected, density is of major importance for the determination of the end point [77], and less important or without importance when the decimal reduction time is determined [19].

The criticism against an inoculum size of $1.25\text{-}5 \times 10^5$ organisms per milliliter of preparation, as prescribed in the *USP XVIII*, stems from this number being considered an exaggeration which is not relevant to the size of accidental contaminations occurring in reality [47,48]. In fact, if one considers that the test is only a very rough model of what happens in reality, the choice of the inoculum density should preferably be adapted to the counting technique of the survivors. Thus fewer organisms, such as 10^3 per milliliter [1,40] or $10^4\text{-}10^5$ per container [78], are recommended and often used in combination with determination of the end point by subsequent inoculation in liquid medium or by using membrane filtration technique. On the other hand, higher densities of $10^4\text{-}10^6$ or more per milliliter of preparation are generally used with the plate count technique for assaying the degree of reduction [6,51,59,62].

The way the inoculum is prepared can also have an influence on the results. Only a few factors among those like the mode of maintaining the strains, the properties of the culture media, the state of the preliminary cultures [51], the suspension medium, and so on, have been systematically investigated with respect to their influence on the test. Also described are several more or less sophisticated procedures with the aim of obtaining an improvement of the reproducibility of the test [5,35,55]. Some of these, however, do not seem to be in accord with real conditions, as, for example, repeated washings of the organisms to eliminate every trace of culture medium. It is well known that the addition of a small amount of peptone to a suspension medium, for example distilled water or physiological or buffered sodium chloride solution, is decisive for the ability of the bacteria to survive in the suspension.

For a general test the inoculum must be easy to prepare, must give reproducible results, and must be accessible to most laboratories. The recommended procedure cannot take into account all the special techniques like lyophilization and spray-drying which have been especially developed for particular products, for instance oils [16,24,54]. Such special techniques are, of course, useful for laboratories that have had experience with them, but the results should in any case be compared with those obtained with the test described in Section IV.

The proposed method requires, in principle, the use of freshly prepared young cultures made from strains maintained on media or stock freezed (in liquid nitrogen, for example) or lyophilized. The strains should be suspended and diluted in physiological solution supplemented with protective peptone to suitable density based on counts or photometric measurements. This is then verified by plate count simultaneously with the test of the preparation.

Antimicrobial Preservation

D. The Preparation in the Test

Several test conditions depend on the kind of preparation—galenical form, container, phase of development, category as regarding application.

Some authors recommend that the test be performed in the original container [51,52,77,78]. Moreover, if stated in a pharmacopoeia, as in the *USP*, that the test apply only to the products in their intact original container, then the test can be performed either in the original containers, if they are suitable, or in better-adapted laboratory vessels. This position adopted for the basic test does not exclude the necessity, in some developmental stage, to submit the product to a more elaborated test in the original containers, especially if there are particular risks involved.

Other important factors are the quantitative inoculum-preparation ratio and the inoculation procedure. The quantitative ratio should be low in order to prevent an appreciable change in composition of the preparation, but large enough to ensure a homogeneous distribution of the contamination: 1:100 seems to be a convenient size and is often used.

The inoculation and the homogeneous repartition of the inoculum pose no technical problems for liquid preparations, where shaking is sufficient.

For semisolid water-oil or oil-water emulsions, it is necessary to use a spatula, mortar, or an emulsifying apparatus like Brown's, for example [32]. In fact, the inoculum is added to the aqueous phase, and the effectiveness of the repartition depends on the precision and perseverance of the experimenter. For water-free fat systems, the question of the suitability of supplying aqueous inoculum arises. This problem is often resolved for oils and ointments by preparation of a dried inoculum through spray-drying [16] or lyophilization [24,54], or on a carrier such as silk thread [26], filter paper, or filter membrane [17,53]. If these techniques are valuable for studying the intrinsic properties of oils or fats, the inoculum in the form of an aqueous suspension, on the other hand, is of value as a model for the spontaneous contamination of a fat product by condensed water, airborne droplets, or mucus.

E. Other Test Conditions

Many conditions have already been discussed. It is essentially the temperature and the contact times which remain to be considered.

It seems natural to keep the contaminated preparations under test at room temperature, as this is the normal way of storing. However, as it has been proved that differences in temperature, for instance, between 15 and 25°C, give various results, depending upon the contaminating organism and the preservative used [19], it seems necessary to be more accurate. In some cases the temperature should correspond to special requirements for storage and use. The desirability to perform the test at several temperatures should normally not take precedence over other factors, so as to avoid an increase in the number of tests [51].

The duration of the test depends on its intended aim. Determination of the microbial reduction after one single contact time [75] is useful for the direct comparison of the microbiocidal activity of various preservatives. Apart from the case where there is a rapid and almost complete destruction of the germs, this technique, however, prevents evaluation of the safety of the system in practical use. It is therefore necessary to test the contaminated preparations at several contact times chosen with regard to what is required of the preparations in practice. This will obviously depend both on the category of the preparation and on the importance of achieving the effect rapidly or slowly. Various authors, depending on whether they give more consideration to one or the other of these two aspects, may suggest different testing procedures: (1) short contact times for ophthalmic preparations combined with complete [55] or 99.9% [19] destruction of the test organisms, which is a severe requirement; (2) short contact times for biological multidose injectables [75], and occasionally for cosmetic lotions combined with the stringent requirement for complete destruction of the test organisms [39]; or (3) short contact times for routine batch testing of manufacturer's products, from a practical viewpoint [41].

A technique involving long contact times and the moderate requirement of 99.9% bacterial destruction within 14 days for the ophthalmic preparations does not seem to be very stringent, whereas this requirement seems more suitable for the investigation of topical and oral products, where it is important to follow the germs surviving where there is slow destruction or perhaps even new proliferation, especially in emulsions [85]. For a basic test method, differentiated systems [42,62] must be taken into consideration in such a way that two fundamental and sometimes incompatible questions may be answered by a satisfactory compromise:

> How long should it take in practice to obtain killing of the germs?
> What rate of decrease or stabilization of the number of organisms can really be obtained with the usual preservatives?

There are great differences in the required degrees of efficacy between injectables, ophthalmics, and the high-risk local preparations withdrawn from a multidose container for several ambulatory treatments on the one hand, and oral preparations on the other hand. For the first-mentioned groups it is desirable to achieve killing within a contact time of a few minutes, while for oral preparations several weeks may suffice. For the sake of uniformity and simplicity, we choose for all categories of preparations the same contact times and test intervals, supplemented with a prolonged testing time for the category of preparations where only a moderate antimicrobial effect is required. A first count made during the first hour after contamination will give information on whether the microbiocidal effect is very rapid or not, and furthermore it gives information on the amount of initial contamination.

The initial contamination is directly comparable to the initial microbial number obtained by counting the microorganisms in a blank of buffered sodium chloride-peptone solution contaminated concurrently with the preparation. Contact times of 1 day and 1, 2, 3, 4, and 6 weeks are long enough to permit, on the one hand, the differentiation in effect of various preparations, and on the other hand, to give a certain assurance that surviving microorganisms do not grow again.

F. Detection of Germs

Except for a few papers using direct inoculation of samples in a liquid medium — methods which give only qualitative results by the presence or absence of growth [55,68,84] — most authors use the plate count [5,6,23,42,59,61,62,69,75,76] or, with some advantages, the membrane filtration technique [35,46]. In the test described here, preference is given to quantitative methods where incorporation in or spreading on agar plates, depending on the strain, is used, or to membrane filtration, according to laboratory experience or to the kind of preparation and its properties. Decisive for the degree of dilution before counting are, on the one hand, the desirability of a low colony count (ideally 20-100 per plate, according to the nature of microorganisms, with variations in practice between 10 and 300 on 50-mm membranes and 90-mm agar plates), and, on the other hand, the assurance of elimination of a residual antimicrobial effect. This last condition determines the threshold for germ detection which must be taken into account for the required germ reduction.

The addition of more or less specific inactivating agents can compensate for insufficient dilution effects and thus eliminate residual action of the preservative. After much experimental work the problems of inactivation of antimicrobial agents has recently been treated in a comprehensive study for cosmetics [80], a great deal of which is also of value for pharmaceuticals, and in a study on many different antimicrobial substances [67].

G. Results and Interpretation

The ratio of the number of surviving germs in the preparation after a certain contact time and the initial number added gives the reduction factor expressed, for instance, in percent or in decimal logarithms. This ratio can also be derived from a killing curve in a semilogarithmic system. Interpretation is obtained by comparing the results and requirements for the preparation tested in the form of the combination of a minimal reduction factor and the maximum time to achieve it. The requirement of total reduction, corresponding to reduction factors of 4, 5, or 6 log [13,76], has already been rejected. A reduction of 2 or 3 log phases has the advantage of being located in the linear part of the reduction curve. This does in fact often show a tail [19] and sometimes a shoulder [20]. Experience shows that this rate of decrease is valid at least for nonsporulating bacteria. Extension of this to many groups of microorganisms in different types of preparations within a week [85] seems to be based on particularly favorable cases. The hazards of such a generalization could lead to a search for a very efficacious preservative even when this is not absolutely necessary or even desirable. This might be the situation, for instance, when other factors, like aluminum tubes, inhibit the microorganisms in creams or ointments [3,79]. Addition of preservatives is sometimes unsuitable because of the risk of allergy [4,25,83]. In such cases one must often be satisfied when microorganisms like yeasts, and especially mold spores, do not proliferate in the preparation.

Therefore the interpretation of Table 1 must be considered only as a guide for evaluating antimicrobial effectiveness. This implies, however, only one aspect of the microbiological quality of the preparations, beginning with the initial purity and ending with a satisfactory behavior during use. Concerning this last-mentioned point, correct handling of the product during use is of great importance [79], because improper handling by the consumer would in fact constitute a too difficult challenge to the preservation effect.

IV. METHOD OF THE BASIC TEST

A. Strains and Inoculum

The following strains are maintained in a way which does not modify their identity and properties:

    A. *Staphylococcus aureus* ATCC 6538
    B. *Escherichia coli* ATCC 8739

C. *Pseudomonas aeruginosa* ATCC 9027
D. *Candida albicans* ATCC 10231
E. *Aspergillus niger* ATCC 16404

Immediately prior to the test, taking into account the necessary delay, strains A, B, and C are first cultured for 24 hr at 30-35°C in casein soybean digest broth (medium 1), and then 24 hr on a casein soybean digest agar slant (medium 2). Strain D is cultured for 3 days on a slant of Sabouraud dextrose agar slant (medium 3) at 30-35°C, and strain E is cultured on the same medium for 5 days at 20-25°C or until suitable sporulation is obtained.

To harvest the slant cultures, use 5 ml of saline with peptone (solution 4) supplemented with 0.1% polysorbate 80 for strain E. Shake the "mother suspensions" in order to dissociate the germ agglomerates; then dilute them using the same solution 4 so that an inoculum of 0.2 ml added to 20 ml or g of preparation gives a germ count between 1 and 5×10^5 per milliliter or gram.

B. Test with Liquid Preparations

Liquid preparations are injectable or ophthalmic solutions, drops, lotions, syrups, suspensions, and so on. Using the inoculum described above, the test is performed with each strain in 20 ml of the preparation, and simultaneously in 20 ml of buffered sodium chloride-peptone solution (solution 5). The latter is used to determine the initial count at time zero. Shake to ensure even distribution of the microorganisms. Maintain the test preparations at 20-25°C protected from light, except if other indications are given for the preparation concerned.

At time zero, that is to say, immediately following the inoculation and then after 1 day and 1, 2, 3, 4, and 6 weeks, the last time depending on the category of the preparation, remove 1-ml samples and dilute them suitably in solution 5 in order to perform a germ count, either with the plate count technique or with the membrane filtration procedure if this is considered preferable, also in regard to elimination of a residual antimicrobial effect.

C. Test with Semisolid Preparations

Semisolid preparations are gels, creams, ointments, and so on. In the same way as for liquid preparations, inoculate 20 g of the preparation with 0.2 ml of the inoculum, and simultaneously 20 ml of solution 5 to determine the initial count at time zero. A suitable homogenization of the inoculated preparation is obtained with an adequate mechanical device. Maintain test preparations at 20-25°C protected from light, except if other indications are given for the product concerned.

At time zero and after 1 day and 1, 2, 3, 4, and 6 weeks, the last time depending on the category of preparation, remove 1-g samples, dilute them in solution 5, possibly supplemented with 0.1-1.0% polysorbate 20 or 80, homogenize by means of an appropriate device, and make a plate count.

D. Count of Microorganisms

For the plate count of strains A and B, transfer 1 ml of the suitable dilution in each of two Petri dishes; then pour 15 ml of medium 2, melted and cooled to

45°C, and mix. For strain C add 0.1 ml of the suitable dilution in each of two plates of medium 2 and spread properly. For strains D and E put 1 ml in each of two Petri dishes; then pour 15 ml of medium 3, melted and cooled to 45°C, and mix.

In the case of the membrane filter technique, a quantity of a suitable dilution of the inoculated preparation is filtered through a membrane of 0.45-µm nominal porosity in a device connected to a vacuum pump. The membrane is rinsed to eliminate the effect of antimicrobial substances, with several portions of 100 ml of solution 5. The membranes are put on agar plates of medium 2 for strains A, B, and C, and on agar plates of medium 3 for strains D and E. For strains A, B, C, and D, incubate 2-5 days at 30-35°C, and for strain E incubate 2-5 days at 20-25°C. Determine the number of colonies at the most favorable state and calculate the viable germ count per gram or milliliter of preparation.

The elimination of the antimicrobial effect is often achieved by the dilutions assigned for the test. This is especially true for alcohols, benzoic and sorbic acids, and, to a lesser extent, phenolic compounds and esters of p-hydroxybenzoic acid. In other cases, suitable inactivating agents must be added or, if this is not possible, the membrane filtration must be used. Recommended inactivating agents are, for instance, the combination of polysorbate 80 and lecithin for quaternary ammonium compounds and chlorhexidine, polysorbate 20 or 80 for phenols and esters of p-hydroxybenzoic acid, and sodium thioglycollate for mercurials.

E. Interpretation

The interpretation of the test results is based upon the ratio of the number of organisms in the preparation after various contact times and the initial count in the control tube with solution 5. Indications for the evaluation of preparations from various categories are summarized in Table 1, which gives the reduction to be obtained within a certain contact time and to be maintained at least until the end of the test period.

In any case one must ensure through the application of good manufacturing procedures that the preparation of categories 2 and 3, with or without preservative, comply with the norms for microbiological purity, and that this last during the whole product's validity.

F. Culture Media and Solutions

1. Casein Soybean Digest Broth

Pancreatic digest of casein	17.0 g
Papaic digest of soybean	3.0 g
Sodium chloride	5.0 g
Dipotassium phosphate	2.5 g
Dextrose	2.5 g
Water (distilled or demineralized)	1000 ml
Autoclave at 120°C for 15 min	
Final pH 7.3 ± 0.2	

TABLE 1 Indications for the Interpretation of the Antimicrobial Preservative Test in Regard to the Category of the eutical Pharmaceutical Preparations

Category and type of preparations[a]	Test organisms	Inoculum per ml or g	Evolution of the microbial count after contact times of					
			1 day	1 week	2 weeks	3 weeks	4 weeks	6 weeks
1a, 1b: Injectable and ophthalmic preparations and preparations to be applied in body cavities normally free from microorganisms or on severe burns and ulcerations (all in multiple-dose containers)	A,B,C	$1\text{-}5 \times 10^5$	[b]	3 log reduction	c	c	No increase	
	D,E	$1\text{-}5 \times 10^5$	c	c	No increase[d]	c	No increase	
2: Preparations for topical applications, e.g., on skin lesions, in the nose, throat, ear, etc.	A,B,C	$1\text{-}5 \times 10^5$	c	c	3 log reduction	c	No increase	
	D,E	$1\text{-}5 \times 10^5$	c	c	No increase[d]	c	No increase	
3: Other preparations, mostly for oral applications	A,B,C	$1\text{-}5 \times 10^5$	c	c	c	2 log reduction	c	No increase
	D,E	$1\text{-}5 \times 10^5$	c	c	c	No increase[d]	c	No increase

[a]Source: Ref. 31.
[b]A marked reduction of, for instance, 2 log in 24 hr is desirable and should be striven for whenever possible.
[c]Microbial counts are also valuable between the main recommended time intervals to follow the evolution of the contamination.
[d]The absence of proliferation during the whole testing period is a minimal requirement, in particular when the preparation, owing to its composition, the type of package, and dispensation, does not necessarily need the addition of a preservative. This is the case, for instance, for creams and ointments in aluminum tubes, where they are better protected than in jars. In general, however, a certain reduction of at least 1 log is desirable at the indicated time.

2. Casein Soybean Digest Agar

Pancreatic digest of casein	15.0 g
Papaic digest of soybean	5.0 g
Sodium chloride	5.0 g
Agar	15.0 g
Water (distilled or demineralized)	1000 ml
Autoclave at 120°C for 15 min	
Final pH 7.3 ± 0.2	

3. Sabouraud Dextrose Agar

Peptones (meat and casein)	10.0 g
Dextrose	40.0 g
Agar	15.0 g
Water (distilled or demineralized)	1000 ml
Autoclave at 120°C for 15 min	
Final pH 5.6 ± 0.2	
(Prolonged heating affects the solidification of the agar)	

4. Physiological Saline with Peptone

Sodium chloride	8.5 g
Peptone	1.0 g
Water (distilled or demineralized)	1000 ml
Optional addition of 0.1% polysorbate 80	
Autoclave at 120°C for 15 min	

5. Buffered Sodium Chloride Solution with Peptone (pH 7)

Monopotassium phosphate	3.56 g
Disodium phosphate · $2H_2O$	7.23 g
Sodium chloride	4.3 g
Peptone (meat or casein)	1.0 g
Water (distilled for demineralized)	1000 ml
Optional addition of 0.1 or 1.0% polysorbate 20 or 80	
Autoclave at 120°C for 15 min	

REFERENCES

1. N. J. van Abbé, Bacteriostasis in solutions for injections, *Pharm. J.*, *159*:146 (1947).
2. K. Anderson and D. Crompton, A test for the bactericidal activity of eyedrops, *Lancet*, *2*:968 (1967).
3. H. Asche, M. Gay, A. Kabay, H. Lehner and S. Urban, Einfluss des Behälters auf die Anfälligkeit von Dermatika Grundlagen gegenüber mikrobieller Kontamination, *Pharm. Ind.*, *40*:1212 (1978).
4. H. J. Bandmann, Die Kontaktallergie durch Konservierungsmittel (A. P. V. Symposium, Berlin, über Konservierung flüssiger und halbfester Arzneizubereitungen), *Dtsch. Apoth. Ztg.*, *116*:483 (1976).
5. R. Barberot, R. Pinzon, and A. Mirimanoff, Contribution à l'étude d'un test de contrôle microbiologique de la conservation de certaines formes galéniques (sirops, gouttes, onguents), 1ère communication, *Pharm. Acta Helv.*, *39*:556 (1964).
6. R. Barberot, R. Pinzon, and A. Mirimanoff, Contribution à l'étude d'un test de contrôle microbiologique de la conservation de certaine formes galéniques (sirops, gouttes, onguents), 2e communication, *Pharm. Acta Helv.*, *40*:229 (1965).
7. M. Barr and L. F. Tice, The preservation of aqueous preparations containing nonionic surfactants. I. Growth of microorganisms in solutions and dispersions of nonionic surfactants, *J. Am. Pharm. Assoc. Sci. Ed.*, *46*:442 (1957).
8. M. Barr and L. F. Tice, The preservation of aqueous preparations containing nonionic surfactants. II. Preservative studies in solutions and products containing nonionic surfactants, *J. Am. Pharm. Assoc. Sci. Ed.*, *46*:445 (1957).
9. M. Barr and L. F. Tice, A study of the inhibitory concentrations of glycerin-sorbitol and propylene glycol-sorbitol combinations on the growth of microorganisms, *J. Am. Pharm. Assoc. Sci. Ed.*, *46*:217 (1957).
10. M. Barr and L. F. Tice, A study of the inhibitory concentrations of various sugars and polyols on the growth of microorganisms, *J. Am. Pharm. Assoc. Sci. Ed.*, *46*:219 (1957).
11. H. S. Bean and S. M. Herman-Ackah, Influence of oil:water ratio on the activity of some bactericides against *Escherichia coli* in liquid paraffin and water dispersions, *J. Pharm. Pharmacol.*, *16*:58T (1964).
12. H. S. Bean and R. C. Farrel, The persistance of *Pseudomonas aeruginosa* in aqueous solutions of phenols, *J. Pharm. Pharmacol.*, *19*:183S (1967).
13. *British Pharmacopoeia 1973*, Injections, Her Majesty's Stationary Office, London, p. 242.
14. M. R. W. Brown and R. M. E. Richards, Effect of polysorbate 80 on the resistance of *Pseudomonas aeruginosa* to chemical inactivation, *J. Pharm. Pharmacol.*, *16*:51T (1964).
15. X. Bühlmann, M. Gay, H. U. Gubler, H. Hess, A. Kabay, F. Knüsel, W. Sackmann, I. Schiller, and S. Urban, Mikrobiolgische Reinheit von Arzneimitteln. Richtlinien und Methoden für eine Pharmakopöe, *Pharm. Acta Helv.*, *46*:321 (1971). [Microbiological quality of pharmaceutical preparations, *Am. J. Pharm.*, *144*:165 (1972).]
16. K. Bullock and W. G. Keepe, Bacterial survival in systems of low moisture content, Part III. Bacteria in fixed oils and fats, *J. Pharm. Pharmacol.*, *3*:700 (1951).

17. H. J. Burkardt and K. H. Müller, Zur Beurteilung der mikroMiziden Eigenschaften lipophiler Salben, *Pharm. Acta Helv.*, *51*:30 (1976).
18. B. Croshaw, M. J. Groves, and B. Lessel, Some properties of "Bronopol," a new antimicrobial agent active against *Pseudomonas aeruginosa*, *J. Pharm. Pharmacol.*, *16*:127T (1964).
19. D. J. G. Davies, N. E. Richardson, and Y. Anthony, Design of a standard protocol for the challenge testing of antimicrobial preservative solutions, *J. Pharm. Pharmacol.*, *28*:49P (1976).
20. D. J. G. Davies, Agents as preservatives in eye-drops and contact lens solutions, *J. Appl. Bacteriol.*, *44*:Sxix (1978).
21. D. J. G. Davies, Conclusions of Bath Symposium on Antimicrobial Efficiency of Preservatives in Sterile Pharmaceuticals, unpublished (1977).
22. P. P. DeLuca and H. B. Kostenbauder, Interaction of preservatives with macromolecules. IV. Binding of quaternary ammonium compounds by nonionic agents, *J. Am. Pharm. Assoc. Sci. Ed.*, *49*:430 (1960).
23. A. Desvignes, J. Bernard, F. Sebastien, and M. Frottier, Etude de quelques agents conservateurs. Contrôle de leur efficacité antimicrobienne dans une préparation émulsionnée, *Ann. Pharm. Fr.*, *32*:37 (1974).
24. J. Dony and B. L'Hoest, Contrôle de stérilité des préparations huileuses, *J. Pharm. Belg.*, *16*:66 (1961).
25. L. P. Durocher, Réaction allergique aux médicaments topiques, *Can. Med. Assoc. J.*, *118*:162 (1978).
26. P. C. Eisman, D. Jaconia, and R. L. Mayer, The preservation of parenteral vegetable oil by chemical agents, *J. Am. Pharm. Assoc. Sci. Ed.*, *42*:659 (1953).
27. P. C. Eisman, E. Ebersold, J. Weerts, and L. Lachman, Microbiological turbidimetric assay method for preservative content, *J. Pharm. Sci.*, *52*:183 (1963).
28. K. Eriksson, Loss of organomercurial preservatives from medicaments in different kinds of containers, *Acta Pharm. Suec.*, *4*:261 (1967).
29. C. K. von Fenyes, Alcohols as preservatives and germicides, *Am. Perfum. Cosmet.*, *85*:91 (1970).
30. FIP, Pureté microbiologique des formes pharmaceutiques non obligatoirement stériles (Rapport commun du Comité des laboratoires et services officiels de contrôle des médicaments et de la Section des pharmaciens de l'industrie), *J. Mond. Pharm.*, *15*:88 (1972). [*Zbl. Pharm.*, *111*:675 (1972); *Pharm. Ind.*, *34*:932 (1972); *WHO/Pharm.*, *74*:477.]
31. FIP, Méthodes d'examen (2e Rapport commun), *Pharm. Acta Helv.*, *50*:285 (1975). [ger.: *Pharm. Acta Helv.*, *51*:41 (1976); engl.: *Pharm. Acta Helv.*, *51*:33 (1976).]
32. C. D. Fox and R. F. Shangraw, Water/oil emulsions prepared by low pressure capillary homogenization. I. Effects of emulgator and composition variables in mannide mono-oleate stabilized systems, *J. Pharm. Sci.*, *55*:318 (1966).
33. M. Gay, A. Kabay, S. Urban, and X. Bühlmann, Agents conservateurs antimicrobiens — test de conservation antimicrobien, projet pour Pharmacopoeia Helvetica (1973).
34. J. R. Haines and M. Alexander, Microbial degradation of polyethylene glycols, *Appl. Microbiol.*, *29*:621 (1975).
35. H. Hess and P. Speiser, Comparative efficacy of bactericidal compounds in buffer solutions, Part I, *J. Pharm. Pharmacol.*, *11*:650 (1959).

36. H. Hess and P. Speiser, Comparative efficacy of bactericidal compounds in buffer solutions, Part II, *J. Pharm. Pharmacol.*, *11*:694 (1959).
37. P. G. Hugbo, Additive and synergistic actions of equipotent admixture of some antimicrobial agents, *Pharm. Acta Helv.*, *51*:284 (1976).
38. W. B. Hugo and J. H. S. Forster, Growth of *Pseudomonas aeruginosa* in solutions of p-hydroxybenzoic acid, *J. Pharm. Pharmacol.*, *16*:209 (1964).
39. G. Jacobs, S. M. Henry, and V. F. Cotty, The influence of pH, emulsifier and accelerated aging upon preservative requirements of O/W emulsions, *J. Soc. Cosmet. Chem.*, *26*:105 (1975).
40. G. Kedvessy and K. Bognár, Anwendung von Phenylquecksilberborat in parenteralen Lösungen, *Pharm. Zentralhalle*, *97*:574 (1958).
41. D. S. Kellog, Preservative testing as applied to quality control analysis, *Bull. Parenter. Drug Assoc.*, *26*:216 (1972).
42. D. S. Kenney, W. E. Grundy, and R. H. Otto, Spoilage and preservative tests as applied to pharmaceuticals, *Bull. Parenter. Drug. Assoc.*, *18*:10 (1964).
43. A. K. Kristensen and J. R. Mortensen, Untersuchungen über die Absorption von Phenylquecksilbernitrat in Verschlussmitteln für Injektionsfläschchen, *Dansk Tidsskr. Farm.*, *42*:132 (1968).
44. L. Lachman, S. Weinstein, G. Hopkins, S. Slack, P. Eisman, and J. Cooper, Stability of antibacterial preservatives in parenteral solutions. I. Factors influencing the loss of antimicrobial agents from solutions in rubber-stopped containers, *J. Pharm. Sci.*, *51*:224 (1962).
45. L. Lachman, The instability of antimicrobial preservatives, *Bull. Parenter. Drug Assoc.*, *22*:127 (1968).
46. S. Lambin, F. Sebastien, and J. Bernard, Détermination de l'activité bactéricide des substances antiseptiques par la technique de filtration sur membrane. Intérêt et limites de cette technique, *Ann. Pharm. Fr.*, *22*:641 (1964).
47. M. Leitz, Critique of USP microbiological test, *Bull. Parenter. Drug Assoc.*, *26*:212 (1972).
48. J. Lingnau, Die Möglichkeiten zur Konservierung pharmazeutischer Zubereitungen, *Pharm. Ind.*, *32*:266 (1970).
49. J. Meyer-Rohn, Gedanken und Untersuchungen zum Problem der chemischen Konservierung von Salben und Cremes in Medizin und Kosmetik, *Fette Seifen Anstrichm.*, *69*:536 (1967).
50. G. M. Miyawaki, N. K. Patel, and H. B. Kostenbauder, Interaction of preservatives with macromolecules. III. Parahydroxybenzoic acid esters in the presence of some hydrophilic polymers, *J. Am. Pharm. Assoc. Sci. Ed.*, *48*:315 (1959).
51. K. E. Moore, Evaluating preservative efficacy by challenge testing during the development stage of pharmaceutical products, *J. Appl. Bacteriol.*, *44*:Sxliii (1978).
52. K. E. Moore and J. E. Taylor, Microbiological standards for nasal solutions, *J. Appl. Bacteriol.*, *41*:379 (1976).
53. M. Muftic, U. Redmann, and H. Richter, Untersuchungsverfahren für Konservierungsmittel in pharmazeutischen Salben, Cremes und Lotionen, *Pharmazie.*, *29*:987 (1967).
54. G. E. Myers, Survival of pathogenic bacteria in pharmaceutical oils, *Can. J. Pharm. Sci.*, *4*:75 (1969).

55. D. A. Norton, D. J. G. Davies, N. E. Richardson, B. J. Meakin, and A. Keall, The antimicrobial efficiencies of contact lens solutions, J. Pharm. Pharmacol., 26:841 (1974).
56. O. H. Paetzold, Untersuchungen über die bakteriellen Verunreinigungen von Externa, Therapiewoche, 1971:2840 (1971).
57. O. H. Paetzold, Experimentelle Untersuchungen über das Verhalten von Bakterien und Candida in Externagrundlagen, Parfuem. Kosmet., 53:244 (1972).
58. M. S. Parker, The preservation of cosmetic and pharmaceutical creams, J. Appl. Bacteriol., 44:Sxxix (1978).
59. H. D. Piersma, Microbiological preservative test for parenteral and ophthalmic solutions (Proposal by PMA Biol. Section Committee, April 5, 1967, to USP), unpublished.
60. F. D. Pisano and H. B. Kostenbauder, Interaction of preservatives with macromolecules. II. Correlation of binding data with required preservative concentration of p-hydroxybenzoates in the presence of Tween 80, J. Am. Pharm. Assoc. Sci. Ed., 48:310 (1959).
61. P. S. Prickett, H. L. Murray, and N. H. Mercer, Potentiation of preservatives (parabens) in pharmaceutical formulations by low concentrations of propylene glycol, J. Pharm. Sci., 50:316 (1961).
62. E. J. Rdzok, W. E. Grundy, F. J. Kirchmeyer, and J. C. Sylvester, Determining the efficacy of preservatives in pharmaceutical products, J. Am. Pharm. Assoc. Sci. Ed., 44:613 (1955).
63. R. M. E. Richards, Inactivation of resistant *Pseudomonas aeruginosa* by antibacterial combinations, J. Pharm. Pharmacol., 23:136S (1971).
64. R. M. E. Richards and R. J. McBride, Cross-resistance in *Pseudomonas aeruginosa* resistant to phenylethanol, J. Pharm. Sci., 61:1075 (1972).
65. R. M. E. Richards and R. J. McBride, Effect of 3-phenylpropan-1-ol, 2-phenylethanol and benzylalcohol on *Pseudomonas aeruginosa*, J. Pharm. Sci., 62:585 (1973).
66. A. Royce and G. Sykes, Losses of bacteriostats from injections in rubber-closed containers, J. Pharm. Pharmacol., 9:814 (1957).
67. A. D. Russell, I. Ahonkai, and D. T. Rogers, Microbiological applications of the inactivation of antibiotics and other antimicrobial agents, J. Appl. Bacteriol., 46:207 (1979).
68. J. Schmid, Herstellung steriler ophthalmologischer Lösungen, Pharmazie, 12:748 (1957).
69. G. Schuster and H. Modde, Untersuchung über die Wirksamkeit von Konservierungsmitteln in anionaktiven Tensiden. I. Fette Seifen Anstrichm., 70:169 (1968).
70. F. S. Skinner, Konservierungsmittelverluste durch Behältereinfluss, Informationsdienst APV, 18:256 (1972).
71. M. T. Suggs, Summary and conclusions of the Symposium in "Preservatives in vaccines" (San Francisco, 1973) Session II. Interaction of products and preservatives, J. Biol. Stand., 2:79 (1974).
72. G. Sykes, The basis for "sufficient of a suitable bacteriostatic" in injections, J. Pharm. Pharmacol., 10:40T (1958).
73. W. J. Tillman and R. Kuramoto, A study of the interaction between methylcellulose and preservatives, J. Am. Pharm. Assoc. Sci. Ed., 46:211 (1957).
74. U.S. Pharmacopoeia XVIII (1970), Antimicrobial Agents — Effectiveness, p. 845-846. U.S.P. Convention, Inc., Rockville, Maryland.

75. R. W. Van der Wyk and A. E. Granston, A bacteriological study of ophthalmic ointments, *J. Am. Pharm. Assoc. Sci. Ed.*, *47*:193 (1958).
76. K. H. Wallhäusser, Prüfung auf ausreichende Konservierung bei Arzneimitteln, *Aerztl. Lab.*, *16*:305 (1970).
77. K. H. Wallhäusser, Antimicrobial preservatives in biologics, *Pharm. Ind.*, *36*:716 (1974).
78. K. H. Wallhäusser, Preservation and sterility of ophthalmic preparations and devices, in *The Quality Control of Medicines* (P. B. Deasy and R. F. Timoney, eds.), Elsevier, Amsterdam (1976).
79. K. H. Wallhäusser, Die mikrobielle Reinheit von Arzneimitseln in der Hand des Verbrauchers, *Dtsch. Apoth. Ztg.*, *118*:1510 (1978).
80. K. H. Wallhäusser, Die Inaktivierung von Konservierungsmitteln durch unverträgliche Bestandteile bei der Formulierung, durch Enthemmungsmittel beim Belastungstest und durch mikroorganismen, *Parfuem. Kosmet.*, *60*:1 (1979).
81. G. Wewalka, T. Gergely, and W. Koller, Eine neuartige Salbengrundlage — Untersuchungen über deren antimikrobiellen Eigenschaften, *Wien. Klin. Wschr.*, *89*:632 (1977).
82. B. Wickliffe and D. N. Entrekin, Relation of pH to preservative effectiveness. II. Neutral and basic media, *J. Pharm. Sci.*, *53*:769 (1964).
83. K. Winkler, Die Kortikoid-Externa in der täglichen Praxis, *Forsch. Praxis, Fortbild.* *18*:36 (1967).
84. R. Woodford and E. Adams, The effect of ethanol and propylene glycol, and a mixture of potassium sorbate with either on *Pseudomonas aeruginosa* contamination of an oil-in-water cream, *Am. Cosmet. Perfum.*, *87*:53 (1972).
85. J. I. Yablonski, Fundamental concepts of preservation, *Bull. Parenter. Drug Assoc.*, *26*:220 (1972).
86. M. Yanagi and G. Onishi, Assimilation of selected cosmetic ingredients by microorganisms, *J. Soc. Cosmet. Chem.*, *22*:851 (1971).
87. R. T. Yousef, The combined bactericidal activity of chlorhexidine diacetate and benzalkonium chloride against *Pseudomonas aeruginosa*, *Acta Pharm. Suec.*, *3*:141 (1966).
88. R. T. Yousef, M. A. El Nakeeb, and S. Salama, Effect of some pharmaceutical materials on the bactericidal activities of preservatives, *Can. J. Pharm. Sci.*, *8*:54 (1973).

24
A PRESERVATIVE EVALUATION PROGRAM FOR DERMATOLOGICAL AND COSMETIC PREPARATIONS

O. J. LORENZETTI *Alcon Laboratories, Inc., Fort Worth, Texas*

I. Introduction 441
II. Development of Test Procedures for Determining Preservative Effectiveness 443
III. Selection and Evaluation of Prototype Formulations 459
IV. Determination of Preservative "Capacity" 459
V. Preservative Evaluation of First-Manufactured Batches 460
VI. Establishment of a Preservative Stability Test System 460
VII. "In-Use" Testing of Preservative Effectiveness 461
VIII. Summary and Conclusions 461
References 462

I. INTRODUCTION

Antimicrobial preservatives are included in topical formulations to protect the health of the consumer, as well as maintain the potency and stability of product formulations. The preservative system acts as a safeguard mechanism to either destroy or inhibit the growth of microorganisms which may be inadvertently introduced into the product subsequent to manufacturing. The incorporation of preservatives into product formulations is never used as a substitute for good manufacturing practices.

What constitutes an adequate preservative system for topical formulations? This question may best be answered by considering the desired properties or attributes of an "ideal" or "utopic" preservative system. The ideal preservative system described by Gladhard et al. (1955), Gottfried (1962), and Grundy (1968) must be (1) effective at relatively low concentrations against a broad spectrum or wide variety of microorganisms which could cause disease or decomposition of the product; (2) soluble in the product formulation at the desired or required concentration; (3) nontoxic and nonsensitizating at in-use concentrations; (4) compatible with the individual ingredients of the formulation as well as packaging components; (5) free of objectionable odors and colors which affect product elegance; (6) capable of maintaining long-term stability over a wide range of pH values and temperatures; and (7) relatively inexpensive, based on the total cost of the product. To date, no preservative system has been developed which totally approaches the ideal. Each preservative system has its attributes and limitations. It is the responsibility of the formulating chemist and microbiologist to choose the preservative system which must be based on an in-depth knowledge of the formulation coupled by an acute awareness of the antimicrobial capabilities and limitations of the preservative or preservative system. The selection of prototype formulations based on sound scientific information results in a significant reduction of the experimental time required for the development of a new product formulation.

Following selection of prototype formulations, the microbiologist must address the "ways and means" of determining preservative efficacy. From the microbiologist's point of view, the overall objective of preservative testing is the determination of the degree of consumer and product protection afforded by the product preservative system. To attain this objective, a *preservative evaluation program* must be designed. Such a program must represent a comprehensive evaluation of all significant factors which affect the preservative effectiveness of product formulations. It not only includes the evaluation of prototype formulations, but is expanded to include evaluation of all facets of product development, manufacturing, and final product marketing.

It is the purpose of this report to present a review of a preservative evaluation program and recommend a select test program including the following:

1. Development and implementation of adequate test methodologies/procedures for evaluating the effectiveness of preservative systems
2. Evaluation of prototype formulations for selection of specific product candidates
3. Evaluation of the level of activity (capacity) of the product candidate preservative system
4. Establishment of a stability program to determine the long-term effectiveness of the product preservative system
5. Evaluation of the first-manufactured batch to ascertain the effects of the manufacturing process on the effectiveness of the product preservative system
6. Establishment of a testing program to evaluate the "in-use" effectiveness of the product preservative system.

A Preservative Evaluation Program

II. DEVELOPMENT OF TEST PROCEDURES FOR DETERMINING PRESERVATIVE EFFECTIVENESS

The test methodologies employed and criteria used in assessing activity are perhaps the most critical aspects of developing a preservative testing program. Numerous chemical (McCarthy, 1974; Wilson, 1970) and biological (Greenberg and Naubert, 1970; Grundy, 1968; Manowitz, 1969; Parker, 1972; Rosen and Berke, 1973) methods have been proposed to evaluate the efficacy of preservative systems. At the present time the "organism challenge" approach is the most widely used and accepted scheme. Basically, the procedure involves the inoculation of product formulations with known types and levels of microorganisms followed by a specific sampling regimen. Activity is assessed on the ability of the preservative system to kill or inhibit the propagation of microorganisms interjected into the formulation. The most critical parameters of this approach are the following: (1) choice of challenge microorganisms; (2) concentration or level of microorganisms in the inoculum; (3) sampling regimens, and (4) data interpretation or standards of activity.

During the past 10 years, a significant amount of energy from all facets of the scientific community has been expended in an effort to design an "optimal" preservative challenge system. In Tables 1-7 we review the various preservative challenge systems which have been proposed as guidelines by compendiary, governmental, and industrial organizations, as well as those which are operational at various cosmetic and pharmaceutical research facilities.

In 1970 the *U.S. Pharmacopoeia (USP) XVIII* established the first official guidelines for determining the preservative effectiveness of pharmaceutical products (Table 1). The major objections to these guidelines as applied to topical formulations have included the following: (1) use of standardized ("pedigree" or "purebred") challenge organisms rather than manufacturing, product, or recent clinical isolates; (2) the length of the test procedure and exclusion of a rechallenge procedure; and (3) the lack of more strenuous standards of effectiveness [cidal (kill) versus static (inhibitory) activity]

TABLE 1 USP XVIII Antimicrobial Preservative Effectiveness Test Procedures

Challenge microorganisms	*Staphylococcus aureus* ATCC 6538 *Escherichia coli* ATCC 8739 *Pseudomonas aeruginosa* ATCC 9027 *Candida albicans* ATCC 10231 *Aspergillus niger* ATCC 16404
Organism inoculum level	$1.25 \times 10^5 - 5 \times 10^5$ cells/ml or g of product formulation
Test formulation sampling	At least two observations (7 days apart) during the 28-day test period
Effectiveness standards	Vegetative cells — 0.1% survival at two sampling periods, 7 days apart
	Yeast/fungi — No significant increase in counts during the 28-day test period

Source: USP XVIII, 1970.

TABLE 2 *USP XIX* Antimicrobial Preservative Effectiveness Test Procedures

Challenge microorganisms	*Staphylococcus aureus* ATCC 6538 *Escherichia coli* ATCC 8739 *Pseudomonas aeruginosa* ATCC 9027 *Candida albicans* ATCC 10231 *Aspergillus niger* ATCC 16404
Organism inoculum level	1.0×10^5-1.0×10^6 cells/ml or g of product formulation
Test formulation sampling	0, 7, 14, 21, and 28 days subsequent to inoculation
Effectiveness standards	Bacteria — <0.1% survival by the 14th day
	Yeast and molds — at or below initial concentrations during first 14 days
	No increase in organism counts for remainder of 28-day test period

Source: USP XIX, 1975.

TABLE 3 International Guidelines for Evaluation of the Preservative Effectiveness of Topical Pharmaceutical Preparations

Challenge microorganisms	*Staphylococcus aureus* ATCC 6538 *Escherichia coli* ATCC 8739 *Pseudomonas aeruginosa* ATCC 9027 *Candida albicans* ATCC 10231 *Aspergillus niger* ATCC 16404
Organism inoculum level	1.0×10^5-5.0×10^5 cells/ml or g of product formulation
Test formulation sampling	1 hr, 1 day, and 1, 2, 3, 4, and 6 weeks postinoculation
Effectiveness standards	Topicals applied to compromised skin:
	Bacteria — minimum reduction of 99.9% after 2-week contact time
	Fungi — minimum reduction of 90.0% after 2-week contact time
	Other topical formulations:
	Bacteria — minimum reduction of 99% after 3-week contact time
	Fungi — no increase over 6-week test period

Source: Buhlmann, 1972.

TABLE 4 Investigative FDA Procedures for Evaluation of the Preservative Effectiveness of Topical Formulations

Challenge microorganisms	*Staphylococcus aureus* ATCC 6538 *Escherichia coli* ATCC 8739 *Pseudomonas aeruginosa* ATCC 9027 *Pseudomonas putida* *Pseudomonas multivorans* *Klebsiella* sp. *Serratia marcescens* *Candida albicans* ATCC 10231 *Aspergillus niger* ATCC 16404
Organism inoculum level	$0.8\text{-}1.2 \times 10^6$ cells/ml or g of product formulation
	Rechallenge — $1.0\text{-}2.0 \times 10^5$ vegetative cells/ml or g of product formulation
Test formulation sampling	Weekly intervals
Effectiveness standards	Vegetative cells — <0.01% survival by 28 days
	Candida albicans — <1% survival
	Aspergillus niger — <10% survival
	Rechallenge — 0.1% survial in 28 days

Source: Bruch, 1972.

TABLE 5 CTFA Guidelines for the Determination of Adequacy of Preservation of Cosmetic and Toiletry Formulations

Challenge microorganisms	*Staphylococcus aureus* ATCC 6538 *Escherichia coli* (may be substituted by *Enterobacter*, *Klebsiella*, or *Proteus*) *Pseudomonas aeruginosa* ATCC 15442 or ATCC 13388 *Candida albicans* *Aspergillus niger* ATCC 9642 *Penicillium luteum* ATCC 9644 *Bacillus subtilis* (not routinely) Formulation isolates (not routinely)
Organism inoculum level	1.0×10^6 cells/g or ml of product formulation
Test formulation sampling	0, 1-2, 7, 14, and 28 days
	In some instances greater than 28 days per rechallenge
Effectiveness standards	Based on the intended use of the formulation

Source: Halleck et al., 1970.

TABLE 6 Colgate-Palmolive Test Procedures for Evaluating the Preservative Effectiveness of Cosmetic Formulations

Challenge microorganisms	Group I *Escherichia coli* ATCC 10536 *Staphylococcus aureus* *Staphylococcus albus* *Escherichia coli* (product isolates) *Pseudomonas aeruginosa* *Enterobacter aerogenes* *Alcaligenes viscosus* Yeasts (2) Group II *Bacillus subtilis* ATCC 6051 *Bacillus* sp. (product isolates) Group III *Aspergillus niger* (product isolates) *Aspergillus glaucus* (2) *Penicillium* sp. (2)
Organism inoculum level	>1.0 × 10^6 cell of individual groups (I-III)/gm or ml of product formulation
Test formulation sampling	18, 24, and 48 hr and 5, 7, 9, 12, and 14 days (streaking procedure)
Effectiveness standards	100% kill of vegetative organisms within 7 days

Source: Owen, 1969.

(Kellogg, 1972; Leitz, 1972; Yablonski, 1972). Nevertheless, the procedures established by the *USP XVIII* and their subsequent revision in the *USP XIX* (Table 2) represent a scientifically valid approach which must be modified when applied to the testing of topical formulations.

International guidelines (Buhlmann et al., 1972) (Table 3) closely parallel those established by the *USP*. The only significant difference is in the length of testing period (42 versus 28 days).

In an effort to strengthen the *USP* guidelines, the Food and Drug Administration (FDA) (Bruch, 1972) and Cosmetic, Toiletry, and Fragrance Association (CFTA) (Halleck et al., 1970) have proposed and established specific guidelines for the testing of topical and cosmetic formulations (Tables 4 and 5). These procedures include the following modifications:

1. Inclusion of additional challenge microorganisms (the FDA placing an emphasis on organisms of clinical significance, and the CTFA emphasizing the use of organisms isolated from contaminated products and/or manufacturing environments)
2. Incorporation of a rechallenge system (rechallenging the formulation to determine if it can withstand repeated insults or challenges which may occur in an "in-use" situation)
3. Inclusion of more strenuous standards of effectiveness [requiring cidal (kill) activity rather than static (inhibitory) activity].

TABLE 7 Revlon Preservatives Evaluation Test Procedures for Cosmetic Formulations

Challenge microorganisms	*Staphylococcus aureus* *Staphylococcus* sp. *Staphylococcus* sp.	group I
	Escherichia coli *Pseudomonas aeruginosa* *Pseudomonas* sp. *Pseudomonas* sp.	group II
	Aspergillus niger *Aspergillus* sp. Yeasts (two strains)	group III
	Product isolates	group IV
Organism inoculum level	5.0×10^5 cells from each group (I-IV)/ml or g of product formulation	
Test formulation sampling	Weekly intervals for 13 weeks	
	Rechallenge at sixth week — continue test for additional 8-12 weeks	
Effectiveness standards	"Almost self-sterilizing" survival of <0.2% of the inocula	

Source: Lanzet, 1972.

The preservative testing procedures (for topical formulations and cosmetics) currently in use at Colgate-Palmolive (Owen, 1969) and Revlon (Lanzet, 1972) are summarized in Tables 6 and 7. These procedures utilize still another modification of the *USP* approach. The use of mixed challenge organism suspensions allows the microbiologist to determine preservative effectiveness against a broader spectrum of microorganisms without significantly increasing manpower expenditures.

In Table 8 we review a typical test procedure used to determine the preservative effectiveness of cosmetic/dermatological formulations. The procedures parallel those described by the *USP XX*, with three major exceptions: (1) A mixed challenge organism suspension is utilized, (2) a product isolate (*Pseudomonas multivorans*) is used as a challenge organisms, and (3) a 1-day sampling of the formulation is taken.

During the past 3 years we have been engaged in studies designed to evaluate the adequacy of the typical procedure for evaluating the preservative efficacy of topical formulations. These studies attempted to evaluate four basic parameters of an existing preservative test system: (1) the use of mixed versus pure organism challenges, (2) the use of product or manufacturing isolates as challenge microorganisms, (3) the significance of utilizing a rechallenge system, and (4) the standards upon which effectiveness is determined.

There has been a point of controversy relative to the use of mixed organism challenges. In a study conducted at Avon (Yablonski, 1972), 19 of 119 (16%) unacceptably preserved samples were missed using a mixed-culture inoculation. However, in a CTFA-supported study (Rodgers et al., 1973),

TABLE 8 A Typical Preservative Test Procedure Currently Utilized for Evaluating Topical Formulations and Cosmetics

Challenge microorganisms	*Staphylococcus aureus* ATCC 6538
	Escherichia coli ATCC 8739 (mixed culture)
	Pseudomonas aeruginosa ATCC 9027
	Candida albicans ATCC 10231
	Aspergillus niger ATCC 16404
	Pseudomonas mulativorans (*Bacillus* sp.)
Organism inoculum level	1.0×10^5-1.0×10^6 cells/ml or g of product formulation
Test formulation sampling	0, 1, 7, 14, 21, and 28 days subsequent to inoculation
Effectiveness standards	Bacteria − <0.1% survival by the 14th day
	Yeast and molds − at or below initial levels during first 14 days
	No increase for remainder of 28-day test period

no differences between pure and mixed organism challenges were detected. In Tables 9-11, we describe data generated on typical formulations in an effort to determine the significance of utilizing mixed organism challenges. We were unable to detect significant differences in the responses of the three formulations 9398, 9403, and 9407 to mixed and pure organism challenges. Therefore, based on these experimental data, previous experience, and published data, it is our opinion that the use of mixed organism challenges does not adversely affect the validity of the test system. Additionally, the use of a mixed organism challenge approach will allow us to determine preservative efficacy against a larger number and broader spectrum of microorganisms without significantly increasing manpower expenditures. If specific or detailed information is required for a specific formulation, a pure-culture challenge will be utilized.

The second phase of these experimental studies was directed toward evaluating the significance of utilizing "isochallenge" microorganism (i.e., organisms isolated from contaminated products and/or the manufacturing environment) in preservative testing. In Tables 12-16 we present the results of studies designed to compare the responses of five product formulations 9398, 9403, 9411, and 9412 to challenges by strains of "standard" and "isochallenge" microorganisms.* No significant differences were recorded in the

---

*The "standard" organisms were obtained from the American Type Culture Collection (ATCC). The "isochallenge" organisms were *Pseudomonas multivorans* (TM-1); isolated from contaminated PiSec; *Aspergillus niger* (TM-4), *Aspergillus clavatus* (TM-7), and *Penicillium* sp. (TM-3), isolated from a contaminated cream; and *Pseudomonas* sp. (TM-6), isolated from the water supply.

A Preservative Evaluation Program

TABLE 9 Comparative Evaluation of the Responses to "Mixed" and "Pure" Organism Challenges: Formulation 9398

Test microorganism	Organism count (per ml formulation)					
	Initial	Day 1	Day 7	Day 14	Day 21	Day 28
Mixed culture[a]	3.2×10^5	0	0	—	—	—
Escherichia coli	6.2×10^5	0	0	—	—	—
Pseudomonas aeruginosa	6.7×10^5	0	0	—	—	—
Staphylococcus aureus	3.0×10^5	0	0	—	—	—
Candida albicans	1.1×10^5	0	0	—	—	—

[a]The mixed culture contained Escherichia coli, Pseudomonas aeruginosa, Staphylococcus aureus, and Candida albicans.

TABLE 10 Comparative Evaluation of the Responses to "Mixed" and "Pure" Organism Challenges: Formulation 9403

Test microorganism	Organism count (per ml formulation)					
	Initial	Day 1	Day 7	Day 14	Day 21	Day 28
Mixed culture[a]	4.0×10^5	3.4×10^3	0	0	—	—
Escherichia coli	1.0×10^6	7.0×10^3	0	0	—	—
Pseudomonas aeruginosa	7.3×10^5	1.0×10^1	0	0	—	—
Staphylococcus aureus	3.6×10^4	0	0	—	—	—
Candida albicans	1.7×10^4	1.1×10^2	0	0	0	—

[a]The mixed culture contained Escherichia coli, Pseudomonas aeruginosa, Staphylococcus aureus, and Candida albicans.

TABLE 11 Comparative Evaluation of the Responses to "Mixed" and "Pure" Organism Challenges: Formulation 9407

Test microorganism	Organism count (per ml formulation)					
	Initial	Day 1	Day 7	Day 14	Day 21	Day 28
Mixed culture[a]	4.0×10^5	5.8×10^4	1.6×10^3	3.7×10^1	1.5×10^1	0
Escherichia coli	1.0×10^6	2.2×10^5	1.2×10^2	0	—	—
Pseudomonas aeruginosa	7.3×10^5	0	0	—	—	—
Staphylococcus aureus	3.6×10^4	4.9×10^4	0	0	—	—
Candida albicans	1.7×10^4	6.3×10^3	0	0	—	—

[a]The mixed culture contained Escherichia coli, Pseudomonas aeruginosa, Staphylococcus aureus, and Candida albicans.

TABLE 12 Comparative Responses to "Isochallenge" and "Standard" Microorganisms: Formulation 9398

Test microorganism	Organism count (per ml formulation)					
	Initial	Day 1	Day 7	Day 14	Day 21	Day 28
Pseudomonas aeruginosa (ATCC 9027)	6.7×10^5	0	0	—	—	—
Pseudomonas multivorans (TM-1)	7.3×10^5	0	0	—	—	—
Aspergillus niger (ATCC 16404)	5.1×10^4	2.2×10^3	0	—	—	—
Aspergillus niger (TM-4)	1.1×10^5	1.1×10^2	0	—	—	—
Penicillium sp. (TM-3)	2.4×10^5	4.4×10^4	0	—	—	—

TABLE 13 Comparative Responses to "Isochallenge" and "Standard" Microorganisms: Formulation 9403

Test microorganism	Organism count (per ml formulation)					
	Initial	Day 1	Day 7	Day 14	Day 21	Day 28
Pseudomonas aeruginosa (ATCC 9027)	7.3×10^5	1×10^1	0	—	—	—
Pseudomonas multivorans (TM-1)	5.1×10^5	0	0	—	—	—
Pseudomonas sp. (TM-6)	9.4×10^5	0	0	—	—	—
Aspergillus niger (ATCC 16404)	2.4×10^4	1.0×10^1	0	—	—	—
Aspergillus niger (TM-4)	3.3×10^5	2.2×10^2	0	—	—	—

TABLE 14 Comparative Responses to "Isochallenge" and "Standard" Microorganisms: Formulation 9407

Test microorganism	Organism count (per ml formulation)					
	Initial	Day 1	Day 7	Day 14	Day 21	Day 28
Pseudomonas aeruginosa (ATCC 9027)	7.3×10^5	0	0	—	—	—
Pseudomonas multivorans (TM-1)	5.1×10^5	2.3×10^5	4.4×10^4	0	—	—
Pseudomonas sp. (TM-6)	9.4×10^5	6.6×10^4	0	0	—	—
Aspergillus niger (ATCC 16404)	2.4×10^4	7.4×10^4	3.3×10^4	$>10^3$	1.7×10^3	9.0×10^2
Aspergillus niger (TM-4)	3.3×10^5	4.5×10^4	2.5×10^2	0	0	—

TABLE 15 Comparative Responses to "Isochallenge" and "Standard" Microorganisms: Formulation 9411

Test microorganism	Organism count (per ml formulation)					
	Initial	Day 1	Day 7	Day 14	Day 21	Day 28
Aspergillus niger (ATCC 16404)	9.0×10^4	2.7×10^4	$>10^4$	1.8×10^4	$>10^4$	1.2×10^3
Aspergillus niger (TM-4)	2.0×10^6	$>10^5$	1.8×10^5	$>10^3$	1.0×10^3	0
Aspergillus clavatus (TM-7)	1.1×10^6	$>10^5$	1.0×10^5	8.0×10^4	4.5×10^4	1.4×10^4

TABLE 16 Comparative Responses to "Isochallenge" and "Standard" Microorganisms: Formulation 9412

Test microorganism	Organism count (per ml formulation)					
	Initial	Day 1	Day 7	Day 14	Day 21	Day 28
Aspergillus niger (ATCC 16404)	9.0×10^4	3.3×10^4	8.0×10^2	0	—	—
Aspergillus niger (TM-4)	2.0×10^6	$>10^5$	9.0×10^4	4.5×10^1	$>10^1$	0
Aspergillus clavatus (TM-7)	1.1×10^6	$>10^5$	6.6×10^5	1.6×10^5	9.0×10^4	6.7×10^4

responses of 9398 and 9403 to the three strains of *Pseudomonas* (Tables 12 and 13). However, the "isochallenge" strains of *Pseudomonas* appeared to be somewhat more resistant to the 9407 formulation than the "standard" (ATCC 9027) strain (Table 14). The most dramatic and significant differences in response to fungal challenges were noted with the 9412 formulation (Table 16). By current standards, the product formulation would be considered adequately preserved by virtue of its activity against *Aspergillus niger* (ATCC 16404). However, inclusion of the TM-7 strain of *Aspergillus clavatus* (isolated from a contaminated sample of Nutracort) demonstrated the significant deficiencies of the preservative system of the formulation. Additionally, inclusion of *A. niger* (TM-4) also demonstrated that the preservative system has specific limitations of activity against fungi which were not adequately detected by using a single strain of a "standard" organism. The reverse of this situation occurred with the 9407 formulation (Table 14). This formulation was less effective against the ATCC ("standard") strain and more effective against the "isochallenge" strain of *A. niger*. These data reinforce the conviction of other investigators who stress the need to incorporate "isochallenge" organisms into a preservative test system. It is our opinion that a number of "isochallenge" *and* "standard" organisms must be included in any Owen-Mahdeen preservative evaluation system. Use of multiple strains of these organisms will allow us to determine preservative efficacy against a broad spectrum of microorganisms.

As previously stated, two of the major objections of the *USP* preservative system as applied to the testing of topical formulations are the inadequacy of the length of the test and the absence of a rechallenge system. It is our opinion that a 28-day test period is of adequate length for evaluating the preservative efficacy of cosmetic/dermatological formulations. However, we do concede that a rechallenge of specific formulations is needed. The rechallenge concept was examined/evaluated experimentally. In Tables 17-19 we present data obtained from the rechallenge* of the three product formulations 9398, 9403, and 9407. Rechallenging these formulations did not significantly alter prior judgments as to their acceptability. In the case of the 9407 formulation (Table 19), the data did strengthen our previous statement that the formulation is inadequately preserved. It is our opinion that rechallenging every product formulation would be a significant expenditure of time, energy, and money, without the provision of significant data which would be critical in selecting product candidates. Therefore rechallenging of formulations will be reserved for those formulations which show marginal activity following initial challenge studies.

The criteria or standards used in assessing preservative efficacy is a critical part of the test system. As previously stated, many investigators feel that the standards of effectiveness set by the *USP* preservative system are inadequate when applied to topical formulations. We have evaluated the standards of preservative studies conducted in industry as well as reviewed the compendiary, governmental, and industrial standards and recommend the

---

*The test formulations were subjected to a routine preservation evaluation. At the end of the 28-day test period the test formulations were rechallenged with each of the microorganisms. Samples were taken at time 0 and at 7-day intervals.

TABLE 17 Responses of Product Formulations to Organism Rechallenges: Formulation 9398

Test microorganism	Organism count (per ml formulation)				
	Initial	Day 7	Day 14	Day 21	Day 28
Mixed culture	1.6×10^5	0	—	—	—
Escherichia coli (ATCC 8739)	5.5×10^5	0	—	—	—
Pseudomonas aeruginosa (ATCC 9027)	5.8×10^5	0	—	—	—
Pseudomonas multivorans (TM-1)	4.6×10^5	0	—	—	—
Staphylococcus aureus (ATCC 6538)	1.2×10^4	0	—	—	—
Candida albicans (ATCC 10231)	6.8×10^4	0	—	—	—
Bacillus sp. (TM-2)	1.0×10^5	8.0×10^3	4.5×10^3	1.7×10^3	9.0×10^2
Aspergillus niger (ATCC 16404)	9.0×10^4	0	—	—	—
Penicillium sp. (TM-3)	1.8×10^5	0	—	—	—

TABLE 18 Responses of Product Formulations to Organism Rechallenges: Formulation 9403

Test microorganism	Initial	Organism count (per ml formulation)			
		Day 7	Day 14	Day 21	Day 28
Mixed culture	3.4×10^5	0	—	—	—
Escherichia coli (ATCC 8739)	8.9×10^5	0	—	—	—
Pseudomonas aeruginosa (ATCC 9027)	5.7×10^5	0	—	—	—
Pseudomonas multivorans (TM-1)	7.5×10^5	0	—	—	—
Pseudomonas sp. (TM-6)	3.6×10^5	0	—	—	—
Staphylococcus aureus (ATCC 6538)	2.5×10^5	0	—	—	—
Candida albicans (ATCC 10231)	1.6×10^4	0	—	—	—
Bacillus sp. (TM-2)	6.2×10^4	1.1×10^5	1.2×10^5	1.9×10^4	1.0×10^4
Bacillus sp. (TM-5)	5.8×10^4	3.4×10^4	1.0×10^4	1.5×10^4	1.3×10^5
Aspergillus niger (ATCC 16404)	3.5×10^4	0	—	—	—
Aspergillus niger (ATCC 16404)	3.5×10^4	0	—	—	—
Aspergillus niger (TM-4)	7.0×10^3	0	—	—	—

TABLE 19 Responses of Product Formulations to Organism Rechallenges: Formulation 9407

Test microorganism	Organism count (per ml formulation)		
	Initial	Day 7	Day 14
Mixed culture	3.4×10^4	1.7×10^4	3.5×10^3
Escherichia coli (ATCC 8739)	8.9×10^5	1.2×10^4	2.8×10^2
Pseudomonas aeruginosa (ATCC 9027)	5.7×10^5	0	0
Pseudomonas multivorans (TM-1)	7.5×10^5	2.4×10^4	1.8×10^2
Pseudomonas sp. (TM-6)	3.6×10^5	0	0
Staphylococcus aureus (ATCC 6538)	2.5×10^5	0	0
Candida albicans (ATCC 10231)	1.6×10^4	3.1×10^1	0
Bacillus sp. (TM-2)	6.2×10^4	5.5×10^5	6.6×10^5
Bacillus sp. (TM-5)	5.8×10^4	2.3×10^5	2.0×10^5
Aspergillus niger (ATCC 16404)	3.5×10^4	1.2×10^4	1.0×10^3
Aspergillus niger (ATCC 16404)	7.0×10^3	0	0
Aspergillus niger (TM-4)	7.0×10^3	0	0

TABLE 20 A Modified Preservative Test Procedure

Challenge microorganisms[a]	Group I *Staphylococcus aureus* (ATCC 6538) *Staphylococcus aureus* (Cl) *Candida albicans* (ATCC 10231) *Candida* sp. (Cl) Other gram-positive bacteria yeasts Group II *Escherichia coli* (ATCC 8739) *Pseudomonas aeruginosa* (ATCC 9027) *Pseudomonas multivorans* (PEI) *Pseudomonas* sp. (PEI) Other gram-negative bacteria Group III *Aspergillus niger* (ATCC 9197) *Aspergillus* sp. (PEI) *Penicillium luteum* (ATCC 9624) *Penicillium* sp. (PEI) Other molds (fungi) Group IV *Bacillus subtilis* (ATCC 6633) *Bacillus* sp. (PEI)
Organism inoculum level	1.0×10^5-1.0×10^6 cells per ml or g of product formulation
Test formulation sampling	0, 1, 7, 14, 21, and 28 days subsequent to inoculation (rechallenge product formulations which show marginal activity)
Effectiveness standards	Vegetative cells — minimum 99.9% reduction within first 14 days, no increase during remainder of the 28 day test Spore formers — stasis throughout 28-day test period Fungi/molds — minimum 90.0% reduction during 28-day test period Rechallenge — same minimum standards applied

[a]PEI, product or environmental isolates; Cl, clinical isolates.

following standards of effectiveness for all formulations undergoing preservative effectiveness testing:

1. Activity against vegetative bacterial cells. A minimum reduction of 99.9% must take place within the first 14 days of the test. During the subsequent 14 days of the test there must be no increase in cell counts.

2. Activity against spore-forming bacterial cells. There must be no increase in cell counts over the 28-day test period (static activity).
3. Activity against yeast. The same minimum standards apply here as applied to vegetative bacteria.
4. Activity against molds and fungi. A minimum cell count reduction of 90.0% must take place during the 28-day test period.
5. Rechallenge. Upon rechallenge the same minimum standards of effectiveness will apply.

It has been the purpose of this presentation to review those significant and critical parameters which must be considered when developing and establishing a preservative test system. Our recommendation for an effective preservative test program is outlined in Table 20.

III. SELECTION AND EVALUATION OF PROTOTYPE FORMULATIONS

The initial step in the development of a new product is the selection and testing of candidate or prototype formulations. Selection of the preservative systems for prototype formulations should be a joint endeavor between the formulating chemist and microbiologist (Goldenberg, 1975). In support of this approach, the microbiologist must continue to develop a preservative reference file which will aid in maintaining and further developing preservative/microbiological expertise. This effort will provide the formulating chemist with a source of information upon which scientifically valid judgments can be made when selecting preservative systems for prototype formulations.

Once the prototype formulations have been selected, adequate testing must be conducted to determine which single formulation will be carried forward through the various other testing regimens of product development. The test procedure established in our earlier discussions (see Sec. 11) will be utilized to evaluate prototype formulations.

IV. DETERMINATION OF PRESERVATIVE "CAPACITY"

The preservative evaluation scheme routinely used to evaluate prototype formulations does not provide the spectrum of data required to determine whether a product formulation is adequately preserved. Primary challenge test data do not reflect the degree or capacity of preservative effectiveness. In many instances the data generated on a formulation with the preservative concentration at a minimum effective level will be similar to data obtained for a formulation in which the preservative concentration is in sufficient excess to provide an adequate safety margin (Henry et al., 1975). Consequently, without these data, products are released which are either "over"- or "under"-preserved. Overpreservation can result in reduced product efficacy, needless sensitization of the consumer, and so on. The most significant adverse effect of underpreservation is the false sense of security obtained relative to the product's ability to maintain its microbiological integrity in the hands of the consumer.

Consequently, we recommend as part of the total preservative evaluation program a testing scheme for evaluating the preservative capacity of product formulations. Briefly, the scheme involves diluting the product formulation (Ramp and Witkowski, 1975). The data generated in these tests will allow one to determine the degree of preservation of any product formulation, as well as to gain an insight as to the specific limitations of the preservative system.

V. PRESERVATIVE EVALUATION OF FIRST-MANUFACTURED BATCHES

Prototype formulations are prepared on a relatively limited scale under controlled laboratory conditions. In scaling up from prototype to manufacturing conditions, a number of factors are encountered which can adversely affect preservative efficacy, including bulk raw material, manufacturing environment, changes in manufacturing procedures, and packaging components. Consequently, conducting preservative evaluations exclusively on prototype formulations does not allow the microbiologist to assess the critical factors encountered in manufacturing which can effect the effectiveness of the product preservative system (Yablonski and Goldman, 1975). It is evident from the foregoing statements that conducting preservative evaluations on actual manufacturing batches is an integral part of a total preservative evaluation program.

VI. ESTABLISHMENT OF A PRESERVATIVE STABILITY TEST SYSTEM

An effective preservative system must maintain its microbiological stability during the shelf life of the product. There are numerous factors which may affect the long-term stability of the preservative system of pharmaceutical and cosmetic products, including pH, temperature, interaction with individual components of the formulations, and interaction with packaging components.

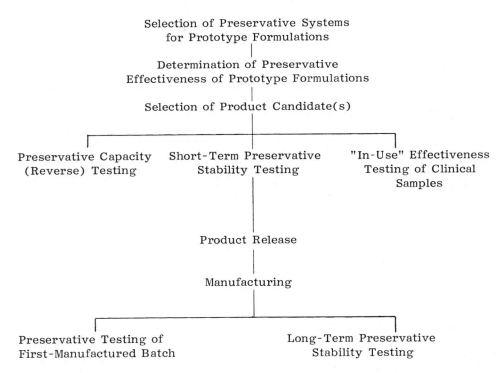

FIGURE 1 Diagrammatical Representation of a Recommended Preservative Evaluation Program

Our approach to establishing a preservative stability system as part of a total preservative evaluation program utilizes both short-term (artificial aging) and long-term (natural aging) stability testing. First, data will be generated on a short-term basis by artificially aging formulation samples. There are numerous systems which can be utilized to achieve this goal. The test system we recommend is a modification of the system currently in use at Bristol-Meyers (Henry, 1975). Formulation samples are exposed to temperatures of 125 and 105°F for 1- and 3-month periods. The formulations are then subjected to preservative challenge testing using the procedures described earlier for prototype formulations. The data generated from these studies will provide us with significant information relative to the 1-year accelerated stability of the preservative system for a specific product formulation (Figure 1).

Our long-term approach involves the testing of stability samples which have naturally aged over prescribed periods of time under ambient conditions. The long-term preservative-stability system will evaluate stability samples from three manufactured batches per year on each manufactured product. Samples will be evaluated at 6 months and 1, 2, and 3 years* post manufacturing. This testing will allow determination of the microbiological stability of product over a 3-year or longer shelf-life period.

VII. "IN-USE" TESTING OF PRESERVATIVE EFFECTIVENESS

As stated previously, the purpose of a preservative system is to protect the consumer and product formulation from microbial contamination. The most significant product contamination can occur due to consumer abuse (repeated insults of multiple-dose containers) (Rdzok et al., 1955). Therefore the ultimate test of a preservative system is to have it evaluated in the hands of the consumer.

For this purpose we recommend a test approach designed to evaluate the effectiveness of product preservative systems following actual consumer use. Formulation samples submitted to patients for clinical evaluation are returned for evaluation. In most instances a *USP XIX* Microbial Limits Test will be conducted on the formulation to determine if the product has become contaminated in the hands of the patient (consumer). In specific product cases, or if a product formulation shows microbial growth, a preservative challenge test is conducted.

The data generated in these studies allow the microbiologist to ascertain whether a specific product formulation can maintain its microbiological safety and stability in the hands of the consumer (under actual "in-use" conditions).

VIII. SUMMARY AND CONCLUSIONS

It has been the purpose of this report to present a review of the various aspects of a typical preservative evaluation program. Each of the segments of the program is designed to evaluate a specific critical factor which could affect the preservative efficacy of a product formulation.

Additionally, utilizing these test procedures, the program establishes testing schemes to evaluate (1) the effectiveness of prototype formulations,

---

*Some specific products may require longer stability testing regimens.

(2) the level or degree (capacity) of preservation of experimental formulations, (3) the effects on preservative efficacy of scaling up to manufacturing conditions, (4) the short- and long-term microbiological stability of preservatives within product formulations, and (5) the actual "in-use" effectiveness of product preservative systems.

Overall, it is our opinion that this program represents a scientifically valid and economically reasonable approach for defining the preservative efficacy of cosmetic/dermatological product formulations.

ACKNOWLEDGMENT

Mr. T. Werner provided valuable input in these studies. This review is correct up to 1980.

REFERENCES

Bruch, C. W. (1972), Possible modifications of USP microbial limits and tests, Drug Cosmet. Ind., 112:32-37, 116-121.

Buhlmann, X., et al. (1972), Microbiological quality of pharmaceutical preparations — guidelines and methods, Am. J. Pharm., 144:165-186.

Gladhard, W. R., Wood, R. M., and Purdum, W. A., (1955). An evaluation of certain bacteriostatic agents used in multiple dose vials sterilized by autoclaving, Bull. Am. Soc. Hosp. Pharm., 12:534-539.

Goldenberg, R. L., (1975), Compounder's corner, Drug Cosmet. Ind., 117: 28, 100.

Gottfried, N. S., (1962), Alkyl p-hydroxybenzoate esters as pharmaceutical preservatives — a review of the parabens, Am. J. Hosp. Pharm., 19: 310-314.

Greenberg, L., and Naubert, J., (1970), Quantitative analysis of preservatives in drug preparations by microbiological assay, Bull. WHO, 43: 663-668.

Grundy, W. E., (1968), Antimicrobial preservatives in pharmaceuticals, in Disinfection, Sterilization and Preservation (C. A. Lawrence and S. S. Block, eds.), Lea and Febiger, Philadelphia, pp. 566-568.

Halleck, F. E., et al., (1970), A guideline for the determination of adequacy of preservation of cosmetics and toiletry formulations, TGA Cosmet. J., 2:20-23.

Henry, S. M., (1975), History, present status, and proposed approaches of preservation testing, presented at the 75th Annual Meeting of the American Society for Microbiology, April 1975, New York City.

Henry, S. M., et al., (1975), Microbiology forum — minimum effective levels of preservatives, CTFA Cosmet. J., 7:32.

Kellogg, D. S., (1972), Critique of preservative challenge systems. II. Preservative testing as applied to quality control analysis, Bull. Parenter. Drug Assoc., 26:216-220.

Lanzet, M., (1972), The development and evaluation of cosmetic preservation systems, CTFA Cosmet. J., 4:4-6.

Leitz, M., (1972), Critique of preservative challenge systems: I. Critique of U.S.P. microbiological test, Bull. Parenter. Drug Assoc., 26:212-216.

Lorenzetti, O. J., and Wernet, T. C., (1977), Topical parabens: benefits and risks, Dermatologica, 154:244-250.

McCarthy, T. J., (1974), Determination of preservative availability from creams and emulsions, Pharm. Weekbl., 109:85-91.

Manowitz, M., (1969), Cosmetic preservatives, in *Disinfection, Sterilization, and Preservation* (C. A. Lawrence and S. S. Block, eds.), Lea and Febiger, Philadelphia, pp. 556-557.

Owen, E. M., (1969), A method for the evaluation of preservative systems in cosmetic formulations, *TGA Cosmet. J.*, 1:12-15.

Parker, M. S., (1972), The rapid screening of preservatives for pharmaceutical and cosmetic preparations, *Soap Perfum. Cosmet.*, 45:103-109.

Ramp, J. A., and Witkowski, R. J., (1975), Microbiological parameters and preservative testing, *Dev. Ind. Microbiol.*, 16:48-56.

Rdzok, E. J., et al., (1955), Determining the efficacy of preservatives in pharmaceutical products, *J. Am. Pharm. Assoc.*, 44:613-616.

Rodgers, J. A., et al., (1973), Evaluation of methods for determining preservative efficacy, *CTFA Cosmet. J.*, 5:2-7.

Rosen, W. E., and Berke, P. A., (1973), Modern concepts of cosmetic preservation, *J. Soc. Cosmet. Chem.*, 24:663-675.

U.S. *Pharmacopoeia XVIII*, (1970), Antimicrobial agents — effectiveness, in *United States Pharmacopoeia*, 18th rev., Mack Publishing, Easton, Pa., pp. 845-846.

U.S. *Pharmacopoeia XIX*, (1975), Antimicrobial preservatives -- effectiveness, in *United States Pharmacopoeia*, 19th rev., Mack Publishing, Easton, Pa., pp. 587-588.

Wilson, C. H., (1970), Some notes on the analysis of preservatives in cosmetics, *CTFA Cosmet. J.*, 4:8-15.

Yablonski, J. I., (1972), Critique of preservative challenge systems. III. Fundamental concepts of preservation, *Bull. Parenter. Drug Assoc.*, 26:220-227.

Yablonski, J. I., and Goldman, C. L., (1975), Microbiology of shampoos, *Cosmet. Perfum.*, 90:45-59.

25
MICROBIAL CHALLENGE AND IN-USE STUDIES OF PERIOCULAR AND OCULAR PREPARATIONS

DONALD JOSEPH REINHARDT *Georgia State University, Atlanta, Georgia*

I. Introduction 465
II. Microbes 467
III. Media 467
IV. Preparation of Inocula 468
V. Challenge Studies 468
VI. In-Use Studies 468
VII. Results 469
VIII. Aquatic Bacteria 473
IX. Comparison of Microbial Contamination of In-Use Contact Lens Solutions and Eye Shadows 475
X. Eye Shadow Challenge Studies 475
XI. Contact Lens Challenge Studies 476
XII. Direct Dilution Sampling of In-Use Solutions 477
XIII. Conclusion 477
References 477

I. INTRODUCTION

The relative susceptibility of the human eye to microbial disease under normal conditions and especially after minor or major trauma suggests that preparations used around or on the eye (e.g., mascaras, eye shadows, and eyedrops) should initially be free of microorganisms and should remain so for as long as possible. Well-documented reports have appeared of infectious diseases of the eye caused by periocular/ocular (P/O) commercial preparations contaminated with microorganisms [9,11,24]. Other reports clearly show that microbial contamination of P/O preparations may occur [3,4,7,18,24]. The ability of these P/O preparations to prevent the survival and growth of microorganisms incidentally introduced into them is an important character that must be evaluated prior to marketing. Furthermore, the quantitation, isolation, and identification of the various types of microorganisms from products in actual use is another important aspect of the ongoing evaluation of P/O commercial preparations. In-use evaluations serve to either corroborate or deny the data obtained under laboratory conditions. These in-use microbial evaluations suggest the inherent capacity of the product to withstand multiple challenges with microorganisms under conditions that cannot be exactly duplicated in controlled laboratory experiments. Laboratory challenge experiments should be designed to assure that the laboratory and the subsequent actual in-use situations more or less resemble one another. The following guidelines can be offered at this point:

1. The challenge microorganisms used for experiments should be a cross section of the potential pathogens of interest and include gram-positive and gram-negative bacteria and possibly some representative fungi such as yeasts (*Candida*) and molds. Ideally the strains should be readily available to other investigators. The American Type Culture Collection (ATCC) is a useful resource.
2. The numbers of microorganisms also should reflect several different densities of microbes likely to intrude into the formulation being tested. A relatively high density of 10^7-10^8 viable cells or colony-forming units (CFUs) should always be a part of the experiment. Intermediate densities of 10^4-10^5 CFUs and low densities of 10^2-10^3 CFUs can also be used for challenge.
3. The microbes and formulations should be in intimate contact to test the full potential of the product to inhibit or kill the potential pathogens.
4. Samples should be withdrawn at reasonable preselected intervals to enable one to determine whether growth, stasis, or death of the original inoculum has occurred.
5. The samples should be mixed thoroughly (vortexing, vibration) and plated in a manner that gives consistent, reproducible CFUs between or among replicate samples.
6. Experiments should be done in duplicate or triplicate and repeated at least once to establish the accuracy of the initial experiments.

With these ideas in mind, I would like to review some of the materials and methods of value for challenge and in-use surveillance studies and some of the results obtained by ourselves and other investigators, and couple these results with relevant discussion comments. My own experiences reflect studies on mascaras, eye shadows, and contact lens solutions. The testing and evaluation of these formulations provide a reasonably broad spectrum of products that can serve as prototypes for the evaluation of similar and related formulations.

II. MICROBES

Gram-positive and gram-negative bacteria and yeasts are the main types of microbes used as challenge organisms. *Staphylococcus aureus* and *Staphylococcus epidermidis* strains are both valuable for challenge studies [2,3]. Inclusion of streptococci is probably not warranted, since they are nutritionally fastidious and usually die rapidly in most preparations with or without inhibitors. On the other hand, staphylococci are recalcitrant gram-positive bacteria that are resistant to drying and a number of inhibitors, and these microbes are common contaminants of in-use preparations.

Acinetobacter calcoaceticus var. *anitratus*, *Pseudomonas aeruginosa*, and either *Klebsiella pneumoniae* or *Enterobacter cloacae* are good choices for challenges with the gram-negative bacteria [1,6,7,8,18]. The latter two genera are less fastidious members of the *Enterobacteriaceae* that are commonly isolated from the environment, as well as from patients. The former two genera are significant oxidative gram-negative bacteria that are common in moist environments and are medically important. The organisms can be maintained on nutrient agar plates or tubes for 2-3 months at 4°C. We have used gram-negative bacterial isolates obtained from cosmetics, clinical sources, and stock reference strains such as the Bact-Chek (Roche Diagnostics, Nutley, N.J.). The latter strains were ATCC strains that were prepared in gelatin and dehydrated and sold as disks. These disks contained viable cells for as long as several years beyond their normal expiration times (previously unreported observations). Although the Bact-Chek disks are no longer available, a similar product with comparable ATCC strains is available (Bactrol, Difco Labs, Detroit). It should be noted that in reporting any of the strains used for research purposes the commercial product origin should be mentioned as the primary source of the ATCC strain. *Candida albicans*, *Candida parapsilosis*, *Aspergillus*, *Penicillium*, and *Fusarium* are some of the fungi that may be considered for challenge studies. A suggested scheme using Food and Drug Administration guidelines for evaluation of fungi in contact lens solutions has recently been published [16].

III. MEDIA

The bacteria can be maintained as stock cultures in 15 × 150 mm screw-capped test tubes with soft nutrient agar (nutrient broth supplemented with 0.3% final concentration of agar). Nutrient agar plates inoculated with each strain of bacteria can serve as sources of the organisms for routine cultures. The organisms are grown overnight at 35°C and refrigerated until needed. Fresh, 18- to 24-hr-old cultures of bacteria are routinely used for challenge studies. The organisms may be grown in liquid culture such as trypticase soy broth, trypticase soy agar, or blood agar. Organisms prepared on agar may be picked as single or multiple colonies and added to sterile water, saline, or broth to form a suspension of cells [7]. Yeasts can be grown on Sabouraud's dextrose agar or on 5% sheep blood agar, with either trypticase soy agar or Columbia agar as a base. The blood agar is prepared by the addition of 50 ml of defibrinated sheep's blood added to 450 ml of the basal medium at 48-50°C followed by mixing and distribution. The yeasts are incubated at 25-30°C for 48 hr prior to their use in challenge experiments.

IV. PREPARATION OF INOCULA

Challenge experiments with microbes from stock strains may be done with washed or unwashed suspensions containing one or more of the microbes of interest. The use of triple centrifuged and washed cells suspended in physiological saline (0.85% NaCl) minimizes carryover of nutrients from the growth medium and permits an evaluation of the ability of a strain or mixture of strains to survive or grow in the test menstruum. The use of unwashed cells permits the study of the microbes in a setting with a carryover of nutrients.

Inocula can be quantitated by the serial dilution technique using sterile 0.85% saline as a diluent and blood agar plates for quantitation of the growth. A bent glass rod dipped in alcohol, flamed, and cooled between each sample can be used to spread the dilution samples.

V. CHALLENGE STUDIES

Liquid or solid menstrua may be used in challenge studies. Ideally the inoculum should be mixed evenly throughout the test sample or menstruum and be recoverable in a uniform and consistent manner, as is usually possible in liquid systems. Semisolid and solid test substrates present a more difficult inoculation and sampling problem, but with careful planning and preliminary evaluatory experiments a suitable experimental system can be designed [7]. In an earlier challenge study we took solid mascaras and scraped them into a fine powder, which we then placed in 380-mg amounts in small, sterile 10 × 100 mm test tubes. One should determine the percentage of retrieval of input organisms immediately, or very soon, after addition to a challenged test system. A zero-time analysis can help determine the efficacy of recovery from noninhibitor systems or the effectiveness of inhibition in inhibitor systems based on the formula

$$\text{Percentage recovery} = \frac{\text{CFU/ml from sampled menstruum (adjusted for dilution)}}{\text{CFU/ml in the initial inoculum}} \times 100$$

(see Table 4 in Sec. VII). This zero-time sampling permits one to analyze the effectiveness of recovery of the added organisms. The ideal situation is to assay the challenge system with and without an inhibitor, for example. The same inhibitor may behave somewhat differently in liquid versus semisolid and solid systems. Usually, in liquid systems the activity of an inhibitor will be more rapid. It is possible that when a solid or semisolid menstruum is tested with microbes, recoveries attempted within the first few minutes will indicate large numbers of survivors. However, platings made at 1, 4, and 24 hr will indicate more clearly the ability of challenge microbes to survive in the menstruum.

VI. IN-USE STUDIES

The sampling, quantitation, and identification of microbes from products that have been used or still are being used by consumers provide an additional opportunity to evaluate the product under uncontrolled (nonlaboratory) circumstances. For the sampling of eye shadows we used sterile Dacron swabs rotated and rubbed over the surface of the eye shadows. In liquid settings, such as with solutions used for soaking, cleaning, and wetting contact lenses, direct dilution samples may be taken. A 0.1-ml sample obtained with a 1.0-ml pipet and 0.01- and 0.001-ml samples taken with sterile calibrated loops permit a simple, rapid, and reliable procedure for the quantitation of microbes in liquid [17-20]. Samples can be plated onto plates of 5% blood agar, eosin-methylene blue agar and Sabouraud's dextrose agar to yield information on the total number of bacteria (on blood agar), gram-negative bacteria (on eosin-methylene blue agar), and possible yeasts and filamentous fungi (on Sabouraud's dextrose agar). Identification of these microorganisms can be done using a combination of standardized [12] and rapid procedures [18]. Some useful and reliable miniaturized biochemical test systems are available for the identification of yeasts (Uni-Yeast Tek, Corning Medical Diagnostics, Roslyn, N.Y.), gram-negative nonfermentative bacteria (Oxiferm, Roche Diagnostics, Nutley, N.J.; Uni-N/F-Tek, Corning Medical Diagnostics, Roslyn, N.Y.), and gram-negative fermentative bacteria (API-20E, Analytab Products, Inc., Plainview, N.Y.; Enterobube, Roche Diagnostics, Nutley, N.J.).

VII. RESULTS

Tables 1 and 2 summarize the incidence of microbial contamination from in-use contact lens solutions, and Table 3 shows the incidence and types of microorganisms obtained from eye shadow testers in retail stores. Note that in both instances the frequency of contamination was high in these in-use P/O preparations. Soft lens solutions showed a generally greater incidence of contamination than hard lens solutions; nevertheless, in both soft and hard lens solutions large numbers of gram-negative bacteria predominated. Species of the genus *Pseudomonas* were most common. Other gram-negative nonfermenters of glucose that were isolated were species of *Alcaligenes*, *Achromobacter*, and *Acinetobacter*. Fermentative gram-negative bacteria that were isolated included *Serratia liquefaciens*, *Serratia marcescens*, and *E. cloacae*. Species of the gram-positive bacteria, *Staphylococcus*, *Micrococcus*, and *Bacillus*, were occasionally isolated from the contact lens solutions; however,

TABLE 1 Incidence of Microbial Contamination from Sampled In-Use Contact Lens Solutions

Type of solution	Number of different brands	Number of contaminated solutions per total sampled	Percentage contaminated
Soft lens			
Soaking	1	1/5	20
Cleaning	3	3/7	43
Saline	2	12/13	92
Distilled water	2	1/2	50
Hard lens			
Soaking	11	0/19	0
Cleaning	6	5/15	33
Wetting	4	11/20	55

Source: Adapted from Ref. 18.

TABLE 2 Types and Frequencies of Microorganisms Isolated from Various Contaminated Contact Lens Solutions

Solution and number of samples thereof	Microorganism	Number of isolates per total number of isolates	Microbial density range (CFU/ml)
Soft lens, saline (12)	*Pseudomonas fluorescens* (3), *P. cepacia* (2), *P. aeruginosa* (1)	13/28	10^3–10^6
	Alcaligenes sp.	4/28	2.5×10^3–10^5
	Acinetobacter calcoaceticus	3/28	1×10^3–3×10^5
	Enterobacter cloacae	3/28	6×10^4
	Serratia liquefaciens	1/28	4×10^2–7.5×10^4
	Gram-positive bacteria/yeasts	4/28	
		28/28	4×10^2–10^6
Soft lens, cleaning/ soaking (4)	*Micrococcus* sp.	3/8	10^2–2×10^3
	Staphylococcus sp.	3/8	2×10^2–6×10^3
	Alcaligenes sp.	2/8	2×10^2–8×10^3
		8/8	10^2–8×10^3

Hard lens, wetting (11)	Pseudomonas cepacia (4), P. maltophilia (2), P. putida (1), P. stutzeri (1), P. aeruginosa (1)	9/18	10^3–3×10^5
	Enterobacter cloacae	3/18	4×10^3–6×10^5
	Alcaligenes denitrificans	1/18	10^3
	Achromobacter xylosoxidans	1/18	2×10^6
	Micrococcus sp.	2/18	10^3–3×10^3
	Bacillus sp.	2/18	10^3
		18/18	
Hard lens, cleaning (5)	Pseudomonas putida (3), P. fluorescens (2), P. maltophilia (1)	6/13	10^2–3.5×10^5
	Alcaligenes sp.	3/13	10^4–3×10^5
	Serratia liquefaciens	2/13	9.5×10^4
	Serratia marcescens	1/13	10^5
	Achromobacter sp.	1/13	6×10^3

Source: Adapted from Ref. 18.

TABLE 3 Types and Frequencies of Microorganisms Isolated from 1345 Eye Shadow Testers in Retail Stores

Microorganism	Total number of colonies isolated	Number of positive wells	Percentage of positive wells	Percentage of all colonies
Micrococcus sp.	3564	524	40.0	41.0
Corynebacterium sp.	1701	272	20.2	19.6
Staphylococcus epidermidis	1513	241	17.9	17.4
Bacillus sp.	1033	419	31.2	11.9
Molds	646	228	17.0	7.4
Staphylococcus aureus	87	31	2.3	1.0
Uncategorized bacteria	45	9	0.7	0.5
Actinomyces	42	15	1.1	0.5
Gram-negative rods	41	12	0.9	0.5
Yeasts	10	7	0.5	0.1
Neisseria sp.	9	8	0.6	0.1

Source: Adapted from Ref. 7.

their numbers averaged about 10^3 CFUs per milliliter. In contrast, the gram-negative bacteria were not only more common, but also generally found in concentrations 10-1000 times greater than the gram-positive bacteria (Table 2). The quantitative data were obtained by direct dilution sampling procedures (see Sec. XII). Table 3 clearly indicates that in eye shadow testers with less available water, the situation is reversed from that in contact lens solutions. In the testers, gram-positive bacteria predominated, while gram-negative bacteria accounted for less than 1% of the total microbial population isolated from 1345 eye shadow testers. In Figure 1 inocula of *S. aureus* were added to and incubated in talcum and an eye shadow powder in polypropylene test tubes. The staphylococci apparently died off in a similar fashion in both talcum and eye shadow powder. Table 4 shows the behavior of *A. calcoaceticus* var. *anitratus* added to two different eye shadows and talcum. The results clearly indicate that in the talcum control, growth of all three different-sized inocula (10,000, 950, 300 CFUs per milliliter) reached high densities within 48-96 hr after introduction into the menstruum. In contrast, two of the three challenge inocula added to the first lot of brand A eye shadow and all three inocula added to brand B were controlled by the eye shadow. However, a second lot of brand A did not effectively prohibit the growth of *Acinetobacter* in any of three different challenge inocula (9500, 1050, and 130 CFUs per milliliter), whereas brand B still manifested disinfecting potency.

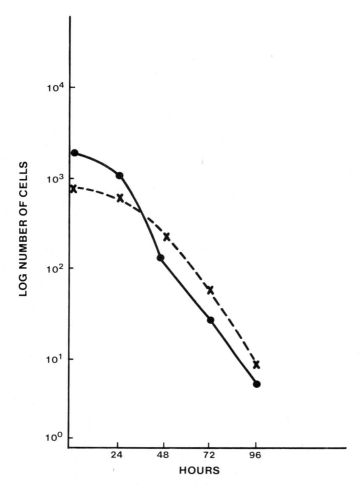

FIGURE 1 Fate of triple washed and centrifuged *Staphylococcus aureus* added to brand A eye shadow and sterile talcum (●————●, talcum; x-------x, brand A).

VIII. AQUATIC BACTERIA

The ubiquity of gram-negative oxidative (nonfermentative) bacteria in aqueous and aquatic environments is well established [5,6,8,18]. In our quantitative and qualitative studies of water in respiratory therapy equipment such as nebulizers and humidifiers, we were impressed by the high percentages of microbial contamination. Often bacterial populations exceeded 10^5 CFUs per milliliter. The microbial contamination of aqueous contact lens solutions was similar in kind and amount to that contamination seen in respiratory therapy equipment. Stanier et al. [22] have shown that *Pseudomonas cepacia* could utilize over 100 of 146 different organic carbon compounds, fluorescent pseudomonads about 75, and *Pseudomonas alcaligenes* and *Pseudomonas maltophilia* about 25 of the 146 compounds. The compounds metabolized included carbohydrates, fatty acids, dicarboxylic and hydroxy acids, miscellaneous organic acids, polyalcohols, glycols, nonnitrogenous aromatic and other cyclic

TABLE 4 Population Changes When Different A. calcoaceticus subsp. antitratus Inocula Were Added to Two Different Eye Shadows and Talcum

Eye shadow	Inoculum (CFU/tube)	Recovery (CFU/0.1 ml) at					
		Zero time	4 h[a]	24 h	48 h	72 h	96 h
Brand A (first lot)	5,500	360	310	10	60	130	3,530
	550	50	40	<1	2	0	0
	60	4	4	0	0	0	0
Brand B	5,500	270	200	10	<1	0	0
	550	30	20	1	0	0	0
	60	3	2	0	0	0	0
Talcum	10,000	600	1,700	22,670	251,000	400,000	530,000
	950	200	300	20,000	219,000	285,000	360,000
	300	30	20	880	77,000	99,000	290,000
Brand A (second lot)	9,500	680	TNTC[b]	970,000	530,000	820,000	100,000
	1,050	80	12,720	237,000	670,000	300,000	209,000
	130	10	862	30,000	75,000	59,000	61,000
Brand B (control)	9,500	150	154	1	0	0	0
	1,050	50	20	0	0	0	0
	130	2	3	0	0	0	0

[a]Length of time that cultures were incubated in cosmetic.
[b]Too numerous to count (usually 10^5 CFU/0.1 ml or more).
Source: Adapted from Ref. 7.

compounds, aliphatic and ring amino acids, and amines. This diverse and extensive potential biochemical menu, coupled with contamination of water by one more of these compounds, suggests that even a pure, sterile water source can soon become a breeding ground for one or more bacteria. Oxidative bacteria, such as *Pseudomonas, Alcaligenes, Achromobacter,* and *Acinetobacter,* are not the only genera that become denizens of water; some of the enterics, such as *Serratia, Enterobacter,* and *Klebsiella,* may establish themselves just as readily and reach high numbers. Van der Kooij [23] has suggested that pseudomonads and aeromonads are not adapted to low substrate concentrations in water and that, ozonation of water leads to an increase of the partial oxidation of humic and fulvic acids, converting them to easily assimilable compounds.

IX. COMPARISON OF MICROBIAL CONTAMINATION OF IN-USE CONTACT LENS SOLUTIONS AND EYE SHADOWS

As shown in this review paper, there were usually significant quantitative and qualitative differences between contaminated in-use liquid (contact lens) preparations when compared to solid or semisolid (eye shadow) P/O preparations. In the liquid preparations, gram-negative bacteria, both nonfermenters and fermenters of glucose, were predominant. In solid and semisolid formulations, gram-positive microbes such as micrococci, staphylococci, diphtheroids, yeasts, and molds were dominant. The general rule is that gram-negative bacteria prefer, compete best, and have a selective advantage with much available water. On the other hand, gram-positive bacteria survive and grow better with less available water if there are adequate nutrients. It is interesting to note that from a total of 35 contaminated cleaning, wetting, soaking, and saline solutions we isolated *P. aeruginosa* only twice (2/35 = 6% of the samples), but other species of *Pseudomonas* were detected alone or in combination in 51% (18/35) of the samples. Both *Pseudomonas cepacia* and *P. putida* were very common. These results suggest that such strains, particularly those isolated from contact lens cleaning solutions, may be useful for future laboratory challenge studies. In the laboratory evaluation of liquid or semisolid or solid P/O formulations, as mentioned previously, challenges with both gram-negative and gram-positive microbes are in order.

X. EYE SHADOW CHALLENGE STUDIES

By challenging eye shadows added to sterile polypropylene test tubes with various doses of *S. aureus, P. aeruginosa,* and *A. calcoaceticus,* we were able to demonstrate that *S. aureus* did not survive in talcum or in either of two brands of eye shadow. Small (110 and 300 CFUs per milliliter) as well as large challenge doses of *P. aeruginosa* survived and increased to 10^4-10^6 CFUs per milliliter by 96 hr in both brands of eye shadow and in the talcum. *Acinetobacter*, on the other hand, was controlled by brand B (with small or large challenge inocula) and was partially controlled by one lot of brand A, but not by another lot or by talcum powder (control). These experiments clearly established that with regard to microbial control there was lot-to-lot variation in brand A but not in brand B. Talcum with no known inhibitor served as a control powder. Bigger and Nelson [5] showed that growth of *Escherichia coli* occurred in talcum with distilled water. They found that talcum and other inorganic compounds may adsorb gases such as CO_2 and NH_3 from the air, making the gases available as nutrients for bacterial growth.

XI. CONTACT LENS CHALLENGE STUDIES

In some recent work, Penley et al. [15] isolated *S. marcescens* from several in-use hard and soft contact lens cases, a corneal ulcer of a patient with hard contact lenses, and the dispensor tip of the patient's wetting solution. These workers chose *S. marcescens*, *P. aeruginosa*, and *S. aureus* as their challenge strains for determination of D values and minimal bactericidal concentrations. A heat-killed suspension of *Saccharomyces cerevisiae* served as an "organic soil." The *Serratia* challenge bacteria subjected to a full cleaning-disinfection regimen survived better than *S. aureus* or *P. aeruginosa* in alkyl triethanol ammonium chloride- (TEAC) based soaking solution. Chlorhexidine-based (CX) soaking solutions showed greater disinfection than TEAC solution.

The D values (time required to kill 90% of the treated population) for the test bacteria were shortest (i.e., most rapid kill) in CX soaking solutions and ranged from under 5 sec to 5.5 min; TEAC solutions gave D values of 9-24 min. Comparisons also were made of the bacteria-inoculated contact lenses that either were cleaned or not cleaned and that were added to either glass tubes or polyethylene cases that contained soaking solution. Controls for that study included bacteria inoculated into soaking CX and TEAC solutions in the glass tubes or polyethylene cases. The polyethylene cases and soft contact lenses apparently afforded some measure of protection from the disinfecting solutions. Richardson et al. reported earlier [21] that there was a loss of antibacterial activity of contact lens solutions during storage. The losses occurred in polyethylene containers, polyethylene powders, and polypropylene containers. Losses were less in the polypropylene containers. Thirty-four different contact lens solutions were evaluated, and more than half contained less than 90% of the stated preservative content. The loss in thimerosal-containing solutions was highest, but losses also occurred in solutions of chlorhexidine gluconate, chlorbutol, and benzalkonium chloride. Norton et al. [14] evaluated the same commercially available 34 contact lens solutions by challenge of each solution with 10^6 organisms per milliliter of each of four challenge strains (*P. aeruginosa*, *S. aureus*, *Micrococcus luteus*, and *C. albicans*). By their standardized methods, six solutions allowed one or more of the test organisms to survive, even after 24 hr of contact with the solution. Only 4 of 14 solutions used to soak and disinfect lenses inactivated all test strains within an hour. Holden and Markides [10] evaluated four types of hydrophilic soaking solutions and one hard lens soaking solution. They noted that solutions with 0.005% chlorhexidine were very effective against *P. aeruginosa* and *S. aureus*; experimental solutions with 0.002% thimerosal and 0.10% ethylenediaminetetraacetate were less effective, and 0.001% thimerosal was very ineffective. Diluted challenge suspensions of bacteria (5×10^3-1.5×10^4) were destroyed by effective solutions in 2 min or less, whereas more concentrated inocula (5×10^6-1.5×10^7) were killed in 15 min or less by effective solutions. Lerman and Sapp [13] reported that Hydron lenses did not concentrate either benzalkonium chloride or chlorbutanol when exposed to normal concentrations of the agents for less than 12 hr, but by 24 hr the concentration rose to 0.08% (8 times the original concentration). However, it was not expected that this increased concentration would be harmful to the eye, according to the authors, since the lens was being washed by tears at 6-7 µl/min, which would rapidly dilute the inhibitor released from the lens.

XII. DIRECT DILUTION SAMPLING OF IN-USE SOLUTIONS

Direct dilution sampling procedures with calibrated (0.01- and 0.001-ml iridium-rhodium) loops and 1.0-ml pipets permit the plating of small samples that can provide very useful quantitative data for solutions in laboratory and consumer evaluations. Over the last 7 years, we have shown the reliability and value of this methodology for the analysis of solutions in respiratory therapy equipment [17-20] and for contact lenses. The direct dilution sampling method is rapid, inexpensive, and provides reliable quantitative data when compared with standard dilution procedures. The method has been used for many years for the quantitation of bacteria in milk and urine. Direct dilution sampling with calibrated loops would show decreased reliability with solutions of increasing viscosity. In these instances the size of the more viscous drops held by the loop would tend to be disproportionately greater than that of water, which has a specific gravity and a density of 1.0.

XIII. CONCLUSION

Microbial challenge and in-use evaluations of P/O products can provide useful information on product stability and performance. Cosmetics research microbiologists should rigorously challenge various formulations of the products in the laboratory to obtain reasonable assurance that the products can withstand in-use, consumer microbial challenges that may even exceed reasonable standards. My own limited experience suggests that many contemporary P/O formulations seem satisfactory from the viewpoint of adequate microbial control. Unfortunately, not all variables that a formulation will be subjected to during actual use can be seen or anticipated. Even the best P/O preparations may fail. This is another aspect of man versus microbe that will continue to make the microbiology of the ocular and periocular formulations and the protection of the human eye sometimes frustrating but always challenging and exciting for decades to come. For every new formulation there is a new potential for success or for the genesis of a new problem. Considering the tremendous technical, biological, and biochemical successes that have been made in the last decade in microbiology, we are winning the battle against microbes.

REFERENCES

1. R. Abel, Jr., J. Schutman, G. L. Boyle, M. A. Meltzer, D. L. Mirow, and J. H. Leopold, Herellea vaginicola and ocular infections, Ann. Ophthalmol., 7:1485-1488 (1975).
2. D. G. Ahearn and L. A. Wilson, Microflora of the outer eye and eye area cosmetics, Dev. Ind. Microbiol., 17:23-28 (1976).
3. D. G. Ahearn, J. Sanghvi, and G. J. Haller, Mascara contamination: in use and laboratory studies, J. Soc. Cosmet. Chem., 29:127-131 (1978).
4. D. G. Ahearn, L. A. Wilson, A. J. Julian, D. J. Reinhardt, and G. Ajello, Microbial growth in eye cosmetics: contamination during use, Dev. Ind. Microbiol., 15:211-216 (1974).
5. J. W. Bigger and J. H. Nelson, the metabolism of coliform bacilli in distilled water, J. Pathol. Biol., 55:321-327 (1943).

6. L. A. Carson, M. S. Favero, W. W. Bond, and N. J. Petersen, Morphological, biochemical, and growth characteristics of *Pseudomonas cepacia* from distilled water, *Appl. Microbiol.*, 25:476-483 (1973).
7. N. L. Dawson and D. J. Reinhardt, Microbial flora of in-use, display eye shadow testers and bacterial challenges of unused eye shadows, *Appl. Environ. Microbiol.*, 42:297-302 (1981).
8. M. S. Favero, L. A. Carson, W. W. Bond, and W. J. Peterson, *Pseudomonas aeruginosa*: growth in distilled water from hospitals, *Science*, 173:836-838 (1971).
9. H. Freedman and J. Sugar, *Pseudmonas* keratitis following soft contact lens wear, *Contact Lens J.*, 10:21-25 (1976).
10. B. A. Holden and A. J. Markides, On the desirability and efficacy of chemical "sterilization" of hydrophilic contact lenses, *Aust. J. Optom.*, 325-336 (1971).
11. J. H. Krachmer and J. J. Purcell, Jr., Bacterial corneal ulcers in cosmetic soft contact lens wearers, *Arch. Ophthalmol.*, 96:57-61 (1978).
12. E. H. Lennette, E. H. Spaulding, and J. P. Truant (eds.), *Manual of Clinical Microbiology*, 2nd. Ed., American Society for Microbiology, Washington, D.C. (1974).
13. S. Lerman and G. Sapp, The hydrophilic (Hydron) corneoscleral lens in the treatment of corneal disease, *Can. J. Ophthalmol.*, 6:1-8 (1971).
14. D. A. Norton, D. J. G. Davies, N. E. Richardson, B. J. Mcaken, and A. Keall, The antimicrobial efficiencies of contact lens solutions, *J. Pharm. Pharmacol.*, 26:841-846 (1974).
15. C. A. Penley, R. A. Schlitzer, D. G. Ahearn, and L. A. L. Wilson, Laboratory evaluation of chemical disinfection of soft contact lenses, *Contact Intraocular Lens Med. J.*, 7:101-110 (1981).
16. C. A. Penley, D. G. Ahearn, R. L. Schlitzer, and L. A. Wilson, Laboratory evaluation of chemical disinfection of soft contact lenses. II. Fungi as challenge organisms, *Contact Intraocular Lens Med. J.*, 7:196-204 (1981).
17. D. J. Reinhardt, C. Kennedy, and B. Malecka-Griggs, Quantitative and qualitative analyses of microbial contamination in solutions and wet, in-use equipment, *Dev. Ind. Microbiol.*, 19:377-384 (1978).
18. D. J. Reinhardt, D. Adams, P. Dickson, and V. Traina, Isolation, identification and quantitation of gram-negative nonfermentative bacilli from aqueous and aquatic sources, *Dev. Ind. Microbiol.*, 20:705-721 (1979).
19. D. J. Reinhardt, C. Kennedy, and B. Malecka-Griggs, Selective non-routine microbial surveillance of in-use hospital nebulizers by aerosol entrapment and direct sampling analyses of solutions in reservoirs, *J. Clin. Microbiol.*, 12:199-204 (1980).
20. D. J. Reinhardt, W. Nabors, C. Kennedy, and B. Malecka-Griggs, *Limulus* amebocyte lysate and direct sampling methods for surveillance of operating nebulizers, *Appl. Environ. Microbiol.*, 42:850-855 (1981).
21. N. E. Richardson, D. J. G. Davies, B. J. Meakin, and D. A. Norton, Loss of antibacterial preservatives from contact lens solutions, *J. Pharm. Pharmacol.*, 29:717-722 (1977).
22. R. Y. Stanier, N. J. Palleroni, and M. Doudoroff, The aerobic pseudomonads. A taxonomic study, *J. Gen. Microbiol.*, 43:159-271 (1966).

23. D. Van der Kooij, The occurrence of *Pseudomonas* sp. in surface water and in tap water as determined on citrate media, *Antonie van Leeuwenhoek, J. Microbiol. Serol.*, *43*:187-197 (1977).
24. L. A. Wilson and D. G. Ahearn, *Pseudomonas*-induced corneal ulcers associated with contaminated eye mascaras, *Am. J. Ophthalmol.*, *84*: 112-119 (1977).
25. L. A. Wilson, J. W. Kuehne, S. W. Hall, and D. G. Ahearn, Microbial contamination in ocular cosmetics, *Am. J. Ophthalmol.*, *71*:1298-1302 (1971).
26. L. A. Wilson, A. J. Julian, and D. G. Ahearn, The survival and growth of microorganisms in mascara during use, *Am. J. Ophthalmol.*, *79*:596-601 (1975).

PART VI
SAFETY, TOXICOLOGICAL, AND REGULATORY ISSUES

26
DERMAL AND OCULAR TOXICITY OF ANTISEPTICS
Methods for the Appraisal of the Safety of Antiseptics

F. LAUTIER *Institut d'Hygiène et de Médecine Préventive, Strasbourg, France*

I. Introduction   484
   A. Defining cosmetics   484
   B. Classification and composition of cosmetics   484
   C. Quality control of cosmetics   484
   D. French legislation for cosmetics   485

II. General Principles Concerning Antiseptics   485
   A. Brief history and definitions   485
   B. Properties of an antiseptic   486
   C. Methods for the study of antiseptics   486

III. Developing Tests to Study Skin Irritation Caused Specifically by Antiseptics   487
   A. Tests   487
   B. Antiseptics tested   489
   C. Interpretation of results   490
   D. Why are two types of skin irritation indexes necessary?   494
   E. Appendixes: Experimental protocol for the two skin irritation tests   495

IV. Developing a Test for the Determination of the Ocular Irritancy Index   498
   A. Experimental protocol   498
   B. Results   499

V. Conclusions   500

   References   500

I. INTRODUCTION

A. Defining Cosmetics

Cosmetics are designed to protect and perfume the skin, the exoskeleton, the ducts, and the mucous membranes of the mouth and genitalia. These products enjoy an extremely wide application during our lives. Are they dangerous to our health? By definition, beauty preparations and cosmetics are never harmful to human health under normal conditions of use. Indeed, cosmetics are actually hygienic in that we use them to care for and protect the skin and mucous membranes.

B. Classification and Composition of Cosmetics

Cosmetics are prepared from extremely pure raw materials, which are variable, depending on their application. Attention is presently directed toward the search for nonreactive, nontoxic, and well-tolerated starting materials. Many contain antiseptics or preservatives.

The French Departmental Order of 1977 [1] established an indicatory list of the various categories of cosmetics and products for body hygiene. These include bath soaps, toilet water, deodorants, shaving creams, makeup, sun lotions, products for the mouth and dental hygiene, internal care, and so on, and antiseptics are found in the composition of most of these products.

It is interesting to note that since 1982 a violent controversy exists between the pharmacists and cosmetic manufacturers in Italy. Since the end of 1981, a portion of the law which allows the latter to praise the antiseptic properties of their products has been passed.

C. Quality Control of Cosmetics

Several checks are carried out before a cosmetic or beauty preparation can be commercialized, and these involve the efficiency and safety of the product.

1. Investigations into the Efficiency

These are carried out on in vitro skin biopsies in animals, or in vivo in volunteers; however, these are costly clinical trials and thus usually carried out on a reduced scale. Such trials determine, for example, the hydration of the corneal layer, the roughness of the skin surface, skin elasticity, the appearance of hair, and the index of sebaceous secretion.

2. Safety Measures

Safety measures are established, which include quality control and choice of raw materials, skin tolerance tests performed in animals, and inquiries made into allergy. The research workers are equally aware of the fact that, occasionally, hypercutaneous toxicity reactions may be produced, when the whole organism may react; skin factors play an important role in this case, such as the age of the patient (for talcum powder containing hexachlorphene), the dosage, and whether antiseptics and preservatives are included. Rare accidents may also take place, such as contact skin allergies, phototoxicity, corneal scarring, or infections (trachoma caused by kohl in endemic countries).

D. French Legislation for Cosmetics

Since the decision of April 5, 1971 [2], concerning the "Official Methods of Analyzing Cosmetics and Beauty Preparations" (published in the *Journal Official of the French Republic*, April 21, 1971, pp. 3862-3864, with the appendix appearing in the *Journal Official of the French Republic*, dated June 5, 1973 [3], the French legislation on cosmetics came into existence, and it demands that three tests be carried out:

> Determination of the primary irritation index (over 72 hr)
> Determination of the ocular irritancy index
> Evaluation of the extent of superficial skin attack on repeated applications (over 90 days).

These three tests should be carried out on albino rabbits.

For the last test, that is, evaluation of the extent of superficial skin attack on repeated applications, a new method was published in the *Journal Official of the French Republic*, January 29, 1980, pp. 1181-1182, and the decision was made on December 18, 1979 [4]. This test is still carried out on rabbits, but with the following changes:

> Instead of three animals, 12 are used, 6 of which act as control.
> Instead of 2 g of the product administered per kilogram, 2 g of the product should be given per 2.5 kg.
> Instead of the experiment lasting 90 consecutive days, it should last 6 weeks, with five applications per week.

Thus several indexes are obtained (the weekly mean, daily maximal, and daily total) instead of a single, well-defined index value, with which the scale of indexes could be established and the products could be classified.

In summary, the decision dated April 5, 1971, is maintained for the determination of the primary irritation index (by 24-hr application of the test product) and the ocular irritancy index. The decision dated December 18, 1979, determined that the method used for the evaluation of the extent of superficial skin attack be carried out with repeated applications (during 6 weeks).

II. GENERAL PRINCIPLES CONCERNING ANTISEPTICS

A. Brief History and Definitions

"Antiseptics are the medicaments capable of destroying contaminating flora" is the definition of antiseptics according to Christol [5]. To elucidate this definition, the following points must be stated:

> Sterility signifies free from germs.
> Asepsis is a method to achieve germ-free conditions.
> Antisepsis means the destruction, by various physical or chemical techniques, of germs; however, antisepsis does not necessarily lead to sterility, but may result in what is called "decontamination" (i.e., the destruction of most, but not all, germs).
> Finally, disinfectants are not antiseptics, but corrosive and caustic products used to decontaminate equipment and surfaces.

It must also be pointed out that antiseptics operate at relatively high doses, their activity is only slightly non-specific, and their toxicity threshold is usually similar to their activity threshold; thus they are generally intended for external use (skin applications, washing wounds, and dressing wounds).

A few historical facts are mentioned by Christol [5], such as the empirical use of antiseptic products by the ancient Egyptians, the ritual washings described in the Bible (*Leviticus* and *Deuteronomy*), and even the ancient writings from China and Persia or by Hippocrates. From more modern times, Fracastor (1445) should be quoted, having advocated the usefulness of prevention. Pringle, in 1750, devised the term *antiseptic*, and in 1825 Labarraque suggested the use of his hypochlorite solution in the treatment of infections. The forefather of hygiene was the Austro-Hungarian Semmelweiss (1818-1856), who expected the hands of obstetricians to be washed with a hypochlorite solution between deliveries. Finally, it was Pasteur who discovered the importance of asepsis and antisepsis through his microbiological researches. Lister took spectacular advantage of this during his operations.

B. Properties of an Antiseptic

A useful antiseptic should be the following:

> Soluble in water and organic solvents
> Stable in solution and on aging
> Nonirritating, nontoxic, and nonallergenic
> Of a wide spectrum of antibacterial activity, that is, be a bactericide
> Quick acting
> Able to conserve its properties on contact with organic solvents

It is clear from the previous observations that, before using an antiseptic, either alone or included in a cosmetic or beauty preparation, it must be tested for its antiseptic potency and it must be proven to be *nonirritating* to skin, mucous, and ocular tissues.

C. Methods for the Study of Antiseptics

Without going into details, the essentials of the methods must be clearly indicated. In France, bactericidal activity is measured using two in vitro techniques, among others: the dilution neutralization technique and the membrane filtration technique. Both are recommended, described, and standardized by the *French Standardization Association (AFNOR)* in the AFNOR T 72-150 and T 72-151 regulations [6]. A third regulation, T 72-160 AFNOR, indicates the way to determine the bactericidal activity in the presence of defined interfering compounds [7].

In vivo studies are also carried out on antiseptic activity. A good antiseptic should lower the number of skin microorganisms to a sufficient level to be able to reduce the risk of infections. Furthermore, according to AFNOR, antisepsis is a process giving temporary results on living tissues in the limit of their tolerance levels, by eliminating or killing microorganisms and/or inactivating viruses.

Skin flora is extremely variable; it can change with time, site (scalp, hands, etc.), and the age and sex of the individual. The following criteria are distinguished: resident flora and transitory flora (e.g., originating in the surroundings).

Methods to test antiseptic activity on this flora are by making an *impression* on the culture medium (the hands are brought into direct contact with the medium); by the *sampling* method, that is, indirectly; and by the *washing* method, where the washing water is subsequently analyzed.

Finally, laboratories and the pharmaceutical industries also carry out *general toxicity tests* on products intended for external use. The latter is carried out to determine the LD_{50} in animals.

Localized skin toxicity determinations are carried out generally, but not systematically. So far, in France there is no official text making the testing of antiseptic products obligatory.

III. DEVELOPING TESTS TO STUDY SKIN IRRITATION CAUSED SPECIFICALLY BY ANTISEPTICS

A. Tests

Research carried out on skin irritation provoked by the application of various antiseptics led to the perfection of an experimental protocol for skin tests, in order to be able to evaluate primary irritation indexes and the extent of superficial skin attack. The official analytical methods published for cosmetics and beauty preparations were adapted by us to antiseptics. The main modification contributed is the establishment of a new index scale whereby antiseptics may be classified as usable or not [8-10].

Whereas the overall toxicity tests with antiseptics in animals by oral and intravenous routes and by percutaneous absorption are generally applied correctly, this is not the case for the skin irritation tests.

Now, an official legislation exists in France for cosmetics and beauty preparations. Why should it not be extended to antiseptics? This is, in fact, what we have been doing since 1976 — taking the official methods intended for cosmetics and adapting them to antiseptics.

We are now in a position to state that the determinations of primary irritation and extent of superficial attack allows the following effects to be defined, depending on the product:

A more or less intense erythema
A more or less widespread edema
Acanthosis, or the presence of thickened, wrinkled skin
Hyperkeratosis, or the thickening of the outer layer of skin

According to the official method, the presence of erythema and/or edema should be noted. During our routine examinations, we have been marking these phenomena systematically, acanthosis and hyperkeratosis, as mentioned by Rouaud [11], as well as the appearance of "skin fatigue," a phenomenon named by Beys et al. [12], the term used by the U.S. Food and Drug Administration to qualify a localized effect on repeated application of a product which is harmless on a single application.

1. Determination of the Primary Irritation Index

A primary irritant is a substance provoking a skin reaction after a single application (see Appendix I in Sec. III.E). The irritation depends on the nature and concentration of the substance, as well as on the length of time the skin is exposed.

The official method suggests applying 0.5 ml of the substance, keeping it in position with a suitable dressing during 24 hr, and observing the skin reaction when the dressing is removed, and again 48 hr later (i.e., 24 and 72 hr after application). We used this method on guinea pigs instead of rabbits. In fact, a guinea pig is easier to handle and keep in an animal house, since it is smaller. Furthermore, the skin of this animal is very sensitive. Thus, instead of 6 male albino rabbits, we took 12 young female albino guinea pigs, weighing between 0.700 and 1 kg, so we could always deal with a soft, sensitive skin. Six guinea pigs had the intact skin and the six others were scarred and thus had skin lesions before the product was applied. The back of the animal was chosen as the test part and a control area was shaved near the rear of the animal.

The guinea pigs were put into individual cages during the 24 hr they were bandaged. They did not require a restraining device, as for rabbits. Thus stress levels were diminished in our animals, and consequently it was easier to obtain valid reactions.

The text laid down in the official method for the evaluation of the irritation index was closely followed, but with further observations made on the second day after application. For certain products erythema was absent at 24 hr; however, it could be seen at 48 hr, disappearing again by 72 hr.

2. Evaluation of the Extent of Superficial Skin Attack on Repeated Applications

Guinea pigs were used for this test (see Appendix II in Sec. III.E) also, for the same reasons. Thus we took three young female albino guinea pigs (weighing 0.700-1 kg). The test product was applied to a shaved part on the back of the animal, and a control area was set up at the rear. Stress was reduced again, as we did not require a retention apparatus and the animals could be left in the same cage throughout the experiment. The official method requires that observations be carried out at the beginning of the experiment and then once a month. We preferred to carry out daily observations, which was easily realized just before the daily application of the test substance.

The last point is very important, as some skin reactions appear on the 8th day and last 5-7 days, while others become evident on the 12, 41, or 85th day only, persist for a short time, and are diminished or disappear altogether. Thus it seems that for certain products the skin becomes accustomed to daily applications after a more or less strong initial skin reaction.

3. Calculation and Interpretation of Indexes

Indexes were always calculated following the official methods for cosmetics, from the observations made 24 and 72 hr after application of the substance (for the primary irritation index) and from the readings realized on days 1, 3, 31, 33, 61, 63, 91, and 93 (for the extent of superficial skin attack).

However, the interpretation of indexes according to the official method is not precise enough; for example, a substance is classified as slightly irritating if the index is between 0 and 2. In our opinion, another scale of values is needed, and we suggest a scale with seven levels (see Table 1).

The first four values (indexes from 0 to 1.50) indicate usable antiseptics. Antiseptics are considered useless if the index value is among the last three (indexes from 1.50 to 5.00). The index values are designated by letters, A-G in Table 1.

TABLE 1 Interpretation of Results

I. Official method published for cosmetics

Index	Classification
0	Nonirritating
0-2	Slightly irritating
2-5	Moderately irritating
5-8	Severely irritating

II. Suggested interpretation for antiseptics

	Index	Classification
a.	0.00-0.00	Nonirritating
b.	0.00-0.50	Slightly irritating
c.	0.50-1.00	Moderately irritating
d.	1.00-1.50	Rather irritating
e.	1.50-2.00	Irritating
f.	2.00-3.00	Most irritating
g.	3.00-5.00	Severely irritating

Source: Ref. 8.

B. Antiseptics Tested

We tested several antiseptics of different chemical compositions. In some cases we carried out tests on the existing preparations, but at various dilutions, according to the anticipated usage. For the antiseptics intended for hand washing, we carried out tests with and without rinsing the hands with water, and with and without brushing the hands.

The solutions tested are given in the following table:

Chlorhexidine
  20% solution
    Final dilution, 1/2000 wt/vol
    Final dilution, 1/5000 wt/vol
  5% Hibitane
    Undiluted solution
    Final dilution, 1/2000 wt/vol
    Alcoholic solution
  Hibiscrub (with 4% chlorhexidine)
    Undiluted solution
      With and without rinsing with water
      With and without brushing
  Vantropol*
    0.3% dilution
  20% chlorhexidine polymer
    Final dilution, 1/100 wt/vol

---

*Vantropol is a combination of chlorhexidine and cetrimide; the product sold under this trade name is intended as a disinfectant (for equipment, surfaces, etc.).

Chlorhexidine (Continued)
 Paragerm
  Savogerm with chlorhexidine with and without rinsing in water
  Bactericidal Paragerm with chlorhexidine
 Bactane (a mixture of chlorhexidine and quaternary ammonium)
 Diluted to 1%

Quaternary ammoniums
 Bactane
  1% dilution
 Benzalkonium chloride
  Diluted to 1/100
 Sterlane
  Emulsion
  Solution
   Undiluted
   Diluted to 1/100
 Tegodor
 0.25%

Chlorine derivatives (triclocarban)
 Septivon
  Undiluted solution
  Diluted to 1/100

Iodinated products
 Iodized alcohol (1% solution in alcohol at 60°C)
 Betadine
  Scrub
  Solution

Hexachlorophene
 2% soap

For all these products, we carried out two skin irritation tests (except for the two Paragerm products, for which only the primary irritation was determined). The dilutions we made up for some of the products were the ones used in the general practice of medicine and surgery. For scrubs, soaps, and the solutions intended for washing and disinfecting the hands, we performed tests either with or without rinsing with water (primary irritation tests with Hibiscrub and Paragerm soap) or with or without brushing (test of superficial attack with Hibiscrub). The commercially available form was used for most antiseptics without dilution.

The results are shown in Table 2 for the primary irritation indexes, and in Table 3 for the extent of superficial skin attack.

C. Interpretation of Results

We followed the numerical scale published in the official method for cosmetics to estimate the extent of erythema and the appearance of edemas:

TABLE 2 Results: Primary Irritation Index

Antiseptic	Dilution	With rinsing	Without rinsing	Index	Interpretation according to our criteria
20% chlorhexidine	1/2000 wt/vol			0.70	C
20% chlorhexidine	1/5000 wt/vol			0.54	C
5% Hibitane	Undiluted solution			0.66	C
5% Hibitane	1/2000 wt/vol			0.33	B
5% Hibitane	Alcoholic solution			0.37	B
Hibiscrub	Undiluted solution		+	0.83	C
Hibiscrub	Undiluted solution	+		0.20	B
Vantropol	0.3 per 100			0.16	B
20% chlorhexidine polymer	1/100 wt/vol			0.20	B
Paragerm solution	Undiluted solution			0.00	A
Savogerm	Undiluted solution	+		0.00	A
Savogerm	Undiluted solution		+	1.00	D
Bactane	1% solution			0.33	B
Benzalkonium chloride	1/100 solution			0.29	B
Sterlane emulsion	Undiluted			0.87	C
Sterlane solution	Undiluted solution			0.41	B
Sterlane solution	1/100			0.37	B
Tegodor	0.25%			0.16	B
Iodized alcohol	Codex solution			0.54	C
Betadine scrub	Undiluted solution			0.12	B
Betadine solution	Undiluted solution			0.37	B
Septivon	Undiluted solution			1.70	E
Septivon	1/100			0.25	B
2% hexachlorophene soap	Undiluted solution			0.75	C

Source: Ref. 8.

Numerical Scale: Erythema and Scab Formation	
Without erythema	0
Slight erythema (hardly visible)	1
Clearly visible erythema	2
Moderate to widespread erythema	3
Severe erythema (purple red) with slight scab formation (deep lesions)	4

TABLE 3 Results: Extent of Superficial Skin Attack

Antiseptic	Dilution	With brushing	Without brushing	Index.	Interpretation according to our criteria
20% chlorhexidine	1/2000			0	A
20% chlorhexidine	1/5000			0	A
5% Hibitane	Undiluted solution			0	A
5% Hibitane	1/2000			0	A
5% Hibitane	Alcoholic solution			0	A
Hibiscrub	Undiluted solution			0	A
Hibiscrub	Undiluted solution	+		0.041	A
Vantropol	0.3 per 100			0	A
20% chlorhexidine polymer	1/100 wt/vol			0	A
Bactane	1% solution			0	A
Benzalkolium chloride	1/100 solution			0	A
Sterlane emulsion	Undiluted solution			4	G
Sterlane solution	Undiluted solution			1.62	E
Sterlane solution	1/100			0.20	B
Tegodor	0.25%			0.75	B
Iodized alcohol	Codex solution			0	A
Betadine scrub	Undiluted solution			2.33	F
Betadine solution	Undiluted solution			0.66	B
Septivon	Undiluted solution			4.54	G
2% hexachlorophene soap	Undiluted solution			4.12	G

Source: Ref. 8.

Numerical Scale: Edema Formation	
Without edema	0
Very slight edema (hardly visible)	1
Slight edema (well-defined outline, apparent swelling)	2
Average edema (about 1 mm thick)	3
Severe edema (more than 1 mm thick, outer surface greater than gauze square)	4

1. Evaluation of Erythema

Erythema with occasional scab formation is easy to evaluate by comparison with the control area. From a batch of three guinea pigs, one may show signs of type 1 erythema during 3-4 days, but it is an isolated, individual reaction and may be disregarded. During the whole of our experiments, whether research was carried out on primary irritation index or superficial attacks, out of the batch of 12 or 3 guinea pigs, some erythema, transient or persistent, was observed.

2. Evaluation of Edema

The same conclusions may be drawn for the estimation of values for edema. Comparison with the control area facilitates estimation.

3. Evaluation of Accompanying Reactions by Animals

It is interesting to note the varied reactions exhibited by the animals due to the different products.

Change in Weight: When a particularly aggressive product is applied, the animals automatically lose some weight. This is associated with the loss of appetite and corresponding oligodipsia, or, on the contrary, polydipsia.

General Reactions Showing Fear or Confidence: With daily applications of certain products, whether erythema and edema exists or not, the treatment represents a pleasant experience for the animal, which may fight or accept the process.

Fur Growth: Fur may grow back normally, faster, or slower, depending on the test product. This factor is usually independent of existing erythema or edema, and it is a transitory phenomenon.

Appearance of Acanthosis or Hyperkeratosis: Acanthosis, hyperkeratosis, and even "skin fatigue" are frequently occurring phenomena, are interesting to note, but generally do not interfere in the calculation of indexes.

4. Miscellaneous Experimental Details

Depending on their chemical composition, antiseptics should be stored with care. Some of the most important considerations are the following:

> The container. It should be made of glass or plastic. Certain products react with certain containers.

Distilled water. The choice is important for diluting antiseptics; for example, chlorhexidine can only be diluted in demineralized water.

Storage temperature. Antiseptics must be kept either in a cold (+4°C) or medium temperature (room temperature between +18 and +20°C) to avoid layer formation.

When a product is tested, the same test should always be carried out on the same batch; furthermore, dilutions should always be prepared extemporaneously.

It should finally be emphasized that the addition of an excipient, an additive, or several antiseptics may considerably alter the results. Furthermore, to calculate the indexes, it is useful to determine, at the same time, the irritation index on the undiluted antiseptic, on various dilutions of the antiseptic (depending on the intended use), and on antiseptics with or without rinsing with water and with or without brushing (this should be included specially in the case of products intended for surgical scrubbing).

All these different factors must be taken into consideration before putting an antiseptic on the market so as to avoid any problems consecutive to irritation reactions, whether temporary or permanent.

D. Why Are Two Types of Skin Irritation Indexes Necessary?

In principle it could be imagined that the primary irritation index carried out in 72 hr would give an exact indication of the test product, and that the extent of superficial skin attack after 90 days only confirms the qualities with precision; this is not so. According to Rouaud [4], the aim of the primary irritation index is to "enable the formulation to be defined" and "to act as quality control," whereas the 90-day tests "allow the diagnosis of an acanthosis or a hyperkeratosis, as well as the effects on fur growth." These tests also allow one to see whether the "absorption of the product leads to death."

Other authors, like Bourrinet and Rodde [13], emphasize that "it is the repetition of application that offers the guarantee of the safety for each component." They also mention that "irritation phenomena depend on the ease with which a substance can penetrate the skin," and "the importance of the more or less long-term tests is that it reveals irritative phenomena further to those observed on single application, which become evident only after a time-lapse, as would be the case for substances used at a low concentration."

From our experiments, it is clear that the primary irritation index may be very similar for two different products, whereas the extent of superficial attack is completely different. Consider, for example, the following table:

Product	Primary index	Index of attack
Vantropol, 0.3 per 100	0.16	0.00
Betadine scrub	0.12	2.33
Hibiscrub	0.83	0.00
Sterlane emulsion	0.87	4.00

Either the primary indexes are very low and similar in value (0.16 and 0.12) and the corresponding index of attack is zero or very high (2.33), or the primary indexes are moderately elevated (0.83 and 0.87) and the indexes of attack are unrelated (0 and 4). Thus it is imperative to carry out both tests in order to assure that an antiseptic is useful and to indicate the methods of use and precautions to be taken.

Furthermore, it is interesting to note the appearance of any intervening phenomena, such as changes in pilosity, acanthosis, hyperkeratosis, or general reactions such as loss of weight, loss of appetite, and so on, which, once again, can be temporary but may lead to a serious attack. All these phenomena should be taken into consideration when defining the methods and conditions for using a product.

E. Appendixes: Experimental Protocol for the Two Skin Irritation Tests

1. Appendix I: Primary Irritation Index Determination

Objective: This method is used to determine the primary irritation capacity of a substance. A primary irritant is a product which provokes a skin reaction after a single application. The resulting irritation depends on the nature and concentration of the substance tested and the length of time it is in contact with the skin.

Materials:

1. Young female albino guinea pigs weighing between 0.700 to 1 kg and having been kept in the animal house for at least 1 week
2. Hand shears
3. Electric razor
4. Sterile vaccinostyle
5. Absorbent gauze pad
6. Absorbent cotton-wool
7. Adhesive plaster, 10 cm wide
8. Baby scales

Methods:

1. Preparation of the skin
   a. Clip the fur with the hand shear on the back (experimental part) and rear (control area) of the animal.
   b. Shave the shorn parts with the electric razor.
   c. Leave the shaved skin untouched for 24 hr.
   d. Leave the skin of six guinea pigs intact and make incisions on the backs of the six others (three parallel cuts 3 cm long and 0.5 cm apart).
2. Application of the substance
   a. Imbibe two 4-cm^2 gauze pieces with 0.5 ml of the substance. Place them on the back of the animal.
   b. Keep the substance in place with a small wad of cotton-wool and two crossed gauze strips and cover the whole with a sterile gauze pad.
   c. Wind the adhesive plaster around the trunk of the animal.
   d. Place the animals in individual cages.
   e. Remove the gauze patches 24 hr after application.
3. Evaluation of the primary skin irritation

The primary irritation index should be evaluated immediately after the gauze is removed, and again 48 hr later (i.e., 24 and 72 hr after application).

4. Numerical scale for erythema and scab formation

Without erythema	0
Slight erythema (hardly visible)	1
Clearly visible erythema	2
Moderate to widespread erythema	3
Severe erythema (purple red) with slight scab formation (deep lesions)	4

5. Numerical scale for edema formation

Without edema	0
Very slight edema (hardly visible)	1
Slight edema (well-defined outline, apparent swelling)	2
Average edema (about 1 mm thick)	3
Severe edema (more than 1 mm thick, outer surface greater than gauze square)	4

6. Scoring
   a. Add up the scores obtained for edema and erythema after 24 and 72 hr for the six areas corresponding to the cut surfaces.
   b. Carry out the same calculation for the six areas corresponding to the intact surfaces.
   c. Add up the scores obtained for the totals of the incised and intact surfaces.
   d. Calculate the average figure by dividing the previous total by 24.
   e. The obtained average (always less than 8) is denoted as the primary skin irritation index.

2. *Appendix II: Evaluation of the Extent of Superficial Skin Attack on Repeated Applications*

Objectives and Principles: The repeated application of small doses of a product over a long period makes reactions become apparent which would not be detected during the primary irritation or sensitization tests. This is the method used to discover the effects of a product on skin after daily applications during 90 days. Most of the time the reactions appear between the 50th and 60th days of application and may vary between slight erythema with peeling to strong hyperacanthosis with skin necrosis.

Materials:
1. Young female albino guinea pigs, weighing between 0.700 to 1 kg, having been kept in the animal house for at least 1 week
2. Baby scales

3. Hand shears and electric razor
4. Gauze pieces, 4 cm^2 each.

Methods:

1. Clip and then shave the back and rear of three guinea pigs (reshave every 2 days).
2. Weigh each guinea pig (once a week).
3. Each day, at the same time, apply the gauze square imbibed with the test substance to the shaved back. The application should last about 1 min and the dosage should be 2 g/kg body weight.
4. Place the animals three to a cage, one cage for each test product.
5. Carry out daily applications during 90 days.

Observations:

1. Note the weight of the animals each week, and shave the test area as well as the control parts (twice a week if necessary).
2. Observe the speed of fur growth and other general aspects.
3. Measure the thickness of the skin and possible hyperacanthosis by measuring the thickness of a skin fold with micrometer callipers.
4. Examine and score the formation of erythema and edema as for the primary irritation index.
    a. Numerical scale for erythema and scab formation

Without erythema	0
Slight erythema (hardly visible)	1
Clearly visible erythema	2
Moderate to widespread erythema	3
Severe erythema (purple red) with slight scab formation (deep lesions)	4

   b. Numerical scale for edema formation

Without edema	0
Very slight edema (hardly visible)	1
Slight edema (well-defined outline, apparent swelling)	2
Average edema (about 1 mm thick)	3
Severe edema (more than 1 mm thick, outer surface greater than gauze square)	4

   c. Scoring. Same as the previous test.

IV. DEVELOPING A TEST FOR THE DETERMINATION OF THE OCULAR IRRITANCY INDEX

A test for the ocular irritancy index was perfected from the protocol published in the *Journal Official of the French Republic*, dated April 21, 1971, which is carried out systematically as a quality control for cosmetics and beauty preparations. This test was adapted for the study of antiseptics [14, 15] with the collaboration of Professor A. Brini of the Opthalmological Clinic, Centre Hospitalier Universitaire, Strasbourg, France.

A. Experimental Protocol

1. Test

The test for each product under study is carried out on six albino guinea pigs. A drop of the product is placed in the right eye of the animal, the left eye acting as a control. Two examinations are carried out:

1. Examination of the conjunctiva with the naked eye, to evaluate watering of the eye, the width of the palpebral fissure, redness of the eye, and chemosis.
2. Examination of the cornea with a slit lamp, looking for the Tyndall effect, and the anterior chamber of the eye, after one drop of a 1% solution of fluorescine has been instilled. This examination enables any corneal lesions to be described, examination of the vitreous humour, and a check to be made on the state of the iris.

These examinations are repeated after 24 hr and 5 and 8 days.

2. Interpretation and Marking

Absence of opacity, loss of shine, or fur loss	0
Presence of an irritated area, diffuse or in plaques, visible details of the iris, no edema	1
Presence of a translucent area, with fluorescent dots, details of the iris obscured, edema without vascularization	2
Presence of an opaque area, with fluorescent dots, showing edema and vascularization, not developing into leukoma	3
Presence of an opaque area, with fluorescent dots, ulcerated, showing edema, vascularization, and developing into leukoma	4

Thus a numerical scale from 0 to 4 is worked out according to the severity of the observed lesions. Scores are not based on the presence of a corneal lesion, but only on the extent of opacity and its evolution.

3. Classification

To calculate the ocular irritancy index, add up the scores obtained for the observations carried out on the cornea, iris, and conjunctiva. Indexes vary between 0 and 110 (see Table 4).

TABLE 4 Irritation Index

Classification	Index	Grade
Nonirritant	0	A
Slightly irritant	0-11	B
Moderately irritant	11-22	C
Quite irritant	22-33	D
Irritant	33-44	E
Extremely irritant	44-66	F
Severely irritant	66-110	G

Source: Ref. 14.

When only a mild, temporary conjunctival irritation is produced, the antiseptic is usable. When a corneal irritation is produced with complete recovery (grades A-D), the antiseptic is usable. When the corneal irritation leads to the formation of a leucoma (grades E-G), the antiseptic is useless.

B. Results

The following solutions were tested:

   20% chlorhexidine, diluted to 1/5000
   20% chlorhexidine, diluted to 1/2000
   5% chlorhexidine, diluted to 1/200
   5% chlorhexidine, undiluted
   Sterlane, diluted to 1/100

All iodinated and alcoholic solutions were routinely eliminated, as well as antiseptics which gave a high skin attack index, these products being incompatible with maxillofacial or ocular use.

Only the 5% chlorhexidine, used in the undiluted form (which is not the standard procedure but which could happen by accident or neglect), causes ulceration and indelible leukoma, leading to blindness.

When 5% undiluted chlorhexidine is instilled in the eye the following results appear:

   After 24 hr secretions are seen, with conjunctival congestion, chemosis, rough corneal surface, but no central ulceration.
   After 5 days the eye shows watering, secretions, conjunctival congestion with chemosis, vascularization, edema, and central corneal ulcer formation.
   After 8 days central corneal leukoma is seen (see Table 5).

This is a simple, fast ocular irritancy test, allowing the antiseptics to be classified partly as to whether they may be useful in clinical medicine (facial or ocular operations), alone, or added to cosmetics and beauty preparations, and partly to develop ideal dilutions according to the intended usage.

TABLE 5

Antiseptic	Cornea	Iris	Conjunctiva	Index
20% chlorhexidine				
1/5000	13.3	0.0	2.3	15.6
1/2000	13.3	0.0	4.0	17.3
5% chlorhexidine				
1/2000	0.0	0.0	3.0	3.3
Undiluted	35.6	5.0	7.83	83.5
Sterlane				
1/100	0.0	0.0	0.0	0.0

Source: Ref. 14.

V. CONCLUSIONS

Before a new antiseptic is put on the market, it is imperative that the skin and ocular irritation tests we have developed since 1976 be performed in order to prove the nontoxicity of the product.

These tests are obviously carried out for cosmetics and beauty preparations. The end products are also tested, as they are made up of a mixture of different constituents. The tests should be performed for each constituent, and, above all, for each antiseptic which may be included in the preparation of the cosmetic.

It should be further emphasized that an antiseptic used in small quantities, but daily, could have long-reaching effects on the living organism, mainly by the phenomenon of skin penetration. Furthermore, many products, antiseptics and others, could lead to sensitization and allergy in some people when used alone or as a mixture. Tests for these phenomena, although different from irritation ratings, exist and are carried out. We would like to mention, among others sources, Volume 7 of *Current Problems in Dermatology*, entitled *Skin: Drug application and evaluation of environmental hazards* [16].

REFERENCES

1. Ministerial order giving an indicatory list of the categories of cosmetics and products of personal hygiene, *J. Off. Repub. Fr.*, March 15 (1977).
2. Official methods for the analysis of cosmetics and beauty products, *J. Off. Repub. Fr.*, April 21 (1971), pp. 3862-3864.
3. Official methods for the analysis of cosmetics and beauty products, *J. Off. Repub. Fr.*, June 5 (1973).
4. Official methods for the analysis of cosmetics and beauty products, *J. Off. Repub. Fr.*, January 29 (1980), pp. 1181-1182.
5. D. Christol, Historique et définitions de l'antisepsie et de la désinfection, *Rev. Prat. Tome.*, 30:2159-2165 (1980).
6. AFNOR, Détermination de l'activité bactéricide des antiseptiques et désinfectants (methode par dilution-neutralisation), norme française T 72.150 (1974).

AFNOR, Détermination de l'activité bactéricide des antiseptiques et désinfectants (methode par filtration sur membranes), norme française T 72.151 (1974).

7. AFNOR, Détermination de l'activité bactéricide des antiseptiques et désinfectants en présence de substances interférentes définies, norme française T 72.160 (1977).
8. F. Lautier, D. Razafitsalama, and J. Lavillaureix, Tests d'irritation cutanée pour l'étude de la toxicité des solutions antiseptiques, *Pharm. Hosp. Fr.*, *43*:59-69 (1978).
9. F. Lautier, D. Razafitsalama, and J. Lavillaureix, Hautentzündstests zur Untersuchung des Toxizität antiseptischer Losungen, *Zeutralbl. Bakteriol. Parasitenkd. Infelctionskr. Hyg. Abt. Orig. Reihe, 167*: 193-205 (1978).
10. F. Lautier, D. Razafitsalama, and J. Lavillaureix, Le cobaye animal de choix pour des test d'irritabilité et d'agressivité cutanée, Comte-rendu du Congrès International de la Fondation Mérieux: l'animal de laboratoire au service de l'homme, Collection Fondation Mérieux, Lyon, pp. 97-102, 1980.
11. J. L. Rouaud, Vérification d'innocuité: les tests biologiques, *Labo. Pharma. Probl. Tech.*, *218*:47-52 (1973).
12. B. Beys, J. Dufaux, I. Prohoroff, and M. Millet, Contrôle toxicologique des cosmétiques, *J. Pharm. Belg.*, *30*:195-222 (1975).
13. P. Bourrinet and D. Rodde, Vérification de la tolérance locale de quelques conservateurs et bactéricides, *Labo. Pharma. Probl. Tech.*, *240*:121-124 (1975).
14. R. Pagot, F. Lautier, and A. Brini, Action toxique d'une solution de chlorhexidine à diverses concentrations sur la cornée du cobaye, *Bull. Soc. Ophthalmol. Fr.*, *80*:631-634 (1980).
15. F. Lautier, R. Pagot, and A. Brini, Etude de la toxicité des solutions antiseptiques: mise au point d'un test d'irritation oculaire, Bull. Soc. Pharm. Strasbg. T XXIV, n°1, 15-17, 1981.
16. G. A. Simon, Z. Paster, M. A. Klingberg, and M. Kaye, Current Problems in Dermatology, Vol. 7. Skin: Drug Application and Evaluation of Environmental Hazards, Series Editor J. W. H. Mali, S. Karger, Basel, 209 pages, 1978.

27
SAFETY EVALUATION OF COSMETIC PRESERVATIVES

ROBERT L. BRONAUGH *U.S. Food and Drug Administration, Washington, D.C.*

HOWARD I. MAIBACH *University of California Medical School, San Francisco, California*

I. Introduction 503
II. Adverse Reactions 506
III. Methods for Safety Evaluation 508
    A. Irritation 508
    B. Sensitization 512
    C. Human assay variables 520
    D. Phototoxicity 521
    E. Contact photosensitization 525
    F. Other types of toxicity 527

References 528

I. INTRODUCTION

A recent study [1] conducted for the Food and Drug Administration (FDA) by the North American Contact Dermatitis Group has added greatly to our knowledge of adverse reactions to cosmetic products. Preservative ingredients, as a group, caused the second highest incidence of reactions (less than fragrances) based on patch test results. The dermatitis-causing ingredients identified in the study are listed in Table 1. Many of the ingredients associ-

---

Adapted and modified in part from R. L. Bronaugh and H. I. Maibach, Adverse reactions to cosmetic ingredients, in *Principles of Cosmetics for the Dermatologist* (P. Frost and S. Horwitz, eds.), C. V. Mosby, St. Louis, Mo. (1982), with permission of publisher.

TABLE 1 Dermatitis-Causing Ingredients[a] Identified by Patch Testing

Ingredient	Number of cutaneous reactions	Ingredient	Number of cutaneous reactions
Allantoin	1	Imidazolidinyl urea	15
Amyl cinnamic aldehyde	2	Isoeugenol	10
Amyldimethyl PABA[b]		Jasmine, absolute	2
(Padimate A)	1	Jasmine, synthetic	2
Beeswax	1	Lanolin	11
Benzalkonium chloride	2	Lanolin alcohol	7
Benzocaine	1	Lanolin oil	2
Benzoin	2	Methacrylate monomer,	
Benzophenone, unspecified	1	unspecified	1
Benzophenone-4	2	Microcrystalline wax	1
Benzophenone-8	1	Musk ambrette	8
Benzyl alcohol	2	Neomycin	1
Benzyl benzoate	1	Nitrocellulose	1
Benzyl salicylate	1	Oak moss	1
Butylated hydroxyanisole	2	Oleyl alcohol	1
Bismuth oxychloride	1	Oxyquinoline	1
2-Bromo-2-nitropropane-		PABA	3
1,3-diol	8	Paraben, unspecified	15
Butyl acetate	1	Peru balsam	2
Captan	2	p-Phenylenediamine	24
Cetearyl alcohol	1	Potassium sorbate	1
Cetyl alcohol	1	Propylene glycol	22
Chloroxylenol	1	Quaternium-15	32
Cinnamic alcohol	14	Resorcinol	1
Cinnamal (Cinnamic		Sandalwood oil	2
aldehyde)	5	Sorbic acid	3
Clove oil	1	Stearamido diethylamine	3
Coal tar	1	Stearic acid	1
Coumarin	4	TEA-stearate	1
Dibutyl phthalate	1	Tetrachlorosalicylanilide	1
Diethylene glycol		Tetrahydrofurfuryl	
dimethacrylate	1	methacrylate	1
Disodium monooleamido-		Thioglycolate,	
sulfosuccinate	1	unspecified	1
Ethyl methacrylate	3	Tocopherol	1
Eugenol	4	Toluenesulfonamide/	
Formaldehyde[c]	10	formaldehyde resin	16
Fragrance, unspecified	39	Tribromsalan (tribromo-	
Geraniol	5	salicylanilide)	1
Glyceryl monothioglycolate	3	Triclosan	1
Glyceryl PABA	3	Triethanolamine	3
Hydrolyzed animal protein	1	Ultraviolet absorber,	
Hydroxycitronellal	9	unspecified	1

[a]Ingredients identified according to the Cosmetics, Toiletry, and Fragrance Association's *Cosmetic Ingredient Dictionary*, 2nd Ed., or nomenclature used by Research Institute for Fragrance Materials.
[b]PABA, p-aminobenzoic acid.
[c]In one case identified as paraformaldehyde.

TABLE 2 Preservative Use in Cosmetic Formulas

Preservative	Frequency
Methylparaben	6785
Propylparaben	6174
Imidazolidinyl urea	1684
Quaternium-15	1001
Formaldehyde solution	874
Butylparaben	668
2-Bromo-2-nitropropane-1,3-diol	566
Butylated hydroxyanisole	518
Butylated hydroxytoluene	405
Sorbic acid	393
Sodium dehydroacetate	191
Ethylparaben	159
Phenylmercuric acetate	147
Citric acid	131
Propyl gallate	121
Dehydroacetic acid	107
Benzylparaben	97
Sodium benzoate	89
Triclosan	88
Benzoic acid	84
DMDM hydantoin	79
Potassium sorbate	76
Chloroxylenol	71
Disodium EDTA	68
Sodium sulfite	67
o-Cymen-5-ol	55
Dimethoxane	45
Captan	41
Trisodium EDTA	40
5-Chloro-2-methyl-4-isothiazolin-3-one	38
2-Methyl-4-isothiazolin-3-one	38
Benzophenone-1	36
Benzyl alcohol	35
Zinc pyrithione	29
Chlorhexidine digluconate	28
Benzalkonium chloride	27
Benzethonium chloride	27
Tripotassium EDTA	27
Boric acid	25
Phenoxyethanol	25
Chloroacetamide	24
Fluorosalan	24
EDTA	23
Glutaral	23
Sodium borate	23
Thimerosal	22
Myristalkonium chloride	20
Paraformaldehyde	20
Dipotassium EDTA	17
Tetrasodium EDTA	17

Source: FDA cosmetic file, December 1980; this file contained 19,548 formulations from 902 companies.

ated with the most frequent reactions are preservatives. The Division of Cosmetic Technology at the FDA receives, on a voluntary basis, information about the composition of cosmetic products from participating members of the cosmetic industry; the most frequently used cosmetic preservatives from this data base are given in Table 2 [2]. A number of the preservatives found to cause a dermatitis are included in this list. Sensitization reactions to preservatives are more commonly observed, at least in part because of the lower doses required for this effect.

Some known reactions to cosmetic preservatives will be presented, and methods for safety evaluation will be discussed.

II. ADVERSE REACTIONS

Marzulli and Maibach [3] have reported irritation reactions to three salts of pyrithione and hexachlorophene; they used repeated applications to increase the sensitivity of the testing procedure. The compounds were tested on human and rabbit skin and applied in a petrolatum vehicle (see Table 3). As is often found, rabbit skin was more sensitive to these irritants and exhibited a dose-response reaction at the three concentrations tested. In both human and rabbit skin, 3% hexachlorophene was the strongest irritant. Sodium pyrithione was more reactive than the other pyrithione salts.

TABLE 3 Comparison of Cumulative Irritation Scores for Antimicrobials and Preservatives in Humans and Rabbits, Using the 21-Day Occlusive Test in Human and the 16-Day Uncovered Test in Rabbits

Compound or active ingredient	Vehicle	Concentration (%)	Mean cumulative irritation score	
			Human	Rabbit
Zinc pyrithione	Petrolatum	0.05	0.0	—
		0.5	0.02	0.8
		2.0	0.1	9.5
Sodium pyrithione	Petrolatum	0.05	0.0	2.8
		0.5	0.0	5.1
		2.0	2.9	37.8
MDS pyrithione	Petrolatum	0.05	0.0	—
		0.5	0.0	6.6
		2.0	0.0	12.0
Hexachlorophene	Petrolatum	0.03	0.0	0.6
		0.3	0.4	7.1
		3.0	10.4	44.1

Safety Evaluation

TABLE 4 Sensitization Response in Human Subjects

Compound	Induction concentration (%)	Challenge concentration (%)	Fraction	Percentage sensitized
Sorbic acid[a]	10	5	0/93	0
	20	5	1/33	3
Formalin[b]	10	1	8/102	7.8
	5	1	4/52	7.7
Bronopol[a]	5	2.5	11/93	12
Chloroxylenol[a]	5	5	0/208	0
	20	10	0/110	0
Phenyl mercuric acetate[a]	0.125	0.01	1/56	2
	0.125	0.001	0/56	0
Zinc pyrithione[a]	3	3	0/10	0
Chloroacetamide[b]	1.25	1.25	35/205	17
Captan[c]	1.0	1.0	9/205	4.4
Methyl (M) and propyl- (P) parabens[a] (same concentration for induction and challenge)		0.2 M + 0.05 P	0/102	0
		1.0 M + 0.25 P	0/101	0
		5.0 M + 1.25 P	1/98	1
		10.0 M + 10.0 P	0/74	0

[a]Petrolatum vehicle.
[b]Water vehicle.
[c]Cream vehicle.

Steinberg et al. [4] compared the relative irritancy of 12 chemicals in rabbits and humans. Five of these compounds are cosmetic preservatives: salicylic acid, benzoic acid, formaldehyde, resorcinol monoacetate, and benzalkonium chloride. The rankings of the compounds in the repeat application tests were similar in human and rabbit skin. Irritation on rabbit skin seemed to be a better predictor of the human response when repeat applications of threshold irritation concentrations were used.

The sensitization response for a number of antimicrobial agents has been tested in humans [5]. The incidence of response to preservatives in cosmetics is listed in Table 4. The most reliable test information was obtained from the human Draize test with at least 200 subjects.

Jordan and King [6] compared three frequently used preservatives in cosmetics with regard to frequency of positive results in the patch and use tests. For all three compounds (propyl- or methylparaben, imidazolidinyl urea, and quaternium-15), the incidence of sensitization was higher in the patch test. These results were explained as due to differences in concentration and vehicles. Quaternium-15 had the largest number of positive patch tests in the authors' clinic (3.5%) and in the clinics of the North American Contact Dermatitis Group (3.1%). Product use tests resulted in a sensitization rate of 3 in 325 subjects. It was suggested that more than one vehicle be used for best results. The use of parabens in cosmetic products and reported hypersensitivity reactions have been discussed by Marzulli and Maibach [7].

III. METHODS FOR SAFETY EVALUATION

A. Irritation

1. Draize Test

The standard test for evaluating the irritancy potential of a compound in animals was published by Draize et al. in 1944 [8]. The nature of the test is such that errors in predicting toxicity in humans are conservative and irritant effects sometimes appear to be greater than those actually observed in humans [9,10]. The long exposure time on the very permeable rabbit skin results in a test that will identify many substances which are also irritants on human skin.

The rabbit Draize test is required by law under provisions of the Federal Hazardous Substance Act and is given below as it appears in the Code of Federal Regulations [11].

Primary irritation to the skin is measured by a patch-test technique on the abraded and intact skin of the albino rabbit, clipped free of hair. A minimum of six animals are used in abraded and intact skin tests. The test substance is introduced under a square patch, such as surgical gauze measuring 1 in. by 1 in. and two single layers thick — 0.5 ml in the case of liquids, or 0.5 g in the case of solids or semisolids. Solids are dissolved in an appropriate solvent and solutions are applied as for liquids. The animals are immobilized with patches held in place by adhesive tape. The entire trunk of the animal is then wrapped with an impervious material, such as rubberized cloth, for the 24-hr period of exposure. This material aids in maintaining the test patches in position and retards the evaporation of volatile substances. After 24 hr of exposure, the patches are removed and the resulting reactions are evaluated on the basis of the designated values shown in Table 5. Scoring is repeated at 72 hr. Equal numbers of exposures are made on abraded skin.

Modifications have been made in the Draize procedures to increase the uniformity of results, to test compounds on a less vigorous basis more similar to use conditions, and to meet the specific needs of the investigator. A shorter period of contact with the skin (4 hr) was proposed by the FDA [12], but this change has not yet been officially promulgated. A committee of the National Academy of Sciences [13] recommended that 4 hr and even shorter times be utilized for certain purposes. An example of the type of warnings sometimes developed from modified tests is given in Table 6.

In a study directed by Weil and Scala [15], 25 laboratories provided data on the irritancy of 16 materials to skin and eyes by the Draize procedure and a modified Draize procedure of their choice. Variable results were obtained from the different laboratories. The primary reason for the variability was considered to be differences in the scoring of the reactions by the different personnel involved. It was suggested that, in order to standardize irritancy testing, training courses in methodology should be developed and given to technicians.

TABLE 5 Grading Values for Skin Reactions in Albino Rabbits Following Topical Application of Potential Primary Irritants[a]

Skin reaction	Value [a]
Erythema and eschar formation	
No erythema	0
Very slight erythema (barely perceptible)	1
Well-defined erythema	2
Moderate to severe erythema	3
Severe erythema (beet redness) to slight eschar formation (injuries in depth)	4
Edema formation	
No edema	0
Very slight edema (barely perceptible)	1
Slight edema (edges of area well defined by definite raising)	2
Moderate edema (raised approximately 1 mm)	3
Severe edema (raised more than 1 mm and extending beyond the area of exposure)	4

[a]The value recorded for each reading is the average value for the six or more animals subjected to the test.
Source: Ref. 11.

2. *Testing Variables*

Abrasion: The maximum effect of an irritant would be expected to be observed with abraded skin, since percutaneous absorption should not be a limiting factor. For this reason the Draize test is done, in part, on abraded skin. Draize noted that the stratum corneum should be incised without drawing blood, but did not specify a procedure. McCreesh and Steinberg [16] investigated the possibility that the different methods commonly used to abrade or scarify the skin could influence the observed results. No difference in irritation was detected when abrasion was caused by either a hypodermic needle, the Berkeley Scarifier (Berkeley Biologicals, Berkeley, Calif.), or the abrader of Haley and Hunziger [17]. The value of abrasion in the Draize rabbit test was questioned by Nixon et al. [9], since no correlation was found between the effects of abrasion on animal results and irritation in humans. Frosch and Kligman [18], however, have proposed the use of this procedure to increase the sensitivity in humans.

The clinical relevance of testing a preservative for irritancy may be questioned in that relatively low concentrations are generally used (with the exception of propylene glycol) in the final cosmetic formulations. Yet the cosmetic formulator must choose between many components of a proposed formulation on the basis of relative safety and toxicity. In carefully weighing these factors in a complex balance, knowledge of the *relative inherent irritancy potential* of each preservative should be of value. The only manner of determining this relative irritancy potential is by increasing the concentration to determine at what level the preservative causes irritation in animal and human skin.

TABLE 6 National Institute for Occupational Safety and Health Interpretation of Skin Testing Ratings

	Rating	Interpretation
Intact skin	0.0.9	Nonirritant; probably safe for intact human skin contact
	1-1.9	Mild irritant; may be safe for use, but appropriate protective measures are recommended during contact
	2-4	Too irritating for human skin contact; avoid contact
Abraded skin	0-0.9	Nontoxic to cellular components of abraded skin; probably safe for human skin contact
	1-1.9	Mild cellular toxins; may be safe for abraded skin contact, provided that protective measures are employed
	2-4	Cellular toxins too irritating for abraded skin contact; avoidance of contact is advised

Intact skin	Abraded skin	Mixed reactions
0-0.9	0-0.9	Safe for human skin contact
	1-1.9	Safe for intact human skin contact; may be safe for abraded skin contact when protection is maintained
	2-4	Safe for intact human skin; contact with abraded skin should be avoided
1-1.9	1-1.9	May be safe for intact and abraded skin when protection is maintained
	2-4	May be safe for intact human skin contact when protection is maintained, but contact with abraded skin is to be avoided
2-4	2-4	Unsafe for intact and abraded human skin contact; avoid contact

Source: Ref. 14.

Animal Species: The animal of choice is the albino rabbit because of its sensitivity and because of the extensive past experience with this animal. Other species have been examined because of a desire for a more convenient animal or one whose response more closely resembles that of humans.

Some investigators [9,19] have observed that the rabbit and the guinea pig respond similarly to certain compounds. Brown [20] contends that a different response might be expected with these two species and that the resultant data with the rabbit and guinea pig should be used to complement each other. Davies et al. [21], after conducting irritant tests in the rabbit, guinea pig, mouse, dog, baboon, piglet, and miniature pig, concluded that the best overall predictor of human response was the rabbit. Very little response was obtained from the skin of the beagle dog in tests that also found guinea pig skin to be less reactive than rabbit skin with 5-day dermal exposure [22]. Marzulli and Maibach [23] obtained an insufficient response to irritants from the hairless rat and the hairless hamster.

Application Method: Since an irritant reaction is observed only after a compound has penetrated the barrier layer of the skin, the well-known effect of the vehicle on percutaneous absorption may influence the reaction observed. Little information is currently available in this regard. Occlusion is recognized as a method of enhancing penetration and might therefore be expected to increase the irritant effect. Future work should more fully document the relative merits of occlusion (versus open application) in acute and repetitive application studies.

Repetitive Applications: Because of the lower sensitivity of human skin to irritants, testing in humans is generally performed by repeated-application patch testing such as the 21-day cumulative irritancy assay described by Philips et al. [10].

In the case of strong or moderate irritants, the single application used in the rabbit Draize test is usually sufficient to elicit a response. However, with weak irritants this Draize test has sometimes failed to predict toxicity in humans. When the Draize procedure was utilized for human testing, it also failed to detect weak irritants on human skin [10].

Repeated-application testing with animals has been used successfully by Ingram and Grasso [24], Brown [20], and MacMillan et al. [22]. Marzulli and Maibach [23] tested 60 materials by repeated applications in humans and rabbits. The rabbit test (16 applications) was patterned after the 21-day occlusive test used in humans. Good correlation was generally obtained between results from animals and humans. Steinberg et al. [4] compared the relative irritancy of 12 chemicals in rabbits and humans. The rabbits were given 21 applications for close comparison to the 21-day human patch test, and predetermined threshold concentrations of irritants were utilized to increase sensitivity. The close agreement between results from rabbit and humans is shown in Table 7.

3. General Comments

The single-application test is invaluable because of its efficiency and relative quality of data. It will detect strong and moderate irritants, and its basic value lies in ranking the irritancy potential of raw ingredients. Accurate extrapolation of the data to the use situation can depend on the experience of the investigator. Including standard substances as an internal control, although all too infrequently performed, adds credence in terms of extrapolating the data to humans.

Cumulative irritancy assays in animals and humans allow for more sophisticated rankings because final products and lower concentrations of individual ingredients can be tested. Also, in these tests the inclusion of internal standards in each assay greatly increases the ability to extrapolate to the use situation.

TABLE 7 Comparison of Irritant Rankings for 12 Compounds Applied Daily for 21 Days Under Saran Wrapping (Occluded) or Elastic Cloth (Partially Occluded) at Threshold Irritant Concentrations[a]

Test compound	Irritant ranking[b]		
	A To people (occluded)	B To rabbits (partially occluded)	C To rabbits (occluded)
Salicylic acid	3	2	1
N-Butylsulfonimidocyclohexamethylene	6	4	4
N-Hexadecylpyridinium	12	12	10
Benzoic acid	2	1	2
Formaldehyde	4	10	8
Resorcinol monoacetate	5	8	11
Cyclohexamethylene carbamide	9	11	12
Triethyleneglycol-N-hexyl ether	7	7	9
Resmethrin	10	6	6
Benzalkonium chloride	11	5	5
Diethyltoluamide	1	3	3
Triethanolamine	8	9	7

Ranking analysis[c] relative irritant ranking analysis

    A vs. B   Rank coeff. = 0.61 $p < 0.05$

    A vs. C   Rank coeff. = 0.53 $p < 0.05$

    B vs. C   Rank coeff. = 0.09 $p < 0.0001$

[a]Ref. 4.
[b]A value of 1 = most irritating.
[c]Ref. 25.

B. Sensitization

1. Draize Test

Draize published the first animal test model with a predictive character for determining the allergenicity of new chemical substances [26,27]; this test is a variant of the classic Landsteiner technique.

A total of 40 albino guinea pigs are divided into two groups of 20: an experimental group and a control group. The test substance is injected intradermally as a 0.1% solution, suspension, or emulsion in 0.95% sodium chloride, paraffin oil, or polyethylene glycol. The animals receive an intradermal injection every second day for a total of 10 injections in 3 weeks. Scoring of the skin reaction is first performed on day 2, and then 24 hr after each injection.

On day 35 the final challenge (elicitation) is performed on the contralateral flank of the experimental animal on a skin site corresponding to the site of the first injection. Control animals are treated in the same way simultaneously, with 0.05 ml of the 0.1% test solution injected intradermally per animal. On days 36 and 37 the animals in both groups are shaved, and the intensity of erythema and occurrence and the size of edema of the test reaction are recorded.

To determine whether a chemical is a sensitizer, the reactions of all animals to the first intradermal injection (0.05 ml) are compared with the challenge test reactions among themselves and with the reactions of the control animals. If there are extreme differences between the reactions within the same group, the mean values for the induction and the test phase within the experimental group are compared. The same holds for the mean values for the eliciting tests within both experimental and control groups.

The guinea pig Draize technique is easily performed. The route of application is often unrealistic compared to intended use conditions, and the induction concentration (0.1%) is fixed without considering the use concentration. Quantitating intradermal skin test readings involves significant difficulties. The problem of the excited skin state probably produces a significant number of false positives [28].

The Draize test is not sufficiently sensitive to identify the allergenic potential of many contact allergens [29,30]. It is useful for individual ingredients, but not for final formulations. Because of its minimal discriminatory powers, it is frequently modified or replaced by other testing techniques [31]. Doubling the number of injections and elevating the test concentrations enhances its sensitivity.

2. Freund's Complete Adjuvant Test

Freund's complete adjuvant test (FCAT) is a variant of the intradermal test method for screening chemicals and components of medical, cosmetic, and other products for their immunogenic properties [29]. It is a semiquantitative test method in which a sensitizing effect can be demonstrated.

Two groups of 8-10 guinea pigs are used as the experimental and control groups. For induction by intradermal application, the substance is incorporated in Freund's complete adjuvant so that the final concentration in the emulsion is usually 5-50%, depending on the physicochemical properties and the toxicity of the material tested. The intended use concentration is considered in determining the induction concentration. For the challenge, performed by open (topical) application, the test material is diluted, emulsified, or suspended in the appropriate vehicle (e.g., water, acetone, alcohol, petrolatum, or polyethylene glycol). The challenge concentrations vary in steps of three, so that the test substance can be tested undiluted and in 30, 10, and 3% concentrations or lower.

The test material in FCA (0.1 ml) is injected intradermally into the right anterior flank of the animals in the experimental group every second day for a total of five times. Animals of the control group are injected with 0.1 ml of FCA only. Four guinea pigs are used to determine the threshold toxic concentration after a single topical application by simultaneously applying four concentrations, such as 100, 30, 10, and 3%, to the left flank skin. The highest concentration used for the topical application test is limited by the solubility of the compound. The application site is left uncovered, and skin reactions are scored 24 hr after application. The minimal irritant concentration and the maximal nonirritant concentration (threshold concentrations of each substance) are determined by an all-or-none criterion. The minimal irritant concentration is defined as the lowest one causing mild erythema in at least 25% of the animals tested. The maximal nonirritant concentration is defined as the highest one not causing a reaction in any animal.

The FCAT determines the sensitizing capacity of the material and the minimal concentration eliciting an allergic response. The test material is considered allergenic when one of eight animals of the experimental groups shows a positive reaction to the nonirritant concentrations used for challenge.

The FCAT is technically simple, but not applicable for finished products. The interpretation of the test results depends on how the material is to be used and on its chemical structure (relationship to known contact sensitizers). Because of the adjuvant, it is generally more sensitive than the Draize test and the Buhler test. It is useful for identification of allergenic potential. Its semiquantitative aspects must be considered when attempting extrapolation to humans. Freund's adjuvant may enhance the likelihood of a false positive due to the excited skin state [28].

3. *Guinea Pig Maximization Test*

By determining factors favoring sensitization in the guinea pig and combining them in a single procedure, Magnusson and Kligman [30,32,33] developed the most sensitive procedure for detecting the capacity of a substance to induce contact sensitivity in guinea pigs. By comparing the results from this test with clinical experience and by using known human contact allergens, Kligman [34] demonstrated a high degree of correlation between the results of the guinea pig maximization test and the human maximization test.

Two groups of 20-25 guinea pigs each are used as an experimental group and a control or vehicle group. The test agent is applied intradermally and topically. The injections are made with the test substance given independently and also incorporated in FCA. When incorporated in FCA, water-soluble test agents are dissolved in the water phase before emulsification. Oil-soluble or oil-insoluble materials are dissolved or suspended in FCA before water is added. The solvent system is similar for the test agent without FCA, with paraffin oil, peanut oil, or propylene glycol being used for dissolving or suspending water-insoluble materials. The concentration of the test agent for intradermal injections is adjusted to the highest well-tolerated level.

The topical induction exposure is performed as an occlusive patch. The concentration of the test material is adjusted to the highest well-tolerated level; if the test agent is irritating, a concentration is chosen that causes weak to moderate inflammation. Solids are micronized or reduced to a fine powder and incorporated in a vehicle, usually petrolatum. Water- and oil-soluble substances are dissolved in water, ethanol, or other appropriate vehicles. Liquid materials are applied as such, or diluted if necessary.

There are two stages in the induction phase. The shoulder region is the induction site. Injections are 0.1 ml of FCA alone (adjuvant blended with an equal amount of water), 0.1 ml of test material, and 0.1 ml of test material in FCA. The control group receives 0.1 ml of FCA alone (adjuvant blended with an equal amount of water), 0.1 ml of vehicle alone, and 0.1 ml of vehicle in FCA. This control treatment should minimize false-positive reactions due to nonspecific hyperirritability challenge responses.

On day 7 the test agent in petrolatum is spread on filter paper to saturation. The patch is covered by an impermeable plastic adhesive tape and secured by an elastic adhesive bandage which is wound around the torso and left in place for 48 hr. If the test agent is nonirritating, the area is pretreated with 10% sodium lauryl sulfate in petrolatum 24 hr before the patch

TABLE 8 Allergenicity Rating

Sensitization rate (%)	Grade	Classification
0-8	I	Weak
9-28	II	Mild
29-64	III	Moderate
65-80	IV	Strong
81-100	V	Extreme

is applied. The control animals are exposed to the vehicle without the test agent in the same way as the experimental group.

On day 21 the experimental and control animals receive occlusive patches for 24 hr. By using ready-made test units of smaller size, for instance, an aluminum chamber (Finn-Chamber, Helsinki), challenge with the test agent at different concentrations and the vehicle can be performed on the same flank [35,36]. Scoring is performed on days 23-28. The intensity and duration of the test responses to the test agent and the vehicle are evaluated. If the challenge reactions in the experimental group clearly outweigh those in the control group, the test agent is a sensitizer.

The important data in the guinea pig maximization test are the frequency of sensitization and not the intensity. The test substances can be assigned, according to the percentage of animals sensitized, to one of five classes, ranging from weak (grade I) to extreme (grade V) (see Table 8).

Treatment with FCA, sodium lauryl sulfate, the test chemicals, and occlusive bandage may lower the threshold level for skin irritation. A challenge concentration of the test agent, which in nontreated animals has been found to be below that level, can evoke false-positive responses at the time of challenge of the animals in the experimental group. To exclude false-positive results, the control group should be exposed to the same manipulations as the animals in the experimental group. A more complete discussion of the nonspecific cutaneous hyperexcitability (excited skin state) is provided by Maibach [28].

The adjuvant injection, especially when it is superficial, often causes ulceration that lasts several weeks. These lesions are undesirable, but do not invalidate the test results except for decreasing the threshold level of skin irritation.

The guinea pig maximization test is an excellent procedure for the identification of contact allergens, but is far less adequate for predicting the sensitization potential of finished products. If a chemical in this test is a clear-cut sensitizer but has virtues that merit continuing interest, it can be tested in the form and concentration of actual use, for example, as a cosmetic, topical drug, fabric finisher, or insecticide. Thus the end product, and not the chemical itself, is tested.

4. Split Adjuvant Technique

Maguire [37-39] and Maguire and Chaee [40] utilized an adjuvant technique that magnifies the processes leading to sensitization in the guinea pig, so that even weak and moderately weak contact allergens in humans can be identified [39]. This technique derives from the split adjuvant test procedures [40,41] in which allergen and FCA are administered separately and together with the test material. Maguire utilized the observation by Magnusson and Kligman [30] that intradermal injection of FCA beneath a site where topical allergens have been applied for challenge greatly potentiates sensitization. The technique has not yet found a place in the routine toxicological armamentarium.

5. Buhler Test

Manufacturers of cosmetics, drugs, fragrances, household products, and chemical raw materials recognize the practical importance of "premarketing safety evaluation."

Buhler [42,43] and Griffith and Buhler [44] presented their own predictive test methods, "using an animal model which mimics the human repeat insult patch test (RIPT), allows realistic exaggeration of usage conditions, and is able to avoid unnecessary sensitization of human subjects." Less than 3% of the materials tested that did not sensitize the guinea pig have been found to sensitize humans by the repeated-insult patch test [45]. The Buhler test is sensitive enough to detect moderate to strong sensitizers and at the same time does not produce false-positive reactions that might lead to the rejection of an otherwise safe and useful compound or product.

Three groups of 10-20 guinea pigs each are used as the experimental, vehicle control, and negative control groups. The test substance is diluted, emulsified, or suspended in a suitable vehicle and applied to the flank skin. The concentration used for induction is allowed to provoke only a weak skin irritation and is calculated with respect to the concentration of use. The test material and the vehicle are applied to the skin with an occlusive patch for 6 hr. During the 6-hr exposure, the animals of both groups are immobilized in a restrainer by a piece of rubber band stretched taut over the back to provide for fixation for the occlusive patch. Induction patches are applied at weekly intervals for 3 weeks.

On day 28 the challenge is identical with the 6-hr occlusive patch test of the induction phase. The test concentration is chosen such that the guinea pig skin tolerates it without irritation. The experimental group is challenged with test material and vehicle, the vehicle control group with vehicle alone, and the negative control group (not pretreated in the induction phase) with test material alone. Grading is on a scale of 1 (slight erythema) to 3 (marked erythema). A comparison of the intensity and duration of the eliciting reactions in all animal groups permits identification of sensitization reactions. The results are expressed in terms of the incidence and severity of response.

This model mimics conditions of use: Application is by the topical route, the induction concentration in most cases corresponds to use concentrations, and the solvent or vehicle is chosen to be identical or similar to that of the final formulation. The finished preparations are tested as such or, if appropriate, diluted (e.g., for shampoos). The sensitivity of the method is comparable to that of the repeated-insult patch test for humans. The validity of the results may be restricted, depending on the limited test concentration, which is particularly important for individual chemicals.

Properly interpreted, this technique has many advantages. It produces fewer false positives than the previously described tests, because there is less irritation. Although it is less sensitive than the FCAT and guinea pig maximization test for individual chemicals, it is a convenient first screen for final formulations.

6. Open Epicutaneous Test

By carefully analyzing the factors and by including all of these factors in the same technique, Klecak et al. [29] developed an animal model, the open epicutaneous test (OET), for screening, simulating the conditions of human use, and obtaining quantitative results expressed as inducing and eliciting concentrations. This approach is in contrast to those methods in which compounds are tested at one concentration, and the results are expressed as the degree of intensity of the lesion or the percentage of sensitized animals.

One to six experimental groups and one control group of six to eight guinea pigs are used. Substances are applied topically uncovered, undiluted, and, if possible and relevant, dissolved, suspended, or emulsified, in concentrations of 30, 10, 3, and 1% or lower in ethanol, acetone, water, petrolatum, polyethylene glycol, and/or other suitable vehicles. Finished products are tested as such or are diluted according to use (shampoos). Constant volumes of each concentration are applied with a pipet or syringe on standard areas of the clipped flank of each animal.

Before starting the induction procedure, an experimental group is used to determine the threshold irritant concentration after a single application by simultaneously applying the different concentrations (e.g., 100, 30, 10, and 3%) to the left flank skin. A 0.025-ml portion of each test concentration is homogeneously spread on an area of flank skin. The highest concentration of a compound used in this local application test is determined by its solubility and (for challenge) by its skin-irritating capacity. The application site remains uncovered.

Skin reactions are read 24 hr after application. The minimal irritant and the maximal nonirritant concentrations (threshold concentrations of the substance) are determined by an all-or-none criterion. The minimal irritant concentration is defined as the lowest one causing mild erythema in at least 25% of an animal group. The maximal nonirritant concentration is the highest one not causing a reaction in any animal. Estimation of the threshold concentrations is valuable for evaluations of the allergenic capacity of the test material, based on the end-point determination.

For induction, a 0.1-ml portion of each undiluted compound and/or of its progressive dilutions is applied to the clipped flank skin of six to eight guinea pigs per concentration group, using one to six such groups for each compound. Applications are repeated daily for 3 weeks or five times weekly for 4 weeks, always at the same reaction site. The application site remains uncovered and the reactions, if continuous daily applications are performed, are read 24 hr after each application or at the end of each week. The maximal nonirritant and minimal irritant concentrations after repeated applications are determined by the same all-or-none criterion. When strong skin reactions are provoked, the application site is changed.

To determine whether contact sensitization was induced, all guinea pigs treated for 21 days as described above, as well as six or eight untreated controls for each compound, are tested on days 21 and 35 on the contralateral flank with the same compound at the minimal irritant and some lower concentrations. The minimal irritant concentration of each compound is used to confirm the biological activity determined before starting the induction (day -1) and to exclude false results due to instability of the test material. Reac-

TABLE 9 Number of Compounds Described as Allergenic in Humans and Detected as Allergenic by the Four Animal Tests

	Total number of compounds	Open Epicutaneous Test	Intradermal tests		
			Draize Test	Maximization Test	Freund's Complete Adjuvant Test
Positive					
In OET and intradermal tests	18	18	7	15	17
In OET only	4	4	0	0	0
In intradermal only	3	0	1	3	3
Total positive	25	22	8	18	20
Negative in all tests	7				
Total	32				

tions are read after 24, 48, and/or 72 hr. This procedure permits determination of the minimal sensitizing concentration necessary for inducing allergic contact hypersensitivity and the minimal eliciting concentration necessary to cause a positive reaction. A concentration is considered allergenic when at least one of the animals of the concentration group concerned shows a positive reaction with nonirritant concentrations.

Klecak et al. [29] compared 32 fragrance materials described as allergenic in humans. The results of these comparative studies are summarized in Table 9. Of 32 compounds incriminated as causing allergic contact dermatitis in humans, 25 were detected as allergenic in the four tests. All compounds with well-established allergenicity in humans were detected with the combined procedure (OET plus FCAT), most of them by the OET alone. The seven remaining substances, which lacked confirmative data such as positive history, positive patch test, or reexposure, were negative in all animal tests. Their "allergenicity" might be due to a cross-sensitizing capacity.

Klecak [46] performed comparative animal tests with 21 highly concentrated perfume compositions, all of them ingredients of cosmetics, pharmaceutics, and household products that had been on the market for years and which were known as clinically harmless. The results obtained are shown in Table 10; of the 21 compositions, 19 did not sensitize the guinea pig in the OET, even when applied undiluted, and the remaining two were sensitizing only at full strength, that is, at concentrations 25-300 times higher than those used in the marketed products. Table 10 also shows that eight compositions were inactive in the four tests; 11 of the 21 were allergenic when tested by the FCAT and other intradermal tests. Klecak concluded that clinically innocuous compositions usually do not sensitize experimental animals by the topical route and that the sensitization by intradermal procedures may not be of clinical relevance.

The OET is a suitable method for testing sample chemicals, mixtures, and finished products for their irritating and sensitizing capacity. The re-

TABLE 10 Allergenicity in Four Animal Tests of 21 Perfume Compositions Harmless in Human Use

			Intradermal tests	
Total number of compounds	Open Epicutaneous Test	Draize Test	Maximization Test	Freund's Complete Adjunant Test
Positive				
In all tests	1	1[a]	1	1	1
In OET only	1	1[a]	0	0	0
In intradermal only	11	0	1	5	11
Total positive	13	2	2	6	12
Negative in all cases	8				
Total	21				

[a]Undiluted, positive; at user concentration, negative.

sults obtained with the OET are correlated with the results of both the maximization test in humans (34, 47-49] and the consumer test. Animal experiments are essential for solving the preliminary questions. The test on humans should be regarded as the final stage of the test series.

7. General Comments

These guinea pig assays are invaluable when used properly. The methods vary greatly in their ease of performance and sensitivity, factors that should be considered in the selection of a method for a study. No matter which assay is employed, the results obtained are influenced by the experience of the investigator. There is no substitute for extensive experience to identify which experiment is reliable or which requires repetition or further study.

The assays differ primarily in terms of dose (fixed or graded) and routes of skin contact (topical, intradermal, or both). It would be simplistic to rank them in terms of sensitivity. The Draize test, generally considered the least discriminating, can be made more sensitive by increasing the dose and the number of injections. The Buhler method, when used properly, has an enviable record as a screen for final formulations before human testing. The model with the greatest sensitivity is the guinea pig maximization test, because of the high concentrations and adjuvant. The major conceptual advantage of the OET resides in quantifying the dose required to induce and to elicit.

Thus the toxicologist has a variety of tools and must determine which is the most relevant to his needs. No matter what the assay, the test indicates potential for sensitization. Extrapolation to the use situation in a final formulation requires numerous other calculations, experimental maneuvers, and experience.

C. Human Assay Variables

All human assays are modifications of the guinea pig model of Landsteiner and Chase [50]. None utilize the intradermal method (such as the guinea pig Draize test, FCAT, and guinea pig maximization test) or Freund's adjuvant. They differ mainly as outlined below.

1. Number of Exposures

Draize (Draize repeated-insult patch test) originally required 10 exposures over approximately 3 weeks. (Some laboratories use nine for convenience.) Five exposures are used in the human maximization test.

2. Test Concentration

The Draize (Draize repeated-insult patch test) procedure did not include details of test concentration. Subsequent experience showed that induction (and elicitation) of contact allergy is dose related; hence the higher the test concentration, the greater the ability to identify the allergic potential. With final formulations it is not often practical to increase the concentration. The maximization test, developed for raw ingredients, uses the highest nonirritating concentration, or 25%.

3. Number of Subjects

Draize (Draize repeated-insult patch test) originally suggested 200 subjects. In spite of numerous statistical objections, this number (combined with elevated concentrations for individual ingredients) has empirically been remarkably efficient in identifying allergens. The maximization test employs 25 subjects.

4. Occlusion

It is generally accepted that all techniques are made more sensitive by utilizing an occlusive patch, presumably by increasing penetration.

5. Comparison of Sensitivity

Marzulli and Maibach [51] have summarized the limited available information on sensitivity. A thorough comparison would require not only examining numerous compounds by the two methods, but also a subsequent follow-up of what has happened in actual use. It appears that often the findings of the two methods are in agreement, but exceptions do exist.

6. Conclusions

Current techniques in humans provide sufficient sensitivity to identify most, if not all, strong to moderate contact allergens. Many but not all low-grade allergens may be discriminated; an important exception is the lanolins. It is not clear whether this exception relates to the lack of discrimination of the tests or the minimal sensitization potential of the lanolins.

As in the case of the guinea pig assays, a critical factor is the experience of the investigator performing the test. Similarly, the human tests identify allergenic potential; extrapolation to actual use requires many other types of information.

Safety Evaluation

D. Phototoxicity

Phototoxicity is a nonimmunological light-related irritation that would occur in (almost) everyone after skin exposure to light of sufficient intensity and wavelength together with a light-activated chemical in an adequate amount. In practice this rarely occurs. Bergamot could not have been used for the last century if it had produced universal dermatitis. Because cosmetic formulation knowledge was advanced enough so that relatively few people developed dermatitis, there was minimal impetus to determine the mechanism involved.

The development of suitable animal models followed the development of a satisfactory human testing model [52]. The basic requisites are nonerythrogenic light (320 nm) and percutaneous penetration of the phototoxic agent.

It is convenient that most phototoxic agents produce dermatitis under irradiation conditions (wavelength of 320 nm) that do not ordinarily yield erythema. For this reason, with light sources such as the Woods light, which has a principal emission at 365 nm, the light-irradiated negative control site is expected to be free of dermatitis. This advantage makes for an all-or-none skin reaction, simplifying the evaluation.

This does not mean that the Woods light is a perfect light source. Although ideal for the study of bergapten, it may not be for other chemicals. Inexpensive, currently marketed broader-spectrum light sources such as the Westinghouse BL F40W are useful for screening the phototoxic potential of new chemicals. Testing is simplified when erythema rays (280-320 nm) are not involved, since the need for estimating minimal erythema dose, a more difficult procedure for evaluating the response, is eliminated.

We cannot predict how many topical phototoxic agents may be activated at wavelengths that produce erythema. The first example of this type is one whose general clinical relevance is not fully understood. Breza et al. [53] showed that vinblastine produced a dermatitis in the light-exposed area after intravenous administration. Their testing suggested that light in the erythema range was required to elicit the response. They experimentally produced a similar skin response in subjects who were given the injection intradermally. As the drug is widely used in the treatment of malignancy, it is not clear why most patients receiving intravenous vinblastine do not develop a phototoxic dermatitis. Perhaps they do not receive sufficient sun exposure, or the dynamics of delivery to the skin and metabolism in these patients may be different from those in the ordinary patient. This issue requires study.

1. Test Animals

Under appropriate test conditions, several species have been utilized successfully to reproduce bergapten phototoxicity [52]. The hairless mouse and the rabbit appeared somewhat to be more sensitive to bergapten than the guinea pig. The pig (swine) was less reactive, but "stripping" of the skin with cellophane enhanced responsiveness; the squirrel monkey appeared resistant. The hamster showed histological changes of phototoxicity that were not apparent on gross examination.

Other anatomic test sites and different treatment schedules might alter these relative rankings. The practical point is that the investigator can select from a variety of test species. Unfortunately, the list of chemicals of known relevance to humans is limited. Until validation is more complete and until some of the variables are better understood, it is prudent that new chemicals intended for wide human use be examined on human skin after exploratory work on animals.

With animal models, judgment is needed in defining the test parameters. For example, both the time after application to the skin and the duration of light exposure should be carefully controlled. With bergapten, maximal human responses are obtained when the skin is irradiated about 1 hr after applying the phototoxic chemical, whereas little or no response results if a 24-hr interval intervenes between chemical application and light exposure. This is not explained by what is known about percutaneous penetration in humans. In animals a 5-10 min exposure to the chemical (before light exposure) is usually sufficient; in humans 1 hr is optimum. Recent studies by Zaynoun et al. [54] provided useful quantitative data in this and related areas.

2. Human Testing

Human bioassays can be safely performed because only small test areas are exposed. As we know of no examples in which the arm or back cannot be used, cosmetic sites such as the face can be avoided. The greatest hazard to the experimental subject is the possible development of a small area of dermatitis, which heals promptly. In addition there may be a long period of hyperpigmentation (weeks to months), but this is rarely a serious consideration to the informed subject. Because the dermatitis can be produced in almost every subject, it is not necessary to have a large test panel [52,55]. With sufficient light exposure and percutaneous penetration, dermatitis will occur. Obtaining high-energy light sources should present no problem (see Tables 11 and 12). Penetration may be enhanced in several ways. Testing on a highly permeable anatomic site, such as the scrotum [56], is one method, and decreasing the barriers to penetration by removing the stratum corneum with cellophane tape is another. Increasing the concentration of the test agent many times over what would ordinarily be used will concomitantly increase penetration and the likelihood of detecting or identifying a phototoxic agent.

TABLE 11 Some Special Units for Investigating Ultraviolet Effects on Humans

Product name	Manufacturer	Product type
Blak-Ray B-100A high-intensity black light	Ultraviolet Products, Inc., 5100 Walnut Grove Ave., San Gabriel, CA 91778	High-intensity reflector,[a] mainly 366 nm
Blak-Ray XX-40A fixture, for use with two 40-W tubes, 48 × 6 × 4 in.	Ultraviolet Products, Inc.	Can be used with fluorescence-type bulbs available at electric store
Blak-Ray XX-15 fixture; 18 in. long	Ultraviolet Products, Inc.	Can be used with commercially available black light
Xenon Solar Simulator, 70% of output below 400 nm	Solar Light Co., 6655 Lawnton Ave., Philadelphia, PA 19126	150-W xenon arc lamp
Filter to cut out below 320 nm	Solar Light Co.	1/8-in. window glass

[a]Disadvantage is heat and small area covered.

Safety Evaluation

TABLE 12 Some Characteristics of Marketed Irradiation Sources[a]

Source	Approximate irradiance at 15 in.
Fluorescent lamps with principal spectral range at 320-400 (UVA)	
1. Sylvania FR 40 T12 PUVA (psoriasis lamp)	317 (bare-base bulb without reflector)
2. F 40 W BL (black light); has some visible	
3. F 40 W BLB (black light with black glass); has no visible	
Mercury vapor lamps (require ballast) with principal spectral range at 320-400 nm (UVA)	
1. GE H 400 A33-1	1115
2. Sylvania H 33AR 400	925
Fluorescent lamps with principal spectral range in visible (400-500 nm)	
1. Westinghouse F20T12 BB (special blue)	300

[a]Additional information: For point sources the inverse-square law applies (double distance = quarter of intensity). For extended sources (fluorescent bulbs) the inverse-square law does not apply for short distances used in exposing animals or humans. By adding a reflector, one can increase the intensity about fivefold; by using multiple lamps (two) plus a reflector, increase may be up to tenfold. By moving closer (i.e., $7\frac{1}{2}$ in. rather than 15 in.) the irradiance is doubled.

It is understood that the study of phototoxic agents in this experimental fashion does not duplicate human use conditions; such tests will only disclose phototoxic potential. The dermatotoxicologist must use judgment in establishing the test concentration, frequency, time, and duration of chemical and light exposure, as well as other factors, to aid in extrapolation to the use situation.

It is not entirely clear why stripping the skin with cellophane tape enhances the phototoxic potential of bergapten. Increased skin penetration is one obvious effect, but other factors may be involved. It is sometimes stated that stripping the skin of the stratum corneum produces complete percutaneous absorption, but this is not so. With hydrocortisone, stripping produces approximately a threefold increase in skin penetration [56]. With several pesticides simiarly studied (H. Maibach and R. Feldmann, unpublished observations) a several-fold increase in penetration occurred, but it was never complete.

3. Chemical-Skin Contact Time Before Light Exposure

With many chemicals there is considerable lag time before a significant degree of percutaneous penetration occurs [57]. In bioassays on skin, this delay factor must be taken into consideration; for example, in the vasoconstrictor corticoid assay responses should be measured 18-24 hr after application of the test substance. It might be expected that light exposure could or should be delayed for many hours after application to skin, but this is not the case with bergamot. Animals are exposed within minutes after application and humans 1-2 hr later; by 4 hr animals are less reactive, and at 24 hr light exposure will often produce no response in humans or animals. This time factor must be taken into account for toxicological assays, and it is of considerable interest in terms of the relationship of pharmacokinetics and the site of action in skin. Future experience will determine if the time relationship found optimal for bergamot will hold for other phototoxic chemicals or if it requires alteration.

4. Vehicles

Vehicles may alter percutaneous penetration. A considerable literature defines chemical and vehicle properties that increase or decrease chemical release and penetration. Much of this work has been done with in vitro or other model test systems; the experience with vehicles in animal or human in vivo test systems is limited. Marzulli and Maibach [52] showed that reactions to bergamot in the rabbit were greater with 70% alcohol than with 95% alcohol.

Kaidbey and Kligman [58], in studying the phototoxic properties of coal tar, methoxsalen, and chlorpromazine, found the results strongly influenced by the vehicle chosen. No single base produced optimal effects for the three chemicals; emulsion-type creams were generally more active than petrolatum. Polyethylene glycol was a uniformly poor vehicle.

The intensity of phototoxic reaction produced by topically applied methoxsalen and coal tar was investigated by Suhonen [59] in relation to three test variables. Petrolatum was a suitable vehicle for testing methoxsalen; carbowax appeared useful in testing coal tar. The optimum concentrations were 5% for coal tar and 0.03-0.05% for methoxsalen. Optimum occlusion times were 1-2 hr for methoxsalen and 24 hr for coal tar.

It is likely that the effect of vehicle on phototoxicity (and penetration) is complex; until all the variables are understood, the vehicle intended for human use should be employed in the predictive assay. A more realistic evaluation of potential hazard is obtained by also using an experimental vehicle (such as alcohol) that is likely to release the test compound.

5. Cutaneous Metabolism

Bergamot dermatitis as a model of phototoxicity can be produced in most humans and in several animal species (see above). It is not known whether special cutaneous metabolic activities are needed to convert bergapten to a toxic material, although this is unlikely. It is possible or even likely that some ordinarily nonphototoxic materials can be converted to toxic materials by individuals having the appropriate metabolic machinery. This is a reasonable explanation for the rare person who develops phototoxicity from oral tetracycline. Investigators also need to investigate this possibility when testing topical agents.

Another method to study phototoxicity avoids the complexities of percutaneous penetration by injecting the chemical intradermally. Kligman and Breit [60] showed that this potential could be demonstrated by injecting certain materials intradermally. This method identifies hazards, but care must be used in extrapolating the findings to what will happen with the actual route of exposure.

6. Incidence of Phototoxicity Caused by Cosmetic Materials

The literature on fragrance toxicity is limited, with the exception of the psoralens. Forbes et al. [61], using hairless mice and miniature swine, tested 160 fragrance raw materials for phototoxicity. By their procedure, 21 compounds elicited a phototoxic response; however, 20 of the compounds were in two botanical families, *Rutaceae* and *Umbelliferae*. The great majority of the approximately 10,000 cosmetic materials have not been tested and therefore the incidence of phototoxicity is unclear.

A phototoxic compound in cosmetics has been identified in another setting. Emmett et al. [62] noted that several occupationally exposed (printing ink) workers had clinical phototoxicity which was traced to a screening agent in the ink. This agent, amyl-para-dimethylaminobenzoate (Escalol 506), was also used widely as a sunscreen [63]. This is a paradox — a sun screen that itself produces phototoxicity. When this became known, other patients were identified who had experienced phototoxicity to this compound and had been misdiagnosed as having simply a sunburn. The lesson is clear: If we screen for phototoxicity, we will identify potential hazards and decrease this adverse effect.

7. Determining the Frequency of Cosmetic Phototoxicity Utilizing the Historical Approach

Earlier in this chapter we noted the primitive state of the epidemiology of cosmetic reactions. This situation with phototoxicity is almost as difficult as that with the covert systemic reactions (mutagenicity, central nervous system toxicity, etc.). When a dermatitis is acute and severe, phototoxicity can sometimes be clinically suspected. But often when a patient has an acute reaction, by the time he or she is examined, only hyperpigmentation of the area remains. We have seen several such examples previously misdiagnosed as melasma or cholasma. Because of its often occult nature, toxicological screening of cosmetic materials will be of great value and presumably make this hazard of only historical interest.

8. Phototoxic Potential Versus Hazard

The screening tests determine toxicity potential, not hazard. Presumably even the most potent phototoxic agents will have a threshold at which this effect will not be observed. In determining the safety of a particular agent, consideration must be given to the concentration of compound in the use situation.

E. Contact Photosensitization

1. Animal Testing (Guinea Pig)

Tetrachlorosalicylamide (TCSA) is frequently studied as a model photoallergen [64-67]. The allergic contact photosensitization technique described is based on that of Landsteiner and Chase [50], modified by the addition of light for the induction of contact photosensitivity.

Hartley strain albino guinea pigs have been used in many photobiological studies. They are exceedingly well suited for investigating contact photosensitivity reactions, but one must also be aware of their limitations [68].

To induce contact photosensitization, the nuchal area of guinea pigs is shaved and depilated, and a 2% solution of TCSA, tribromosalicylanilide, or trichlorocarbanilide in acetone is topically applied. The material is applied 30 min before irradiation. The irradiation administered to the treated site consists of two types, used successively:

1. Ultraviolet light B, such as from a fluorescent "sunlamp" tube (Westinghouse); dose, 1×10^7 ergs/cm^2; emission, 285-350 nm.
2. "Black light," from fluorescent tubes (Westinghouse, General Electric, Sylvania, and Toshiba); dose, 3×10^8 ergs/cm^2; emission, 320-450 nm.

This procedure, consisting of topical exposure and irradiation, is performed three times during a 7-day period. Three weeks after the last sensitizing exposure, the back of the guinea pig is shaved and depilated for the first time. This area, not previously exposed to the photosensitizing agent, receives (for challenge or elicitation) a topical application of appropriate dilutions, such as 0.01, 0.1, and 1% solutions (in acetone or ethanol), of the halogenated salicylanilide to two symmetrical skin sites on the dorsal surface. After 30 min one site is exposed to nonerythrogenic (320-450 nm) radiation for the first time and the animal is placed in a dark room. The other site receives no light exposure.

The light source used to elicit the reaction is usually ultraviolet A. The radiation is passed through a 3-mm-thick pane of window glass that has previously been determined to permit no ultraviolet radiation lower than 320 nm to reach the test sites.

All test sites, irradiated and nonirradiated, are scored and interpreted 24 hr later as follows: 0, no erythema; 1, minimal but definite erythema; 2, moderate erythema; 3, considerable erythema; and 4, maximal erythema. The interpretation of the data is made as follows:

	Irradiated	Nonirradiated
Normal	0	0
Contact photosensitivity	1-4	0
Contact sensitivity (nonphotosensitivity)	1-4	1-4
Contact photosensitivity and contact sensitivity	1-4	1-4

Phototoxicity and primary irritant control studies are performed simultaneously by using animals that have received no previous photosensitizing exposure. One would expect no response to be elicited by the concentrations of photosensitizers and solvents used with the irradiation factors noted.

Primary contact photosensitivity to TCSA has been induced in guinea pigs in at least six different laboratories with minor modifications of the preceding technique. The data of Harber and co-workers [65] (see Table 13) are representative of the typical findings.

The guinea pig model is beguilingly simple. What is not clear is the clinical significance and extrapolation to humans. Part of the problem rests in the observation that most photoallergens are also (plain) contact allergens; it is not obvious how the two reactions are related. Most of our animal model studies with photoallergens have been *retrospective*, that is, we identify

Safety Evaluation

TABLE 13 Induction of Contact Photosensitivity to TCSA in 65 Guinea Pigs

Type of sensitivity	Number of animals sensitized
Contact sensitivity	4
Contact photosensitivity	20
Contact sensitivity and contact photosensitivity	12

Source: Ref. 63.

photoallergens in the guinea pig or mouse after they are clinically identified as a hazard in humans. For example, TCSA was extensively studied after its identification as a photoallergen in humans. What remains to be determined is whether we will have many false-positive responses when we start to perform more such tests *before* the compounds are released for human use.

To date, two factors have probably helped protect the consumer and the reputation of the toxicologist working in this area: (1) Since most photoallergens are also contact allergens, they are most easily identified by the contact sensitization assays with which we have much greater experience, and (2) there appear to be far fewer photoallergens than contact allergens; therefore the probability is greater that we will incorrectly identify more contact allergens than photoallergens.

2. *Human Assay*

The human assays of photosensitization are derived from the Draize repeated-insult patch test modified by the addition of appropriate amounts and qualities of ultraviolet light. The test patches are generally removed at 24 hr and the skin is exposed to approximately three minimal erythema doses of ultraviolet B and large amounts of ultraviolet A. Challenge is with ultraviolet A alone to simplify the reading, since the ultraviolet A will not produce erythema at the doses used. In the photomaximization assay [69], 5 exposures (rather than 9 or 10 as in the Draize test) are used.

The human assays readily identify the most potent photoallergens (i.e., TCSA). However, most of the human subjects are also (plain) contact sensitive. Dibromosalicylate is similarly a photoallergen and a contact allergen.

Extrapolation from the assay to the use situation remains to be defined. There is no doubt about the validity of extrapolation with TCSA. However, hexachlorophene, bithional, and 6-methylcoumarin are examples of photoallergens whose meager epidemiological data in humans are difficult to relate with the positive animal or human assays. After two decades of laboratory and use experience with tribromosalicylanilide, the actual allergen and the frequency of sensitization remain confusing and controversial.

F. Other Types of Toxicity

This chapter has considered the more classic forms of dermatotoxicity as they relate to preservatives. For a current view of other hazards from preservatives, including the contact uticaria syndrome, pigmentation dermatitis, subjective irritation, pigmented contact dermatitis, dermatitis from other routes of exposure (oral, etc.), and systemic toxicity, refer to the second edition of *Dermatotoxicology* [(F. N. Marzulli and H. I. Maibach, eds.), Hemisphere, New York (1982)].

REFERENCES

1. H. J. Eiermann, W. Larsen, H. I. Maibach, and J. S. Taylor, Prospective study of cosmetic reactions: 1977-1980, *J. Am. Acad. Dermatol.*, 6:909-917 (1982).
2. E. L. Richardson, Update-frequency of preservative use in cosmetic formulas as disclosed to FDA, *Cosmet. Toiletries*, 96:91-92 (1983).
3. F. N. Marzulli and H. I. Maibach, The rabbit as a model for evaluating skin irritants: a comparison of results obtained on animals and man using repeated skin exposures, *Food Cosmet. Toxicol.*, 13:533-540 (1975).
4. M. Steinberg, W. A. Akers, M. Weeks, A. H. McCreesh, and H. I. Maibach, A comparison of test techniques based on rabbit and human skin responses to irritants with recommendations regarding the evaluation of mildly or moderately irritating compounds, in *Animal Models in Dermatology* (H. I. Maibach, ed.), Churchill Livingstone, New York (1975), pp. 1-22.
5. F. N. Marzulli and M. I. Maibach, Antimicrobials: experimental contact sensitization in man, *J. Soc. Cosmet. Chem.*, 24:399-421 (1973).
6. W. P. Jordan and S. E. King, Human experimental contact dermatitis, in *Safety and Efficacy of Topical Drugs and Cosmetics* (A. M. Kligman and J. J. Leyden, eds.), Grune and Stratton, New York (1982), pp. 193-203.
7. F. N. Marzulli and H. I. Maibach, Status of topical parabens: skin hypersensitivity, *Int. J. Dermatol.*, 13:397-399 (1974).
8. J. H. Draize, G. Woodard, and H. O. Calvery, Methods for the study of irritation and toxicity of substances applied topically to the skin and mucous membranes, *J. Pharmacol. Exp. Ther.*, 82:377-389 (1944).
9. G. A. Nixon, C. A. Tyson, and W. C. Wertz, Interspecies comparisons of skin irritancy, *Toxicol. Appl. Pharmacol.*, 31:481-490 (1975).
10. L. Philips, M. Steinberg, H. I. Maibach, and W. A. Akers, A comparison of rabbit and human skin response to certain irritants, *Toxicol. Appl. Pharmacol.*, 21:369-382 (1972).
11. Anonymous, Method of testing primary irritant substances, *United States Code of Federal Regulations*, 16 CFR, 1500.41 (1979).
12. U.S. Food and Drug Administration, Hazardous substances: proposed revision of test for primary skin irritants, *Fed. Regist.*, 37:27,635-27,636 (1972).
13. Committee for the Revision of NAS Publication 1138, *Principles and Procedures for Evaluating the Toxicity of Household Substances*, National Academy of Sciences, Washington, D.C. (1977), p. 30.
14. K. I. Campbell, E. L. George, L. L. Hall, and J. F. Stara, Dermal irritancy of metal compounds, *Arch. Environ. Health*, 30:168-170 (1975).
15. C. S. Weil and R. A. Scala, Study of intra- and interlaboratory variability in the results of rabbit eye and skin irritation tests, *Toxicol. Appl. Pharmacol.*, 19:276-360 (1971).
16. A. H. McCreesh and M. Steinberg, Skin irritation testing in animals, in *Dermatotoxicology and Pharmacology* (F. N. Marzulli and H. I. Maibach, eds.), Wiley, New York (1977), pp. 193-210.

17. T. Haley and J. Hunziger, Instrument for producing standardized skin abrasion, *J. Pharm. Sci.*, 63:106 (1974).
18. P. J. Frosch and A. M. Kligman, The chamber-scarification test for irritancy, *Contact Dermatitis*, 2:314-324 (1976).
19. R. A. Roudabush, C. J. Terhaar, D. W. Fassett, and S. P. Dziuba, Comparative acute effects of some chemicals on the skin of rabbits and guinea pigs, *Toxicol. Appl. Pharmacol.*, 7:559-565 (1965).
20. V. Brown, A comparison of productive initiation tests with surfactants on human and animal skin, *J. Soc. Cosmet. Chem.*, 22:411-420 (1970).
21. R. E. Davies, K. H. Harper, and S. R. Kynoch, Interspecies variation in dermal reactivity, *J. Soc. Cosmet. Chem.*, 23:371-381 (1972).
22. F. S. MacMillan, R. R. Rafft, and W. B. Cloers, A comparison of the skin irritation produced by cosmetic ingredients and formulations in the rabbit, guinea pig and beagle dog to that observed in the human, in *Animal Models in Dermatology* (H. I. Maibach, ed.), Churchill Livingstone, Edinburgh (1975), pp. 12-22.
23. F. N. Marzulli and H. I. Maibach, The rabbit as a model for evaluating skin irritants: a comparison of results obtained on animals and man using repeated skin exposure, *Food Cosmet. Toxicol.*, 13:533-540 (1975).
24. A. J. Ingram and P. Grasso, Patch testing in the rabbit using a modified patch test method, *Br. J. Dermatol.*, 92:131-142 (1975).
25. S. Siegel, *Nonparametric Statistics for the Behavioral Sciences*, McGraw-Hill, New York (1956).
26. J. H. Draize, Dermal toxicity, *Food Drug Cosmet. Law J.*, 10:722-732 (1955).
27. J. H. Draize, Appraisal of the safety of chemicals in foods, drugs and cosmetics, in *Dermal Toxicity*, Association of Food and Drug Officials of the United States, Texas State Department of Health, Austin, Texas (1959), p. 46.
28. H. Maibach, Excited skin state, in *Cutaneous Allergy* (J. Ring, ed.), Springer-Verlag, Munich (1981), pp. 208-221.
29. G. Klecak, H. Geleick, and J. R. Frey, Screening of fragrance materials for allergenicity in the guinea pig. I. Comparison of four testing methods, *J. Soc. Cosmet. Chem.*, 28:53-64 (1977).
30. B. Magnusson and A. M. Kligman, Allergic contact dermatitis in the guinea pig, in *Identification of Contact Allergens*, Charles C. Thomas, Springfield, Ill. (1970).
31. T. Maurer, Tierexperimentelle Methoden zur pradiktiven Erfassung sensibilisierender Eigenschaften von Kontaktallergenen, unpublished dissertation, University of Basel, Basel (1974).
32. B. Magnusson, The relevance of results obtained with the guinea pig maximization test, in *Animal Models in Dermatology* (H. I. Maibach, ed.), Churchill Livingstone, Edinburgh (1975), pp. 76-83.
33. B. Magnusson and A. M. Kligman, The identification of contact allergens by animal assay. The guinea pig maximization test, *J. Invest. Dermatol.*, 52:268-276 (1969).
34. A. M. Kligman, The identification of contact allergens by human assay. III. The maximization test. A procedure for screening and rating contact sensitizers, *J. Invest. Dermatol.*, 47:393-409 (1966).
35. V. Pirila, Patch testing technique. A new modification of the chamber test, *Excerpta Med. Int. Congr. Ser.*, 235:50 (1971).

36. V. Pirila, Chamber test versus lapptest, *Forh. Nord. Dermatol. Foren.*, *20*:43 (1974).
37. H. C. Maguire, Mechanism of intensification by Freund's complete adjuvant of the acquisition of delayed hypersensitivity in the guinea pig, *Immunol. Commun.*, *1*:239-246 (1973).
38. H. C. Maguire, The bioassay of contact allergens in the guinea pig, *J. Soc. Cosmet. Chem.*, *24*:151-162 (1973).
39. H. C. Maguire, Estimation of the allergenicity of prospective-human contact sensitizers in the guinea pig, in *Animal Models in Dermatology* (H. I. Maibach, ed.), Churchill Livingstone, Edinburgh (1975), pp. 67-75.
40. H. C. Maguire and M. W. Chase, Studies on the sensitization of animals with simple chemical compounds. XIII. Sensitization of guinea pigs with picric acid, *J. Exp. Med.*, *135*:357-374 (1972).
41. H. C. Maguire and M. W. Chase, Exaggerated delayed-type hypersensitivity to simple chemical allergens in the guinea pig, *J. Invest. Dermatol.*, *49*:460-468 (1967).
42. E. V. Buhler, A new method for detecting potential sensitizers using the guinea pig, *Toxicol. Appl. Pharmacol.*, *6*:341 (1964).
43. E. V. Buhler, Delayed contact hypersensitivity in the guinea pig, *Arch. Dermatol.*, *91*:171-177 (1965).
44. J. F. Griffith and E. V. Buhler, *Experimental Skin Sensitization in the Guinea Pig and Man*, Procter and Gamble Co., Cincinnati, Ohio (1969).
45. E. V. Buhler and F. Griffith, Experimental skin sensitization in the guinea pig and man, in *Animal Models in Dermatology* (H. I. Maibach, ed.), Churchill Livingstone, Edinburgh (1975), pp. 56-66.
46. G. Klecak, Identification of contact allergens: predictive tests in animals, in *Dermatotoxicology and Pharmacology* (F. N. Marzulli and H. I. Maibach, eds.), Wiley, New York (1977), pp. 305-339.
47. A. M. Kligman, The identification of contact allergens by human assay. I. A critique of standard methods, *J. Invest. Dermatol.*, *47*:369-374 (1966).
48. A. M. Kligman, The identification of contact allergens by human assay. II. Factors influencing the induction and measurement of allergic contact dermatitis, *J. Invest. Dermatol.*, *47*:375-392 (1966).
49. A. M. Kligman, The SLS provocative patch test, *J. Invest. Dermatol.*, *46*:573-589 (1966).
50. K. Landsteiner and M. W. Chase, Experiments on transfer of cutaneous sensitivity to simple compounds, *Proc. Soc. Exp. Biol. Med.*, *49*:688 (1942).
51. F. N. Marzulli and H. I. Maibach, Contact allergy: predictive testing of fragrance ingredients in humans by Draize and maximization methods, *J. Environ. Pathol. Toxicol.*, *3*:235-245 (1980).
52. F. N. Marzulli and H. I. Maibach, Perfume phototoxicity, *J. Soc. Cosmet. Chem.*, *21*:685-715 (1970).
53. T. Breza, K. Halperin, and J. Taylor, Photosensitivity reaction to vinblastine, *Arch. Dermatol.*, *111*:1168-1170 (1975).
54. S. T. Zaynoun, B. E. Johnson, and W. Frain-Bell, A study of oil of bergamot and its importance as a phototoxic agent. II. Factors which affect the phototoxic reaction induced by bergamot oil and psoralen derivatives, *Contact Dermatitis*, *3*:225-239 (1977).

55. K. Burdick, Phototoxicity of Shalimar perfume, *Arch. Dermatol.*, *93*: 424-425 (1966).
56. R. Feldmann and H. Maibach, Regional variation in percutaneous penetration of hydrocortisone in man, *J. Invest. Dermatol.*, *48*:181-183 (1967).
57. R. Feldmann and H. Maibach, Absorption of some organic compounds through the skin in man, *J. Invest. Dermatol.*, *54*:399-404 (1970).
58. K. Kaidbey and A. Kligman, Topical photosensitizers, influence of vehicles on penetration, *Arch. Dermatol.*, *110*:868-870 (1974).
59. R. Suhonen, Photoepicutaneous testing, *Contact Dermatitis*, *2*:218-226 (1976).
60. A. Kligman and R. Breit, Identity of phototoxic drugs by human assay, *J. Invest. Dermatol.*, *51*:90-99 (1968).
61. P. D. Forbes, F. Urbach, and R. E. Davies, Phototoxicity testing of fragrance raw materials, *Food Cosmet. Toxicol.*, *15*:55-60 (1977).
62. E. A. Emmett, B. R. Taphorn, and J. R. Kominsky, Phototoxicity occurring during the manufacture of ultraviolet-cured ink, *Arch. Dermatol.*, *113*:770-775 (1977).
63. K. H. Kaidbey and A. M. Kligman, Phototoxicity to a sunscreen ingredient, *Arch. Dermatol.*, *114*:547-549 (1978).
64. J. Griffith and R. O. Carter, Patterns of photoreactivity and cross reactivity in persons sensitive to TCSA, *Toxicol. Appl. Pharmacol.*, *12*:304 (1968).
65. L. C. Harber, S. E. Targovnik, and R. L. Baer, Contact photosensitivity patterns to halogenated salicylanilides in man and guinea pigs, *Arch. Dermatol.*, *96*:646 (1967).
66. P. S. Herman and W. M. Sams, Carrier protein specificity in salicylanilide sensitivity, *J. Invest. Dermatol.*, *54*:438 (1970).
67. L. Vinson and V. F. Borselli, A guinea pig assay of the photosensitizing potential of topical germicides, *J. Soc. Cosmet. Chem.*, *17*:123 (1966).
68. L. C. Harber, Use of guinea pigs in photobiologic studies, in *The Biologic Effects of Ultraviolet Radiation* (F. Urbach, ed.), Pergamon, Oxford (1969), pp. 291-299.
69. K. H. Kaidbey and A. M. Kligman, Photomaximization test for identifying photoallergic contact sensitizers, *Contact Dermatitis*, *6*:161-169 (1980).

28
EVALUATION OF CHEMICAL
TOXICOLOGY OF COSMETICS

F. HOWARD SCHNEIDER *Bioassay Systems Corporation, Woburn, Massachusetts*

I. Introduction 533

II. Current Methods of Animal Toxicity Evaluation 535
   A. Acute toxicity screening assays 535
   B. Subchronic toxicity tests 542
   C. Chronic toxicity tests 543
   D. Teratology tests 544
   E. Reproductive toxicity 544
   F. Toxicokinetics 545
   G. Photosensitivity reactions 546

III. In Vitro Toxicity Assays 547
   A. Cytotoxicity 548
   B. Mutagenicity 551
   C. In vitro phototoxicity assay 555

IV. Summary 555

   References 556

I. INTRODUCTION

The number of toxic effects associated with the use of cosmetics is relatively low, although the number of individuals throughout the world exposed to cosmetics is staggering. The apparent safety of cosmetic products may reflect both the relatively low inherent toxicity of most cosmetic ingredients, as well as the effectiveness of the skin in preventing the absorption of exogenous chemicals. However, this is not to say that cosmetics are totally devoid of toxicity; they are not. Even a cursory review of the literature reveals problems of skin irritation, allergic sensitization responses, and various systemic toxicities, especially with products like hair dyes, tints, antibacterial soaps, nail polish, and perfumes [1]. In fact, documented toxic responses are, in part, responsible for regulatory guidelines for the evaluation of cosmetic ingredients.

Toxicological evaluation of cosmetic ingredients is based to a large extent on studies acceptable to the U.S. Food and Drug Administration (FDA) which are designed to reveal both local and systemic toxic effects. Authorization for regulation of cosmetic safety is provided by the Federal Food, Drug, and Cosmetic Act, which states that a cosmetic is "adulterated" if it contains "any poisonous or deleterious substances which may render it injurious to users. . . under such conditions of use as are customary or usual" [2]. The various categories of products manufactured by the cosmetic industry, and the toxicity tests appropriate to these products, are described by Giovacchini [3].

An important source of information on the toxicity evaluation of cosmetic ingredients is the Cosmetic Toiletry and Fragrance Association (CTFA). This group recently established the Cosmetic Ingredient Review, a program which promotes the safety of chemicals to be used in cosmetics. The panel of experts holds open meetings at which results of their literature reviews and study group recommendations are presented. The group is especially concerned about the adequacy of existing animal and human toxicity data, and is willing to recommend additional testing if it is required for the demonstration of safety. The panel is charged with the evaluation of approximately 3000 chemicals, a number which reflects not only the magnitude of the challenge, but also the diversity of cosmetic ingredients. The 1980 Ingredient Review Priority list includes 239 materials, the first of which is formaldehyde. Those scientific literature reviews which have been completed are available from the Cosmetic Ingredient Review. Safety assessment reports have recently been published for nine separate cosmetic ingredients or groups of ingredients, including glycol sterates, octyl and related palmitates, squalane and squalene, stearalkonium chloride, quaternium-18 formulations, decyl and isodecyloleates, cety and myristyl lactates, and various carbomers [4].

The CTFA has published safety testing guidelines which pertain to the tests which would be used most commonly to evaluate the toxicity of cosmetic ingredients such as preservatives [5]. Included in these guidelines are procedures for evaluating skin irritation, eye irritation, contact sensitization, photodermatitis, mucous membrane irritation, percutaneous toxicity, oral toxicity, and inhalation toxicity. Although other guidelines and procedures will be referred to in this review, this document should be central to developing test protocols for the examination of any cosmetic preservative.

Although the FDA and the CTFA provide valuable support, it is the commercial firm profiting from the sale of its products which must assume final responsibility for the safety of its products. It is imperative that cosmetic manufacturers employ or retain individuals who are not only aware of compliance requirements, but who also possess an understanding of product safety evaluation. Expertise of this type can range from a full staff of toxicologists and in-house toxicology laboratories to a single independent consultant; knowledge of the basic principles of the toxicological evaluation of chemicals and experience are the key qualifications. The cosmetic product toxicologist must be able to effectively evaluate new safety questions and toxicity data as they develop, as well as relate toxicity demonstrated in laboratory studies to potential toxicity to humans.

The purpose of this chapter is to provide a summary of the laboratory methods available to the cosmetic industry toxicologist which are or may be useful for evaluating the safety of the preservatives his or her company manufactures or provides for the cosmetic industry.

II. CURRENT METHODS OF ANIMAL TOXICITY EVALUATION

Preservatives which come in contact with the external surfaces of the body can cause two general types of toxicity, local and systemic. Owing to the physical properties of most cosmetics, variables such as formulation ingredient solubilities, and the concentration of active materials, local toxic effects associated with the external surface of the body are of primary concern. However, the potential for systemic toxicity cannot be dismissed, since some chemicals are effectively transported or absorbed across the skin and attain significant blood levels. Furthermore, it is also known that skin cells can metabolize certain chemicals, allowing externally applied chemicals to be converted to one or more metabolites prior to entry into the general circulation [6]. Understanding the mechanisms by which toxic effects are produced will allow the design of not only safer preservatives, but also more meaningful test systems. Systems for the evaluation of product toxicity can be divided into two broad categories: acute or short-term tests, and those designed to evaluate the potential for toxic effects which are generally manifested over a longer period of time, an example of which is the "lifetime" rodent carcinogenicity study. Each of these test categories is discussed in this section.

A. Acute Toxicity Screening Assays

Toxicity is defined here as the capacity to cause injury, and the specific toxic end point must be clearly defined. It is important to realize that toxicity and hazard do not have the same meaning; measures of toxicity are used to estimate potential hazard. Hazard is not only a function of the inherent toxicity, but is also related to the bioavailability of the substance. This distinction between toxicity and hazard, and the importance of information on the bioavailability of the material, must be kept in mind when assessing the safety of any products to which living organisms are exposed.

Acute toxicity assays include tests in which the organism is exposed to the test chemical either once only or more than once in a short period of time, generally 24 hr; the period for observation is usually not more than 14 days.

The primary reasons for subjecting cosmetic preservatives to acute toxicity screening assays are the following:

To obtain data for use in estimating potential hazard to the user
To provide information about the mechanism by which toxicity is produced
To provide information for setting dose levels in subsequent long-term toxicity tests, if they are, in fact, necessary.

Information gained from acute studies is also important in new product development. For example, an acute toxicity battery could be used to evaluate a series of structurally related compounds, any of which could serve as an effective preservative in a cosmetic product. In this case toxicity would provide the discrimination for selection of the preservative for further investigation.

One frustrating aspect of acute toxicity studies is the apparent variability of data obtained for the same test (e.g., oral LD_{50}) conducted in different laboratories. The underlying reason for such variability is the extent to which experimental and biological variables influence the outcomes of tests in which live organisms are used. Variables such as route of administration, species, strain, and sex differences in the test animals, age-related effects, and procedures for the housing and care of the animals can have marked influences. Additional and more subtle influences, which are frequently more difficult to control, include test animal health, stress upon being handled, variations in test substance administration, definition of and measurement of toxic end points, and observer bias. One of the important functions of an experimental toxicologist is to standardize test conditions and minimize variability.

Acute toxicity tests can be divided into two subcategories: evaluation of local effects and evaluation of systemic effects.

1. Evaluation of Local Effects

Local toxic effects are thought of as those actions resulting from contact between the toxicant and the target tissues which are restricted to the immediate physiological area. Topical application to the skin or eyes are the most common routes of administration for cosmetic products. However, exposure by intradermal or subdermal injection, intravaginal and intranasal application, and direct application to the inner surfaces of the oral cavity are occasionally required for special products. Only the four most common assays will be discussed in detail, although less frequently used tests will also be mentioned.

Primary Skin Irritation Test: The most commonly, in fact almost universally, used test procedure for the estimation of the potential of a chemical to cause skin irritation in man is the primary skin irritation test in rabbits as described by Draize and his colleagues [7]. Basically, the method consists of clipping the hair of the skin and applying the test substance either directly, as with solids, or impregnated into an absorbant patch, as with liquids. An occlusive covering is applied over the application patch and the skin reactions are evaluated 24 and 72 hr later with the 0 to +4 scoring system. The final score is an average of the 24- and 72-hr readings for erythema, eschar formation, and localized edema. The amount of sample applied is generally constant, 0.5 g for solids and 0.5 ml for liquids. However, ointments can be applied in a dosage of 4 g/kg body weight. Volatile antimicrobial agents or solvents are tested by the open occlusive method, which includes a 30-min period for evaporation between sample application and putting on the occlusive dressing; this procedure is intended to simulate the intended use.

The rabbit primary skin irritation test is used widely, not only for cosmetic preservatives, but for many types of industrial and consumer chemical products. It is routinely included in most acute toxicity screening batteries and is used to generate data for establishing handling, labeling, shipping, and storage instructions and precautions. Primary skin irritation in this test is considered a local inflammatory response of normal living skin to direct exposure to a chemical agent without the involvement of an immunological mechanism. The test is usually conducted with albino rabbits according to the procedures outlined in any one of several test guidelines, including Section 1500.41 of Chapter II, Title 16, *Code of Federal Regulations* [8], Department of Transportation regulations [9], or Section 163.81 of the Environmental Protection Agency (EPA) FIFRA guidelines [10,11].

An excellent review has been published recently by Steinberg et al. on the correlation of the skin inflammatory response between animals and humans [12]. The authors conclude that animal skin irritation tests can be useful in predicting human skin irritancy if the animal test model is tailored to the special test material and if the technicians conducting the test are carefully trained and routinely monitored.

Eye Irritation Test: The purpose of this toxicity test is to assess the ability of a germicidal or other chemical to cause eye irritation upon physical contact with the eye. This phase of testing is employed in cosmetics for two reasons:

1. Many products containing preservatives are used in close proximity to the eye (mascara, eye shadow).
2. Accidental spraying or splashing into the eye is possible with any liquid preserved product (shampoo, skin lotions).

The methods of Draize [7] and Friedenwald et al. [13], both using albino rabbits, are employed to evaluate injuries to the iris, cornea, conjunctiva, and palpebral mucosa. Ophthalmic parameters which can be evaluated include pupil size, reaction to light, redness, congestion, chemosis, iris swelling and congestion, corneal opacity, and hemorrhage. The rabbits should be housed one to a cage in a clean atmosphere free of extraneous particles which might themselves cause eye irritation.

Test material (0.1 ml of liquid or 100 mg of solid) is placed into the conjunctival sac of one eye of each of six animals; the other eye serves as the control. The eyes are not washed after application of the test substance in the standard protocol; modifications of this procedure are possible if the effects of immediate eye washing subsequent to exposure are to be evaluated. Irritation is assessed in the cornea, iris, palpebral mucosa, and conjunctiva by the Draize scoring scheme on a scale of 0 to +4. Animals are examined 1, 24, 48, and, in some protocols, 72 hr after exposure. Test results are considered positive if four or more of the six test animals exhibit a positive reaction.

Skin Sensitization Test: Evaluation of the potential for a cosmetic product or preservative to cause an allergic or sensitization reaction of the skin is an important step in product qualification. Skin sensitization reactions are reflected in cutaneous reactions similar to those observed with primary skin irritation reactions, although they are produced only after repeated exposure to the sensitizer. This delayed-type hypersensitivity, also referred to as allergic contact dermatitis, is an example of a condition in which the immune system plays a central role, including host defense against infectious organisms, graft-versus-host disease, and tumor immunity. It is especially important to cosmetic producers, since repeated applications enhance the response. Sensitization to cosmetics depends upon individual idiosyncrasy, just as with other types of reactions of the immune system. Skin reactions vary from transient redness and itching to the disabling effects of severe and widespread eczematous eruptions.

The time required for development of these reactions varies from one individual to the next and with the chemical properties of the sensitizing agent. However, the common finding is that repeated applications of a sensitizer, even in increasingly smaller amounts, leads to skin reactions which can vary between barely detectable to severe. In view of the frequency with which skin sensitization reactions occur, and their potential consequences, it is critical that manufacturers of cosmetic products which are used topically obtain reliable data on the possibility of such reactions.

Sensitization tests are conducted in animals and in humans, although animal tests are performed prior to testing with humans in order to avoid serious toxic responses. The most immunologically responsive laboratory animal used for testing allergic potential is the guinea pig, the animal used in most skin sensitization assays. A frequently used procedure is that developed by Buehler in 1965 [14] which consists of three weekly skin applications of 0.5 ml of the preserved and nonpreserved cosmetic product on the closely clipped flanks of the animal under complete occlusive patch conditions. The occlusion enhances the chances of absorption of the test material, a requisite for stimulating an allergic reaction. The skin is rechallenged with similar concentrations of the product 2 weeks following the original application. Skin reactions are evaluated 24 and 48 hr later. The responses are graded for (1) severity, using a scale from 0 to 3, with 3 indicating strong erythema with or without edema, and (2) incidence, which is the number of animals showing a response of 1 or more at either 24 or 48 hr. An important step in this test is the evaluation of any primary skin irritation response to the test cosmetic so that it may be accounted for in the response produced at the time of challenge. It is necessary to use a nonirritating concentration of the test preservative; an initial concentration-response test allows selection of the highest nonirritating concentration for the sensitization study. Another important guinea pig sensitization test is that developed by Magnusson and Kligman [15] in which sample application and frequency differ somewhat from those in the Buehler method, although conceptually the two assays are the same.

In cases in which skin sensitization results are inconclusive and there is suspicion that sensitization has occurred, but the data are too variable or inconsistent to substantiate it, other approaches can be used to induce a response, such as subcutaneous injection of the preservative. The subsequent challenge and scoring procedures are identical to the standard Buehler or Kligman tests.

Human Sensitization Patch Test: Information about the ability of a cosmetic product to cause delayed cutaneous allergic sensitization is critical in the decision to market the product for repeated topical use. For this reason, a reliable predictive human test for evaluating this potential should be used by manufacturers of such products. The first test method was proposed in 1944 by Schwartz and Peck [16], and numerous other procedures of greater complexity have been described and used in the ensuing years. In each, one or more induction exposures are followed by a rest period and then a challenge application. An enhancement of the skin response after challenge over that seen during early induction exposure is considered an indication of delayed contact allergy.

The cosmetic is tested in humans only after it has passed all previous animal tests by showing no signs of irritation or toxicity. Sufficient toxicological information regarding the test sample is obtained prior to initiation of the sensitization study to assure a minimum exposure risk for the human test subjects. Acute animal toxicity tests (oral, dermal), eye irritation, skin irritation, subchronic percutaneous studies, and guinea pig sensitization tests are appropriate sources of this information. Familiarity with recognized sensitizers in the published literature provides useful background information, and previous human experience with similar materials is also useful in forecasting the outcome with the test sample.

A useful variation of the human patch test has been developed by the Proctor and Gamble Company and described recently by Stotts in *Current Concepts in Cutaneous Toxicity* [17]. The test consists of a 3-week induction period during which nine occlusive patches are applied each Monday, Wednesday, and Friday. Each patch is left in place for 24 hr and then removed by the subject. Scoring is done after a 24-hr rest period (48 hr over a weekend) and immediately preceding the next patch application. A rest period permits an accurate determination of the degree of primary irritation developing during induction, since such a response tends to peak between 8 and 24 hr after patch removal. An accurate assessment of primary irritation is important in making concentration adjustments and decisions about subsequent patch placement as well as in obtaining a pattern of reactivity for a given subject. Allergic reactions are also more fully manifested after 24 hr and this aids interpretation when the test is completed. Seventeen days after the last induction application (Monday of the sixth week) duplicate challenge patches are applied for 24 hr. Readings from these patches are taken 48 and 96 hr after application. The timing of the scoring allows for the detection of delayed and persistent responses, which are characteristic of allergic contact sensitivity. A rechallenge of any subject suspected of being sensitized takes place 6-8 weeks following the original challenge and is conducted in essentially the same manner. Repeated induction applications, particularly of mild irritants, can provoke stronger irritant responses at challenge which mimic sensitization. The delay before the rechallenge allows all previous reactivity to subside and permits more conclusive confirmation of the presence or absence of sensitization. More than one rechallenge may be necessary if a mixture needs to be fractionated or other pertinent information is desired.

Stotts also mentions the importance of adequate training of the technicians who read the reactions and of the need for documentation of procedures and results. Both factors must be given adequate consideration in the light of current regulatory and consumer pressure for adequate evaluation of products prior to their entry into the market.

Less Frequently Used Tests: Application to mucosal surfaces such as those found in the mouth, nasal sinuses, and genital tract can result in local toxicities due to irritation, corrosion, and sensitization reactions. In fact, these reactions can be even more severe than on skin owing to lack of protection by the keratin layer, even though fluids produced by local mucous glands may dilute or wash away the toxic substances to some extent.

The most common laboratory animal assays for evaluating toxicity to mucosa are the hamster cheek pouch and rabbit vagina tests. In each test toxic effects are evaluated by gross observations and histological examination. Gross observations consist of visual scoring on a scale of 0 to 4 for erythema, edema, and ulceration; histological evaluation includes ulceration or necrosis, acanthosis, hyperkeratosis and dysplasia of the epithelial layer, hyperemia or hemorrhage, inflammation, eschar and necrosis of the lamina propria, and degeneration or necrosis, inflammation, and eschar of the subcutis. These gross and microscopic observations adequately measure local toxic responses to test materials and provide a warning to the manufacturer that products which elicit a response may pose a danger to humans.

2. Evaluation of Systemic Effects

Preservatives applied directly to the skin, as well as those ingested, can be absorbed and distributed throughout the body via the circulatory system.

For this reason, it is necessary to evaluate systemic toxicity as well as local toxicity when testing preserved cosmetic products. The three most likely means by which products would be absorbed systemically are (1) percutaneous absorption, (2) inhalation of vapors, mists, or powders, and (3) oral ingestion. Accordingly, laboratory toxicity tests are recommended which incorporate each of these routes of test substance administration. The most commonly performed acute tests for systemic toxicity of cosmetic products are oral toxicity in rats or mice, dermal toxicity in rabbits, and inhalation toxicity in rats.

Oral Toxicity: The most commonly used indicator of acute toxicity is the LD_{50}, the dose of a substance which produces death in half the population of a group of test animals. This value is used by regulatory agencies as an overall indicator of the inherent toxicity of a substance, even though the value has shortcomings. Some of the major drawbacks to the use of LD_{50} value by itself include the inability to judge the range of toxic doses (estimated from the slope of the dose-response curve), variation among animal species and age, inconsistent correlation with toxicity to humans, variation between laboratories, lack of information on the nature of the toxic actions responsible for death, and costs relative to benefits. Its advantages include universal acceptance or recognition by government agencies and private companies, its quantitative nature, availability of standard protocols, and relative ease of conducting the LD_{50} assay. The LD_{50} values can be calculated from data derived in a laboratory study by several standard mathematical procedures [18,19].

A number of tests have been recommended as replacements for the standard LD_{50} assay as a means of reducing the number of animals used and the testing costs. These alternative tests, reviewed recently by Rowan [20], include the "limit" test, in which one group of animals receives a single large dose (e.g., 5 g/kg body weight) and if there are no deaths, the substance is presumed to be nontoxic. Another suggested test is the "approximate lethal dose," which is assessed by treatment with graduated doses to a single animal, each dose increasing by 50%, until death results. It is unlikely that these or any other method will completely replace the LD_{50} assay in the near future, although there does appear to be an increase in use of the single-dose oral toxicity assay using one high dose level, such as 5 or 10 g/kg.

When conducting a full LD_{50} assay, or the single-dose or approximate lethal dose assays, there are certain test variables which must be considered. A standard test protocol must address, among others, the following features:

- Duration. A 14-day observation period following a single dose of the test substance.
- Animal. Rats are most frequently used.
- Initial age. Approximately 5 weeks.
- Dose administration. Most commonly by gavage.
- Test groups. At least four groups of five males and five females given different doses selected from a range-finding study.
- Diet. A defined diet, ad libitum, withdrawn from the animal overnight (12 hr) before treatment and reintroduced about 4 hr after treatment.
- Observations. All animals are observed frequently for 8 hr after dosing and at least twice daily thereafter. Signs of toxic effects are recorded together with the time of onset, duration, and their intensity.

Although the critical test end point is death, other signs of toxicity are important in evaluating mechanisms of toxicity and in planning additional tests. Typical signs include weight change, appearance, behavior, and abnormal movements. Gross pathological evaluation of organs and tissues at necropsy and evaluation of histological preparations of tissues are also important in identifying affected tissues.

Usually, a range-finding study (with doses increasing by a factor of 4, for example, 200, 800 mg) is conducted with a few animals to determine the approximate lethal dose of the test chemical. However, practical experience and an adequate data bank on the acute toxicity of a wide variety of chemicals allows the investigator to estimate the order of magnitude of the lethal dose. The doses for the full-scale study are selected so that some of the animals in at least three groups will die within the test period.

Dermal Toxicity: It is well established that skin is permeable to certain types of chemicals, especially lipid-soluble compounds, and many examples of toxicity to specific organs have been reported after skin contact with toxic chemicals [21]. In view of this possibility, it is recommended that new cosmetic preservatives or products be subjected to a standard dermal toxicity assay.

The acute dermal toxicity assay most frequently used involves application of the test material to the skin of rabbits, even though the monkey and domestic swine are closer to humans with regard to the percutaneous penetration of chemicals [21]. The application site is gently cleaned after 3 hr and the animals are observed for 14 days. Either a single dose of 2 g/kg body weight or multiple doses are applied. Generally, if toxicity (indicated by the death of at least one animal out of six) is observed at 2 g/kg, a complete LD_{50} assay is carried out to quantitatively establish the concentration of material which, when applied to the skin, causes death in one-half of the animals. As with the oral LD_{50} value, this figure is valuable in providing a quantitative measure of toxic potential of a chemical with regard to a specific route of exposure.

The signs of toxicity evaluated in dermal toxicity studies are essentially the same as for systemic toxic effects produced by oral administration of the test substance, and, for that matter, similar to those evaluated upon inhalation exposure. Observation of the test animal over a 14-day period, coupled with gross pathology at terminal necropsy, are the minimum criteria used to assess toxic responses. Guidelines for dermal toxicity tests are provided in the EPA FIFRA guidelines [9,10], the Organization for Economic Cooperation and Development (OECD) guidelines [22], and in the procedures outlined by the U.S. government's Interagency Regulatory Liaison Group [23]. Additional criteria which can be included are histopathological evaluation of selected tissues, effects on hematological and blood and urine chemistry parameters, neurological effects as measured by behavioral changes, and mutagenic effects on specific cell types, especially newly formed blood cells. Consideration of the chemical or potential toxic properties of the test preservative will allow inclusion in the study of the most appropriate measures of specific toxic effects.

Inhalation Toxicity: Exposure to cosmetic ingredients by inhalation of vapors or aerosols is possible during manufacture and packaging or during actual use of the end product. For this reason it is sometimes advisable to evaluate potential toxicity by allowing animals to breath air containing the test chemical. Acute inhalation toxicity studies consist of exposing rats for 1-4 hr to air containing specific concentrations of vapors of volatile liquids or aerosols of powdered solid substances. If an ambient air concentration of 2 mg/liter produces no death in the test animals (generally 10 males), the substance is considered safe as far as acute toxic effects are concerned under the *Code of Federal Regulations*, Title 49, Department of Transportation's General Requirements for Shipments and Packagings [8].

If, on the other hand, the material is toxic, a quantitative measure of toxicity can be obtained by determining the concentration which causes death in 50% of the animals, the LC_{50} value (concentration lethal to 50% of the animals). The LC_{50} can be expressed as weight per standard volume of air (e.g., milligrams per liter) or as parts per million. The statistical methods used to obtain this value are the same as for the oral and dermal LD_{50} values. Additional parameters of toxicity are the same as for other acute toxicity studies; however, it should be noted that inhalation studies are more expensive to conduct, since special exposure chambers are required and chemical measurements for confirming the test sample concentration in the chamber air requires sophisticated analytical procedures.

B. Subchronic Toxicity Tests

The term *subchronic* as used in describing toxicity assays generally refers to tests in which animals are treated 5-7 days per week with the test material for more than 20 days, but less than 1 year. Routes of administration include oral gavage, via the feed or water, topical application, inhalation, and various injections. Topical application is naturally of most use in testing cosmetic ingredients. In some instances a single product is tested by more than one route of administration; for example, oral and dermal routes might be used to assess which form of exposure would lead to the lowest toxicity.

The major difference between acute and subchronic studies, aside from the duration, is the degree to which the experimental animals are evaluated for toxic effects. Subchronic studies frequently include, in addition to those observations listed for acute studies, eye examinations, clinical chemistry and hematology on blood samples, urine analyses, food consumption measurements, weekly body weight changes, and complete histopathological evaluation. A subchronic study not only serves to identify the toxic effects of a test substance, especially with regard to target organs, but also provides data necessary for selecting dose levels for a subsequent chronic study. These data are also valuable in understanding the development and mechanism of toxic effects. In view of the expense and time requirement of subchronic studies (and chronic studies as well), it is important to design the protocol to yield as much information as is needed and to include numbers of animals and measurements which will be sufficient for valid statistical analyses. It should be kept in mind that some animals, especially in high-dose groups, will die before the end of the study and the remaining number must be sufficient for end-of-study observations.

The most commonly used subchronic study with cosmetic materials is the 21-day repeated-dose rabbit dermal toxicity assay. This study requires that the test substance be applied to the clipped skin of rabbits 5 or 7 days per week for 21 days either at a single dose of 2 g/kg body weight or in graded doses less than the estimated maximum tolerated dose. At the end of 21 days the remaining animals are evaluated for systemic toxicity, including using those parameters described for evaluating acute systemic toxicity. Histopathological evaluation of major organs and tissues is an important part of the study. It is possible to incorporate additional study parameters into the subchronic study, such as absorption, distribution, and elimination of the test material and evaluation of specific toxic effects. The subchronic dermal study is useful in the evaluation of cosmetic preservatives, since it allows assessment of local skin effects and potential systemic actions with repeated exposure over reasonably long periods of time at a far lesser cost in time and money than for chronic lifetime studies. The EPA FIFRA guidelines [9,10] provide a good foundation for developing a subchronic study protocol.

C. Chronic Toxicity Tests

Chronic toxicity studies, also referred to as long-term or lifetime studies, are designed to last throughout most of the test animals' life-span. With mice this period is 18-24 months, with rats 24-30 months, and with dogs and monkeys up to 5-7 years. Routes of administration are the same as for acute and subchronic studies. Dosing usually begins soon after weaning and acclimation. A lifetime oncogenicity study is intended to evaluate the ability of the test substance to cause benign and malignant tumors and preneoplastic lesions; a chronic toxicity study is designed to detect other toxic effects as well. The two types of studies are generally combined in the form of a chronic toxicity and carcinogenicity assay. Chronic carcinogenicity studies are usually conducted when there is likelihood of widespread exposure to the test substance or if there is reason to suspect that the test substance may be carcinogenic. Types of data important in considering the potential for carcinogenicity include results of in vitro mutagenicity assays, chemical relationship to known or suspected carcinogens, and epidemiological evidence from human exposure. Occasionally evidence of carcinogenicity will show up in subchronic studies, although carcinogenic effects are most reliably confirmed late in the animal's life or after prolonged exposure to the test agent.

Common protocol features are the inclusion of two or three dose levels, vehicle control, and untreated controls. The maximum tolerated dose of the test substance is used as the highest dose group: The other dose groups are fractions of the maximum tolerated dose. Dose levels are selected from subchronic studies, knowledge of relevant pharmacokinetics, known toxic properties of the test substance, and information on potential exposure to humans. Both males and females are included, unless the test chemical will clearly be restricted in use to one sex or the other. A typical study will include 50 animals per sex per dose or control group, totaling 500 animals for three doses and two controls. The complexity of such a study is demonstrated by the fact that body weight changes, food consumption, signs of toxicity, and occurrence of death must be unerringly recorded throughout the course of the study. Additional test parameters include the occurrence of observable tumors, blood and urine chemistry, gross pathology, and extensive histopathological evaluation. As in acute and subchronic assays, special studies can be incorporated to provide valuable information on the test chemical.

The majority of cosmetic ingredients do not require chronic toxicity evaluation, since most materials with known potential for carcinogenicity have been eliminated from use. However, it should be kept in mind that as more data from in vitro mutagenicity assays are accumulated, there may occur cases in which materials now in use are shown to be mutagenic. If so, it may be prudent, or required, either to conduct chronic assays or substitute chemicals with established safety. FIFRA, OECD, and Interagency Regulatory Liaison Group guidelines have been developed for chronic animal toxicology and carcinogenicity studies.

D. Teratology Tests

A teratogen is a substance which causes abnormalities in the developing fetus. Teratogenic effects occur early in the fetal period, during organogenesis, and can alter the structure and function of developing cells and tissues. Such changes in the developing organism should not be confused with direct fetotoxic actions of chemicals, which can occur at any stage in development. Teratogenic effects refer to morphological, biochemical, and functional abnormalities produced by a toxicant before death. Test models for screening teratogenic potential fall into two general categories: animal models for detecting teratogens directly and in vitro models for detecting biochemical or cellular changes which may produce teratogenic effects in vivo. Chemical attack during the critical stages of organogenesis will most likely result in teratogenic effects; the earlier this takes place during organogenesis, the higher the likelihood of a malformed offspring. When the developing organism enters the fetal growth period, changes due to chemical substances are almost certain to be toxicological rather than teratogenic.

Although there is no animal model which duplicates man in all physiological aspects, several systems have been developed as workable laboratory models for screening the teratogenic effects of chemicals [10,11,21,22,24]. The most commonly used are the rat and rabbit teratology tests. In each case the pregnant female is treated with various doses of the test substance daily for a set period of time early in fetal development. As with other toxic effects, it is important to establish dose-response relationships. Multiple exposures and doses are important, since teratogenic effects vary with the dose of the teratogen as well as with the stage of development. Teratogenic effects are detected and quantitated in routine screening assays by enumerating gross malformations on the exterior of the fetus, soft-tissue anomalies as observed by dissection or histological examination, and bone anomalies as seen in specially stained skeletal preparations.

It is important to keep in mind that physical anomalies are detected by these classic teratology procedures. Teratogenic effects in biochemical processes and nervous system functions can be detected only if the newborn animals are allowed to survive and evaluated with a battery of appropriate biochemical, clinical chemistry, and behavioral tests.

E. Reproductive Toxicity

Reproductive toxicity includes all effects resulting from parental exposure to chemicals which interfere with mating, fertilization, implantation, gestation, birth, and subsequent development of offspring. It includes effects on the behavior and fertility of the parents and effects on the offspring (including behavioral impairment, structural malformations, infertility, and cancer) which may not be manifested until they reach adult life. Not generally included are genetic or chromosomal effects in parents, unless there is reason to associate them with specific types of reproductive impairment.

The effects of chemicals on human reproduction, and the risks from exposure are obviously difficult to assess accurately because of the complexity of the reproductive process and the many years required for reproductive maturation. Moreover, it is well recognized that many chemical agents to which an individual may be exposed may be overtly or potentially harmful, causing structural or functional change immediately, after a considerable lapse of time, or after some indeterminate time following exposure.

Chemicals can adversely affect reproduction in mammals at several different stages; the following are some examples:

    Damage to parental gametes, resulting in sterility or abnormal development of the fertilized egg or embryo
    Interference with normal uterine development and the nutrition of the conceptus
    Damage to the embryo or inhibition of embryogenesis
    Toxic effects on the fetus, fetal membrane (yolk sac and amnion), or placenta
    Inhibition of maternal metabolism, causing secondary effects on the fetus
    Inhibition of uterine growth
    Adverse effects on parturition
    Adverse effects on lactation
    Latent effects on the progeny, manifested in later life (e.g., impaired development, infertility, or cancer)

Extensive consideration has been given to these factors in order to optimize different protocols for reproductive toxicity testing. Guidelines have been published by the U.S. Food and Drug Administration [25], the World Health Organization [24], the Organization for Economic Cooperation and Development [22], the U.S. Environmental Protection Agency [8,9], and the National Research Council [26]. In addition, a large number of experimental assay systems for investigating the potential effects of chemicals on the reproductive functions of mammals have been published. A critical review of the different types of studies, end points, and their applicability to cosmetic products is beyond the scope of this review.

F. Toxicokinetics

The skin is an important barrier separating man from toxic chemicals; however, this barrier is not complete and toxicants do enter the body to produce damage; the higher the concentration attained in the body, the greater the damage. The concentration of the toxicant in the body is obviously a function of the amount of toxicant the individuals comes in contact with, but it also depends on the rate and amount absorbed, the distribution of the toxicant within the body, the rate of metabolism, and the rate of excretion of the toxicant [27].

A toxicant is usually absorbed and enters the blood before it produces its undesirable effects, unless it acts topically. The major routes by which toxicants enter the body are the lungs, gastrointestinal tract, and skin. Once the chemical has entered the bloodstream, it can reach the site in the body where it produces its effect. For a chemical to produce a toxic effect in a certain organ, the toxic agent must reach that organ, but the organ in which the toxicant is most highly concentrated is not necessarily the organ where most of the tissue damage occurs. For example, the chlorinated hydrocarbons attain the highest concentration in the fat depots of the body, but produce no toxic effects there. A toxicant is removed from the blood by biotransformation, excretion, and accumulation at various storage sites. The relative importance of these processes is related to the physical and chemical characteristics of the toxicant. Although the kidney plays a major role in the elimination of toxicants, other organs are important. An example is the excretion of volatile chemicals by the lungs. Although the liver is the most active organ in the biotransformation of toxicants, enzymes in other tissues, such as the esterases in the plasma, and enzymes in the kidney, lung, and gastrointestinal tract, may also metabolize foreign chemicals. Biotransformation is often a prerequisite for the renal excretion of a toxicant, because the toxicant may be so lipid soluble that it is reabsorbed by the renal tubules after filtration. After the toxicant is biotransformed, its metabolites may be excreted into the bile, as are the metabolites of DDT, or they may be excreted into the urine, as are the metabolites of organophosphate insecticides.

The science of toxicokinetics involves the quantitative measurement of the absorption, metabolism, distribution, and elimination of toxic substances. Standard analytical chemistry and radioisotope procedures are used to trace the fate of the chemical at the site of application, in body fluids and tissues, and in urine, fecal matter, and expired air. Balance studies can be conducted to ensure recovery of all the applied test material. Sophisticated instruments such as mass spectrophotometers and high-performance liquid chromatographs are used to identify the metabolites of chemicals taken up into the circulation and distributed to various tissues. Metabolic pathways can be different for a chemical for different routes of administration and under different use conditions. For example, metabolism of some chemicals by skin can be influenced by light; photosensitive chemicals will undergo different chemical reactions in the presence of photons than in the dark.

The complexity of toxicokinetic studies is enhanced by the need to combine reliable biochemical and physiological studies with accurate and sensitive analytical chemistry procedures. Even to a greater extent than for other animal toxicology studies, the reliability of the experimental data reflects the quality of experimental design, execution, and equipment.

G. Photosensitivity Reactions

Photosensitivity can be defined most simply as "cell injury by photons" [28]; chemical phototoxicity is described as phototoxicity in the presence of exogenous chromophores. Early work by Epstein [29] showed that photosensitivity reactions can be divided into phototoxicity and photoallergic responses. Phototoxicity is similar to primary skin irritation, since it is not based on immune mechanisms and occurs upon first exposure. Photocontact allergy is analogous to delayed contact dermatitis, although light energy is required for the formation of the effective antigen. It is a classic immunological reaction which occurs only in previously sensitized individuals.

It is known that exposure to certain chemicals can lead to increased sensitivity to light [30]; sulfanilamide, chlorothiazide, tetracyclines, and chlorpromazine are examples of photosensitizing drugs; other examples are many essential oils (e.g., lime, bergamot, rue, angelica root), 7-methoxycoumarin, methylene blue, eosin (on scarified skin), amyldimethylaminobenzoate, coal tars, and furocoumarins such as 8-methoxypsoralen.

The most commonly used parameter to measure phototoxicity is primary skin irritation, especially erythema; albino rabbits, guinea pigs, standard mice, hairless mice, and miniature swine are routinely used for this test. A standard procedure has also been developed by Kaidbey and Kligman for evaluating photosensitivity in humans as well [30]. Intact and scarified skin should be included in these assays.

Although there are no federal regulations requiring the evaluation of the phototoxicity potential of cosmetic ingredients in general and preservatives in particular, it should be considered when new preservatives contain structures which are related to chemicals known to cause photoirritation. Preserved sunscreen products are logical candidates in view of the prolonged exposure to sunlight associated with their use.

The interest of the FDA in phototoxic responses is reflected in its annual Photochemical Toxicity Symposia, at which participants from industry, academia, and federal health services discuss current research and safety regulations pertaining to this topic [31].

III. IN VITRO TOXICITY ASSAYS

Protests by animal rights groups against the use of laboratory animals in toxicology studies have intensified significantly over the past several years, although the movement now actively promotes the development of alternative methods as well as attacking what they interpret as inhumane treatment. In fact, efforts of this type have resulted in the introduction of a bill in Congress (HR 556) which proposes the establishment of a Center for Alternative Research to be under the auspices of the National Institutes of Health, and includes allocating up to 50% of the funds designated for animal-related research to the development of alternative systems. Such a bill would formalize at the government level a trend to reduce the use of animals in testing and biomedical research which has been quietly but steadily developing over the past decade. However, in reality, scientific and economic reasons are the driving forces for seeking nonanimal test systems, rather than pressures from the animal welfare movement.

Animals can be eliminated from toxicity screening only when scientifically sound alternatives are available. In fact, some scientists feel that animals will never be completely replaced, although their use may be significantly reduced by the development of nonanimal adjunct procedures. Procedures which are discussed in this context include mathematical models, isolated organs, tissue and cell cultures, chemical and physical assays, anthropomorphic dummies, simulated tissues and body fluids, mechanical models, computer simulations, and lower organisms. However, it is important to note that for the most part these possibilities are at the level of discussion only, with the exception of in vitro biological assays and, to a limited extent, computer modeling. Considerable research is required for the development of alternative methods; much of the research required to provide alternative procedures could even increase the use of animals, since correlations between data provided by any new method and data obtained from animal (or human) studies is necessary in order to validate the alternate method.

The economics of toxicity testing are of major importance as an impetus to develop alternate test procedures. For example, a lifetime carcinogenicity study in one species, the rat, for example, can cost between $400,000 and $800,000, depending on the route of administration and specific test requirements. A complete toxicological evaluation of a new product can cost as much as $2 million. These costs are put into perspective when the number of new chemicals developed each year is considered. Hundreds of new chemicals are introduced in the United States each year, and there are already approximately thousands in use. A small fraction of these have been tested for carcinogenicity; the combined costs of testing these chemicals would be prohibitive. For economic reasons of this nature, the availability of low-cost reliable screening assays is highly desirable (see Ref. 32).

Enormous progress has been made over the past 30 years in cell, tissue, and organ culture, although only over the past few years has this knowledge been put to use on a routine basis for the evaluation of the toxic properties of chemicals. However, as more applied scientists discover the advantages of in vitro cell systems, and as the procedures become validated by their use, it is inevitable that the number and use of routine in vitro tests will be expanded. Progress in this area will be stimulated by the recent initiation of major research programs directed toward the development of nonanimal testing procedures. Tufts University was awarded funds by the American Fund for Alternatives to Animal Research, The Medical College of Pennsylvania received a grant from the New England Antivivisection Society, Rockefeller University was awarded a grant from Revlon, and Johns Hopkins University received funds from the CTFA and Bristol-Myers. Many private drug firms and independent testing laboratories are also engaged in research of this type, and it is likely that rapid progress will be made over the next decade and that the selection of nonanimal test procedures will be far more extensive. Directions in in vitro toxicity have been reviewed recently in two conferences on this topic convened in 1975 and 1976 by the Tissue Culture Association's Committee on Carcinogenesis, Mutagenesis, and Toxicity Testing In Vitro [33].

A. Cytotoxicity

Cell culture systems can play an important role in toxicology testing; they are sensitive, homogeneous, easy to establish and manage, cost effective, informative, and predictive of toxic effects which may occur in humans. In many cases in vitro systems provide the most realistic and logical approach to screening potential toxic agents prior to their comprehensive investigation in acute, subchronic, and chronic animal studies. Currently, mammalian cell culture systems are being effectively used to evaluate the potential cytotoxic effects of cosmetic ingredients, medical device materials, pharmaceuticals, chemicals, and environmental pollutants. Soluble materials are dissolved in isotonic buffer and added directly to cultures; insoluble materials can be either extracted under standard conditions and the extract added to the cultures, tested as suspensions, or tested directly in the agar overlay test. Cytotoxicity is based on the effects of the test substance on cellular integrity, cell growth, and changes in specific biochemical or physiological parameters. Selected assay systems are described below to illustrate the diversity and use of in vitro cytotoxicity tests.

1. Direct Exposure Test

In this assay, cultures of human diploid fibroblasts are used to evaluate the potential cytotoxicity of test materials. Samples are added to confluent cultures of cells and incubated for a period of 24 hr; at 24 hr the cells are examined microscopically for signs of cytotoxicity such as cell lysis, vauole formation, and nuclear abnormalities. The degree of cytotoxicity is scored on a relative scale of 0 to 4, based on the percentage of affected cells. A score of 1 or greater for either cell lysis or vacuole formation is considered a cytotoxic response. This assay is useful for quality control analysis of medical device materials and screening the biocompatibility of pharmaceuticals and cosmetic ingredients.

2. Agar Overlay Test

This test system makes use of confluent cultures of cells, such as mouse L929 fibroblasts, overlaid with 1.5% Noble agar in cell culture medium [34]. The agar surface provides a flexible system for testing a variety of different solid samples such as plastics, powders, and rubbers, as well as liquid samples impregnated on paper disks. The cells are treated with the vital dye neutral red prior to adding the test samples. The cultures containing samples are incubated for 24 hr, at which time they are examined microscopically for zones of decoloration and cell lysis. The degree of cytotoxicity is based on the percentage of affected cells and is scored according to decoloration and lysis. This test measures the cytotoxicity of water-extractable substances leached from the test sample.

3. Neutral Red Uptake

This test also provides a valuable in vitro method for screening cytotoxicity and is effectively used in combination with the Chinese hamster ovary cell clonal cytotoxicity test [35]. Human diploid fibroblasts in exponential growth are incubated for 48 hr with an extract of the test substance incorporated into the culture medium. After incubation the cells are stained with neutral red for 2 hr, rinsed with a buffer, and the intracellular neutral red extracted in acid-alcohol. The amount of dye extracted from the cells is measured spectrophotometrically at 540 nm. The degree of cytotoxicity is based upon the percentage of neutral red uptake as compared to the uptake in the control cultures. A neutral red uptake score of 85% or less is considered cytotoxic. This assay system measures the ability of water-extractable substances from the test article to affect cellular integrity, cell proliferation, and the uptake and accumulation of neutral red. This assay system is more suitable for screening cytotoxicity than are the direct exposure and the agar overlay tests because of its greater sensitivity and its ability to provide a quantitative indication of cytotoxicity.

4. Chinese Hamster Ovary Cell Clonal Cytotoxicity Test

The Chinese hamster ovary clonal cytotoxicity test provides a sensitive and quantitative procedure for evaluating cytotoxicity [35]. Chinese hamster ovary cells are seeded into 10×100 mm Petri dishes at an inoculum density which ensures the development of colonies from individual cells. Twenty-four hours after seeding the cultures the cells are incubated with at least five concentrations of the test substance for a period of 5-8 days. At this time the number of colonies which have developed are counted using an automatic colony counter. The percentage of colony survival, based on the number of colonies present in the medium control cultures, is obtained for each concentration of the test substance. The cytotoxic effects of the test substance are quantitated by calculating the LC_{50} (the concentration lethal to 50% of the cells) value obtained from the concentration-response curve. This is currently one of the most sensitive in vitro assays available for evaluating the cytotoxicity of medical device materials and environmental pollutants. The high degree of sensitivity is related to the limited nutritional conditions under which the cells are grown. Under these conditions, the growth and survival of the Chinese hamster ovary colonies is solely dependent on the concentration of growth factors present in the culture medium, since the concentration of growth factors synthesized and secreted by the cells are at too low a concentration to be effective.

5. L929 Chromium-51 Release Test

The L929 chromium-51 release test is recommended by the American Dental Association for evaluating the biocompatibility of dental materials [36] and would also be useful for evaluating the toxicity of cosmetic ingredients. L929 mouse fibroblasts previously labeled with the radioisotope chromium-51 are seeded into 24-well tissue culture plates at a density providing a confluent monolayer. One hour after seeding, a period of time sufficient for cell attachment, the medium is aspirated from the wells and the cells are treated with the test and control substances incorporated into culture medium. At 4 and 24 hr of incubation an aliquot of medium is removed from each well and the amount of chromium-51 released into the medium is counted using a radioisotope counter. The degree of cytotoxicity is based on the percentage of chromium-51 released into the medium as compared to that in the medium control cultures. If the percentage of chromium-51 release exceeds twice the mean control value after a 4-hr treatment, the sample is considered cytotoxic. This test measures the ability of a test substance to produce cytotoxic effects on cellular integrity.

6. Rabbit Alveolar Macrophage Assay

The rabbit alveolar macrophage assay provides a sensitive method for analyzing the potential toxicity of particulate matter [37]. Alveolar macrophages provide an excellent indicator cell for this assay, since they normally participate in the detoxification of inhaled particulates, they are easy to obtain in large quantities from the lungs of rabbits, and they can be maintained in a functional state under in vitro cell culture conditions. The basis of the assay is the impairment in the functional status of the cells.

Alveolar macrophages are obtained from rabbit lungs by tracheobronchial lavage with 0.9% sodium chloride. After the cells have been washed several times, a standard number of cells are suspended in culture medium and dispensed into each well of six-well tissue culture plates. Solutions of the test substance prepared at twice their test concentrations are added to an equal volume of suspended cells contained in each well. Five concentrations of the test substance and the medium control are tested in triplicate wells. The culture plates are placed on a rocker platform set at 20 oscillations per minute and incubated in a moist atmosphere of 95% air-5% CO_2 at 37°C for 20 hr. After the incubation period the cultures are prepared for measurement of cell number, viability, and ATP content. Phagocytosis of particulate matter, such as latex beads, for example, can be measured microscopically. These parameters adequately define the impairment in macrophage function, and the relative degree of cytotoxicity is based on the EC_{50} values obtained for the percentage inhibition in cell number, viability, phagocytosis, and for the reduction in cellular ATP content.

7. Target Cell Assays

Ideally, cosmetic ingredients should also be evaluated in in vitro systems consisting of cell types derived from tissue which is exposed to the particular cosmetic ingredient being evaluated. Intensive research efforts in industrial, academic, and government laboratories are directed toward developing such systems. Eye and skin cells are excellent candidates, not only because they can be grown in culture, but also in view of their importance as sites of exposure to cosmetics. The availability of specific target tissues, rather than reliance on only the general toxicity reactions obtained by use of the systems described above, will enhance the predictive value of in vitro systems.

B. Mutagenicity

There are over 2000 diseases which have a genetic basis, and over 100 have been identified as being related to specific enzyme deficiencies, presumably representative of point mutations [38]. Sickle cell anemia, phenylketonuria, and Tay-Sachs disease are examples of diseases arising from point mutations. Other genetic diseases are associated with chromosomal disorders, such as translocation, deletions, and aneuploidy. Down's syndrome, for example, results from an extra #21 chromosome and is one of the leading causes of mental retardation. Although it is difficult to demonstrate that a particular mutation leading to a human genetic disease was the result of exposure to a specific chemical, such relationships are almost certain to exist. Given the similarity of the organization of DNA in mammals, and the evidence that exposure to chemicals can induce heritable mutations in experimental mammals, exposure to chemicals would also increase the incidence of genetic disease in humans. Mutations in somatic cells (cells other than germ cells) can also have a serious effect on human health; there is a great deal of evidence supporting the hypothesis that mutations in somatic cells lead to the formation of cancer [39-42]. This evidence is mainly derived from studies on the inheritance of cancer in humans, clinical investigations of patients with defective DNA repair systems who are also more cancer prone than normal, observation of chromosomal changes in malignant cells, and studies of carcinogenic agents which also cause DNA damage and mutations.

A variety of short-term tests (both in vitro and in vivo) have been developed in which a high percentage of known carcinogens produced a positive response [43]. These tests assess changes in the genetic material in bacteria, animal cells, or whole animals. In order to mimic mammalian metabolism, which is responsible for the conversion of promutagens to their active forms, the in vitro tests can be conducted with microsomal enzymes from rodent livers containing mixed function oxidases which activate promutagens. Since the cause of the induction of most cancers is presumably the interaction of the chemical with the DNA of the chromosome, and the induction of genetic disease also results from the same interaction, these short-term tests have been employed to identify compounds or complex mixtures which are potential human mutagens or carcinogens. The remainder of this section discusses the types of screening tests which are routinely employed for this purpose and describes the tests which would be most useful for assessing the long-term risk to human health posed by cosmetic ingredients and products.

Genetic toxicology tests can be selected to provide a qualitative assessment of mutagenic or carcinogenic potential and a quantitative assessment of actual risk to germ cells. Tests used for a qualitative assessment of mutagenic potential include bacterial and mammalian cell culture assays, chromosomal damage assays, unscheduled DNA synthesis, and assays in yeast. Assays used to assess the effect of a chemical on germ cells include the *Drosophila* sex-linked recessive lethal test, the heritable translocation test, the mouse specific locus assay, the Russell spot test, and the sperm head abnormality test.

In vitro genetic toxicology tests are recommended by the EPA in its FIFRA and Toxic Substance Control Act (TSCA) guidelines and are accepted in support of new product registrations by the FDA or by the OECD. For most new product evaluations, one or two assays are selected from each of several test categories, these test categories being the following:

Point mutations
Chromosome damage
Primary DNA damage and repair
Cell transformation

Such a battery of three to six tests should be adequate to assess the genotoxic potential of new cosmetic preservatives. Negative results in each test would imply that the material has little likelihood for teratogenic or carcinogenic effects. Positive results in each test would indicate a strong probability for genetic damage. Borderline results, always the most difficult to evaluate, raise the possibility of genetic toxicity and will force decisions with regard to further in vitro testing or animal teratology and carcinogenicity assays.

A number of the tests which have been proposed will not be discussed, since they have little applicability to the assessment of risk to humans from cosmetic products. The Russell spot test and the heritable translocation test are both expensive and not used routinely. Owing to the large number of chemicals tested in the Ames/*Salmonella* mutagenesis assay, other bacterial assays, such as those using *Escherichia coli* or *Bacillus subtilis* will not be considered. The remaining tests will be described briefly.

1. Ames/Salmonella Mutagenicity Assay

The *Salmonella*/mammalian microsome mutagenicity assay or "Ames test" measures the ability of a chemical to induce back-mutations in the histidine biosynthetic pathway of five specially constructed *Salmonella typhimurium* mutants [44]. These strains, TA98, TA100, TA1535, TA 1537, and TA1538, are unable to grow in the absence of histidine (his^-). In the presence of mutagenic substances, bacteria undergo a back mutation to his^+ and no longer require histidine for growth. The strains have been specially constructed to detect as many classes of mutagens as possible. To permit entry of large molecules, such as polycyclic hydrocarbons, aflatoxins, and aromatic amines, the cell wall has been modified (*rfa* mutation); an excision repair system (*uvrB* mutation) has been deleted in order to maximize the expression of mutations. Strains TA100 and TA1535 detect mutagens which cause point mutations (substitution of one base for another in the DNA); TA1537, TA98, and TA1538 detect different types of frameshift mutagens (addition or deletion of one or more bases from the DNA). In addition, a plasmid has been introduced into two of the strains, TA100 and TA98, to increase their sensitivity to some mutagens, such as aflatoxin B_1 and benzo[a]pyrene, presumably by increasing error-prone repair. Strains TA100 and TA98 were created by transferring a resistance transfer plasmid, pKM101, into TA1535 and TA1438, respectively. The use of a rat liver homogenate (also called S9 preparation) provides the enzymes found in most mammals needed to metabolize and to convert many potential mutagens to their reactive forms.

2. Yeast Assays

Saccharomyces cerevisiae is the most commonly used yeast strain for mutagenicity assays. Point mutations, mitotic crossing over, and gene conversion can be detected by a standard assay procedure [45]. A mutagenic event is expressed and detected in alterations in colony color, which reflects mutations in the adenine biosynthesis pathway.

3. Cell Culture Mutagenesis Assays

Two forward mutation systems utilizing mammalian cells in culture are used to detect mutagenic substances, the mouse lymphoma L5178Y cells [46] and the hypoxanthine-guanine phosphoribosyl transferase (HGPRT) [47] assays using either Chinese hamster ovary (CHO cell line) or Chinese hamster lung cells (V79). The mouse lymphoma assay uses cells which are heterozygous for thymidine kinase ($TK^+/^-$) and detects substances which cause cells to mutate to $TK^-/^-$. The mutated cells can grow in the presence of a toxic pyrimidine analog (usually trifluorothymidine) and develop colonies in the selective medium. Hypoxanthine-guanine phosphoribosyl transferase is a purine salvage enzyme which provides a scavenger pathway for the synthesis of purine nucleotides. The enzyme converts the preformed purine bases hypoxanthine and guanine to their respective nucleotides, thus bypassing any cellular requirement to synthesize the purines from carbon and nitrogen sources. Mutants deficient in HGPRT activity are selected by growing mammalian cells in medium containing 6-thioguanine (6TG). This purine analog is an HGPRT substrate and is converted to a toxic nucleotide, thus killing the wild-type cells. Mutant cells, those lacking HGPRT activity, cannot convert 6TG to its toxic form and are able to survive the selection step. Biochemical methods have shown that over 98% of the surviving cells have reduced HGPRT activity, validating the selection procedure. Additionally, detailed studies have shown that the reduced HGPRT activity results from a change in the HGPRT enzyme, proving that survival in 6TG medium selects for true HGPRT mutations.

The V79/CHO HGPRT system is a well-characterized mammalian cell culture system which measures the frequency of mutations induced at a specific gene locus. It utilizes optimal selection conditions for such crucial parameters as the mutagenic exposure time, phenotypic expression time (time required for expression of mutation), 6TG concentration, and the cell density which permits maximum mutant recovery. It is an excellent short-term in vitro assay for evaluating the mutagenic activities of compounds and is obviously more closely related to man than the bacterial mutagenicity assays.

4. Drosophila Sex-Linked Recessive Lethal Test

The *Drosophila* (fruitfly) assay detects the induction of lethal mutations in treated male flies [48]. Treated males are crossed with *Basc/Basc* (extreme bar, apricot eye) females, producing first-generation offspring which are $Basc/X^t$ females and *Basc*/Y males; the X^t is the X chromosome from treated males. Individual first-generation females are then mated to their brothers. Each single-female culture is scored for the presence of wild-type males. If the treated X chromosome from the males carried lethal mutations, the class of wild-type males will be missing. Although this assay is time-consuming, it is an excellent quantitative test for mutations in a living organism.

5. Micronucleus Test

The micronucleus assay is utilized primarily for screening test substances for the ability to break chromosomes [49]. After treatment of animals with known chromosome-breaking agents (clastogens), acentric fragments of chromosomes lag behind during cell division and are not included in the nuclei of the daughter cells; they form micronuclei in the cytoplasm of the daughter cells instead. In bone marrow smears the erythroblast is the most useful cell type for detecting micronuclei. A few hours after the completion of mitosis, erythroblasts expel their nuclei. However, the micronuclei remain in the cytoplasm of the young erythrocyte and are easily recognized. Since young erythrocytes (up to 24 hr after expulsion of the nucleus) stain blue and older erythrocytes stain red, it is possible to score for micronuclei only in young cells. This type of scoring thus detects anomalies which must have occurred during the preceding mitosis while the test substance was present in the animal.

6. In Vivo Dominant Lethal Assay in Rodents

The dominant lethal assay in rodents measures the ability of a test substance to induce a lethal mutation in the sperm of the treated males [50]. After treatment, males are mated to untreated, virgin females. Any sperm carrying a lethal mutation will result in a nonviable zygote after fertilization. Usually the treatment must result in gross genetic damage, such as deletions or translocations, to produce nonviable zygotes. In most protocols males are treated for five consecutive days and then mated with two virgin females each week for 7 weeks, a time period which covers the total spermatogenic cycle. After the mating week the females are housed until mid-pregnancy, at which time they are killed and examined for the number of living and dead implantations, thus providing a quantitative evaluation of mutations in the male.

7. In Vitro Chromosomal Damage Assays

The in vitro sister chromatid exchange (SCE) and the chromosomal aberration assays detect substances which cause chromosomal damage in cultured mammalian cells [51]. To detect SCEs, cells are labeled for two cycles with bromodeoxyuridine (BrdU), which results in differentially stained sister chromatids after treatment with fluorescent and Giemsa stains. Following administration of a test substance, chromosomes are analyzed for exchanges between sister chromatids; the exchanges appear as reciprocal interchanges along the differentially stained chromosomes. These SCEs presumably involve DNA breakage and reunion and may represent the expression of a DNA repair system. A number of studies using known alkylating agents have shown that SCEs are induced at concentrations well below those needed to cause chromosomal breaks or aberrations [52].

The in vitro chromosome aberration assay is used to detect chromosome-breaking agents, clastogens. The type and extent of damage inflicted by such agents are usually dose dependent; more breaks, deletions, and complex rearrangements such as quadriradials and triradials are seen at higher doses than at lower doses. In vitro testing allows better control over dosing and exposure time conditions than can be obtained during in vivo (bone marrow) testing.

Introduction of a metabolic activation system (the S9 liver fraction) improved the sensitivity of the in vitro assay, and promutagens such as nitrosamines, aflatoxin B_1, and benzo[a]pyrene produce chromosomal damage after activation.

Chromosomal damage assays can also be performed in vivo. To detect chromosomal aberrations or breaks in vivo, bone marrow cells from rodents are removed following administration of the test substance. To detect SCEs in vivo, the animals must be treated with BrdU, which can be accomplished by injection, the implantation of tablets, or infusion through the tail vein. The detection of SCEs can take place in a variety of tissues, including regenerating liver, bone marrow cells, spleen, and spermatogonia. Sister chromatid exchanges can also be detected in systems other than rodents, such as the mudminnow, chicken, herring gull, chick embryo, and marine worm.

8. Mutagenicity and Ultraviolet Light

A problem unique to the cosmetic industry is the use of compounds which are specifically manufactured to absorb ultraviolet light and be used either as sun blockers or tanning agents. Although these compounds can be used at concentrations which do not produce irritation to human skin, their potential mutagenicity following irradiation with ultraviolet light must be considered. An example of this problem is the use of 5-methoxypsoralen in suntan lotion in Europe. Although it is relatively nontoxic by itself following exposure to ultraviolet light, it induces cell killing, mutations, and sister chromatid exchanges in vitro [53]. This material also induced tumors in mice [54]. A similar compound, 8-methoxypsoralen, used in conjunction with ultraviolet light to treat psoriasis, is known to induce skin cancer and produces tumors in mice [55]. These findings illustrate the need to perform mutagenicity studies as well as phototoxicity studies for additives to cosmetics. Although some mutagenicity testing of cosmetic ingredients is currently being performed, it is equally important to conduct such studies with inclusion of ultraviolet light exposure.

C. In Vitro Phototoxicity Assay

Weinberg and Springer have developed a sensitive in vitro procedure for detecting photoirritation [56]. The assay, using yeast cultured on agar, is based on a microbiological phototoxicity test developed earlier by Daniels [57] and can be easily used to screen cosmetic materials for phototoxic potential prior to human testing. The method employs ultraviolet (UVA) irradiation of the plate during diffusion of the test sample from a paper disk placed upon the agar freshly seeded with Fleischman's baker's yeast. The positive control is 8-methoxypsoralen. Inhibition of yeast growth reflects phototoxicity and can be easily quantitated. The test is inexpensive, easy to perform, and provides results within 72 hr.

Correlations with positive human phototoxicity results have been excellent. Tests such as this illustrate the value of in vitro screening assays with regard to savings in time and money and provide the impetus for further efforts in in vitro assay development.

IV. SUMMARY

This paper presents a discussion of standard animal toxicology tests used to evaluate the safety of cosmetic ingredients, and gives summaries of in vitro assays which, in the author's opinion, are becoming increasingly important for new product testing. However, it is clear that these tests do not provide completely reliable replacements for product evaluation with human volunteers, and, as indicated by Dr. Paul Koehler of Chesebrough-Ponds, Inc., "the 1980's will see a greater demand for, and reliance on, human testing..." [58].

Irrespective of the specific evaluation programs selected for new products, which will most likely include tests from each category, animal, in vitro, and human, the need for testing will increase rather than decrease.

Pressures from consumer groups and regulatory agencies for greater product safety will not abate, and as medical science becomes even more adept at detecting the long-term adverse effects of chemicals, their efforts will even intensify.

The higher costs of more extensive product evaluation will force manufacturers to rely more on less expensive initial screening assays for "prequalifying" new product candidates prior to submitting them to expensive and time-consuming tests with animals and humans. For this reason, cosmetic ingredient producers must continue, and expand, their active involvement in product safety evaluation and new assay development.

ACKNOWLEDGMENTS

The author thanks Drs. K. S. Loveday, M. P. Bear, and J. A. Salinas for suggestions and Ms. Chris Ulrickson for her expert assistance in the preparation of the manuscript.

REFERENCES

1. M. A. Barletta, *Drug Cosmet. Ind.*, *120*:44-46 (1977).
2. *Code of Fed. Regulations*, Title 21 361 (a) (1978).
3. R. P. Giovacchini, *CTFA Cosmet. J.*, *8*:7-11 (1976).
4. *J. Am. Coll. Toxicol.*, *1*:1-177 (1982).
5. Cosmetic, Toiletry, and Fragrance Association, *CTFA Safety Testing Guidelines*, Cosmetic, Toiletry, and Fragrance Association, Inc., Washington, D.C. (1981).
6. D. R. Bickers, in *Current Concepts in Cutaneous Toxicity* (V. A. Drill and P. Lazar, eds.), Academic, New York (1980), pp. 95-126.
7. J. H. Draize, G. Woodard, and H. O. Calvery, *J. Pharmacol. Exp. Ther.*, *82*:377-390 (1948).
8. *Code of Fed. Regulations*, Title 16, Part 1500.41 (1973).
9. *Code of Fed. Regulations*, Title 49, Part 173.240 (1972).
10. *Code of Fed. Regulations*, Title 40, Part 772.112 (1978).
11. *Code of Fed. Regulations*, Title 40, Part 163.81 (1973).
12. M. Steinberg, G. W. Wolfe, and A. H. McCreesh, *J. Toxicol. Cutaneous Ocular Toxicol.*, *1*:33-48 (1982).
13. J. S. Friedenwald, W. E. Hughes, Jr., and H. Herrmann, *Arch. Ophthalmol.*, *31*:279-288 (1944).
14. E. V. Buehler, *Arch. Dermatol.*, *91*:171-177 (1965).
15. B. Magnusson and A. M. Kligman, *J. Invest. Dermatol.*, *52*:268-276 (1969).
16. L. Schwartz and S. M. Peck, *Public Health Rep.*, *59*:2-10 (1944).
17. J. Stotts, in *Current Concepts in Cutaneous Toxicity* (V. A. Drill and P. Lazar, eds.), Academic, New York (1980), pp. 41-53.
18. J. T. Litchfield and F. Wilcoxon, *J. Pharmacol. Exp. Ther.*, *96*:99-113 (1949).
19. C. S. Weil, *Biometrics*, *8*:249-263 (1952).
20. A. N. Rowan, *Pharm. Technol.*, April: 65-94 (1981).
21. National Academy of Sciences, *Principles and Procedures for Evaluating the Toxicity of Household Substances*, National Academy of Sciences, Washington, D.C. (1977), pp. 26-27.

22. Organization for Economic Cooperation and Development, *OECD Short Term and Long Term Toxicology Groups*, final report, Paris, December 31 (1979).
23. Interagency Regulatory Liaison Group, *Guidelines for Selected Acute Toxicity Tests*, Washington, D.C. (1979).
24. World Health Organization, *Principles for the Testing of Drugs for Teratogenicity*, ITS Tech. Serv. No. 364, Geneva (1967).
25. *Guidelines for Reproduction Studies for Safety Evaluation of Drugs for Human Use*, U.S. Food and Drug Administration, Washington, D.C. (1966).
26. *Principles and Procedures for Evaluating the Toxicity of Household Substances*, National Academy of Sciences, Washington, D.C. (1977), pp. 99-118.
27. C. D. Klaassen, in *Toxicology: The Basic Science of Poisons* (L. J. Casarett and J. Doull, eds.), MacMillan, New York (1975), pp. 26-44.
28. J. Parrish, in an article by R. Goldenberg, *Drug Cosmet. Ind.*, May: 44, 45 (1981).
29. J. H. Epstein, in *Yearbook of Dermatology* (F. D. Malinson and W. Pearson, eds.), Medical Year Book, Chicago (1939), pp. 5-43.
30. K. H. Kaidbey and A. M. Kligman, in *Current Concepts in Cutaneous Toxicity* (V. A. Drill and P. Lazar, eds.), Academic, New York (1980), pp. 55-68.
31. R. Goldenberg, *Drug Cosmet. Ind.*, May: 44, 45 (1981).
32. M. Hollstein and J. McCann, *Mutat. Res.*, 65:133-226 (1979).
33. *Short Term In Vitro Testing for Carcinogenesis, Mutagenesis and Toxicity* (J. Berky and P. C. Sherrod, eds.), The Franklin Institute Press, Philadelphia (1977).
34. W. L. Guess, S. A. Rosenbluth, B. Schmidt, and J. Autian, *J. Pharm. Sci.*, 54:156-170 (1965).
35. M. P. Bear, D. S. Johnson, and F. H. Schneider, in *Safety Evaluation and Regulation of Chemicals* (F. Homburger, ed.), S. Karger, Basel (1983), pp. 277-283.
36. Adendum to American National Standards Institute/American Dental Association Document 41* for Recommended Standard Practices for Biological Evaluation of Dental Materials, American National Standards Institute, New York (1982).
37. M. D. Waters, D. E. Gardner, and D. L. Coffin, *Toxicol. Pharm.*, 28:253-263 (1974).
38. V. A. McKusick, *Mendelian Inheritance in Man, Catalogs of Autosomal Dominant, Autosomal Recessive, and X-Linked Phenotypes*, 4th Ed., Johns Hopkins University Press, Baltimore (1975).
39. J. Cairns, *Cancer: Science and Society*, Freeman, San Francisco (1978).
40. *Origins of Human Cancer* (H. Hiatt, J. D. Watson, and J. A. Winstein, eds.), Cold Spring Harbor Laboratory, Cold Spring Harbor, New York (1977).
41. *Chromosomes and Cancer* (J. German, ed.), Wiley, New York (1974).
42. A. G. Knudson, *Genetics of Human Cancer* (J. J. Mulvihill, R. W. Miller, and J. F. Fraumeni, eds.), Raven, New York (1977), pp. 391-399.
43. D. Brusick, in *Carcinogens: Identification and Mechanisms of Action* (A. C. Griffin and G. R. Shaw, eds.), Raven Press, New York (1979), pp. 93-105.

44. J. McCann, E. Choi, E. Yamasaki, and B. N. Ames, *Proc. Nat. Acad. Sci. U.S.A.*, 72:5135-5139 (1975).
45. D. J. Brusick and V. W. Mayer, *Environ. Health Perspect.*, 6:83-96 (1973).
46. D. Clive, in *Progress in Genetic Toxicology* (D. Scott, B. A. Bridges, and F. H. Sobels, eds.), Elsevier/North Holland, Amsterdam (1977), pp. 241-247.
47. J. P. O'Neill, D. B. Couch, R. Machanoff, J. R. SanSebastian, P. A. Brimer, and A. W. Hsie, *Mutat. Res.*, 45:91-101 (1977).
48. E. Voegel and F. H. Sobels, in *Chemical Mutagens: Principles and Methods for Their Detection*, Vol. 4 (A. Hollaender, ed.), Plenum, New York (1980), pp. 93-142.
49. D. Wild, *Mutat. Res.*, 56:319-327 (1978).
50. A. J. Bateman, in *Handbook of Mutagenicity Test Procedures* (B. J. Kilbey, M. Legator, W. Nichols, and C. Ramel, eds.), Elsevier, Amsterdam (1977), pp. 325-334.
51. D. G. Stetka, A. V. Carrano, and J. Minkler, in *Proceedings of EPA Symposium on Application of Short-Term Bioassays in the Fractionation and Analysis of Complex Environmental Mixtures*, Williamsburg, Va. (1978).
52. S. A. Latt, R. R. Schreck, K. L. Loveday, C. P. Dougherty, and C. F. Shuler, in *Advances in Human Genetics*, Vol. 10 (H. Harris and K. Hirschhorn, eds.), Plenum, New York (1980), pp. 267-331.
53. M. D. Ashwood-Smith, G. A. Poutton, M. Barker, and M. Mildenbergh, *Nature*, 285:407-409 (1980).
54. F. Zajdela and E. Bisagni, *Carcinogenesis*, 2:121-127 (1981).
55. R. A. Stern, L. A. Thibodeau, A. B. Kleinerman, J. A. Parrish, T. B. Fitzpatrick, and 22 participating investigators, *N. Engl. J. Med.*, 300:809-813 (1979).
56. E. H. Weinberg and Springer, *J. Soc. Cosmet. Chem.*, 32:303-315 (1981).
57. F. Daniels, Jr., *J. Invest. Dermatol.*, 44:259-270 (1965).
58. P. B. Koehler, *J. Soc. Cosmet. Chem.*, 31:213-218 (1980).

29
COSMETIC PRODUCT PRESERVATION
Safety and Regulatory Issues

HEINZ J. EIERMANN *U.S. Food and Drug Administration, Washington, D.C.*

I. Introduction 559
II. The Regulation of Cosmetics 560
III. Regulatory Activities 562
IV. Preservation Effectiveness During Product Use 563
References 567

I. INTRODUCTION

Cosmetic product preservation and its associated issues have not always received the kind of attention they are getting today. Years ago, manufacturers of cosmetics viewed microbial contamination problems mostly as problems of product stability. As long as creams, lotions, or similar products were not covered with blotches of mold or did not have a foul odor, microbiological problems were not perceived to exist.

When *Aspergillus* was discovered on the surface of a product, the cosmetic chemist simply increased the formulation's paraben concentration, and the problem was considered resolved. Sanitary manufacturing practices usually meant avoidance of chemical contaminants or filth. The practice of sanitizing utensils and equipment and examining cosmetic contact surfaces for microbial contamination received only Spartan attention. Equally rare was the practice of sampling and microbiological testing of raw materials, packaging components, and finished products. Once a cosmetic had been tested for preservation as part of the product development process, its microbiological fate usually was left to chance, and there was little interest in the safety aspect of microbial contamination.

In the mid-1960s, Nobel and Savin [1], Morse and colleagues [2,3], and others reported about staphylococcal infections in hospitals in the United States and abroad from the use of contaminated hand creams and hand lotions, and the cosmetic industry's attitude toward microbial preservation issues changed drastically. Microbiology became a major topic of scientific meetings and the cosmetic literature [4-8].

II. THE REGULATION OF COSMETICS

A discussion of the regulatory issues associated with cosmetic product preservation requires review of some of the statutory provisions affecting cosmetics.

Cosmetics are regulated primarily under the authority of the Federal Food, Drug, and Cosmetic Act [9]. The other law applicable to cosmetics is the Fair Packaging and Labeling Act [10]. The Food, Drug, and Cosmetic Act was passed in 1938, and the Fair Packaging and Labeling Act in 1966. Both laws are enforced by the Food and Drug Administration (FDA), an agency of the Department of Health and Human Services.

Congress enacted the Food, Drug, and Cosmetic Act in order to protect consumers from unsafe or deceptively labeled or packaged products by prohibiting the movement in interstate commerce of adulterated or misbranded food, drugs, devices, and cosmetics. The purpose of the Fair Packaging and Labeling Act is to ensure that packages and their labels provide consumers with accurate information about the quantity of contents and facilitate value comparisons.

The Food, Drug, and Cosmetic Act defines cosmetics as articles intended to be applied to the human body or any part thereof for cleansing, beautifying, promoting attractiveness, or altering the appearance without affecting the body's structure or its functions. Soap is exempted from the term *cosmetic* [11]. Articles intended to cure, mitigate, treat, or prevent disease and articles intended to affect the structure or any function of the body are drugs [12]. Articles intended to cleanse, beautify, or promote attractiveness as well as treat or prevent disease or otherwise affect the structure or any function of the human body are drugs and cosmetics and must comply with the regulatory requirements of both product categories [13].

Typical examples of products which, by law, are drugs as well as cosmetics are hormone creams, antidandruff shampoos, antiperspirants that are also deodorants, and tanning preparations that also claim to protect against sunburn. However, sunscreen formulations marketed as tanning products without making reference to their sunscreening ability are considered cosmetics, and so are deodorants that contain antimicrobial agents but do not claim in labeling to have antibacterial action.

The Food, Drug, and Cosmetic Act prohibits the introduction into or the receipt in interstate commerce of any food, drug, device, or cosmetic that is adulterated or misbranded. Also prohibited is their adulteration or misbranding while in interstate commerce [14].

Contrary to popular belief, there is no statutory requirement that cosmetic products or ingredients be proved safe, as is the case with food additives or new drugs, or that the truthfulness of cosmetic product labeling be substantiated before cosmetics are introduced into interstate commerce. Furthermore, cosmetic manufacturers or distributors are not required by law to register their manufacturing establishments, product formulations, or consumer reports of adverse reactions with the FDA, or make available other information on their products.

The burden of proof that a product is unsafe or deceptively packaged or labeled rests with the agency. With the exception of color additives, a cosmetic manufacturer may use essentially any ingredient or market any cosmetic. If the FDA needs to remove a cosmetic from the market for lack of safety or for deception, it can do so only by showing that the product is harmful or that its labeling or packaging is false or misleading. If the FDA needs to prohibit the further use of an ingredient, it must demonstrate that the ingredient may be poisonous or deleterious under the intended conditions of use.

The statutory provisions for drugs are in many ways identical to those applicable to cosmetics unless a drug is not generally recognized by experts as safe and effective under the conditions of intended use or it has become so recognized but has not been used to a material extent or for a material time under such conditions. A drug not generally recognized as safe and effective or a safe and effective drug not used to a material extent or for a material time is considered a new drug and must be proven safe and effective prior to its introduction into the marketplace [15]. The long-standing review of marketed nonprescription drugs by expert panels appointed by the FDA serves to determine which of these drugs are safe and effective for their intended use and may continue to be distributed as "over-the-counter" drugs and which are not so recognized and thus legally become new drugs [16].

Most drugs which are also cosmetics are over-the-counter drugs; several are new drugs for which safety and effectiveness had to be proved to the agency before they could be marketed.

A cosmetic is deemed adulterated if it bears or contains a poisonous or deleterious substance which may render it injurious to users under the conditions of prescribed, customary, or normal use; if it consists in whole or in part of a filthy, putrid, or decomposed substance; if it has been prepared, packed, or held under insanitary conditions whereby it may have become contaminated with filth, or whereby it may have been rendered injurious to health; if its container is composed in whole or in part of a poisonous or deleterious substance which may render the contents injurious to health; or if it is not a hair dye and it is, or bears or contains, a nonpermitted color additive. Coal-tar hair dyes which bear the label Caution prescribed by law and provide directions for preliminary testing are exempted from the adulteration provision [17].

A cosmetic is deemed misbranded if its labeling is false or misleading in any particular; if the container is made, formed, or filled in a misleading manner; if it does not comply with the labeling requirements applicable to color additives; or if its packaging or labeling is in violation of regulations issued under the authority of the Poison Prevention Packaging Act of 1970 [18].

For its enforcement, the Food, Drug, and Cosmetic Act authorizes the FDA to promulgate regulations, conduct examinations and investigations of products, inspect establishments in which products are manufactured, packed, or held, and proceed against adulterated or misbranded products introduced into interstate commerce [19].

To prevent further shipment of a hazardous or misbranded product, the agency may request a federal district court to issue a restraining order against a manufacturer or distributor of a violative cosmetic. The agency may also initiate criminal proceedings against a person violating the provisions of the Food, Drug, and Cosmetic Act [20].

Regulations are published for two purposes, namely, to interpret with specificity broadly stated statutory provisions as, for example, certain statutory labeling requirements; and to permit or prohibit certain acts or activities regulated under the authority of the Food, Drug, and Cosmetic Act as, for example, the listing of a color additive, the prohibition of use of a substance as an ingredient or, in the case of drugs, the manufacture of products in compliance with good manufacturing practices.

The promulgation of a regulation is a rather complex and time-consuming procedure involving publication of a proposed regulation for public comment, careful review and consideration of the submitted arguments, data, and other information, and publication of the final rule. In some instances, a person adversely affected by a regulation may file an objection with the agency and request a public hearing. Furthermore, any adversely affected person may seek judicial review of a regulation by the federal courts [21].

The Fair Packaging and Labeling Act authorizes the FDA to publish regulations which provide that consumer commodities, including cosmetics, bear label statements of product identity and net quantity of contents as specified by law. The agency also has the authority to promulgate regulations preventing the deception of consumers or facilitating value comparisons, including the listing of ingredients other than trade secrets in order of descending predominance [22].

III. REGULATORY ACTIVITIES

The limited statutory authority of the FDA to regulate cosmetics has had a corresponding effect on the agency's regulatory activities associated with microbial preservation. It is difficult to detect preservation-related hazards and initiate enforcement action when there is no requirement for registration of manufacturers and product formulations and when the burden is on the agency to document harmfulness or deception in order to seize adulterated or misbranded products instead of on the producer to conduct safety testing and submit data to the agency. In turn, this shortcoming may have contributed to the slowness in the evolution and industrial application of cosmetic preservation.

The number of recall or seizure actions involving microbial contamination of cosmetics has been relatively small in relation to the size of the cosmetic industry. For example, in 1968 Dunnigan reported that during 1966, 1967, and the early part of 1968, the FDA was involved in 25 recalls of contaminated cosmetics [23]. During fiscal years 1974-1980, recalls or seizures of cosmetics because of microbial hazards averaged about five products per year [24].

With regard to the publication of regulations addressing cosmetic product preservation, the agency published in 1977 in the *Federal Register* a notice of intent to propose regulations; however, none has been proposed to date [25]. This notice will be discussed in more detail later.

In 1973 a regulation was published concerning the safety of mercurial preservatives. Because of the topical effect of mercury compounds as skin irritants or sensitizers, their absorption through the skin, and their chronic systemic toxicity, mercury compounds may be used only in eye area cosmetics and only if an effective and safe nonmercurial substitute is not available. If used, the mercury concentration of the cosmetic, calculated as the metal, may not exceed 65 ppm (0.0065%). The continued use of mercurial preservatives in eye area cosmetics is considered warranted because mercury compounds have been found to be highly effective in preventing *Pseudomonas* contamination [26].

During the early 1970s, great strides were made by the cosmetic industry toward implementation of sanitary manufacturing practices and the development and production of better-preserved and hence safer cosmetic products. In a nationwide survey in 1969 of topical drug and cosmetic preparations used in hospitals, almost 20% of the samples examined by the FDA were found to be contaminated [27]. In a survey conducted during the same year by Wolven, 61 of 250 marketed cosmetics, that is, 24%, were determined to be contaminated with various microorganisms [28]. In both surveys several products contained more than one type of microorganism. In the FDA survey 3.6% of the cosmetics were contaminated with pseudomonads; in the Wolven survey *Pseudomonas* contamination was 6.4%. Wolven's second survey of cosmetics in 1972 uncovered only 8 contaminated samples among the 223 tested [29]. In 1975 the FDA conducted a survey involving 340 domestic cosmetic products and found only 17, mostly facial makeup preparations and creams, that were contaminated with microorganisms, and only a few of these presented cause for concern [30].

These data and the information obtained from the agency's inspectional activities suggest that microbial contamination of cosmetics during manufacture is no longer a major regulatory issue. Judicious use of preservatives, concern for sanitary manufacturing conditions, and rigorous microbiological control of finished products have caused most products entering interstate commerce to have low microbial counts and to be free of pathogenic microorganisms. Largely unresolved, however, is the question of whether or not these products remain in that condition when used by consumers.

IV. PRESERVATION EFFECTIVENESS DURING PRODUCT USE

The hazard associated with inadequately preserved cosmetics which become contaminated during consumer use is twofold, namely, the effect of contamination of the product, that is, spoilage, product separation, and formation of harmful microbial metabolites; and the direct effect of the microorganisms on human health. The hazard of inadequately preserved cosmetics to human health has been amply demonstrated by the reports of staphylococcal infections in hospitals from the use of contaminated hand creams and hand lotions [2,3] and the FDA-funded studies conducted by Wilson and Ahearn [31].

The Wilson-Ahearn studies were initiated in 1971 following Wilson's observation that some eye injuries appeared to be associated with contaminated mascaras [32,33]. The initial studies involved the identification of microorganisms in new and used mascaras and on the eyelids and fingers of consumers. The results were quite specific: New mascaras were rarely contaminated and the microbial densities of the contaminated mascaras were usually low. Used mascaras, however, were frequently found to be contaminated and many contained a variety of microorganisms [34].

The most widely encountered microorganisms were *Staphylococcus epidermidis* and molds. Others found were diptheroids, *Micrococcus* sp. and *Bacillus* sp., *Pseudomonas aeruginosa* and *Klebsiella pneumoniae* were also encountered, though rarely [35]. In most instances, the microorganisms isolated from used samples of mascaras were also those found on the outer eye, the fingers, or the saliva of the users [34].

Subsequent studies involved controlled-use experiments with commercial brands of mascaras. The purpose of these studies was to determine the frequency and pattern of contamination as well as the survival of the organisms after the products had been withdrawn from users. The samples were given to small study groups for normal use over a 9-week period. At the beginning of the test, the mascaras and the users were examined for microbial flora, the latter around the eye and the fingers. Only mascaras that were free of microorganisms were permitted in the study. During the test, the mascaras were sampled about every 2 weeks. When a test sample yielded high populations in two successive microbial examinations, it was withdrawn from use. The withdrawn sample was stored and tested periodically for changes in microbial count.

The results of the tests showed that by the time the products were used 25 times, as many as 96% of the samples of individual brands yielded microorganisms. About half of the cosmetics which had become contaminated during use remained contaminated over a 30-day storage period after the samples had been withdrawn from use. In some instances, the microbial level increased significantly during storage. The contaminated products which contained an organic mercurial preservative in addition to parabens showed a relatively rapid decrease in microbial population levels after withdrawal from use [35].

At the time the controlled-use experiments were underway, Wilson cultured the eyes of over 150 female patients with eye infections for aerobic microorganisms. He also examined women with healthy eyes and, where warranted by clinical findings, the eye cosmetics of the patients with infections.

The lid margins, and to a lesser extent the conjunctivae, of normal eyes were found to harbor high populations of microorganisms. The most common organisms isolated were *S. epidermidis* and diphtheroids. Infected eyes showed a higher incidence and higher numbers of *S. epidermidis*, and a decreased incidence of diphtheroids [36].

About 10% of the patients with eye infections were found to be using eye area cosmetics contaminated with *S. epidermidis* or *Fusarium* sp. The contaminated products were suggested as a factor in the eye infections. At least four of the cases of staphylococcal blepharitis or conjunctivitis were determined to be definitely associated with the use of contaminated cosmetics, because the infections subsided when the products were withdrawn from use [36].

In addition to the 32 blepharoconjunctival infections encountered during the course of the 7-year study, Wilson also became aware of 16 cases of corneal ulceration. Of these, 12 were associated with mascaras contaminated with *P. aeruginosa*, 3 with *Fusarium* sp., and 1 with *Bacillus circularis*. All infections appeared to have resulted from the use of the contaminated mascaras and inadvertent abrasion of the patient's cornea with the applicator wand, or other corneal trauma. The ulcers occurred within 24 hr of the injury and resulted in loss of useful vision in the ulcerated eye [37,38].

Although the healthy eye is known to be quite resistant to microbial infection, these cases of temporary or permanent eye injury confirm that the cornea and outer eye, nevertheless, may become infected, particularly when traumatized. A heavily contaminated eye or facial makeup, therefore, constitutes a potentially serious health hazard, and equally hazardous to the eye could be a shampoo or other hair product attacking the conjunctiva or cornea and containing pathogenic microorganisms. Additionally, products applied to other parts of the human body also may present a health risk when contaminated with pathogenic microorganisms. Cosmetics are used not only by healthy consumers, but also by persons with abrasions or burns, inflamed mucosa, or irritated or diseased skin.

The FDA expressed its concern about the adequacy of preservation, particularly eye area cosmetics, in a *Federal Register* notice published in 1977 [39]. The notice made reference to the reported corneal ulcerations associated with *P. aeruginosa* and announced the agency's intention to propose regulations regarding microbial preservation of cosmetics that may come in contact with the eye. The notice also advised manufacturers that it considered inadequately preserved cosmetics to be in violation of the Food, Drug, and Cosmetic Act [40]:

> Under section 601 of the act, a cosmetic is considered adulterated if it is prepared under conditions whereby it may have been rendered injurious to health, as well as if it bears any poisonous or deleterious substance that may render it injurious to users under the conditions of use. Furthermore, under sections 201(n), 601, and 602 of the act and 21 CFR 740.10, the label must bear any warning statements that are necessary or appropriate to prevent a health hazard that may be associated with the product. Manufacturers and distributors should be advised that FDA intends to take whatever regulatory action is necessary to remove from the market any cosmetic that poses an unreasonable risk of injury because of inadequate preservation to withstand contamination under customary conditions of use. FDA does not intend to await the completion of the rule making proceeding announced in this notice of intent before taking needed regulatory action.

The need for effective preservation of products during consumer use has been recognized for many years. In 1970 Halleck recommended on behalf of the Preservation Subcommittee of the Toilet Goods Association Microbiology Committee that, in carrying out preservation studies, consideration be given not only to the product formulation, manufacturing conditions, packaging, or product stability, but also to the "continued effectiveness of the preservation system" and to the "intended use of the product by the consumer" [41]. In 1971 the Cosmetic, Toiletry, and Fragrance Association (CTFA) Subcommittee on Quality Assurance advocated that "cosmetics should be adequately preserved for their intended use and packaging by the judicious use of appropriate biocidal and biostatic agents" [42]. Tenenbaum stated in 1973 that "it is desirable that nonsterile preparations be essentially free of microorganisms when delivered to the public and remain so after repeated withdrawals by the consumer" [43].

In the tentative final order regarding the regulation of over-the-counter antimicrobial drug products, the FDA characterized the term *effectively preserved* as "preservation sufficient to prevent spoilage or prevent growth of inadvertently added microorganisms" [44]. Since the Antimicrobial I Panel made specific reference in its monograph to protection "from the growth of microorganisms introduced as a result of customer use," the term *effectively preserved* must be understood to include also effective preservation during use by the consumer [45]. The Antimicrobial I Panel was the advisory review panel appointed by the FDA for the evaluation of data on the use of antimicrobial ingredients in topical over-the-counter drugs. Its recommendations served as a basis for the subsequently published proposed and tentative final regulation regarding these products.

The literature contains various laboratory methods and performance standards for measuring the preservative activity and adequacy of preservation of cosmetics. Olson, for example, favored a titration technique, that is, inoculation of a product with increasing concentrations of selected test organisms and determination of the highest microbial density at which antimicrobial activity occurred [46]. The *U.S. Pharmacopoeia XX* recommends (for multiple-dose parenteral, otic, nasal, and ophthalmic products) single inoculation with cultures of *Candida albicans*, *Aspergillus niger*, *Escherichia coli*, *P. aeruginosa*, and *Staphylococcus aureus* of appropriate concentrations to obtain microbial densities in the test preparations ranging between 100,000 and 1,000,000 microorganisms per milliliter of product. A preservative system is considered effective if "(a) the concentrations of viable bacteria are reduced to not more than 0.1 percent of the initial concentrations by the fourteenth day, (b) the concentrations of viable yeasts and molds remain at or below the initial densities during the first 14 days, and (c) the concentration of each test microorganism remains at or below these designated levels during the remainder of the 28-day test period" [47]. The methods proposed by the Preservation Subcommittee of the CTFA Microbiological Committee [48] and the Council of the Society of Cosmetic Chemists of Great Britain [49] are based on the same concept of measuring the reduction of the concentrations of viable microorganisms; however, they also recommend rechallenge of the test samples. A further modification of these tests is the procedure suggested by Orth [50].

The problem with preservative efficacy testing is not so much a matter of selecting an appropriate microbiological testing procedure as it is one of establishing a rational performance standard. No matter how sensitive, precise, and reliable a laboratory testing procedure may be, its usefulness remains academic as long as the data obtained from such testing lack a practical basis for meaningful interpretation, that is, the given performance standard is not supported by actual consumer exposure experiences.

The performance standards recommended in connection with the published preservative efficacy testing procedures have not been reported to be validated by such consumer experiences. The Preservation Subcommittee of the CTFA Microbiology Committee acknowledged this shortcoming in its guidelines for preservation testing. The section on performance criteria for aqueous eye cosmetic formulations is prefaced by the caveat, "Little has been published regarding performance of preservatives in aqueous eye cosmetics. The correlation of preservative effectiveness in actual consumer use has not been adequately demonstrated. ...It is the responsibility of the manufacturer to select appropriate methodology and criteria" [51].

Of special concern is preservation of products that are used daily and may last for several weeks or months. The recommended preservative efficacy tests require either none, one, or an unspecified number of rechallenges, and the microorganisms used for rechallenge are the same as those used initially. Wilson and Ahearn [37] and Ahearn et al. [34] demonstrated that repeated challenge of mascaras with *S. epidermidis* increased the incidence of microbial contamination of marginally preserved products. Mascaras first challenged with *S. epidermidis* were shown to become more susceptible to *Pseudomonas* sp. contamination.

The absence of validated performance standards also affected the FDA's rule-making efforts. In 1977, instead of being able to propose regulations regarding microbial preservation of cosmetics that may come into contact with the eye, the FDA could only express its intent to propose them. The cosmetic industry and others were requested to submit information on "microbial testing methods and standards of performance suitable to assure that such cosmetics do not become contaminated with microorganisms during manufacturing, subsequent storage and/or use by consumers" [52]. It was hoped that the agency would receive the data necessary to correlate laboratory test results with preservation experience under customary use conditions and thus be able to propose meaningful microbiological performance standards. However, no useful information was received in response to this request, and rule-making plans had to be postponed. Apparently, no comparative studies have ever been conducted (or at least none has been reported in the scientific literature) comparing the results of microbiological challenge tests with those obtained from testing of the same product under conditions of consumer use.

In 1978 the San Francisco Medical Center was awarded a contract to provide the data base for proposing meaningful performance standards. The project was to encompass comparative testing of experimental products in the laboratory and under simulated use conditions. The test products contained preservative systems of various efficacies. Some experimental formulations were expected to become contaminated under both testing conditions, while others were not to show microbial growth. The performance standard for predictive laboratory testing was to be the antimicrobial activity level of the experimental product that had been found to be adequately preserved during simulated use testing [53].

The contract expired before completion of the study. The data obtained from a new study are expected to be published in the scientific literature. They may stimulate others to build on this experience in the interest of better consumer protection.

REFERENCES

1. W. C. Noble and J. A. Savin, Steroid cream contaminated with *Pseudomonas aeruginosa*, Lancet, 1:347-349 (1966).
2. L. J. Morse, H. L. Williams, F. P. Grenn, Jr., E. E. Eldridge, and J. R. Rotta, Septicemia due to *Klebsiella pneumoniae* originating from a hand cream dispenser, N. Engl. J. Med., 277:472-473 (1967).
3. L. J. Morse and L. F. Schornbeck, Hand lotions — a potential nosocomial hazard, N. Engl. J. Med., 278:376-378 (1968).
4. Scientific conferences of the Toilet Goods Association, New York, N.Y., May 9 and December 1, 1969.
5. Workshop: The Preservation of Cosmetic Products, Cosmetic, Toiletry, and Fragrance Association, 1971, Scientific Conference, Atlanta, Ga., October 1971.
6. S. J. Tenenbaum, The microbial content of cosmetics and toiletries, TGA Cosmet. J., 1:19-21 (1969).
7. F. E. Halleck, A. guideline for the determination of adequacy of preservation of cosmetics and toiletry formulations, TGA Cosmet. J., 2:20-23 (1970).
8. S. M. Henry, What constitutes a microbiologically acceptable cosmetic?, CTFA Cosmet. J., 3:28-32 (1971).
9. Federal Food, Drug, and Cosmetic Act of 1938, as amended, Secs. 201-702, 21 U.S.C. 321-392.
10. Fair Packaging and Labeling Act, Pub. L. 89-755, Secs. 2-13, 15 U.S.C. 1451-1460 (November 3, 1966).
11. Sec. 201(i), 21 U.S.C. 321(i) (see also Ref. 9).
12. Sec. 201(g), 21 U.S.C. 321(g).
13. Sec. 509, 21 U.S.C. 359.
14. Sec. 301, 21 U.S.C. 331.

15. Sec. 505, 21 U.S.C. 355 [see also Sec. 201(p), 21 U.S.C. 321(p)].
16. *Code of Federal Regulations*, 21, Part 330, Over-the-counter (OTC) human drugs which are generally recognized as safe and effective and not misbranded, Secs. 330.1 and 330.10.
17. Sec. 601, 21 U.S.C. 361.
18. Sec. 602, 21 U.S.C. 362.
19. Secs. 701, 702, and 704, 21 U.S.C. 371, 372, and 374.
20. Secs. 302, 303, and 304, 21 U.S.C. 332, 333, and 334.
21. Sec. 710, 21 U.S.C. 371.
22. Secs. 4 and 5, 14 U.S.C. 1453 and 1454 (see also Ref. 10).
23. A. P. Dunnigan, Microbial control of cosmetics, *Drug Cosmet. Ind.*, *102*:43-45, 152-158 (1968).
24. Unpublished annual reports of the evaluation of FDA cosmetic compliance programs.
25. Preservation of cosmetics coming in contact with the eye, *Federal Register*, *42*:54837-54838 (1977) (hereinafter cited as 42 Fed. Reg. 54837).
26. Use of mercury compounds in cosmetics including use of skin-bleaching agents in cosmetic preparations also regarded as drugs, *Federal Register*, *38*:853-854 (1973).
27. A. P. Dunnigan, Report of a special survey: microbiological contamination of topical drugs and cosmetics, *TGA Cosmet. J.*, *2*:39-41 (1970).
28. A. Wolven and I. Levenstein, Cosmetics – contaminated or not, *TGA Cosmet. J.*, *1*:34-37 (1969).
29. A. Wolven and I Levenstein, Microbial examination of cosmetics, *Am. Cosmet. Perfum.*, *87*:63-65 (1972).
30. Unpublished report on a sampling survey conducted in fiscal year 1975 involving microbiological analysis of 2541 units of 501 domestic and imported cosmetic products.
31. Food and Drug Administration, Microbial contamination in ocular cosmetics, Contracts No. 71-74 et seq., D. Ahearn and L. Wilson, principal investigators.
32. L. A. Wilson, D. G. Ahearn, D. B. Jones, and R. R. Sexton, Fungi from the outer eye, *Am. J. Ophthalmol.*, *67*:52-56 (1969).
33. L. A. Wilson, W. Kuehne, S. W. Hall, and D. G. Ahearn, Microbial contamination in ocular cosmetics, *Am. J. Ophthalmol.*, *71*:1298-1302 (1971).
34. D. G. Ahearn, L. A. Wilson, A. J. Julian, D. J. Reinhardt, and G. Ajello, Microbial growth in eye cosmetics: contamination during use, *Dev. Ind. Microbiol.*, *15*:211-216 (1974).
35. L. A. Wilson, A. J. Julian, and D. G. Ahearn, The survival and growth of microorganisms in mascara during use, *Am. J. Ophthalmol.*, *79*: 596-601 (1975).
36. D. G. Ahearn and L. A. Wilson, Microflora of the outer eye and eye area cosmetics, *Dev. Ind. Microbiol.*, *17*:23-25 (1976).
37. L. A. Wilson and D. G. Ahearn, *Pseudomonas*-induced corneal ulcers associated with contaminated eye mascaras, *Am. J. Ophthalmol.*, *84*: 112-119 (1977).
38. L. A. Wilson, Report No. 29 and Summary Report, Laboratory Studies, FDA Contract No. 223-74-2016 Mod. 13, submitted September 1978.
39. *Federal Register*, *42*:54837 (1977) (see also Ref. 25).
40. *Federal Register*, *42*:54838.
41. F. E. Halleck, A guideline for the determination of adequacy of preservation of cosmetics and toiletry formulations, *TGA Cosmet. J.*, *2*: 20-23 (1970).

42. Microbiology Subcommittee on Quality Assurance of the CTFA Microbiology Committee, Microbiological Aspects of Quality Assurance, *CTFA Technical Guidelines*, The Cosmetic, Toiletry, and Fragrance Association, Inc., Washington, D.C. (1971).
43. S. Tenenbaum, Microbial content of cosmetics and non-sterile drugs, *Cosmet. Perfum.*, 88:49-53 (1973).
44. Over-the-counter drugs generally recognized as safe, effective and not misbranded, OTC antimicrobial products, *Federal Register*, 43: 1210-1249 (1978).
45. Over-the-counter topical antimicrobial products and drug and cosmetic products, establishment of monograph and use of certain halogenated salicylanilides as active or inactive ingredients, *Federal Register*, 39:33102-33141 (1974).
46. S. W. Olson, The application of microbiology to cosmetic testing, *J. Soc. Cosmet. Chem.*, 18:191-198 (1967).
47. U.S. Pharmacopoeia, Antimicrobial preservatives — effectiveness, *The United States Pharmacopoeia*, 20th rev., United States Pharmacopoeial Convention, Inc., Mack, Easton, Pa. (1980), pp. 873-874.
48. Preservation Subcommittee of the CTFA Microbiological Committee, A guideline for the determination of adequacy of preservation of cosmetics and toiletry formulations, *CTFA Technical Guidelines*, The Cosmetic, Toiletry, and Fragrance Association, Inc., Washington, D.C. (1973).
49. The Council of the Society of Cosmetic Chemists of Great Britain, The hygienic manufacture and preservation of toiletries and cosmetics, *J. Soc. Cosmet. Chem.*, 21:754 (1970).
50. D. S. Orth, Linear regression method for rapid determination of cosmetic preservative efficacy, *J. Soc. Cosmet. Chem.*, 30:321-332 (1979).
51. Preservation Subcommittee of the CTFA Microbiological Committee, A guideline for preservation testing of aqueous liquid and semi-liquid eye cosmetics, *CTFA Technical Guidelines*, The Cosmetic, Toiletry, and Fragrance Association, Inc., Washington, D.C. (1975).
52. *Federal Register*, 42:54837 (see also Ref. 25).
53. Food and Drug Administration, Development and validation of preservative efficacy testing method for cosmetic products, Contract No. 223-78-2015, R. Aly, principal investigator (1978).

**PART VII
APPENDIXES**

APPENDIX A
MICROBIOLOGICAL METHODS FOR COSMETICS

JOSEPH M. MADDEN *U.S. Food and Drug Administration, Washington, D.C.*

I. Introduction 574
II. Equipment and Materials 575
III. Media and Reagents 575
IV. Sampling and Sample Handling 575
V. Preliminary Sample Preparation 576
VI. Microbiological Evaluations 576
   A. Screening test 576
   B. Screening test for anaerobes 576
   C. Aerobic plate count 577
   D. Anaerobic plate count 577
   E. *Staphylococcus aureus* plate count 578
   F. Yeast and mold plate count 578
   G. Enrichment cultures 579
   H. Identification of bacteria 579
VII. Appendix I: Media for the Enumeration of Microorganisms in Cosmetics and the Identification of Gram-Positive Bacteria 592
   A. Anaerobic agar 592
   B. Bile esculin agar 592
   C. Brain-heart infusion broth and agar 592
   D. MacConkey agar 592
   E. Malt extract or potato dextrose agar 592
   F. Mannitol salt agar 592
   G. Modified letheen agar 593
   H. Modified letheen broth 593

       I. Oxidative-fermentative test medium 593
       J. Sabouraud dextrose broth 593
       K. Sheep blood agar 593
       L. Starch agar 593
       M. Thioglycollate medium 135c 593
       N. Trypticase soy agar and tryptic soy agar and broth 594
       O. Vogel-Johnson agar 594

VIII. Appendix II: Media for *Enterobacteriaceae* Identification 594

       A. Andrade carbohydrate broths 594
       B. Lysine iron agar 594
       C. Malonate broth 594
       D. Motility-indole-ornithine medium 594
       E. MR-VP broth 595
       F. Simmon's citrate agar 595
       G. Triple sugar iron agar 595
       H. Urea agar (Christensen's) 595

IX. Appendix III: Media for Identification of Gram-Negative Nonfermentative Bacilli 596

       A. Acetamide medium 596
       B. Clark's flagellar stain 596
       C. Esculin agar, modified 597
       D. Nutrient gelatin 598
       E. Indole medium 598
       F. Indole medium 599
       G. King's B medium 599
       H. Lysine decarboxylase medium 599
       I. Motility nitrate medium 600
       J. Nitrate broth, enriched 601
       K. King's oxidative-fermentative basal medium 601
       L. Oxidase test strips 602

       References 602

I. INTRODUCTION

The ability of microorganisms to grow and reproduce in cosmetic products has been known for many years [2]. The cosmetic industry is well aware that microorganisms may cause spoilage or chemical changes in their products or possible injury to the user [2,3,8-11]. Injury to the user is the basis of the cosmetic provisions of the Food, Drug, and Cosmetic Act, as amended in 1976, delegating authority to the Food and Drug Administration for the enforcement of its provisions.

The isolation of microorganisms from cosmetic products is based on enrichment culture, direct dilution, and plating methods. Products that are not water soluble are initially treated to render them miscible before enrichment culture or dilution procedures. Media that partially inactivate preservative systems commonly found in cosmetic products are used for all enrichment and bacterial plating procedures. The isolated microorganisms are identified by common microbiological methods.

Microbiological Methods for Cosmetics

II. EQUIPMENT AND MATERIALS

1. Pipets, sterile, 1, 5, and 10 ml, graduated
2. Gauze pads, sterile, 4 × 4 in.
3. Sterile instruments: forceps, scissors, scalpel and blades, spatulas, and microspatulas
4. Test tubes, screw cap, 13 × 100, 16 × 125, and 20 × 150 mm
5. Dilution bottles, screw cap
6. Balance, sensitivity of 0.01 g
7. Petri dishes, sterile, plastic, 15 × 100 mm
8. Bent glass rods, sterile
9. Incubators, 30 ± 2°C, 35 ± 2°C
10. BBL anaerobic GasPak envelopes, indicator strips and jars, or anaerobic incubator, 35 ± 2°C, or anaerobic glove box, 35 ± 2°C
11. Candle jars or CO_2 incubator, 35 ± 2°C

III. MEDIA AND REAGENTS

1. Aqueous solution of 80% ethanol and 1% hydrochloric acid (vol/vol)
2. Tween 80 (polysorbate 80)
3. 95% ethanol
4. Lyophilized rabbit coagulase plasma with ethylenediaminetetraacetic acid (EDTA)
5. 3% aqueous solution of hydrogen peroxide (H_2O_2)
6. Gram stain and endospore stain [7]
7. Media and reagents listed in Sections VII-IX

IV. SAMPLING AND SAMPLE HANDLING

The following are guidelines to be followed in handling cosmetic samples for microbiological analysis:

1. Analyze samples as soon as possible after arrival. If storage is necessary, store samples at room temperature. Do not incubate, refrigerate, or freeze samples before or after analysis.
2. Inspect samples carefully before opening and note any irregularities of the sample container.
3. Disinfect the surface of the sample container with an aqueous mixture of 80% ethanol (vol/vol) and 1% HCl (vol/vol) before opening and removing the contents. Dry the surface with sterile gauze before opening.
4. A representative portion of the contents must be used for microbial analysis. If possible, use a 10-g (ml) portion of the sample. For products whose weight is less than 10 g (ml), analyze the entire contents.
5. If only one sample unit is available and multiple analyses are requested, that is, microbial, toxicological, and chemical, the subsample for microbiological examination must be taken before those for other analyses. In this situation, the amount of the subsample used for microbial analysis will depend on the other analyses to be performed. For example, if the total sample content is 5 ml, use a 1- or 2-ml portion for microbial analysis.

V. PRELIMINARY SAMPLE PREPARATION

1. Liquids. No special sample preparation is required. The liquid may be added directly to the modified letheen broth (see Sec. VII. H).
2. Solids and powders. Aseptically remove and weigh 1 g of the sample into a 20 × 150 mm screw-cap test tube containing 1 ml of sterile Tween 80. Disperse the product in the Tween 80 with a sterile spatula. Add 8 ml of sterile modified letheen broth and mix thoroughly.
3. Cream and oil-based products. Aseptically remove and weigh 1 g of the sample into a 20 × 150 mm screw-cap test tube containing 1 ml of sterile Tween 80. Dispense the product in the Tween 80 with a sterile spatula. Adjust the total volume to 10 ml with sterile modified letheen broth (8 ml of broth).
4. Aerosols of powders, soaps, liquids, and other materials. After decontaminating the nozzle of the spray can, expel an appropriate amount of the product into a tared dilution bottle, for example, 1 g of product into 9-ml sterile modified letheen broth. Thoroughly mix the product and broth and reweigh.

VI. MICROBIOLOGICAL EVALUATIONS

A. Screening Test

Screen cosmetic products for microbial contaminants by inoculating 1 ml (g) of sample directly into a screw-cap bottle containing 90 ml of modified letheen broth (see Sec. VII.H) and another 1-ml (g) portion into a bottle containing 90 ml of Sabouraud dextrose broth (see Sec. VII.J). If the product is water immiscible, for example, creams, oils, and powders, treat the product as described in Section V before inoculating the broths.

Incubate at 30 ± 2°C for 7 days. Confirm growth at 2 and 7 days (or at any time growth becomes evident) by plating on modified letheen agar (see Sec. VII.G) and either malt extract agar containing 40 ppm chlortetracycline or potato dextrose agar (see Sec. VII.E).

From the bottle or Sabouraud dextrose broth transfer a 0.5-ml portion to prepared malt extract or potato dextrose agar plate, and from the modified letheen broth transfer 0.5 ml to a prepared modified letheen agar plate. Spread the inoculum over the surface of the agar plates, using a sterile bent glass streaking rod. Incubate 4 days at 30 ± 2°C. Record the results of the screening test.

B. Screening Test for Anaerobes

Talcum and other powders should be tested for both aerobic and anaerobic microorganisms by inoculating a dilution bottle containing 90 ml of thioglycollate medium 135c (see Sec. VII.M) with sample prepared as described in Section V, Item 2, above. Do not shake the bottle after adding the cosmetic preparation. Incubate at 35 ± 2°C for a minimum of 7 days. At the end of the 7-day incubation period, or sooner if growth is suspected, plate 0.1 ml of the thioglycollate medium 135c onto modified letheen agar and prereduced anaerobe agar (see Sec. VII.A) plates. Spread the inoculum with a sterile bent glass rod. Anaerobic agar plates are prereduced before inoculation by placing them in an anaerobic atmosphere overnight (12-16 hr). Incubate the anaerobic agar plates in an anaerobic atmosphere (GasPak jars, anaerobic incubator, or anaerobic glove box) for 4 days at $35 \pm 2°C$; incubate the modified letheen agar plates aerobically for 4 days at $35 \pm 2°C$. Observe the plates for growth at the end of the incubation period.

If growth appears on any of the plates after the procedures above are performed carry out plate counts on the cosmetic sample. If no growth is observed on any of the plates, report the cosmetic sample as containing no viable microorganisms.

C. Aerobic Plate Count

1. Prepare and label duplicate Petri dishes containing modified letheen agar for dilutions of 10^{-1}-10^{-6}.
2. Make serial dilutions of samples in modified letheen broth by adding either 5 or 10 ml of the prepared cosmetic preparation to 45 or 90 ml, respectively, of modified letheen broth, yielding a 1:10 dilution. Dilute the samples decimally until a dilution series from 10^{-1} to 10^{-6} is obtained.
3. After thoroughly mixing the preparations in the dilution bottles, pipet 0.5 ml of each dilution onto the surface of the labeled Petri dishes. Spread the inoculum over the entire surface of the plate with a bent glass rod that has first been sterilized in 95% ethanol and quickly flamed to remove alcohol traces.
4. Let the plates absorb the inoculum before inverting and incubating them for 48 hr at $30 \pm 2°C$.
5. Count all colonies on plates containing 30-300 colonies, and record results per dilution counted. If plates do not contain 30-300 colonies, record the dilution counted and note the number of colonies found.
6. Average the counts obtained, multiply by the appropriate dilution factor, and report the results as the aerobic plate count per gram (milliliter) of sample.

D. Anaerobic Plate Count

This is to be performed only on talcs and powders. Perform as described above for the aerobic plate count, using anaerobe agar and 5% defibrinated sheep blood agar (see Sec. VII.K) for plating. Incubate the blood agar plates in a 5-10% carbon dioxide atmosphere (candle jar or CO_2 incubator), and the anaerobe agar plates in GasPak jars. Incubate both for 48 hr before counting.

Strict anaerobes will grow only in the GasPak jars, and a minimum amount of exposure to oxygen is essential for good recovery of these organisms. It is thus recommended that a small amount of inoculum be utilized (e.g., 0.1 ml) and that the inoculated plates be placed in an anaerobic atmosphere within minutes after inoculation. Suspected anaerobic organisms must be subcultured aerobically (under CO_2) as well as anaerobically to establish their true oxygen relationship.

E. *Staphylococcus aureus* Plate Count

1. Transfer 0.5-ml aliquots from each of the dilutions prepared for the aerobic plate count to appropriately labeled duplicate plates of Vogel-Johnson agar (see Sec. VII.O).
2. Distribute the inoculum over the plate surface with a sterile, bent glass spreader rod.
3. When the inoculum has been completely absorbed, invert the plates and incubate 48 hr at $35 \pm 2°C$.
4. Count plates at the dilution having 30-300 well-distributed colonies that are convex, shiny, and black, with or without a yellow zone surrounding the colony. Count colonies after 48 hr of incubation.
5. Plates having more than 300 colonies may be selected when plates at a greater dilution do not contain the colonial types described above. Plates from minimal dilutions having less than 30 colonies may also be used if necessary.
6. From each plate demonstrating growth, pick one or more of the above-described colonies. Transfer colonies to agar slants containing any suitable maintenance medium, for example, trypticase soy or tryptic soy agars (see Sec. VII.N) or brain-heart infusion agar (see Sec. VII.C). Incubate slants until growth is evident.
7. Inoculate a small amount of growth from the maintenance slant into a 13 × 100 mm test tube containing 0.2 ml of brain-heart infusion broth (see Sec. VII.C). Incubate 18-24 hr at $35 \pm 2°C$. Then add 0.5 ml of reconstituted rabbit coagulase plasma (with EDTA) and mix thoroughly. Incubate at $35 \pm 2°C$.
8. All strains that yield a positive coagulase reaction may be considered as *Staphylococcus aureus*.
9. Calculate the number of *S. aureus* organisms present by determining the fraction of colonies tested that are coagulase positive. Multiply this fraction by the average number of colonies appearing on the Vogel-Johnson plates represented by the characteristics previously described for *S. aureus* colonies on Vogel-Johnson agar medium. Multiply the number obtained by the appropriate dilution factor and report as the number of *S. aureus* organisms per gram (milliliter) of sample.

F. Yeast and Mold Plate Count

1. Transfer 0.5-ml aliquots of the dilution series to appropriately labeled duplicate plates of either malt extract agar containing 40 ppm chlortetracycline or potato dextrose agar. Spread the inoculum over the surface of the medium with a sterile glass spreader rod.
2. After the inoculum has been absorbed by the medium, invert the plates, incubate at $30 \pm 2°C$, and observe daily for 7 days.
3. Average the counts obtained on the duplicate plates and report as the yeast or mold count per gram (milliliter) of sample.
4. The identification of the isolated yeasts or molds should be performed by an experienced mycologist.

G. Enrichment Cultures

Incubate all dilution bottles at $30 \pm 2°C$ for a minimum of 7 days. Examine the broths daily for growth. If growth is suspected, as well as at the end of the 7-day incubation period, subculture all bottles onto both modified letheen agar and MacConkey agar (see Sec. VII.D) plates. Incubate plates 48 hr at $30 \pm 2°C$.

H. Identification of Bacteria

Examine all plates and streak morphologically dissimilar colonial types onto MacConkey and modified letheen agar media. Prepare a Gram stain of all morphologically dissimilar colonial types once obtained in pure culture.

1. Gram-Positive Rods

Report aerobic gram-positive rods as either spore forming or non-spore-forming. To enhance sporulation, inoculate a starch agar (see Sec. VII.L) plate with the isolate and incubate 48 hr at room temperature. Prepare either a Gram stain or an endospore stain (see Sec. III, Item 6) from an isolated colony and note the position of the endospore within the vegetative cell (central, terminal, or subterminal), the shape of the endospore (round or ellipsoidal), and the morphology of the sporulating cell's sporangium (swollen or not swollen). All aerobic spore-forming rods should also be tested for motility either by inoculation into a motility test medium or motility-indole-ornithine medium (see Sec. VIII.D) or by microscopic examination.

Cultural Method:

1. Inoculate a tube of either motility (see Sec. VIII.D) or motility-indole-ornithine medium by stabbing.
2. Incubate aerobically 18-24 hr at room temperature.
3. Growth from the line of the stab (indicated by turbidity of the medium around the stab) constitutes a positive test.

Microscopic Examination:

1. Inoculate an isolated colony into a suitable broth.
2. Incubate aerobically 18-24 hr at room temperature.
3. Place one drop of the broth culture on a clean microscope slide and cover with a cover slip.
4. Motility is indicated by the observance of individual bacterial cells moving in random directions. Observation may be performed either at a magnification of 400× or under oil immersion.

Further characterization of gram-positive rods is generally unnecessary. Consult Ref. 7 if further characterization of these organisms is required.

2. Gram-Positive Cocci

1. Streak a modified letheen agar plate, incubate 18-24 hr at $35 \pm 2°C$, and test the resultant growth for catalase production either by adding a drop of 3% H_2O_2 to an isolated colony or by adding a drop of

3% H_2O_2 to a clean microscope slide and placing a platinum loop containing the isolate into the drop. The reaction is positive if there is a rapid evolution of oxygen gas (bubble formation). A Nichrome wire loop may give a false-positive reaction, and it should be noted that if the H_2O_2 is placed onto a medium, the bacteria will be rendered nonviable. A positive control (*Staphylococcus* species or an enteric) and a negative control (*Streptococcus* species) should be run daily to ensure the quality of the H_2O_2 solution.

2. If no catalase is produced, inoculate a bile esculin agar (see Sec. VII.B) slant, a tube of trypticase soy or tryptic soy broth (see Sec. VII.N) containing 6.5% NaCl, and a 5% sheep blood agar plate. Incubate 18-24 hr at 35 ± 2°C.

   a. If the organism produces a blackening of the bile esculin medium and will grow in the presence of 6.5% NaCl, report the organism as a group D Enterococcus.
   b. If the organism produces a blackening of the bile esculin medium but will not grow in the presence of 6.5% NaCl, report the organism as a group D *Streptococcus* sp., not Enterococcus.
   c. If the organism does not produce a blackening of the bile esculin medium, report the organism as either an alpha, beta, or gamma hemolytic *Streptococcus* sp., not group D.
   d. Additional speciation of streptococci may be performed if required after procedures outlined in Ref. 7 or by the use of serological kits commercially available for this purpose (e.g., Phadebact from Pharmacia Diagnostics, Piscataway, N.J.).

3. If catalase is produced, inoculate the following media with a freshly isolated colony: mannitol salt agar (see Sec. VII.F), oxidative-fermentative medium (see Sec. VII.I) with dextrose (overlay one tube with sterile vaspar or mineral oil; leave one tube loosely capped with no overlay), and an enriched agar slant for use in the coagulase test (see Sec. VI.E for this method). Report the organism as follows:

   a. *Staphylococcus aureus* if it is coagulase positive and/or will ferment mannitol.
   b. *Staphylococcus epidermidis* if it is fermentative as well as oxidative on oxidative-fermentative dextrose, is coagulase negative, and will ferment mannitol.
   c. *Micrococcus* species if it is only oxidative on oxidative-fermentative dextrose.

3. Gram-Negative Rods

1. Inoculate a triple sugar iron agar (see Sec. VIII.G) slant, a MacConkey agar plate, and a modified letheen agar plate with all gram-negative rods. Incubate 18-24 hr at 35 ± 2°C.
2. Reactions that may be observed in triple sugar iron agar slants are shown in Figure 1. If triple sugar iron reactions are masked by hydrogen sulfide production, lactose and glucose carbohydrate broths should be inoculated and incubated 18-24 hr at 35 ± 2°C. Figure 1 refers the investigator to Schemes A-D utilized for the identification of the *Enterobacteriaceae*.

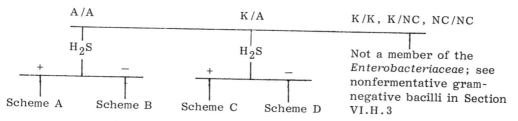

FIGURE 1 Reactions (slant/butt) in triple sugar iron medium (A, acid; K, alkaline; NC, no change; +, positive; −, negative).

These schemes, based on Table 1, are meant only as an aid in the identification for those unfamiliar with this group of microorganisms. Once an isolate is identified by reference to the flow charts, it is strongly recommended that additional biochemical tests listed in Table 1 be performed to confirm the identity of the isolate. All biochemical reactions in Table 1 are based on results obtained in the media listed in Section VIII and on schemes and biochemical reactions in Refs. 1, 4, and 7. These references should be consulted if an organism cannot be identified through the use of the schemes in this chapter.

3. If the organism is identified as a nonfermenter, Tables 2-5 may be used for identification. Figure 2 initially divides the nonfermentative gram-negative bacilli into groups based on the aerobic production of acid from either glucose, sucrose, xylose, or mannitol, fluorescence, and the production of nitrogen gas from an inorganic nitrogen source. Figure 2 refers the user to Schemes E-H, which are meant only as an aid in the identification of these bacilli for those unfamiliar with the nonfermentative gram-negative bacilli. The tables from which these schemes were constructed were devised by Lucy Anne Otto (Los Angeles District, Food and Drug Administration), and employ the media and techniques described in Section IX. Ms. Otto has based these tables primarily on the Elizabeth O. King scheme of identification. When an organism cannot be identified from these tables, refer to the full King tables [6].

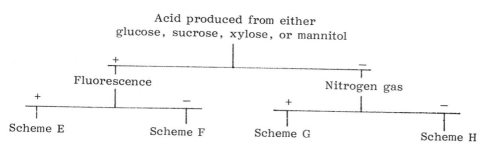

FIGURE 2 Preliminary separation of gram-negative nonfermentative bacilli (+, positive; −, negative).

TABLE 1 Biochemical Reactions[a] of the *Enterobacteriaceae*

Enterobacteriaceae	Triple sugar iron slant/butt	LYDC	LYDA	MOT	IND	ORN	URE
Citrobacter freundii	K(A)/A	−	−	+	−	V	V
C. diversus	K(A)/A	−	−	+	+	+	V
Enterobacter aerogenes	A/A	+	−	+	−	+	−
E. agglomerans	A/A	+	−	+	V	−	V
E. cloacae	A/A	−	−	+	−	+	V
E. gergoviae	A/A	+	−	+	−	+	+
E. sakazaki	A/A	−	−	+	−	+	−
Escherichia coli	A(K)/A	V	−	V	+	V	−
Hafnia alvei	K/A	+	−	+	−	+	−
Klebsiella pneumoniae	A/A	+	−	−	−	−	+
K. oxytoca	A/A	+	−	−	+	−	+
K. ozaenae	A/A	V	−	−	−	−	V
K. rhinoscleromatis	A/A	−	−	−	−	−	−
Morganella morganii	K/A	−	V	V	+	+	+
Proteus mirabilis	K(A)/A	−	+	+	−	+	V
P. vulgaris	A(K)/A	−	−	−	−	−	−
Providencia stuartii	K/A	−	+	V	+	−	V
Serratia liquefaciens	A-K/A	+	−	+	−	+	−
S. marcesans	A-K/A	+	−	+	−	+	V
S. rubidaea	A/A	+	−	+	−	−	−

[a]LYDC, lysine decarboxylase; LYDA, lysine deaminase; MOT, motility; IND, indole; ORN, ornithine decarboxylase; URE, urease; CTR, citrate; MAL, malonate; TSI, triple sugar iron; MR, methyl red; VP, Voges-Proskauer; ARAB, arabinose; RHAM, rhamnose; SORB, sorbitol; MANN, mannitol; +, 90% of strains positive; −, 90% of strains negative; V, 10.1-89.9% of strains positive; A, acid reaction; K, alkaline reaction.

CTR	MAL	H₂S (TSI)	MR	VP	ARAB	RHAM	SORB	MANN
+	V	V	+	−	+	+	+	+
+	+	−	+	−	+	+	+	+
+	V	−	−	+	+	+	+	+
V	V	−	V	V	+	+	V	+
+	V	−	−	+	+	+	+	+
+	+	−	V	+	+	+	−	+
+	−	−	−	+	+	+	−	+
V	−	−	+	−	−	V	+	+
V	V	−	V	V	+	V	−	+
+	+	−	V	+	+	+	+	+
+	+	−	V	+	+	V	+	+
V	−	−	+	−	V	V	V	+
−	V	−	+	−	+	V	V	+
−	−	−	+	−	−	−	−	−
V	−	+	+	V	−	−	−	−
V	−	−	−	−	−	−	−	−
+	−	−	+	−	−	V	+	V
+	−	−	V	V	+	V	+	+
+	−	−	−	V	−	−	+	+
V	V	−	−	V	+	−	−	+

TABLE 2 Identification of Gram-Negative Nonfermentative Bacilli: Glucose-Positive Saccharolytic Nonfermentative Bacilli[a]

Criteria	Pseudomonas aeruginosa	Pseudomonas fluorescens	Pseudomonas putida	Pseudomonas cepacia	Pseudomonas maltophilia	CDC group Vd-2	CDC group Va-1
Oxidase	+	+	+	v	−	+	+
Motility	+	+	+	+	+	+	v
Pigment	v	−	−	v	v	−	−
Fluorescence	+	+	+	−	−	−	−
NO$_3^-$ → NO$_2^-$	+	v	−	−	−	+	+
NO$_2^-$ → N$_2$	+	v	−	−	v	+	v
42°C growth	+	−	−	−	v	−	v
Lysine decarboxylase	−	−	−	+	+	−	−
Indole	−	−	−	−	−	−	−
Gelatinase	v	+	−	−	+	+	+
Urease	v	v	v	v	v	−	−
Esculin hydrolysis	−	−	−	−	v	−	−
Acetamide	+	−	−	−	−	+	+
Glucose	+	+	+	+	+	+	+
Xylose	v	+	+	(+)	−	+	−
Mannitol	v	v	v	+$_w$	−	+	−
Lactose	−	v	v	(+)	−	−	+
Sucrose	−	v	v	+	−	+	−
Maltose	−	v	v	+$_w$	+	+	+
Flagella	Polar	Polar	Polar	Polar	Polar	Peritrichous	Polar

[a]+, positive; −, negative; v, variable; (+), delayed positive; +$_w$, weakly positive.

TABLE 3 Identification of Gram-Negative Nonfermentative Bacilli: Glucose-Positive Pigmented Nonfermentative Bacilli[a]

Criteria	CDC group IIK-1	CDC group IIK-2	Flavobacterium meningosepticum	Flavobacterium IIb	CDC group Vc-1	CDC group Ve-2	Pseudomonas stutzeri
Oxidase	v	+	+	+	−	−	+
Motility	v	−	−	−	+	+	+
Pigment	y	v	y	y	y	y	y
Fluorescence	−	−	−	−	−	−	−
$NO_3^- \longrightarrow NO_2^-$	−	−	−	−	v	−	+
$NO_2^- \longrightarrow N_2$	−	−	−	−	−	−	+
42°C growth	−	−	−	v	v	−	+
Lysine decarboxylase	−	−	+	−	−	−	−
Indole	−	−	+	+	−	−	−
Gelatinase	−	v	+	v	v	−	−
Urease	−	+	−	v	(+)	(+)	v
Esculin hydrolysis	+	+	(+)	v	+	−	−
Glucose	+	+	+	+	+	+	+
Xylose	+	+	−	v	+	+	+
Mannitol	−	−	+	v	+w	+w	+w
Lactose	+	+	v	−	−	v	−
Sucrose	+	+	−	v	−	v	−
Maltose	+	+	+	v	+	+	v
Salicin	v	+	ND	ND	+	−	ND
Sorbitol	−	−	ND	ND	−	+	ND
Flagella	Polar	Polar	−	−	Polar tuft	Polar	Polar

[a]y, nondiffusible yellow; +, positive; −, negative; v, variable; (+), most reactions occur within 1 or 2 days, and some are delayed; +w, weakly positive; ND, not determined.

TABLE 4 Identification of Gram-Negative Nonfermentative Bacilli: Weakly Saccharolytic Nonfermentative Bacilli[a]

Criteria	Pseudomonas vesicularis	Pseudomonas acidovorans	Achromobacterium xylosoxidans IIIa	Achromobacterium xylosoxidans IIIb	CDC group Vd-1	Achromobacterium anitratus	Pseudomonas pickettii	Pseudomonas putrefaciens group 1
Oxidase	+	+	+	+	+	−	+	+
Motility	+	+	+	+	+	−	+	+
Pigment	Orange	−	−	−	−	−	−	Brown
Fluorescence	−	−	−	−	−	−	−	−
$NO_3^- \rightarrow NO_2^-$	−	+	+	+	+	−	+	−
$NO_2^- \rightarrow N_2$	−	−	−	−	−	−	v	−
42°C growth	−	−	+	+	−	+	v	−
Indole	−	−	−	−	−	−	−	−
Gelatinase	−	+	−	−	−	v	−	+
Urease	−	−	−	−	+	v	v	−
Esculin hydrolysis	+	−	−	−	ND	−	−	−
Acetamide	−	+	−	−	−	−	−	ND
Glucose	+	+	+$_w$	−	+	+	+	v
Xylose	+	−	+	+	+	+	+	−
Mannitol	−	+	−	−	−	−	−	−
Lactose	−	−	−	−	−	(+)	−	−
Sucrose	−	−	−	−	−	−	−	+
Maltose	+	−	−	−	−	v	−	+
Fructose	−	+	−	−	ND	ND	+	v
Flagella	Polar	Polar tuft	Peri-trichous	Peri-trichous	Peri-trichous		Polar	Polar

[a]H_2S positive; Group 2 is asaccharolytic; +, positive; −, negative; (+), delayed positive; +$_w$, weakly positive; v, variable;

TABLE 5 Identification of Gram-Negative Nonfermentative Bacilli: Asaccharolytic Nonfermentative Bacilli[a]

Criteria	Alcaligenes faecalis	Alcaligenes odorans	Alcaligenes denitrificans	Borditella bronchianis	Pseudomonas testosteroni	Pseudomonas diminuta	Pseudomonas alcaligenes	Pseudomonas pseudoalcaligenes	Flavobacterium IIf	Moraxella	Acinetobacter lwoffii
Oxidase	+	+	+	+	+	+	+	+	+	+	−
Motility	+	+	+	+	+	+	+	+	−	−	−
Pigment	−	−	−	−	−	−	−	−	y	−	−
Fluorescence	−	−	−	−	−	B	−	−	−	−	−
$NO_3^- \rightarrow NO_2^-$	v	−	+	+	+	−	(+)	(+)	−	v	−
$NO_2^- \rightarrow N_2$	−	+	+	−	−	−	−	+	+	−	−
42°C growth	+	+	+	(+)	+	v	−	+	+	v	v
Indole	−	−	−	−	−	−	−	−	+	−	−
Gelatinase	−	−	−	−	+	−	−	−	+	−	−
Urease	−	−	−	+	v	−	−	−	−	−	v
Esculin hydrolysis	−	−	−	−	−	−	−	−	−	−	v
Acetamide	+	+	+	−	−	−	−	−	−	−	−
Flagella	Peritrichous	Peritrichous	Peritrichous	Peritrichous	Polar	Polar	Polar	Polar	−	−	−

[a]B, soluble brown pigment; y, nondiffusible yellow; +, positive; −, negative; v, variable; (+), delayed positive.

SCHEME A *Enterobacteriaceae* Identification (Lysine Deaminase)

I. Negative: *Citrobacter freundii*
II. Positive: Indole
    A. Negative: *Proteus mirabilis*
    B. Positive: *Proteus vulgaris*

SCHEME B *Enterobacteriaceae* Identification (Lysine Decarboxylase)

I. Negative: Ornithine decarboxylase
    A. Negative: *Citrobacter freundii, Enterobacter agglomerans*, or *Escherichia coli*. See biochemical reactions in Table 1 for the separation of these microorganisms.
    B. Positive: Indole
        1. Positive: Malonate
            a. Positive: *Citrobacter diversus*
            b. Negative: *Escherichia coli*
        2. Negative: Sorbitol
            a. Negative: *Enterobacter sakazaki*
            b. Positive: Methyl red
                i. Positive: *Citrobacter freundii*
                ii. Negative: *Enterobacter cloacae*
II. Positive: Urease
    A. Positive: Motility
        1. Positive: *Enterobacter gergoviae*
        2. Negative: Indole
            a. Positive: *Klebsiella oxytoca*
            b. Negative: Malonate
                i. Positive: *Klebsiella pneumoniae*
                ii. Negative: *Klebsiella ozaenae*
    B. Negative: Indole
        1. Positive: *Escherichia coli*
        2. Negative: Motility
            a. Negative: *Klebsiella ozaenae* or *Klebsiella rhinoscleramatis*. See biochemical reactions in Table 1 for separation of these organisms.
            b. Positive: Ornithine decarboxylase
                i. Positive: *Enterobacter aerogenes* or *Serratia liquifaciens*. See biochemical reactions in Table 1 for separation.
                ii. Negative: *Serratia rubidaea*

SCHEME C *Enterobacteriaceae* Identification (Arabinose)

I. Positive: *Citrobacter freundii*
II. Negative: Ornithine decarboxylase
   A. Negative: *Proteus vulgaris*
   B. Positive: *Morganella morganii*

SCHEME D *Enterobacteriaceae* Identification (Malonate)

I. Positive: Indole
   A. Positive: *Citrobacter diversus*
   B. Negative: Sorbitol
      1. Positive: *Citrobacter freundii*
      2. Negative: *Hafnia alvei*
II. Negative: Indole
   A. Positive: Arabinose
      1. Positive: *Escherichia coli*
      2. Negative: Ornithine decarboxylase
         a. Positive: *Morganella morganii*
         b. Negative: *Providencia stuartii*
   B. Negative: Arabinose
      1. Negative: *Serratia marcescans*
      2. Positive: Sorbitol
         a. Negative: *Hafnia alvei*
         b. Positive: Lysine decarboxylase
            i. Positive: *Serratia liquifaciens*
            ii. Negative: *Citrobacter freundii*

SCHEME E Nonfermentative Gram-Negative Bacilli Identification (Acetamide Utilization and Growth at 42°C)

I. Positive: *Pseudomonas aeruginosa*
II. Negative: Gelatin liquefication
   A. Positive: *Pseudomonas fluorescens*
   B. Negative: *Pseudomonas putida*

SCHEME F Nonfermentative Gram-Negative Bacilli Identification (Oxidase)

I. Positive: Pigment production
    A. Pigment produced: Lysine decarboxylase
        1. Positive: Sucrose
            a. Positive: *Pseudomonas cepacia*
            b. Negative: *Pseudomonas vesicularis*
        2. Negative: Urease
            a. Positive: Nitrogen gas
                i. Produced: *Pseudomonas stutzeri*
                ii. Not produced: Indole
                      (1) Positive: *Flavobacterium* II-b, *Flavobacterium meningosepticum*. See biochemical reactions in Table 1 for differentiation of these organisms.
                      (2) Negative: CDC (Centers for Disease Control) Group IIK-2
            b. Negative: Indole
                i. Positive: *Flavobacterium* II-b
                ii. Negative: Nitrogen gas
                      (1) Produced: *Pseudomonas stutzeri*
                      (2) Not produced: Brown pigment
                          (a) Produced: *Pseudomonas putrefaciens*
                          (b) Not produced: Lactose
                              (i) Positive: CDC Group IIK-1
                              (ii) Negative: *Pseudomonas vesicularis*
    B. Pigment not produced: Sucrose
        1. Positive: Lysine decarboxylase
            a. Positive: *Pseudomonas cepacia*
            b. Negative: Motility
                i. Motile: CDC Group VD-2
                ii. Nonmotile: CDC Group IIK-2
        2. Negative: Acetamide
            a. Positive: *Pseudomonas acidovorans*
            b. Negative: Urease
                i. Positive: Lactose
                      (1) Positive: CDC Group Va-1
                      (2) Negative: Fructose
                          (a) Positive: *Pseudomonas pickettii*
                          (b) Negative: CDC Group Vd-1
                ii. Negative: Nitrogen gas
                      (1) Produced: Fructose
                          (a) Positive: *Pseudomonas pickettii*
                          (b) Negative: *Acinetobacter xylosoxidans* IIIb
                      (2) Not produced: Fructose
                          (a) Positive: *Pseudomonas pickettii*
                          (b) Negative: *Acinetobacter xylosoxidans* IIIa

II. Negative: Sucrose
    A. Positive: Lysine decarboxylase
        1. Positive: *Pseudomonas cepacia*
        2. Negative: Esculin
            a. Positive: CDC Group IIK-1
            b. Negative: CDC Group VE-2
    B. Negative: Esculin

SCHEME F Continued

1. Positive: Lysine decarboxylase
   a. Positive: *Pseudomonas maltophilia*
   b. Negative: CDC Group VE-1
2. Negative: Motility
   a. Motile: Lysine decarboxylase
      i. Positive: *Pseudomonas maltophilia*
      ii. Negative: CDC Group VE-2
   b. Non-motile: *Acinetobacter anitratus*

SCHEME G Nonfermentative Gram-Negative Bacilli Identification

Nitrite production
A. Positive
   Growth at 42°C
   1. Positive
      Acetamide
      a. Positive
         Alcaligenes denitrificans
      b. Negative
         Pseudomonas pseudoalcaligenes
   2. Negative
      Pseudomonas alcaligenes
B. Negative
   Alcaligenes odorans

SCHEME H Nonfermentative Gram-Negative Bacilli Identification (Urease)

I. Positive: Gelatinase
   A. Positive: Motility
      1. Motile: *Pseudomonas testosteroni*
      2. Nonmotile: *Acinetobacter lwoffii*
   B. Negative: Motility
      1. Motile: *Bordetella bronchieanis*
      2. Nonmotile: *Acinetobacter lwoffii*
II. Negative: Gelatinase
   A. Positive: Motility
      1. Motile: *Pseudomonas testosteroni*
      2. Nonmotile: *Acinetobacter lwoffii*
   B. Negative: Motility
      1. Motile: Acetamide
         a. Positive: *Alcaligenes faecalis*
         b. Negative: Brown pigment
            i. Produced: *Pseudomonas diminuta*
            ii. Not produced: Indole
               (1) Positive: *Flavobacterium* IIf
               (2) Negative: *Pseudomonas alcaligenes*
      2. Nonmotile: Oxidase
         a. Positive: *Moraxella* sp.
         b. Negative: *Acinetobacter lwoffii*

VII. APPENDIX I: MEDIA FOR THE ENUMERATION OF MICROORGANISMS IN COSMETICS AND THE IDENTIFICATION OF GRAM-POSITIVE BACTERIA

A. Anaerobe Agar

Trypticase soy agar or tryptic soy agar	40 g
Agar (bacteriological grade)	5 g
Yeast extract	5 g
Hemin solution	0.5 ml
Vitamin K_1 solution	1.0 ml
L-Cysteine (dissolved in 5 ml of 1 N NaOH)	0.4 g
Distilled water	1000 ml

Dissolve all ingredients in medium by boiling. Adjust pH to 7.5; autoclave for 15 min at 121°C. Dispense into 15 × 100 mm Petri plates. Medium must be reduced before inoculation (24 hr under anaerobic conditions in either an anaerobic glove box, incubator, or GasPak jar).

Stock hemin solution. Dissolve 1 g of hemin (Eastman Chemical No. 2203) in 100 ml of distilled water. Autoclave 15 min at 121°C. Refrigerate.

Stock vitamin K_1 solution. Dissolve 1 g of vitamin K_1 (Sigma Chemical Co. No. V-3501) in 100 ml of 95% ethanol. Solution may require 2-3 days with intermittent shaking. Refrigerate.

From: V. R. Dowell, Jr., and G. L. Lombard, Presumptive identification of anaerobic non-sporforming gram-negative bacilli, Center for Disease Control, Atlanta, Ga. (1977).

B. Bile Esculin Agar

Commercially available in dehydrated form.

C. Brain-Heart Infusion Broth and Agar

Commercially available in dehydrated form.

D. MacConkey Agar

Commercially available in dehydrated form.

E. Malt Extract or Potato Dextrose Agar

Commercially available in dehydrated form.

F. Mannitol Salt Agar

Commercially available in dehydrated form.

G. Modified Letheen Agar

Letheen agar (Difco or AOAC from BBL)	32 g
Trypticase peptone	5 g
Thiotone peptone	10 g
Yeast extract	2 g
NaCl	5 g
Sodium bisulfite	5 g
Agar-agar	5 g
Distilled water	1000 ml

Dissolve ingredients by boiling. Autoclave 15 min at 121°C and dispense into 15 × 100 mm Petri Plates.

H. Modified Letheen Broth

Prepared from either Difco or BBL commercially available medium with the same supplementation as for modified letheen agar, omitting only the additional NaCl and agar-agar. Dispense in appropriate amounts into dilution bottles.

I. Oxidative-Fermentative Test Medium

Commercially available in dehydrated form.

J. Sabouraud Dextrose Broth

Commercially available in dehydrated form.

K. Sheep Blood Agar

Trypticase soy agar, tryptic soy agar, blood agar base, *Brucella* agar, or any medium that will support the growth of fastidious organisms prepared according to manufacturer's directions and cooled to 45-50°C in a water bath. Add 5% defibrinated sheep blood, mix, and dispense into 15 × 100 mm Petri plates or any other sterile container desired.

L. Starch Agar

Commercially available in dehydrated form.

M. Thioglycollate Medium 135c

Prepare thioglycollate medium 135c without indicator according to manufacturer's (BBL) instructions. Before autoclaving or dispensing, supplement with 0.5 ml of stock hemin solution and 0.5 ml of stock vitamin K_1 solution (see Sec. VII.A) per liter. Mix and dispense into dilution bottles. Autoclave 15 min at 121°C. After cooling, tighten the screw caps of the dilution bottles and store in the dark. Do not refrigerate the bottles. If possible, use only freshly prepared medium.

N. Trypticase Soy Agar and Tryptic Soy Agar and Broth

Commercially available in dehydrated form.
   Add 65 g of NaCl per liter to make a 6.5% NaCl-TSB solution for group D *Streptococcus* differentiation.

O. Vogel-Johnson Agar

Commercially available in dehydrated form.

VIII. APPENDIX II: MEDIA FOR ENTEROBACTERIACEAE IDENTIFICATION

A. Andrade Carbohydrate Broths

Beef extract	3 g
Bacto peptone, gelysate, or protease pancreatic digest of gelatin	10 g
Sodium chloride	10 g
Andrade indicator	10 ml
Distilled water	1000 ml

   Adjust pH to 7.2 with 1 N NaOH and autoclave 15 min at 121°C.
   Andrade indicator.

Acid fushsin	0.2 g
Distilled water	100 ml
1 N Sodium hydroxide	16 ml

   Let decolorize before use. Add an additional 1 or 2 ml of alkali if necessary.
   Add 1.0% dextrose, lactose, sucrose, and mannitol. Add 0.5% dulcitol, salicin, and other carbohydrates. Add dextrose, mannitol, dulcitol, salicin, adonitol, and inositol before sterilization. All other carbohydrates must be added after autoclaving the basal medium. Carbohydrates may be prepared in 10% concentrations and sterilized by autoclaving.

B. Lysine Iron Agar

Commercially available in dehydrated form.

C. Malonate Broth

Commercially available in dehydrated form.

D. Motility-Indole-Ornithine Medium

Commercially available in dehydrated form.
   Modified Kovac's reagent (formula in Sec. IX) should be used for the indole test. Individual tubes containing motility test medium, 1% tryptone or

trypticase, and ornithine test medium (all commercially available) may be used instead of motility-indole-ornithine medium.

E. MR-VP Broth

Commercially available in dehydrated form.

Methyl red test. Inoculate MR-VP medium lightly with the growth from a young agar culture. Incubate 48 hr at 35-37°C. Tests should not be performed on cultures that have been incubated for less than 48 hr. If the results are equivocal, the cultures should be incubated for 4-5 days at room temperature and at 35-37°C. Add 5-6 drops of methyl red indicator (see below) to 5 ml of medium. Read reactions immediately: bright red is positive; yellow is negative.

Methyl red indicator.

Methyl red	0.1 g
Ethanol (95%)	300 ml

Dissolve the dye in the alcohol and add sufficient distilled water to make 500 ml of solution.

Voges-Proskauer test. Add 0.6 ml of solution A and 0.2 ml of solution B (see below) to 1 ml of MR-VP broth that has been inoculated and incubated for 48 hr at 35-37°C. Shake the tube well after the introduction of each of the two reagents. Positive reactions occur within 5 min and are indicated by the development of a red color. A copper color is a negative test.

Solution A.

α-Naphthol	5 g
Ethanol (absolute)	100 ml

Solution B.

Potassium hydroxide	40 g
Distilled water	100 ml

F. Simmon's Citrate Agar

Commercially available in dehydrated form.

G. Triple Sugar Iron Agar

Commercially available in dehydrated form.

H. Urea Agar (Christensen's)

Commercially available in dehydrated form.

Perform all biochemical tests after incubation at 35-37°C for 18-24 hr, with the exceptions of malonate broth (incubation at 35-37°C for 48 hr) and MR-VP broth (incubation at 35-37°C for 48 hr or longer).

IX. APPENDIX III: MEDIA FOR IDENTIFICATION OF GRAM-NEGATIVE NONFERMENTATIVE BACILLI

A. Acetamide Medium

Stock basal medium.

KH_2PO_4, 0.5 M	14 ml
K_2HPO_4, 0.5 M	6 ml
PR-CV, 500×	1 ml
Agar	0.5 g
Distilled water, to make	400 ml

Stock acetamide (1%).

Acetamide	1.0 g
Distilled water	100 ml

Store over chloroform in a screw-cap container. The solution is stable indefinitely at room temperature.

PR-CV, 500×.

Phenol red	2.0 g
Crystal violet	0.2 g
Distilled water	200 ml

Add concentrated alkali until both ingredients are dissolved.

Preparation. Add 0.8 ml of basal medium to a 13-mm tube. Add 0.2 ml of acetamide solution to the tube. Steam (100°C) 10 min. Cool.

Inoculation. Add 0.1 ml of a heavy suspension of young (24-hr) cells. Suspension should be opaque, and as heavy as possible. Shake well.

Incubation. Incubate at 35°C for 4 days. Read daily.

Reactions. Positive is indicated by red; negative is indicated by yellow.

B. Clark's Flagellar Stain

Solution A.

Basic fuchsin, special	1.2 g
95% ethyl alcohol	100 ml

Mix and let stand overnight at room temperature.

Solution B.

Tannic acid	3.0 g
NaCl	1.5 g
Water	200 ml

Mix solutions A and B. Adjust pH to 5.0 with 1 N NaOH or 1 N NCl if necessary.

Refrigerate 2-3 days before use. The stain is stable 1 month at 4°C or may be stored frozen indefinitely (50-ml aliquots). To use, thaw stain, remix, and store at 4°c.

Each batch will vary in optimum staining time from 5 to 15 min. Determine the staining time after the stain has ripened 2-3 days at 4°C; prepare three or more cleaned slides for staining, using a known flagellated organism. Stain the slides for various times (e.g., 5, 10, 15 min) and examine for the best staining time. Mark the staining time on all containers.

IMPORTANT! Stain will *not* work unless slides are clean. Clean slides by soaking them 4 days at room temperature in cleaning solution (either acid dichromate or 3% concentrated HCl in 95% ethyl alcohol). Rinse the slides 10 times in fresh tap water. Rinse twice in distilled water. Air-dry at room temperature and store in a covered container.

Staining procedure.

Suspension preparation. Pick a small amount of growth from an 18-24 hr plate (equivalent to a 1-mm colony). Do not pick up agar. Suspend gently in 3 ml of distilled water. Flagella can be knocked off. The suspension should be faintly opalescent.

Slide preparation. Pass a cleaned slide through the blue part of a burner flame several times until very hot to remove any residual dirt from the cleaning procedure. Cool the slide on a paper towel on the bench with the flamed side up. Mark a wax line across the slide to give an area of 2.5 × 4.5 cm. Place a large loopful of suspension in the center of the slide adjacent to the wax line. Tilt the slide and let the drop run down the center of the slide to the end. (If the drop does not run evenly, the slide is dirty. Discard it.) Air-dry the slide on a level surface.

Staining. Add 1 ml of stain to the slide. This will cover the marked off area. Let stand 5-15 min as determined above. Wash with tap water, air-dry, and examine.

C. Esculin Agar, Modified

Esculin	1.0 g
Ferric citrate	0.5 g
Heart infusion agar	40 g
Distilled water	1 liter

Heat to dissolve, cool to 55°C, and then adjust to pH 7.0. Dispense 4-ml portions into 13-mm tubes. Autoclave 15 min at 121°C. Slant.

Inoculation. Streak the slant.
Incubation. Incubate 24 hr at 35°C.
Reactions. Positive is indicated by a dark brown color in the medium; negative is indicated by absence of color change.

D. Nutrient Gelatin

Heart infusion broth	25 g
Gelatin	120 g
Distilled water	1 liter

Heat to boiling to dissolve. Cool to 55°C and adjust to pH 7.4. Dispense 4-ml portions into 13-mm screw-cap tubes. Autoclave 15 min at 121°C.
Inoculation. Use 1-2 drops of a heavy suspension of 24-hr cells.
Incubation. Incubate 48 hr at 35°C.
Reading the test. Chill the tube about 30 min in the refrigerator. Tilt the tube and observe liquidity.
Reactions. Positive is indicated by any amount of liquidity in upper portion; negative is indicated when the entire deep is solid.

E. Indole Medium

Tryptone	20 g
Distilled water	1 liter

Adjust to pH 7.3. Dispense 4-ml portions into 13-ml tubes. Autoclave 15 min at 121°C.
Inoculation. Use loop or needle.
Incubation. Incubate 48 hr at 35°C.
Assay. Add 1 ml of xylene to the tube. Shake (or mix by vortex) vigorously to extract indole. Let stand 1-2 min for separation of the two phases. Xylene will layer on top. Add 0.5 ml of Ehrlich's reagent (see below) down the side of the tube so that it forms a layer between the two phases DO NOT SHAKE.
Reactions. Positive is indicated when a red ring develops just below the xylene layer; negative is indicated by the absence of color.
Erhlich's reagent.

Ethanol (95%)	95 ml
p-Dimethylaminobenzaldehyde	1 g
HCl, concentrated	20 ml

Dissolve the aldehyde in the alcohol; then slowly add HCl. Store in refrigerator.

F. Indole Medium

L-Tryptophan	1.0 g
NaCl	1.0 g
K_2HPO_4, 0.5 M	7.2 ml
KH_2PO_4, 0.5 M	192 ml

Dispense 1.0-ml portions into 13-mm screw-cap tubes. May be sterilized in bulk (e.g., 100 ml per bottle). Autoclave 15 min at 121°C.

Inoculation. Use 0.1 ml of a heavy cell suspension to 1 ml of medium in a 13-mm tube.

Incubation. Incubate 18-24 hr at 35°C.

Assay. Add 5 drops of modified Kovac's reagent (see below) to the tube.

Reactions. Positive is indicated by a red color in the reagent; negative is indicated by absence of color change in the reagent.

Modified Kovac's reagent.

Butanol	80 ml
HCl	20 ml
p-Dimethylaminobenzaldehyde	1 g

Dissolve the aldehyde in the alcohol and slowly add HCl. Store in refrigerator.

G. King's B Medium

Proteose peptone no. 3	20 g
Glycerol, C.P.	10 ml
K_2HPO_4	1.5 g
$MgSO_4$	15 g
Distilled water	1 liter

Adjust to pH 7.2. Dispense 4-ml portions into 13-mm tubes. Autoclave 15 min at 121°C. Slant to give half butt and half slant.

Inoculation. Streak the slant with a loop.

Incubation. Incubate 24 hr at 35°C, and then for 48 hr at room temperature.

Reactions. Positive is indicated by a fluorescent yellow-green pigment when observed under ultraviolet light at 366 nm; negative is indicated by absence of fluorescence.

H. Lysine Decarboxylase Medium

L-Lysine, HCl	0.5 g (DL-lysine, 1.0 g)
D-Glucose	0.5 g

KH_2PO_4	0.5 g
Distilled water	100 ml

Adjust to pH 4.6. Autoclave 15 min at 121°C. Aseptically dispense 1.0-ml portions into sterile 13-mm tubes.

Inoculation. Use 0.1 ml of a heavy cell suspension.

Incubation. Incubate overnight at 35°C.

Assay. Add 1 drop of 40% (wt/vol) KOH. Shake. Add 1 ml of lysine decarboxylase reagent (see below). DO NOT SHAKE. Let stand 15 min for extraction and color development.

Reactions. Positive is indicated by a purple color in the bottom layer; negative is indicated by absence of color in the bottom layer.

Lysine decarboxylase reagent.

Ninhydrin (hydantoin or 1,2,3-triketohydrindene)	0.1 g
Chloroform	100 ml

Store tightly stoppered in hood.

I. Motility Nitrate Medium

Tryptose	10 g
Heart infusion agar	8 g
KNO_3 (nitrite free)	1 g
Distilled water	1 liter

Heat to dissolve. Dispense 4-ml portions into 13-mm screw-cap tubes. Autoclave 15 min at 121°C. Store at room temperature. If refrigerated, the tubes *must* be brought to room temperature before inoculation to avoid false-positive gas readings. The solution is stable indefinitely if no drying occurs.

Inoculation. Stab about 5-10 mm deep.

Incubation. Incubate 24 hr at room temperature.

Motility reading. After 6 hr of incubation a hazy balloon appears around the stab. This may not be visible with weakly motile organisms. Always check negative motility by wet mount or hanging drop from a 24-hr room temperature broth culture.

Nitrite assay. Add 0.5 ml of nitrite reagent (see below) and let stand 1-2 min.

Reactions. Positive is indicated by a dark red color; negative is indicated by the absence of color. Confirm by adding zinc dust: A red color should appear.

Nitrite reagent.

N,N-dimethyl-1-naphthylamine	0.2 g
Sulfanilic acid	0.3 g

Succinic acid	2.0 g
Distilled water	100 ml

Warm gently to dissolve. The finished reagent is pale pink. Store in a refrigerator. The solution is stable indefinitely.

J. Nitrate Broth, Enriched

Heart infusion broth	25 g
KNO_3 (nitrite free)	2 g
Distilled water	1 liter

Dispense 4-ml portions into 13-mm tubes with gas trap inserts. Autoclave 15 min at 121°C.

Inoculation. Use loop or needle.

Incubation. Incubate 48 hr at 30-35°C.

Assay. Add 0.5 ml of reagent no. 1 and 0.5 ml of reagent no. 2 to the tube. (Reagents may be combined and then added immediately to the broth.)

Reactions. Positive is indicated by a dark red color; negative is indicated by absence of color change. Confirm negative reactions by adding a small amount of zinc.

Reagent no. 1.

Sulfanilic acid	8 g
Acetic acid, 5 N	1 liter

Reagent no. 2.

N,N-Dimethyl-1-naphthylamine	6 ml
Acetic acid, 5 N	1 liter

Acetic acid, 5 N.

Glacial acetic acid (17.2 N)	100 ml
Distilled water	250 ml

K. King's Oxidative-Fermentative Basal Medium

Trypticase (or casitone)	2 g
1.5% phenol red solution	2 ml
Distilled water	1 liter

Warm to dissolve. Adjust to pH 7.3, add 3.0 g of agar, and dissolve by heating. Autoclave 15 min at 121°C. Aseptically add sterile carbohydrates (10% solution) to give a final concentration of 1% (10 ml of 10% solution + 100 ml basal medium = 1%). Dispense 3.0-ml portions into 13-mm sterile tubes.

Inoculation. Use 1-2 drops of a heavy cell suspension
Incubation. Incubate 48 hr at 35°C.
Stock carbohydrates (10%).

Carbohydrate	10 g
Distilled water	100 ml

Sterilize by either of the following methods:

1. Filter through a 0.22-μm membrane filter. Store in a screw-cap sterile bottle or tube.
2. Put the solution in a screw-cap container (bottle or tube). Add a few drops of chloroform so that a bubble is visible in the bottom of the container. Let stand at room temperature for 48 hr for sterilization to occur. Replenish chloroform if the bubble disappears. Store under refrigeration; the solution is stable indefinitely.

L. Oxidase Test Strips

Tetramethyl-p-phenylenediamine-dihydrochloride	1.0 g
Ascorbic acid	0.1 g
Distilled water	100 ml

Cut filter paper (Whatman No. 40) into small strips of about 10 × 40 mm. Soak in reagent. Drain. Spread strips on paper towels on a tray. Cover with paper towels. Light degrades the reagent. Dry in a 35°C incubator. (Reagent degrades at higher temperature.) When dry, store in a brown bottle at room temperature. Strips must be protected from light and moisture; they should be white. The strips are stable indefinitely.

Assay. Use a *platinum* loop to smear a mass of cells on a portion of the strip. (Nichrome wire gives false-positive reactions.) Read at 10 sec, *no longer*.

Reactions. Positive is indicated by a deep purple color; negative is indicated by the absence of color or when a purple color appears *after* 10 sec.

REFERENCES

1. D. J. Brenner, J. J. Farmer, F. W. Hickman, M. A. Asbury, and A. G. Steigerwalt, Taxonomic and nomenclature changes in *Enterobacteriaceae*, Center for Disease Control, Atlanta, Ga. (1977).
2. M. G. DeNavarre, *The Chemistry and Manufacture of Cosmetics*, D. van Nostrand Co., Inc., New York (1941).

3. A. P. Dunnigan, Microbiological control of cosmetics, *Drug Cosmet. Ind.*, *102*:43-45, 152-158 (1968).
4. P. R. Edwards and W. H. Ewing, *Identification of Enterobacteriaceae*, Burgess, Minneapolis (1972).
5. Food and Drug Administration, Bureau of Foods, Division of Microbiology, *Bacteriological Analytical Manual*, 5th ed., Association of Official Analytical Chemists, Washington, D.C. (1978).
6. E. O. King, *The Identification of Unusual Pathogenic Gram Negative Bacteria*, Center for Disease Control, Atlanta, Ga. (1964) (rev. 1972).
7. E. H. Lennette, E. H. Spaulding, and J. P. Truant (eds.), *Manual of Clinical Microbiology*, American Society for Microbiology, Washington, D.C. (1974).
8. L. J. Morse, H. L. Williams, F. P. Grenn, Jr., E. E. Eldridge, and J. R. Rotta, Septicemia due to *Klebsiella pneumoniae* originating from a hand-cream dispenser, *N. Engl. J. Med.*, *277*:472-473 (1967).
9. R. Smart and D. F. Spooner, Microbiological spoilage in pharmaceuticals and cosmetics, *J. Soc. Cosmet. Chem.*, *23*:721-737 (1972).
10. L. A. Wilson and D. G. Ahearn, *Pseudomonas* induced corneal ulcers associated with contaminated eye mascaras, *Am. J. Ophthalmol.*, *84*:112-119 (1977).
11. L. A. Wilson, A. J. Jilian, and D. G. Ahearn, The survival and growth of microorganisms in mascara during use, *Am. J. Ophthalmol.*, *79*:596-601 (1975).

APPENDIX B
ANTIMICROBIAL PRESERVATIVES USED BY THE COSMETIC INDUSTRY

KARL H. WALLHÄUSSER *Hoechst A.G., Frankfurt, West Germany*

I. History and Background Information 605
II. Antimicrobial Preservatives Used in Cosmetics 609
III. Monographs for the Antimicrobial Preservatives 620
References 744

I. HISTORY AND BACKGROUND INFORMATION

Personal hygiene already played an important part in the advanced civilizations of antiquity, particularly among the privileged classes. Thus the manufacture of soap is described [1] on a Sumerian clay tablet from the period around 2500 B.C.; according to this tablet 5.5 qa (liters) of potash was required in addition to 1 qa oil for the manufacture of soap. Later on the Romans enhanced the importance of hygienic measures. To this day the ruins of Roman baths, often gigantic and spread throughout the former Roman empire, bear witness to these endeavors. The great religions with a mass following also ensured that washing before prayers became common practice.
By contrast, the necessary hygienic conditions were often lacking in the public baths of the Middle Ages, so that as a rule infectious diseases, above all, venereal diseases, were transmitted particularly in these places. It was not until the rapid development of microbiology in the second half of the nineteenth century that the preconditions were created for combating pathogenic microorganisms by suitable chemical means. The first "disinfectants" for the prevention of infections were successfully used in hospitals by men such as Lister and Semmelweis (see Table 1).
  The antimicrobial activity of phenol and several phenol derivatives was studied exhaustively by Koch and Pasteur. Toward the end of the nineteenth

TABLE 1 Discovery and Application of the Most Important Disinfectants

Antimicrobial agent	First description		First application	
Chlorine	1774		Semmelweis	1847
Sodium hypochlorite	1789	Berthollet	Alcock	1827
Cl⁻-releasing compounds	1915	Dakin	Dakin	1916
Chlorine dioxide	1925			1946
Iodine	1812	Courtors	Prout	1816
Iodophors	1949	Shelanski		1956
Formaldehyde	1867	Hofmann		1894
Phenol	1834		Lister	1867
Cresol	1842			1890
Chlorocresol	1906	Pechhold		1908
bis-Phenols	1906	Pechhold and Ehrlich		1927
Quaternary ammonium compounds	1856		Jacobs	1916
			Domagk	1933
Amphoteric compounds	1952	Schmitz	Wegener	1954
Chlorhexidine	1946	Curd and Rose	Davies	1954
Peracetic acid	1955			
Organotins	1954	Kerk		
Mercury compounds				
Sublimate	2637	B.C.		
Merthiolate	1931	Jamieson		
Alcohols		B.C.		

Source: Ref. 2.

century the first industrially produced vaccines were marketed, and since they were readily attacked by microbes (degradable), they had to be "preserved." In those days phenol and cresol in concentrations of 0.3-0.5% were used as preservatives against spoilage due to microbes. In the case of the industrially produced "injection preparations," which were still being supplied as "bulk material" to pharmacies around the turn of the century and passed on from there as a "dispensable preparation" to doctors, there were also frequent incidents, since these products were often highly contaminated with microorganisms, depending on the chemical constitution of the active ingredient and on further auxiliaries, but depending above all on the type of preparation concerned. It only became gradually possible to remedy this situation by means of sterilization in the final container (ampule) or, in the case of heat-sensitive products, by adding a suitable preservative. If one views the list of preservatives [3] in Table 2, it will be noted, in comparison with Table 1, that many "disinfectants" are also used as preservatives. The only difference lies in the concentration applied, which is usually lower by a power of 10 in preservation than in disinfection.

Antimicrobial Preservatives

TABLE 2 Discovery and Application of the Most Important Antimicrobial Preservatives

Antimicrobial preservative	Introduced for preservation		First description of the substance	
Sublimate			2637 B.C.	
Phenol	1890		1834	Runge
Cresol	1890		1842	
Acetone-chloroform	1903		1903	
Phenylethyl alcohol	1904			
Benzyl alcohol			1853	
Sodium benzoate	1875	Fleck		
Formaldehyde	1900		1867	Hofmann
Chlorocresol	1908		1906	Pechhold
PHB esters	1924	Sabalitschka	1860	Graebe
Thimerosal	1928			
Phenylmercuric compounds	1928-1936			
Quaternary ammonium compounds	1933	Domagk	1856	Hofmann
Sorbic acid	1939	Müller	1859	Hofmann
Sodium thimerfonate			1931	
Chlorhexidine	1954	Davies	1946	Curd and Rose

Together with the hygienic measures that were instituted in hospitals, the general awareness of hygiene developed particularly after the First World War. Although in those days, as had been the case for over 4500 years, soap was still the main cleansing agent both in personal hygiene and in laundering, the first detergents and shampoos — the latter still in powder form — appeared on the market in the 1920s. Skin care products, creams, sun protection oils, and beauty products manufactured on an industrial scale increasingly gained acceptance in the "cosmetic field." However, it was not until the second half of this century that the cosmetic-hygienic measures, which have become common practice meanwhile, reached a hardly surpassable peak in the United States and Europe. These facts are manifest particularly in the daily water consumption. Whereas the daily per capita consumption of water was 10-30 liters at the beginning of this century, it had already risen to 130-200 liters in the 1970s.

This increased awareness of hygiene is to be welcomed, as it certainly provides a number of advantages, but one should not fail to notice the problems this engenders. First of all there is the anxiety about our water resources, which we must use economically. More and more sources are drying up and the groundwater table is falling. In addition, the waste water produced has to be purified in ever more elaborate and expensive processes in order that pollution of our surface waters, streams, rivers, and lakes, which constitute our drinking water potential, does not continue. The change with respect to the base products of our detergents and cleansing agents has led to numerous problems. With the discovery of the fatty alcohol sulfates as early as 1928, the first step was made in the direction of the synthetic and partially synthetic surfactants, but it was not until the period between 1950 and 1960 that the breakthrough was made throughout the world which was

TABLE 3 Production of Cosmetics in the Federal Republic of Germany in 1979[a]

Cosmetic	Production (tons/year)	Percentage of total production
Eau de cologne, lavender water and scents	4,750	2.45
Face lotions	1,400	0.72
Shaving lotions	2,500	1.29
Perfumes	200	0.10
Sun care products	4,300	2.22
Toothpastes	23,800	12.30
Shampoos	42,000	21.70
Bubble baths	79,500	41.10
Creams, lotions, and other skin care products	34,850	18.00

[a]More than 96% of the total production has to be preserved in order to prevent microbial spoilage. Only in the case of products with an alcohol content of over 20% (this limit is fixed at an alcohol content of 25% in the United States) may the addition of a preservative be dispensed with.
Source: Ref. 6.

soon noticeable by the white crests and mountains of foam in our rivers, especially at barrage weirs and dams. This was attributable to the insufficient biodegradability of these detergents, particularly tetrapropylenebenzene sulfonate (TPS).

Thanks to the rapid and judicious cooperation between governmental institutions and the detergent-manufacturing industry it soon became possible to change to types which are more readily biodegradable, for example, the linear alkylbenzene sulfonates (LAS). In the European Economic Community (EEC) countries, for example, a biodegradability of at least 80% is required at present for anionic detergents, this having to be proved in accordance with compulsory standard test methods. In 1978 the consumption of anionic surfactants in the Federal Republic of Germany (see Table 3) was 140,000 tons/year and accounted for 60% of the total surfactants consumption (225,000 tons/year) [4]. Since the preponderant part of detergents and cleansing agents is marketed in powder form (in Europe the share of these products is 95%, in Japan 90%, and in the United States 80%) [5], the disadvantages of the biodegradability of these surfactants do not have the same drastic effect on the stability of the final product as, for example, in the cosmetics field, where, especially with respect to liquid shampoos, bubble baths, toothpastes, and so on, the biological time bomb ticks from the start, that is to say, from the production process to the consumer. Fatty alcohol ether sulfates are mainly used in these products, whose share is about 10% of the anionic surfactants produced. The addition of suitable preservatives is imperative here to ensure that biodegradation does not already set in on the way to the consumer or in his hand, but only when further dilution with water in the biological waste water treatment plant takes place. In addition, radical changes have made themselves felt during the last 30 years in the field of skin care products. The general trend was away from the water-oil ointments, stable from the microbiological aspect, to the oil-water creams and lotions, which are considerably more prone to microbial attack. Preservatives are unavoidable nowadays, both in shampoos and creams, in order to prevent spoilage due to microorganisms and any injury this may cause to the consumer.

Over 96% of the total production of cosmetics listed in Table 3 has to be preserved against microbial attack nowadays. Only in the case of products with an alcohol content of more than 20% (in the United States more than 25%) can the addition of preservatives be dispensed with.

However, this necessity to add preservatives poses further problems with regard to skin compatibility — mainly allergies. The most widely used preservatives too — foremost among them the parabens — are no exception. What is aggravating in addition is that they are widely used and rank first both in the cosmetic and pharmaceutical fields, let alone the fact that they are also used for the protection of foods (meat salads, mayonnaise, etc.).

It is not surprising that dermatologists have repeatedly suggested that the fields of application of preservatives be more strictly limited and that certain products, for example, be reserved exclusively for the cosmetic sector. Unfortunately, this can hardly be carried out in practice, owing to the lack of really good preservatives. So far only the "expensive" products, for example, Germall 115, meet this requirement. And even this product has the drawback of only being effective against bacteria, and its spectrum of action has to be complemented by the addition of an antimycotic component.

II. ANTIMICROBIAL PRESERVATIVES USED IN COSMETICS

While the pharmacopoeias, for example, the *U.S. Pharmacopoeia XX (USP XX)* [7] or the *European Pharmacopoeia* [3,8] do not yet include "positive lists" of antimicrobial preservatives, there has been more progress in this respect in the cosmetics field, particularly in Europe. As many as three lists were submitted by three bodies in this field in 1978-1979, namely, the following:

The Comité de Liaison des Syndicats Européen de l'Industrie de la Parfumerie et des Cosmétiques [9] (COLIPA), CL 38-78/AMR/gp, January 27, 1978; (COLIPA) CL 50-78/AMR/gp, February 1, 1978, and CL 51-78/AMR/gp.

The Council of Europe, "Cosmetic products and their ingredients," Strasbourg 1978 [10] PA/SG (78) 1; 52.421; 03.5, January 1978.

The Commission of the European Economic Community (EEC) (82/368/EWG), definitive, Brussels, May 1, 1982 [11].

On comparing the three lists (see Table 4) the large number of "preservatives" mentioned, namely, 76, 68, and 70, respectively, is conspicuous, while hardly a dozen compounds are in use in the pharmaceutical field [12]. Of the three lists, that of the EEC is finding increasing acceptance. Consequently I do not want to go into the details of the other two. Only two groups of active ingredients are distinguished in the EEC list: those that are definitively accepted and those that are provisionally accepted.

Table 4 shows that the EEC list contains only 12 definitively accepted preservatives, thus proceeding with considerably greater caution than the other two bodies. What is more, only six of these compounds or classes of compounds may really be used without restrictions [13] (PHB esters, without PHB benzyl ester; benzoic acid, propionic acid, sorbic acid, o-phenylphenol, in organic sulfites and bisulfites), while the remaining six preservatives (form-

TABLE 4 Preservatives for the Cosmetic Industry

Organization	Total number	Total number of preservatives		
		Definitively accepted	Provisionally accepted	For which files are needed
Colipa [9]	76	30	27	19
Council of Europe [10]	68	32	28	8
European Economic Community (EEC) [11]	70	12	58	

TABLE 5 Definitively Permitted Substances

EEC reference number	Substances	Maximum authorized concentration	Limitations and requirements	Conditions of use and warnings which must appear on the label
1	Benzoic acid, salts, and esters	0.5%		
2	Propionic acid and its salts	2% (acid)		
3	Salicylic acid and its salts	0.5% (acid)	Not to be used in preparations for children, except in shampoos	
4	Sorbic acid and its salts	0.6% (acid); not to be cumulated with the concentration specified for esters		
5	Formaldehyde	0.2% (except for products for oral hygiene) 0.1% (products for oral hygiene); expressed as free formaldehyde	Prohibited in aerosol dispensers, except for foams	Contains formaldehyde if concentration exceeds 0.05%
	Paraformaldehyde	0.2% (except for products for oral hygiene) 0.1% (products for oral hygiene) Concentrations calculated as theoretically available formaldehyde; not to be cumulated with the concentration specified for formaldehyde	Prohibited in aerosol dispensers, except for foams	Contains formaldehyde if concentration exceeds 0.05%
6	2,2'-Dihydroxy-3,3',5,5',6,6'-hexachlorodiphenylmethane (hexachlorophene)	0.1%	Prohibited in products for children and intimate hygiene	Not to be used for babies; contains hexachlorophene
7	o-Phenylphenol and its salts	0.2% expressed as phenol		

TABLE 5 Continued

EEC reference number	Substances	Maximum authorized concentration	Limitations and requirements	Conditions of use and warnings which must appear on the label
8	Pyridine-1-oxide-2-thiol (zinc pyrithione)	0.5%	Only in products rinsed off after use	
9	Inorganic sulfites and bi-sulfites	0.2% expressed as free SO_2		
10	Sodium iodate	0.1%	Only in products rinsed off after use	
11	Chlorobutanol	0.5%	Prohibited in aerosols	Contains chlorobutanol
12	p-Hydroxybenzoic acid, salts, and esters except benzyl ester (parabens)	0.4% (acid) for one ester 0.8% (acid) for mixtures of esters		

Source: Ref. 1.

TABLE 6 Test for the Innocuousness of Cosmetic Starting Products[a]: Determination of the Toxicological Profile for Assessing Health Risk

1. Acute toxicity (LD_{50})
2. Subacute toxicity
   Regularly repeated applications for 28-90 days
3. Chronic toxicity
   Long-term study of more than 6 months, 2 years as a rule
4. Toxicokinetic studies in regard to the following:
   Absorption
   Distribution
   Metabolism
   Excretion
5. Studies with respect to the influence on reproduction:
   Teratogenicity
   Fertility
   Perinatal and postnatal development
   Mutagenicity
   Carcinogenicity

[a]According to recommendations of the German Federal Ministry of Health of March 1981 [15].

aldehyde, hexachlorophene, chlorobutanol; see Table 5) require a declaration or warning notice, for example, "Not in Children's Care Products" or "Only for Rinsing Off Products."

The preservatives named in the second part of the EEC list [11] are only provisionally accepted, that is to say, until December 31, 1985. Further data, particularly in regard to toxicity, teratogenicity, mutagenicity, and carcinogenicity, still have to be provided with respect to these preservatives (see Table 6). The demands made here correspond largely to those in the United States.

Antimicrobial preservatives are also counted among the starting products. The costs of such a test procedure are considerable. Including the tests for chronic toxicity, which normally take 2 years to complete, several hundred thousand dollars have to be reckoned with. In addition, further dermatological and allergological tests, partly in experimental animals or in man, have to be carried out (see Table 7).

The monographs in Section III for the individual antimicrobial active agents show that there is a considerable backlog of demands in this respect in regard to most compounds. At the same time the question arises as to whether such a large number of active agents is really required and whether it would not be more appropriate to limit oneself to a few compounds [14].

The lists by Richardson [16], published at intervals of several years, of the preservatives most frequently used in cosmetics in the United States provide an answer. It is clearly shown that fewer than 10 preservatives are used in more than 85% of cosmetic products for protection against microbial attack. The situation is similar in Europe. Thus, according to a representative survey among several leading manufacturers of cosmetics in the Federal Republic of Germany [17], about 80% of the cosmetic creams and lotions are

TABLE 7 Required Dermatological and Allergological Tests Needed When Substances are Used Whose Toxicological Properties are Sufficiently Well Known

1. Animal experiments
   Direct irritant effect
   Sensitizability
   Phototoxic properties
   Photoallergic properties
   Acne-producing properties
   Mucosal compatibility

2. Tests in human beings (should only be carried out if the animal experiments do not provide translatable results)
   Primary irritant effect on single application
   Repetitive epicutaneous test (irritant effect on prolonged, repeated application on the same site, e.g., three times weekly for 3 months)
   Special tests, e.g., test for phototoxicity

3. Test carried out with the product under the actual conditions of use.
   Recommended area of application: on the introduction of a product containing novel substances.
   Test procedure: testing the cosmetic in a large number of probands using the intended concentration and frequency on the same skin area over a prolonged period. Collection of information via skin irritations and subjective tolerability.

preserved with PHB esters, although in 50% of the cases a combination with Germall 115 or a similar product, such as Euxyl or Biopure 100, is used. In the case of the "rinse off" products, for example, shampoos and bubble bath concentrates, the preservatives formaldehyde, Bronopol, Bronidox, and Dowicil 200 predominate (see Table 8).

Ethanol as well as several more recent compounds, which were produced particularly in the last 2 years in the United States, were included in this list (distinguished by a minus sign). Ethanol is not classified as a preservative by the EEC. The EEC [11] defines preservatives as substances which may be added to cosmetics in amounts up to the "permissible" concentrations in order to prevent the development of microorganisms in these products. It is pointed out, however, that other substances which are used in the production of cosmetics, for example, essential oils and various alcohols, may also make a contribution toward the preservation of these products. In the case of alcohols, usually ethanol, n-propanol, and isopropanol, considerably higher concentrations than those of the usual preservatives are required for protection against microbial attack, namely, 20-25%. However, these concentrations are only reached in face lotions, shaving lotions, and hair lotions and are then sufficient for preservation (see Table 9).

TABLE 8 Antimicrobial Preservatives Approved Definitively (o) or Provisionally and Some Newer Compounds Not Listed by the EEC for Use in Cosmetics (-)

		Spectrum of activity (MIC in ppm) against				Optimal pH range	Percentage in U.S. cosmetics
		Bacteria		Fungi and Yeasts	Molds		
Preservative	Concentration used (%)	Gram positive	Gram negative			4 5 6 7 8 9	
Ethanol	20.0	+	+	+	+		
Isopropanol	20.0	+	+	+	+		
o Chlorobutanol	0.5	800	1000	2000	5000		0.02
Benzyl alcohol	1.0	25	2000	2500	5000		0.18
2,4-Dichlorobenzyl alcohol	0.15	1000	1000	500	500		
2-Phenoxyethanol	1.0	2000	4000	5000	5000		0.13
Phenoxyisopropanol		2500	4000	4000	4000		0.02
Phenylethyl alcohol	0.5	2500	5000	5000	5000		0.03
Chlorophenesin	0.5	1250	2500	1250	2500		
Bronopol	0.1	70	70	50	50		2.60
5-Bromo-5-nitrodioxane	0.2	75	50	25	25		
o Formaldehyde	0.2	125	125	500	500		4.02
o Paraformaldehyde	0.2	125	125	500	500		0.09
Hexamethylenetetramine	0.5	125	125	500	500		
Monomethyloldimethylhydantoin	0.2	1000	1250	2000	2000		0.07
Dimethyloldimethylhydantoin	0.2	800	1000	1250	1500		0.36
Glutaraldehyde	0.2	250	500	500	1000		0.11

TABLE 8 Continued

Preservative	Concentration used (%)	Spectrum of activity (MIC in ppm) against				Optimal pH range	Percentage in U.S. cosmetics
		Bacteria		Fungi and Yeasts	Molds		
		Gram positive	Gram negative				
Chloracetamide	0.3	2000	3000	500	500	4–9	0.11
Quaternium-15	0.3	200	1000	3000	3000	4–9	4.60
Imidazolidinyl urea	0.6	1000	2000	8000	8000	4–9	7.75
Germal II diazolidinyl urea	0.2	250	1000	8000	4000	4–9	
o Inorganic sulfites	0.2	100	200	200	400		0.31
Boric acid	3.0	5000	10,000	5000	1250	4–6	0.22
Formic acid	0.5	1400	800	1600	3000	4–6	
o Propionic acid	2.0	2000	3000	2000	2000	4–6	0.01
o Undecylenic acid	0.2			1250	2500	4–6	2.16
o Sorbic acid	0.6	100	300	500	300	4–6	0.79
o Benzoic acid	0.5	1000	1500	750	750	4–6	0.04
o Salicylic acid	0.5	2000	3000	3000	5000	4–6	1.37
Dehydroacetic acid	0.5	8000	8000	300	1000	4–6	
Usnic acid	0.2	10	4000	500			
o Methylparaben	0.2	1250	1250	5000	5000	4–9	31.22
o Ethylparaben	0.15	625	625	2500	5000	4–9	0.73
o Propylparaben	0.02	180	625	625	2500	4–9	28.40
o Butylparaben	0.02	160	320	625	1250	4–9	3.07

616

Compound						
Benzylparaben	0.006	120				
○ Blend of PHB esters	0.8	500				
Parachlor-m-cresol	0.2	625	160	250	1000	0.45
Chloroxylenol	0.5	500	500	400	300	0.33
Chlorothymol	0.1	500	1250	2500	2500	0.25
o-Cymen-5-ol	0.1	200	1000	2000	2000	
Isopropyl-o-cresol	0.1	1000	1000	1500	2000	0.02
Tetrabromo-o-cresol	0.3	5	100	100	100	
Dichloro-m-xylenol	0.1	10	1500	1000	1500	0.12
○ o-Phenylphenol	0.2	100	1250	1250	1250	
Chlorophene	0.2	5	1000	1000	1000	0.04
Dichlorophene	0.1	2	500	500	500	
Bromochlorophene	0.1	10	100			
○ Hexachlorophene	0.1	1	50			0.04
Triclosan	0.5	10	1000		300	0.40
Trichlorocarbanilide	0.2	1	200	1000	900	0.005
Dichlorofluorocarbanilide	0.3	5	100	500	2000	0.009
Propamidine isethionate	0.1	16	5	1000	1500	
Dibromopropamidine	0.15	4	500	500	1000	
Hexamidine isethionate	0.1	10	256	500	1000	
Pyrithione sodium	0.5	1	256	500	1000	0.03
○ Zinc pyrithione	0.5	16	256	4	2	0.13
Hexetidine	0.2	10	500	1	2	
			10,000	5000	10	

TABLE 8 Continued

Preservative	Concentration used (%)	Spectrum of activity (MIC in ppm) against				Optimal pH range	Percentage in U.S. cosmetics
		Bacteria		Fungi and Yeasts	Molds		
		Gram positive	Gram negative				
Tris(hydroxyethyl)hexahydrotriazine	0.3	160	320	640	640		0.19
Captan	0.2	200	1000	1000	2000		
Dimethoxane, dioxin	0.2	1250	625	2500	1250		
- Oxadine A	0.2	500	500	1000	1000		
Kathon CG	0.1	150	300	300	300		0.17
Hydroxy-8-quinoline	0.3	4	128	256	512		0.21
Chlorhexidine	0.1	1	50	20	200		0.17
Polyhexamethylenebiguanide	0.3	2	128	64	256		
Alkyltrimethylammonium bromide	0.1	2	20	50	100		0.03
Benzalkonium chloride	0.5	2	10	20	50		0.12
Benzethonium chloride	0.1	1	100	50	100		0.12
Piroctone olamine	0.1	32	625	40	625		
Thiomersal	0.003	1	8	32	128		0.1
Phenylmercuriacetate	0.003	1	4	8	16		0.68
o Sodium iodate	0.1						
- Glyceryl monolaurate	0.5	250	—	500	250		

TABLE 9 Antimicrobial Activity of Ethereal Oils in Emulsions[a]

	Staphylococcus aureus		Escherichia coli		Pseudomonas aeruginosa		Candida albicans		Aspergillus niger	
Initial germ count	10^4		10^4		10^4		10^4		10^4	
Ethereal oil concentration (%)	0.1	1.0	0.1	1.0	0.1	1.0	0.1	1.0	0.1	1.0
Oil of eucalyptus	3	3	–[b]	3	–	10	–	10	–	–
Oil of lemon	3	3	1	1	–	–	3	3	3	3
Oil of cloves	3	1 hr	3	1 hr	3	1	3	1	3	1 hr
Oil of orange	3	3	10	10	–	10	3	3	–	–
Oil of sage	3	1	3	1	3	1	–	3	–	–
Oil of rosemary	3	3	10	10	–	3	–	3	–	–
Oil of pineneedle	3	3	3	3	–	10	–	10	–	–

[a]The killing time for the reduction of the initial germ count (10^4 microorganisms per milliliter) to less than 10^2 per milliliter is given in days; for oil of cloves this time is given in hours.
[b]–, no reduction quota to less than 10^2 microorganisms per milliliter in 20 days.
Source: Ref. 17.

III. MONOGRAPHS FOR THE ANTIMICROBIAL PRESERVATIVES

In the following section all the available information with respect to the preservatives listed in Table 8 is compiled in the form of monographs. Surprisingly, the smallest amount of toxicological data is available on the preservatives that have been in use for the longest periods of time, for example, chlorobutanol (used since the turn of the century for the preservation of injectables and ophthalmics). A large backlog of demands certainly exists here. However, the question as to who is to bear the costs for testing these products which have been "tried and proven for over 80 years" is still unanswered.

Table 8 is an aid to the practitioner in selecting the suitable preservative during the development of a new cosmetic. The most widely used preservatives can be recognized by their percentage of use in the United States. Further important criteria for selection are the optimum pH range and the antimicrobial spectrum. It is seen in Table 8 that the use of alcohols and acids is restricted to the acidic pH range. This means that only if the product to be preserved has a pH of less than 6 does it make sense to use this preservative. The spectrum of activity is not always complete and often has to be complemented by means of a combination.

Following this "rough" preselection, all further data on solubility and compatibility or incompatibility with other additives must be derived from the monographs, where all the available information about toxicity, skin tolerability, and so on, has been compiled. The latter data are more for the purpose of orientation.

Of an official list of suggestions one ought really to expect that the agents contained have overcome the toxicological hurdle. And this is the very reason why only the agents named here should be used. In the case of new products caution is advised; one ought to wait until sufficient data are available and "registration" has been granted. It is true that in the case of a novel product decomposition products may form owing to incompatibility with other starting materials, which may result in skin irritation (allergization). To avoid any surprises, the performance tests required by governmental agencies [15] should be carried out. After introduction on the market, "compatibility" in particular should be continually kept under surveillance by means of "surveys" among distributors and consumers.

CONTENTS OF SECTION III

Acetone chloroform	625
Adermykon	634
Alkyltrimethylammonium Bromide	730
Benzalkonium Chloride	731
Benzethonium Chloride	734
Benzoic Acid	670
Benzyl Alcohol	627
Benzyl Carbinol	632
Bisulfites	659
Boric Acid	661
Bromochlorophene	697
5-Bromo-5-nitro-1,3-dioxane	638
2-Bromo-2-nitropropane-1,3-diol	635
Bromophene	697
Bronidox L	638

Antimicrobial Preservatives — 621

Bronopol	635
CA 24	651
Captan	716
Carvacrol	687
Cetrimonium Bromide	730
Chloretone	625
Chlorhexidine	726
Chlorobutanol	625
p-Chloro-m-cresol	683
Chloroacetamide	650
Chlorophene	694
Chlorophenoxypropanediol	634
Chlorothymol	686
Chloroxylenol	685
4-Chloro-3,5-xylenol	684
p-Chloro-m-xylenol	684
Chlorphenesin	634
Cloflucarban	704
Cosmocil CQ	728
o-Cymenol	693
p-Cymenol	687
Dantoin 685	645
Dehydroacetic Acid	694
Diaminodiphenoxypropane	705
Diazolidinyl Urea	657
Dibromopropamidine	707
2,4-Dichlorobenzyl Alcohol	629
Dichlorophene	695
Dichloro-m-xylenol	690
Dimethoxane	718
Dimethyloldimethylhydantoin	647
4,4-Dimethyl-1,3-oxazolidine	720
Dioxin CO	719
Dowicil 200	653
Ethanol	623
Ethylmercurithiosalicylate	735
Formaldehyde	640
Formic Acid	663
Germall II	657
Germall 115	655
Glutaral	649
Glutaraldehyde	649
Glyceryl Monolaurate	740
Glydant	647
Hexachlorophene	698
Hexadienoic Acid	668
Hexahydrotriazin	714
Hexamethylenetetramine	644
Hexamidine Isethionate	708
Hexetidine	713
Hibitane	726

o-Hydroxybenzoic Acid	672
p-Hydroxybenzoic Acid Benzyl Ester	680
p-Hydroxybenzoic Acid Esters	678
8-Hydroxyquinoline	725
Imidazolidinyl Urea	655
Inorganic Sulfites	659
Irgasan CF 3	704
Irgasan DP 300	700
Isothiazoline (Mixture) (Kathon CG)	722
Isopropyl-o-cresol	687
4-Isopropyl-3-methylphenol	693
Kathon CG	722
Lauricidin	740
Liquapar	682
Methylacetic Acid	665
Monoglyceride	740
Monomethyloldimethylhydantoin	645
Mycil	634
Octopirox	742
Oxadine A	720
Parabens	678
Paraform	643
Paraformaldehyde	643
1,5-Pentandial	649
PHB Esters	678
Phenonip	682
Phenoxetol	630
2-Phenoxyethanol	630
Phenylcarbinol	627
Phenethyl Alcohol	632
Phenylmercuric Acetate	737
o-Phenylphenol	691
Piroctone Olamine	742
Polyhexamethylenebiguanide Hydrochloride	728
Propamidine Isethionate	705
Propionic Acid	665
Quaternium-15	653
Quaternary Compounds	730
Quinoline Compounds	725
Salicylic Acid	672
Sodium Iodate	739
Sodium Omadine	709
Sodium Pyrithione	709
Sorbic Acid	668
Tetraazatricyclodecane	644
Tetrabromo-o-cresol	688

Antimicrobial Preservatives 623

Thimerosal	735
Trichlorocarbanilide	702
Trichlorohydroxydiphenyl Ether	700
Trichloro-2-methyl-2-propanol	625
Triclocarban	702
Triclosan	700
Tris(hydroxyethyl)-s-triazine	714
Undecenoic Acid	667
Undecylenic Acid	667
Urotropin	644
Usnic Acid	676
Zinc Omadine	711
Zinc Pyrithione	711

ETHANOL

Chemical names:	Ethyl alcohol, ethanol. The term *ethanol* means absolute ethyl alcohol; the term *alcohol* is used for alcohol at 95% (vol/vol). Where other strengths are intended, the term *alcohol* is used followed by precision of the strength (*European Pharmacopoeia*).
Cosmetic, Toiletry, and Fragrance Association (CTFA) adopted name:	Alcohol
Type of compound:	Alcohol by fermentation of sugar or starch

1. Structure and Chemical Properties

Appearance:	Clear, colorless liquid; mobile
Odor:	Characteristic odor; burning taste
Solubility:	Miscible with water, and acetone and glycerol
Optimum pH:	Acidic pH
Stability:	Absorbs water; volatile
Compatibility and inactivation:	May be inactivated by nonionics; alcohol at 95% is incompatible with acacia, albumin, bromium, chlorine, chromic acid, permanganate
Structural formula:	C_2H_5OH Mol. wt. 46.07; bp 78-79°C
Synthesis:	Fermentation (biosynthetic)
Assay:	Gas chromatography [7]

2. Antimicrobial Spectrum

Test organisms (~10^6 CFU[b]/ml)	Minimal killing time (sec) by concentration (suspension test; contact imes of 24 and 72 hr)			Minimal inhibitory concentration (%) (serial dilution test; incubation times of 24 and 72 hr)[a] [7]		
	60%	70%	80%	5%	10%	15%
Staphylococcus aureus	15	15	15	+	−	−
Escherichia coli	60	30	30	+	−	−
Klebsiella pneumoniae		10		+	−	−
Pseudomonas aeruginosa		10		+	−	−
Pseudomonas fluorescens		10		+	−	−
Pseudomonas cepacia		10		+	−	−
Candida albicans				+	−	−
Aspergillus niger				+	−	−
Penicillium notatum				+	−	−

[a]+, growth; −, inhibition.
[b]CFU *Colony Forming Unit*.

As a preservative, only satisfactory in concentrations above 15%, better at 20%. As a disinfectant, used in concentrations of 60-70%; quick bactericidal action (less than 1 min). Inhibits some enzyme systems. FDA's Over-the-Counter (OTC) Miscellaneous External Panel for category I (safe and effective): only concentrations greater than 60% [5].

3. Toxicity

Acute oral toxicity [4]

   Rat: LD_{50}, 13.7 g/kg [8]
   Guinea pig: LD_{50}, 5.5 g/kg
   Rabbit: LD_{50}, 9.5 g/kg

Acute intravenous toxicity

   Guinea pig: [2] LD_{50}, 2.3 g/kg
   Rat: LD_{50}, 4.2 g/kg.

Primary skin irritation: In a predictive skin sensitization test a 50% alcohol solution provoked a delayed allergic skin reaction [3]

Human chronic toxicity: Daily tolerable dose, 80 g

Acceptable daily intake: 7 g/kg per day [6]

4. Cosmetic Applications	Self-protecting agent against microbiological contamination in some cosmetic products in concentrations greater than 15% (skin lotions)
5. Other Applications	Topical anti-infective, antiseptic, evaporating skin lotions, disinfectants
6. Suppliers	Union Carbide Corporation, USI Chemicals

7. References

1. K. H. Wallhäusser, *Sterilisation, Desinfektion, Konservierung.* Vol. 2, Aufl. Thieme, Stuttgart (1978) p. 391.
2. P. L. Epée, et al., *J. Med. Legal. Criminol. Police Sci. Toxicol.*, 42:141 (1962).
3. J. Stotts and W. J. Ely, *J. Invest. Dermatol.*, 69:219 (1977).
4. H. P. Fiedler, *Lexikon der Hilfsstoffe*, Editio Cantor, Aulendorf (1981).
5. *FDC Reports*, Pink Sheet *41*, No. 33, T&G, 7 August 13 (1979).
6. U. Rydberg and S. Skerfving, *Adv. Exp. Med. Biol.*, 85B:403 (1977).
7. M. Elefant and J. M. Talmage, *J. Pharm. Sci.*, 56:133 (1967).
8. Smyth, et al., *J. Ind. Hyg. Toxicol.*, 23:259 (1941).

CHLOROBUTANOL

Chemical name:	1,1,1-Trichloro-2-methylpropan-2-ol
EEC No.:	I/11
CAS No.:	57-15-8
Nonproprietary names:	Chlorobutanol, acetonechloroform
CTFA adopted name:	Chlorobutanol
Registered names:	Chloreton, chlorbutol, Methaform, Sedaform
Type of compound:	Alcohol

1. Structure and Chemical Properties

Appearance:	Colorless crystals
Odor:	Camphorlike odor
Solubility:	Slightly soluble in water (about 0.8%); very soluble in ethanol (1 g in 1 ml), propylene glycol, and liquid paraffin; soluble in glycerin (1 g in 10 ml)

Optimum pH:	Acidic pH (up to 4.0)
Stability:	Decomposed by alkalis and heat
Compatibility and inactivation:	Incompatible with some nonionics and alkalis; partial inactivation by Tween 80 and polyvinylpyrrolidone; unstable in polyethylene containers (eyedrops) [1]
Structural formula:	$C_4H_7Cl_3O$ Mol. wt. 177.47; mp 97°C

$$Cl_3C - \underset{\underset{CH_3}{|}}{\overset{\overset{OH}{|}}{C}} - CH_3$$

Assay:	Colorimetric assay [2], gas chromatography [4], USP XX (1980, p. 915), infraredspectrometry [3]

2. Antimicrobial Spectrum

Test organisms ($\sim 10^6$ CFU/ml)	Minimal germicidal concentration (μg/ml) (suspension test; contact times of 24 and 72 hr; pH 6.0)
Staphylococcus aureus	625
Escherichia coli	625
Pseudomonas aeruginosa	1000
Candida albicans	2500
Aspergillus niger	5000

3. Toxicity		Acute oral toxicity
Dog:		LD, 238 mg/kg
4. Cosmetic Applications		Use concentration up to 0.5%; prohibited in aerosol dispensers, except for foams; warning must appear on the label: Contains Chlorobutanol
5. Other Applications		Preservative in oral and cutaneous pharmaceuticals with acidic pH; use concentration, 0.3-0.5%
6. Supplier		Stauffer Chemical Company

7. References

1. W. T. Friesen and M. Elmer, *Am. J. Hosp. Pharm.*, 28:507 (1971).
2. L. Chaftez and R. W. Mahoney, *J. Pharm. Sci.*, 54:1805 (1965).
3. D. C. Healton and A. W. Davidson, *J. Assoc. Off. Anal. Chem.*, 49:850 (1966).
4. A. W. Davidson, *J. Assoc. Off. Anal. Chem.*, 50:669 (1967).

BENZYL ALCOHOL

Chemical names:	Benzyl alcohol, benzenemethanol, phenylcarbinol, phenylmethanol
EEC No.:	51
CAS No.:	100-51-6
CTFA adopted name:	Benzyl alcohol
Type of compound:	Natural alcohol, constituent of jasmine and other plants

1. Structure and Chemical Properties

Appearance:	Liquid; sharp burning taste
Odor:	Faint aromatic odor
Solubility:	1 g dissolves in about 25 ml of water; 1 vol in about 1.5 vols 50% ethanol; miscible with absolute and 94% alcohol
Optimum pH:	Above 5
Stability:	Slowly oxidizes to benzaldehyde; dehydrates at low pH
Compatibility and inactivation:	Inactivated by nonionics; partial inactivation by Tween 80
Structural formula:	C_7H_8O Mol. wt. 108.13 ⌬—CH$_2$OH
Assay:	Gas chromatography [3], *USP XX* (1980, p. 915)

2. Antimicrobial Spectrum [4]

Test organisms (~10^6 CFU/ml)	Minimal germicidal concentration (μg/ml) (suspension test; contact times of 24 and 72 hr)
Staphylococcus aureus	25
Escherichia coli	2000
Pseudomonas aeruginosa	2000
Candida albicans	2500
Aspergillus niger	5000

3. Toxicity

Acute oral toxicity

   Rat: LD_{50}, 1.23 g/kg
   Mouse: LD_{50}, 1.58 g/kg
   Rabbit: LD_{50}, 1.94 g/kg

Acute dermal toxicity

   Guinea pig: LD_{50}, 5.0 ml/kg
   High percutaneous toxicity [2]

Human toxicity: By dermal application; only a small amount resorbed by the dermis [1]

Toxicokinetic data: Metabolized to hippuric acid

4. Cosmetic Applications

Use concentration, 1.0-3.0%; EEC guideline, 1.0%

5. Other Applications

Preservative in injectable drugs, ophthalmic products, and oral liquids; use concentration, 0.5-2.0%; and solvent for cellulose acetate and shellac

6. Supplier

Norda, Inc.

7. References

1. E. Menczel and H. J. Maibach, *Acta Derm. Venereol.*, 52:38 (1972).
2. H. Wollmann et al., *Pharmazie*, 22:455 (1967).
3. C. J. Lindemann and A. Rosolia, *Pharm. Sci.*, 58:118 (1969).
4. K. H. Wallhäusser, *Sterilisation-Desinfektion-Konservierung*, Vol. 2, Aufl. Thieme, Stuttgart (1978), p. 395.

2,4-DICHLOROBENZYL ALCOHOL

Chemical names:	2,4-Dichlorobenzyl alcohol, 2,4-DCBA
EEC No.:	24
Registered names:	Dybenal, Myacide SP
Type of compound:	Alcohol

1. Structure and Chemical Properties

Appearance:	White to yellowish crystalline powder
Solubility:	In water (20°C), 0.1%; in propylene glycol, 73.0%; in polypropylene glycol, 95.0%; soluble in ethanol and isopropanol
Optimum pH:	Wide pH range (3.0-9.0)
Stability:	Oxidation in aqueous solutions
Compatibility and inactivation:	Incompatible with some anionics and some nonionics
Structural formula:	$C_7H_6Cl_2O$ Mol. wt. 177.04

2. Antimicrobial Spectrum

Test organisms (~10^6 CFU/ml)	Minimal inhibitory concentration (μg/ml) (serial dilution test; incubation times of 24 and 72 hr)
Staphylococcus aureus	1000
Escherichia coli	500
Serratia marcescens	500
Klebsiella pneumoniae	500
Pseudomonas aeruginosa	1000
Pseudomonas fluorescens	1000
Pseudomonas cepacia	1000
Candida albicans	500
Saccharomyces cerevisiae	500
Aspergillus niger	500
Penicillium notatum	250

Broad spectrum of activity, especially against yeasts and molds; combination with Bronopol or Germall 115 (see compatibility).

3. Toxicity

Acute oral toxicity

  Mouse: LD_{50}, 2.3 g/kg
  Rat: LD_{50}, 3.0 g/kg

Subchronic toxicity

  Rat: 98-day test, 7.2 and 14.4 ppm daily with the diet — no toxic effect

Primary skin irritation

  Rabbit: A 0.5% solution to the shaved flanks of rabbits for 5 consecutive days was not irritating.

Local dermal reactions

  Guinea pig: A 1% solution in contact with the shaved abdomen of guinea pigs for 4-5 hr caused little or no damage, but higher concentrations produced swelling, inflammation, and blistering.

Rabbit eye irritation: 0.8% in aqueous solution had no effect.

Guinea pig sensitization: No sensitization

Mutagenicity: Negative Ames test

Environmental toxicity

  Daphnia magna: LC_{50} (24 hr), 22.0 ppm, (48 hr) 13.1 ppm
  Rainbow trout: LC_{50} (24 hr), 18.9 ppm, (48 hr) 14.4 ppm, (72 hr) 13.3 ppm
  Mallard duck: Acute oral LD_{50}, 2.5 mg/kg

4. Cosmetic Applications   Use concentration, 0.15%; preservative in aqueous solutions, skin lotions, soft creams, and gel formulations

5. Other Applications   Antiseptic, disinfectant, antiseptic cream, antiseptic mouthwash and gargle, and sore mouth pastilles (0.06%) in combination with amyl-m-cresol (0.03%)

6. Supplier   Boots Co., Ltd., Nottingham, England

7. Reference   1. Boots Co., Ltd., July 1980.

PHENOXETOL

Chemical names: 2-Phenoxyethanol, ethylene glycol monophenyl ether

EEC No.:   43

CAS No.:   122-99-6

CTFA adopted name:	Phenoxyethanol
Registered names:	Dowanol EPH (Dow); Phenyl Cellosolve (Union Carbide), Phenoxethol, Phenoxetol, Arosol, Phenonip
Type of compound:	Phenolic derivative

1. Structure and Chemical Properties

Appearance:	Oily liquid; burning taste
Odor:	Faint aromatic odor
Solubility:	2.67% in water; miscible with alcohol (freely soluble) and propylene glycol; 2% in olive oil
Optimum pH:	Wide pH tolerance
Stability:	Fully stable
Compatibility and inactivation:	May be inactivated somewhat by nonionics; compatible with anionic and cationic detergents
Structural formula:	$C_8H_{10}O_2$ Mol. wt. 138.16 ⌬—O—CH_2CH_2OH
Synthesis:	Treating phenol with ethylene oxide in an alkaline medium

2. Antimicrobial Spectrum

Test organisms (~10^6 CFU/ml)	Minimal inhibitory concentration (μg/ml) (serial dilution test; incubation times of 24 and 72 hr)
Staphylococcus aureus	2000
Escherichia coli	4000
Klebsiella pneumoniae	4000
Pseudomonas aeruginosa	4000
Candida albicans	5000
Aspergillus niger	5000

Especially effective against gram-negative bacteria, mostly in combination with parabens to give broad-spectrum activity; active against *Pseudomonas* in high concentrations.

3. Toxicity

Acute oral toxicity

Rat: LD_{50}, 1.3 g/kg

Primary skin irritation: No irritation, in contact with wounds the daily maximum dose should be less than 40 ml (for a solution with 2.2% active ingredient).

4. Cosmetic Applications

Use concentration, 0.5-2.0%; in combination with parabens (→ Phenonip), dehydroacetic acid, or sorbic acid; mostly addition of propylene glycol for better water solubility of Phenoxetol.

5. Other Applications

Bactericide in conjunction with quaternary ammonium compounds; insect repellant; topical antiseptic [1].

6. Suppliers

Nipa Laboratories, Emery Industries

7. References

1. *Merck Index*, 9th ed., Merck, Rahway, N.J. (1976).
2. S. M. Henry and G. Jacobs, *Cosmet. Toiletries*, 96:34 (1981).

PHENETHYL ALCOHOL

Chemical names: 2-Phenylethanol, β-phenylethyl alcohol

Nonproprietary name: Benzyl carbinol

Type of compound: Natural alcohol, found in natural essential oils such as rose, hyacinth, Aleppo pine

1. Structure and Chemical Properties

Appearance: Colorless liquid; density, 1.0222-1.0230

Odor: Floral odor; rose character

Solubility: 2 ml dissolves in 100 ml water after thorough shaking; miscible with alcohol (1 part in 1 part 50% alcohol).

Optimum pH: Acidic pH preferred

Stability: Instable with oxidants

Compatibility and inactivation: Partially inactivated by nonionics and Tween 80

Structural formula: $C_8H_{10}O$
Mol. wt. 122.16

Antimicrobial Preservatives

$$\text{C}_6\text{H}_5-\overset{\beta}{\text{CH}_2}\overset{\alpha}{\text{CH}_2}\text{OH}$$

Synthesis: Reduction of ethyl phenylacetate with sodium in absolute alcohol

Assay: Gas chromatography [3]

2. Antimicrobial Spectrum

Test organisms (~10^6 CFU/ml)	Minimal inhibitory concentration (μg/ml) (serial dilution test; incubation times of 24 and 72 hr; pH 6.2)
Staphylococcus aureus	1250
Escherichia coli	2500
Klebsiella pneumoniae	2500
Pseudomonas aeruginosa	2500-5000
Pseudomonas fluorescens	2500
Candida albicans	2500
Aspergillus niger	5000
Penicillium notatum	5000

Mostly against gram-positive and gram-negative bacteria; only small activity against yeasts and molds. Phenylethyl alcohol enhances the permeability of the bacterial cell wall [2]. Used in combination with benzalkonium chloride, chlorhexidine, parabens, chlorobutanol, and chlorocresol.

3. Toxicity

Acute oral toxicity

  Rat: LD_{50}, 1.79 g/kg [1]

Acute dermal toxicity

  Guinea pig: LD_{50}, 5-10 ml/kg

Human eye irritation: A concentration of 0.75% provoked irritation.

Guinea pig sensitization: No sensitization in concentrations of 1-2%

Teratogenicity: No effect

4. Cosmetic Applications

Eye makeup (1%)

5. Other Applications

Antimicrobial agent in pharmaceuticals; preservative in eyedrops, 0.3% in combination with 0.01% benzalkonium chloride: cutaneous antiseptic; oral pharmaceutical products (0.3-0.5%)

6. References

1. Jenner et al., *Food Cosmet. Toxicol.*, 2:327 (1964).
2. S. Silver and L. Wendt, *J. Bacteriol.*, 93:560 (1967).
3. C. J. Lindemann and A. Rosolia, *J. Pharm. Sci.*, 58:118 (1969).

CHLORPHENESIN

Chemical name:	3-(4-Chlorophenoxy)-1,2-propanediol
EEC No.:	3
CAS No.:	104-29-0
Registered names:	Mycil, Adermykon, Chlorphenesin, Geophen
Type of compound:	Alcohol

1. Structure and Chemical Properties

Appearance:	White crystalline powder, mp 77-79°C
Odor:	Odorless
Solubility:	Soluble in water less than 1%; ethylurea and propylene glycol act as solubilizers; soluble in alcohol
Optimum pH:	4-6
Structural formula:	$C_9H_{11}ClO_3$ Mol. wt. 202.64

$$Cl-\langle\text{C}_6\text{H}_4\rangle-O-CH_2\overset{OH}{\underset{|}{CH}}-CH_2OH$$

Synthesis: Prepared by condensing equimolar amounts of p-chlorophenol and glycidol in the presence of a tertiary amine or a quaternary ammonium salt as catalyst

2. Antimicrobial Spectrum

Test organisms ($\sim 10^6$ CFU/ml)	Minimal inhibitory concentration (μg/ml) (serial dilution test; incubation times of 24 and 72 hr)
Staphylococcus aureus	1250
Escherichia coli	1250
Klebsiella pneumoniae	1250
Pseudomonas aeruginosa	2500
Candida albicans	1250
Aspergillus niger	2500

Antimicrobial Preservatives

Chlorphenesin is a potent antibacterial and antifungal agent, including the common dermatophytes with low toxicity [1,2].

3. Toxicity — Low toxicity [2]

4. Cosmetic Applications — EEC use concentration, 0.5%

5. Other Applications — Topical antifungal

6. References
   1. W. B. Hugo, *Inhibition and Destruction of the Microbial Cell*, Academic, London (1971), p. 665.
   2. F. Q. Hartley, *J. Pharm. Pharmacol.*, 20:388 (1947).

BRONOPOL

Chemical name:	2-Bromo-2-nitropropane-1,3-diol
EEC No.:	19
CAS No.:	52-51-7
CTFA adopted name:	2-Bromo-2-nitropropane-1,3-diol
Registered name:	Bronopol
Type of compound:	Alcohol

1. Structure and Chemical Properties

Appearance:	White crystalline powder; can be irritant in concentrated form
Odor:	Faint characteristic odor
Solubility:	Percentages (wt/vol) at 22-25°C: water, 25; ethanol, 50; isopropanol, 25; glycerol, 1; polyethyleneglycol 300, 11.0; propylene-glycol, 14.0; liquid paraffin, <0.5; cottonseed oil, <0.5; olive oil, <0.5
Optimum pH:	5.0-7.0
Stability:	Can be stored at room temperature in the dark for up to 2 years with no decomposition; stable in aqueous solution at low pH; decomposition by increasing pH or temperature.

	Percentage Bronopol remaining (microbiological assay)
Effect of temperature (°C; 10 weeks; pH 6.0)	
30	87
40	38
50	8

	Percentage Bronopol remaining (microbiological assay)
Effect of pH (10 weeks; 40°C)	
pH 4	94
pH 6	37
pH 8	5

Under alkaline conditions and when exposed to light, Bronopol may become yellow or brown. In contact with iron and aluminum, deactivation, discoloration, and corrosion follow; stable with stainless steel and tin. Bronopol decomposes under these conditions to formaldehyde and bromide ion. In the presence of certain amines, nitrite (Bronopol can liberate nitrite on decomposition) and nitrosamine can be formed which may be carcinogenic.

Compatibility and inactivation: Bronopol is not adversely affected by anionic, cationic, or nonionic surfactants or proteins. A 10% serum or milk has little effect on the activity of Bronopol, whereas a 50% serum reduces the activity 4-8 times, and horse blood 32-64 times. Polysorbate 80 (1%), 5% Tween 80 (nonionic), 5% Span 80 (nonionic), 5% Empicol MD (anionic), and 0.1% lecithin have no significant effect on the bacteriostatic activity of Bronopol. Sulfhydryl compounds (cysteine, thioglycolate) thiosulfate, and metabisulfite are markedly antagonistic; decrease 8- to 64-fold. Cystine hydrochloride at 0.1% has been used as an inactivating agent in recovery medium for the bactericidal test.

Structural formula: $C_3H_6BrNO_4$
Mol. wt. 169.97

$$HOCH_2-\underset{NO_2}{\overset{Br}{C}}-CH_2OH$$

Assay: Thin-layer chromatography, polarography, microbiological assay, gas-liquid chromatography

Antimicrobial Preservatives

2. Antimicrobial Spectrum

Test organisms (~10^6 CFU/ml)	Minimal inhibitory concentration (μg/ml) (serial dilution test; incubation times of 24 and 72 hr)
Staphylococcus aureus	62.5
Escherichia coli	31.25
Klebsiella pneumoniae	62.5
Pseudomonas aeruginosa	31.25
Pseudomonas fluorescens	31.25
Pseudomonas cepacia	31.25
Candida albicans	200-1000
Aspergillus niger	200-1000
Penicillium notatum	200-1000

Bronopol is bactericidal in 24 hr at 37°C at concentrations only slightly higher (two- to fourfold) than bacteriostatic levels. It is more active against gram-negative bacteria than against gram-positive bacteria. It demonstrates low sporicidal activity. Bronopol forms disulfide bonds from thiol groups and these may account for the observed inhibition of dehydrogenase activity.

3. Toxicity

Acute oral toxicity

    Mouse (male): LD_{50}, 374 mg/kg
    Rat (male): LD_{50}, 307 mg/kg

Acute intraperitoneal toxicity

    Rat (male): LD_{50}, 22 mg/kg

Acute dermal toxicity: Rat (acetone solution) death at 160 mg/kg

Primary skin irritation: A 0.5% solution in acetone is nonirritant; 2.5% aqueous methylcellulose solution is nonirritant; 5.0% solution in polyethyleneglycol 300 is nonirritant.

Rabbit eye irritation: 0.5% in normal saline is nonirritant when used once daily for 4 successive days; 2% in polyethylene glycol 400 is nonirritant (single application), but 5% is irritant.

Guinea pig sensitization (Magnusson-Kligman test): Three challenges were necessary before 2 out of 10 animals became sensitized.

Human sensitization: A volunteer study showed that Bronopol is slightly irritant to human skin at 1% in soft paraffin and at 0.25% in aqueous buffer at pH 5.5 (closed-patch test). On patients 0.25% Bronopol in soft yellow paraffin is a mild irritant.
Maibach [2] confirmed that Bronopol caused direct irritation to human skin at concentrations above 1%.

Chronic toxicity (90-day test):	Daily oral doses of 20 mg/kg administered to male and female rats were well tolerated. Doses of 80 and 160 mg/kg caused gastrointestinal lesions, respiratory distress, and some deaths.
Carcinogenicity:	Application of 0.3 ml of aqueous acetone solutions containing 0.2 or 0.5% Bronopol to the shaved backs of mice three times weekly for 80 weeks: Bronopol did not alter the spontaneous tumor profile, either locally or systemically. A 2-year toxicity and tumorgenicity test, in which rats received 10, 40, or 160 mg/kg daily in drinking water, provided no evidence to suggest that Bronopol affected tumor incidence.
Mutagenicity:	No mutagenic activity (Ames-type assay and host-medicated assay in mice)
4. Cosmetic Applications	Concentrations of 0.01-0.1% in hand and face creams, shampoos, hair dressings, mascaras, and bath essence formulations
5. Other Applications	Pharmaceutical products, household products (fabric conditioners and washing detergents)
6. Supplier	Boots Co., Ltd., Nottingham, England
7. References	1. Bronopol, Boots Technical Bulletin Issue 3, September 1979. 2. H. J. Maibach, *Contact Dermatitis*, 3: 99 (1977).

BRONIDOX L

Chemical name:	5-Bromo-5-nitro-1,3-dioxane
EEC No.:	18
CTFA adopted name:	Bronidox L
Registered name:	Bronidox L
Type of compound:	o-Acetal, o-formal

1. Structure and Chemical Properties [1]

Appearance:	Solution, 10% (wt/vol) active ingredient in propylene glycol
Solubility (20°C):	Above 25% in ethanol, above 10% in isopropanol, 0.46% (wt/vol) in water, above 10.0% (wt/vol) in propyleneglycol; soluble in vegetable oils; insoluble in paraffin oil

Antimicrobial Preservatives

Optimum pH: 5-7

Stability: Unstable at pH <5 and temperature above 50°C; corrosive to metal containers

Compatibility and inactivation: Incompatible with cysteine; compatible with nonionics; in the presence of proteins a higher concentration (0.5%) is required

Structural formula:

   $C_4H_6BrNO_4$
Mol. wt. 212.0

2. Antimicrobial Spectrum

Test organisms (~10^6 CFU/ml)	Minimal inhibitory concentration (μg/ml) (serial dilution test; incubation times of 24 and 72 hr)
Staphylococcus albus	50
Staphylococcus aureus	75
Streptococcus faecalis	75
Escherichia coli	50
Proteus vulgaris	50
Serratia marcescens	25
Pseudomonas aeruginosa	50
Pseudomonas fluorescens	50
Candida albicans	25
Saccharomyces cerevisiae	10
Aspergillus niger	10
Penicillium camerun	25

The antimicrobial activity of Bronidox L is probably based on the oxidation of thiol groups in enzyme systems and mercapto amino acids. Under normal conditions (pH 5-7, below 40°C) no formaldehyde splits off. Bronidox L has a broad spectrum of antimicrobial activity.

3. Toxicity [1]

Acute oral toxicity

Mouse: LD_{50}, 590 mg/kg
Rat: LD_{50}, 455 mg/kg

Subacute dermal toxicity

Rat: After 15 weeks, 10, 50, and 100 mg/kg per day had no effect; with 200 mg/kg per day some deaths occurred after a few doses.

Primary skin irritation: In concentrations above 0.5%, skin irritation occurs; a 0.5% suspension in olive oil provoked no reaction.

Human skin:	Patch test at 0.1%, no irritation; 0.5% in suspension and 0.25% in Vaseline showed irritation. Partial resorption (cutaneous), some metabolites in urine.
Rabbit eye irritation:	Ten applications of 0.05% produced no reaction; irritation threshold about 0.1%.
Guinea pig sensitization:	No sensitization

4. Cosmetic Applications — In use concentration, 0.1% in EEC guideline; only for rinse-off products

5. Other Applications — Preservative for technical products

6. Supplier — Henkel KGA Düsseldorf

7. Reference
1. M. Potokar et al., *Fette Seifen Anstrichm.*, 78:269 (1976).

FORMALDEHYDE

Chemical names:	Formaldehyde; solution: Formalin 37% by weight of formaldehyde gas in water, usually 10-15% methanol added to prevent polymerization
EEC No.:	I/5
CAS No.:	50-00-0
CTFA adopted name:	Formaldehyde
Registered names:	Formalin, Formol
Type of compound:	Aldehyde

1. Structure and Chemical Properties

Appearance (Formalin):	Colorless liquid; powerful reducing agent, especially in the presence of alkali; keep container well closed; density, 1.12
Odor:	Pungent odor
Solubility:	Freely soluble in water
Optimum pH:	Broad range; pH 3-10; Formalin pH 2.8-4.0
Stability:	In the cold it may become cloudy; at very low temperatures trioxymethylene is formed (→ paraformaldehyde); in the air it slowly oxidizes to formic acid.
Compatibility and inactivation:	Incompatible with ammonium, alkali, H_2O_2, iodine, potassium permanganate, tannin, iron, gelatin, heavy metals-salts; compatible with anionics, cationics, and nonionics; inactivation by proteins

Structural formula: CH_2O

Mol. wt. 30.03

Assay: Photometry [3], gas chromatography [4], chromatography with Draeger tube [6]

2. Antimicrobial Spectrum

Test organisms (~10^6 CFU/ml)	Minimal germicidal concentration (μg/ml) (suspension test; contact times of 24 and 72 hr)	Minimal inhibitory concentration (μg/ml) (serial dilution test; incubation times of 24 and 72 hr)
Staphylococcus aureus	62.5	125
Escherichia coli	31.25	125
Klebsiella pneumoniae	31.25	125
Pseudomonas aeruginosa	62.5	125
Pseudomonas fluorescens	62.5	125
Pseudomonas cepacia	62.5	125
Candida albicans	250	500
Aspergillus niger	500	500
Penicillium notatum	500	500

3. Toxicity [1,5]

Acute oral toxicity

  Rat: LD_{50}, 800 mg/kg
  Guinea pig: LD_{50}, 260 mg/kg

Acute subcutaneous toxicity

  Mouse: LD_{50}, 300 mg/kg
  Dog: LD, 800 mg/kg

Inhalation toxicity

  Rat: 250 ppm (4 hr)
  Cat: LD, 800 mg/kg
  Man: LD, 36 mg/kg

Human toxicity: Vapor intensely irritating to mucous membranes. The current permissible exposure limit for formaldehyde is 3 ppm, as set by the OSHA (in Germany 1 ppm). NIOSH has recommended a limit of 1 ppm. Irritation effects on the eye, nose, and throat between 0.03 and 4 ppm up to 35 min. Ef-

Human toxicity: (Continued)	fects on human skin are irritation (nonimmunological by mediated dermatitis), allergic contact dermatitis, contact urticaria. Allergy caused by 1-2% solutions and through inhalation exposure. The threshold for this effect is in the range of 0.3% solutions. 1-4% of all people will give a positive patch test to 2% formaldehyde. Four out of nine people sensitized to formaldehyde will show a positive patch test to a 30 ppm solution [2].
Mutagenicity:	Ames test, negative; mouse lymphoma, positive; sister chromatid exchange, positive; cell transformation 10 T 1/2, negative; with TPA positive. Chromosomal aberrations in bone marrow, negative; *Drosophila*, positive [2].
Carcinogenicity:	Rats, inhalation of formaldehyde 6 hr/day, 5 days/week for 24 months: 2 ppm, no effect; 6 and 15 ppm, 1.5% and 43.2% tumor frequency (most of the tumors were squamous cell carcinomas, a few were adenomatous polyps). In the same experiments with mice there was no effect due to 2 and 6 ppm, and a tumor frequency of 2.4% by 15 ppm formaldehyde [2]. These studies failed to demonstrate any significant relationship between exposure to formaldehyde and excess cancer among morticians and plant workers.
4. Cosmetic Applications	Preservative for shampoos (0.1-0.2% formaldehyde) only with warning on the label: Contains Formaldehyde (if the concentration is greater than 0.05%) in EEC.
5. Other Applications	Disinfectant (rooms, plants), antiseptic, preservative (detergents, emulsions, hides, rubber, Latex), steam formaldehyde disinfection in hospitals (beds).
6. Supplier	J. T. Baker Chemical Co., Tenneco Chemicals
7. References	1. *Merck Index*, 9th ed., Rahway, N.J. (1976). 2. CIIT, Conference on Formaldehyde Toxicity, November 20 (1980). 3. G. Durand et al., *Anal. Biochem.*, *61*:232 (1974). 4. H. P. Harke, *Zentralbl. Bakteriol. Abt. I Orig. B*, *164*:279 (1977).

Antimicrobial Preservatives

7. References (Continued)

5. H. P. Fiedler, *Lexikon der Hilfsstoffe*, Vol. 1, Editio Cantor, Aulendorf (1981).
6. Quick test with Draeger-tube (Dräger-Röhrchen) Dräger-Werk, Lübeck, Germany.

PARAFORMALDEHYDE

Chemical name:	Polyoxymethylene
EEC No.:	I/5
CAS No.:	9002-81-7
Nonproprietary name:	Paraform
CTFA adopted name:	Polyoxymethylene
Registered names:	Triformol, Formagene, Foromycen
Type of compound:	Polymerized formaldehyde

1. Structure and Chemical Properties

Appearance:	White crystalline powder
Odor:	Formaldehyde odor
Solubility:	Slowly soluble in cold; more readily soluble in hot water, with evolution of formaldehyde; insoluble in alcohol; soluble in fixed alkali solutions; keep container tightly closed!
Optimum pH:	4-8
Stability:	Instable in alkali solutions
Compatibility and inactivation:	Compatible with anionics and nonionic detergents; incompatible with proteins, NH_3, oxidants, and heavy metals
Structural formula:	$(CH_2O)_n$
Assay:	See Ref. 1.

2. Antimicrobial Spectrum

See formaldehyde.

3. Toxicity

Acute oral toxicity (see formaldehyde)

Rat:	LD_{50}, 800 mg/kg

4. Cosmetic Applications

Preservative, maximum concentration of 0.2%

5. Other Applications

Disinfectants for sickrooms

6. Supplier   Degussa

7. References   1. B. Schilling et al., *Pharmazie*, *33*:103
                                         (1978).

HEXAMETHYLENETETRAMINE

Chemical name: 1,3,5,7-Tetraazatricyclo[3,3,1,1^{3,7}]decane

EEC No.:   44

CAS No.:   100-97-0

Nonproprietary names: Methenamine, hexamethylenetetramine,
                                      HMT, Urotropin

Registered names: Aminoform, Formin, Uritone, Cystamin

Type of compound: Formaldehyde donor, N-acetal

1. Structure and Chemical Properties

Appearance: Crystals, granules, or powder; hygro-
                                      scopic; volatile at low temperature

Odor: Odorless

Solubility: 1 g dissolves in 1.5 ml of water or 12.5 ml
                                      of alcohol

Optimum pH: pH of 0.2 M aqueous solution 8.4

Stability: Split formaldehyde at acidic pH

Compatibility and inactivation: Compatible with anionics, cationics, non-
                                      ionic detergents and proteins

Structural formula: $C_6H_{12}N_4$
                                      Mol. wt. 140.19

Synthesis: Reaction of formaldehyde with ammonia in
                                      aqueous solution

Assay: Photometric test: reaction of the released
                                      formaldehyde with chromotropic acid (1,8-
                                      dinaphthalene-3,6-disulfonic acid) [1]

2. Antimicrobial Spectrum

Hexamethylenetetramine itself shows no
antimicrobial activity; responsible for this
is the formaldehyde split under hydrolytic
conditions in an acidic milieu.

Antimicrobial Preservatives

3. Toxicity

Acute subcutaneous toxicity

Rat:	LD_{50}, 200 mg/kg
Subchronic toxicity:	90-day test (rat) with 0.4 g/day produced no toxic effect [2]; only the animals' coats were yellow.
Chronic toxicity:	Daily addition of 1% hexamethylenetetramine to the drinking water over 60 weeks showed no effect on mice and rats [3].
Acceptable daily intake:	0-0.15 mg/kg per day
Carcinogenicity:	No carcinogenic effect [3] by oral application; repeated injections provoked sarcoma [4].
4. Cosmetic Applications	Preservation of lotions and creams (0.2%); if the concentration is higher than 0.05% free formaldehyde, warning Contains Formaldehyde must appear on the label.
5. Other Applications	Preservation of hides; agent for hardening phenol-formaldehyde resin; corrosion inhibitor for steel; antibacterial (urinary)
6. Suppliers	Merck, Eastman

7. References

1. E. Bremanis, Die photometrische Bestimmung des Formaldehyds mit Chromotropsäure, *Z. Anal. Chem.*, 130:44-47 (1949).
2. R. Brendel, *Arzneim. Forsch.*, 14:51-53 (1964).
3. G. Della Porta, M. J. Colnaghi, and G. Parmiani, *Food Cosmet. Toxicol.*, 6:707-715 (1968).
4. F. Watanabe and S. Sugimoto, *Gan No Rinsho*, 46:365-366 (1955).

MONOMETHYLOLDIMETHYLHYDANTOIN

Chemical names:	1-(Hydroxymethyl)-5,5-dimethylhydantoin, 1-(hydroxymethyl)-5,5-dimethyl-2,4-imidazolidinedione
EEC No.:	39
CAS No.:	28453-33-0
CTFA adopted name:	MDM hydantoin
Registered name:	Dantoin 685
Type of compound:	Nonionic cyclic, formaldehyde donor, N-acetal

1. Structure and Chemical Properties

Appearance:	Crystals; mp 100°; contains about 19% formaldehyde
Odor:	Odorless
Solubility:	Water soluble; soluble in ethanol or methanol
Optimum pH:	4.5-9.5
Stability:	Stable at temperatures below 85°C. Formaldehyde is split off at pH 6 in aqueous solutions.
Structural formula:	$C_6H_{10}N_2O_3$ Mol. wt. 158.16

$$\underset{HN}{\overset{CH_3}{\underset{CH_3}{>}}}\!\!\!\!\!\!\!\!\!\!\underset{O}{\overset{O}{\bigcirc}}\!\!\!\!N\text{-}CH_2OH$$

2. Antimicrobial Spectrum

Test organisms ($\sim 10^6$ CFU/ml)	Minimal inhibitory concentration ($\mu g/ml$) (serial dilution test; incubation times of 24 and 72 hr)
Staphylococcus aureus	512
Escherichia coli	1024
Klebsiella pneumoniae	1024
Pseudomonas aeruginosa	2048
Pseudomonas fluorescens	1024
Pseudomonas cepacia	2048
Candida albicans	2048
Aspergillus niger	2048
Penicillium notatum	1024

3. Cosmetic Applications

Use concentration, 0.25%; EEC, 0.2% preservative for shampoos, active compound in deodorants; 0.2% expressed as free formaldehyde or theoretically available formaldehyde; only for rinse-off products

4. Other Applications

Preservative for technical products (cutting oils)

5. Suppliers	Glyco Inc., General Aniline, Rex Campbell
6. References	1. H. P. Fiedler, *Lexikon der Hilfsstoffe*, Vol. 1, Editio Cantor, Aulendorf (1981).

DIMETHYLOLDIMETHYLHYDANTOIN

Chemical names:	Dimethyloldimethylhydantoin, 1,3-dimethylol-5,5-dimethylhydantoin, 1,3-bis-(hydroxymethyl)-5,5'-dimethyl-2,4-imidazolidinedione
EEC No.:	50
CAS No.:	6440-58-0
CTFA adopted name:	DMDM hydantoin
Registered names:	Glydant, Dantoin 55% solution, DMDMH-55
Type of compound:	Nonionic cyclic, formaldehyde donor

1. Structure and Chemical Properties

Appearance:	Glydant 55: total formaldehyde, 17.7%; free available, 2.12%; at the in-use concentration of 0.2%, only 0.0042% formaldehyde available
Odor:	Formaldehyde odor
Solubility:	Freely soluble in water (>50%)
Optimum pH:	3.5-10.0; pH of the 55% solution; 6.5-7.5
Stability:	Stable over wide pH range and temperature conditions (<90°C)
Compatibility and inactivation:	Compatible with anionics, cationics, nonionics, and proteins
Structural formula:	$C_7H_{12}O_4N_2$ Mol. wt. 188.12

2. Antimicrobial Spectrum

Test organisms (~10^6 CFU/ml)	Minimal germicidal concentration (μg/ml) (suspension test; contact times of 24 and 72 hr)	Minimal inhibitory concentration (μg/ml) (serial dilution test; incubation times of 24 and 72 hr)
Staphylococcus aureus	4000	250-800
Escherichia coli	4000	500
Proteus mirabilis	4000	800
Klebsiella pneumoniae		500
Pseudomonas aeruginosa	4000	800-1000
Pseudomonas fluorescens		800-1000
Candida albicans	5000	725-1250
Aspergillus niger		750-1500
Penicillium notatum		750-1500

Broad spectrum of activity, but preferable against bacteria and only in high concentrations against fungi; combination with antifungal components necessary, such as parabens, Kathon CG, and formaldehyde.

3. Toxicity

Acute oral toxicity [2]

    Rat (female): LD_{50}, 3.8 g/kg
    Rat (male): LD_{50}, 2.7 g/kg

Acute dermal toxicity

    Rabbit: LD_{50}, >20 g/kg

Subacute dermal toxicity (28-day test):

    Rabbit: Daily dermal applications of aqueous solutions at 4000 and 400,000 ppm concentration levels: no evidence of abnormalities

Primary skin irritation: No effect (rabbit)

Human patch test: 4000 ppm diluted in tap water during 9- and 24-hr occluded induction patch applications provoked no irritation [3,4].

Rabbit eye irritation: No irritation using a 1.0% wt/vol solution

Phototoxicity: None; not a photoallergic agent

Human patch test: No sensitization with 4000 ppm (50 persons) [1]

Mutagenicity: Ames Salmonella/microsome plate test, not mutagenic under these test conditions.

4. Cosmetic Applications	Use concentration, 0.15-0.4%; EEC, 0.2%; shampoos, henna cream conditioner, in creams (hand cream) sometimes a lag against yeast and molds at the highest concentration (0.4%); only to be used with a warning on the label (Contains Formaldehyde).
5. Other Applications	Preservative agent in detergents
6. Suppliers	Glyco Inc., Narden (Jan Dekker)
7. References	1. R. J. Schanno et al., *J. Soc. Cosmet. Chem.*, 31:85 (1980). 2. H. P. Fiedler, *Lexikon der Hilfsstoffe*, Vol. 1, Editio Cantor, Aulendorf (1981). 3. Warf Institute Report No. 40-69 7 IK-419-7, September 30, 1976, No. 605-1229. 4. J. H. Kay and J. C. Calandra, *J. Soc. Cosmet. Chem.*, 281-289 (1962).

GLUTARALDEHYDE

Chemical names:	Glutaraldehyde, glutardialdehyde, 1,5-pentandial, dioxopentane
CTFA adopted name:	Glutaral
Registered names:	Ucarcide, Alhydex
Type of compound:	Dialdehyde

1. Structure and Chemical Properties [4]

Appearance:	Oily liquid; commercially a 25% solution, stabilized with ethanol or alkaline pH, stabilized by 0.1-0.25% hydroquinone
Odor:	Pungent odor
Solubility:	Slightly soluble in water
Optimum pH:	Broad pH range; optimal at pH 5; for bactericidal activity pH 7.5-8.5 [5]
Stability:	Polymerization in the presence of water; stabilization by the addition of ethanol
Compatibility and inactivation:	Inactivated by ammonia or primary amines at neutral to basic pH values
Structural formula:	$C_5H_8O_2$ Mol. wt. 100.12

$$OHC-(CH_2)_3-CHO$$

Assay:	Quantitative method [3]

2. Antimicrobial Spectrum

Compared with 4% aqueous formaldehyde, 2% aqueous glutaraldehyde is 10 times as effective as a bactericidal and sporicidal agent; 0.2% kills 10^4 gram-positive or gram-negative vegetative bacteria within 20 min. As with formaldehyde, sodium bisulfite is an effective neutralizing agent in bactericidal testing. The activity of glutaraldehyde is not diminished in the presence of 10% serum. The antimicrobial activity is better at neutral to alkaline pH. Bactericidal only when rendered alkaline to pH 7.5-8.5 [5].

3. Toxicity

Acute oral toxicity

Rat:	LD_{50}, 60 mg/kg; LD_{50} (25% solution), 2.38 mg/kg

Inhalation toxicity

Rat:	8 hr in saturated glutaraldehyde atmosphere caused no deaths.
Primary skin irritation:	Human contact dermatitis observed [1]; 550 ppm without effect (706 persons)
Mutagenicity (mouse):	Not mutagenic [2]
4. Cosmetic Applications	Use concentration, 0.02-0.2% (of 50% solution)
5. Other Applications	Disinfectant for instruments and equipment; preservative at 0.2%, pH about 5; corrosive to metals
6. Supplier	Union Carbide Corporation

7. References

1. J. E. Weaver and H. J. Maibach, *Contact Dermatitis*, 3:65 (1977).
2. M. Tamada et al., *Bokin Bobai*, 6:62 (1978).
3. J. Haidu and P. Friedrich, *Anal. Biochem.*, 65:273 (1975).
4. H. P. Fiedler, *Lexikon der Hilfsstoffe*, Vol. 1, Editio Cantor, Aulendorf, (1981).
5. A. A. Stonehill et al., *Am. J. Hosp. Pharm.*, 20:458 (1963).

CHLOROACETAMIDE

Chemical names:	2-Chloroacetamide, CA 24 (70% chloroacetamide + 30% sodium benzoate)
EEC No.:	22

Antimicrobial Preservatives

CAS No.: 79-07-2
Registered name: CA 24
Type of compound: Acidamide

1. Structure and Chemical Properties

Appearance:	White crystalline powder; mp 119-120°C
Odor:	Odorless
Solubility:	Soluble in 20 parts water and in 10 parts absolute alcohol
Optimum pH:	4-8
Stability:	Stable
Compatibility and inactivation:	Compatible with anionics, cationics, and nonionics; incompatible with strong acids and alkalis (saponification)
Structural formula:	C_2H_4ClNO
	Mol. wt. 93.51

$$Cl-CH_2-\overset{\overset{O}{\|}}{C}-NH_2$$

Synthesis:	Preparation from ethyl chloroacetate and ammonia
Assay:	See E. Calvet, *J. Chim. Phys.*, 30:160 (1933).

2. Antimicrobial Spectrum

Test organisms ($\sim 10^6$ CFU/ml)	Minimal germicidal concentration (μg/ml) (suspension test; contact times of 24 and 72 hr)	Minimal inhibitory concentration (μg/ml) (serial dilution test; incubation times of 24 and 72 hr)
Staphylococcus aureus	5000	2000
Escherichia coli	3000	3000
Klebsiella pneumoniae	3000	3000
Pseudomonas aeruginosa	5000	2500
Candida albicans	2500	500
Aspergillus niger	5000	500
Penicillium notatum	2500	500

The antimicrobial activity of chloroacetamide is increased in the presence of detergents, such as lauryl sulfate.

3. Toxicity

Acute oral toxicity

  Mouse: LD_{50}, 150 mg/kg [1]

  Rat: LD_{50}, 138 mg/kg

Subacute oral toxicity

  Rat: 12.5 and 50 mg/kg per day over 13 weeks, no external reaction but testicular atrophy [11]

Subchronic dermal toxicity

  Rabbit: 30 times 50 mg/kg per day, no pathological reaction [3], no testicular atrophy observed [3]

Primary skin irritation: No irritation up to 10%; with rabbits, no irritation with 5% solution [7]

Human skin test: 0.1% solution in water, 200 persons with allergic reactions after 24, 48, and 72 hr, no irritation [2]

Rabbit eye irritation: No irritation with 5% solution [8]

Guinea pig sensitization: 1% solution applied nine times produced no effect; Buehler test, 0.3% solution applied three times over 3 weeks, no sensitization [9]

Mutagenicity: Ames test, negative [4,10], 1000 µg per agar plate; Dominant lethal test, 114 mg/kg i.p., no mutagenicity [5]

Carcinogenicity: Micronucleus test twice 50 mg/kg: no increase of structural and numeric chromosome aberrations and micronuclei [6]

4. Cosmetic Applications

Preservative EEC, 0.3%; shampoos, bath lotions; only with warning on the label (Contains Chloroacetamide)

5. Other Applications

Preservatives for emulsions, cutting oils, hides, paintings

6. Suppliers

Hoechst A.G., Werk Gersthofen; CA 24, Biochema Schwaben

7. References

1. Abteilung für Gewerbetoxikologie der Hoechst A.G., September 4, 1964.
2. Report of Professor Röckl, Würzburg, November 16, 1970.
3. IBR (International Bioresearch Institute), Hannover Project, March 1971.

7. References (Continued)

4. IBR Report, Project, March 14, 1979.
5. IBR Report, Project No. 2-1-183-79, June 28, 1979.
6. IBR Report, October 1979/Kn.
7. IBR Report, Project No. 1-3-418-81, June 1981.
8. IBR Report, Project No. 1-3-417-81, 1981.
9. IBR Report, Project No. 2-5-419-81, August 1981.
10. Hoechst A.G. Report, *Arbeitsgr. Molekularbiologie*, May 28, 1979.
11. Dr. Leuchner, *Labor für Pharmakologie und Toxikologie*, Hamburg, Report March 12, 1970.

QUATERNIUM -15

Chemical names:	Cis isomer of 1-(3-chloroallyl)-3,5,7-triaza-1-azoniaadamantane-chloride, N-(3-chloroallyl)-hexammonium chloride
EEC No.:	48
CAS No.:	4080-31-3
CTFA adopted name:	Quaternium 15
Registered names:	Dowicil 200; Dowicide Q, Preventol D1
Type of compound:	Quaternary adamantane

1. Structure and Chemical Properties [2,3]

Appearance:	Cream-colored powder; hygroscopic
Odor:	Odorless
Solubility:	127 g soluble in 100 ml water (freely soluble); concentrated solutions are yellow; 18.7% in propylene glycol, 12.6% in glycerol, 1.85% in ethanol, 0.25% in isopropanol, 0.1% in paraffin oil
Optimum pH:	Broad range, 4-10
Stability:	Unstable above 60°C, coloration (yellowish) in cream formulations
Compatibility and inactivation:	Compatible with anionics, nonionics, cationics and proteins
Structural formula:	$C_9H_{16}Cl_2N_4$ Mol. wt. 251.17

$$\left[\begin{array}{c}\text{CH}_2\text{-CH=CHCl}\\ \overset{\oplus}{\text{N}}\\ \text{N} \quad \text{N}\\ \text{N}\end{array}\right] \text{Cl}^{\ominus}$$

2. Antimicrobial Spectrum

Test organisms ($\sim 10^6$ CFU/ml)	Minimal inhibitory concentration (μg/ml) (serial dilution test; incubation times of 24 and 72 hr)
Staphylococcus aureus	200
Escherichia coli	500
Klebsiella pneumoniae	750
Pseudomonas aeruginosa	1000
Pseudomonas fluorescens	1000
Pseudomonas cepacia	1000
Candida albicans	>3000
Aspergillus niger	>3000
Penicillium notatum	>3000

More effective against bacteria (minimal inhibitory concentration, 200-1000 ppm) than against molds and yeasts (>3000 ppm) [1].

3. Toxicity

Acute oral toxicity

Rat:	LD_{50}, 0.94-1.5 g/kg
Rabbit:	LD_{50}, 40-80 g/kg
Primary skin irritation:	Not a primary skin irritant (human skin)
Guinea pig sensitization:	Nonsensitizing at concentrations up to 2%
Mutagenicity:	Nonmutagenic (Ames test and unscheduled DNA synthesis)
4. Cosmetic Applications	Does not release gaseous formaldehyde; in-use concentration, 0.02-0.3%; EEC guideline; 0.2% in shampoos (0.1-0.2%), hair care products, and lotions
5. Other Applications	Preservative for technical products
6. Supplier	Dow Chemical Co.
7. References	1. K. H. Wallhäusser, *Parfuem. Kosmet.*, 60:1 (1979).

7. References
   (Continued)

2. *Merck Index*, 9th ed., Rahway, N.J. (1976).
3. S. M. Henry and G. Jacobs, *Cosmet. Toiletries*, *96*:31 (1981).

IMIDAZOLIDINYL UREA

Chemical name:	N,N'-methylene-bis-[N'-[1-(hydroxymethyl)-2,5-dioxo-4-imidazolidinyl]urea
EEC No.:	36
CAS No.:	39236-46-9
Nonproprietary name:	Imidazolidinyl urea
CTFA adopted name:	Imidazolidinyl urea
Registered names:	Germall 115, Biopure 100, Euxyl K 200
Type of compound:	Heterocyclic substituted urea

1. Structure and Chemical Properties [1]

Appearance:	Stable white powder; tasteless; of neutral pH; decomposes at temperatures above 160°C
Odor:	Odorless
Solubility (g/100 g solvent at 25°C):	Water, 200; glycerol, 100; isopropanol, 0.05; ethyleneglycol, 150; methanol, 0.05; sesame oil, 0.05; propyleneglycol, 120; ethyl alcohol, 0.05; mineral oil, 0.05
Optimum pH:	Wide range
Stability:	Stable; over 10°C Germall 115 releases formaldehyde (decomposition)
Compatibility and inactivation:	Compatible with all ionics, nonionics and protein
Structural formula:	$C_{11}H_{16}N_8O_8$ Mol. wt. 388.31

Assay:

1. Thin-layer chromatography (TLC) silica gel plates; flow solvent, chloroform:methanol:acetic acid:water (50:30:10:10). Spray reagent is ninhydrin. The plate is heated at 150°C,

Assay (Continued):

    cooled, and viewed under ultraviolet light at 366 nm. Two pale yellow fluorescent zones appear of 0.27 and 0.35.

2. TLC with polyamid plates. Reagent $K_3[Fe(CN)_6] - Na_2[Fe(CN)_5NO] \cdot 2H_2O$ or phenylhydrazine-4-sulfonic acid. Reflectance densitometry at 550 nm to measure the concentration of Germall 115 on the plate (Gottschalk and Oelschläger, 1977).
3. Fluorometric method (Sheppard and Wilson, 1974).
4. Colorimetric assay for Germall 115 incorporated in cosmetic emulsions, measured at 520 nm. The absorbance is compared with the absorbance of a standard solution.

2. Antimicrobial Spectrum

Test organisms ($\sim 10^6$ CFU/ml)	Minimal germicidal concentration ($\mu g/ml$) (suspension test; contact times of 24 and 72 hr)	Minimal inhibitory concentration ($\mu g/ml$) (serial dilution test; incubation times of 24 and 72 hr)
Staphylococcus aureus	2000	1000
Escherichia coli	8000	2000
Klebsiella pneumoniae	8000	2000
Pseudomonas aeruginosa	8000	2000
Pseudomonas fluorescens	8000	2000
Pseudomonas cepacia	4000	2000
Candida albicans	8000	8000
Aspergillus niger	8000	8000
Penicillium notatum	8000	8000

Imidazolidinyl urea has good bacteriostatic activity with a concentration of 0.2%, a bactericidal effect only against gram-positive bacteria, and with a concentration of 0.5%, a bactericidal effect also gainst gram-negative bacteria. With in-use concentrations (0.2-0.5%), there is no antifungal activity. It is therefore necessary to combine imidazolidinyl urea with another antifungal agent, such as PHB esters, or in formulations with an acid pH sorbic acid or dehydroacetic acid. A synergistic behavior has been observed in combination with parabens, sorbic acid, dehydroacetic acid, quaternary ammonium compounds, and Triclosan. The antimicrobial effect is apparently increased by the presence of surfactants, and other cosmetic additives.

Antimicrobial Preservatives

3. Toxicity

Acute oral toxicity

Rat:	5.2 g/kg
Mouse:	7.2 g/kg
Subacute toxicity (90-days test):	Rat, oral, 6-600 mg/kg; all animals survived without any toxic effect.
Acute dermal toxicity:	Rabbit, LD_{50}, >8 g/kg (Sutton, 1973)
Subacute dermal toxicity:	Rabbit, skin contact with 20, 45, 90, or 200 mg of powder of Germall 115/kg per day — no irritation
Primary skin irritation (50% solution in water):	Rabbit, no edemas; with preirritated skin, severe edemas and erythemas
Patch test:	Repeated patch tests with 10% Germall 115 on 200 human subjects showed no irritation or sensitization.
Rabbit eye irritation:	5, 10, and 20% solutions produced no irritation.
Teratogenicity (mouse):	Not detectable (oral doses of 30, 95, and 300 mg/kg); it is slightly fetotoxic.

4. Cosmetic Applications

Use concentrations, 0.1-0.5% in combination with parabens or other antifungal preservatives; used for lotions, creams, hair conditioners, shampoos, and deodorants

5. Other Applications

Preservative for technical products, but limited by costs

6. Supplier

Sutton Laboratories, Inc.

7. References

1. R. L. Elder (ed.), Cosmetic ingredients, *J. Environ. Pathol. Toxicol.*, 4: 133-146 (1980).
2. D. S. Ryder, *J. Soc. Cosmet. Chem.*, 25:535 (1974).
3. H. Gottschalk and T. Oelschläger, *J. Soc. Cosmet. Chem.*, 28:497-520 (1977).

GERMALL II

Chemical name:	N-(Hydroxymethyl)-N-(1,3-dihydroxymethyl-2,5-dioxo-4-imidazolidinyl)-N'-(hydroxymethyl) urea
CAS No.:	78491-02-8
CTFA adopted name:	Diazolidinyl urea
Type of compound:	Heterocyclic imidazolidinyl urea

1. Structure and Chemical Properties [1]

Appearance:	Fine, white, free-flowing powder
Odor:	None or characteristically mild
Solubility:	Water soluble
Optimum pH:	Wide range
Stability:	Stable
Compatibility and inactivation:	Like Germall 115
Structural formula:	$C_8H_{14}N_4O_2$ Mol. wt. 278.23
Assay:	Colorimetric assay for incorporated Germall II. The absorbance at 520 nm is measured in a photometer and compared with the absorbance of a standard solution.

2. Antimicrobial Spectrum [2]

Test organisms ($\sim 10^6$ CFU/ml)	Minimal germicidal concentration (μg/ml) (suspension test; contact times of 24 and 72 hr)	Minimal inhibitory concentration (μg/ml) (serial dilution test; incubation times of 24 and 72 hr)
Staphylococcus aureus	1000	250
Escherichia coli	4000	1000
Klebsiella pneumoniae	4000	500
Pseudomonas aeruginosa	4000	1000
Pseudomonas fluorescens	2000	1000
Pseudomonas cepacia	2000	1000
Candida albicans	8000	8000
Aspergillus niger	8000	4000
Penicillium notatum	8000	4000

The antimicrobial activity of Germall II is much better than that of Germall 115. In the range of the in-use concentrations (0.2-0.5%), gram-negative and gram-positive bacteria are inhibited and often killed. The antifungal activity is a little bit better than that of Germall 115, but not good enough to work without the addition of another antifungal agent, such as PHB esters.

3. Toxicity [1]

Acute oral toxicity

Rat: LD_{50}, 2.57 g/kg

Acute dermal toxicity
  Rabbit: LD_{50}, >2.0 g/kg

Primary skin irritation
  Rabbit: A 1.0 or 5.0% solution is not a primary skin irritant.

Rabbit eye irritation: A 1.0 or 5.0% solution is not an eye irritant.

Guinea pig sensitization: No sensitization

4. Cosmetic Applications — 0.1-0.5 in combination with parabens or other antifungal preservatives

5. Other Applications — None

6. Supplier — Sutton Laboratories, Inc.

7. References
  1. Sutton Laboratories, Inc., Germall II.
  2. K. H. Wallhäusser, *Parfuem. Kosmet.*, 62:379 (1981).

INORGANIC SULFITES AND BISULFITES

Chemical name: Inorganic sulfites and bisulfites
EEC No.: I/9
Type of compound: Inorganic acid

1. Structure and Chemical Properties

Appearance: The sulfites are white powders; the bisulfites are only available in solutions.

Odor: Characteristic SO_2 odor

Solubility: The powders, without $CaSO_3$, are soluble in water.

Optimum pH: Below pH 4

Stability: Instable in solutions

Structural formula:

$Na_2SO_3 \cdot 7H_2O$	Mol. wt. 252.15	Sulfite
K_2SO_3	Mol. wt. 158.27	
$NaHSO_3$	Mol. wt. 104.06	Bisulfite
$KHSO_3$	Mol. wt. 120.16	
$Na_2S_2O_5$	Mol. wt. 190.10	Pyrosulfite
$K_2S_2O_5$	Mol. wt. 222.34	
$CaSO_3 \cdot 2H_2O$	Mol. wt. 156.17	

Assay: Titration with iodine

2. Antimicrobial Spectrum

Test organisms ($\sim 10^6$ CFU/ml)	Minimal inhibitory concentration (μg/ml) (serial dilution test; incubation times of 24 and 72 hr)	
	pH 4	pH 6
Staphylococcus aureus		80
Escherichia coli		100-200
Klebsiella pneumoniae		100-200
Pseudomonas aeruginosa		100-200
Pseudomonas fluorescens		50-100
Pseudomonas cepacia		100-200
Candida albicans	100-200	
Aspergillus niger	200-400	
Penicillium notatum	200-400	

The inhibition and destruction of microorganisms by an acid depends on both the hydrogen ion concentration and dissociation constant of the acid. The antimicrobial activity is much higher in the un-ionized state and is dependent on the percentage of undissociated acid, which is 6% at pH 3, 0.6% at pH 4, and 0.01% at pH 6. These compounds are thus more active at acidic pH values. Inhibition of enzyme systems takes place with SH groups.

3. Toxicity

Acute oral toxicity

Rat: LD_{50}, 1000-2000 mg SO_2/kg
Rabbit: LD_{50}, 600-700 mg SO_2/kg

Subchronic toxicity

Rat: 0.6% sodium bisulfite as feed additive causes vitamin deficiency and diarrhea [2]

Chronic toxicity

Rat: 0.5-2% sodium bisulfite as feed additive [3] over 1 year showed injuries in the nervous system, but a dose of 0.25% resulted in no toxic effects, only slight diarrhea [3] if the dose is >0.1% [3]; 0.12% potassium pyrosulfite to drinking water (30-90 mg SO_2/kg) over 20 months results in no abnormal growth of the rats [4].

Human toxicity: Very different reactions; some people can consume up to 4 g of sulfite per day (about 50 mg/kg body weight) without trouble; others suffer headaches, nausea, or diarrhea at much lower doses [1].

Carcinogenicity: Sulfite shows no carcinogenic effect [5].

Mutagenicity: Mutagenic against bacteria [6].

4. Cosmetic Applications   In-use concentration, 0.2% (EEC)

5. Other Applications   Food preservative and disinfectant in the food industry, especially the wine industry; disinfectants, 1-2%

6. References

1. K. Lang, *Schriftreihe des Bundes für Lebensmittelrecht und Lebensmittelkunde*, Heft 31, Behr's Verlag, Hamburg (1960).
2. B. Bhagat and M. F. Lockett, *Food Cosmet. Toxicol.*, 2:1 (1964).
3. O. G. Fitzhugh et al., *J. Pharmacol. Exp. Ther.*, 86:37 (1946).
4. R. Cluzan et al., *Ann. Biol. Anim. Biochim. Biophys.*, 5:267 (1965).
5. M. F. Lockett and J. L. Natoff, *J. Pharm. Pharmacol.*, 12:488 (1960).
6. H. Hayatus and A. Miura, *Biochem. Biophys. Acta*, 39:156 (1970).

BORIC ACID

Chemical names: Boric acid, boracic acid, orthoboric acid

EEC No.:   2

CAS No.:   10043-35-3

Type of compound: Mineral acid

1. Structure and Chemical Properties

Appearance: White crystalline powder; mp about 171°C

Odor: Odorless

Solubility: 1 g dissolves in 18 ml of cold water, 4 ml of boiling water, 18 ml of cold alcohol, 6 ml of boiling alcohol, and 4 ml of glycerol

Optimum pH: 5.1 (0.1 M)

Stability: Volatile with steam

Compatiblity and inactivation: Incompatible with alkali carbonates and hydroxides

Structural formula: H_3BO_3
Mol. wt. 61.84

2. Antimicrobial Spectrum

Test organisms ($\sim 10^6$ CFU/ml)	Minimal inhibitory concentration (μg/ml) (serial dilution test; incubation times of 24 and 72 hr; pH 6.1)
Staphylococcus aureus	5000
Escherichia coli	10,000
Klebsiella pneumoniae	10,000
Pseudomonas aeruginosa	10,000
Pseudomonas fluorescens	10,000
Pseudomonas cepacia	10,000
Candida albicans	5000
Aspergillus niger	2500
Penicillium notatum	1250

Small selective antimicrobial spectrum, mainly against yeasts, only in high concentrations (1-3%) against molds and bacteria. Responsible for the antimicrobial activity is the undissociated part of the boric acid. But the dissociation constant, which is, at 7.3×10^{-10}, much lower than that of all other acids used as preservatives, allows boric acid to have antimicrobial activity up to pH 7.

3. Toxicity

Acute oral toxicity

Mouse:	LD_{50}, 3.45 g/kg
Rat:	LD_{50}, 5.14 g/kg
Dog:	LD_{50}, 1-4 g/kg
Mouse:	Subcutaneous, LD_{50}, 1.74 g/kg
Human toxicity:	The concentration dangerous to human life is 1-3 g for babies, 5 g for children, and 15-20 g [1] for adults. Heavy intoxications in babies and infants after treatment of sore skin with powder. Ingestion or absorption may cause nausea, vomiting, diarrhea, abdominal cramps, erythematous lesions on skin and mucous membranes, circulatory collapse, tachycardia, cyanosis, delirium, convulsions, and coma. Chronic use may cause borism (dry skin, eruptions, gastric disturbances). Death has occurred from <5 g in infants and from 5-20 g in adults [4]. Dangerous to life, 50-80 g boric acid ointment [5]. Boric acid level in blood 16.1 μmol/liter (0.1 mg/100 ml) without toxic effects; these were detectable at 0.8 mmol/liter (5 mg/100 ml). Quick resorption by the intestinal tract, mucous membranes, and wounds, but not by the intact skin [5].

4. Cosmetic Applications	Not for use in baby cosmetics [2,3] and baby powder; EEC maximum dose, 3%; for mouth care only 0.5%.
5. Other Applications	Medical eyedrops, astringents, antiseptics, aqueous solutions, powders for external use, for weatherproofing wood and fireproofing fabrics.
6. References	1. T. Schuppli, *Dermatologica*, *141*:130 (1970). 2. *Br. Med. J.*, July 23, 1966. 3. *Soap Perfum. Cosmet.*, *39*:796 (1966). 4. E. Browning, *Toxicity of Industrial Metals*, 2nd ed., Appleton-Century-Crofts, New York, (1969), pp. 90-97. 5. W. Wirth and C. Gloxhuber, *Toxikologie*, 3rd ed., Thieme, Stuttgart (1981), p. 78.

FORMIC ACID

Chemical name:	Formic acid
EEC No.:	5
CAS No.:	64-18-6
Type of compound:	Organic acid, observed in 1670 by Fischer in the products resulting from the destillation of ants [2].

1. Structure and Chemical Properties

Appearance:	Colorless liquid; avoid contact with skin!
Odor:	Pungent odor; MAK (*Maximale Arbeitsplatz-Konzentratic* threshold limit value) 5 ppm (Germany)
Solubility:	Miscible with water, alcohol, and glycerol
Optimum pH:	3.5
Structural formula:	
HCOOH	CH_2O_2 Mol. wt. 46.02 Sodium formate Mol. wt. 68.01 Calcium formate Mol. wt. 130.11
Assay:	Photometric assay [1]

2. Antimicrobial Spectrum

Test organisms ($\sim 10^6$ CFU/ml)	Minimal inhibitory concentration (μg/ml) (serial dilution test; incubation times of 24 and 72 hr)	
	pH 6	pH 3
Staphylococcus aureus	1400	
Escherichia coli	100-500	
Klebsiella pneumoniae	100-500	
Pseudomonas aeruginosa	200-800	
Pseudomonas fluorescens	200-800	1600
Candida albicans		
Aspergillus niger	1500-6000	
Penicillium notatum	1500-6000	

Formic acid is the strongest of the fatty acids, with the highest dissociation constant. This means the undissociated part is 98.3% at pH 2, 36.1% at pH 4, and only 0.56% at pH 6; it is therefore only active at acidic pH values.

3. Toxicity

Acute oral toxicity

  Rat: LD_{50}, 1.25 g/kg
  Mouse: LD_{50}, 1.1 g/kg [3]

Acute intravenous toxicity

  Rabbit: Minimum lethal dose, 239 mg/kg

Subchronic toxicity: 0.5-1.0% to the drinking water of rats leads to organic injuries [5]

Chronic toxicity

  Rat: 0.2% calcium formate to drinking water (about 150-200 mg/kg body weight over 2 years showed no effect [6]

Primary skin irritation: Dangerously caustic to skin! Chronic absorption has been reported to cause albuminuria and hematuria.

Human toxicity: 10-g dose dangerous; 50-60 g lethal dose [4] in concentrated form; irritation of skin and mucous membranes

Acceptable daily intake: 0-3 mg/kg per day

Mutagenicity: Mutagen for *Drosophila*

Carcinogenicity: Not detectable [4]

Teratogenicity: Not detectable [4]

Antimicrobial Preservatives

4. Cosmetic Applications   EEC maximum dose, 0.5%

5. Other Applications   Preservative for food since 1865 [2], mostly in the form of sodium, potassium or calcium formate; not allowed in the United States

6. References
   1. H. Tanner, *Schweiz. Z. Obst. Weinbau*, *112*:38 (1976).
   2. F. H. Jodin, *C. Rend.*, *61*:1179 (1865).
   3. G. Malorny, *Z. Ernaehrungswiss.*, *9*: 332 (1969).
   4. Tracor Jitco, Inc., Scientific literature reviews on generally recognized as safe (GRAS) food ingredients, PB-228 558, Springfield, Ill., National Technical Information Services, U.S. Department of Commerce (1974).
   5. A. Sporn et al., *Igiena*, *11*:507 (1962).
   6. G. Malorny, *Z. Ernaehrungswiss.*, *9*: 332 (1969).

PROPIONIC ACID

Chemical names:	Methylacetic acid, ethylformic acid
EEC No.:	I/2
CAS No.:	79-09-4
Nonproprietary name:	Propionic acid
Registered name:	Mycoban
Type of compound:	Organic acid

1. Structure and Chemical Properties

Appearance:	Oily liquid; slightly pungent; disagreeable; mp 21.5°C
Odor:	Rancid odor
Solubility:	Miscible with water; can be salted out of water solutions by the addition of $CaCl_2$ or other salts
Optimum pH:	3.5-4.5; limit pH 6
Stability:	Stable
Structural formula:	$C_3H_6O_2$ Mol. wt. 74.08 CH_3CH_2COOH
Synthesis:	Fermentation product using bacteria of the genus *Propionibacterium*

Assay: Gas chromatography [1]

2. Antimicrobial Spectrum

Test organisms (~10^6 CFU/ml)	Minimal inhibitory concentration (μg/ml) (serial dilution test; incubation times of 24 and 72 hr; pH 3.9)
Staphylococcus aureus	2000
Escherichia coli	2000
Klebsiella pneumoniae	1250
Pseudomonas aeruginosa	3000
Pseudomonas fluorescens	1250
Pseudomonas cepacia	3000
Candida albicans	2000
Aspergillus niger	2000
Penicillium notatum	2000

The antimicrobial activity is dependent on the degree of undissociated acid; 88% at pH 4, 6.7% at pH 6, and 0.71% at pH 7. The spectrum of activity is very indifferent, preferable molds are inhibited; but there are some *Penicillium* species with a high propionic acid tolerance (more than 5%). The same is true for yeasts; some *Torula* species need propionic acid for their metabolism. The acid is more active against gram-negative bacteria than against gram-positive organisms. In food preservation it is mainly used against ropiness in bread, which is caused by *Bacillus mesentericus*.

3. Toxicity

Acute oral toxicity

Rat:	LD_{50}, 2.6 g/kg
Subchronic toxicity:	Addition of 1-3% sodium or calcium propionate to animal feed (rats) over some weeks showed no effect [2]
Chronic toxicity:	3.75% addition to animal feed (rats) over 1 year without negative effects
Primary skin irritation:	In concentrated form causes irritation of skin and mucous membranes.
Acceptable daily intake:	No limit

4. Cosmetic Applications

2% (EEC)

5. Other Applications

Preservative for food; against ropiness in bread (0.15-0.4%).

6. References

1. E. Lück et al., *Z. Lebensm. Unters. Forsch.*, 158:27 (1975).
2. K. E. Harshbarger, *J. Dairy Sci.*, 25: 168 (1942).

Antimicrobial Preservatives

UNDECYLENIC ACID

Chemical name:	10-Undecenoic acid
EEC No.:	13
CAS No.:	112-38-9
Registered names:	Declid, Renselin, Sevinon
Type of compound:	Organic acid, appears in sweat

1. Structure and Chemical Properties

Appearance:	Liquid or crystals
Odor:	Suggestive of perspiration
Solubility:	Insoluble in water; soluble in alcohol
Optimum pH:	4.5-6.0; above pH 6.0 the calcium salt is inactivated to a greater extent than the free acid.
Compatibility and inactivation:	Compatible with boric acid and salicylic acid
Structural formula:	$C_{11}H_{20}O_2$ Mol. wt. 184.27

$$CH_2=CH(CH_2)_8COOH$$

Zinc salt: $C_{22}H_{38}O_4Zn$; amorphous white powder; mp 115-116°C Undecylenic acid monoethanol-amide-di-sodium-sulfosuccinate

$$CH_2=CH(CH_2)_8\underset{O}{\overset{}{C}}-NH-C_2H_4O_2C-CH_2-CH\underset{COONa}{\overset{SO_3Na}{\diagup\!\!\!\diagdown}}$$

2. Antimicrobial Spectrum

Undecylenic acid is a powerful antifungal agent against a variety of pathogenic fungi. It is used alone or in combination with other antifungal substances such as boric acid, salicylic acid, and methyl salicylate as a local application in concentrations of 2-15% in emulsions, ointments, or dusting powders for the treatment of tinea pedis, capitis, and cruris, moniliasis, mycotic vulvovaginitis, and related conditions. Antifungal activity is enhanced when the acid is combined with its zinc salt and with other fatty acids; antifungal activity is greatest at acidic pH.

3. Toxicity

Acute oral toxicity

Rat:	LD_{50}, 2.5 g/kg
Primary skin irritation:	Applied to mucous membranes, undecylenic acid is irritant in concentrations exceeding 1%.

4. Cosmetic Applications	EEC, 0.2%. Undecylenic acid monoethanol-amide-di-sodium-sulfosuccinate is used as an antidandruff agent (1%) in shampoos. Stable only in the pH range 5.0-6.5.
5. Other Applications	Pharmaceutical products as a topical antifungal; 2-15% in emulsions, ointments, dusting powders; often in combination with other antifungal agents (boric acid, salicylic acid)
6. Supplier	Riedel de Haen
7. References	1. W. B. Hugo, *Inhibition and Destruction of the Microbial Cell*, Academic, London (1971), p. 668. 2. K. H. Wallhäusser and H. Schmidt, *Sterilisation-Desinfektion-Konservierung-Chemotherapie*, Thieme, Stuttgart (1967), p. 162.

SORBIC ACID

Chemical name:	2,4-Hexadienoic acid, 2-Propenyl acrylic acid
EEC No.:	I/4
CAS No.:	110-44-1
Nonproprietary name:	Sorbic acid
CTFA adopted names:	Sorbic acid, potassium sorbate
Registered name:	Sentry
Type of compound:	Organic acid
History:	In berries of the mountain ash *Sorbus aucuparia* as the lactone parasorbic acid, isolated in 1859 by von Hofmann [1].

1. Structure and Chemical Properties

Appearance:	Crystalline (needles); mp 134.5°C
Odor:	Faint characteristic odor
Solubility:	Water (30°C), 0.25%, (100°C) 3.8%; propylene glycol at 20°C, 5.5%; absolute ethanol, 0.29%; glacial acetic acid, 11.5%; glycerol, 0.31%; isopropanol, 8.4%; oils, 0.5-1.0%.
Potassium sorbate:	Mol. wt. 150.22; white powder or granulate; soluble in water at 25°C, 138 g/100 ml water
Calcium sorbate:	White powder; soluble in water, 1.2%

Optimum pH: Up to 6.5

Compatibility and inactivation: Slightly incompatible with nonionics

Structural formula: $C_6H_8O_2$
Mol. wt. 112.12

$$CH_3-CH_2=CH-CH=CH-COOH$$

Synthesis: Condensing crotonaldehyde and malonic acid in pyridine solution

Assay:
1. Oxidation of sorbic acid with calcium dichromate and reaction with 2-thiobarbiturate produces a reddish color; colorimetry [2]
2. Spectrophotometry using the absorbance maximum at 260 nm [3]

2. Antimicrobial Spectrum

Test organisms ($\sim 10^6$ CFU/ml)	Minimal inhibitory concentration (μg/ml) (serial dilution test; incubation times of 24 and 72 hr; pH 6.0)
Staphylococcus aureus	50-100
Clostridium sporogenes	100-500
Escherichia coli	50-100
Klebsiella pneumoniae	50-100
Pseudomonas aeruginosa	100-300
Pseudomonas fluorescens	100-300
Pseudomonas cepacia	50-100
Candida albicans	25-50
Saccharomyces cerevisiae	200-500
Aspergillus niger	200-500
Penicillium notatum	200-300

Fungistatic activity (yeasts and molds), but only little activity against bacteria. Mode of action is by inhibition of different microbial enzyme systems, such as enolase, lactate dehydrogenase and other dehydrogenase systems, and catalase. For inhibition it is necessary that sorbic acid traverse the cell membrane; this is done by the undissociated part independent from the pH At pH 3.15 this undissociated quota is about 40%, and at pH 7.0 less than 1%.

3. Toxicity

Acute oral toxicity

Rat: LD_{50}, 7.36 g/kg (acid); LD_{50}, 5.94 g/kg (sodium sorbate)

Chronic toxicity

Rat: Addition of 5% sorbic acid to animal feed over a lifetime had no negative effects [4], while 10% over 2 years showed reduced increase in body weight, but an increase of thyroid gland, liver, and kidney [5].

Carcinogenicity: No carcinogenic effect [6]

4. Cosmetic Applications

Antifungal preservation in creams and lotions

5. Other Applications

Preservation in pharmaceutical oral dosage forms, in food and wine

6. Supplier

Hoechst A.G.

7. References

1. A. W. Hofmann, *Ann. Chem. Pharmac.*, *34*:129-140 (1859).
2. H. Schmidt, *Z. Anal. Chem.*, *178*:173-184 (1960).
3. F. H. Luckmann and D. Melnick, *Food Res.*, *20*:649 (1955).
4. K. Lang, *Arzneim. Forsch.*, *10*:997 (1960).
5. J. F. Gaunt et al., *Food Cosmet. Toxicol.*, *13*:31 (1975).
6. F. Dickens et al., *Br. J. Cancer*, *22*:762 (1968).

BENZOIC ACID

Chemical name: Benzene carboxylic acid

EEC No.: I/1

CAS No.: 65-85-0

Nonproprietary name: Benzoic acid

CTFA adopted names: Benzoic acid, sodium benzoate

Type of compound: Organic acid, free in nature in berries

History: First description by H. Fleck in 1875 [1]

1. Structure and Chemical Properties

Appearance: Monoclinic tablets, plates, leaflets, or white powder; mp 122°C

Solubility: In water (20°C), 0.29%; ethanol (20°C), 1 g in 2.3 ml; 1 g in 1.5 boiling ethanol; sodium salt in water (20°C), 1 g in 1.8 ml

Antimicrobial Preservatives

Optimum pH:	2-5
Stability:	Stable at low pH
Compatibility and inactivation:	Loss of activity in the presence of proteins and glycerol; incompatible with nonionics, quaternary compounds and gelatin
Structural formula:	$C_7H_6O_2$ Mol. wt. 122.12 Sodium benzoate mol. wt. 162

Assay:	1. Polarography [2] 2. Gas chromatography [3]

2. Antimicrobial Spectrum

Test organisms (~10^6 CFU/ml)	Minimal germicidal concentration (μg/ml) (suspension test; contact times of 24 and 72 hr; pH 6)	Minimal inhibitory concentration (μg/ml) (serial dilution test; incubation times of 24 and 72 hr; pH 6)
Staphylococcus aureus	20	50-100
Escherichia coli	160	100-200
Klebsiella pneumoniae	160	100-200
Pseudomonas aeruginosa	160	200-500
Pseudomonas fluorescens	160	200-500
Pseudomonas cepacia	160	
Candida albicans	1200	500-1000
Aspergillus niger	1000	500-1000
Penicillium notatum	1000	500-1000

Only the undissociated form has antimicrobial activity, 60% at pH 4.0, 1.5% at pH 6, and 0.15% at pH 7; this means benzoic acid is only effective in an acidic milieu. In oil-water emulsions benzoic acid migrates from the water to the oil phase; only the amount solved in water is effective.

3. Toxicity

Acute oral toxicity

Mouse:	LD_{50}, 2.37 g/kg
Rat:	LD_{50}, 1.7 g/kg

Subchronic toxicity

Mouse:	80 mg/kg per day in feed over 3 months results in higher mortality.

Chronic toxicity:	40 mg/kg per day in feed (mice and rats) over 17 and 18 months, respectively, was growth inhibiting [4].
Acceptable daily intake (human):	0-5 mg/kg body weight per day
Human skin:	Toxic dose, 6 mg/kg

4. Cosmetic Applications — Use concentration, 0.1-0.2%; EEC maximum concentration, 0.5%

5. Other Applications — Preservative agent in food and pharmaceuticals (oral dosage forms)

6. Suppliers — Pfizer Inc., Kalama Chemical

7. References

1. B. Strahlmann, *Mitt. Geb. Lebensmittelunters. Hyg.*, 65:96 (1974).
2. J. Davidek et al., *Z. Lebensm. Unters. Forsch.*, 129:370 (1966).
3. J. Vogel and J. Deshusses, *Mitt. Gebiete Lebensmittelunters. Hyg.*, 56:35 (1965).
4. A. J. Shtenberg, A. D. Ignatev, *Food Cosmet. Toxicol.*, 8:369 (1970).

SALICYLIC ACID

Chemical names:	o-Hydroxybenzoic acid, 2-hydroxybenzoic acid
EEC No.:	I/3
CAS No.:	69-72-7
CTFA adopted name:	Salicylic acid
Type of compound:	Organic acid, occurs in the form of esters in several plants
History:	First synthesis in 1874 by Kolbe [1]

1. Structure and Chemical Properties

Appearance:	Acicular crystals or crystalline powder, mp 157-159°C, sublimates at 76°C; when rapidly heated at atmospheric pressure, decomposition into phenol and CO_2
Odor:	Odorless
Solubility:	1 g in 460 ml of water, 15 ml of boiling water, 2.7 ml of alcohol, 60 ml of glycerol, and about 80 ml of fats or oils; solubility in water increased by addition of sodium phosphate, borax, alkali acetates or citrates.

Optimum pH:	4.0-6.0
Stability:	Gradually discolors in sunlight; keep protected from light in well-closed container; discolors with iron salts
Compatibility and inactivation:	Incompatible with iron salts, spirit of nitrous ether, lead acetate, iodine
Structural formula:	$C_7H_6O_3$ Mol. wt. 138.12 [benzene ring with COOH and OH substituents in ortho position]
Synthesis:	By heating sodium phenolate with carbon dioxide under pressure
Assay:	1. Colorimetric assay with Fe^{+++}-chloride [2] 2. Spectrophotometric assay for salicylic acid in pharmaceutical preparations [3]

2. Antimicrobial Spectrum

Test organisms (~10^6 CFU/ml)	Minimal inhibitory concentration (μg/ml) (serial dilution test; incubation times of 24 and 72 hr; pH 3.2)
Staphylococcus aureus	1250
Escherichia coli	1250
Klebsiella pneumoniae	1250
Pseudomonas aeruginosa	2500
Pseudomonas fluorescens	1250
Pseudomonas cepacia	2500
Candida albicans	2500
Aspergillus niger	2500
Penicillium notatum	2500

Salicylic acid is more active against yeasts and molds than against bacteria, but the antibacterial activity is better than that of benzoic acid. As in the case of all fatty acids, only the undissociated part is active; this is 90% at pH 2, 8.6% at pH 4, and 0.09% at pH 6. Salicylic acid therefore should be used as an antimicrobial preservative only in the pH range 2-5. Salicylic acid attacks the plasma membrane of bacteria and inhibits some enzyme systems.

3. Toxicity

Acute oral toxicity

Rat:	LD_{50}, 891 mg/kg
Rabbit:	LD_{min}, 1300 mg/kg
Toxicokinetic data:	Salicylic acid is quickly resorbed but not metabolized. The secretion is slow, so it may cumulate.

4. Cosmetic Applications

Use concentration; 0.025-0.2%; EEC guideline, 0.5%, not to be used in preparations for children (<3 years)

5. Other Applications

As preservative of food products, but now forbidden in some countries; in pharmaceutical preparations with topical keratolytic activity

6. Supplier

J. T. Baker Chemical Company

7. References

1. H. Kolbe, *J. Prakt. Chem.*, 10:89 (1874).
2. *European Pharmacopeia*, Vol. 3, Maisonneuve, 57-Sainte-Ruffine (France), p. 137 (1975).
3. S. Adams and J. H. M. C. B. Miller, *J. Pharm. Pharmacol.*, 30:81 (1978).

DEHYDROACETIC ACID

Chemical names:	3-Acetyl-6-methyl-2H-pyran-2,4(3H)-dione; 2-acetyl-5-hydroxy-3-oxo-4-hexenoic acid δ lactone
EEC No.:	4
CAS No.:	520-45-6
Nonproprietary name:	DHA
CTFA adopted name:	Dehydroacetic acid
Type of compound:	Organic acid

1. Structure and Chemical Properties

Appearance (Sodium salt):	Hydrate; colorless, tasteless powder; mp 109-111°C (sublimates)
Odor:	Odorless
Solubility:	Acid; wt/wt at 25°C: ethanol, 3%; glycerol, <0.1%; olive oil, <0.1%; water, <0.1%
Sodium salt (wt/wt at 25°C):	Water, 33%; propylene glycol, 48%; ethanol, 1%; olive oil, <0.1%

Optimum pH:	5-6.5; the activity decreases with higher pH.
Structural formula:	$C_8H_8O_4$ Mol. wt. 168.15 (acid) Mol. wt. 201.8 (sodium salt)

$$H_3C-C\underset{\underset{O}{\parallel}}{\overset{O}{\diagdown}}\underset{CH}{C}\underset{}{\overset{O}{\diagup}}C\overset{\nearrow O}{\diagdown}CH-\underset{\underset{O}{\parallel}}{C}-CH_3$$

Synthesis:	Polymerization product of ketene
Assay:	1. Fluorometry [4] 2. Spectrophotometry [5]; absorbance maximum at 312 nm

2. Antimicrobial Spectrum

Test organisms (~10^6 CFU/ml)	Minimal germicidal concentration (μg/ml) (suspension test; contact times of 24 and 72 hr)	Minimal inhibitory concentration (μg/ml) (serial dilution test; incubation times of 24 and 72 hr; pH 6.0)
Staphylococcus aureus	20,000	10,000
Bacillus subtilis	10,000	5,000
Escherichia coli	20,000	10,000
Pseudomonas aeruginosa	20,000	>20,000
Candida albicans		200
Aspergillus niger		200
Penicillium notatum		200

Only the undissociated part of the dehydroacetic acid is active against microorganisms, and this is strictly dependent on the pH of the medium.

3. Toxicity

Acute oral toxicity

Rat:	LD_{50}, 1.0 g/kg (sodium salt)
Human toxicity:	Causes impaired kidney function; large doses can cause vomiting, ataxia, and convulsions.

Chronic toxicity

Rat:	Daily dose with food, 0.1%, over 2 years showed no toxic effect; no-effect level, >50 mg/kg [1]

Primary skin irritation:	No primary skin irritation
Human application:	6-13 mg/kg body weight over 150 days had no effect [3,6]; human skin, no irritation or sensitization [3]; quick resorption by the human body; serum levels detectable (10-15 mg/100 ml)

4. Cosmetic Applications

Use concentration, 0.02-0.2%; EEC maximum, 0.6%

5. Other Applications

Antimicrobial preservative; in the United States used for the preservation of pumpkins; maximal dose, 65 mg/kg; in Europe not allowed for food preservation

6. Supplier

Gaines Chemical Works, Inc.

7. References

1. G. M. Cramer et al., *Food Cosmet. Toxicol.*, *16*:255 (1978).
2. L. A. Woods et al., *J. Pharmacol. Exp. Ther.*, *99*:84 (1950).
3. H. C. Spencer et al., *J. Pharmacol. Exp. Ther.*, *99*:57 (1950).
4. T. Shibazaki, *J. Pharmacol. Soc. J.*, *88*:601 (1968).
5. R. Gren, *Dtsch. Apoth. Ztg.*, *111*:219 (1971).
6. L. A. Woods et al., *J. Pharmacol. Exp. Ther.*, *99*:84 (1950).

USNIC ACID

Chemical name:	2,6-Diacetyl-7,9-dihydroxy-8,9b-dimethyl-1,3(2H,9bH)-dibenzofurandione
EEC No.:	14
CAS No.:	125-46-2
Nonproprietary name:	Usnic acid
Type of compound:	Antibiotic produced by *Cladonia stellaris*, *Usnea barbata*, and other lichens

1. Structure and Chemical Properties

Appearance:	Yellow crystalline powder; mp 204°C
Solubility:	Soluble in water, 0.01%; ethanol, 0.02%; isopropanol, 0.28%; oils, 0.1-0.3%; acetone, 0.77%
Stability:	Heat stable (250°C); stable against ultraviolet light; unstable in aqueous solutions

Structural formula:

$C_{18}H_{16}O_7$
Mol. wt. 344.31

[Structural diagram of the compound]

Assay [3]:
1. Spectrophotometry in methyl ethyl ketone with p-phenylenediamine
2. Gas chromatography

2. Antimicrobial Spectrum

Test organisms ($\sim 10^6$ CFU/ml)	Minimal inhibitory concentration (μg/ml) (serial dilution test; incubation times of 24 and 72 hr)
Staphylococcus aureus	4-10
Streptococcus faecalis	6-8
Escherichia coli	>1000
Pseudomonas aeruginosa	>1000
Mycobacterium tuberculosis	10-50

Only small-spectrum activity; microbiostatic, mostly against gram-positive bacteria, especially *mycobacteria* and *propionibacteria (P. acne)*; no resistance observed [1].

3. Toxicity [4]

Acute intravenous toxicity

Rat: LD_{50}, 30 mg/kg
Mouse: LD_{50}, 25 mg/kg

Acute subcutaneous toxicity

Mouse: LD_{50}, 700 mg/kg

Sensitization: No sensitization

4. Cosmetic Applications

Used in antiacne formulations, 0.1-0.3%, and deodorants; EEC, 0.2%

5. Other Applications

Local chemotherapeutic agent

6. Supplier

H. Passek GmbH

7. References

1. J. Möse, *Arzneim. Forsch.*, 5:508 (1955).
2. Klosa, *Pharmazie*, 8:436 (1953).
3. R. Fischer and G. Mikula, *Arzneim. Forsch.*, 19:2025 (1969).
4. *Antibiotics*, Vol. 1, (D. Gottlieb and P. Shaw, eds.), Springer-Verlag, New York (1967), p. 611.

PARABENS (PHB ESTERS)

Chemical name:	Esters of p-hydroxybenzoic acid, PHB esters
EEC No.:	I/12
CTFA adopted name:	Parabens
Registered names:	Solbrol, Nipagin, Nipasol, Nipakombin
Type of compound:	Benzoic acid esters
History:	Sabalitschka first used these compounds as preservative agents in 1924 [3]

1. Structure and Chemical Properties

Appearance:	White crystalline powders
Odor:	Odorless

Solubility (in g/100 ml solvent):

	Water				Ethanol, 25°C	Peanut oil, 25°C
	10°C	25°C	50°C	80°C		
Methyl ester	0.13	0.25	0.75	3.2	52	0.1
Ethyl ester	0.06	0.11	0.30	0.86	70	1.0
Propyl ester	0.018	0.04	0.12	0.45	95	14.0
Butyl ester	0.012	0.02	0.065	0.15	210	5.0
Benzyl ester		0.006			72	0.5

Optimum pH:	3-9.5
Stability:	Stable
Compatibility:	Incompatible with anionics, nonionics, and proteins

Antimicrobial Preservatives

Structural formula:

	COOCH$_3$	COOC$_2$H$_5$	COOC$_3$H$_7$	COOC$_4$H$_9$
	⟨phenol⟩	⟨phenol⟩	⟨phenol⟩	⟨phenol⟩
	OH	OH	OH	OH
	Nipagin M	Nipagin A	Nipasol	Nipabutyl
Mol. wt.	152.14	166.17	180.20	194.23
Mol. wt. of NA Salt	174.12	188.15	202.18	
mp	125-128	116-119	95-98	68-72

Assay: Spectrophotometry (destination as PHB at 255 nm) [1]

2. Antimicrobial Spectrum of the Different PHB Esters

Test organisms (~10^6 CFU/ml)	Minimal germicidal concentration (µg/ml) (suspension test; contact times of 24 and 72 hr)				Minimal inhibitory concentration (µg/ml) (serial dilution test; incubation times of 24 and 72 hr)			
	Methyl	Ethyl	Propyl	Butyl	Methyl	Ethyl	Propyl	Butyl
Staphylococcus aureus	1250	625	180	160	800	500	150	120
Escherichia coli	1250	1250	360	160	800	600	300	150
Klebsiella pneumoniae	1250	625	360	160	800	600	300	150
Pseudomonas aeruginosa	1250	625	625	160	1000	800	400	175
Pseudomonas fluorescens	1250	625	625	160	1000	800	400	175
Pseudomonas cepacia	1250	1250	625	320	1000	800	400	175
Candida albicans	5000	2500	625	625	1000	800	250	125
Aspergillus niger	5000	5000	2500	1250	600	400	200	150
Penicillium notatum	5000	2500	1250	1250	500	250	125	100

3. Toxicity

Acute oral toxicity

  Dog: Methyl LD_{50}, 3.0; ethyl LD_{50}, 5.0, propyl LD_{50}, 6.0; butyl LD_{50}, 6.0 (in g/kg)

Subchronic toxicity

  Rabbit: 500 mg PHB methyl ester per kilogram body weight per day for 6 days had no effect; 3000 mg shows toxic effect; cats and dogs are more sensitive. The data for the propyl ester are in the same range [2].

Human toxicity: 2 g methyl and propyl PHB ester per day over 1 month showed no effect [2]

4. Cosmetic Applications

In-use concentration, 0.4% for the single ester and 0.8% for a mixture of esters; mostly used, 0.18% methyl + 0.02% propyl PHB ester

5. Other Applications

Preservative for pharmaceuticals and food

6. Suppliers

Inolex Personal Care Products, Mallinckrodt Inc., Napp Chemicals, Protameen Chemicals, Tenneco Chemicals, Tri-K Industries

7. References

1. W. Lorenzen and R. Sieh, *Z. Lebensm. Unters. Forsch.*, *118*:222 (1962).
2. K. Schübel and J. Manger, *Naunyn Schmiedebergs Arch. Exp. Pathol. Pharmakol.*, *146*:208 (1929).
3. T. Sabalitschka and R. Neufeld-Crzellitzer, *Arzneim. Forsch.*, *4*:575 (1954).

p-HYDROXYBENZOIC ACID BENZYL ESTER

Chemical name:	p-Hydroxybenzoic acid benzyl ester
EEC No.:	6
CTFA adopted name:	Parabens
Registered name:	Nipabenzyl
Type of compound:	Benzoic acid esters

1. Structure and Chemical Properties

Appearance: White crystalline powder

Odor: Odorless

Solubility: In water (20°C), 0.006%; ethanol, 42%; peanut oil, 0.5%

Antimicrobial Preservatives

Optimum pH: 3-9.5

Stability: Stable

Compatibility and inactivation: Inactivated in the presence of nonionic emulsifiers, Tween, saccharose esters, polyoxy-40-stearate, polyvinylpyrrolidone, methylcellulose

Structural formula: Mol. wt. 228.24; mp 108-113°C

$$\text{HO-C}_6\text{H}_4\text{-COOCH}_2\text{-C}_6\text{H}_5$$

2. Antimicrobial Spectrum

Test organisms (~10^6 CFU/ml)	Minimal germicidal concentration (μg/ml) (suspension test; contact times of 24 and 72 hr)	Minimal inhibitory concentration (μg/ml) (serial dilution test; incubation times of 24 and 72 hr)
Staphylococcus aureus	50	120
Escherichia coli	125	160
Klebsiella pneumoniae	125	160
Pseudomonas aeruginosa	175	160
Pseudomonas fluorescens	175	160
Pseudomonas cepacia	175	160
Candida albicans	100	250
Aspergillus niger	125	1000
Penicillium notatum	125	500

3. Toxicity (see Parabens)

4. Cosmetic Applications In-use concentration, 0.1% (EEC)

5. Other Applications (see Parabens)

6. Suppliers Inolex Personal Care Products, Nipa Lab.

7. References (see Parabens)

PHENONIP

Chemical composition:	Mixture of 14.5% methyl-; 5.7% ethyl-; 2.4% propyl-; 2.4% isobutyl-, and 2.4% n-butyl-parabens in 69.6% 2-phenoxyethanol (Phenoxetol)
Registered name:	Phenonip
Type of compound:	Blend of p-hydroxybenzoic acid esters

1. Structure and Chemical Properties

Appearance:	Clear solution
Odor:	Odorless
Solubility:	In water, 0.5%; miscible with ethanol, isopropanol, propylene glycol, glycerin (for all other data, see Parabens)
Optimum pH:	3.0-9.5
Stability:	Stable
Compatibility and inactivation:	(see PHB esters)

2. Antimicrobial Spectrum
(see PHB esters)

3. Toxicity
(see PHB esters)

4. Cosmetic Applications
In-use concentration, 0.25-0.35%

5. Other Applications
(see PHB esters)

6. Supplier
Nipa Lab.

7. References

1. H. P. Fiedler, *Lexikon der Hilfsstoffe*, Vol. 2, Editio Cantor, Aulendorf (1981), p. 708.
2. J. E. Lucas et al., *Acta Pharm. Suec.*, 7:149 (1970).

LIQUAPAR

Chemical composition:	Mixture of N-butyl-, isobutyl-, and isopropylparabens
CTFA adopted name:	Mixture of N-butyl-, isobutyl-, and isopropylparabens
Registered name:	Liquapar
Type of compound:	Blend of p-hydroxybenzoic acid esters

1. Structure and Chemical Properties

Appearance:	White emulsion containing 50% total parabens

Antimicrobial Preservatives

Odor:	Odorless
Solubility:	In water, 0.06%; miscible with alcohol and propylene glycol
Optimum pH:	Wide pH range
Stability:	Relatively stable
Compatibility and inactivation:	Incompatible with nonionics (for all other data, see Parabens)

2. Antimicrobial Spectrum (see PHB esters)

3. Toxicity (see PHB esters)

4. Cosmetic Applications Use concentration, 0.05-0.3%

5. Other Applications (see PHB esters)

6. Supplier Mallinckrodt, Inc.

p-CHLORO-m-CRESOL

Chemical names:	4-Chloro-m-cresol, PCMC, 4-chloro-3-methylphenol
EEC No.:	26
CAS No.:	59-50-7
CTFA adopted name:	p-Chloro-m-cresol
Registered names:	BP, PCMC, Preventol CMK
Type of compound:	Halogenated phenolic

1. Structure and Chemical Properties

Appearance:	Dimorphous crystals; mp 55.5°C; volatile with steam; white powder
Odor:	Odorless when very pure
Solubility:	1 g dissolves in 260 ml of water at 20°C; more soluble in hot water; freely soluble in alcohols, fixed oils, terpenes, aqueous alkaline solutions
Optimum pH:	More active in acid than in alkaline solutions
Stability:	Aqueous solutions turn yellow in light and in contact with air
Compatibility and inactivation:	Partial inactivation in the presence of nonionics; discoloration with iron salts
Structural formula:	C_7H_7ClO Mol. wt. 142.58

Synthesis: By chlorination of m-cresol

2. Antimicrobial Spectrum

Test organisms (~10^6 CFU/ml)	Minimal inhibitory concentration ($\mu g/ml$) (serial dilution test; incubation times of 24 and 72 hr)
Staphylococcus aureus	625
Escherichia coli	1250
Klebsiella pneumoniae	625
Pseudomonas aeruginosa	1250
Pseudomonas fluorescens	1250
Candida albicans	2500
Aspergillus niger	2500

Broad-spectrum activity at acidic pH values.

3. Toxicity

Acute oral toxicity

Mouse: LD_{50}, about 4 g/kg

Guinea pig sensitization: No sensitization

4. Cosmetic Applications

Use concentration, 0.1-0.2% in protein shampoos and baby cosmetics; EEC maximum, 0.2%

5. Other Applications

Topical antiseptic, disinfectant, preservative in pharmaceutical products

6. Suppliers

Gerbstoffchemie Margold Griesheim, Bayer Leverkusen

4-CHLORO-3,5-XYLENOL

Chemical names: 4-Chloro-3,5-xylenol, 4-chloro-3,5-dimethylphenol, p-chloro-m-xylenol, PCMX

EEC No.: 32

CAS No.: 88-04-0

CTFA adopted name: Chloroxylenol
Registered name: Ottasept
Type of compound: Halogenated phenolic

1. Structure and Chemical Properties [1,2]

Appearance: Crystalline powder; volatile with steam

Odor: Phenolic odor

Solubility: 1 g dissolves in 3 liters of water at 20°C; more soluble in hot water; in 1 part of 95% alcohol, terpenes, fixed oils; soluble in alkaline solutions

Optimum pH: Wide pH range

Compatibility and inactivation: Incompatible with many cationics and non-ionics

Structural formula: C_8H_9ClO
Mol. wt. 156.61

$$\underset{\underset{Cl}{}}{\underset{H_3CCH_3}{\overset{OH}{\bigcirc}}}$$

Synthesis: By treating 3,5-dimethylphenol with Cl_2 or SO_2Cl_2

2. Antimicrobial Spectrum

Test organisms (~10^6 CFU/ml)	Minimal inhibitory concentration ($\mu g/ml$) (serial dilution test; incubation times of 24 and 72 hr)
Staphylococcus aureus	250
Escherichia coli	1000
Klebsiella pneumoniae	500
Pseudomonas aeruginosa	1000
Pseudomonas fluorescens	1000
Pseudomonas cepacia	1000
Candida albicans	2000
Aspergillus niger	2000

About 60 times as potent as phenol, greater antimicrobial activity than p-chloro-m-cresol.

3. Toxicity

Primary skin irritation:	Less irritating than phenol or cresol
Guinea pig sensitization:	No sensitization

Listed in *The British Pharmacopoeia*, 1976 edition, registered with the EPA and FDA, Category I (OTC Panel) for safety.

4. Cosmetic Applications

Preservative in protein solutions, hair conditioners, silicone emulsions, children's cosmetics; EEC, 0.5%. Maximum concentration in toilet soaps and deodorant soaps 2%; for 2,4-dichloro-m-xylenol also 2%.

5. Other Applications

Topical and urinary antiseptic, preservative for pharmaceutical products, constituent in disinfectants

6. Supplier

Ottawa Chemical Division of Ferro Corp.

7. References

1. H. P. Fiedler, *Lexikon der Hilfsstoffe*, Vol. 1, 2nd ed., Editio Cantor, Aulendorf (1981).
2. S. M. Henry and G. Jacobs, *Cosmet. Toiletries*, 96:29 (1981).

CHLOROTHYMOL

Chemical names:	4-Chloro-5-methyl-2-(1-methylethyl)-phenol, 6-chloro-4-isopropyl-1-methyl-3-phenol, 6-chlorothymol (4-chlorothymol)
CAS No.:	89-68-9
Registered name:	KM 6
Type of compound:	Phenolic compound

1. Structure and Chemical Properties

Appearance:	Crystalline powder
Odor:	Characteristic odor
Solubility:	1 g dissolves in 1000 ml of water, 0.5 ml ethanol
Optimum pH:	4-8
Stability:	Stable
Compatibility and inactivation:	Incompatible with nonionics
Structural formula:	$C_{10}H_{13}ClO$ Mol. wt. 184.66

Antimicrobial Preservatives

[Chemical structure: 4-chloro-5-methyl-2-isopropylphenol — benzene ring with Cl, CH₃, OH, and CH(CH₃)₂ substituents]

Synthesis: Made by the action of sulfuryl chloride on thymol in CCl_4

2. Antimicrobial Spectrum

Test organisms (~10^6 CFU/ml)	Minimal inhibitory concentration (μg/ml) (serial dilution test; incubation times of 24 and 72 hr)
Staphylococcus aureus	500
Escherichia coli	1000
Klebsiella pneumoniae	1000
Pseudomonas aeruginosa	1000
Candida albicans	1500
Aspergillus niger	2000

Thymol and chlorothymol are both used as local antifungal agents, mainly as a dusting powder in the treatment of superficial fungal infections. Chlorothymol is intensely irritating to mucous membranes. From the halothymols, the 6-iodothymol is the most effective agent, followed by the corresponding bromo compound, and finally the chloro and fluorocompounds. Propylene glycol solutions have good antifungal activity against several pathogenic fungi such as *Trichophyton* spp., *Microsporum* spp., and *Candida albicans*.

3. Toxicity

Primary skin irritation: Irritating to mucous membranes

4. Cosmetic Applications

Use concentration, 0.1%; preservative

5. Other Applications

Germicide

6. Supplier

Gerbstoffchemie Margold Griesheim

CARVACROL

Chemical names: Isopropyl-o-cresol, 1-methyl-2-hydroxy-4-isopropylbenzene, p-cymenol

CAS No.: 499-75-2

Type of compound: Phenolic

1. Structure and Chemical Properties

Appearance:	Liquid; volatile with steam
Odor:	Thymol odor
Solubility:	Practically insoluble in water; freely soluble in alcohol
Structural formula:	$C_{10}H_{14}O$ Mol. wt. 150.21

$$\underset{(H_3C)_2CH}{\overset{OH\quad CH_3}{\text{[structure]}}}$$

Synthesis:	Chlorination of α-pinene with t-butyl hypochlorite

2. Antimicrobial Spectrum

Broad antimicrobial spectrum

3. Toxicity

Acute oral toxicity

Rabbit:	LD, 100 mg/kg

4. Cosmetic Applications

In-use concentration, 0.1%

5. Other Applications

Disinfectant; pharmaceutical antifungal; chlorocarvacrol is used instead of iodine for skin disinfection.

3,4,5,6-TETRABROMO-o-CRESOL

Chemical name:	3,4,5,6-Tetrabromo-o-cresol
EEC No.:	21
CAS No.:	576-55-6
Registered name:	Rabulen-TL
Type of compound:	Phenolic

1. Structure and Chemical Properties

Appearance:	White to buff crystalline powder; mp 205-208°C
Odor:	Cresol-like

Antimicrobial Preservatives

Solubility:	Practically insoluble in water; soluble in alcohol, alkali hydroxides, 1% in ethanol 75%, 1% in isopropanol 60%
Optimum pH:	4-4.5
Stability:	Stable in the absence of light under normal temperatures; short period of heating up to 50°C is possible
Compatibility and inactivation:	Incompatible with nonionics and iron salts
Structural formula:	$C_7H_4Br_4O$ Mol. wt. 423.76

Synthesis:	Bromination of o-cresol
Assay:	Determination of brome (75.43%); titration [1]

2. Antimicrobial Spectrum

Test organisms (~10^6 CFU/ml)	Minimal inhibitory concentration (μg/ml) (serial dilution test; incubation times of 24 and 72 hr)
Staphylococcus aureus	2.5
Escherichia coli	1250
Klebsiella pneumoniae	1250
Pseudomonas aeruginosa	2500
Pseudomonas fluorescens	1250
Pseudomonas cepacia	1250

3. Toxicity

Acute oral toxicity

Rat:	LD_{50}, 1.1 g/kg [1]
Mouse:	LD_{50}, 0.8 g/kg [2]

Subchronic toxicity

Mouse:	Oral administration of 0.6 ml of a 3% solution in oil over 9 days showed no toxic effect [2].
Rats:	2 ml of a 3% solution in oil over 6 days showed no toxic effects [2].

Primary skin irritation:	Irritating to skin and mucous membranes; in-use concentrations, no irritation [1]
Rabbit eye irritation:	No irritation [1]
Human skin test:	No irritation [2]

4. Cosmetic Applications — EEC maximum concentration, 0.3%; deodorant, 0.5-1.0%; shampoos

5. Other Applications — Fungicide

6. Supplier — Dr. Kurt Richter

7. References
 1. Dr. K. Richter, GMBH Berlin, Information and product specification, "Deodorant Richter/K", Berlin, October 24, 1979.
 2. Dr. K. Richter, GMBH Berlin, Information, "Wirkstoff CLR."

DICHLORO-m-XYLENOL

Chemical names:	2,4-Dichloro-3,5-xylenol, DCMX
EEC No.:	33
CAS No.:	133-53-9
CTFA adopted name:	Dichloro-m-xylenol
Type of compound:	Halogenated phenolic

1. Structure and Chemical Properties

Appearance:	White crystalline powder; volatile with steam
Solubility:	1 g soluble in 5 liters of water
Compatibility and inactivation:	Incompatible with nonionics, quaternary compounds, and proteins
Structural formula:	$C_8H_8Cl_2O$ Mol. wt. 191.06

2. Antimicrobial Spectrum

Test organisms ($\sim 10^6$ CFU/ml)	Minimal inhibitory concentration (μg/ml) (serial dilution test; incubation times of 24 and 72 hr)
Staphylococcus aureus	10
Escherichia coli	1000
Klebsiella pneumoniae	1000
Pseudomonas aeruginosa	1000
Candida albicans	1000
Aspergillus niger	1000

3. Cosmetic Applications — In baby cosmetics and soaps; EEC guideline, 0.1%

4. Other Applications — Antiseptic, preservative

o-PHENYLPHENOL

Chemical names:	o-Phenylphenol, 2-phenylphenol, o-hydroxydiphenyl, OPP, o-hydroxybiphenyl
EEC No.:	I/7
CAS No.:	90-43-7
Registered names:	Dowicide 1, Preventol O
Type of compound:	Phenolic

1. Structure and Chemical Properties

Appearance:	White, flaky crystals
Odor:	Mild characteristic odor
Solubility:	Practically insoluble in water; soluble in fixed alkali hydroxide; sodium salt tetrahydrate (in g/100 g of solvent) in water, 122; methanol, 138; propylene glycol, 28
Optimum pH:	Sodium salt, saturated water solution, pH 12.0-13.5
Compatibility and inactivation:	Incompatible with nonionics, carboxymethylcellulose, polyethylene glycols, quaternary compounds, and proteins
Structural formula:	$C_{12}H_{10}O$ Mol. wt. 170.20

[Structure: 2-hydroxybiphenyl (o-phenylphenol)]

Assay: Photometric [2]

2. Antimicrobial Spectrum

Test organisms (~10^6 CFU/ml)	Minimal germicidal concentration (μg/ml) (suspension test; contact times of 24 and 72 hr)	Minimal inhibitory concentration (μg/ml) (serial dilution test; incubation times of 24 and 72 hr)
Staphylococcus aureus	625	100
Escherichia coli	1250	500
Klebsiella pneumoniae	1250	500
Pseudomonas aeruginosa	1250	1000
Candida albicans	125	50
Aspergillus niger	125	50

3. Toxicity

Acute oral toxicity

  Rat: LD_{50}, 2.48 g/kg [1]

Subchronic toxicity

  Rat: 200 mg o-phenylphenol per kilogram per day over 32 days produced no toxic effects [3].

Chronic toxicity

  Rat: 0.2% o-phenylphenol added to feed over 2 years produced no toxic effects [4].

4. Cosmetic Applications

Maximum use concentration, 0.2% (EEC)

5. Other Applications

Germicide and fungicide preparations, preservative for citrus fruits

6. Suppliers

Gerbstoffchemie Margold Griesheim, Bayer Leverkusen

7. References

1. H. C. Hodge et al., *J. Pharmacol. Exp. Ther.*, *104*:202 (1952).
2. H. Böhme and G. Hofmann, *Z. Lebensm. Unters. Forsch.*, *114*:97 (1961).
3. F. C. Macintosh, *Analyst*, *70*:334 (1945).

7. References (Continued)

4. H. C. Hodge et al., *J. Pharmacol. Exp. Ther.*, 104:202 (1952).

4-ISOPROPYL-3-METHYLPHENOL

Chemical names:	3-Methyl-4-(1-methylethyl)-phenol, 4-isopropyl-m-cresol, 4-isopropyl-3-methylphenol, o-cymen-5-ol
EEC No.:	37
CAS No.:	3228-20-2
Registered names:	Biosol (Osaka Kasei, Japan)
Type of compound:	Substituted phenol

1. Structure and Chemical Properties

Appearance:	Colorless or white powder, white needlelike crystal, mp 110-113°C
Odor:	Odorless
Solubility (at 26-28°C):	In water, 0.03-0.04%; ethanol, 36%; isopropanol, 50%; ethylene glycol, 3.5%; propylene glycol, 8%; glycerin, 0.1%
Compatibility and inactivation:	Incompatible with nonionics and quaternary compounds
Structural formula:	$C_{10}H_{14}O$ Mol. wt. 150.22
Assay:	Ultraviolet absorbance at about 272 nm (Japanese Standards of Cosmetic Ingredients, Yakyi Nippon Ltd., 1979)

2. Antimicrobial Spectrum

Test organisms (~10^6 CFU/ml)	Minimal inhibitory concentration (µg/ml) (serial dilution test; incubation times of 24 and 72 hr)
Bacillus subtilis	200
Staphylococcus aureus	200
Streptococcus haemolyticus	100
Escherichia coli	60

2. Antimicrobial Spectrum
(Continued)

Test organisms ($\sim 10^6$ CFU/ml)	Minimal inhibitory concentration (μg/ml) (serial dilution test; incubation times of 24 and 72 hr)
Klebsiella pneumoniae	100
Pseudomonas aeruginosa	200
Pseudomonas fluorescens	100
Pseudomonas cepacia	200
Candida albicans	100
Saccharomyces cerevisiae	100
Aspergillus niger	100
Aspergillus oryzae	50
Microsporium japonicum	100

Good antifungal activity, used locally, mainly as a dusting powder; see also chlorothymol.

3. Toxicity [1]

Acute oral toxicity

  Mouse: LD, 5 g/kg

Acute intraperitoneal toxicity

  Mouse: LD_{50}, 0.47 g/kg

Primary skin irritation: No irritation or inflammation to skin

4. Cosmetic Applications
EEC, 0.1%

5. Other Applications
Active agent in local pharmaceutical products

6. Suppliers
Jan Dekker by Naarden International Chemicals Division, Wormerveer

7. References
1. Information Letter, Naarden International, Biosol 3-52040.

CHLOROPHENE

Chemical names:	2-Benzyl-4-chloro-phenol, 4-chloro-2-(phenylmethyl)phenol, chlorophene
EEC No.:	17
CAS No.:	120-32-1

Antimicrobial Preservatives

Registered names: Santophen 1, Septiphene, Chlorophen, Ketolin

Type of compound: Phenolic

1. Structure and Chemical Properties [1]

Appearance: Crystals; mp 48.5°C

Odor: Characteristic odor

Solubility: Water (25°C), 0.007%; propylene glycol, 80%; isopropanol, 85%; ethanol (70%), 87%

Compatibility and inactivation: Incompatible with nonionics, quaternary compounds, and proteins

Structural formula: $C_{13}H_{11}ClO$

Mol. wt. 218.69

Synthesis: Chlorination of o-benzylphenol

2. Antimicrobial Spectrum

Like the higher chlorinated compounds, only a very small spectrum of antimicrobial activity, especially against gram-positive bacteria.

3. Toxicity

Primary skin irritation: No irritation

4. Cosmetic Applications

EEC in-use concentration, 0.2%; preservative

5. Other Applications

Disinfectant, preservatives

6. References

1. H. P. Fiedler, *Lexikon der Hilfsstoffe*, Vol. 1, Editio Cantor, Aulendorf (1981), p. 173.

DICHLOROPHENE

Chemical names: 5,5'-Dichloro-2,2'-dihydroxydiphenylmethane, 2,2'-methylene-bis-(4-chlorophenol)

EEC No.: 29

CAS No.: 1215-74-3

Registered names: G-4, Preventol GD, Dichlorophen, DCP

Type of compound: Phenolic

1. Structure and Chemical Properties

Appearance: Crystals, mp 177-178°C

Solubility: Practically insoluble in water; 1 g soluble in 1 g ethanol (95%)

Optimum pH: Bactericidal effect better at pH 5-6; inhibitory activity at pH 6 or 7 is less than that at pH 8 [2]

Compatiblity and inactivation: Tween 80 is a neutralizing agent. Blood, serum, and milk markedly depress the bactericidal efficiency of diphenyl derivatives, but bacteriostasis can persist at very low concentrations.

Structural formula: $C_{13}H_{10}Cl_2O_2$
Mol. wt. 269.12

2. Antimicrobial Spectrum

Test organisms ($\sim 10^6$ CFU/ml)	Minimal inhibitory concentration (μg/ml) (serial dilution test; incubation times of 24 and 72 hr)
Staphylococcus aureus	3
Escherichia coli	3
Proteus vulgaris	3
Klebsiella pneumoniae	3
Pseudomonas aeruginosa	300
Candida albicans	30
Aspergillus niger	100
Penicillium notatum	100

3. Toxicity [1]

Acute oral toxicity

Mouse: LD_{50}, 1.2 g/kg
Rat: LD_{50}, 2.69 g/kg

Antimicrobial Preservatives

4. Cosmetic Applications	In-use concentration (EEC), 0.2%; in soaps maximum concentration, 1%; warnings on the label (Contains Dichlorophen).
5. Other Applications	Preservative
6. Suppliers	Givaudan, Bayer
7. References	1. H. P. Fiedler, *Lexikon der Hilfsstoffe*, Vol. 1, Editio Cantor, Aulendorf (1981). 2. G. Sykes, *Disinfection and Sterilization*, 2nd ed., Spon, London (1965), p. 319.

BROMOCHLOROPHENE

Chemical names:	3,3'-Dibromo-5'5'-dichloro-2,2'-dihydroxy-diphenylmethane, Bromophene
EEC No.:	20
CAS No.:	15435-29-7
Registered names:	Bromochlorophen, Bromophen
Type of compound:	Phenolic

1. Structure and Chemical Properties

Appearance:	White powder; mp 188-191°C
Odor:	Faint odor
Solubility:	In 95% ethanol, 5.5%; n-propanol, 7%; isopropanol, 4%; 1,2-propylene glycol, 2.5%; paraffin oil, 0.5%; glycerol, <0.1%; water, <0.1%
Optimum pH:	Bactericidal activity optimal at pH 5-6, bacteriostasis better at alkaline pH values (pH 8)
Stability:	Unstable in light
Compatiblity and inactivation:	Incompatible with Tween, blood, serum, milk
Structural formula:	$C_{13}H_8Br_2Cl_2O_2$ Mol. wt. 426.93

2. Antimicrobial Spectrum

Test organism	Minimum inhibitory concentration ($\mu g/ml$)
Staphylococcus aureus	1000
Escherichia coli	1000
Pseudomonas aeruginosa	10

Moderate fungistatic activity

3. Toxicity

Acute oral toxicity

  Rat: LD_{50}, 3.7 g/kg
  Mouse: LD_{50}, 1.55 g/kg

Acute dermal toxicity

  Rat: LD_{50}, >10 g/kg

Primary skin irritation: No effect, irritation, or sensitization

4. Cosmetic Applications

In-use concentration (EEC), 0.1%; deodorants maximum concentration, 1%

5. Other Applications

Antibacterial agent in pharmaceutical preparations for local use

6. Supplier

Merck Darmstadt

HEXACHLOROPHENE

Chemical names:	2,2'-Dihydroxy-3,3',5,5',6,6'-hexachlorodiphenylmethane, 2,2'-methylenebis[3,4,6-trichlorophenol]
EEC No.:	I/6
CAS No.:	70-30-4
Nonproprietary names:	Hexachlorophene, Hexachlorophane
Registered names:	Hexachlorophene, G-11, Hexosan, Gamophen
Type of compound:	Phenolic

1. Structure and Chemical Properties

Appearance:	White crystals; mp 164-165°C; anionic compound
Solubility:	Practically insoluble in water; soluble in alcohol, propylene glycol, polyethylene glycols, olive oil, cottonseed oil

Antimicrobial Preservatives

Optimum pH: Optimal for bactericidal effect pH 5-6, for bacteriostasis very alkaline pH (pH 8)

Compatiblity and inactivation: Incompatible with Tween

Structural formula: $C_{13}H_6Cl_6O_2$
Mol. wt. 406.92

Assay: Gas chromatography USP XX (1980)

2. Antimicrobial Spectrum

Test organisms (~10^6 CFU/ml)	Minimal inhibitory concentration (μg/ml) (serial dilution test; incubation times of 24 and 72 hr)
Staphylococcus aureus	0.5
Escherichia coli	12.5
Klebsiella pneumoniae	12.5
Pseudomonas aeruginosa	250
Pseudomonas fluorescens	250
Pseudomonas cepacia	250
Candida albicans	1000
Aspergillus niger	300
Penicillium notatum	300

Optimal pH for bactericidal activity is pH 5-6; 1000 ppm kills *Staphylococcus aureus* within 1-10 min. The inhibitory effect is greater at pH 8 than under acidic conditions.

3. Toxicity

Acute oral toxicity

Rat: LD_{50}, 59 mg/kg

Chronic toxicity: See Ref. 2

Influence on skin: High absorption-resorption coefficient; accumulation in stratum corneum

Human toxicity: Potential neurotoxicity in humans, the FDA has regulated its use [1]

Mutagenicity: Negative Ames test

Teratogenicity: No effect [2]

4. Cosmetic Applications	EEC guideline maximum dose for cosmetics, 0.1%; not to be used in preparations for children and intimate hygiene; warnings on the label: Not To Be Used for Babies. Contains Hexachlorophene; not for children under 3 years of age
5. Other Applications	Topical anti-infective, pHisohex disinfection
6. Suppliers	Givaudan, Analabs
7. References	1. J. D. Lockhart, *Pediatrics*, 50:229 (1972). 2. B. P. Vaterlaus and J. J. Hostynek, *J. Soc. Cosmet. Chem.*, 24:291 (1973).

TRICLOSAN

Chemical name:	2,4,4'-Trichloro-2'-hydroxydiphenylether
EEC No.:	28
CAS No.:	3380-34-5
Registered names:	Irgasan DP 300, Triclosan
Type of compound:	Diphenyl ether

1. Structure and Chemical Properties

Appearance:	White crystalline powder
Odor:	Faint aromatic odor; mp 60-61°C
Solubility (20°C):	In water, 0.001%; ethanol (70%), ~100%; propylene glycol, 100%; methyl cellosolve, 100%; vegetable oils, 60-90%; glycerol, 0.15%; poor solubility in alkali hydroxyde solution; soluble in different organic solvents
Optimum pH:	4-8
Stability:	Thermostable up to 280°C; labile in the presence of bleaching agents with active chlorine
Compatibility and inactivation:	Neutralized by Tween and lecithin
Structural formula:	$C_{12}H_7O_2Cl_3$ Mol. wt. 289.5

Assay: Gas chromatography [2]; thin-layer chromatography [3]

2. Antimicrobial Spectrum

Test organisms (~10^6 CFU/ml)	Minimal germicidal concentration (μg/ml) (suspension test; contact times of 10 min)	Minimal inhibitory concentration (μg/ml) (serial dilution test; incubation times of 24 and 72 hr)
Staphylococcus aureus	25	0.1
Streptococcus faecalis		10
Escherichia coli	500	5
Klebsiella pneumoniae		5
Proteus vulgaris		1
Pseudomonas aeruginosa		>300
Candida albicans	25	10
Saccharomyces cerevisiae		10
Aspergillus niger		100
Trichophyton spp.		10
Penicillium notatum		50
Chaetomium globosum		100

3. Toxicity

Acute oral toxicity [1]

  Rat: LD_{50}, 4.53 to >5 g/kg
  Dog: LD_{50}, >5 g/kg

Subchronic toxicity

  Rat: Oral administration of 0.45 g/kg per day over 4 weeks produced no toxic reactions

Primary skin irritation

  Rat: Cutaneous application over 4 weeks produced no irritation

Rabbit eye irritation: No irritation

Guinea pig sensitization: No sensitization

Human sensitization: No effect (epicutaneous test)

Photosensitization: No effect

Carcinogenicity (mouse): Skin application 3 times per week over 18 months produced no carcinogenic reaction

4. Cosmetic Applications

In-use concentration, 0.1-0.3% in deodorants, shampoos, soap; EEC, 0.3%

5. Other Applications

Skin disinfectants, preservative

6. Supplier   Ciba Geigy GmbH

7. References   1. H. P. Fiedler, *Lexikon der Hilfsstoffe*, Editio Cantor, Aulendorf (1981).
                                    2. M. König, Z. Anal. Chem., 266:119 (1973).
                                    3. C. H. Wilson, J. Soc. Cosmet. Chem., 26:75 (1975).

TRICLOCARBAN

Chemical names: 3,4,4'-Trichlorocarbanilide, TCC, N-(4-chlorophenyl)-N'-(3,4-dichlorophenyl)urea

EEC No.:   25

CAS No.:   101-20-2

Nonproprietary name: Trichlorocarbanilide

Type of compound: Carbanilide

1. Structure and Chemical Properties

Appearance: Fine white plates; mp 255.2-256°C [2]

Solubility: Sparingly soluble in water; soluble in acetone and polyethylene glycol

Compatibility and inactivation: Incompatible with nonionics (Tween 80), phosphatides, and proteins; compatible with anionic and cationic tensides

Structural formula: $C_{13}H_9Cl_3N_2O$
                                    Mol. wt. 315.59 [1]

$$Cl-\!\!\langle\!\!\!\!\bigcirc\!\!\!\!\rangle\!\!-NH-CO-HN-\!\!\langle\!\!\!\!\bigcirc\!\!\!\!\rangle\!\!-Cl\ \ (Cl)$$

Synthesis: Preparation from 3,4-dichloraniline and 4-chlorophenylisocyanate

Assay: Thin-layer chromatography [4].

2. Antimicrobial Spectrum

Test organisms ($\sim 10^6$ CFU/ml)	Minimal inhibitory concentration (μg/ml) (serial dilution test; incubation times of 24 and 72 hr)
Staphylococcus aureus	0.5
Escherichia coli	10
Proteus vulgaris	1

2. Antimicrobial Spectrum
   (Continued)

Test organisms ($\sim 10^6$ CFU/ml)	Minimal inhibitory concentration (μg/ml) (serial dilution test; incubation times of 24 and 72 hr)
Klebsiella pneumoniae	500
Pseudomonas aeruginosa	>1000
Candida albicans	500
Aspergillus niger	500
Penicillium notatum	500
Trichophyton gypseum	50
Chaetomium globosum	100

3. Toxicity

Acute oral toxicity

  Mouse: LD_{50}, 0.6 g/kg
  Rat: LD_{50}, 3.6 g/kg

Rabbit eye irritation: 2% solution shows no irritation

Human skin toxicity: Cutaneous application results in only low blood levels (<25 ppb) [3], most of the TCC remained in the stratum corneum; only a small amount penetrated through the epidermis.

4. Cosmetic Applications

Bacteriostatic and antiseptic in soaps, antiperspirants, and other cleansing compositions; in-use concentration (EEC), 0.2%; should not be used by those dealing with newborn infants (some causes of methemoglobinemia have been reported) [5].

5. Other Applications

Disinfectant; should not be used in maternity units and by those dealing with newborn infants [5]

6. Suppliers

Monsanto, Ferak

7. References

1. *Merck Index*, 9th ed., Rahway, N.J. (1976).
2. H. P. Fiedler, *Lexikon der Hilfsstoffe*, Editio Cantor, Aulendorf (1981).
3. D. Howes and J. G. Black, *Toxicology*, 6:67 (1976).
4. C. H. Wilson, *J. Soc. Cosmet. Chem.*, 26:75 (1975).
5. F. Hackenberger, Antiseptic drugs, in *Meyler's Side Effects of Drugs*, 9th

7. References
(Continued)

ed., Excerpta Medica, Amsterdam (1980), p. 402.

CLOFLUCARBAN

Chemical names:	N-(4-Chlorophenyl)-N'-[4]chloro-3-(trifluoromethyl)phenyl]urea, 4,4'-dichloro-3-(3-fluoromethyl)-carbanilide, halocarban
EEC No.:	27
CAS No.:	369-77-7
CTFA adopted name:	
Registered name:	Irgasan CF 3
Type of compound:	Carbanilide

1. Structure and Chemical Properties [1]

Appearance:	White crystalline solid; mp 214-215°C
Solubility:	Insoluble in water; soluble in organic solvents such as ethanol, isopropanol, polyethylene glycol 400, Tween 20
Stability:	Good; unstable in strong alkaline solutions
Compatibility and inactivation:	Incompatible with proteins and strong alkalis; compatible with nonionics, anionics, and cationic compounds
Structural formula:	$C_{14}H_9Cl_2F_3N_2O$ Mol. wt. 349.15

$$Cl-\underset{CF_3}{\underset{|}{C_6H_3}}-NHCNH-C_6H_4-Cl$$
$$\quad\quad\quad\quad\quad\overset{\|}{O}$$

Assay:	High-power liquid chromatography [2]

2. Antimicrobial Spectrum

Test organisms (~10^6 CFU/ml)	Minimal inhibitory concentration (μg/ml) (serial dilution test; incubation times of 24 and 72 hr)
Staphylococcus aureus	5
Escherichia coli	250
Klebsiella pneumoniae	250
Pseudomonas aeruginosa	500
Pseudomonas fluorescens	500
Pseudomonas cepacia	500

2. Antimicrobial Spectrum (Continued)

Test organisms ($\sim 10^6$ CFU/ml)	Minimal inhibitory concentration (μg/ml) (serial dilution test; incubation times of 24 and 72 hr)
Candida albicans	1000
Aspergillus niger	2000
Penicillium notatum	1000

3. Cosmetic Applications   In-use concentration (EEC), 0.3%; maximum concentration in aerosols, 0.2%; deodorants and soaps, 1.5%

4. Other Applications   Disinfectant

5. Suppliers   Ciba Geigy, Cyanamid

6. References
   1. *Merck Index*, 9th ed., Rahway, N.J. (1976).
   2. T. Wolf and D. Semionow, *J. Soc. Cosmet. Chem.*, 24:363 (1973).

PROPAMIDINE ISETHIONATE

Chemical names: 4,4-Diaminodiphenoxypropane, p,p'-(trimethylenedioxy)dibenzamidine-bis-(β-hydroxyethane sulfonate), 4,4'-diamidino-α,ω-diphenoxypropane isethionate

Type of compound: Benzamidine

History: First described by Ashley et al. [2]

1. Structure and Chemical Properties

Appearance: Bitter crystals or granular powder; mp 235°C; hygroscopic

Odor: Odorless

Solubility: In water, 1 part in 5 parts; glycerol, 95%; in alcohol, about 1 part in 32 parts; pH of a 5% solution in water 4.5-6.5

Optimum pH: 6.3-7.6, with increasing pH the activity of propamidine increases.

Stability: Precipitated by PO_4^{3-} (0.1 M); phospholipids (soy bean lecithin), polyamines (triethylenetetramine) and amino and oxo acids antagonized antibacterial activity [1].

Compatibility and inactivation: 10% serum or blood decreases the sensitivity of *Staphylococcus aureus* and *Escherichia coli*, but not that of *Pseudomonas aeruginosa*.

Structural formula: $C_{21}H_{32}N_4O_{10}S_2$
Mol. wt. 564.63

$$\text{OCH}_2\text{–CH}_2\text{–CH}_2\text{O}\; \left\langle \text{Ar–C(=NH)NH}_2 \right\rangle_2 \cdot \left[\begin{array}{c} CH_2OH \\ CH_2SO_3H \end{array} \right]_2$$

2. Antimicrobial Spectrum

Test organisms (~10^6 CFU/ml)	Minimal germicidal concentration (μg/ml) (suspension test; contact times of 24 and 72 hr)	Minimal inhibitory concentration (μg/ml) (serial dilution test; incubation times of 24 and 72 hr)
Staphylococcus aureus	4	2
Streptococcus faecalis		25
Escherichia coli	128	64
Proteus vulgaris	128	256
Klebsiella pneumoniae		256
Salmonella enteriditis		256
Pseudomonas aeruginosa	256	256
Clostridium perfringens		32
Trichophyton tonsurans		100
Epidermophyton floccosum		250

Active against a number of pathogenic fungi (glucose-agar dilution test). Soy bean lecithin will antagonize its antifungal activity.

3. Toxicity

Acute intravenous toxicity

Mouse: LD_{50}, 42 mg/kg

Acute subcutaneous toxicity

Mouse: LD_{50}, 55 mg/kg

4. Cosmetic Applications

Use concentration, 0.1%; only for topical use

5. Other Applications

Antimicrobial, antiseptic

6. References

1. W. B. Hugo, *Inhibition and Destruction of the Microbial Cell*, Academic, London (1971), pp. 123-129.
2. J. N. Ashley et al., *J. Chem. Soc.*, 103-116 (1942).

DIBROMOPROPAMIDINE

Chemical name:	4,4'-(Trimethylenedioxy)-bis-(3-bromobenzamidine)diisethionate
EEC No.:	9
CAS No.:	614-87-9
Registered name:	Brolene
Type of compound:	Benzamidine

1. Structure and Chemical Properties

Appearance:	White, crystalline; mp 226°C
Solubility:	Freely soluble in water, 50%; 1 g dissolves in 60 g of 95% ethanol; soluble in glycerol
Optimum pH:	5.5-8; if the pH increases, the antimicrobial activity increases too.
Stability:	100°C for 30 min; aqueous solution may decompose slightly after prolonged storage.
Compatibility and inactivation:	Incompatible with chlorides, sulfates, anionics; blood and serum reduce the antimicrobial activity (factor of 2-4) [2]
Structural formula [1]:	$C_{21}H_{30}Br_2N_4O_{10}S_2$ Mol. wt. 722.45

$$H_2N-C(=NH)-C_6H_3(Br)-O(CH_2)_3O-C_6H_3(Br)-C(NH_2)=NH \cdot \left[\begin{array}{c} CH_2OH \\ CH_2SO_3H \end{array} \right]_2$$

2. Antimicrobial Spectrum

Test organisms (~10^6 CFU/ml)	Minimal germicidal concentration (μg/ml) (suspension test; contact times of 24 and 72 hr)	Minimal inhibitory concentration (μg/ml) (serial dilution test; incubation times of 24 and 72 hr)
Staphylococcus aureus		1
Streptococcus pyogenes		1
Escherichia coli	32	4

2. Antimicrobial Spectrum
(Continued)

Test organisms ($\sim 10^6$ CFU/ml)	Minimal germicidal concentration (μg/ml) (suspension test; contact times of 24 and 72 hr)	Minimal inhibitory concentration (μg/ml) (serial dilution test; incubation times of 24 and 72 hr)
Proteus vulgaris	256	128
Pseudomonas aeruginosa	64	32
Clostridium perfringens		512
Trichophyton tonsurans		25

As with propamidine, dibromopropamidine is more active against gram-positive non-spore-forming organisms than against gram-negative bacteria or *Clostridia*. Halogenation produces an increase in antibacterial activity, but not general enhancement of antifungal activity. Cross-resistance was demonstrated to other diamidines.

3. Toxicity

Acute intravenous toxicity

  Mouse: LD, 10 mg/kg

Acute subcutaneous toxicity

  Mouse: LD, 300 mg/kg

Primary skin irritation

  Guinea pig: Minimum toxic concentration, 0.05 g/100 ml

4. Cosmetic Applications
In-use concentration (EEC), 0.1%

5. Other Applications
Antiseptic

6. Supplier
May and Baker

7. References
1. *Merck Index*, 9th ed., Rahway, N.J. (1976).
2. W. B. Hugo, *Inhibition and Destruction of the Microbial Cell*, Academic, London (1971), p. 129.

HEXAMIDINE ISETHIONATE

Chemical names: 1,6-Di(4-amidinophenoxy)-n-hexane, p,p'-(hexamethylenedioxy)dibenzamidine-bis-(β-hydroxyethane sulfonate), 4,4'-hexamethylene dioxybenzamidine and salts

Antimicrobial Preservatives

EEC No.:	7
CAS No.:	3811-75-4
Registered names:	Hexamidin, Desomedine, Esomedina, Hexomedine
Type of compound:	Benzamidine

1. Structure and Chemical Properties

Appearance:	(Compare propamidine)
Solubility:	Salts are soluble in water and alcohol; insoluble in oils
Compatibility and inactivation:	Incompatible with chloride and sulfate ions, anionics, and proteins
Structural formula:	$C_{29}H_{38}N_4O_{10}S_2$ Mol. wt. 606.72

$$\text{OCH}_2\text{-CH}_2\text{-CH}_2\text{-CH}_2\text{-CH}_2\text{-CH}_2\text{O}$$

(structure: two para-substituted benzamidine groups linked by a hexamethylenedioxy chain)

2. Antimicrobial Spectrum

(See Propamidine)

3. Cosmetic Applications

Use concentration (EEC), 0.1%

4. Other Applications

Topical antiseptic

5. Supplier

May and Baker

SODIUM PYRITHIONE

Chemical names:	Pyridine-1-oxide-2-thiol-sodium salt, sodium-2-pyridinethiol-1-oxide
EEC No.:	40
CAS No.:	3811-73-2
CTFA adopted name:	Sodium pyrithione
Registered names:	Sodium Omadine, Pyrion-Na
Type of compound:	Cyclic thiohydroxamic acid (salt), pyridine derivative
History:	First synthesized by Olin Chemical, Stamford, in 1948 [3]

1. Structure and Chemical Properties [2]

Appearance:	White to yellowish powder; mp 250°C
Odor:	Mild odor
Solubility:	Soluble in water, 53%; ethanol, 19%; PEG 400, 12%
Optimum pH:	7-10; 2% solution has a pH of 8.0
Stability:	Unstable to light and oxidizing agents and strong reducing agents
Compatibility and inactivation:	Slight inactivation by nonionics, chelating reaction with heavy metal ions
Structural formula:	C_5H_4NOSNa
	Mol. wt. 149.2

2. Antimicrobial Spectrum

Test organisms (~10^6 CFU/ml)	Minimal inhibitory concentration ($\mu g/ml$) (serial dilution test; incubation times of 24 and 72 hr)
Staphylococcus aureus	1
Streptococcus faecalis	2
Escherichia coli	8
Salmonella typhimurium	64
Pseudomonas aeruginosa	512
Candida albicans	4
Trichophyton mentagrophytes	0.5
Aspergillus niger	2
Penicillium notatum	2

3. Toxicity [1,3]

Acute oral toxicity

Rat:	LD_{50}, 875 mg/kg
Mouse:	LD_{50}, 1172 mg/kg

Subchronic toxicity

Rat:	Oral administration of 75 mg/kg per day over 30 days and of intraperitoneal administration 40 mg/kg per day, produced no pathological effects

Rabbit eye irritation:	10% solution, pH 7.8-7.9, no effect |"

4. Cosmetic Applications

Use concentration, 250-1000 ppm (active basis); EEC guideline, 0.5%; only for rinse-off products

5. Other Applications

Preservative for cutting oils, 0.005%

6. Suppliers

Olin Chemical, available as a powder or a 40% aqueous solution; Riedel de Haen

7. References

1. H. P. Fiedler, *Lexikon der Hilfsstoffe*, Vol. 2, Editio Cantor, Aulendorf (1981), p. 681.
2. S. M. Henry and G. Jacobs, *Cosmet. Toiletries*, 96:37 (1981).
3. Olin Chemical, Omadine, Information No. 735-026 (1980).

ZINC PYRITHIONE

Chemical names:	Zinc bis-(2-pyridinethiol-1-oxide)bis-(2-pyridylthio)zinc-1,1'-dioxide, bis-[1-hydroxy-2(1H)-pyridinethionato-O,S]-(T-4) zinc
EEC No.:	I/8
CAS No.:	13463-41-7
CTFA adopted name:	Zinc pyrithione
Registered names:	Zinc Omadine, Vancide
Type of compound:	Cyclic thiohydroxamic acid (zinc complex)
History:	First synthesized in 1948 by Olin Chemical [1]

1. Structure and Chemical Properties

Appearance:	White to yellowish crystalline powder
Odor:	Mild odor
Solubility:	Soluble in water, 15 ppm, at pH 8, 35 ppm; ethanol, 100 ppm; PEG 400, 2000 ppm
Optimum pH:	4.5-9.5; 10% dispersion, pH 3.6
Stability:	Forms an insoluble product with some cationics and amphoterics; unstable to light and oxidizing agents, unstable in acid or alkaline solutions at higher temperatures
Compatibility and inactivation:	Incompatible with EDTA, slight inactivation by nonionics; in the presence of heavy metal ions chelation or transchelation will occur; some complexes are very insoluble.

Structural formula: $C_{10}H_8N_2S_2Zn$
Mol. wt. 317.7

2. Antimicrobial Spectrum

Test organisms (~10^6 CFU/ml)	Minimal inhibitory concentration (μg/ml) (serial dilution test; incubation times of 24 and 72 hr)
Staphylococcus aureus	4
Streptococcus faecalis	16
Escherichia coli	16
Salmonella typhimurium	16
Pseudomonas aeruginosa	512
Candida albicans	0.25
Aspergillus niger	2
Penicillium vermiculatum	1
Trichophyton mentagrophytes	0.25

3. Toxicity

Acute oral toxicity

Rat: LD_{50}, 200 mg/kg
Mouse: LD_{50}, 300 mg/kg

Subchronic toxicity

Rat: 10 ppm (about 1 mg/kg per day) as food additive over 30 days shows no effect; higher doses are toxic.

Chronic toxicity: Ointments with different concentrations of zinc pyrithione (use concentration) produces no irritation to rabbits and human skin.

Primary skin irritation: 48% dispersion and powder are irritating to the skin and extremely irritating to rabbits eyes.

Guinea pig sensitization: Not an allergic sensitizer

Antimicrobial Preservatives

4. Cosmetic Applications — Use concentration, 250-1000 ppm (active basis) in gels, creams, heavy lotions, talcum powder, shampoos; EEC guideline; 0.5% for rinse-off products only

5. Other Applications — Sanitized paper goods, deodorized woolens, sanitary hospital goods

6. Suppliers — Olin Chemical (Zinc Omadine), Ruetgers-Nease (Zinc Pyrion)

7. References
   1. Olin Chemical, Bulletin No. 735-026 (1980).

HEXETIDINE

Chemical names: 5-Amino-1,3-bis-(2-ethylhexyl)-5-methyl-hexahydropyrimidine, 1,3-bis-(2-ethylhexyl)hexahydro-5-methyl-5-pyrimidine

EEC No.: 15

CAS No.: 141-94-6

Registered names: Hexetidine, Hextril, Hexatidine

Type of compound: N-Acetal

1. Structure and Chemical Properties [1]

Appearance: Colorless liquid

Odor: Aminelike odor

Solubility: Soluble in methanol; insoluble in water; freely soluble in ethanol and acetone

Stability: Thermostable

Structural formula: $C_{21}H_{45}N_3$
Mol. wt. 339.59

2. Antimicrobial Spectrum

Test organisms ($\sim 10^6$ CFU/ml)	Minimal inhibitory concentration (μg/ml) (serial dilution test; incubation times of 24 and 72 hr)
Staphylococcus aureus	5
Escherichia coli	1250
Klebsiella pneumoniae	>10,000
Pseudomonas aeruginosa	>10,000
Candida albicans	5000
Aspergillus niger	10
Penicillium notatum	5

3. Toxicity

Acute oral toxicity

  Rat: LD, 1 g/kg

4. Cosmetic Applications

In-use concentration (EEC), 0.2%; widely advertised as an oral disinfectant (1 mg/ml in mouthwashes), antibacterial and antifungal effects and also mild local anesthetic properties when applied to the mucosa [2]; preservative with good fungistatic activity

5. Other Applications

In pharmaceuticals 100 mg hexetidine in 100 ml of solution for mouthwashes against *Candida albicans* infections

6. Suppliers

Commercial Solvents Corp., Gödecke A.G., Woelm-Pharma

7. References

1. H. P. Fiedler, *Lexikon der Hilfsstoffe*, Vol. 1, Editio Cantor, Aulendorf (1981), p. 464.
2. F. Hackenberger, Antiseptic drugs, in *Meyler's Side Effects of Drugs*, 9th ed. (M. N. G. Dukes, ed.), Excerpta Medica, Amsterdam (1980), p. 394.

TRIS(HYDROXYETHYL)-s-TRIAZINE

Chemical name:	Tris-hydroxyethylhexahydrotriazine, Hexahydrotriazine
EEC No.:	35
CAS No.:	4719-04-4
Registered names:	Bacillat 35, Grotan BK, Bakzid 80 (mixture of different derivatives), KM 200

Antimicrobial Preservatives

Type of compound: N-Acetal

1. Structure and Chemical Properties

Appearance: Yellowish solution

Odor: Aminelike

Solubility: Soluble in water, alcohol, and acetone

Optimum pH: Alkaline range (pH 8-10)

Stability: Stable in alkaline solutions; changes by heating to oxazolidine

Compatibility and inactivation: Compatible with anionics and nonionics

Structural formula [1]: $C_9H_{21}O_3N_3$
Mol. wt. 219.10

R: $-CH_2-CH_2OH$

2. Antimicrobial Spectrum

Test organisms ($\sim 10^6$ CFU/ml)	Minimal germicidal concentration (μg/ml) (suspension test; contact times of 24 and 72 hr)		Minimal inhibitory concentration (μg/ml) (serial dilution test; incubation times of 24 and 72 hr; pH 8.8)
Staphylococcus aureus			156-312
Escherichia coli			312
Proteus mirabilis			156-312
Klebsiella pneumoniae			312
Pseudomonas aeruginosa	200	50	312
Pseudmonoas fluorescens			312
Pseudmonas cepacia			312
Candida albicans			625-1250
Candida spp.	200	100	
Aspergillus niger	500	100	625-1250

3. Toxicity

Acute oral toxicity

Rat: LD_{50}, 1200 mg/kg (Bakzid 80)
Mouse: LD_{50}, 950 mg/kg (Bakzid 80)

Acute dermal toxicity

Rat: LD_{50}, >5 ml/kg

Subacute dermal toxicity

    Rabbit: 1 and 5% Bakzid 80 solution, 2 ml/kg daily, no effect (21 days) [4]

Human skin irritation: In-use concentration, no irritation; with 0.5% solution, 97 out of 100 probands had no irritation; 3 formaldehyde sensitive. Contact dermatitis observed in some cases; 4 positive [2] out of 300 persons.

Rabbit eye irritation: Irritation in the concentrated form

Guinea pig sensitization: (Magnusson-Kligman test): Induction 1 and 10%, provocation 10%, no sensitization (Bakzid 80) [4]

Mutagenicity: Nonmutagenic [3]

Environmental toxicity

    Rainbow trout: LC_{50}, 69.5 ml liter (96 hr) (Bakzid 80) [4]

4. Cosmetic Applications — Use concentration (EEC), 0.3%

5. Other Applications — (Bakzid 80): Cutting oils, emulsions, suspensions as preservative

6. Suppliers — Gerbstoffchemie Margold, Pennwalt Corp.; Bakzid 80, Dr. Bode and Co.

7. References
   1. H. P. Fiedler, *Lexikon der Hilfsstoffe*, Vol. 1, Editio Cantor, Aulendorf (1981).
   2. K. Keczkes and P. M. Brown, *Contact Dermatitis*, 2:212 (1977).
   3. C. Urwin et al., *Mutat. Res.*, 40:43 (1976).
   4. Information Bacillolfabrik, Dr. Bode, Hamburg, January 1982.

CAPTAN

Chemical name: N-(Trichloromethylthio)-4-cyclohexene-1,2-dicarboximide

EEC No.: 30

CAS No.: 133-06-2

Nonproprietary name: Captan

CTFA adopted name:

Registered names: Vancid 89 RE, Advacide TMP

Type of compound: Phthalimid derivative

Antimicrobial Preservatives

1. Structure and Chemical Properties

Appearance:	Crystals; mp 172-173°C
Odor:	Odorless
Solubility:	Practically insoluble in water; in ethanol, 0.29%
Optimum pH:	Acidic pH
Stability:	Stable in acidic pH
Compatibility and inactivation:	Compatible with nonionics, anionics, and cationic compounds at pH <7
Structural formula:	$C_9H_8Cl_3NO_2S$ Mol. wt. 300.61

<chemical structure: cyclohexene fused with imide ring, N–S–CCl₃> [3]

Assay: Colorimetry [1]

2. Antimicrobial Spectrum

Test organisms (~10^6 CFU/ml)	Minimal germicidal concentration ($\mu g/ml$) (suspension test; contact times of 24 and 72 hr)	Minimal inhibitory concentration ($\mu g/ml$) (serial dilution test; incubation times of 24 and 72 hr) (agar dilution test)
Staphylococcus aureus	50	200
Proteus spp.	250	50-400
Escherichia coli	1000	50-200
Shigella spp.		200-400
Salmonella spp.		100-200
Pseudomonas aeruginosa	1000	200-2000
Pseudomonas fluorescens		1000
Pseudomonas cepacia		1000
Candida albicans		1000
Aspergillus niger		2000
Penicillium notatum		2000

Dilutions prepared in 3:1 acetone:water mixture

3. Toxicity

Acute oral toxicity
- Rat: LD_{50}, 9 g/kg
- Rabbit: LD_{50}, 2 g/kg

Acute intraperitoneal toxicity
- Rat: LD_{50}, 50-100 mg/kg

Human toxicity: Ingestion of large quantities may cause vomiting and diarrhea.

Primary skin irritation
- Rat: Application of 30 ml of a 20% suspension 1-2 times per week on 15% of the skin shows no toxic effects, only slight flush

Chronic toxicity
- Rat: Oral application of 0.025, 0.25, and 1% daily. Only with the high concentration was growth suppressed, but no other toxic effects from all other concentrations.

Human patch test: Aqueous paste with 50% active compound contact with skin for 24 hr shows no irritation

Rabbit eye irritation: 0.1 ml of 0.18% solution, no effect

Mutagenicity (mice): Nonmutagenic [2]

Teratogenicity: No effect

4. Cosmetic Applications: In-use concentration (EEC), 0.5%; bacteriostatic in soap and shampoos

5. Other Applications: Agricultural fungicide

6. Suppliers: Vanderbilt, Ciba-Geigy, Lehman and Voss

7. References
1. A. R. Kittleson, *Anal. Chem.*, *24*:1173 (1952).
2. G. L. Kennedy et al., *Food Cosmet. Toxicol.*, *13*:55 (1975).
3. W. B. Hugo, *Inhibition and Destruction of the Microbial Cell*, Academic, London (1971), p. 670

DIMETHOXANE

Chemical names: 6-Acetoxy-2,4-dimethyl-m-dioxane, 2,6-dimethyl-1,3-dioxan-4-ol-acetate

EEC No.: 1

CAS No.: 828-00-2

Antimicrobial Preservatives

Nonproprietary name: GIV-GARD DXN
CTFA adopted name: Dimethoxane
Registered name: Dioxin CO
Type of compound: o-Acetal

1. Structure and Chemical Properties

Appearance:	Colorless liquid
Odor:	Mustardlike odor
Solubility:	Miscible with water
Optimum pH:	Broad range
Stability:	Hydrolyzes in aqueous solutions; stabilization with alkali
Compatibility and inactivation:	Discoloration in the presence of amines and amides; compatible with nonionic emulsifiers
Structural formula [1]:	$C_8H_{14}O_4$ Mol. wt. 174.2

2. Antimicrobial Spectrum

Test organisms ($\sim 10^6$ CFU/ml)	Minimal inhibitory concentration (μg/ml) (serial dilution test; incubation times of 24 and 72 hr)
Staphylococcus aureus	1250
Bacillus subtilis	625
Escherichia coli	625
Salmonella typhosa	625
Aerobacter aerogenes	625
Pseudomonas aeruginosa	625
Pseudomonas fluorescens	625
Candida albicans	1250
Saccharomyces cerevisiae	2500
Aspergillus niger	1250
Aspergillus oryzae	1250
Penicillium piscarium	625

3. Toxicity

Acute oral toxicity

Rat: LD_{50}, 1.9 ml/kg

Primary skin irritation:	1% solution, no effect (human skin); no sensitization
Rabbit eye irritation:	1% solution, no effect
Carcinogenicity:	No oncogenic effect when applied biweekly to the intact skin of male and female mice for 80 consecutive weeks (~0.1 ml of GIV-2-0494 1% in H_2O and acetone) [3]

4. **Cosmetic Applications**   0.1%; EEC proposal, 0.2%

5. **Other Applications**   Preservative for cutting oils, resin emulsions, water-based paints, 0.03-0.1%, gasoline additive

6. **Supplier**   Givaudan Corp.

7. **References**
   1. H. P. Fiedler, *Lexikon der Hilfsstoffe*, Vol. 1, Editio Cantor, Aulendorf (1981).
   2. Givaudan Information.
   3. Food and Drug Research Laboratories, Inc., Lab. Report No. 2459, March 15, 1977.

OXADINE A

Chemical name:	4,4-Dimethyl-1,3-oxazolidine
EEC No.:	—
CAS No.:	51 200-87-4
Registered name:	Oxadine A
CTFA adopted name:	Dimethyl oxazolidine
Type of compound:	Cyclic substituted amine, oxazolidine

1. Structure and Chemical Properties

Appearance:	Colorless liquid
Odor:	Penetrating aminelike odor
Solubility:	Completely water soluble
Optimum pH:	6.0 (6.0-11.0)
Stability:	Unstable below pH 5.0
Compatibility and inactivation:	Compatible with cationic, anionic, and nonionic systems over the pH range 5.5-11

Antimicrobial Preservatives

Structural formula:

```
      CH3
H3C\  |
     C—CH2
HN /    \
    C    O
    |
    H2
```

2. Antimicrobial Spectrum

Test organisms ($\sim 10^6$ CFU/ml)	Minimal germicidal concentration (μg/ml) (suspension test; contact times of 24 and 72 hr)[a]	Minimal inhibitory concentration (μg/ml) (serial dilution test; incubation times of 24 and 72 hr)	
		Ref. 2	Ref. 1
Staphylococcus aureus	500	500	125-250
Escherichia coli	500	500	250-500
Klebsiella pneumoniae	500	500	
Pseudomonas aeruginosa	500	500	250-500
Pseudomonas fluorescens	250	500	250-500
Pseudomonas cepacia	500	500	
Candida albicans	1000	>1000	500-1000
Aspergillus niger	500	>1000	250-500
Penicillium notatum	1000	>1000	

[a]See Ref. 2.

Effective against a broad spectrum of microorganisms over a wide pH-range.

3. Toxicity

Acute oral toxicity
  Male rat: LD_{50}, 950 mg/kg

Acute dermal toxicity
  Rabbit: LD_{50}, 1400 mg/kg

Inhalation test
  Rat: LC_{50}, 11.7 mg/liter

Rabbit eye irritation: 5000 ppm in water has no discernible effect upon the eyes.

Mutagenicity: Ames test, nonmutagenic

Human sensitization: At use levels, no sensitization

4. Cosmetic Applications

Protein shampoos, hand creams; 0.05-0.2% use concentration; not in EEC guideline

5. Other Applications	Antimicrobial preservative for cutting oils
6. Supplier	International Minerals and Chemical Corp.
7. References	1. E. S. Demers, Oxadine A — The new alternative, *Cosmet. Toiletries, 96*:79 (1981). 2. K. H. Wallhäusser, *Parfuem. Kosmet., 62*:379 (1981).

KATHON CG

Chemical names:	Mixture of 5-chloro-2-methyl-4-isothiazoline-3-one and 2-methyl-4-isothiazoline-3-one
EEC No.:	45
CAS No.:	26172-55-4
CTFA adopted names:	Mixture of methyl, chloromethyl isothiazolinone, and methyl isothiazolinone
Type of compound:	Two isothiazolinones and inorganic magnesium salts in water

1. Structure and Chemical Properties

Appearance:	Nonviscous liquid; light amber; 1.5% active substance; specific gravity 1.2 at 20°C; concentrations >20%; Kathon CG cause primary irritation to the eyes, skin, and mucous membranes.
Solubility:	Highly soluble in water, lower alcohols, and glycols
Optimum pH:	Effective over a wide pH range; stability reduced in systems of increasing alkaline pH (pH >8).
Stability:	Optimum pH 4-8; loses activity upon storage in some formulations at elevated temperatures
Compatibility and inactivation:	Compatible with anionic, cationic, and nonionic surfactants and emulsifiers of all ionic types; inactivated by bleach and high pH.
Structural formula:	Mol. wt. 148.5 + 114

Structural formula: (Continued)	Bactericidal activity; fungicidal activity. Active ingredients: 5-chloro-2-methyl-4-isothiazolon-3-one, 1.15%, and 2-methyl-4-isothiazolin-3-one, 0.35%
Assay:	Gas chromatography using a reference standard (Spring House Research Lab., January 5, 1981) [1].

2. Antimicrobial Spectrum

Test organisms (~10^6 CFU/ml)	Minimal germicidal concentration (μg/ml) (suspension test; contact times of 24 and 72 hr)	Minimal inhibitory concentration (μg/ml) (serial dilution test; incubation times of 24 and 72 hr)
Staphylococcus aureus	500-1000	150
Escherichia coli	1500	300
Klebsiella pneumoniae	1000	200
Proteus mirabilis	1000	200
Pseudomonas aeruginosa	1500-2000	300
Pseudomonas fluorescens	1500	150
Pseudomonas cepacia	1500	300
Candida albicans	125	300
Aspergillus niger		200-600

3. Toxicity [1]

Acute oral toxicity

Rat (male):	LD_{50}, 3.35 g/kg
Rat (female):	LD_{50}, 2.63 g/kg

Acute dermal toxicity

Rabbit (NZW, male):	LD_{50}, >5.0 g/kg

Subchronic toxicity

Rat:	3-month dietary study, 2 g/kg per day nontoxic, no pathological findings; no-effect level, 633 mg/kg per day
Rabbit:	3-month percutaneous 1 application per day, 5 days a week, at 0.40%, nontoxic; no pathological findings
Primary skin irritation:	Primary skin irritant (rabbits), product diluted at 3.7% nonirritant
Human patch test:	0.37% Kathon CG (10 × minimum dose) shows no primary irritation or sensitization

Human irritation studies

Nonionic ointment
3 applications per week,
 5 weeks, occluded: 0.37%, 3/10 sensitized, moderate-severe irritation

3 applications per week,
 occluded: 0.18%, no sensitization, slight-moderate irritation

3 applications per week,
 occluded: 0.18%, 0/10 sensitized, nonirritant

Anionic lotion
3 applications per week,
 3 weeks, occluded: 0.37%, 4/50 sensitized

Aqueous solution
5 applications per week,
 4 weeks, occluded arm dip: 0.37%, 0/10 sensitized

Repeated insult (patch test): 0.17%, 1/18 contact sensitization, no primary irritation

Guinea pig sensitization: Skin sensitization (Magnusson-Kligman Test), 0.37%; Kathon CG shows no skin sensitization.

Phototoxicity/photo-
 sensitization: Neither phototoxic nor photosensitizing

Rabbit eye irritation: 18% corrosive to the eye, 3.7% no irritation

Vapor inhalation (rats): Greater than 13.7 mg/liter

Teratology (rats): No fetus toxicity; no teratogenicity in the range 100-1000 mg/kg

Mutagenicity: Ames test, variable; cytogenetic study on rats (male), no effect at 19-190 and 1900 mg/kg

4. Cosmetic Applications   Use concentrations, 0.035-0.15%; shampoos, hair conditioners, hair and body gels, color dye solutions, bubble baths, skin creams and lotions, mascaras [1,2]

5. Other Applications   Cutting oils

6. Supplier   Rohm and Haas Co.

7. References
 1. Rohm and Haas, Technical Bulletin, May 1981.
 2. *Manuf. Chem. Aerosol News*, 49:44, 48 (1978).

Antimicrobial Preservatives

8-HYDROXYQUINOLINE

Chemical names:	8-Hydroxyquinoline; 8-quinolinol; phenoxypyridine
EEC No.:	34
CAS No.:	1130-05-8
Nonproprietary name:	8-Quinolinol
Registered names:	Bioquin, Chinosol, Quinosol
Type of compound:	Quinoline

1. Structure and Chemical Properties [1]

Appearance:	White crystals or crystalline powder; mp 76°C; sulfate, yellow crystalline powder
Solubility:	Almost insoluble in water, freely soluble in alcohol; sulfate freely soluble in water; 1% in glycerol
Compatibility and inactivation:	Forms normal chelate compounds with metals
Structural formula:	C_9H_7NO

Mol. wt. 145.16
Potassium sulfate
Mol. wt. 281.34
Sulfate
Mol. wt. 406.42

$$\left[\begin{array}{c} \text{OH} \\ \bigcirc\!\!\!\bigcirc_N \end{array} \right]^+ \cdot \begin{array}{c} KSO_4^- \\ (H_2SO_4 \cdot H_2O) \end{array}$$

2. Antimicrobial Spectrum

Test organisms (~10^6 CFU/ml)	Minimal inhibitory concentration (μg/ml) (serial dilution test; incubation times of 24 and 72 hr)
Staphylococcus aureus	4
Escherichia coli	64
Klebsiella pneumoniae	64
Pseudomonas aeruginosa	128
Pseudomonas fluorescens	128
Pseudomonas cepacia	128
Candida albicans	128-256
Aspergillus niger	256-512
Penicillium notatum	128-256

Activity of 8-hydroxyquinoline against gram-positive bacteria is given when either Cu^{2+}, Fe^{2+}, or Fe^{3+} is present. The compound is only weakly active against gram-negative bacteria, and no clear requirement is shown for any particular metal. Yeasts are damaged by 8-hydroxyquinoline only in the presence of cupric ions, and toxicity to molds also requires cupric ions. Iron and 8-hydroxyquinoline together catalyze the oxydation of $-SH$ groups in nucleoproteins and enzyme systems.

3. Toxicity

Acute intraperitoneal toxicity [1]

Mouse: LD_{50}, 48 mg/kg

4. Cosmetic Applications

Fungistatic preservative; EEC use concentration, 0.3%; 0.001-0.02% in lotions for the treatment of mycotic skin conditions and in bacterially infected eczemas [2]; not in products for children under 3 years, warning on the label; not for sun protection products.

5. Other Applications

Disinfectant and preservative, antiseptic (topical); potassium hydroxyquinoline sulfate applied to the skin as a cream or lotion at a concentration of 0.05-0.5%.

6. Supplier

Riedel de Haen

7. References

1. *Merck Index*, 9th ed., Rahway, N.J. (1976).
2. W. B. Hugo, *Inhibition and Destruction of the Microbial Cell*, Academic, London (1971), p. 115.

CHLORHEXIDINE

Chemical name:	Bis(p-chlorophenyldiguanido)hexane
EEC No.:	31
CAS No.:	55-56-7
Nonproprietary name:	n-Hexane
CTFA adopted name:	Chlorhexidine
Registered names:	Hibitane; Novalsan; Rotersept, Sterilon, Hibiscrub, Arlacide
Type of compound:	Cationic compound

1. Structure and Chemical Properties

Appearance:	White crystalline powder
Odor:	Odorless

Antimicrobial Preservatives

Solubility: Digluconate, water, >70%; Diacetate, water, 1.8%; dihydrochloride in water, 0.06%; soluble in alcohol, glycerol, propylene glycol, polyethylene glycols

Optimum pH: Neutrality test; 5-8; 0.2% solution in water, pH 6.5-7.5

Stability: Unstable at high temperature (above 70°C)

Compatibility and inactivation: Compatible with cationics and nonionics; incompatible with anionics, various gums, soap, sodium alginate; partial inactivation by Tween 80

Structural formula: Dihydrochloride $C_{22}H_{30}Cl_2N_{10}$
Mol. wt. 505.48
Diacetate $C_{26}H_{38}Cl_2N_{10}O_4$

Chlorhexidine

$$Cl-\underset{}{\bigcirc}-NH-\underset{\underset{NH}{\|}}{C}-NH-\underset{\underset{NH}{\|}}{C}-NH$$
$$Cl-\underset{}{\bigcirc}-NH-\underset{\underset{NH}{\|}}{C}-NH-\underset{\underset{NH}{\|}}{C}-NH$$
$$(CH_2)_6$$

2. Antimicrobial Spectrum [2]

Test organisms ($\sim 10^6$ CFU/ml)	Minimal germicidal concentration (μg/ml) (suspension test; contact times of 24 and 72 hr)	Minimal inhibitory concentration (μg/ml) (serial dilution test; incubation times of 24 and 72 hr)
Staphylococcus aureus	100	0.5-1.0
Escherichia coli	100	1.0
Klebsiella pneumoniae	100	5-10
Pseudomonas aeruginosa	400	5-60
Pseudomonas fluorescens	200	10-20
Pseudomonas cepacia	200	10-20
Candida albicans	20-40	10-20
Aspergillus niger	400	200
Penicillium notatum	200	200

Bacteriostatic at low concentrations, bactericidal at about 10-fold concentration; but under practical conditions there is a lag against Staphylococcus aureus. At low concentrations no effect against fungi; activated by ethanol (7%) or isopropanol 4%; combination with parabens, chlorocresol, and β-phenoxyethanol.

3. Toxicity

Acute oral toxicity
  Mouse: LD_{50}, 2 g/kg (diacetate) [1,4]

Chronic toxicity
  Rat: 0.05% with the drinking water over 2 years without toxic effects

Primary skin irritation: No effect

Human sensitization: Allergic tendency [3]

Mutagenicity: Positive Ames test; positive DNA repair test

4. Cosmetic Applications

Use concentrations 0.01-0.1%, nonionic creams (0.05%), in toothpaste, against film on the teeth and caries, deodorants, antiperspirants (0.1%); EEC maximum, 0.3%

5. Other Applications

Disinfectants (skin); preservative in eyedrops, 0.01%; topical and uterine antiseptic (0.02%); powder, creams 0.1-1.0%

6. Supplier

ICI Americas, Inc.

7. References

1. Davies et al., *Br. J. Pharmacol.*, 9:192 (1954).
2. K. H. Wallhäusser, *Praxis der Sterilisation-Desinfektion-Konservierung*, 3, Auflage Thieme, Stuttgart (1984), p. 485.
3. Dolder and Skinner, *Ophthalmica*, Vol. 2, Wiss. Verlagsges., Stuttgart (1978), p. 276.
4. N. Senior, 1. *International Atlas Symposium*, Düsseldorf, May 1974.

COSMOCIL CQ

Chemical name: Polyhexamethylene biguanide hydrochloride
EEC No.: 42
CTFA adopted name: Cosmocil CQ
Registered names: Cosmocil 20% solution, Vantocil IB
Type of compound: Cationic

1. Structure and Chemical Properties

Appearance: Clear yellowish liquid
Odor: Odorless
Solubility: Water soluble, alcohol soluble

Optimum pH: 4-8 (3.5-10.0)

Stability: Stable below 80°C; nonvolatile

Compatibility and inactivation: Incompatible with anionics; compatible with nonionics and cationic products; no corrosive action

Structural formula:

$$[-(CH_2)_6-NH-\underset{\underset{NH}{\|}}{C}-NH-\underset{\underset{NH}{\|}}{C}-NH-]_n \cdot HCl$$

2. Antimicrobial Spectrum

Test organisms ($\sim 10^6$ CFU/ml)	Minimal inhibitory concentration (μg/ml) (serial dilution test; incubation times of 24 and 72 hr)
Staphylococcus aureus	20
Streptococcus faecalis	5
Escherichia coli	20
Enterobacter cloacae	20
Proteus vulgaris	250
Salmonella typhi	5
Pseudomonas aeruginosa	100
Saccharomyces cerevisiae	25
Aspergillus niger	375
Penicillium notatum	1250

3. Toxicity [2]

Acute oral toxicity

Rat: LD_{50}, 5 g/kg

Chronic toxicity: 90-day feeding test, 3.1 and 6.2 ppm over 90 days, no toxicity or abnormalities with the lower dose. At a dietary level of 6.2 ppm there was a slight retardation of growth and a lower level of food intake. Apart from this, there were no abnormalities.
2½-year feeding test with pathogen-free Wistar rats, dietary levels 1000, 5000, and 10,000 ppm. During the first 3 months evident reduction in body weight and food intake; no-effect level, 5000 ppm.

Primary skin irritation: The concentrated form is a strong irritant; 50,000 ppm was tolerated by rats, with no irritation; for mice the no-effect level was 100 mg/kg per day; in the concentrated form it was irritating.

Rabbit eye irritation:	0.1 ml of a 2000 ppm dilution produces no irritant effect.
Photoirritation:	Even at high concentrations, no significant photoirritancy
Guinea pig sensitization:	No sensitization
Environmental toxicity	
Rainbow trout:	LC_{50} (96 hr), 10 ppm |//

4. Cosmetic Applications — Use concentration, 0.2-1.0% (of 20% solution); EEC guideline, 0.3%

5. Other Applications — Disinfectants, preservative for technical products

6. Supplier — ICI Americas, Inc.

7. References
 1. S. M. Henry and G. Jacobs, *Cosmet. Toiletries*, 96:30 (1981).
 2. ICI Technical Information, D 1534, Cosmocil CQ.

ALKYLTRIMETHYLAMMONIUM BROMIDE

Chemical name:	Alkyltrimethylammonium bromide
EEC No.:	55
CAS No.:	57-09-0
Nonproprietary names:	Cetrimonium bromide, cetrimide
Registered names:	Arquad, Cetavlon, Cetab, Micol, Dodigen 5594
Type of compound:	Quaternary compound

1. Structure and Chemical Properties

Appearance:	Crystals; mp 237-243°C
Solubility:	In water, 10%; freely soluble in alcohol
Optimum pH:	4.0-10.0
Stability:	Stable in acidic solution
Compatibility and inactivation:	Incompatible with anionics, soap, nitrates, heavy metals, oxidants, rubber, proteins, blood
Structural formula:	$C_{19}H_{42}BrN$ Mol. wt. 364.48

$$\left[\text{Alkyl}\overset{\oplus}{-}\text{N}\begin{array}{c}\diagup\text{CH}_3\\-\text{CH}_3\\\diagdown\text{CH}_3\end{array} \right] \text{Br}^-$$

2. Antimicrobial Spectrum

Test organisms ($\sim 10^6$ CFU/ml)	Minimal germicidal concentration (μg/ml) (suspension test; contact times of 24 and 72 hr)	Minimal inhibitory concentration (μg/ml) (serial dilution test; incubation times of 24 and 72 hr)
Staphylococcus aureus	8	4
Escherichia coli	32	16
Klebsiella pneumoniae	32	16
Pseudomonas aeruginosa	256-1024	64-128
Pseudomonas fluorescens	256	64
Pseudomonas cepacia	256	64
Candida albicans	25	12.5
Aspergillus niger	100	50
Penicillium notatum	250	100

3. Toxicity

Chronic toxicity

Rat: Addition of 10, 20, and 45 mg/kg per day to the drinking water over 1 year showed no effect with 10- and 20-mg doses; only the highest concentration caused loss of body weight

Mutagenicity: Negative Ames test

Teratogenicity: Not teratogenic

4. Cosmetic Applications

EEC guideline, 0.1% (maximum); additive to deodorants (maximum, 0.05-0.1%).

5. Other Applications

Cationic detergent, antiseptic, disinfectant, preservative

6. Supplier

Riedel de Haen

BENZALKONIUM CHLORIDE

Chemical names: Mixture of alkyldimethylbenzylammonium chloride, N-dodecyl-N,N-dimethylbenzyl-ammonium chloride

EEC No.: 54

CAS No.: 139-07-1

Registered names: Zephirol, Roccal, Dodigen 226, Barquat MB-50

Type of compound: Quaternary (cationic) compound

1. Structure and Chemical Properties

Appearance: White or yellowish white, amorphous powder, mostly in 50% solution; very bitter taste

Odor: Aromatic odor

Solubility: Very soluble in water (about 50%) and alcohols

Optimum pH: 4.0-10.0; 1% solution pH 6.0-8.0

Stability: Good, stable at 121°C for 30 min

Compatibility and inactivation: Incompatible with anionics, soap, nitrates, heavy metals, citrates, sodium tetraphosphate, sodium hexymetaphosphate, and oxidants, rubber, proteins, blood; adsorbed by plastic materials [2]

Structural formula:

$$\left[\text{Alkyl}-\overset{\oplus}{\text{N}}\underset{\text{CH}_3}{\overset{\text{CH}_3}{<}}\text{C}_6\text{H}_5 \right] \text{Cl}^-$$

Represents a mixture of the alkyls C_8H_{17} to $C_{18}H_{37}$

Assay:
1. Gas chromatography [1]
2. Thin layer chromatography

2. Antimicrobial Spectrum [3]

Test organisms ($\sim 10^6$ CFU/ml)	Minimal germicidal concentration (μg/ml) (suspension test)	
	24 hr	5 min
Staphylococcus aureus	4-10	50
Escherichia coli	10	80
Klebsiella pneumoniae	10	
Pseudomonas aeruginosa	10-100	200
Pseudomonas fluorescens	10-100	
Pseudomonas cepacia	10-100	

2. Antimicrobial Spectrum [3]
   (Continued)

Test organisms ($\sim 10^6$ CFU/ml)	Minimal germicidal concentration (μg/ml) (suspension test)	
	24 hr	5 min
Candida albicans	10	160
Aspergillus niger	100-200	
Penicillium notatum	100-200	

Activation by addition of 0.1% EDTA sodium or nonionic detergents; reduced activity if pH <5.

3. Toxicity

Acute oral toxicity

  Mouse: LD_{50}, 300 mg/kg
  Rat: LD_{50}, 450-750 mg/kg

Subacute oral toxicity

  Rat: Daily oral application of 550 ppm over 3 months without toxic effects.

Chronic toxicity

  Rat: Dietary study with addition of 0.25% over 2 years showed no toxic effects.

Primary skin irritation: 0.1% solution shows no effect on the skin of animals and human volunteers; 0.5% caused irritation.

Rabbit eye irritation: 1:3000 dilution is tolerated.

Mutagenicity: Negative Ames test

4. Cosmetic Applications

Hair conditioner with cationics; addition as preservative or antibacterial compound; eye preparations; EEC maximum, 0.5%; addition to deodorants (0.05-0.1%) and hair conditioners

5. Other Applications

Preoperative antiseptic for skin or for wounds, burns; disinfectants, technical germicide; preservative in pharmaceutical preparations; eyedrops, 0.01-0.1%

6. Suppliers

Onyx Chemical, Lonza Inc., Hoechst Corp.

7. References

1. H. Mitchner and E. C. Jennings, J. Pharm. Sci., 56:1595 (1967).

7. References
   (Continued)

2. H. P. Fiedler, *Lexikon der Hilfsstoffe*, Editio Cantor, Aulendorf (1981), p. 169.
3. K. H. Wallhäusser, *Praxis der Sterilisation-Desinfektion-Konservierung*, Thieme, 3. Auflage Stuttgart (1984), p. 473.

BENZETHONIUM CHLORIDE

Chemical names:	N,N-Dimethyl-N-[2-[2-[4-(1,1,3,3-tetramethylbutyl)phenoxy]ethoxy]ethyl]benzenemethane ammonium chloride; diisobutylphenoxyethoxyethyldimethylbenzylammonium chloride
EEC No.:	53
CAS No.:	121-54-0
Registered names:	Hyamine 1622, Solamine, Phemeride
Type of compound:	Quaternary compound

1. Structure and Chemical Properties [1,2]

Appearance:	Monohydrate; thin, hexagonal plates from chloroform; sinters slightly at 120°C; mp 164-166°C
Solubility:	Very soluble in water, producing a foamy, soapy solution; soluble in alcohol; pH of a 1% aqueous solution, 4.8-5.5
Optimum pH:	Wide range, pH 4-10
Compatibility and inactivation:	Incompatible with soap and anionic detergents; precipitation with mineral acids and many salts
Structural formula:	$C_{27}H_{42}ClNO_2$ Mol. wt. 448.1

$$\left[(CH_3)_3C{-}CH_2{-}C(CH_3)_2{-}C_6H_4{-}OCH_2CH_2OCH_2CH_2\overset{+}{N}(CH_3)_2{-}CH_2{-}C_6H_5 \right] Cl^-$$

2. Antimicrobial Spectrum [2]

Test organisms ($\sim 10^6$ CFU/ml)	Minimal germicidal concentration (μg/ml) (suspension test; contact time of 10 min, at 20°C)	Minimal inhibitory concentration (μg/ml) (serial dilution test; incubation times of 24 and 72 hr)
Staphylococcus aureus	40	0.5
Streptococcus pyogenes	33	0.5
Escherichia coli	50	32
Proteus vulgaris		64
Pseudomonas aeruginosa	800	250
Pseudomonas fluorescens		250
Pseudomonas cepacia		250
Candida albicans		64
Aspergillus niger		128
Penicillium notatum		64

3. Toxicity

Acute oral toxicity

  Mouse: LD_{50}, 500 mg/kg
  Rat: LD_{50}, 420 mg/kg; LD_{50}, 765 mg/kg

Human toxicity: Ingestion may cause vomiting, collapse, convulsions, coma

4. Cosmetic Applications

In-use concentration (EEC), 0.1% as a preservative in deodorants

5. Other Applications

Topical anti-infective, antiseptic, disinfectants, wound powders

6. Supplier

Rohm and Haas

7. References

1. Merck Index, 9th ed., Rahway, N.J. (1976).
2. K. H. Wallhäusser and H. Schmidt, Sterilisation-Desinfektion-Konservierung, Chemotherapie, Thieme, Stuttgart (1967), p. 226.

THIMEROSAL

Chemical name:	Ethylmercurithiosalicylate
EEC No.:	10
CAS No.:	54-64-8
Nonproprietary names:	Thiomersal, Thimerosal, Thiomersalate

CTFA adopted name: Thiomersal
Registered names: Merfamin, Merthiolat, Merzonin
Type of compound: Organic mercurial (anionic)

1. Structure and Chemical Properties

Appearance: Cream-colored crystalline powder; discoloration in light

Solubility: 1 g soluble in 1 ml of water or 8 ml of ethanol

Optimum pH: 7-8; 1% solution, pH 6.7

Stability: Aqueous solutions are stabilized with EDTA; unstable in presence of potassium iodide

Compatibility and inactivation: Incompatible with nonionics, lecithin, thioglycolate, and proteins

Structural formula: $C_9H_9HgNaO_2S$
Mol. wt. 404.84

$$\underset{\text{benzene ring}}{\text{C}_6H_4}(COONa)(S-Hg-C_2H_5)$$

Assay:
1. Polarography [2], USP XX (1980, p. 916).
2. Thin-layer chromatography [3].

2. Antimicrobial Spectrum

Test organisms (~10^6 CFU/ml)	Minimal germicidal concentration (μg/ml) (suspension test; contact times of 24 and 72 hr)	Minimal inhibitory concentration (μg/ml) (serial dilution test; incubation times of 24 and 72 hr)
Staphylococcus aureus	16	0.2
Escherichia coli	128	4
Klebsiella pneumoniae	64	4
Pseudomonas aeruginosa	128	8
Pseudomonas fluorescens	128	4
Pseudomonas cepacia	128	8
Candida albicans	128	32
Aspergillus niger	4096	128
Penicillium notatum	2048	128

Antimicrobial Preservatives

Broad antimicrobial spectrum, especially against gram-negative bacteria; sterilizing effect in ophthalmic solutions, starting with 10^8 cells of *Pseudomonas aeruginosa* per milliliter and using concentrations between 0.005 and 0.1% thiomersalate in 3-48 hr. Recovery medium: fluid thioglycollate or lecithin-polysorbate 80-thioglycollate medium; no sporicidal activity.

3. Toxicity

Subcutaneous toxicity

Rat:	LD_{50}, 98 mg/kg
Human sensitization:	High percentage in Sweden (3.7%) [1]

4. Cosmetic Applications

EEC, 0.007% (of Hg) as maximum concentration, also if mixed with other mercurial compounds. The Hg content of thiomersal is 49.56%. The EEC has limited its use to eye makeup only. The warning Contains Ethylmercurithiosalicylate must appear on the label.

5. Other Applications

Preservative for ophthalmic preparations, eyedrops, and biologics, preferably at acidic pH (about 6.0); topical anti-infective

6. Supplier

7. References

1. B. Magnusson and H. Möller, *Arch. Dermatol. Res.*, 258:235 (1977).
2. J. Birner and J. R. Garnet, *J. Pharm. Sci.*, 53:1264 (1964).
3. H. Möller, *Dermatologica*, 145:280 (1972).

PHENYLMERCURIC ACETATE

Chemical names:	Phenylmercuric acetate, PMA, acetoxyphenylmercury
EEC No.:	11
CAS No.:	62-38-4
CTFA adopted name:	Phenylmercuric acetate
Registered names:	Advacide PMA 18, Cosan PMA, Mergal A 25, Metasol 30, Nildew AC 30, Nuodex PMA 18, Nylmerate, Troysan
Type of compound:	Organic mercurial, cationic

1. Structure and Chemical Properties

Appearance:	Small, lustrous prisms from ethanol; mp 149°C

Solubility:	Poorly soluble in water (1 part in 600 parts water); soluble in hot ethanol
Optimum pH:	Neutral pH
Compatibility and inactivation:	Incompatible with iodine compounds sulfides, thioglycollates, anionics, halogenics, ammonia; compatible with nonionic emulsifiers
Structural formula:	$C_6H_6HgOOCCH_3$, $C_8H_8HgO_2$ Mol. wt. 336.75 ⟨C$_6$H$_5$⟩—Hg—O—CO—CH$_3$
Synthesis:	Heating benzene with mercuric acetate
Assay:	Polarography [4], *USP XX* (1980, p. 916)

2. Antimicrobial Spectrum

Test organisms (~10^6 CFU/ml)	Minimal inhibitory concentration (μg/ml) (serial dilution test; incubation times of 24 and 72 hr)
Staphylococcus aureus	0.1
Escherichia coli	0.5
Klebsiella pneumoniae	0.5
Pseudomonas aeruginosa	1-5
Pseudomonas fluorescens	0.5
Pseudomonas cepacia	1
Candida albicans	8
Aspergillus niger	16
Penicillium notatum	16

3. Toxicity

Acute oral toxicity

Rat:	LD_{50}, 30 mg/kg (acetate); LD_{50}, 60 mg/kg (nitrate) [2]
Chronic toxicity:	0.1 mg Hg/kg animal feed over 1 year shows no effect in rats; 0.5 mg Hg/kg, some animals showed kidney effects [1]
Primary skin irritation:	0.1% solution is a skin irritant; 0.01%, no effect [3].

4. Cosmetic Applications

Limited to eye cosmetics and shampoos which cannot be stabilized with other preservatives; EEC maximum concentration, 0.003%, limited by the EEC to creams with

4. Cosmetic Applications (Continued)	nonionic bases and concentrated shampoos where other preservatives are ineffective; must be labeled Contains Phenylmercurial Compounds
5. Other Applications	Pharmaceutical eyedrops, 0.002-0.005%; herbicides, fungicides
6. Suppliers	Riedel de Haen, Nuodex, Heyden Newport, Cosan, Naftone
7. References	1. Fitzhugh et al., *Arch. Ind. Hyg. Occup. Med.*, 2:433 (1950). 2. W. S. Spector, *Handbook of Toxicology*, Vol. 1, Saunders, Philadelphia (1956). 3. N. Hjorth and C. Trolle-Lassen, *Arch. Pharm. Chem.*, 69:16 (1962). 4. T. M. Hopes, *J. Assoc. Off. Anal. Chem.*, 49:840 (1966).

SODIUM IODATE

Chemical name:	Sodium iodate
EEC No.:	I/10
CAS No.:	7681-55-2
Type of compound:	Iodine compound

1. Structure and Chemical Properties

Appearance:	White crystalline powder
Solubility:	1 g soluble in about 11 parts water, 9 parts boiling water; insoluble in alcohol; the aqueous solution is neutral.
Structural formula:	$INaO_3$ Mol. wt. 197.90

2. Antimicrobial Spectrum

Broad antimicrobial spectrum

3. Toxicity

Acute intravenous toxicity

Dog:	LD, 200 mg/kg [1]

4. Cosmetic Applications

Use concentration (EEC), 0.1%; only for rinse-off products

5. Other Applications

Antiseptic for mucous membranes

6. References

1. W. S. Spector (ed.), *Handbook of Toxicology*, Vol. 1, Saunders, Philadelphia (1956), pp. 274-275.

LAURICIDIN

Chemical name:	Glyceryl monolaurate (alpha and beta forms)
CAS No.:	27215-38-9
Nonproprietary name:	Monoglyceride
CTFA adopted name:	Glyceryl monolaurate
Registered name:	Lauricidin
Type of compound:	Distilled monoglyceride (>90% mono content) (fatty acid composition: C_{10} maximum, 2%; C_{12}, maximum, 90%; C_{14}, maximum, 8%)

1. Structure and Chemical Properties

Appearance:	Waxy solid powder or pastelike solid with an off-white color, mp 56°C; saponification value, 200-206
Odor:	Faint characteristic odor
Solubility:	At 25°C, in water, <0.1%; methanol, 250%; ethanol, 80%; isopropanol, 60%; propylene glycol, 4.5%; glycerin, 0.2%; mineral oil, 0.2%
Optimum pH:	Unaffected in the pH range 3.5-8.0; optimal at pH 7.0-7.5
Stability:	Stable at room temperature, in formulations heated to 68°C until Lauricidin is completely dispersed in water
Compatibility and inactivation:	Compatible with most emulsifiers; inactivated by sodium lauryl sarcosine and ethoxylated and propoxylated nonionics (Tween 80)
Structural formula:	$C_{15}H_{30}O_4$ Mol. wt. 274.4

$$\begin{array}{cc}
\text{H}_2\text{C}-\text{O}-\overset{\text{O}}{\overset{\|}{\text{C}}}-(\text{CH}_2)_{10}-\text{CH}_3 & \text{H}_2\text{C}-\text{OH} \\
| & | \\
\text{HC}-\text{OH} & \text{HC}-\text{O}-\overset{\text{O}}{\overset{\|}{\text{C}}}-(\text{CH}_2)_{10}-\text{CH}_3 \\
| & | \\
\text{H}_2\text{C}-\text{OH} & \text{H}_2\text{C}-\text{OH} \\
\alpha\text{-form} & \beta\text{-form}
\end{array}$$

Assay:	Thin-layer chromatography [1]; gas-liquid chromatography of the acetylated material

2. Antimicrobial Spectrum

Test organisms (~10^6 CFU/ml)	Minimal germicidal concentration (μg/ml) (suspension test; contact times of 24 and 72 hr)	Minimal inhibitory concentration (μg/ml) (serial dilution test; incubation times of 24 and 72 hr)	
Staphylococcus aureus	17-250	78	156
Bacillus subtilis	17		
Escherichia coli	Not affected	5,000	5,000
Klebsiella pneumoniae		>10,000	>10,000
Proteus mirabilis		>10,000	>10,000
Pseudomonas aeruginosa	Not affected	5,000	5,000
Candida albicans	500	78	312
Candida utilis	69		
Saccharomyces cerevisiae	137-250		
Aspergillus niger	137		250
Penicillium notatum			39
Microsporum audouini	12.5		
Trichophyton mentagrophytes (interdigitale)	6.25		

High activity against gram-positive bacteria, but not against gram-negative bacteria except in the presence of EDTA, lactic acid, etc. Therefore Lauricidin should be formulated with other additives (butylhydroxyanisol, EDTA, parabens, etc.). There is good antifungal activity against molds and yeasts; also against lipid-coated viruses [3], such as herpes I and II; for broad-spectrum activity, combination with parabens [1].

3. Toxicity

Acute oral toxicity

  Rat: LD_{50}, 50 g/kg; no toxicity
  Mouse: LD_{50}, 25 g/kg

Subchronic toxicity

  Rat: Feed diets containing 25% for a period of 10 weeks; no histological abnormalities

Primary skin irritation

  Rabbit (patch test [2]):   0.5 ml of 20% emulsions; moderate irritant according to *U.S. Federal Register*, 1973 skin test

  Rabbit eye irritation: A 20% solution, 0.1-ml sample, produced positive reactions in only 1 of 6 rabbits; classified as a negative eye irritant (*U.S. Federal Register* 1973 eye test)

  Guinea pig sensitization: According to Magnusson's method, classified as a mild (grade II) sensitizer, having

Guinea pig sensitization: (Continued)	a sensitization rate of 9-28% (test material, for injection 10% in ethanol; topical application 20% in ethanol)
4. Cosmetic Applications	In-use concentration, 0.5%, in deodorant, soaps, and powders, medicated shampoos, hand and foot care products, preservative, nontoxic surfactant, dental and gum care products
5. Other Applications	Food-grade antimicrobial agent which is approved as an emulsifier in foods by the FDA (21 CFR GRAS 182.4505); antifoaming and wetting agent; base lotions for pharmaceuticals (gel)
6. Supplier	Lauricidin Inc.
7. References	1. Technical Information of Med-Chem Labs. 2. *U.S. Federal Register*, 38, No. 187, Section 1500:41 (1973). 3. S. C. Hienholzer and J. Kabara, *J. Food Safety*, 4:1 (1982). 4. J. J. Kabara, *J. Soc. Cosmet. Chem.*, 29:735 (1980).

PIROCTONE OLAMINE

Chemical name:	1-Hydroxy-4-methyl-6-(2,4,4-trimethylpentyl)-2(1H)pyridone ethanolamine salt
EEC No.:	57
Nonproprietary name:	Piroctone olamine
Registered name:	Octopirox
Type of compound:	Pyridone derivative

1. Structure and Chemical Properties

Appearance:	White to slightly yellowish fine powder; mp 130-135°C (decomposition)
Odor:	Odorless
Solubility:	In water, 0.2% (wt/wt); in oils, 0.05-0.1%; in alcohols, 10% (wt/wt); in water containing surfactants, 1-10% (wt/wt)
Optimum pH:	5-9
Stability:	In the pH range 5-9 stable (incorporated in a shampoo base) at 80°C for at least 14 days and is therefore stable at temperatures normally used in formulating hair care products.

Compatibility and inactivation: Compatible with ingredients commonly used in hair care products, e.g., anionic, cationic, and amphoteric surfactants; incompatibility with some components of fragrance might occasionally occur

Structural formula: Mol. wt. 298.4

2. Antimicrobial Spectrum

Test organisms ($\sim 10^6$ CFU/ml)	Minimal inhibitory concentration (μg/ml) (serial dilution test; incubation times of 24 and 72 hr)
Staphylococcus aureus	32
Escherichia coli	64
Klebsiella pneumoniae	32
Pseudomonas aeruginosa	625-1250
Candida albicans	64
Penicillium notatum	625

3. Toxicity [1]

Acute oral toxicity

Rat: LD_{50}, 8.1 g/kg (practically no toxicity)
Mouse: LD_{50}, 5.0 g/kg

Acute dermal toxicity: Studies did not show any symptomatology which gives evidence for a specific organ toxic effect in subacute and subchronic studies; no-effect level for rats and dogs, 100 mg/kg body weight per day.

Primary skin irritation: Good skin tolerance in animals and human volunteers under practical conditions of use

Rabbit eye irritation: Good eye mucous membrane tolerance

Human sensitization: No sensitization

Guinea pig sensitization: Did not induce any signs of sensitization

Teratogenicity test: Not teratogenic

Mutagenicity test:	Not mutagenic (point mutation test, chromosomal mutation test)
4. Cosmetic Applications	Antidandruff hair tonics, transparent shampoos; concentration in rinse-off products, 0.5-1.0%; in formulations remaining on the hair and scalp after application, 0.05-0.1% (wt/wt); piroctone olamine exhibits considerable substantivity to skin. The amount adsorbed onto keratin is dependent on time, temperature, and concentration; independent of pH (5-8).
5. Other Applications	Preservative for technical products
6. Supplier	Hoechst A.G.
7. References	1. Hoechst A.G., Information, No. 80-10-20.

REFERENCES

1. F. Bertrich, *Kulturgeschichte des Waschens*, Econ-Verlag, Düsseldorf (1966), p. 6.
2. K. H. Wallhäusser, Disinfectants as an aid for good manufacturing practice in the pharmaceutical industry. *J. Pharm. Belg.*, 36:283-297 (1981).
3. K. H. Wallhäusser, Die Bedeutung der Konservierung für die Stabilisierung von Arzneimitteln und Kosmetika gegen einen mikrobiellen. *Verderb. (1981) Pharm. Ind.*, 43:472.
4. L. Dinkloh and I. Aur, Die Detergentiengesetzgebung in nationaler und internationaler Sicht, Symposium on Detergents and Water, Essen, April 1980.
5. P. Berth et al., Symposium of Detergents and Water, Essen, April 1980.
6. Industrieverband Körperpflege und Waschmittel, Information, Frankfurt (1980).
7. *U.S. Pharmacopoiea*, 20th rev., Mack, Easton, Pa. (1980).
8. *European Pharmacopoiea*, Maisunneuve, 57-Sainte-Ruffine (France), Vol. 3.
9. Comité de Liaison des Syndicats Européens de l'Industrie de la Parfumerie et des Cosmétiques, Brussels. (1) Preservatives definitively accepted, CL 38-78/AMR/gp, January 27, 1978; (2) preservatives provisionally accepted, CL 50-78/AMR/gp, February 1, 1978; and (3) preservatives for which files are needed, CL 51-78/AMR/gp, February 1, 1978.
10. Council of Europe, Strasbourg, PA/SG (78) 1; 52.421; 03.5, January 1978.
11. *Amtsblatt der Europäischen Gemeinschaften* (EEC), Nr. L 167/1 Richtlinie des Rates, May 17, 1982 (82/368/EWG), 15.6 1982, p.L 167/2, L 167/28.
12. K. H. Wallhäusser, Antimicrobial Preservatives in Europe: experience with preservatives used in pharmaceuticals and cosmetics, International

Symposium on Preservatives in Biological Products, San Francisco, 1973, *Dev. Biol. Stand.*, *24*:9-28 (1974).
13. K. H. Wallhäusser, Die mikrobielle Reinheit von Kosmetika. Situationsbericht, *Parfuem. Kosmet.*, *61*:121-134 (1980).
14. K. H. Wallhäusser, Praxis der Sterilisation, Desinfektion, Konservierung, Thieme Stuttgart, 1984, 3. Auflage.
15. Empfehlunge des Bundesgesundheitsamtes zur Prüfung der gesundheitlichen, Unbedenklichkeit von kosmetischen Mitteln (1981) *Bundesgesundheitsblatt (BGA) Bundesgesundheitsant*, *24*: No. 6, March 20, 1981.
16. E. L. Richardson, Update — frequency of preservative use in cosmetic formulas as disclosed to FDA, *Cosmet. Toiletries*, *96*, March (1981); *Cosmet. Toiletries*, *92*:85 (1977).
17. K. H. Wallhäusser, Konservierung von Kosmetika, *Seifen Oele Fette Wachse*, *107*:173-179 (1981).

INDEX

Abrasion 509
Acceptance criteria 419
Accompanying reactions by animals 493
Acetamide medium 596
Acetone-Chloroform 607, 625
Acute oral LD50 183, 186
Acute oral toxicity 148, 149, 187
Acute toxicity screening assays 535-542
Adapted organisms 417, 418
Adermydon 634
Adsorption-Complexation 370-374
Adulterated products 561
Advacide PMA 737
Advacide TMP 716
Aerobic plate count 577
Agar diffusion test 425
Agar overlay test 549
Alcohols 265-267, 606, 614
  as preservative 608, 609, 614, 615
Aldehyde 265-267
Alhydex 649
Alkylbenzene Sulfonates 608
Alkyltrimethylammoniumbromide 618, 730
Allergic contact dermititis 537
  allergies 609
  applications 607
  conditions of use 611

[Allergic contact dermititis]
  costs for toxicological tests 613
  declaration 613
  definition (EEC) 614
  discovery 607
  history 605
  limitations 611
  most widely used 609, 613
    number of compounds used 619, 610
  optimal pH-range 615-617, 620
  positive lists 609
    COLIPA 610, 629
    council of Europe 609, 610
    EEC 609, 610, 615-617
    in use concentration 611
  selection in development phase 615-617, 620
  spectrum of activity 615-617
  toxicological profile 613
  used in cosmetics 605, 609, 610
  warning notice 613
  warnings, on the label 611
Alveolar macrophage assay 550
Ames test 552
Aminoform 644
Amphoteric compounds 606
Anaerobe agar 592
Anaerobes 576
Anaerobic plate count 577

Andrade carbohydrate broths 594
Animal oils and fats 227
Animal species 510
Animal studies 139, 140
Animal testing 525
Anionic hand lotion 135
Anionic preservatives 328-329
Anionic protein shampoo 136
Anionic surfactants 608
  biodegradability 608
  consumption 508
Antibacterial activity 41-48, 285
Antibacterial effects 310
Antifungal activity 48-50, 285
Antifungal effects 310, 311
Antimicrobial preservatives
  history and background information 605-609
  "Positive Lists" 609
Antioxidants 230
  effect of chain length 347
  Tert-Butylhydroxyanisole 345, 346
  Tert-Butylhydroxytoluene 345, 346
Antiperspirant 180
Antiseptics 245, 246
  dermal toxicity
    evaluation of erythema 493
    primary irritation index determination 495
    repeated application, evaluation of 496
    two types of skin irritation indices 494
    varied reaction in animals 493
  history 485
  iodinated products 490
  methods for the study of 486
  ocular irritancy index 498-500
    classification 497
    interpretation and marking 498
    testing for 498
  paragerm 490
  primary irritation indices 491
  properties of 486
  skin irritation
    toxicity of 483-501
  triclocarban 490
  vantropol (Chlorhexidine + Cetrimide) 489
Antiviral effects 309
Aquatic bacteria 473
Aquatic toxicity 184, 185

Arlacide 726
Aroma preservatives 237-270
  antimicrobial activity 240-264
  chemical constituents
    alcohol, aldehyde, ketone and esters 265-267
    aromatic derivatives 269
    chain length 269
    hexavalent aliphatic terpene 268
    isoprene unit 265
    phenolic fractions 268
  essential oils
    antiseptics 245, 246
    inhibitory activity of combinations 242-245
    oil-water lotion 247, 248
    phenol coefficients 240
    zones of inhibition 241
  fragrances
    deosafe agents 264
    effects on hand-degerming test 263
    effects on petri plate cultures 263
    zone diameter vs. MIC values 250-262
  history 237-239
  toxicology 270
Arquad 730
Artificial contamination test 426
Assessment of preservative system 389-400
  biodeterioration 395
  classification of preservative 392, 393
  combination of preservative 394-396
  in-use tests 396, 397
  interaction of microorganisms 391, 392
  nutritive status of cosmetics
    nutrient content 391
    pH 391
    water availability 390
  rapid evaluation techniques 397-399
    microcalorimetry 399
    microelectrophoresis 399
    radiometric methods 399
    spore swelling measurement 398
    turbidimetric assays 397
  role of formulation 393-395
  screening procedures 396

Bacillat 35: 714
Bacteria vs. Fungi 23

Index

Bakzid 80: 714
Barquat MB-50: 732
Benzalconium chloride 618, 731
Benzene carbocyclic acid 670
Benzenemethanol 627
Benzethonium chloride 618, 734
Benzoic acid 609, 611, 616, 670
Benzyl alcohol 607, 615, 627
Benzyl carbinol 632
2-Benzyl-4-Chloro-Phenol 694
Benzylparaben 617, 678, 680
Biodegradability of detergents 608
Biodegradation 125
Biological time bomb 608
Biomembrane 8, 9, 12
Biopure 100: 614
Bioquin 725
Biotransformation 72, 73
Bird wildlife 184, 186
Bis (1-Hydroxy 2 (1H) Pyridinethio (see pyrithione) 115-126
Bis-Phenols 606
Bisulfites 609, 612, 659
BNPD (see bronopol)
Boric acid 616, 651
B-Phenoxethanol
  (see phenoxyethanol)
B-Propiolactone
  antimicrobial activity 211
Brain-heart broth and agar 592
Brolene 707
Bromochlorophene 617, 697
5-Bromo-5-Nitrodioxane 615, 638
2-Bromo-2-Nitropropane-1, 3-Diol
  (see Bronopol) 31-61, 635
Bromophene 697
Bronidox L 614, 615, 638
Bronopol 31-61, 614, 615, 629, 635
  analytical methods 38-41
  antibacterial spectrum 42, 43
    cidal 45, 48
    effect of cysteine 46, 48
    effect of organic matter 44
    effect of pH 42
    effect of surfactants 44
    resistance 48
    static 41
  antifungal activity 48-50
    effect of cysteine 49
  background information 32-33
  cosmetic applications
    advantage 57
    limitations 58

[Bronopol]
  metabolism 51
  microbiological activity 41-50
    antibacterial activity 41-48
    antifungal activity 48-50
    mode of action 48
  names and synonyms 32
  partition coefficients 34
  physical properties 33
  regulatory status 59
  solubility 33
  stability 34-38
  structure and chemical properties 33-41
  synthesis 33
  toxicity
    acute 51
    carcinogenicity 55
    chronic 52
    decomposition product and impurities 55
    irritancy and contact sensitivity 52-54
    mutagenicity 55
    phototoxicity 54
    reproduction, effect on 54
Buhler test 516
Butylparaben 616, 679

CA 24: 650
Captan N-(Trichloromethylthio)-4-cyclohexene-1,2-dicarboximide) 618, 716
Carcinogenicity studies 55, 73, 543
Carvacrol 687
Cationic preservatives 328
Cell culture assays 553
Cell wall 25
  bacteria vs. fungi 23
Cell wall-membrane 22, 23
  composition of membrane 22
Cetab 730
Cetavlon 730
Cetrimide 730
Cetrimonium bromide 730
CDTA (see Cyclohexanediaminetetra-acetic 331)
CFU (colony forming units) 624
Challenge inocula added to eye shadow 474
Challenge inocula added to talc 474

Challenge studies
  percentage recovery 468
  zero-time analysis 468
Challenge testing 88-100
Challenges microbes 467
Chelating agents
  chemistry of 324, 325
  ethylenediaminetetraacetate (EDTA) 323-337
  naturally occurring 346
  preservative functions 325, 326
Chemical reactivity 123
Chemical-skin contact time 523
Chinosol 725
Chlorbutol 625
Chlorhexidine 489 606, 607, 618, 726
Chlorine 606
Chlorine dioxide 606
Chloroacetamide 616, 650
Chlorobutanol 612, 613, 615, 625
Chlorocresol 606, 607
p-Chloro-m-cresol 683
p-Chlorom-xylenol (PCMX) 684
Chlorophene 637, 694
Chlorophenesin 615, 634
Chlorophenoxypropanediol 634
Chlorothymol 617, 686
Chloroxylenol 617, 685
4-Chloro-3,5-xylenol 684
Chromium-51 release test 550
Chromosomal damage assays 554
Chronic toxicity 151, 152
Clark's flagellar 596
Classification of preservative 392, 393
Cloflucarban 704
Clonal cytotoxicity test 549
Cobalt-60 225
Code of Federal Regulations 536
Colgate-Palmolive test 446, 447
Colony forming unit (CFU) 624
Combination of preservatives 394-396
Combination with other agents 196, 198, 199
Combination with other preservatives 125
Commonly used preservatives 411
Compatibility problems 425
Composition of bacterial membrane 22
Contact growth index 264

Contact intervals 429-431, 433, 434
Contact lens challenge studies 476
  d-values 476
Container effects 493
Contamination of basic material 224
Contamination of contact lens solution 470
Contamination of eye shadow testers 472
Contamination of topical ointments 362
Control procedures 216, 218
Corrosion 169
Cosan PMA 737
Cosmetic ingredient review 534
Cosmetic preservation
  problems and solutions 3-6
Cosmetics 342, 350-354
  classification and composition of 484
  definition of 484
  quality control
    French legislation requirements 485
    investigation into the efficiency 484
  safety measures 484
Cosmetics, production in Germany 608
  percentage of preserved products 608, 609
Cosmocil CQ 728
Counts of organisms 430-432
Cream rinse 10, 104, 180
Creams
  microbial attack 609
Cresol 111, 606
Crystalline formation 306
CTFA guidelines 445
CTFA method 405
Culture media 433, 435
Cutaneous metabolism 524
o-Cymen-5-01: 713
p-Cymenol 687
Cystamin 644
Cytotoxicity 548

Dantoin 685: 645
Declaration of preservatives 613
Declid 667
Decomposition product and impurities 55
Detection of contamination 430-433

Dehydroacetic acid (DHA) 616, 674
Delayed-type hypersensitivity 537
Deosafe agents 274
Dermal sensitization 316, 317
Dermal toxicity 184, 186, 188, 561
Dermatitis causing ingredients 504
Dermatological preparations 441
  test procedures 441-463
    Colgate-Palmolive test 446, 447
    CTFA guidelines 445
    FDA procedures 445
    ideal preservative system 442
    in-use testing effectiveness 461
    international guidelines 444
    isochallenge microorganism 448-453
    mixed organism challenge 447
    Owen-Mahdeen evaluation system 454
    preservative capacity 459
    preservative stability tests 460, 461
    prototype formulations, selection 459
    rechallenge test 454-457
    Revlon evaluation test 447, 448
    standards of effectiveness 458, 459
    USP Pharmacopoeia CIC 444
    USP Pharmacopoeia XVIII 443
Desomedine 709
Detergents 607
  biodegradability 608
  soap 605, 607
  water pollution 607
Diaminodiphenoxypropane 705
Diazolidinyl urea (Germall II) 616, 657
Dibromopropamidine 617, 707
Dichlorbenzyl alcohol 615, 629
Dichlorofluorocarbanilide 617
Dichloro-m-xylenol 617, 695
Diethylenetriaminepentaacetic acid (DTPA) 324, 325, 331
Differential scanning calorimetry 170, 172, 175
Diluents and plating media 412
Dimethoxane 618, 718
Dimethyloldimethylhydantoin (DMDMH) 165-188, 615, 647
  analysis 170-177
    differential scanning colorimetry 170, 172, 175, 176

[Dimethyloldimethylhydantoin]
    gas chromatography 170, 172, 173, 176
    infrared 170-172
    nuclear magnetic resonance 170, 174, 175, 177
    ultraviolet absorption 170, 173, 174
  cosmetic applications 177-183
    antiperspirant 180
    cream rinse 180
    face cream 180
    night cream 180
  formaldehyde 166, 169-170, 173, 176-179
  MIC values 178, 179
  physical properties 168, 169
    acute oral LD50 183, 186
    corrosion 169
    stability 168, 169
  structure 167, 168
  synonyms 167
  synthesis 166, 170
  toxicology
    aquatic toxicity 184, 185
    bird wildlife 184, 185
    dermal toxicity 184, 186
    eye irritation 184
    inhalation 186
    phytosensitization 186, 187
    skin irritation 183
4,4-Dimethyl-1,3-Oxazolidine 720
Dioxin CO 618, 719
Dioxopentane 649
Diphenyl ether 700
Direct exposure test 548
Discoloration 148
Disinfectants 606
  application 606
  discovery 606
DMDMH
  (see Dimethyloldimethylhydantoin 165-188)
Dodigen 732
  5594: 730
Dominant lethal assay 554
Dowicide 1: 691
Dowicide Q 653
Dowicil 200 (1-(3-Chloroally 1)-3,5,7-Triaza-1-azoniaadamantane chloride) 143-163, 614, 653
  analysis 156-163
  calculations 159, 160
  safety precautions 146

[Dowicil 200]
  antimicrobial properties 145, 146
  formulation techniques
    discoloration 148
    method of addition 146
    pH 147
    preservation levels 147
    stock solutions 148
    temperature 147
  solubility 145
  stability 148
  toxicology
    acute oral toxicity 148, 149
    chronic toxicity 151, 152
    eye irritation 148, 149
    irritation (human) 154
    mutagenicity 155
    N-Nitrosamine analysis 155, 156
    percutaneous absorption 150, 151
    phototoxicity and photosensitization 153
    sensitization 152
    sensitization on Formaldehyde – sensitized subjects 154
    skin irritation 149, 150
Draize test 508, 512, 536
DTPA
  (see diethylenetraiminepentaacetic acid) 324, 325, 331
D values 476
  graphical method 406
  mathematical method 406
Dybenal 629

EEC 608
EEC – Preservative – List 609
  definitively accepted 609, 610, 611
  provisionally accepted 610, 613
Edema, evaluation of 493
Effect on preservatives
  effect of containers 367, 378, 380
  effect of Cysteine 46, 48, 49
  effect of heat 123, 381
  effect of light 123, 381
  effect of media 432, 433
  effect of organic matter 44
  effect of pH 42, 123, 364-366
  effect of radiation 382-384
  effect of storage 366-369, 371, 372, 374, 380
  effect of surfactants 44

[Effect of preservatives]
  effect of temperature 367, 368, 369, 429, 432
  effect of temperature and concentration 212, 213
  effect of water vapor 213, 214
Effective preservation 565
Effects of formulations 86-88
  effects of gels 373-376
  effects of hand-degerming test 263
Effects on Petri Plate cultures 263
Efficacy in formulations 122, 123
Efficacy test results 418
  acceptance criteria 419
Efficacy testing
  CTFA method 405
  D values-graphical method 406
  D values-mathematical method 406
  linear regression method 406
  U.S. Pharmacopoeia method 405
  preservative death time curve 408-410
  recovery system
    diluents and plating media 412
    reliability of the system 412
    stressed organisms 413
  selection of the preservative system
    commonly used preservatives 411
    ideal preservatives 411
    single vs. combined 411
  test organisms selection 414, 415
    adapted organisms 417, 418
    pure vs. mixed culture 415
    repeated – challenge testing 415, 416
Electron accelerators 225
Emeressence 1160 Rose Ether
  (see Phenoxyethanol) 79-108
Emulsifiers 229
Enrichment cultures 579
Enterobacteriaceae 582, 588, 589
EPA FIFRA guidelines 541, 543, 551
Erythema, evaluation 493
Esculin agar, modified 597
Esomedina 709
Essential oils
  antiseptics 245, 246
  inhibitory activity of combinations 242-245
  oil-water lotion 247, 248
  phenol coefficients 240
  zones of inhibition 241
Essential oils and fats 228

Esters, mono 282-286
　antimicrobial activity 341, 343, 344
　mechanism of action 286-290
　monoglycerides 282-284
　　antibacterial activity 285
　　antifungal activity 285
　polyglyceride 341
　polyglycerol esters 284-286
　sucrose caprate 341
　sucrose esters 284
　toxicology 293-296
Esters of P-Hydroxybenzoic acid (see parabens)
Ethanol 614, 615, 623
Ethereal oils 619
　antimicrobial activity 619
Ethylenediaminetetraacetate (EDTA) 323-337
Ethylenediaminetetraaceticacid
　anionic preservatives 328-329
　cationic preservatives 328
　monionic preservatives 329, 330
　titrimetric estimation 333
Ethyleneglycolmonophenyl ether 630
Ethylene glycolphenyl ether (see phenoxyethanol)
Ethylene oxide
　antimicrobial activity 210
　　effect of temperature and concentration 212, 213
　　effect of water vapor 213, 214
　　mechanism of action 215
　　spectrum activity 214
　control procedures 216
　physical properties 210
　regulatory agencies 216
　residual effects 215
Ethylmercurithiosalicylate 735
Ethylparaben 616, 679
European economic community 608
Euxyl K 200: 614, 655
Eye irritation 73, 148, 149, 316, 537
Eye shadow challenge studies 475

Face cream 180
Face lotion 614
　preservation 614
Factors adversely affecting preservatives
　adsorption — complexation 370-374
　effect of chemical reaction 381
　　effect of containers 367, 378, 380

[Factors adversely affecting preservatives]
　effect of heat 381
　effect of light 381
　effect of pH 364-366
　effect of radiation 382-384
　effect of storage 366-369, 371, 372, 374, 380
　effect of temperature 367, 368, 369
　effect of gels 373-376
　phase distribution 377-379
　solubilization — complexation 376, 377
　sorbic acid degradation 368, 369
Fair packaging and labeling act 562
Fatty acids
　effect of geometric isomers 277
　effect of hydroxyl group 280
　effect of saturation 276, 277
　effect of sulfur group 281
　effect of unsaturation 277
　fluoro acids 278
　mechanism of action 286-290
　MIC values 279, 282
　structure — function relationships 282
　toxicology
　　LD50 290, 291
　　patch test for corrosivity 292
　　skin irritation 292
Fatty acids, alcohols and esters 228
Fatty acids and esters 275-297
　history 275, 276
Fatty alcohol ether sulfates 608
Fatty alcohol sulfates 607
FDA procedures 445
FIFRA guidelines 537
Fluoro acids 278
Food, drug and cosmetic act 560, 561
Food grade chemicals
　preservatives 347
　see antioxidants
　see chelating agents
　see monolaurin
　see systems approach 339-354
Food systems 342-345
Formagene 643
Formaldehyde 166, 169-170, 173, 176-179, 188, 606, 607, 609, 611, 614, 615, 640
　antimicrobial activity 211, 216-219
　applications 218

[Formaldyhyde]
  concentration used 611
  control procedures 218
  physical properties 216
  reaction with melamine 217
    urea 217
Formalin 640
Formic acid 616, 663
Formin 644
Formol 640
Formulating properties
  chemical reactivity 123
  compatibility 124
  effect of heat 123
  effect of light 123
  effect of pH 123
  solubility 124
Formulation effects
  formulations 100-102
Formulation principles 359-385
  contamination of topical ointments 362
  factors adversely affecting preservatives
    adsorption − complexation 370-374
    effect of chemical reaction 381
    effect of containers 367, 378, 380
    effect of gels 373-376
    effect of heat 381
    effect of light 381
    effect of pH 364-366
    effect of radiation 382-384
    effect of storage 366-369, 371, 372, 374, 380
    effect of temperature 367, 368, 369
    phase distribution 377-379
    solubilization − complexation 376, 377
    sorbic acid degradation 368, 369
    microbial contamination 360, 362, 363
    occurrence of microorganisms 361, 362
    ocular contamination 362
    spoilage organism 361
Formulation techniques
  discoloration 148
  method of addition 146
  pH 147
  preservation levels 147

[Formulation techniques]
  stock solutions 148
  temperature 147
Foromycen 643
Fragrances
  deosafe agents 264
  effects of hand-degerming test 263
  effects of petri plate cultures 263
  zone diameter vs. MIC values 250-262
French legislation requirements 485
Freund's complete adjuvant test 513
Fungal contamination 467
Fungus
  cell wall 25

G-4: 696
G-11: 698
Gamma radiation 225
Gamophen 698
Gas chromatography 170, 172, 173, 176
Gases, sterilant 209-219
  mechanism of action 219
General principles 424-426
Genotoxicity 140
Geophen 634
Germall II: 616, 657
Germall 115: 609, 614, 629, 655
  (see imidazolidinyl urea 191-203)
GIV-GARD DXN 719
Glutaral 649
Glutaraldehyde 615, 649
Glyceryl monolaurate 618, 740
  (see monolaurin 305-319)
Glycol monophenyl ether
  (see phenoxyethanol)
Glydant 647
  (see dimethyloldimethylhdantoin 165-188)
Glydant 55: 647
Gram-negative 23
Gram-negative nonfermentative Bacilli 584-594
Gram-negative rods 580-591
Gram-positive 24, 25
Gram-positive Cocci 579
Gram-positive rods 579
Grotan BK 714
Guidelines for challenges 466

Hair lotion 614
  preservation 614
Hair-styling lotion 348-350
Halocarban 704
Hamster cheek pouch test 539
HEDDTA
  (see hydroxethyl ethylenediamine-
    triacetic 324, 325, 331, 333)
Hexachlorophene (G-11) 611, 613,
    617, 698
Hexadienoic acid 668
Hexahydrotriazinc 618, 714
Hexamethylenetetramine 615, 644
Hexamidine 617, 708
n-Hexane 726
Hexatidine 713
Hexavalent aliphatic terpene 268
Hexetidine 703
Hexosan 698
Hextril 713
Hibiscrub 726
Hibitane 726
Human assay 527
Human assay variables 519
  number of exposures 520
  number of subjects 520
  occlusion 520
  sensitivity 520
  test concentrations 520
Human sensitization patch test 538
Human studies 140
Human testing 532
Hyamine 1622: 734
Hydrophilic-Lipophilic Balance (HLB)
  temp. 15, 16
0-Hydroxybenzoic acid 672
m-Hydroxybenzoic acid 672
p-Hydroxy-Benzoic acid, benzyl
  ester 680
p-Hydroxy-Benzoic acid esters 607,
    609, 612, 614, 678-683
0-Hydroxybiphenyl 691
Hydroxy-8-Quinoline 618, 725
Hypochlorites 109

Ideal preservative system 442
Imidazolidinyl urea (Germall 115) 191-
    203, 616, 655
  analysis 194, 195
  antimicrobial action
    combinations with other agents
      196, 198, 199

[Imidazolidinyl urea]
  pseudomonas aeruginosa 199
  reinocularion tests 197
  spectrum of action 200-202
  typical formulations 198
  history 192
  physical properties 193
  solubilities 196
  regulatory status 203
  synonyms 191, 193
  synthesis 196
  toxicology 202, 203
Inactivating agents 71
Incidence of microbial contamination
    469
Indole medium 598, 599
Infrared 170-172
Inhalation 186
Inhalation toxicity 541
Inhibitory activity of combinations
    242-245
Inoculum size 429, 432
Inorganic sulfites 609, 612, 616, 659
In raw material 132, 137
Interagency regulatory liaison group
    541, 543
Interfacial tension 8
International guidelines 444
Interpretation and marking 498
Interpretation of indexes 488, 489
In-use contamination 475
In-use testing effectiveness 461
Investigation into the efficiency 484
In vitro toxicity assays 547-555
  cytotoxicity 548
Iodinated products 490
Iodine 109, 606
Iodophors 606
Irgasan CF 3: 704
Irgasan DP 300: 700
Irradiation
  effect on preservatives 384
Irradiation sources 522, 523
Irritancy and contact sensitivity 52-
    54
Irritation
  abrasion 509
  animal species 510
  application methods 511
  draize test 508
  human 154
Isochallenge microorganism 448-453
Isoprene unit 265

Iso-propanol 614, 615
4-Isopropylbenzene 687
Isopropyl-o-cresol 617, 687
4-Isopropyl-3-methylphenol 693
Isothiazoline one (see Kathon CG) 722

Kathon CG: 618, 722
Ketolin 695
King's B medium 689
King's oxidative-fermentative basal medium 601
KM 200: 701

LAS (linear alkylbenzene sulfonates) 618
Lauricidin
  (see monolaurin 305-319, 740)
Lauricidin 112
  (see monolaurin 311)
Lauricidin 802
  (see monolaurin 311)
Lauricidin 812
  (see monolaurin 311)
LD_{50} 290, 291
Lens solution
  fungal contamination 467
Letheen agar or broth 593
Linear regression method 406
Lipopolysaccharide
  gram-negative 23
  gram-positive 24, 25
  structure 24
Liquapar 682
Liquid crystal 9, 11, 12, 13
  optically anisotropic 11
Liquid preparations 432
Liquid soap 103
Lysine decarboxylase medium 599
Lysine iron agar 594

MacConkey agar 592
Malonate broth 594
Mannitol salt agar 592
Maximization test 514
MDMH (see Glydant 165-188)
  Monomethyloldimethylhydantoin 186, 187

Mechanism of action 85-86, 120-122, 215
Media and reagents 575
Media for identification of Enterobacteriaceae
  andrade carbohydrate broths 594
  lysine iron agar 594
  malonate broth 594
  Motility-Indole-Ornithine 594
  MR-VP broth 595
  triple sugar iron agar 595
  urea agar (Christensen's) 595
Media for identification of gram (−) Nonfermentative Bacilli
  acetamide medium 596
  Clark's flagellar 596
  esculin agar, modified 597
  indole medium 599, 618
  King's B medium 599
  King's Oxidative-Fermentative Basal medium 601
  lysine decarboxylase medium 600
  motility nitrate medium 600
  nitrate broth, enriched 601
  nutrient gelatin 598
  oxidase test strips 602
Media for identification of gram (+) bacteria
  anaerobe agar 592
  bile esculin agar 592
  brain-heart broth and agar 592
  letheen agar or broth 593
  MacConkey agar 592
  mannitol salt agar 592
  oxidative-fermentative test media 593
  potato dextrose agar or malt extract 592
  sabouraud dextrose broth 593
  starch agar 593
  thioglycollate media 135C 593
  tryptic soy agar and broth 594
  trypticase soy agar 594
  Vogel-Johnson agar 514
Media for P/O organisms 467, 468
Melamine 217
Mercury 109
Method of addition 146
Methyl bromide
  antimicrobial activity 211
MIC data 425
MIC values 84, 85, 131, 133
Microbial control problems 477

Microbial testing
  challenge testing 88-100
  effects of formulations 86-88
  mixed-culture 99, 100
Microbiological methods 573-603
  enrichment cultures 579
  equipment and materials 575
  identification of bacteria
    enterobacteriaceae 582, 588, 589
    gram-negative nonfermentative bacilli 584-591
    gram-negative rods 580-591
    gram-positive cocci 579
    gram-positive rods 579
  media and reagents 575
  media for identification of enterobacteriaceae
    andrade carbohydrate broths 594
    lysine iron agar 594
    malonate broth 594
    motility-indole-ornithine 594
    MR-VP broth 595
    triple sugar iron agar 595
    urea agar (Christensen's) 595
  media for identification of gram (−) nonfermentative bacilli
    acetamide medium 596
    Clark's flagellar 596
    esculin agar, modified 597
    indole medium 598, 599
    King's B medium 599
    King's oxidative-fermentative basal medium 601
    lysine decarboxylase medium 599
    motility nitrate medium 600
    nitrate broth, enriched 601
    nutrient gelatin 598
    oxidase test strips 602
  media for identification of gram (+) bacteria
    anaerobe agar 592
    bile esculin agar 592
    brain-heart broth and agar 592
    letheen agar or broth 593
    MacConkey agar 592
    mannitol salt agar 592
    oxidative-fermentative test media 593
    potato dextrose agar or malt extract 592

[Microbiological methods]
    sabouraud dextrose broth 593
    starch agar 593
    thioglycollate media 135C 593
    tryptic soy agar and broth 594
    trypticase soy agar 594
    Vogel-Johnson agar 594
  sample preparation 596
  sampling methods 595
  screening test
    aerobic plate count 577
    anaerobes 576
    staphylococcus aureus 578
    yeast and mold 578
Microcalorimetry 399
Microelectrophoresis 399
Microemulsions 7-20
  biomembrane 8, 9, 12
  droplet size 7
  hydrophilic-lipophilic balance temp. 15, 16
  interaction with microbes 12, 13
  interfacial tension 8
  liquid crystal 9, 11-13
    optically anisotropic 11
  nonionic surfactants 8, 9, 15
  oil-water emulsion 15, 16
  stabilization of 8
  temperature-dependent 14
  transparent emulsions 7, 8, 9, 12, 13
  water-oil emulsion 15, 16
  water-oil system 9, 10
Micronucleus 553
Microorganism
  cell wall
    bacteria vs. fungi 23
  cell wall/membrane 22, 23
    composition of membrane 22
  composition and structure 21-26
  identification of 469
  prokaryotic 21, 23
  types and frequencies 472
Mineral oils and fats 228
Misbranded products 561
Mixed-culture 99, 100
Mixed organism challenge 447
Mode of contamination, multiple 427
Mode of contamination, single 427
Monoglycerides 282-284
  antibacterial activity 285
  antifungal activity 285
Monohydric alcohols 282

Monolaurin (Lauricidin) 305-319
  analysis of 307, 308
  crystalline formation 306
  effects on spores 342
  effects on viruses 741
  history 305, 306
  microbiological data 308, 312
    antibacterial effects 310
    antifungal effects 310, 311, 344
    antiviral effects 309
  oil-water system 319
  phase diagram 307
  preservative effects
    cosmetics 342, 350-354
    food systems 342-345
  solubility of 308
  structure 306
  synthesis 306
  toxicology 312-317
    dermal sensitization 316, 317
    eye irritation 316
    oral chronic studies 317
    oral LD50 317
    regulatory status 318
    skin irritation 313, 316
Monomethyloldimethydantoin 186, 187
  toxicology
    acute oral 187
    dermal 188
    skin irritation 187
  uses 188
Motility-Indole-Ornithine 594
Motility nitrate medium 600
MR-VP broth 595
Mutagenicity 55, 155, 551-555
  ames test 552
  cell culture assays 553
  chromosomal damage assays 554
  dominant lethal assay 554
  micronucleus 553
  phototoxicity 555
  sex-linked recessive lethal test 553
  yeast assays 552
Myacide AS
  (see Bronopol 31-61, 635)
Myacide BT
  (see Bronopol 31-61, 635)

Neutral red uptake 549

Night cream 180
Nildev AC 30, 737
Nipabenzyl 680
Nipabutyl 679
Bipakombin 678
Nipasol 679, 698
Nitrate broth, enriched 601
N-Nitrosamine analysis 155, 156
No. Amer. contact dermatitis 503
Nonionic preservative 329, 330
Nonionic ointment 134
Nonionic surfactants 8, 9, 15
Novalsan 726
Nuclear magnetic resonance 170, 174, 175, 177
Nuodex PMA 737
Nutrient gelatin 598
Nutritive status of cosmetics — nutrient content 391
Nutritive status of cosmetics — pH 391
  water availability 390
Nylmerate 737

Occlusion 500
Occurrence of microorganisms 361, 362
Octopirox 742
Ocular contamination 362
Ocular irritancy index 498-500
  interpretation and marking 498
  testing for 498
Ocular preparations 465-479
  aquatic bacteria 473
  challenge inocula added to eye shadow 474
  challenge inocula added to talc 474
  challenge studies
    percentage recovery 468
    zero-time analysis 468
  challenge microbes 467
  contact lens challenge studies 476
  D-values 476
  contamination of contact lens solution 470
  contamination of eye shadow testers 472
  eye shadow challenge studies 475
  guidelines for challenges 466
  in-use contamination 475
  incidence of microbial contamination 469

[Ocular preparations]
  lens solution
    fungal contamination 467
    media for P/O organisms 467, 468
    microbial control problems 467, 468
  microorganisms
    identification of 469
    types and frequencies 472
  preparation of inocula 468
  sampling of in-use solutions 477
Oil of cloves 619
  eucalyptus 619
  lemon 619
  orange 619
  pineneedle 619
  rosemary 619
  sage 619
Oil-water creams 619
Oil-water emulsion 15, 16
Oil-water lotion 105, 247, 248, 348-350, 353
Oil-water system 319
Omadine, Na and Zn (see Pyrithione 617)
Open epicataneous test 517
O-Phenylphenol 112 (see also Dowicide 1 or A; see also Topane)
O-Phenylphenols 609, 611, 617
  OTC 624
Oral chronic studies 317
Oral LD50: 317
Oral toxicity 540
Organization for Econ Coop and Devel (DECD) 541, 543, 551
Ottasept 685
  over the counter 624
Owen-Mahdeen evaluation system 454
Oxadine A 638, 720
Oxidase test strips 602
Oxidative-Fermentative test media 593

Parabens 599, 612, 616, 617, 678-683
  analysis and determination 74
  antimicrobial activity 65, 67, 68
  antimicrobial activity, effects of inactivating agents 71
  partition coefficients 70

[Parabens]
  pH 70, 71
  surfactants 69
  application and method of incorporation 71, 72
  blend of PHB esters 617, 682
  effect of chain length 345
  physical and chemical properties 65, 66
  toxicity
    acute 72
    biotransformation 72, 73
    carcinogenicity 73
    chronic 72
    eye irritation 73
    skin irritation studies 73
    teratogenicity 73
Parachlor-m-cresol 617, 683
Paraform 643
Paraformaldehyde 611, 615, 643
Paragerm 490
Partition coefficients 34, 70
Patch test for corrosivity 292
PCMX (see p-chloro-m-xylenol 684)
1,5-Pentandial 649
Peracetic acid 606
Percutaneous absorption 150, 151
Percutaneous penetration 541
Periocular and ocular preparations 465-479
Periocular preparations (see ocular preparations) 465-479
Personal hygiene 605
pH 70, 71, 147, 391
Pharmaceuticals
  antimicrobial preservation 423-440
    agar diffusion test 425
    analytical methods 425
    artificial contamination test 426
    compatibility problems 425
    contact intervals 429-431, 433, 434
    counts of organisms 430-432
    culture media 433, 435
    detection of contamination 430-433
    effect of media 432, 433
    effect of temperature 429, 432
    general principles 424-426
    inoculum size 429, 432
    liquid preparations 432
    MIC data 425
    mode of contamination, multiple 427

[Pharmaceuticals]
    mode of contamination, single 427
    resistance of organisms 425
    semisolid preparations 432
    strains and inoculum 431
    test conditions 429
    test organism 426-427, 431-432
    test results and interpretation 431, 433, 434
Phase diagram 307
Phase distribution 377-379
PHB esters 607, 609, 614, 678-693
Phenethyl alcohol 532
Phenol and derivatives 605, 607
Phenol coefficients 240
Phenolic fractions 268
Phenonip 631, 682
Phenolic lipophilic agents (see antioxidants)
Phenolic lipophilic agents (see parabens)
Phenols 109-112
    history 109-110
    mode of action 110
    regulatory constraints 110
    structure/function relationships 110
Phenoxetol (see phenoxyethanol)
2-Phenoxyethanol (see phenoxyethanol)
Phenoxyethanol 79-108, 615, 630, 631
    formulation effects
        formulations 100-102
    history 79, 84
    microbial testing
        challenge testing 88-100
        effects of formulations 86-88
        mixed culture 99, 100
    microbiology
        mechanism of action 85-86
        MIC values 84-85
    physical and chemical properties 80-82
    solubilities and partition coefficients 82
    synthesis 83
    toxicology 106, 107
    viscosity and stability testing 103-106
    cream rinse 10, 104

[Phenoxyethanol]
    liquid soap 103
    oil-water lotion 105
    water-oil cream 105, 106
Phenoxyethyl alcohol (see phenoxyethanol)
Phenoxyisopropanol 615
Phenylcarbinol 627
Phenyl cellosolve 631
Phenylethyl alcohol 607, 615, 632
Phenylmercuriacetate 618, 737
Phenylmercuric compounds 607, 618
Phenylmethanol 627
Phenylphenols 609, 617, 691
Photoallergens
    tetrachlorosalicylamide 525
    tribromosalicylanilide 525
    trichlorocarbanilide 526
Photosensitivity reactions 546
Photosensitization 186, 187
    animal testing 525
    human assay 527
    tetrachlorosalicylamide (TCSA) 525
Phototoxicity 54, 520-525, 555
    chemical-skin contact time 523
    cutaneous metabolism 524
    human testing 522
    incidence and frequency 525
    irradiation sources 522, 523
    skin stripping 523
    test animal 521
    vehicles effects 524
Phototoxicity and photosensitization 153
Phototoxicity in vitro 555
Piroctone olamine 618, 742
P/O
    periocular and ocular preparations 465-479
P/O
    preparations (see ocular preparations) 465-479
Polyglyceride 341
Polyglycerol esters 284-286
Polyhexamethylenebiguanide 618, 728
Polyoxymethylene 643
Potato dextrose agar or malt extract 592
Powders 227
Preparation of inocula 468
Preservation Effectiveness in-use 563

Index

Preservation levels 147
Preservative capacity 459
Preservative death time curve 408-410
Preservative effects
  food systems 342-345
Preservative stability tests 460, 461
Preservative systems 340
  hair styling lotion 348-350
  in cosmetics 347-354
  oil-water lotion 348-350, 353
  optimization 354
  protein hair treatment 348-350
  shampoo 348-350
  water-oil skin cream 348-350
Preventol CMK 683
  0: 691
Primary irritation index 487
Primary irritation index determination 495
Primary irritation indices 491
Prokaryotic 21, 23
Propamidine 617, 705
n-Propanol 614
Propionic acid 609, 611, 616, 665
Propylene oxide
  antimicrobial activity 211
Propylparaben 616
Protein hair treatment 348-350
Prototype formulation, selection of 459
2-Pyridinethiol-1-Oxide (see pyrithione 115-126)
Pyrion-Sodium 709
Pyrithione 115-126
  analysis 120
  antimicrobial action
    efficacy in formulations 122-123
    mechanism of action 120-122
    MIC values
    spectrum 120
  biodegradation 125
  chemicals properties 118-119
  combination with other preservatives 125
  formulating properties
    chemical reactivity 123
    compatibility 124
    effect of heat 123
    effect of light 123
    effect of pH 123
    solubility 124
  history 117

[Pyrithione]
  physical properties 117-118
  preparation 117
  solubilities 118
  specific gravity and particle size 124
  toxicology 125
Pyrithione zinc 613, 617
  sodium 617

Quality control
  French legislation requirements 485
  investigations into the efficiency 484
  safety measures 484
Quaternary ammonium compounds 606, 607, 730
Quaternium-15 646, 653 (see also Dowicil 200)
Quinoline compounds 725
8-Quinolinol 725
Quinosol 725

Rabulen — TL 688
Radiation, gamma 224-231
Radiation ionizing
  cesium 137, 225
  cobalt-60 225
  effects on
    animal oils and fats 227
    antioxidants 230
    emulsifiers 229
    essential oils and fats 228
    fatty acids, alcohols and esters 228
    mineral oils and fats 228
    powders 227
    preservatives 229
    solvents 226
    synthetic oils and fats 228
    thickening agents 226
    vegetable oils 227
    vitamins 230
    waxes 227
  electron accelerators
  gamma
  regulatory status 230
  sources of 225
  units and dose 225

Radiometric methods 399
Rapid evaluation techniques 397-399
  microcalorimetry 399
  microelectrophoresis 399
  radiometric methods 399
  spore swelling measurement 398
  turbidimetric assays 397
Rechallenge test 454-457
Recovery system
  diluents and plating media 412
  reliability of the system 412
  stressed organisms 413
Regulatory agencies 216
Regulatory constraints 110
Regulatory status 59, 318
Reinoculation tests 197
Reliability of the system 112
Reneselin 667
Repeated inoculation, effect of 488, 496
Reproductive toxicity 544
Resistence of organisms 48, 425
Revlon evaluation test 447, 448
Roccal 732
Rotersept 726

Sabouraud dextrose broth 593
Safety and regulatory issues 559, 569
  effective preservation 565
  preservation effectiveness in-use 563
  regulation of cosmetics 560-562
    adulterated products 561
    fair packaging and labeling act 562
    food, drug and cosmetic act 560, 561
    misbranded products 561
  regulatory activities 562
  toilet goods association 565
Safety evaluation
  dermatitis-causing ingredients 504
  methods 503-527
Safety evaluation methods
  human assay variables 519
    number of exposures 520
    number of subjects 520
    occlusion 520
    sensitivity 520
    test concentrations 520

[Safety evaluation methods]
  irritation
    abrasion 509
    animal species 510
    application methods 511
    draize test 508
  photosensitization
    animal testing 525
    human assay 527
    tetrachlorosalicylamide (TCSA) 525
  phototoxicity 520-525
    chemical-skin contact time 523
    cutaneous metabolism 524
    human testing 522
    incidence and frequency 525
    irradiation sources 522, 523
    skin stripping 523
    test animal 521
    vehicles effects 524
  sensitization 512-519
    Buhler tests 516
    draize test 512
    Freund's complete adjuvant test 513
    maximization test 514
    open epicutaneous test 517
    split adjuvant technique 515
Safety measures 484
Safety precautions 156
Salicylic acid 601, 616, 672
Sample preparation 576
Sampling of in-use solutions 477
Sampling methods 575
Santophen 1: 695
Sedaform 625
Selection of the preservative system
  commonly used preservative 411
  ideal preservative 411
  single vs. combined 411
Semisolid preparations 432
Sensitivity 520
Sensitization
  Buhler test 516
  draize test 512
  Freund's complete adjuvant test 513
  maximization test 514
  open epicutaneous test 517
  split adjuvant technique 515
Sentry 668
Septiphene 695
Sevinon 667
Sex-lined recessive lethal test 553

Index

Shampoos 348-350, 607-609, 614
  preservation 614
Single vs. combined inoculation 411
Skin care products 608
  compatibility 609
Skin irritation 149, 150, 183, 187, 291, 292, 313, 316, 536
  interpretation of indexes 488, 489
  primary irritation index 487
  repeated application, effect of 488
Skin irritation studies 73
Skin sensitization test 537
Skin stripping 523
Soap manufacture
  by the sumerians 605
Sodium benzoate 607
Sodium iodate 612, 618, 739
Sodium omadine 729 (see pyrithione 115-126)
Sodium pyrithione 700 (see pyrithione 115-126)
Sodium thimerfonate 607
Solamine 734
Solubility 124, 196
Solubilization-complexation 376, 377
Solvents 226
Sorbic acid 607, 609, 611, 616, 668
Sorbic acid degradation 368, 369
Split adjuvant technique 515
Spoilage organism 361
Spore swelling measurement 398
Standards of effectiveness 458, 459
Staphylococcus aureus 578
  toxin 3
Starch agar 593
Static effects 41
Sterilants (see gases and/or radiation)
Sterilon 726
Strains and inoculum 431
Stressed organisms 413
Sublimate 606, 607
Sucrose caprate 341
Superficial skin attack 492
Surfactants 69
Sucrose esters 284
Synthetic oils and fats 228
Synthetic surfactants 607

Target cell assays 550
Temperature 147
Teratogenicity 73
Tert-Butylhydroxyanisole 345, 346
Tert-Butylnydroxytoluene 345, 346
Test organism selection 414, 415
  adapted organisms 417, 418
  pure vs. mixed culture 415
  repeated-challenge testing 415, 416
Test procedures 441-463
  Colgate-Palmolive test 446, 447
  CTFA guidelines 445
  FDA procedures 445
  ideal preservative system 442
  in-use testing effectiveness 461
  international guidelines 444
  isochallenge microorganism 448-453
  mixed organism challenge 447
  Owen-Mahdeen evaluation system 454
  preservative capacity 459
  preservative stability tests 460, 461
  prototype formulation, selection of 459
  rechallenge test 454-457
  Revlon evaluation test 447, 448
  standards of effectiveness 458, 459
  USP Pharmacopoeia XIX 444
  USP Pharmacopoeia XVIII 443
Test results and interpretation 431, 433, 434
Tetraazatricyclodecane 644
Tetrabromo-o-cresol 617, 698
Tetrachlorosalicylamide (TCSA) 525
Tetrapropylenebenzene sulfonate (TPS) 608
Thickening agents 226
Thimerosal 607, 735
Thioglycollate media 135, 593
Thiomersal 618, 736
Thymol 112
Titrimetric estimation 333
Tomicide Z-50 (see pyrithione 115-126)
Toxicity evaluation 533-558
  acute toxicity screening assays 535-542
  allergic contact dermititis 537
  carcinogenicity studies 543
  chronic toxicity test 543
  cosmetic ingredient review 534
  cytotoxicity
    agar overlay test 549
    alveolar macrophage assay 550
    chromium-51 release test 550
    clonal cytotoxicity test 549
    direct exposure test 548
    neutral red uptake 549
    target cell assays 550

[Toxicity evaluation]
  delayed-type hypersensitivity 537
  EPA fifra guidelines 541, 543, 551
  federal food, drug and cosmetic
      act 534
  interagency regulatory liaison
      group 541, 543
  in vitro toxicity assays 547-555
      cytotoxicity 548
  LD-50: 536, 540
  local effects
      code of federal regulations 536
      draize test 536
      eye irritation 537
      hamster cheek pouch test 539
      human sensitization patch test
          538
      skin irritation effects 536
      skin sensitization test 537
      vagina test 539
  mutagenicity 551-555
      Ames test 552
      cell culture assays 553
      chromosomal damage assays 554
      dominant lethal assay 554
      micronucleus 553
      phototoxicity 555
      sex-linked recessive lethal test
          553
      yeast assays 552
  organization for Econ Coop. and
      Devel. (OECD) 541, 543, 551
  percutaneous penetration 541
  photosensitivity reactions 546
  phototoxicity in vitro 555
  reproductive toxicity 544
  subchronic toxicity test 542
  systemic effects
      dermal toxicity 541
      inhalation toxicity 541
      oral toxicity 540
  teratology tests 544
  toxicokinetics 545
Toxicological properties 132, 138-141
  animal studies 139, 140
  genotoxicity 140
  human studies 140
Toxicology 106, 107, 125, 202, 203,
      270, 293-296, 290-292, 312-317,
      332
  acute oral 187
  acute oral toxicity 148, 149
  aquatic toxicity 184, 185

[Toxicology]
  bird wildlife 184, 186
  chronic toxicity 151, 152
  dermal 188
  dermal sensitization 316, 317
  dermal toxicity 184, 186
  eye irritation 148, 149, 184, 316
  inhalation 186
  irritation (human) 154
  LD50 290, 291
  mutagenicity 155
  N-nitrosamine analysis 155, 156
  oral chronic studies 317
  oral LD50: 317
  patch test for corrosivity 292
  percutaneous absorption 150, 151
  photosensitization 186, 197
  phototoxicity and photosensitization
      153
  regulatory status 318
  sensitization 152
  sensitization of formaldehyde-sensi-
      tized subj. 154
  skin irritation 149, 150, 183, 187,
      291, 292, 313, 316
Toxin 3
TPS (tetrapropylenebenzene sulfate)
    608
Transparent emulsions 7, 8, 9, 12, 13
Tribromosalicylanilide 525
Trichlorocarbanilide 526, 617, 702
Trichloro-2-hydroxydiphenyl ether 700
Trichloro-2-methyl-2-propanol 625
Triclocarban 490, 702
Triclosan 617, 700
Triformol 613
Triple sugar iron agar 595
Tryptic soy agar and broth 594
Trypticase soy agar 594
Tris-(hydroxyethyl)-hexahydrotria-
    zinc 618, 714
Turbidimetric assays 397
Typical formulations 198

Ucaricide (see Glutaraldehyde) 649
Ultraviolet absorption 170, 173, 174
10-Undecenoic acid 667
Undecylenic acid 616, 667
Urea 217
Urea agar (Christensen's) 595
Uriton 649

Index

Usnic acid 616, 697
U.S. Pharmacopoeia method 405
USP Pharmacopoeia XIX 444
USP Pharmacopoeia XVIII 443

Vancid 89 RE
Vantocil IB 728
Vantropol (Chlorhexidine + Cetrimide 469
Vegetable oils 227
Vehicles effects 524
Virus, effects on 741
Viscosity and stability testing 103-106
  cream rinse 10, 104
  liquid soap 103
  oil-water lotion 105
  water-oil cream 105, 106
Vitamins 230
Vogel-Johnson agar 514

Water availability 390

Water consumption 607
  pollution 607
Water-oil cream 105, 106
Water-oil emulsion 15, 16
Water-oil skin cream 348-350
Water-oil system 9, 10
Water-oil ointments 609
Waxes 227

Yeast and mold 578
Yeast and assays 562

Zephirol 732
Zero time analysis 468
Zinc omadine (see pyrithione 115-126) 711
Zinc pyrion (see pyrithione 115-126)
Zinc pyrithione 115-126, 612, 711
Zone diameter vs MIC values 250-262
Zones of inhibition 241